Table of Problem-Solving Strategies

Note for users of the two-volume edition:
Volume 1 (pp. 1–547) includes chapters 1–16.
Volume 2 (pp. 548–1043) includes chapters 17–30.

Table of Math Relationship Boxes

ActivPhysics™ OnLine Activities

www.aw-bc.com/knightjonesfield

COLLEGE PHYSICS

A STRATEGIC APPROACH

VOLUME 1

RANDALL D. KNIGHT

California Polytechnic State University, San Luis Obispo

BRIAN JONES

Colorado State University

STUART FIELD

Colorado State University

PEARSON

Addison
Wesley

San Francisco   Boston   New York
Cape Town   Hong Kong   London   Madrid   Mexico City
Montreal   Munich   Paris   Singapore   Sydney   Tokyo   Toronto

Vice President and Editorial Director:	Adam Black, Ph.D.
Development Editor:	Alice Houston, Ph.D.
Project Editor:	Martha Steele
Executive Producer:	Claire Masson
Associate Media Producer:	Deb Greco
Editorial Assistant:	Grace Joo
Director of Marketing:	Christy Lawrence
Senior Market Development Manager:	Josh Frost
Market Development Coordinator:	Jessica Lyons
Managing Editor:	Corinne Benson
Production Supervisor:	Shannon Tozier
Production Service and Composition:	WestWords, Inc.
Illustrations:	Precision Graphics
Copyeditor:	Kevin Gleason
Text Design:	Patrick Devine Design and Side by Side Studios
Cover Design:	Yvo Riezebos Design
Manufacturing Manager:	Pam Augspurger
Director, Image Resource Center:	Melinda Patelli
Manager, Rights and Permissions:	Zina Arabia
Photo Research:	Cypress Integrated Systems, Inc.
Printer and Binder:	RR Donnelley, Willard
Cover Printer:	Phoenix Color
Cover Image:	Ken Wilson, Papilio/CORBIS

Library of Congress Cataloging-in-Publication Data

Knight, Randall Dewey.
   College physics : a strategic approach / Randall D. Knight, Brian Jones, Stuart Field. — 1st ed.
      p. cm.
  ISBN 0-8053-0634-X
1.   Physics—Textbooks. I. Jones, Brian. II. Field, Stuart. III. Title.
    QC23.2.K55 2006
    530—dc22

                                  2006032583

ISBN 0-8053-0629-3 (Student edition)

2 3 4 5 6 7 8 9 10—DOW—10 09 08 07
www.aw-bc.com

Brief Contents

About the Authors

Randy Knight has taught introductory physics for 25 years at Ohio State University and California Polytechnic University, where he is currently Professor of Physics and Director of the Minor in Environmental Studies. Randy received a Ph.D. in physics from the University of California, Berkeley and was a postdoctoral fellow at the Harvard-Smithsonian Center for Astrophysics before joining the faculty at Ohio State University. It was at Ohio State, under the mentorship of Professor Leonard Jossem, that he began to learn about the research in physics education that, many years later, led to *Five Easy Lessons: Strategies for Successful Physics Teaching, Physics for Scientists and Engineers: A Strategic Approach,* and now to this book. Randy's research interests are in the field of lasers and spectroscopy. When he's not in the classroom or in front of a computer, you can find Randy hiking, sea kayaking, playing the piano, or spending time with his wife Sally and their seven cats.

Brian Jones has won several teaching awards at Colorado State University during his 17 years teaching in the Department of Physics. His teaching focus in recent years has been the College Physics class, including writing problems for the MCAT exam and helping students review for this test. Brian is also Director of the *Little Shop of Physics,* the Department's engaging and effective hands-on outreach program, which has merited coverage in publications ranging from the *APS News* to *People* magazine. Brian has been invited to give workshops on techniques of science instruction throughout the United States and internationally, including Belize, Chile, Ethiopia, Azerbaijan and Slovenia. Previously, he taught at Waterford Kamhlaba United World College in Mbabane, Swaziland, and Kenyon College in Gambier, Ohio. Brian and his wife Carol have dozens of fruit trees and bushes in their yard, including an apple tree that was propagated from a tree in Isaac Newton's garden, and they have traveled and camped in most of the United States.

Stuart Field has been interested in science and technology his whole life. While in school he built telescopes, electronic circuits, and computers. After attending Stanford University, he earned a Ph.D. at the University of Chicago, where he studied the properties of materials at ultralow temperatures. After completing a postdoctoral position at the Massachussetts Institute of Technology, he held a faculty position at the University of Michigan. Currently at Colorado State University, Stuart teaches a variety of physics courses, including algebra-based introductory physics, and was an early and enthusiastic adopter of Knight's *Physics for Scientists and Engineers.* Stuart maintains an active research program in the area of superconductivity. His hobbies include woodworking; enjoying Colorado's great outdoors; and ice hockey, where he plays goalie for a local team.

BUILT FROM THE GROUND UP
FOR MORE EFFECTIVE LEARNING

- **An NSF-funded educational research program**

- **Input from an unprecedented 4,500 student**

- **Detailed reviews by 250 instructors**

This is just some of the unusual history that shaped *Physics for Scientists and Engineers* by Randy Knight into the most widely adopted new physics text published in more than 30 years. In *College Physics: A Strategic Approach*, Randy Knight is joined by Brian Jones and Stuart Field to apply the best solutions from educational research to the algebra-based introductory physics course and the particular needs of these students.

Built from the ground up on a wealth of research into how students learn physics, and how they can be taught more effectively, *College Physics*:

- **builds** students' problem-solving abilities and confidence

- **integrates** relevant examples and interesting topics

- **promotes** deeper understanding by explicitly addressing students' preconceptions, misconceptions, and common stumbling-blocks

- **provides** a highly readable text and research-based visual pedagogy

The following pages provide an overview of how *College Physics* is designed to achieve these goals based on input from thousands of students and hundreds of instructors to date.

"Visual pedagogy is a critical feature of physics textbooks because so much of physics is expressed in graphics. I am extremely impressed with the care and research-based ideas that have gone into the unique instructional design of the graphics in Knight's textbooks. His physics textbooks are leaders in applying research-based principles of visual pedagogy to support student learning."

—Richard E. Mayer, *Professor of Psychology, University of California, Santa Barbara*

"No text I have seen has done more to lead students to thinking than Knight/Jones/Field."

—Dieter R. Brill, *University of Maryland*

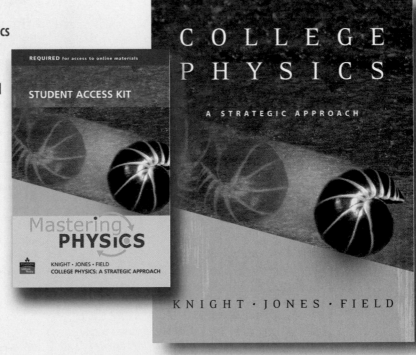

Clear, consistent instruction

Students build confidence and success through a consistent 3-step approach that encourages them to **Prepare** the problem, before trying to **Solve it** and **Assess** their answer.

Topic-specific *Problem-Solving Strategies* follow the same 3-step framework and provide more detailed guidance.

1 PREPARE reinforces the value of gathering information, drawing figures, making assumptions, and planning—key steps research shows students otherwise skip.

2 SOLVE carefully works through the mathematical steps of the solution, explaining algebraic manipulations and use of key information.

3 ASSESS reminds the student to verify whether their answer makes sense—numerically and in context.

(MP) PROBLEM-SOLVING
STRATEGY 24.1 **Magnetic–field problems**

PREPARE Because current-carrying wires do not lie in the same plane as the fields they produce, you'll need to prepare an especially careful drawing. Generally, you should choose the plane of your drawing so that the magnetic field vectors lie either in the plane of the paper or perpendicular to it.

- Str
  the
- Us
  fiel
- Sol
  of t
- Ma
  bes
  Th

If you
syste

SOLVE
noid)
each f
solend
additi

ASSESS
reason

(MP) PROBLEM-SOLVING
STRATEGY 10.1 **Conservation of energy problems**

PREPARE Choose what to include in your system (see Tactics Box 10.1). Draw a before-and-after visual overview, as outlined in Tactics Box 9.1. Note known quantities, and determine what quantity you're trying to find. If the system is isolated and if there is no friction, your solution will be based on Equation 10.7, otherwise you should use Equation 10.6.
  Identify which mechanical energies in the system are changing:

- If the *speed* of the object is changing, include K_i and K_f in your solution.
- If the *height* of the object is changing, include $(U_g)_i$ and $(U_g)_f$.
- If the *length* of a spring is changing, include $(U_s)_i$ and $(U_s)_f$.
- If kinetic friction is present, ΔE_{th} will be positive. Some kinetic or potential energy will be transformed into thermal energy.

If an external force acts on the system, you'll need to include the work W done by this force in Equation 10.6.

SOLVE Depending on the problem, you'll need to calculate initial and/or final values of these energies and insert them into Equation 10.6 or 10.7. Then you can solve for the unknown energies, and from these any unknown speeds (from K), positions (from U), or displacements or forces (from W).

ASSESS Check the signs of your energies. Kinetic energy, as we'll see, is always positive. In the systems we'll study in this chapter, thermal energy can only increase, so that its change is positive. In Chapters 11 and 12 we'll study systems for which the thermal energy can decrease.

(MP) TACTICS BOX 4.3 Drawing a free-body diagram 📝 Exercises 17–22

❶ **Identify all forces acting on the object.** This step was described in Tactics Box 4.2.
❷ **Draw a coordinate system.** Use the axes defined in your visual overview (Tactics Box 2.2). If those axes are tilted, for motion along an incline, then the axes of the free-body diagram should be similarly tilted.
❸ **Represent the object as a dot at the origin of the coordinate axes.** This is the particle model.
❹ **Draw v**
describe
❺ **Draw a**
diagram
check t
on your

TACTICS BOX 24.1 Right-hand rule for fields 📝 Exercises 7–12

❶ Point your *right* thumb in the direction of the current.

❷ Curl your fingers around the wire to indicate a circle.

❸ Your fingers point in the direction of the magnetic field lines around the wire.

Tactics Boxes provide step-by-step procedures that build key skills the student will use over and over—such as drawing free-body diagrams, and using ray tracing.

"The Prepare, Solve, and Assess steps are right on, not too many steps to confuse the students, but enough to thoroughly discuss the problem. The strategies are well thought-out and expertly executed."

—Toby Moleski, *Muskegon Community College*

Math Relationship Boxes ensure the student is confident with the key mathematical relationships most common in this course. Each relationship is consolidated in words, math, and graphics, along with tips on reasoning with limiting cases and scaling. Icons in the text refer back to these boxes to help students see connections in the math they use.

📐 Quadratic relationships 📝 Exercises 14, 19 (MP)

Two quantities are said to have a **quadratic relationship** if y is proportional to the square of x. We write the mathematical relationship as

$$y = Ax^2$$

y is a quadratic function of x

The graph of a quadratic relationship is a parabola.

Doubling x causes y to change by a factor of 4.

$y = Ax^2$

SCALING Suppose we double x:

$$x_f = 2x_i$$

This will change y as well:

$$y_f = Ax_f^2 = A(2x_i)^2 = 2^2Ax_i^2 = 2^2y_i$$
$$= 4y_i$$

Increasing x by a factor of 2 causes y to increase by a factor of 2^2, or 4.
  Generally, we can say that

Changing x by a factor C changes y by a factor C^2.

Explicit, guided practice

Worked Examples implement the Strategies and follow the same PREPARE/SOLVE/ASSESS framework as part of developing good problem-solving habits. They carefully walk the student through the underlying reasoning and pitfalls to avoid.

EXAMPLE 2.9 Calculating the minimum length of a runway

A fully-loaded 747 with all engines at full thrust accelerates at 2.6 m/s². Its minimum takeoff speed is 70 m/s. How much time will the plane take to reach its takeoff speed? What minimum length of runway does the plane require for takeoff?

PREPARE The visual overview of Figure 2.34 summarizes the important details of the problem. We set x_i and t equal to zero at the starting point of the motion, when the plane is at rest and the acceleration begins. The final point of the motion is when the plane achieves the necessary takeoff speed of 70m/s. The plane is accelerating to the right, so we will compute the time for the plane to reach a velocity of 70m/s and the position of the plane at this time, giving us the minimum length of the runway.

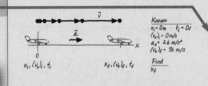

FIGURE 2.34 Visual overview for an accelerating plane.

SOLVE First we solve for the time required for the plane to reach takeoff speed. We can use the first equation in Table 2.4 to compute this time:

$$(v_x)_f = (v_x)_i + a_x \, \Delta t$$
$$70 \text{ m/s} = 0 \text{ m/s} + (2.6 \text{ m/s}^2) \, \Delta t$$
$$\Delta t = 26.9 \text{ s}$$

We keep an extra significant figure here as we will use this result in the next step of the calculation.

Given the time that the plane takes to reach takeoff speed, we can compute the position of the plane when it reaches this speed using the second equation in Table 2.4:

$$x_f = x_i + (v_x)_i \, \Delta t + \tfrac{1}{2} a_x (\Delta t)^2$$
$$= 0 \text{ m} + (0 \text{ m/s})(26.9 \text{ s}) + \tfrac{1}{2}(2.6 \text{ m/s}^2)(26.9 \text{ s})^2$$
$$= 940 \text{ m}$$

Our final answers are thus that the plane will take 27 s to reach takeoff speed, with a minimum runway length of 940 m.

ASSESS Think about the last time you flew; 27 s seems like a reasonable time for a plane to accelerate on takeoff. Actual runway lengths at major airports are 3000 m or more, a few times greater than the minimum length, because they have to allow for emergency stops during an aborted takeoff. (If we had calculated a distance greater than 3000 m, we would know we had done something wrong!)

EXAMPLE 4.3 Forces on an upward-accelerating elevator

An elevator, suspended by a cable, speeds up as it moves upward from the ground floor. Draw a free-body diagram of the elevator.

PREPARE Figure 4.26 illustrates the steps outlined in Tactics Box 4.3.

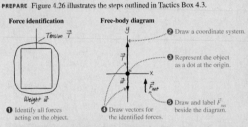

Force identification

- ❶ Identify all forces acting on the object.

Free-body diagram

- ❷ Draw a coordinate system.
- ❸ Represent the object as a dot at the origin.
- ❹ Draw vectors for the identified forces.
- ❺ Draw and label $\vec{F}_{net}$ beside the diagram.

FIGURE 4.26 Free-body diagram of an elevator accelerating upward.

Pencil Sketches provide students with an explicit and accessible example of what to draw in solving a problem—a key step research indicates students often skip—often following steps outlined in Tactics Boxes.

Conceptual Examples target students' qualitative reasoning skills. Since no math is involved, they follow a Reason and Assess approach.

CONCEPTUAL EXAMPLE 24.2 Canceling a magnetic field

A current loop carries a counter clockwise current of 3 A. A second current-carrying loop with twice the radius is placed outside the first loop, as shown in Figure 24.25. What current in the outer loop will exactly cancel the field in the center of the first loop?

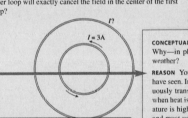

FIGURE 24.25 Two current loops.

REASON The inner loop produces a magnetic field at its center of magnitude B. By the right-hand rule for fields, the direction of this field is out of the page. To make the total field at the center zero, the outer loop must produce a field at the center that has the same magnitude B but pointing in the opposite direction, into the page. Equation 24.2 shows that the field at the center of a loop is *inversely proportional* to its radius. So if the outer loop also carried 3 A, its field at the center would be only

CONCEPTUAL EXAMPLE 11.4 Energy transfers and the body

Why—in physics terms—is it more taxing on the body to exercise in very hot weather?

REASON Your body continuously converts chemical energy to thermal energy, as we have seen. In order to maintain a constant body temperature, your body must continuously transfer heat to the environment. This is a simple matter in cool weather when heat is spontaneously transferred to the environment, but when the air temperature is higher than your body temperature, your body cannot cool itself this way and must use other mechanisms to transfer this energy, such as perspiring. These mechanisms require additional energy expenditure.

ASSESS Strenuous exercise in hot weather can easily lead to a rise in body temperature if the body cannot exhaust heat quickly enough.

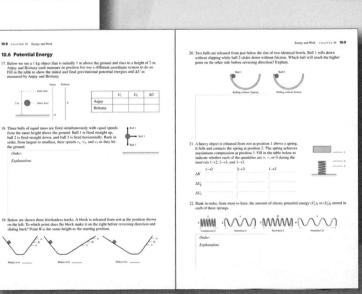

ENERGY WORKSHEET Name _____ Problem _____

PREPARE
- Identify the system
- Sketch "before and after"
- Define coordinates
- List knowns/unknowns

Known

Find

• Is the system isolated?

SOLVE
What law or concept is most relevant for...

ASSESS

The **Student Workbook** provides straightforward confidence- and skill-building exercises—bridging the gap between worked examples and end-of-chapter problems. Also included are worksheets—research-based templates that help students structure their approach to solving problems. Workbook activities are referenced throughout the text by ✐.

INTEGRATES INTERESTING AND RELEVANT TOPICS

An active, inductive approach

Drawing from students' majors and the world around them, relevant examples and interesting topics are carefully woven into the text. These provide students with motivation, a means to consolidate their understanding, and a clear context for how physics is valuable to them.

> "The writing is clear and engaging. The coverage is well thought-out. *College Physics* is better than any other text I have encountered for algebra-based physics."
> —Bruce A. Schumm, *University of California, Santa Cruz*

Worked Examples incorporates scenarios from everyday life and the world around us.

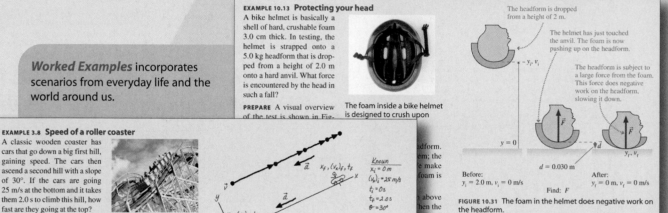

EXAMPLE 10.13 Protecting your head
A bike helmet is basically a shell of hard, crushable foam 3.0 cm thick. In testing, the helmet is strapped onto a 5.0 kg headform that is dropped from a height of 2.0 m onto a hard anvil. What force is encountered by the head in such a fall?

PREPARE A visual overview of the test is shown in Fig-

The foam inside a bike helmet is designed to crush upon

...adform. ...em; the ...e make ...foam is

...n above ...hen the

The headform is dropped from a height of 2 m.

The helmet has just touched the anvil. The foam is now pushing up on the headform.

The headform is subject to a large force from the foam. This force does negative work on the headform, slowing it down.

$y = 0$

$d = 0.030$ m

Before:
$y_i = 2.0$ m, $v_i = 0$ m/s

After:
$y_f = 0$ m, $v_f = 0$ m/s

Find: F

FIGURE 10.31 The foam in the helmet does negative work on the headform.

EXAMPLE 3.8 Speed of a roller coaster
A classic wooden coaster has cars that go down a big first hill, gaining speed. The cars then ascend a second hill with a slope of 30°. If the cars are going 25 m/s at the bottom and it takes them 2.0 s to climb this hill, how fast are they going at the top?

PREPARE We start with the visual overview in Figure 3.24, which includes a motion diagram, a pictorial representation, and a list of values. Notice how the motion diagram of Figure 3.24 differs from that of the previous example: The velocity decreases as the car moves up the hill, so the acceleration vector is opposite the direction of the velocity vector. The motion is along the *x*-axis, as before, but the acceleration vector points in the negative-*x* direction, so the component a_x is negative.

Known
$x_i = 0$ m
$(v_x)_i = 25$ m/s
$t_i = 0$ s
$t_f = 2.0$ s
$\theta = 30°$

Find
$(v_x)_f$

FIGURE 3.24 The coaster's speed decreases as it goes up the hill.

SOLVE To determine the final speed, we need to know the acceleration. We will assume that there is no friction or air

---

Less of a drag At the high speeds attained by racing cyclists, air drag can become very significant. The world record for the farthest distance traveled in one hour on an ordinary bicycle is 56.38 km, set by Chris Boardman in 1996. But a bicycle with an aerodynamic shell has a much lower drag force, allowing it to attain significantly higher speeds. The bike shown here was pedaled 84.22 km in one hour by Sam Whittingham in 2004, for an amazing average speed of 52.3 mph!

Free-standing Applications, provided in the margin with photographs and a self-contained caption, connect the physical principles with the real world.

TRY IT YOURSELF

Getting the ketchup out The ketchup stuck at the bottom of the bottle is initially at rest. If you hit the bottom of the bottle, as in (a), the bottle suddenly moves down, taking the ketchup on the bottom of the bottle with it, so that the ketchup just stays stuck to the bottom. But if instead you hit *up* on the bottle, as in (b), you force the bottle rapidly upward. By the first law, the ketchup that was stuck to the bottom stays at rest, so that it separates from the upward-moving bottle: the ketchup has moved forward with respect to the bottle!

TRY IT YOURSELF

Noisy magnets The striped pole structure of refrigerator magnets can be demonstrated by using two *identical* such magnets. Place them together, back to back, then pull them quickly across each other as shown by the arrows. The magnets alternately attract and repel as they bump past each other. If you pull them quickly enough, you will actually hear an audible tone as the rapidly passing magnets are alternately pulled together and pushed apart.

Try It Yourself Activities throughout the text provide students with simple real-world experiments designed to quickly reinforce a key idea through direct experience.

> "Extremely readable and instructive."
> —Tiffany Landry, *Folsom Lake College*

Engaging life-science students

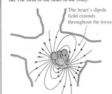

(a) The electric dipole of the heart.

A cross section of the heart showing muscle tissue.

Depolarized tissue

This line separates cells that have depolarized and those that have not.

Tissue not yet depolarized

The charge separation at the line between the two regions creates an electric dipole.

(b) The field of the heart in the body.

The heart's dipole field extends throughout the torso.

FIGURE 20.35 The beating heart generates a dipole electric field.

- The electric field is created by charges. Field lines start on a positive charge and end on a negative charge.

We can combine the above information into a technique for drawing electric field lines for an arrangement of charges. For example, Figure 20.34 pictures the electric field of a dipole using electric field lines. You should compare this to Figure 20.29b, which illustrates the field with field vectors.

The Electric Field of the Heart

Nerve and muscle cells have a prominent electrical nature. As we will see in detail in Chapter 23, a cell membrane is an insulator that encloses a conducting fluid and is surrounded by conducting fluid. While resting, the membrane is *polarized* with positive charges on the outside of the cell, negative charges on the inside. When a nerve or a muscle cell is stimulated, the polarity of the membrane switches; we say that the cell *depolarizes*. Later, when the charge balance is restored, we say that the cell *repolarizes*.

All nerve and muscle cells generate an electrical signal when depolarization occurs, but the largest electrical signal in the body comes from the heart. The rhythmic beating of the heart is produced by a highly coordinated wave of depolarization that sweeps across the tissue of the heart. As Figure 20.35a shows, the surface of the heart is positive on one side of the boundary between tissue that is depolarized and tissue that is not yet depolarized, negative on the other. In other words, the heart is a large electric dipole. The orientation and strength of the dipole change during each beat of the heart as the depolarization wave sweeps across it.

The electric dipole of the heart generates a dipole electric field that extends throughout the torso, as shown in Figure 20.35b. As we will see in Chapter 21, an electrocardiogram measures the changing electric field of the heart as it beats. Measurement of the heart's electric field can be used to diagnose the operation of the heart.

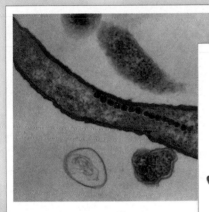

Magnetotactic Bacteria BIO Several organisms use the earth's magnetic field to navigate. The clearest example of this is *magnetotactic bacteria*. The dark areas in this image are small pieces of iron; each piece is a single domain and hence a very strong magnet. Such a bacterium possesses a very strong magnetic moment: It acts like a bar magnet, lining up with the earth's magnetic field! In the temperate regions where such bacteria live, the earth's field has a large vertical component. The bacteria use their alignment with this vertical field component to navigate up and down.

Calf muscle

Achilles tend[on]

On each stride, the t[endon] stretches, storing ab[out] of energy.

Spring in your step BIO As you run, lose some of your mechanical energy time your foot strikes the ground; this is transformed into unrecoverable ther[mal] energy. Luckily, about 35% of the decr[ease in] your mechanical energy when your foo[t] is stored as elastic potential energy in t[he] stretchable Achilles tendon of the lowe[r] On each plant of the foot the tendon is stretched, storing some energy. The ten[don] springs back as you push off the groun[d] again, helping to propel you forward. T[his] recovered energy reduces the amount [of] internal chemical energy you use, increasing your efficiency.

Dinner at a distance BIO A chameleon's tongue is a powerful tool for catching prey. Certain species can extend the tongue to a distance of over 1 ft in less than 0.1 s! A study of the kinematics of the motion of the chameleon tongue, using techniques like those in this chapter, reveals that the tongue has a period of rapid acceleration followed by a period of constant velocity. This knowledge is a very valuable clue in the analysis of the evolutionary relationships between chameleons and other animals.

Passage Problems

Kangaroo Locomotion BIO

Kangaroos have very stout tendons in their legs that can be used to store energy. When a kangaroo lands on its feet, the tendons stretch, transforming kinetic energy of motion to elastic potential energy. Much of this energy can be transformed back into kinetic energy as the kangaroo takes another hop. The kangaroo's peculiar hopping gait is not very efficient at low speeds but is quite efficient at high speeds.

Figure P11.61 shows the energy cost of human and kangaroo locomotion. The graph shows oxygen uptake (in mL/s) per kg of body mass, allowing a direct comparison between the two species.

FIGURE P11.61 Oxygen uptake (a measure of energy use per second) for a running human and a hopping kangaroo.

For humans, the energy used per second (i.e., power) is proportional to the speed. That is, the human curve nearly passes through the origin, so running twice as fast takes approximately twice as much power. For a hopping kangaroo, the graph of energy use has only a very small slope. In other words, the energy used per second changes very little with speed. Going faster requires very little additional power. Treadmill tests on kangaroos and observations in the wild have shown that they do not become winded at any speed at which they are able to hop. No matter how fast they hop, the necessary power is approximately the same.

61. | A person runs 1 km. How does his speed affect the total energy needed to cover this distance?
   A. A faster speed requires less total energy.
   B. A faster speed requires more total energy.
   C. The total energy is about the same for a fast speed and a slow speed.

62. | A kangaroo hops 1 km. How does its speed affect the total energy needed to cover this distance?
   A. A faster speed requires less total energy.
   B. A faster speed requires more total energy.
   C. The total energy is about the same for a fast speed and a slow speed.

63. | At a speed of 4 m/s,
   A. A running human is more efficient than an equal-mass hopping kangaroo.
   B. A running human is less efficient than an equal-mass hopping kangaroo.
   C. A running human and an equal-mass hopping kangaroo have about the same efficiency.

64. | At approximately what speed would a human use half the power of an equal-mass kangaroo moving at the same speed?
   A. 3 m/s   B. 4 m/s   C. 5 m/s   D. 6 m/s

65. | At approximately what speed would a human use twice the power of a kangaroo of half the mass moving at the same speed?
   A. 3 m/s   B. 5 m/s   C. 7 m/s   D. 9 m/s

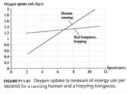

PROMOTES DEEPER UNDERSTANDING

A structured learning path

This text incorporates many subtle but powerful techniques from cognitive science that improve student learning and retention, including a self-evident and structured learning path and a unique visual pedagogy.

"Clearly written by experienced, knowledgeable teachers who have spent time in the trenches... *College Physics* sets a new standard."

—Marllin L. Simon, *Auburn University*

"The text is exceptionally easy to read... and takes special care to explain the fundamentals and debunk common misconceptions..."

—Michael Pravica, *University of Nevada–Las Vegas*

Looking Ahead ▶▶

The goals of Chapter 3 are to learn more about vectors and to use vectors as a tool to analyze motion in two dimensions. In this chapter, you will learn to:

▶ Work with vectors, coordinate systems, and components.
▶ Find acceleration vectors on motion diagrams.
▶ Use vectors to understand motion on a ramp and relative velocity.
▶ Solve problems involving projectiles following parabolic paths.
▶ Understand the basic ideas of motion in a circle.

Looking Back ◀◀

This chapter will build naturally on the material from the previous two chapters. Please review:

◀ Sections 1.1–1.2 Basic motion concepts.
◀ Section 1.5 Vectors.
◀ Sections 2.4–2. Acceleration and free fall.

---

3 VECTORS AND MOTION IN TWO DIMENSIONS

With astonishing precision, an archer fish secures a meal by spitting drops of water at an unsuspecting insect. What factors determine how high and how far the water drops travel?

Looking Ahead ▶▶

The goals of Chapter 3 are to learn more about vectors and to use vectors as a tool to analyze motion in two dimensions. In this chapter, you will learn to:

▪ Work with vectors, coordinate systems, and components.
▪ Find acceleration vectors on motion diagrams.
▪ Use vectors to understand motion on a ramp and relative velocity.
▪ Solve problems involving projectiles following parabolic paths.
▪ Understand the basic ideas of motion in a circle.

Looking Back ◀◀

This chapter will build naturally on the material from the previous two chapters. Please review:

◀ Sections 1.1–1.2 Basic motion concepts.
◀ Section 1.5 Vectors.
◀ Sections 2.4–2. Acceleration and free fall.

O nce the drops of water leave the archer fish's mouth and fly through the air, their motion follows a gently curving path. This path, a parabola, is the same type of path followed by a basketball shot toward the basket, a high jumper leaping over a bar, or a skier going off a jump. These are all examples of motion in two dimensions, as the objects are moving horizontally and vertically at the same time.

In order to extend our study of motion to two dimensions, we will need to develop techniques for working with vectors. We will first use vector techniques to analyze motion on a ramp and relative velocity, continue on to look at the parabolic motion of projectiles, then close the chapter by describing the motion of objects moving in circles. Along the way we'll use numerous examples drawn from sports—from skiers gliding down slopes to balls flying through the air.

70

Each chapter begins with a motivational roadmap—*Looking Ahead* provides learning goals to keep the student focused on the key ideas, and *Looking Back* consolidates connections with earlier material.

Stop to Think Questions at the end of a section allow the student to quickly check their understanding. Using powerful ranking-task and graphical techniques, they are designed to efficiently probe key misconceptions and encourage active reading. (Answers are provided at the end of the chapter.)

STOP TO THINK 10.4 Rank in order, from largest to smallest, the gravitational potential energies of identical balls 1 to 4.

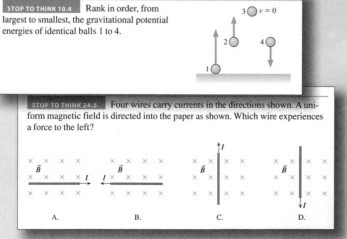

STOP TO THINK 24.5 Four wires carry currents in the directions shown. A uniform magnetic field is directed into the paper as shown. Which wire experiences a force to the left?

A.   B.   C.   D.

---

20   CHAPTER 24 · Magnetic Fields and Forces

will not speed up or slow down. Suppose a positively charged particle is moving perpendicular to a uniform magnetic field $\vec{B}$ as shown in Figure 24.32. In Chapter 7, we looked at the motion of objects subject to a force that was always perpendicular to the velocity. The net result was *circular motion at constant speed*. For a mass moving in a circle at the end of a string, the tension force is always perpendicular to $\vec{v}$. For a satellite moving in a circular orbit, the gravitational force is always perpendicular to $\vec{v}$. Now, for a charged particle moving in a magnetic field, it is the magnetic force that is always perpendicular to $\vec{v}$ and so causes the particle to move in a circle. Thus *a particle moving perpendicular to a uniform magnetic field undergoes uniform circular motion at constant speed.*

NOTE ▶ Since the right-hand rule for forces specifies the orientation of the force on a positive charge, the force on a negative charge will be in the opposite direction. A negatively charged particle will orbit in the opposite sense from that shown in Figure 24.32 for a positive charge. ◀

FIGURE 24.32 A charged particle moving perpendicular to a uniform magnetic field.

Figure 24.33 shows a particle of mass m moving at a speed v in a circle of radius r. We found in Chapter 7 that this motion requires a force directed toward the center of the circle with magnitude

$$F = \frac{mv^2}{r}$$ (24.7)

NOTE ▶ Since the right-hand rule for forces specifies the orientation of the force on a positive charge, the force on a negative charge will be in the opposite direction. A negatively charged particle will orbit in the opposite sense from that shown in Figure 24.32 for a positive charge. ◀

FIGURE 24.34 A charged particle in a magnetic field follows a helical trajectory.

$$v_\perp = \frac{qBr}{m}$$ (24.9)

The radius depends on the ratio of the mass of the particle to its charge, a fact we will use later. The radius also depends on the speed and the magnetic field. Increasing the speed will increase the radius of the circular motion, while increasing the field will decrease the radius.

So a particle moving *perpendicular* to a magnetic field will move in a circle. In the table at the start of this section, we saw that a particle moving *parallel* to a magnetic field experiences no magnetic force, and so will continue in a straight line. A more general situation in which a charged particle's velocity $\vec{v}$ is neither parallel to nor perpendicular to the field $\vec{B}$ is shown in Figure 24.34. The net result is a circular motion due to the perpendicular component of the velocity coupled with a constant velocity parallel to the field. The charged particle spirals around the magnetic field lines in a helical trajectory.

High-energy particles and radiation streaming out from the sun at high wind create ions and electrons as they strike molecules high in the earth's atmosphere. Some of these charged particles become trapped in the earth's magnetic field, as Figure 24.35 shows, the electrons spiral in helical trajectories along the earth's magnetic field lines. Some of these particles enter the atmosphere near the north and south poles, ionizing gas and creating the ghostly glow of the aurora.

NOTE Paragraphs throughout guide students away from known preconceptions, around common sticking points, and highlight many math-related issues that cause difficulties.

Proven visual pedagogy

Figures are carefully streamlined in detail and color so students focus on the physics—for instance, the object of interest in mechanics—using an instructional approach well documented to improve learning.

The proven technique of annotation—adding chalkboard-like dialog onto the figure (rather than providing as a lengthy caption beneath)—is used to more effectively guide students in:

- interpreting a graph or figure
- understanding a process
- translating between text, math, graphs, and figures
- grasping a difficult concept through a visual analogy

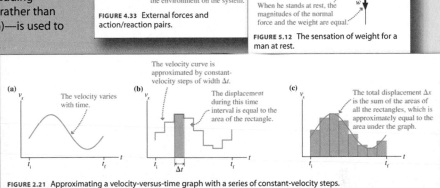

FIGURE 4.33 External forces and action/reaction pairs.

FIGURE 5.12 The sensation of weight for a man at rest.

FIGURE 2.21 Approximating a velocity-versus-time graph with a series of constant-velocity steps.

"The summaries are very effective in helping students to identify the key ideas and uniting principles of the chapter."
—Kristi D. Concannon, *King's College*

Critically acclaimed **Visual Chapter Summaries** consolidate understanding by providing each concept in words, math, and figures and organizing these into a coherent hierarchy—from General Principles to Applications.

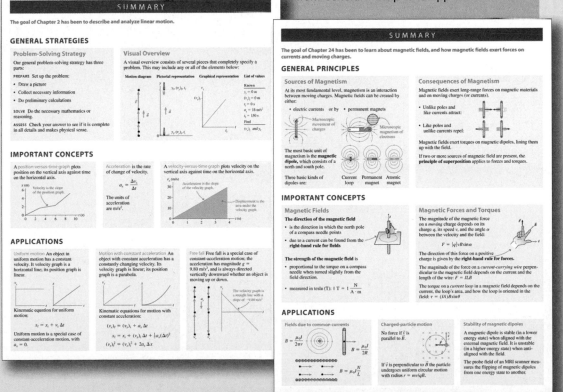

THE BEST-TESTED PROBLEMS

Enhanced by analysis

End-of-chapter problems of superior quality and diversity have been carefully crafted to exercise and test the full range of qualitative and quantitative problem-solving skills. The wealth of 3,000 problems in *College Physics: A Strategic Approach* are the first problems to be pre-tested for reliability, difficulty, and duration on thousands of students nationally before publication in a new textbook. *(See fold-out grid at front of instructor copy for more details).*

Simpler subproblems are provided upon request when a student gets stuck.

Conceptual Questions

Multiple-Choice Questions

Conceptual Questions require careful reasoning and can be used for either group discussions or individual work. The **Multiple-Choice Questions** use carefully chosen distractors to elicit common misconceptions and misunderstandings.

Problems

Problems, keyed to sections, allow students to build up their skills, often drawing on real-world or life-science applications to provide motivational examples.

Students select only the specific help they need. Thousands of solution paths are allowed.

More advanced **General Problems** include many context-rich problems which require students to simplify and model more complex real-world situations. Those labeled INT integrate significant material from earlier chapters.

General Problems

*See the **MasteringPhysics™** brochure for more details on how it enables instructors to more quickly build assignments of just the right coverage, difficulty, and duration, and the time-saving and probing diagnostics it provides (including most-common wrong answers by a class at each step, the step-by-step work of each student, even national average comparison data).*

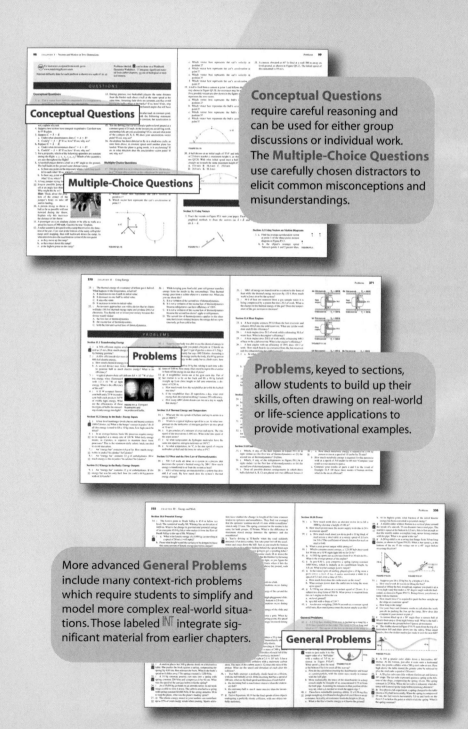

INTEGRATED WITH MASTERINGPHYSICS™

Unparalleled teaching diagnostics and individualized tutoring

MasteringPhysics™ is the most widely used and educationally proven physics homework and tutorial system in the world. NSF-sponsored research and independent studies at a wide range of schools show that students who use MasteringPhysics™ significantly improve their scores on final exams and conceptual tests such as the Force Concept Inventory. Drawing from the world's largest metadatabase of the step-by-step work of how real students solve problems, accumulated over eight years of research, MasteringPhysics™ offers unparalleled individualized guidance.

Instantaneous feedback is provided specific to the most common wrong answers.

MasteringPhysics™ provides Instructors with more than 3,000 problems from which to build wide-ranging weekly assignments that include:

- end-of-chapter textbook problems
- tutorials for practicing all the Problem-Solving Strategies, Tactics Boxes, and Math Relationship Boxes in the textbook
- skill-building tutorials for mastering key concepts, and self-tutoring problems for guided practice
- math-specific skill-building tutorials, ranking task problems, even essay questions

Numerical problems provide randomized numbers and sig-fig feedback, and tutorial problems provide wrong-answer specific feedback and simpler subproblems upon request.

A **vector-drawing tool** provides grading and feedback on student-drawn free-body diagrams embedded in the tutorials.

Partial credit is provided for the method, not just the final numerical answer.

A **graph-drawing tool** provides similar grading and feedback in tutorials that develop students' graph-reading skills.

MasteringPHYSICS

www.masteringphysics.com

Real-World Applications

Applications of biological or medical interest are marked BIO in the list below. MCAT-style Passage Problems are marked BIO below. Other end-of-chapter problems of biological or medical interest are marked BIO in the chapter. "Try It Yourself" experiments are marked **TIY**.

Preface to the Instructor

College Physics: A Strategic Approach is a new algebra-based physics textbook for students majoring in the biological and life sciences, architecture, natural resources, and other disciplines. Students entering this course typically have strong academic abilities and are highly motivated to do well. However, their success and enjoyment of physics is often hampered by a number of obstacles. First, these students do not see the relevance of physics to their major; they tend to view this course as a hurdle to overcome rather than an opportunity to broaden their knowledge of science. Second, they have often learned a rather linear approach to problem solving in other courses and this differs substantially from the more open and wide-ranging approach required in physics. Finally, even if they have taken previous physics courses, students tend to arrive with a wealth of strongly held preconceptions, inconsistent with the way the physical world works.

Objectives

Our primary goal in writing *College Physics: A Strategic Approach* has been to tackle the learning obstacles listed above head on and to thereby provide instructors and students with a more effective teaching and learning resource. Our axioms have been:

- To provide students with a textbook that's a more manageable size, less encyclopedic in its coverage, and better designed for learning.
- To integrate proven techniques from physics education research into the classroom in a way that accommodates a range of teaching and learning styles.
- To help students develop both quantitative reasoning skills and solid conceptual understanding, with special focus on concepts well documented to cause learning difficulties.
- To help students develop problem-solving skills and confidence in a systematic manner using explicit and consistent tactics and strategies.
- To motivate students by integrating real-world examples relevant to their majors—especially from biology, sports, medicine, the animal world—and that build upon their everyday experiences.
- To utilize proven techniques of visual instruction and design from educational research and cognitive psychology that improve student learning and retention and address a range of learner styles.
- To support an active-learning environment.

A more complete explanation of these goals and the rationale behind them can be found in Randy Knight's paperback book, *Five Easy Lessons: Strategies for Successful Physics Teaching* (Addison Wesley, 2002). Please request a copy from your local Addison Wesley sales representative if it would be of interest to you (ISBN 0-8053-8702-1).

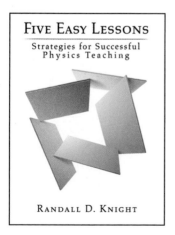

Textbook Organization

College Physics: A Strategic Approach is a 30-chapter text intended for use in a two-semester course. The textbook is divided into seven parts: Part I: *Force and Motion,* Part II: *Conservation Laws,* Part III: *Properties of Matter,* Part IV:

- **Complete edition,** with MasteringPhysics™ (ISBN 0-8053-9062-6): chapters 1–30.
- **Volume 1** with MasteringPhysics™ (ISBN 0-8053-1695-7): chapters 1–16.
- **Volume 2** with MasteringPhysics™ (ISBN 0-8053-1696-5): chapters 17–30.
- **Complete edition,** without MasteringPhysics™ (ISBN 0-8053-6582-6): chapters 1–30.
- **Volume 1** without MasteringPhysics™ (ISBN 0-8053-4662-7): chapters 1–16.
- **Volume 2** without MasteringPhysics™ (ISBN 0-8053-7209-1): chapters 17–30.

Oscillations and Waves, Part V: *Optics,* Part VI: *Electricity and Magnetism,* and Part VII: *Modern Physics.*

Part I covers Newton's laws and their applications. We cover the two fundamental conserved quantities, momentum and energy, in Part II for two reasons. First, the way that problems are solved using conservation laws—comparing an *after* situation to a *before* situation—differs fundamentally from the problem-solving strategies used in Newtonian dynamics. Second, the concept of energy has a significance far beyond mechanical (kinetic and potential) energies. In particular, the key idea in thermodynamics is energy, and moving from the study of energy in Part II into thermal physics in Part III allows the uninterrupted development of this important idea.

Optics (Part V) is covered directly after oscillations and waves (Part IV), and *before* electricity and magnetism (Part VI). Further, we treat wave optics before ray optics. Our motivations for this organization are twofold. First, wave optics is largely an extension of the general ideas of waves; in a more traditional organization, students have forgotten much of what they learned about waves by the time they get to wave optics. Second, optics as it is presented in introductory physics makes no use of the properties of electromagnetic fields. The documented difficulties that students have with optics are difficulties with waves, not difficulties with electricity and magnetism. There's little reason other than historical tradition to delay optics. However, the optics chapters can be deferred easily until after Part VI for instructors who prefer that ordering of topics.

The Student Workbook

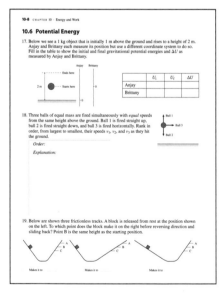

A key component of *College Physics: A Strategic Approach* is the accompanying *Student Workbook.* The workbook bridges the gap between textbook and homework problems by providing students the opportunity to learn and practice skills prior to using those skills in quantitative end-of-chapter problems, much as a musician practices technique separately from performance pieces. The workbook exercises, which are keyed to each section of the textbook, focus on developing specific skills, ranging from identifying forces and drawing free-body diagrams to interpreting field diagrams.

The workbook exercises, which are generally qualitative and/or graphical, draw heavily upon the physics education research literature. The exercises deal with issues known to cause student difficulties and employ techniques that have proven to be effective at overcoming those difficulties. The workbook exercises can be used in-class as part of an active-learning teaching strategy, in recitation sections, or as assigned homework. More information about effective use of the *Student Workbook* can be found in the *Instructor's Guide.*

Available versions: Volume 1 (ISBN 0-8053-8209-7): chapters 1–16, and Volume 2 (ISBN 0-8053-0626-9): chapters 17–30.

Instructor Supplements

- The **Instructor Guide for** *College Physics: A Strategic Approach* (ISBN 0-8053-0622-6), a comprehensive and highly acclaimed resource written by the textbook authors, provides chapter-by-chapter creative ideas and teaching tips for using *College Physics: A Strategic Approach* in your class. In addition, it contains an extensive review of what has been learned from physics education research, and provides guidelines for using active-learning techniques in your classroom.

- The **Instructor Solutions Manuals, Chapters 1–16** (ISBN 0-8053-0495-9), and **Chapters 17–30** (ISBN 0-8053-0638-2), written by Professor Larry Smith, Snow College; Professor Pawan Kahol, Missouri State University; Professor Marllin Simon, Auburn University; and Walter Polkosnik, University of California-Santa Barbara, provide *complete* solutions to all the end-of-chapter questions and problems. All solutions follow the Prepare/Solve/Assess problem-solving strategy used in the textbook for quantitative problems, and Reason/Assess strategy for qualitative ones. They are available as editable Word documents and as pdf files on the *Media Manager CD-ROMs* for your own use or for posting on your password-protected course website.

- The cross-platform **Media Manager CD-ROMs** (ISBN 0-8053-0306-5) provide invaluable and easy-to-use resources for your class, organized by textbook chapter. The contents include a comprehensive library of more than 220 applets from **ActivPhysics OnLine**™, as well as all figures, tables, summaries, and key equations from the textbook in JPEG format. In addition, all the Problem-Solving Strategies, Math Relationship Boxes, Tactics Boxes, and Key Equations are provided in editable Word format. The **Instructor Guide** and **Instructor Solutions Manuals** are also included as pdfs and editable Word files, along with pdfs of answers to the **Student Workbook** exercises, and **Lecture Outlines (with Classroom Response System "Clicker" Questions)** in PowerPoint® written by the textbook authors.

- The online **Instructor Resource Center** (www.aw-bc.com/irc) provides updates to files on the Media Manager CD-ROMs, and also includes the lecture outlines in Keynote®. To obtain a Login Name and Password, contact your Pearson Addison-Wesley sales representative.

- **MasteringPhysics**™ (www.masteringphysics.com), is the most advanced, educationally effective, and widely used physics homework and tutorial system in the world. It is designed to assign, assess, and track each student's progress using a wide diversity of extensively pre-tested problems. Icons throughout the book indicate that *MasteringPhysics*™ offers specific tutorials for all the textbook's Tactics Boxes, Problem-Solving Strategies, and Math Relationship Boxes, as well as all the End-of-Chapter problems, Test Bank items, and Reading Quizzes. *MasteringPhysics*™ provides instructors with a fast and effective way to assign uncompromising, wide-ranging online homework assignments of just the right difficulty and duration. The powerful post-assignment diagnostics allow instructors to assess the progress of their class as a whole or to quickly identify individual students' areas of difficulty.

- **ActivPhysics OnLine**™ (www.aw-bc.com/knightjonesfield) provides a comprehensive library of more than 420 tried and tested *ActivPhysics* applets. In addition, it provides a suite of highly regarded applet-based tutorials developed by education pioneers Professors Alan Van Heuvelen and Paul D'Alessandris. The *ActivPhysics* icons that appear throughout the book direct students to specific interactive exercises that complement the textbook discussion.

  The online exercises are designed to encourage students to confront misconceptions, reason qualitatively about physical processes, experiment quantitatively, and learn to think critically. They cover all topics from mechanics to electricity and magnetism and from optics to modern physics. The highly acclaimed *ActivPhysics OnLine* companion workbooks (see Student Supplements) help students work through complex concepts and understand them more clearly. More than 220 applets from the *ActivPhysics OnLine* library are also available on the Instructor *Media Manager CD-ROMs*.

- The **Printed Test Bank** (ISBN 0-8053-1697-3) and cross-platform **Computerized Test Bank** (ISBN 0-8053-0618-8), prepared by Sen-Ben Liao, University of California, provides more than 2,000 high-quality problems, with a range of multiple-choice, true/false, short-answer, and regular homework-type questions. The computerized version includes all of the questions from the Printed Test Bank, more than half of which can be assigned with randomized values.
- The **Transparency Acetates** (ISBN 0-8053-1689-2) provide more than 200 key figures from *College Physics: A Strategic Approach* for classroom presentation.

Student Supplements

- The **Student Solutions Manuals, Chapters 1–16** (ISBN 0-8053-0632-3) and **Chapters 17–30** (ISBN 0-8053-0631-5), written by Professor Larry Smith, Snow College; Professor Pawan Kahol, Missouri State University; Marllin Simon, Auburn University; and Walter Polkosnik, University of California-Santa Barbara, provide *detailed* solutions to more than half of the odd-numbered end-of-chapter problems. Following the problem-solving strategy presented in the text, thorough solutions are provided to carefully coach students through both the qualitative (Reason/Assess) and quantitative (Prepare/Solve/Assess) steps in the problem-solving process.

- **MasteringPhysics**™ (www.masteringphysics.com) is the most advanced, widely used, and educationally proven physics tutorial system in the world. It is the result of eight years of detailed studies of how real students work physics problems and of precisely where they need help. Studies show that students who use MasteringPhysics™ significantly improve their scores on final exams and conceptual tests such as the Force Concept Inventory. MasteringPhysics™ achieves this improvement by providing students with instantaneous feedback specific to their wrong answers, simpler subproblems upon request when they get stuck, and partial credit for their method(s) used. This individualized, 24/7 tutor system is recommedned by 9 out of 10 students to their peers as the most effective and time-efficient way to study.

- **ActivPhysics**™ **OnLine** (www.aw-bc.com/knightjonesfield) provides students with a suite of highly regarded applet-based tutorials (see above). The accompanying workbooks help students work though complex concepts and understand them more clearly. The *ActivPhysics* icons throughout the book direct students to specific interactive exercises that complement the textbook discussion. The following workbooks provide a range of tutorial problems designed to use the ActivPhysics OnLine simulations:
- **ActivPhysics OnLine Workbook Volume 1: Mechanics • Thermal Physics • Oscillations & Waves** (ISBN 0-8053-9060-X)
- **ActivPhysics OnLine Workbook Volume 2: Electricity & Magnetism • Optics • Modern Physics** (ISBN 0-8053-9061-8)
- The **Addison-Wesley Tutor Center** (www.aw.com/tutorcenter) provides one-on-one tutoring via telephone, fax, email, or interactive website during evening hours and on weekends. Qualified college instructors answer questions and provide instruction for *Mastering Physics*™ and for the examples, exercises, and problems in *College Physics: A Strategic Approach*.

Acknowledgments

We have relied upon conversations with and, especially, the written publications of many members of the physics education community. Those who may recognize their influence include Arnold Arons, Uri Ganiel, Fred Goldberg, Ibrahim Halloun, Richard Hake, David Hestenes, Leonard Jossem, Jill Larkin, Priscilla

Laws, John Mallinckrodt, Lillian McDermott, Edward "Joe" Redish, Fred Reif, John Rigden, Rachel Scherr, Bruce Sherwood, David Sokoloff, Ronald Thornton, Sheila Tobias, and Alan Van Heuleven.

We are grateful to Marty Gelfand for taking on the significant task of editing and organizing the end-of-chapter problem sets; Larry Smith, Pawan Kahol, Marllin Simon, and Walt Polkosnik for the difficult task of writing the *Instructor Solutions Manuals;* to Jim Andrews for coauthoring the Student Workbook (and literally writing out all its answers); to Wayne Anderson, Jim Andrews, Nancy Beverly, David Cole, Karim Diff, Jim Dove, Marty Gelfand, Kathy Harper, Charlie Hibbard, Robert Lutz, Matt Moelter, Kandiah Manivannan, Ken Robinson, and Cindy Schwarz-Rachmilowitz for their contributions to the end-of-chapter questions and problems; and to Steven Vogel for his careful review of the biological content of many chapters and for helpful suggestions.

We especially want to thank our editor Adam Black, development editor Alice Houston, project editor Martha Steele, production supervisor Shannon Tozier, and all the other staff at Addison Wesley for their enthusiasm and hard work on this project. Jared Sterzer and the team at Westwords, Inc., copy editor Kevin Gleason, and photo researcher Brian Donnelly get much credit for making this complex project all come together. In addition to the reviewers and classroom testers listed below, who gave invaluable feedback, we are particularly grateful to Charlie Hibbard for his close scrutiny of every word, symbol, number, and figure.

Randy Knight: I would like to thank my Cal Poly colleagues, especially Matt Moelter, for many valuable conversations and suggestions. I am endlessly grateful to my wife Sally for her love, encouragement, and patience, and to our many cats for nothing in particular other than being cats.

Brian Jones: I would like to thank my fellow AAPT and PIRA members for their insight and ideas, the creative and wonderful students who are my partners in the Little Shop of Physics, the students in my College Physics classes who help me become a better teacher, and my wife Carol, my best friend and gentlest editor, whose love makes the journey worthwhile.

Stuart Field: I would like to thank my friends and colleagues for their support and insights, both professional and personal, that helped me with the writing of this book. I would especially like to thank my wife Julie for her love, support, and encouragement; and my children Sam and Ellen, who grew up before me (or behind me) as I sat in front of my computer screen.

Reviewers and Classroom Testers

Susmita Acharya, *Cardinal Stritch University*
Ugur Akgun, *University of Iowa*
Ralph Alexander, *University of Missouri-Rolla*
Donald Anderson, *Ivy Tech*
Steve Anderson, *Montana Tech*
James Andrews, *Youngstown State University*
Charles Ardary, *Edmond Community College*
Charles Bacon, *Ferris State University*
John Barry, *Houston Community College*
David H. Berman, *University of Northern Iowa*
Richard Bone, *Florida International University*
Jeff Bodart, *Chipola College*
James Borgardt, *Juniata College*
Daniela Bortoletto, *Purdue University*

Don Bowen, *Stephen F. Austin State University*
Asa Bradley, *Spokane Falls Community College*
Elena Brewer, *SUNY at Buffalo*
Dieter Brill, *University of Maryland*
Hauke Busch, *Augusta State University*
Kapila Castoldi, *Oakland University*
Michael Cherney, *Creighton University*
Lee Chow, *University of Central Florida*
Song Chung, *William Paterson University*
Alice Churukian, *Concordia College*
Kristi Concannon, *Kings College*
Teman Cooke, Georgia *Perimeter College at Lawrenceville*
Jesse Cude, *Hartnell College*
Melissa H. Dancy, *University of North Carolina at Charlotte*

Loretta Dauwe, *University of Michigan-Flint*
Mark Davenport, *San Antonio College*
Carlos Delgado, *Community College of Southern Nevada*
David Donovan, *Northern Michigan University*
Archana Dubey, *University of Central Florida*
Andrew Duffy, *Boston University*
Taner Edis, *Truman State University*
Ralph Edwards, *Lurleen B. Wallace Community College*
Steve Ellis, *University of Kentucky*
Paula Engelhardt, *Tennessee Technical University*
Davene Eryes, *North Seattle Community College*
Gerard Fasel, *Pepperdine University*
Luciano Fleischfresser, *OSSM Autry Tech*
Cynthia Galovich, *University of Northern Colorado*
Bertram Gamory, *Monroe Community College*
Delena Gatch, *Georgia Southern University*
Martin Gelfand, *Colorado State University*
Terry Golding, *University of North Texas*
Robert Gramer, *Lake City Community College*
William Gregg, *Louisiana State University*
Paul Gresser, *University of Maryland*
Robert Hagood, *Washtenaw Community College*
Heath Hatch, *University of Massachusetts*
Carl Hayn, *Santa Clara University*
James Heath, *Austin Community College*
Greg Hood, *Tidewater Community College*
Sebastian Hui, *Florence-Darlington Technical College*
Joey Huston, *Michigan State University*
David Iadevaia, *Pima Community College-East Campus*
Erik Jensen, *Chemeketa Community College*
Todd Kalisik, *Northern Illinois University*
Ju H. Kim, *University of North Dakota*
Armen Kocharian, *California State University Northridge*
J. M. Kowalski, *University of North Texas*
Laird Kramer, *Florida International University*
Chrisopher Kulp, *Eastern Kentucky University*
Richard Kurtz, *Louisiana State University*
Kenneth Lande, *University of Pennsylvania*
Tiffany Landry, *Folsom Lake College*
Todd Leif, *Cloud County Community College*
John Levin, *University of Tennessee-Knoxville*
John Lindberg, *Seattle Pacific University*
Rafael López-Mobilia, *The University of Texas at San Antonio*
Robert W. Lutz, *Drake University*
Lloyd Makorowitz, *SUNY Farmingdale*
Eric Martell, *Millikin University*
Mark Masters, *Indiana University-Purdue*
Denise Meeks, *Pima Community College*
Henry Merrill, *Fox Valley Technical College*

Mike Meyer, *Michigan Technological University*
Tobias Moleski, *Nashville State Tech*
April Moore, *North Harris College*
Gary Morris, *Rice University*
Charley Myles, *Texas Tech University*
Fred Olness, *Southern Methodist University*
Charles Oliver Overstreet, *San Antonio College*
Paige Ouzts, *Lander University*
Russell Palma, *Minnesota State University-Mankato*
Richard Panek, *Florida Gulf Coast University*
Joshua Phiri, *Florence-Darling Technical College*
Iulia Podariu, *University of Nebraska at Omaha*
David Potter, *Austin Community College*
Promod Pratap, *University of North Carolina-Greensboro*
Michael Pravica, *University of Nevada, Las Vegas*
Earl Prohofsky, *Purdue University*
Marilyn Rands, *Lawrence Technological University*
Andrew Rex, *University of Puget Sound*
Andrew Richter, *Valparaiso University*
Phyliss Salmons, *Embry-Riddle Aeronautical University*
Michael Schaab, *Maine Maritime Academy*
Bruce Schumm, *University of California, Santa Cruz*
Mizuho Schwalm, *University of Minnesota Crookston*
Cindy Schwarz, *Vassar College*
Natalia Semushkhina, *Shippensburg University*
Khazhgery (Jerry) Shakov, *Tulane University*
Bart Sheinberg, *Houston Community College*
Marllin Simon, *Auburn University*
Kenneth Smith, *Pennsylvania State University*
Michael Smutko, *Northwestern University*
Jon Son, *Boston University*
Noel Stanton, *Kansas State University*
Donna Stokes, *University of Houston*
Chuck Stone, *North Carolina A&T*
Chun Fu Su, *Mississippi State University*
William Tireman, *Northern Michigan University*
Negussie Tirfessa, *Manchester Community College*
Rajive Tiwari, *Belmont Abbey College*
Herman Trivilino, *College of the Mainland*
Douglas Tussey, *Pennsylvania State University*
James Vesenka, *University of New England*
Christos Valiotis, *Antelope Valley College*
Stamatis Vokos, *Seattle Pacific University*
James Wanliss, *Embry-Riddle Aeronautical University*
Henry Weigel, *Arapahoe Community College*
Courtney Willis, *University of Northern Colorado*
Katherine Wu, *University of Tampa*
Ali Yazdi, *Jefferson State Community College*
David Young, *Louisiana State University*
Hsiao-Ling Zhou, *Georgia State University*
Ulrich Zurcher, *Cleveland State University*

Preface to the Student

The most incomprehensible thing about the universe is that it is comprehensible.
—Albert Einstein

If you are taking a course for which this book is assigned, you probably aren't a physics major or an engineering major. It's likely that you aren't majoring in a physical science. So why are you taking physics?

It's almost certain that you are taking physics because you are majoring in a discipline that requires it. Someone, somewhere, has decided that it's important for you to take this course. And they are right. There is a lot you can learn from physics, even if you don't plan to be a physicist. We regularly hear from doctors, physical therapists, biologists and others that physics was one of the most interesting and valuable courses they took in college.

So, what can you expect to learn in this course? Let's start by talking about what physics is. Physics is a way of thinking about the physical aspects of nature. Physics is not about "facts." It's far more focused on discovering *relationships* between facts and the *patterns* that exist in nature than on learning facts for their own sake. Our emphasis will be on thinking and reasoning. We are going to look for patterns and relationships in nature, develop the logic that relates different ideas, and search for the reasons *why* things happen as they do.

Once we've figured out a pattern, a set of relationships, we'll look at applications to see where this understanding takes us. Let's look at an example. Part (a) of the figure shows an early mechanical clock. The clock uses a *pendulum,* a mass suspended by a thin rod free to pivot about its end, as its timekeeping element. When you study oscillatory motion, you will learn about the motion of a pendulum. You'll learn that the *period* of its motion, the time for one swing, doesn't depend on the amplitude, the size of the swing. This makes a pendulum the ideal centerpiece of a clock.

But there are other systems that look like pendulums too. The gibbon in part (b) of the figure is moving through the trees by swinging from successive handholds. The gibbon's mass is suspended below a point about which it is free to pivot, so the gibbon's motion can be understood as pendulum motion. You can then use your knowledge of pendulums to describe the motion, explaining, for example, why this gibbon is raising its feet as it swings.

Like any subject, physics is best learned by doing. "Doing physics" in this course means solving problems, applying what you have learned to answer questions at the end of the chapter. When you are given a homework assignment, you may find yourself tempted to simply solve the problems by thumbing through the text looking for a formula that seems like it will work. This isn't how to do physics; if it was, whoever required you to take this course wouldn't bother. The folks who designed your major want you to learn to *reason,* not to "plug and chug." Whatever you end up studying or doing for a career, this ability will serve you well. And that's why someone, somewhere, wants you to take physics.

(a) Pendulum clock

(b) Gibbon locomotion

How do you learn to reason in this way? There's no single strategy for studying physics that will work for all students, but we can make some suggestions that will certainly help:

- **Read each chapter *before* it is discussed in class.** Class attendance is much less effective if you have not prepared. When you first read a chapter, focus on learning new vocabulary, definitions, and notation. You won't understand what's being discussed or how the ideas are being used if you don't know what the terms and symbols mean.
- **Participate actively in class.** Take notes, ask and answer questions, take part in discussion groups. There is ample scientific evidence that *active participation* is far more effective for learning science than is passive listening.
- **After class, go back for a careful rereading of the chapter.** In your second reading, pay close attention to the details and the worked examples. Look for the *logic* behind each example, not just at what formula is being used. We have a three-step process by which we solve all of the worked examples in the text. Most chapters have detailed Problem-Solving Strategies to help you see how to apply this procedure to particular topics, and Tactics Boxes that explain specific steps in your analysis.
- **Apply what you have learned to the homework problems at the end of each chapter.** By following the techniques of the worked examples, applying the tactics and problem-solving strategies, you'll learn how to apply the knowledge you are gaining. In short, you'll learn to reason like a physicist.
- **Form a study group with two or three classmates.** There's good evidence that students who study regularly with a group do better than the rugged individualists who try to go it alone.

And we have one final suggestion. As you read the book, take part in class, and work through problems, step back every now and then to appreciate the big picture. You are going to study topics that range from motions in the solar system to the electrical signals in the nervous system that let you order your hand to turn the pages of this book. You will learn quantitative methods to calculate things such as how far a car will move as it brakes to a stop and how to build a solenoid for an MRI machine. It's a remarkable breadth of topics and techniques that is based on a very compact set of organizing principles. It's quite remarkable, really, well worthy of your study.

Now, let's get down to work.

Detailed Contents

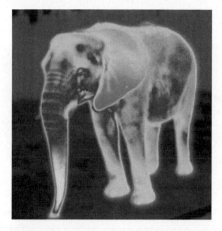

PART I

FORCE AND MOTION

Why Things Change

Each of the seven parts of this book opens with an overview that gives you a look ahead, a glimpse of where your journey will take you in the next few chapters. It's easy to lose sight of the big picture while you're busy negotiating the terrain of each chapter. In Part I, the big picture is, in a word, *change.*

Simple observations of the world around you show that most things change. Some changes, such as aging, are biological. Others, such as sugar dissolving in your coffee, are chemical. We will look at changes that involve *motion* of one form or another—running and jumping, throwing balls, lifting weights.

There are two big questions we must tackle to study how things change by moving:

- **How do we describe motion?** How should we measure or characterize the motion if we want to analyze it mathematically?
- **How do we explain motion?** Why do objects have the particular motion they do? Why, when you toss a ball upward, does it go up and then come back down rather than keep going up? Are there "laws of nature" that allow us to predict an object's motion?

Two key concepts that will help answer these questions are *force* (the "cause") and *acceleration* (the "effect"). Our basic tools will be three laws of motion elucidated by Isaac Newton. Newton's laws relate force to acceleration, and we will use them to explain and explore a wide range of problems. As we learn to solve problems, we will learn basic techniques that we can apply in all of the parts of this book.

Simplifying Models

Reality is extremely complicated. We would never be able to develop a science if we had to keep track of every detail of every situation. Suppose we analyze the tossing of a ball. Is it necessary to analyze the way the atoms in the ball are connected? Do we need to analyze what you ate for breakfast and the biochemistry of how that was translated into muscle power? These are interesting questions, of course. But if our task is to understand the motion of the ball, we need to simplify!

We can do a perfectly fine analysis if we treat the ball as a round solid and your hand as another solid that exerts a force on the ball. This is a *model* of the situation. A model is a simplified description of reality—much as a model airplane is a simplified version of a real airplane—that is used to reduce the complexity of a problem to the point where it can be analyzed and understood.

Model building is a major part of the strategy that we will develop for solving problems in all parts of the book. We will introduce different models in different parts. We will pay close attention to where simplifying assumptions are being made, and why. Learning *how* to simplify a situation is the essence of successful modeling—and successful problem solving.

The cheetah is the fastest land animal, able to run at speeds exceeding 60 miles per hour. Nonetheless, the rabbit has an advantage in this chase. It can *change* its motion more quickly, and will likely escape. How can you tell, by looking at the picture, that the cheetah is changing its motion?

1 CONCEPTS OF MOTION AND MATHEMATICAL BACKGROUND

As this snowboarder moves in a graceful arc through the air, the direction of his motion, and the distance between each of his positions and the next, is constantly changing. What language should we use to describe this motion?

Looking Ahead ▶▶

The goal of Chapter 1 is to introduce the fundamental concepts of motion and to review the related basic mathematical principles. In this chapter you will learn to:

- ▶ Draw and interpret motion diagrams.
- ▶ Apply the particle model of motion.
- ▶ Describe motion in terms of distance, displacement, time, and velocity.
- ▶ Express quantities with appropriate units and the correct number of significant figures.
- ▶ Describe motion using vectors, and learn how to add vectors.
- ▶ Understand how different physical quantities can often be expressed using the same mathematical form.

Socrates: *The nature of motion appears to be the question with which we begin.*

Plato, 375 BCE

The universe in which we live is one of change and motion. This snowboarder was clearly in motion in the series of photos that make up the image. In the course of a day you probably walk, run, bicycle, or drive your car, all forms of motion. The clock hands are moving inexorably forward as you read this text. The pages of this book may look quite still, but a microscopic view would reveal jostling atoms and whirling electrons. The stars look as permanent as anything, yet the astronomer's telescope reveals them to be ceaselessly moving within galaxies that rotate and orbit yet other galaxies.

Motion is a theme that will appear in one form or another throughout this entire book. Although we all have intuition about motion, based on our experiences, some of the important aspects of motion turn out to be rather subtle. So rather than jumping immediately into a lot of mathematics and calculations, this first chapter focuses on *visualizing* motion and becoming familiar with the *concepts* needed to describe a moving object. We will use mathematical ideas when needed, because they increase the precision of our thoughts, but we will defer actual calculations until Chapter 2. Our goal is to lay the foundations for understanding motion.

The quest to understand motion dates to antiquity. The ancient Babylonians, Chinese, and Greeks were especially interested in the celestial motions of the night sky. The Greek philosopher and scientist Aristotle wrote extensively about the nature of moving objects. However, our modern understanding of motion did not begin until Galileo (1564–1642) first formulated the concepts of motion in mathematical terms. And it took Newton (1642–1727) to put the concepts of motion on a firm and rigorous footing. This connection between motion and mathematics was the breakthrough that allowed the growth of the science of physics.

One key difference between physics and other sciences is how we set up and solve problems. We'll often use a two-step process to solve motion problems. The first step is to develop a simplified representation of the motion so that key elements stand out. For example, the above photo allows us to observe the position of the snowboarder at several successive times. It is precisely by considering this sort of picture of motion that we will begin our study of this topic. The second step is to analyze the motion with the language of mathematics. The process of putting numbers on nature is often the most challenging aspect of the problems you will solve. In this chapter, we will explore the steps in this process as we introduce the basic concepts of motion.

1.1 Motion: A First Look

As a starting point, let's define **motion** as the change of an object's position or orientation with time. Examples of motion are easy to list. Bicycles, baseballs, cars, airplanes, and rockets are all objects that move. The path along which an object moves, which might be a straight line or might be curved, is called the object's **trajectory.**

Figure 1.1 shows four basic types of motion that we will study in this book. In this chapter, we will start with the first type of motion in the figure, motion along a straight line. In later chapters, we will learn about circular motion, which is the motion of an object along a circular path; projectile motion, the motion of an object through the air; and rotational motion, the spinning of an object about an axis.

Straight-line motion

Circular motion

Projectile motion

Rotational motion

FIGURE 1.1 Four basic types of motion.

The fundamental question we want to ask is: What *concepts* are needed to give a full and accurate description of motion?

FIGURE 1.2 Several frames from the video of a car.

The same amount of time elapses between each image and the next.

FIGURE 1.3 A motion diagram of the car shows all the frames simultaneously.

Making a Motion Diagram

An easy way to study motion is to record a video of a moving object. A video camera takes images at a fixed rate, typically 30 images every second. Each separate image is called a *frame.* As an example, Figure 1.2 shows a few frames from a video of a car going past. Not surprisingly, the car is in a somewhat different position in each frame.

Suppose we now edit the video, layering the frames on top of each other, and then look at the final result. We end up with the picture in Figure 1.3. This composite image, showing an object's position at several *equally spaced instants of time,* is called a **motion diagram.** As simple as motion diagrams seem, they will turn out to be powerful tools for analyzing motion.

NOTE ▶ It's important to keep the camera in a *fixed position* as the object moves by. Don't "pan" it to track the moving object. ◀

Now let's take our camera out into the world and make a few motion diagrams. The following table illustrates how a motion diagram shows important features of different kinds of motion.

Examples of motion diagrams

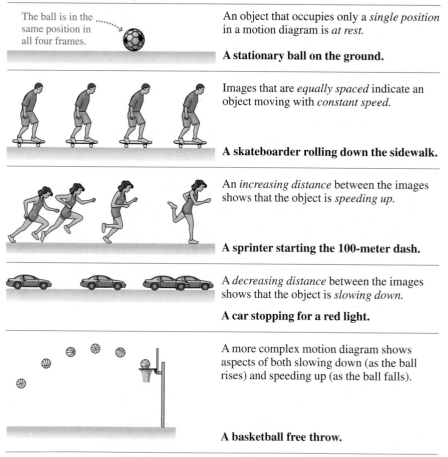

The ball is in the same position in all four frames.

An object that occupies only a *single position* in a motion diagram is *at rest.*

A stationary ball on the ground.

Images that are *equally spaced* indicate an object moving with *constant speed.*

A skateboarder rolling down the sidewalk.

An *increasing distance* between the images shows that the object is *speeding up.*

A sprinter starting the 100-meter dash.

A *decreasing distance* between the images shows that the object is *slowing down.*

A car stopping for a red light.

A more complex motion diagram shows aspects of both slowing down (as the ball rises) and speeding up (as the ball falls).

A basketball free throw.

We have defined several concepts (at rest, constant speed, speeding up, and slowing down) in terms of how the moving object appears in a motion diagram. These are called **operational definitions,** meaning that the concepts are defined in terms of a particular procedure or operation performed by the investigator. For example, we could answer the question "Is the airplane speeding up?" by checking whether or not the images in the plane's motion diagram are getting farther apart. Many of the concepts in physics will be introduced as operational definitions. This reminds us that physics is an experimental science.

STOP TO THINK 1.1 Which car is going faster, A or B? Assume there are equal intervals of time between the frames of both videos.

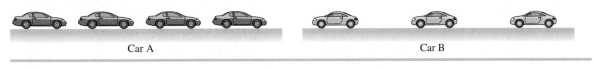

Car A   Car B

NOTE ▶ Each chapter in this textbook will have several *Stop to Think* questions. These questions are designed to see if you've understood the basic ideas that have been presented. The answers are given at the end of the chapter, but you should make a serious effort to think about these questions before turning to the answers. If you answer correctly, and are sure of your answer rather than just guessing, you can proceed to the next section with confidence. But if you answer incorrectly, it would be wise to reread the preceding sections carefully before proceeding onward. ◀

The Particle Model

For many objects, such as cars and rockets, the motion of the object *as a whole* is not influenced by the "details" of the object's size and shape. To describe the object's motion, all we really need to keep track of is the motion of a single point: You could imagine looking at the motion of a dot painted on the side of the object.

In fact, for the purposes of analyzing the motion, we can often consider the object *as if* it were just a single point, without size or shape. We can also treat the object *as if* all of its mass were concentrated into this single point. An object that can be represented as a mass at a single point in space is called a **particle.** A particle has no size, no shape, and no distinction between top and bottom or between front and back.

If we treat an object as a particle, we can represent the object in each frame of a motion diagram as a simple dot rather than having to draw a full picture. Figure 1.4 shows how much simpler motion diagrams appear when the object is represented as a particle. Note that the dots have been numbered 0, 1, 2, . . . to tell the sequence in which the frames were exposed. These diagrams are more abstract than the pictures, but they are easier to draw and they still convey our full understanding of the object's motion.

Treating an object as a particle is, of course, a simplification of reality. Such a simplification is called a **model.** Models allow us to focus on the important aspects of a phenomenon by excluding those aspects that play only a minor role. The **particle model** of motion is a simplification in which we treat a moving object as if all of its mass were concentrated at a single point. This might seem like an oversimplification, but if all we are concerned with is the motion of an object, it may not be. Using the particle model may allow us to see connections that are very important. Consider the motion of the two objects shown in Figure 1.5. These two very different objects have exactly the same motion diagram! As we will see, all objects falling under the influence of gravity move in exactly the same manner if no other forces act. The simplification of the particle model has revealed something about the physics that underlies both of these situations.

Not all motions can be reduced to the motion of a single point. Consider a rotating gear. The center of the gear doesn't move at all, and each tooth on the gear is moving in a different direction. Rotational motion is qualitatively different from motion along a line, and we'll need to go beyond the particle model later when we study rotational motion.

(a) Motion diagram of a car stopping

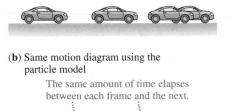

(b) Same motion diagram using the particle model

The same amount of time elapses between each frame and the next.

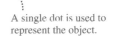

0   1   2   3

Numbers show the order in which the frames were taken.

A single dot is used to represent the object.

FIGURE 1.4 Simplifying a motion diagram using the particle model.

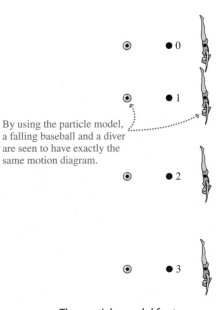

By using the particle model, a falling baseball and a diver are seen to have exactly the same motion diagram.

FIGURE 1.5 The particle model for two falling objects.

STOP TO THINK 1.2	Three motion diagrams are shown. Which is a dust particle settling to the floor at constant speed, which is a ball dropped from the roof of a building, and which is a descending rocket slowing to make a soft landing on Mars?	A. 0● 1● 2● 3● 4● 5●	B. 0● 1● 2● 3● 4● 5●	C. 0● 1● 2● 3● 4● 5●

1.2 Position and Time: Putting Numbers on Nature

To develop our understanding of motion further, we need to be able to make quantitative measurements. That is, we need to use numbers. As we look at a motion diagram, it would be useful to know where the object is (its *position*) and when the object was at that position (the *time*). We'll start by considering the motion of an object that can move only along a straight line. Examples of this **one-dimensional** or "1D" motion would be a bicyclist moving along the road, a train moving on a long straight track, or an elevator moving up and down a shaft.

Position and Coordinate Systems

Suppose you are driving along a long, straight country road, as in Figure 1.6, and your friend calls and asks where you are. You might reply that you are four miles east of the post office, and your friend would then know just where you were. Your location at a particular instant in time (when your friend phoned) is called your **position.** Notice that to know your position along the road, your friend needed three pieces of information. First, you had to give her a reference point (the post office) from which all distances are to be measured. We call this fixed reference point the **origin.** Second, she needed to know how far you were from that reference point or origin—in this case, four miles. Finally, she needed to know which side of the origin you were on: You could be four miles to the west of it, or four miles to the east.

We will need these same three pieces of information in order to specify any object's position along a line. We first choose our origin, from which we measure the position of the object. The position of the origin is arbitrary, and we are free to place it where we like. Usually, however, there are certain points (such as the well-known post office) that are more convenient choices than others.

In order to specify how far our object is from the origin, we lay down an imaginary axis along the line of the object's motion. Like a ruler, this axis is marked off in equally spaced divisions of distance, perhaps in inches, meters, or miles, depending on the problem at hand. We place the zero mark of this ruler at the origin, allowing us to locate the position of our object by reading the ruler mark where the object is.

Finally, we need to be able to specify which side of the origin our object is on. To do this, we imagine the axis extending from one side of the origin with increasing, positive markings; on the other side, the axis is marked with increasing *negative* numbers. By reporting the position as either a positive or a negative number, we know on what side of the origin the object is.

These elements—an origin and an axis marked in both the positive and negative directions—can be used to unambiguously locate the position of an object. We call this a **coordinate system.** We will use coordinate systems throughout this

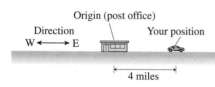

Origin (post office)

Direction
W ←—→ E

Your position

4 miles

FIGURE 1.6 Describing your position.

Sometimes measurements have a very natural origin. This snow depth gauge has its origin set at road level.

book, and we will soon develop coordinate systems that can be used to describe the position of objects moving in more complex ways than just along a line. Figure 1.7 shows a coordinate system that can be used to locate various objects along the country road discussed earlier.

Although our coordinate system works well for describing the positions of objects located along the axis, our notation is somewhat cumbersome. We need to keep saying things like "the car is at position +4 miles." A better notation, and one that will become particularly important when we study motion in two dimensions, is to use a symbol such as x or y to represent the position along the axis. Then we can say "the cow is at $x = -5$ miles." The symbol that represents a position along an axis is called a **coordinate.** The introduction of symbols to represent positions (and, later, velocities and accelerations) also allows us to work with these quantities mathematically.

Figure 1.8 shows how we would set up a coordinate system for a sprinter running a 50-meter race (we use the standard symbol "m" for meters). For horizontal motion like this we usually use the coordinate x to represent the position.

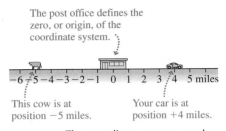

The post office defines the zero, or origin, of the coordinate system.

This cow is at position −5 miles.   Your car is at position +4 miles.

FIGURE 1.7 The coordinate system used to describe objects along a country road.

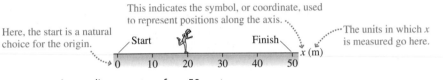

This indicates the symbol, or coordinate, used to represent positions along the axis.

Here, the start is a natural choice for the origin.

The units in which x is measured go here.

FIGURE 1.8 A coordinate system for a 50-meter race.

Motion along a straight line need not be horizontal. As shown in Figure 1.9, a rock falling vertically downward and a skier skiing down a straight slope are also examples of straight-line or one-dimensional motion.

Time

The pictures in Figure 1.9 show the position of an object at just one instant of time. But a full motion diagram represents how an object moves as time progresses. So far, we have labeled the dots in a motion diagram by the numbers 0, 1, 2, ... to indicate the order in which the frames were exposed. But to fully describe the motion, we need to indicate the *time,* as read off a clock or a stopwatch, at which each frame of a video was made. This is important, as we can see from the motion diagram of a stopping car in Figure 1.10. If the frames were taken one second apart, this motion diagram shows a leisurely stop; if 1/10 of a second apart, it represents a screeching halt.

For a complete motion diagram, we thus need to label each frame with its corresponding time (symbol t) as read off a clock. But when should we start the clock? That is, which frame should be labeled $t = 0$? This choice is much like that of choosing the origin $x = 0$ of a coordinate system: You can pick any arbitrary point in the motion and label it "$t = 0$ seconds." This is simply the instant you decide to start your clock or stopwatch, so it is the origin of your time coordinate. A video frame labeled "$t = 4$ seconds" means it was taken 4 seconds after you started your clock. We typically choose $t = 0$ to represent the "beginning" of a problem, but the object may have been moving before then.

To illustrate, Figure 1.11 shows the motion diagram for a car moving at a constant speed, and then braking to a halt. Two possible choices for the frame labeled $t = 0$ seconds are shown; our choice depends on what part of the motion we're interested in. Each successive position of the car is then labeled with the clock reading in seconds (abbreviated by the symbol "s").

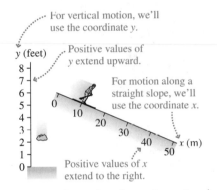

For vertical motion, we'll use the coordinate y.

Positive values of y extend upward.

For motion along a straight slope, we'll use the coordinate x.

Positive values of x extend to the right.

FIGURE 1.9 Examples of one-dimensional motion.

FIGURE 1.10 Is this a leisurely stop or a screeching halt?

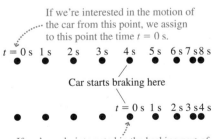

If we're interested in the motion of the car from this point, we assign to this point the time $t = 0$ s.

Car starts braking here

If we're only interested in the braking part of the motion, we would assign $t = 0$ s here.

FIGURE 1.11 The motion diagram of a car that travels at constant speed and then brakes to a halt.

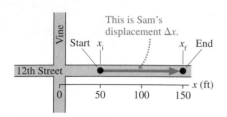

FIGURE 1.12 Sam undergoes a displacement Δx from position x_i to position x_f.

The size and the direction of the displacement both matter. Roy Riegels (pursued above by teammate Benny Lom) found this out in dramatic fashion in the 1928 Rose Bowl when he recovered a fumble and ran 69 yards—toward his own team's end zone. An impressive distance, but in the wrong direction!

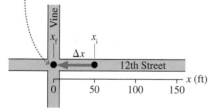

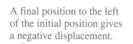

FIGURE 1.13 A displacement is a signed quantity. Here Δx is a negative number.

Changes in Position and Displacement

Now that we've seen how to measure position and time, let's return to the problem of motion. To describe motion we'll need to measure the *changes* in position that occur with time. Consider the following:

> Sam is standing 50 feet (ft) east of the corner of 12th Street and Vine. He then walks to a second point 150 ft east of Vine. What is Sam's change of position?

Figure 1.12 shows Sam's motion on a map. We've placed a coordinate system on the map, using the coordinate x. We are free to place the origin of our coordinate system wherever we wish, so we have placed it at the intersection. Sam's initial position is then at $x_i = 50$ ft. The positive value for x_i tells us that Sam is east of the origin.

> **NOTE** ▶ We will label special values of x or y with subscripts. The value at the start of a problem is usually labeled with a subscript "i," for *initial,* and that at the end is labeled with a subscript "f," for *final.* For cases having several special values, we will usually use subscripts "1," "2," etc. ◀

Sam's final position is $x_f = 150$ ft, indicating that he is 150 feet east of the origin. You can see that Sam has changed position, and a *change* of position is called a **displacement.** His displacement is the distance labeled Δx in Figure 1.12. The Greek letter delta (Δ) is used in math and science to indicate the *change* in a quantity. Thus Δx indicates a change in the position x.

> **NOTE** ▶ Δx is a *single* symbol. You cannot cancel out or remove the Δ in algebraic operations. ◀

To get from the 50 ft mark to the 150 ft mark, Sam clearly had to walk 100 ft, so the change in his position—his displacement—is 100 ft. We can think about displacement in a more general way, however. **Displacement is the *difference* between a final position x_f and an initial position x_i.** Thus we can write

$$\Delta x = x_f - x_i = 150 \text{ ft} - 50 \text{ ft} = 100 \text{ ft}$$

> **NOTE** ▶ A general principle, used throughout this book, is that the change in any quantity is the final value of the quantity minus its initial value. ◀

Displacement is a *signed quantity.* That is, it can be either positive or negative. If, as shown in Figure 1.13, Sam's final position x_f had been at the origin instead of the 150 ft mark, his displacement would have been

$$\Delta x = x_f - x_i = 0 \text{ ft} - 50 \text{ ft} = -50 \text{ ft}$$

The negative sign tells us that he moved to the *left* along the x-axis, or 50 ft *west.*

Change in Time

A displacement is a change in position. In order to quantify motion, we'll need to also consider changes in *time,* which we call **time intervals.** We've seen how we can label each frame of a motion diagram with a specific time, as determined by our stopwatch. Figure 1.14 on the next page shows the motion diagram of a bicycle moving at a constant speed, with the times of measured points indicated. The displacement between the initial position x_i and the final position x_f is

$$\Delta x = x_f - x_i = 120 \text{ ft} - 0 \text{ ft} = 120 \text{ ft}$$

Similarly, we define the time interval between these two points to be

$$\Delta t = t_f - t_i = 6 \text{ s} - 0 \text{ s} = 6 \text{ s}$$

A time interval Δt measures the elapsed time as an object moves from an initial position x_i at time t_i to a final position x_f at time t_f. Note that unlike Δx, Δt is always positive because t_f is always greater than t_i.

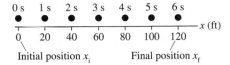

FIGURE 1.14 The motion diagram of a bicycle moving to the right at a constant speed.

STOP TO THINK 1.3   Sarah starts at a positive position along the x-axis. She then undergoes a negative displacement. Her final position

A. Is positive.
B. Is negative.
C. Could be either positive or negative.

1.3 Velocity

We all have an intuitive sense of whether something is moving very fast or just cruising slowly along. To make this intuitive idea more precise, let's start by examining the motion diagrams of some objects moving along a straight line at a *constant* speed, objects that are neither speeding up nor slowing down. This motion at a constant speed is called **uniform motion.** As we saw for the skateboarder in Section 1.1, for an object in uniform motion successive frames of the motion diagram are *equally spaced.* We know now that this means that the object's displacement Δx is the same between successive frames.

To see how an object's displacement between successive frames is related to its speed, consider the motion diagrams of a bicycle and a car, traveling along the same street as shown in Figure 1.15. Clearly the car is moving faster than the bicycle: In any one-second time interval, the car undergoes a displacement $\Delta x = 40$ ft, while the bicycle's displacement is only 20 ft.

The distances traveled in one second by the bicycle and the car are a measure of their speeds. The greater the distance traveled by an object in a given time interval, the greater its speed. This idea leads us to define the speed of an object as

During each second, the car moves twice as far as the bicycle. Hence the car is moving at a greater speed.

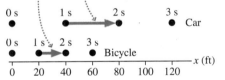

FIGURE 1.15 Motion diagrams for a car and a bicycle.

$$\text{speed} = \frac{\text{distance traveled in a given time interval}}{\text{time interval}} \tag{1.1}$$

Speed of a moving object

For the bicycle, this gives

$$\text{speed} = \frac{20 \text{ ft}}{1 \text{ s}} = 20 \, \frac{\text{ft}}{\text{s}}$$

while for the car we have

$$\text{speed} = \frac{40 \text{ ft}}{1 \text{ s}} = 40 \, \frac{\text{ft}}{\text{s}}$$

The speed of the car is twice that of the bicycle, which seems reasonable.

NOTE ▶ The division gives units that are a fraction: ft/s. This is read as "feet per second," just like the more familiar "miles per hour." ◀

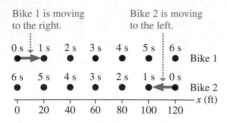

FIGURE 1.16 Two bicycles traveling at the same speed, but with different velocities.

To fully characterize the motion of an object, it is important to specify not only the object's speed but also the *direction* in which it is moving. For example, Figure 1.16 shows the motion diagrams of two bicycles traveling at the same speed of 20 ft/s. The two bicycles have the same speed. but something about their motion is different—the *direction* of their motion.

The problem is that the "distance traveled" in Equation 1.1 doesn't capture any information about the direction of travel. But we've seen that the *displacement* of an object does contain this information. We can then introduce a new quantity, the **velocity,** as

$$\text{velocity} = \frac{\text{displacement}}{\text{time interval}} = \frac{\Delta x}{\Delta t} \tag{1.2}$$

Velocity of a moving object

The velocity of bicycle 1 in Figure 1.16, computed using the one-second time interval between the $t = 2$ s and $t = 3$ s positions, is

$$v = \frac{\Delta x}{\Delta t} = \frac{x_3 - x_2}{3\text{ s} - 2\text{ s}} = \frac{60\text{ ft} - 40\text{ ft}}{1\text{ s}} = +20\,\frac{\text{ft}}{\text{s}}$$

while the velocity for bicycle 2, during the same time interval, is

$$v = \frac{\Delta x}{\Delta t} = \frac{x_3 - x_2}{3\text{ s} - 2\text{ s}} = \frac{60\text{ ft} - 80\text{ ft}}{1\text{ s}} = -20\,\frac{\text{ft}}{\text{s}}$$

NOTE ▶ We have used x_2 for the position at time $t = 2$ seconds, and x_3 for the position at time $t = 3$ seconds. The subscripts 2 and 3 serve the same role as before—identifying particular positions—but in this case the positions are identified by the time at which each position is reached. ◀

The two velocities have opposite signs because the bicycles are traveling in opposite directions. **Speed measures only how fast an object moves, but velocity tells us both an object's speed** *and its direction.* A positive velocity indicates motion to the right or, for vertical motion, upward. Similarly, an object moving to the left, or down, has a negative velocity.

NOTE ▶ Learning to distinguish between speed, which is always a positive number, and velocity, which can be either positive or negative, is one of the most important tasks in the analysis of motion. ◀

The velocity as defined by Equation 1.2 is actually what is called the *average* velocity. On average, over each 1 s interval bicycle 1 moves 20 ft, but we don't know if it sped up or slowed down a little during that 1 s. In Chapter 2, we'll develop the idea of *instantaneous* velocity, the velocity of an object at a particular instant in time. Since our goal in this chapter is to *visualize* motion with motion diagrams, we'll somewhat blur the distinction between average and instantaneous quantities, refining these definitions in Chapter 2, where our goal will be to develop the mathematics of motion.

The "Per" in Meters Per Second

The units for speed and velocity are those of a distance (feet, meters, miles) divided by a unit of time (seconds, hours). Thus we could measure velocity in units of m/s or mph, pronounced "meters *per* second" and "miles *per* hour." The word "per" will often arise in physics when we consider the ratio of two quantities. What do we mean, exactly, by "per"?

If a car moves with a speed of 23 m/s, we mean that it travels 23 meters *for each* 1 second of elapsed time. The word "per" thus associates the number of units in the numerator (23 m) with *one* unit of the denominator (1 s). We'll see many other examples of this idea as the book progresses. You may already know a bit about *density;* you can look up the density of gold and you'll find that it is 19.3 g/cm³ ("grams *per* cubic centimeter"). This means that there are 19.3 grams of gold *for each* 1 cubic centimeter of the metal. Thinking about the word "per" in this way will help you better understand physical quantities whose units are the ratio of two other units.

1.4 A Sense of Scale: Significant Figures, Scientific Notation, and Units

Physics attempts to explain the natural world, from the very small to the exceedingly large. And in order to understand our world, we need to be able to *measure* quantities both minuscule and enormous. A properly reported measurement has three elements. First, we can only measure our quantity with a certain precision. To make this precision clear, we need to make sure that we report our measurement with the correct number of *significant figures.*

Second, writing down the really big and small numbers that often come up in physics can be awkward. To avoid writing all those zeros, scientists use *scientific notation* to express numbers both big and small.

Finally, we need to choose an agreed-upon set of *units* for the quantity. For speed, common units include meters per second and miles per hour. For mass, the kilogram is the most commonly used unit. Later, we'll study more esoteric quantities such as magnetic fields, which have the units of "tesla." Every physical quantity that we can measure has an associated set of units.

Measurements and Significant Figures

When we measure any quantity, such as the length of a bone or the weight of a specimen, we can do so only with a certain *precision.* The digital calipers in Figure 1.17 can make a measurement to within ±0.001 in, so they have a precision of 0.001 in. If you used the tape shown to make a measurement, you probably couldn't do so to better than about ±1 mm, so the precision of the tape measure is about 1 mm. The precision of a measurement can also be affected by the skill or judgment of the person performing the measurement. A stopwatch might have a precision of 0.001 s, but, due to your reaction time, your measurement of the time of a sprinter would be much less precise.

It is important that your measurement be reported in a way that reflects its actual precision. Suppose, for example, that you use a ruler to measure the length of a particular specimen of a newly discovered species of frog. You judge that you can make this measurement with a precision of about 1 mm, or 0.1 cm. In this case, the frog's length should be reported as, say, 6.2 cm. We interpret this to mean that the actual value falls between 6.15 cm and 6.25 and thus rounds to 6.2 cm. Reporting the frog's length as simply 6 cm is saying less than you know; you are withholding information. On the other hand, to report the number as 6.213 cm is wrong. Any person reviewing your work would interpret the number 6.213 cm as meaning that the actual length falls between 6.2125 cm and 6.2135 cm, thus rounding to 6.213 cm. In this case, you are claiming to have knowledge and information that you do not really possess.

This example shows that one way to state your knowledge precisely is through the proper use of **significant figures.** You can think of a significant figure as being a digit that is reliably known. A measurement such as 6.2 cm has *two* significant figures, the 6 and the 2. The next decimal place—the one-hundredths—is

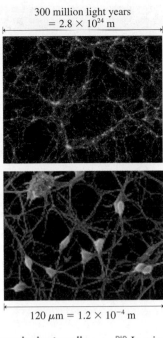

300 million light years = 2.8×10^{24} m

$120 \; \mu m = 1.2 \times 10^{-4}$ m

From galaxies to cells . . . BIO In science, we need to express numbers both very large and very small. The top image is of a computer simulation of the structure of the universe. Bright areas represent regions of clustered galaxies. The bottom image is of cortical nerve cells. Nerve cells relay signals to each other through a complex web of dendrites. These images, though similar in appearance, differ in scale by a factor of about 2×10^{28}!

These calipers have a precision of 0.001 in.

A tape measure has a precision of about 1 mm.

FIGURE 1.17 The precision of a measurement depends on how the measurement is made.

Tatyana Lebedeva won the 2004 Olympic gold medal with a long jump reported as 7.07 m. This number implies that the long jump is measured to an accuracy of about 0.01 m, or 1 cm.

not reliably known, and is thus not a significant figure. Similarly, a time measurement of 34.62 s has four significant figures, implying that the 2 in the hundredths place is reliably known.

When we perform a calculation such as adding or multiplying two or more measured numbers, we can't claim more accuracy for the result than was present in the initial measurements. Calculations with measured numbers follow the "weakest link" rule. The saying, which you probably know, is that "a chain is only as strong as its weakest link." If nine out of ten links in a chain can support a 1000 pound weight, that strength is meaningless if the tenth link can support only 200 pounds. Similarly, nine out of the ten numbers used in a calculation might be known with a precision of 0.01%; but if the tenth number is poorly known, with a precision of only 10%, then the result of the calculation cannot possibly be more precise than 10%. The weak link rules!

Determining the proper number of significant figures is straightforward, but there are a few definite rules to follow. We will often spell out such technical details in what we call a "Tactics Box." A Tactics Box is designed to teach you particular skills and techniques. Each Tactics Box will use the ✐ icon to designate exercises in the *Student Workbook* that you can use to practice these skills.

MP   **TACTICS BOX 1.1 Using significant figures**   ✐ Exercise 16

❶ When multiplying or dividing several numbers, or when taking roots, the number of significant figures in the answer should match the number of significant figures of the *least* precisely known number used in the calculation:

Three significant figures

$$3.73 \times 5.7 = 21$$

Two significant figures

Answer should have the *lower* of the two, or two significant figures.

❷ When adding or subtracting several numbers, the number of decimal places in the answer should match the *smallest* number of decimal places of any number used in the calculation.

$$\begin{array}{r} 18.54 \\ +106.6 \\ \hline =125.1 \end{array}$$

18.54 — Two decimal places
+106.6 — One decimal place
=125.1 ◂···· Answer should have the *lower* of the two, or one decimal place.

❸ **Exact numbers** have no uncertainty and, when used in calculations, do not change the number of significant figures of measured numbers. Examples of exact numbers are π and the number 2 in the relation $d = 2r$ between a circle's diameter and radius.

There is one notable exception to these rules:

■ It is acceptable to keep one or two extra digits during *intermediate* steps of a calculation. The goal here is to minimize round-off errors in the calculation. But the *final* answer must be reported with the proper number of significant figures.

TRY IT YOURSELF

How tall are you really? If you measure your height in the morning, just after you wake up, and then in the evening, after a full day of activity, you'll find that your evening height is *less* by as much as 3/4 inches. Your height decreases over the course of the day as gravity compresses and reshapes your spine. If you give your height as 66 3/16 in, you are claiming more significant figures than are truly warranted; the 3/16 in isn't really reliably known, as your height can vary by more than this. Expressing your height to the nearest inch is plenty!

EXAMPLE 1.1 Measuring the velocity of a car

To measure the velocity of a car, clocks A and B are set up at two points along the road, as shown in Figure 1.18. Clock A is precise to 0.01 s, while B is precise only to 0.1 s. The distance between these two clocks is carefully measured to be 124.5 m. The two clocks are automatically started when the car passes a trigger in the road; each clock stops automatically when the car passes that clock. After the car has passed both clocks, clock A is found to read $t_A = 1.22$ s, and clock B to read $t_B = 4.5$ s. The time from the less-precise clock B is correctly reported with fewer significant figures than that from A. What is the velocity of the car, and how should it be reported with the correct number of significant figures?

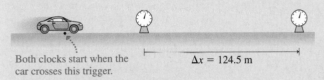

Both clocks start when the car crosses this trigger. $\Delta x = 124.5$ m

FIGURE 1.18 Measuring the velocity of a car.

SOLVE We've already determined the car's displacement $\Delta x = 124.5$ m as it moves between the two clocks. To calculate its velocity, we need to determine the time interval Δt.

This is

This number has one decimal place.   This number has two decimal places.

$$\Delta t = t_B - t_A = (4.5 \text{ s}) - (1.22 \text{ s}) = 3.3 \text{ s}$$

By rule 2 of Tactics Box 1.1, the result should have *one* decimal place.

We can now calculate the velocity with the displacement and time interval from above:

The displacement has four significant figures.

$$v = \frac{\Delta x}{\Delta t} = \frac{124.5 \text{ m}}{3.3 \text{ s}} = 38 \text{ m/s}$$

The time interval has two significant figures.   By rule 1 of Tactics Box 1.1, the result should have *two* significant figures.

ASSESS Our final value has two significant figures. Suppose you had been hired to measure the speed of a car this way, and you reported 37.72 m/s. It would be reasonable for someone looking at your result to assume that the measurements you used to arrive at this value were correct to four significant figures and thus that you had measured time to the nearest 0.01 second. Our correct result of 38 m/s has all of the accuracy that you can claim, but no more!

Scientific Notation

It's easy to write down measurements of ordinary-sized objects: your height might be 1.72 meters, the weight of an apple 0.34 pounds. But the radius of a hydrogen atom is 0.000 000 000 053 m and the distance to the moon is 384 000 000 m. Writing and keeping track of all those zeros is quite cumbersome.

NOTE ▶ Scientists usually write numbers with many digits by arranging the digits in groups of three, with the groups separated by spaces instead of commas. This makes it easier to read long numbers. ◀

Beyond requiring you to deal with all the zeros, writing quantities this way makes it unclear how many significant figures are involved. In the distance to the moon given above, how many of those digits are significant? Three? Four? All nine?

Writing numbers using scientific notation avoids both these problems. A value in scientific notation is a number with one digit to the left of the decimal point and zero or more to the right of it, multiplied by a power of ten. This solves the problem of all the zeros and makes the number of significant digits immediately apparent. In scientific notation, writing the distance to the sun as 1.50×10^{11} m implies that three digits are significant; writing it as 1.5×10^{11} m implies that only two digits are.

Even for smaller values, scientific notation can clarify the number of significant figures. Suppose a distance is measured as 1200 m. If this distance is known to within 1 m, we could write it as 1.200×10^3 m, showing that all four digits are significant; if it were accurate to only 100 m or so, we would report it as 1.2×10^3 m, indicating two significant figures.

Tactics Box 1.2 shows how to convert a number to scientific notation, and how to correctly indicate the number of significant figures.

Exercises 17, 18

(MP) TACTICS BOX 1.2 Using scientific notation

To convert a number into scientific notation:

❶ For a number greater than 10, move the decimal point to the left until only one digit remains to the left of the decimal point. The remaining number is then multiplied by 10 to a power; this power is given by the number of spaces the decimal point was moved. Here we convert the diameter of the earth to scientific notation:

We move the decimal point until there is only one digit to its left, counting the number of steps.

Since we moved the decimal point 6 steps, the power of ten is 6.

$$6\,370\,000 \text{ m} = 6.37 \times 10^6 \text{ m}$$

The number of digits here equals the number of significant figures.

❷ For a number less than 1, move the decimal point to the right until it passes the first digit that isn't a zero. The remaining number is then multiplied by 10 to a negative power; the power is given by the number of spaces the decimal point was moved. For the diameter of a red blood cell we have

We move the decimal point until it passes the first digit that is not a zero, counting the number of steps.

Since we moved the decimal point 6 steps, the power of ten is −6.

$$0.000\,007\,5 \text{ m} = 7.5 \times 10^{-6} \text{ m}$$

The number of digits here equals the number of significant figures.

Proper use of significant figures is part of the "culture" of science. We will frequently emphasize these "cultural issues" because you must learn to speak the same language as the natives if you wish to communicate effectively! Most students "know" the rules of significant figures, having learned them in high school, but many fail to apply them. It is important that you understand the reasons for significant figures and that you get in the habit of using them properly.

Units

As we have seen, in order to measure a quantity we need to give it a numerical value. But a measurement is more than just a number—it requires a *unit* to be given. You can't go to the grocery and ask for "three-and-a-half of flour." You need to use a unit—here, one of weight, such as pounds—in addition to the number.

In your daily life, you generally tend to use the English system of units, in which distances are measured in inches, feet, and miles. These units are well adapted for daily life, but they are rarely used in scientific work. Given that science is an international discipline, it is also important to have a system of units that is recognized around the world. For these reasons, scientists use a system of units called *le Système Internationale d'Unités,* commonly referred to as **SI units.** In casual speaking we often refer to these as *metric units,* because the meter is the basic standard of length.

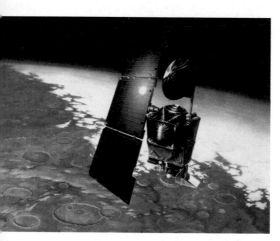

The importance of units In 1999, the $125 million Mars Climate Orbiter burned up in the Martian atmosphere instead of entering a safe orbit from which it could perform observations. The problem was faulty units! An engineering team had provided critical data on spacecraft performance in English units, but the navigation team assumed these data were in metric units. As a consequence, the navigation team had the spacecraft fly too close to the planet, and it burned up in the atmosphere.

The three basic SI quantities, shown in Table 1.1, are time, length (or distance), and mass. Other quantities needed to understand motion can be expressed as combinations of these basic units. For example, speed or velocity are expressed in meters per second or m/s. This combination is a ratio of the length unit (the meter) to the time unit (the second).

The SI units have a long and interesting history. SI units were originally developed by the French in the late 1700s as a way of standardizing and regularizing numbers for commerce and science. Some of their other innovations of the time did not survive (such as the 10-day week), but their units did.

TABLE 1.1 Common SI units

Quantity	Unit	Abbreviation
time	second	s
length	meter	m
mass	kilogram	kg

Using Prefixes

We will have many occasions to use lengths, times, and masses that are either much less or much greater than the standards of 1 meter, 1 second, and 1 kilogram. We will do so by using *prefixes* to denote various powers of ten. For instance, the prefix "kilo" (abbreviation k) denotes 10^3, or a factor of 1000. Thus 1 km equals 1000 m, 1 MW equals 10^6 watts, and 1 μV equals 10^{-6} V. Table 1.2 lists the common prefixes that will be used frequently throughout this book. Memorize it! Few things in physics are learned by rote memory, but this list is one of them. A more extensive list of prefixes is shown inside the cover of the book.

Although prefixes make it easier to talk about quantities, the proper SI units are meters, seconds, and kilograms. Quantities given with prefixed units must be converted to base SI units before any calculations are done. Thus 23.0 cm must be converted to 0.230 m before starting calculations. The exception is the kilogram, which is already the base SI unit.

TABLE 1.2 Common prefixes

Prefix	Abbreviation	Power of 10
mega-	M	10^6
kilo-	k	10^3
centi-	c	10^{-2}
milli-	m	10^{-3}
micro-	μ	10^{-6}
nano-	n	10^{-9}

Unit Conversions

Although SI units are our standard, we cannot entirely forget that the United States still uses English units. Even after repeated exposure to metric units in classes, most of us "think" in the English units we grew up with. Thus it remains important to be able to convert back and forth between SI units and English units. Table 1.3 shows a few frequently used conversions that will come in handy.

One effective method of performing unit conversions begins by noticing that since, for example, 1 mi = 1.609 km, the ratio of these two distances—*including their units*—is equal to one, so that

$$\frac{1 \text{ mi}}{1.609 \text{ km}} = \frac{1.609 \text{ km}}{1 \text{ mi}} = 1$$

A ratio of values equal to one is called a **conversion factor.** The following Tactics Box shows how to make a unit conversion. It uses the example of converting 60 mi into the equivalent distance in km.

TABLE 1.3 Useful unit conversions

1 inch (in) = 2.54 cm
1 foot (ft) = 0.305 m
1 mile (mi) = 1.609 km
1 mile per hour (mph) = 0.447 m/s
1 m = 39.37 in
1 km = 0.621 mi
1 m/s = 2.24 mph

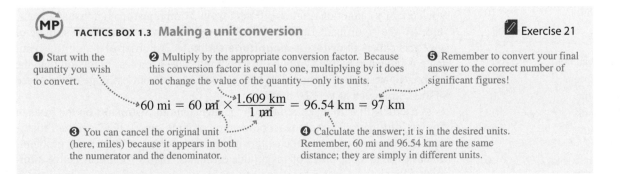

TACTICS BOX 1.3 **Making a unit conversion** ✎ Exercise 21

❶ Start with the quantity you wish to convert.

❷ Multiply by the appropriate conversion factor. Because this conversion factor is equal to one, multiplying by it does not change the value of the quantity—only its units.

❺ Remember to convert your final answer to the correct number of significant figures!

$$60 \text{ mi} = 60 \text{ mi} \times \frac{1.609 \text{ km}}{1 \text{ mi}} = 96.54 \text{ km} = 97 \text{ km}$$

❸ You can cancel the original unit (here, miles) because it appears in both the numerator and the denominator.

❹ Calculate the answer; it is in the desired units. Remember, 60 mi and 96.54 km are the same distance; they are simply in different units.

Note that we've rounded the answer to 97 kilometers because the distance we're converting, 60 miles, has only two significant figures.

More complicated conversions can be accomplished with several successive multiplications of conversion factors, as we see in the next example.

EXAMPLE 1.2 Can a bicycle go that fast?

In Section 1.3, we calculated the speed of a bicycle to be 20 ft/s. Is this a reasonable speed for a bicycle?

SOLVE In order to determine whether or not this speed is reasonable, we will convert it to more familiar units. For speed, the unit you are most familiar with is likely miles per hour.

We first collect the necessary unit conversions:

$$1 \text{ mi} = 5280 \text{ ft} \qquad 1 \text{ hour (1 h)} = 60 \text{ min} \qquad 1 \text{ min} = 60 \text{ s}$$

We then multiply our original value by successive factors of 1 in order to convert the units:

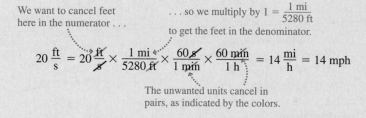

We want to cancel feet here in the numerator . . .   . . . so we multiply by $1 = \dfrac{1 \text{ mi}}{5280 \text{ ft}}$ to get the feet in the denominator.

$$20 \,\frac{\text{ft}}{\text{s}} = 20\,\frac{\text{ft}}{\text{s}} \times \frac{1 \text{ mi}}{5280 \text{ ft}} \times \frac{60 \text{ s}}{1 \text{ min}} \times \frac{60 \text{ min}}{1 \text{ h}} = 14\,\frac{\text{mi}}{\text{h}} = 14 \text{ mph}$$

The unwanted units cancel in pairs, as indicated by the colors.

Our final result of 14 miles per hour (14 mph) is a very reasonable speed for a bicycle, which gives us confidence in our answer. If we had calculated a speed of 140 miles per hour, we would have suspected that we had made an error, as this is quite a bit faster than the average bicyclist can travel!

Estimation

When scientists and engineers first approach a problem, they may do a quick measurement or calculation to establish the rough physical scale involved. This will help establish the procedures that should be used to make a more accurate measurement—or the estimate may well be all that is needed.

Suppose you see a rock fall off a cliff and would like to know how fast it was going when it hit the ground. By doing a mental comparison with the speeds of familiar objects, such as cars and bicycles, you might judge that the rock was traveling at "about" 20 mph. This is a one-significant-figure estimate. With some luck, you can probably distinguish 20 mph from either 10 mph or 30 mph, but you certainly cannot distinguish 20 mph from 21 mph just from a visual appearance. A one-significant-figure estimate or calculation, such as this estimate of speed, is called an **order-of-magnitude estimate.** An order-of-magnitude estimate is indicated by the symbol $\sim$ which indicates even less precision than the "approximately equal" symbol $\approx$. You would report your estimate of the speed of the falling rock as $v \sim 20$ mph.

A useful skill is to make reliable order-of-magnitude estimates on the basis of known information, simple reasoning, and common sense. This is a skill that is acquired by practice. Several chapters in this book will have homework problems that ask you to make order-of-magnitude estimates. Tables 1.4 and 1.5 have information that will be useful for doing estimates.

Later in the book, we will do some analysis of locomotion and look at the walking and running speeds of different animals. To help put things in perspective, it might be useful to have an estimate of how fast a person walks.

How many jellybeans are in the jar? Some reasoning about the size of one bean and the size of the jar can give you a one-significant-figure estimate.

TABLE 1.4 Some approximate lengths

	Length (m)
Circumference of the earth	4×10^7
New York to Los Angeles	5×10^6
Distance you can drive in 1 hour	1×10^5
Altitude of jet planes	1×10^4
Distance across a college campus	1000
Length of a football field	100
Length of a classroom	10
Length of your arm	1
Width of a textbook	0.1
Length of your little fingernail	0.01
Diameter of a pencil lead	1×10^{-3}
Thickness of a sheet of paper	1×10^{-4}
Diameter of a dust particle	1×10^{-5}

TABLE 1.5 Some approximate masses

	Mass (kg)
Large airliner	1×10^5
Small car	1000
Large human	100
Medium-size dog	10
Science textbook	1
Apple	0.1
Pencil	0.01
Raisin	1×10^{-3}
Fly	1×10^{-4}

EXAMPLE 1.3 How fast do you walk?

Estimate how fast you walk, in meters per second.

SOLVE In order to compute speed, we will need a distance and a time. If you walked a mile to campus, how long would this take? You'd probably say 30 minutes or so—a half an hour. Let's use this rough number in our estimate.

We have

$$\text{speed} = \frac{\text{distance}}{\text{time}} \sim \frac{1 \text{ mile}}{1/2 \text{ hour}} = 2 \frac{\text{mi}}{\text{h}}$$

But we want this in meters per second. Since our calculation is only an estimate, we use an approximate form of the conversion factor from Table 1.3:

$$1 \frac{\text{mi}}{\text{h}} \approx 0.5 \frac{\text{m}}{\text{s}}$$

This gives an approximate walking speed of about 1 m/s.

ASSESS Is this a reasonable value? Let's do another estimate. Your stride is probably about one yard long—about one meter. And you take about one step per second; next time you are walking, you can count and see. So a walking speed of 1 meter per second sounds pretty reasonable.

This sort of estimation is very valuable. We will see many cases in which we need to know an approximate value for a quantity before we start a problem or after we finish a problem, in order to assess our results.

STOP TO THINK 1.4 Rank in order, from the most to the fewest, the number of significant figures in the following numbers. For example, if B has more than C, C has the same number as A, and A has more than D, you would give your answer as B > C = A > D.

A. 0.43   B. 0.0052   C. 0.430   D. 4.321×10^{-10}

1.5 Vectors and Motion: A First Look

Many physical quantities, such as time, mass, and temperature, can be described completely by a number with a unit. For example, the mass of an object might be 6 kg and its temperature 30° C. When a physical quantity is described by a single number (with a unit), we call it a **scalar quantity.** A scalar can be positive, negative, or zero.

Many other quantities, however, have a directional quality and cannot be described by a single number. To describe the motion of a car, for example, you must specify not only how fast it is moving, but also the *direction* in which it is moving. A **vector quantity** is a quantity that has both a *size* (the "How far?" or "How fast?") and a *direction* (the "Which way?"). The size or length of a vector is called its **magnitude.** The magnitude of a vector can be positive or zero, but it cannot be negative.

Some examples of vector and scalar quantities are given at right.

We graphically represent a vector as an *arrow,* as illustrated for the velocity and force vectors in the table at right. The arrow is drawn to point in the direction of the vector quantity, and the *length* of the arrow is proportional to the magnitude of the vector quantity. If we choose to draw an arrow 2 cm long to represent a velocity with magnitude 23 m/s, we will draw an arrow 4 cm long to represent a velocity with magnitude 46 m/s. This graphical notation for representing vectors

Vectors and scalars

Scalars

Time, temperature, and weight are all *scalar* quantities. To specify your weight, only one number—150 pounds—need be given. The temperature is reported by a single number, such as 70° F, and the time of day is really just the number of seconds after midnight, although we break it into hours and minutes for convenience.

Vectors

The velocity of the race car is a *vector.* To fully specify a velocity, we need to give its magnitude (e.g., 120 mph) *and* its direction (e.g., west).

The force with which the boy pushes on his friend is another example of a vector. To completely specify this force, we must know not only how hard he pushes (the magnitude) but also in which direction.

is so useful that we often think of the arrow as being the vector itself. Thus we might say, "draw a vector pointing to the right," and you would draw an arrow pointing to the right.

When we want to represent a vector quantity with a *symbol,* we need somehow to indicate that the symbol is for a vector rather than for a scalar. We do this by drawing an arrow over the letter that represents the quantity. Thus $\vec{r}$ and $\vec{A}$ are symbols for vectors, whereas r and A, without the arrows, are symbols for scalars. In handwritten work you *must* draw arrows over all symbols that represent vectors. This may seem strange until you get used to it, but it is very important because we will often use both r and $\vec{r}$, or both A and $\vec{A}$, in the same problem, and they mean different things! Without the arrow, you will be using the same symbol with two different meanings and will likely end up making a mistake. Note that the arrow over the symbol always points to the right, regardless of which direction the actual vector points. Thus we write $\vec{r}$ or $\vec{A}$, never $\overleftarrow{r}$ or $\overleftarrow{A}$.

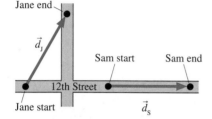

FIGURE 1.19 Two displacement vectors.

Displacement Vectors

For motion along a line, we found in Section 1.2 that the displacement is a quantity that specifies not only how *far* an object moves, but also the *direction*—to the left or to the right—that the object moves. Since displacement is a quantity that has both a magnitude ("How far") and a direction, it can be represented by a vector, the **displacement vector.** Figure 1.19 shows the displacement vector for Sam's trip that we discussed earlier. We've simply drawn an arrow—the vector—from his initial to his final positions and assigned it the symbol $\vec{d}_S$. Because $\vec{d}_S$ has both a magnitude and a direction, it is convenient to write Sam's displacement as $\vec{d}_S = (100 \text{ ft, east})$. The first value in the parentheses is the magnitude of the vector (i.e., the size of the displacement) and the second value specifies its direction.

Also shown in Figure 1.19 is the displacement vector $\vec{d}_J$ for Jane, who started on 12th Street and ended up on Vine. As with Sam, we draw her displacement vector as an arrow from her initial to her final position. In this case, $\vec{d}_J = (100 \text{ ft, } 30° \text{ east of north})$.

The boat's displacement is the straight-line connection from its initial to its final position.

Jane's trip illustrates an important point about displacement vectors. Jane started her trip on 12th Street and ended up on Vine, leading to the displacement vector shown. But to get from her initial to her final position, she needn't have walked along the straight-line path denoted by $\vec{d}_J$. If she walked east along 12th Street to the intersection, then headed north on Vine, her displacement would still be the vector shown. **An object's displacement vector is drawn from the object's initial position to its final position, regardless of the actual path followed between these two points.**

Vector Addition

Let's consider one more trip for the peripatetic Sam. In Figure 1.20, he starts at the intersection and walks east 50 ft; then he walks 100 ft to the northeast through a vacant lot. His displacement vectors for the two legs of his trip are labeled $\vec{d}_1$ and $\vec{d}_2$ in the figure.

Sam's trip consists of two legs that can be represented by the two vectors $\vec{d}_1$ and $\vec{d}_2$, but we can represent his trip as a whole, from his initial starting position to his overall final position, with the *net* displacement vector labeled $\vec{d}_{net}$. Sam's net displacement is in a sense the *sum* of the two displacements that made it up, so we can write

$$\vec{d}_{net} = \vec{d}_1 + \vec{d}_2$$

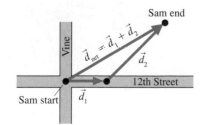

FIGURE 1.20 Sam undergoes two displacements.

Sam's net displacement thus requires the *addition* of two vectors, but vector addition obeys different rules from the addition of two scalar quantities. The directions of the two vectors, as well as their magnitudes, must be taken into account. Sam's trip suggests that we can add vectors together by putting the "tail" of one vector at the tip of the other. This idea, which is reasonable for displacement vectors, in fact is how *any* two vectors are added. Tactics Box 1.4 shows how to add two vectors $\vec{A}$ and $\vec{B}$.

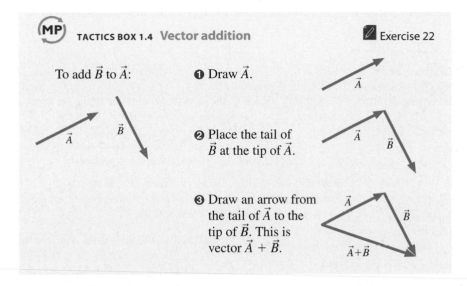

TACTICS BOX 1.4 Vector addition   Exercise 22

To add $\vec{B}$ to $\vec{A}$:   ❶ Draw $\vec{A}$.

                                      ❷ Place the tail of
                                         $\vec{B}$ at the tip of $\vec{A}$.

                                      ❸ Draw an arrow from
                                         the tail of $\vec{A}$ to the
                                         tip of $\vec{B}$. This is
                                         vector $\vec{A} + \vec{B}$.

EXAMPLE 1.4 Finding Sam's net displacement

Find Sam's net displacement in Figure 1.20, writing it in the form $\vec{d}_{net} = $ (magnitude of displacement, direction).

SOLVE We'll solve this graphically, using a ruler and a protractor. As shown in Figure 1.21, we first draw vector $\vec{d}_1$ pointing to the east, or to the right on our paper. We'll choose a scale where 1 cm on our paper represents 25 ft of Sam's neighborhood. Thus the length of $\vec{d}_1$ on the paper is 2 cm, representing the 50 ft magnitude of Sam's first displacement.

We then draw the second vector $\vec{d}_2$ with its tail at the tip of $\vec{d}_1$. Sam walked to the northeast during this leg, so we draw the direction of the vector at 45° to the horizontal; since he walked a distance of 100 ft, we draw the vector with a length of 4 cm.

The net displacement is the vector sum of the two displacements $\vec{d}_1$ and $\vec{d}_2$. It extends from the tail of $\vec{d}_1$ to the tip of $\vec{d}_2$. Using a ruler, we measure its length to be about 5.6 cm, corresponding to 5.6 × 25 ft = 140 ft. We can use a protractor to

find that the angle θ (the Greek letter *theta*) is about 30°. We thus have

$$\vec{d}_{net} = (140 \text{ ft}, 30° \text{ north of east})$$

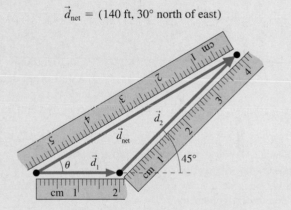

FIGURE 1.21 Graphical addition of two vectors.

Vectors and Trigonometry

Adding two vectors together using a ruler and protractor isn't very precise or practical. We need a more accurate and efficient method that can be used for adding any two vectors. Trigonometry provides us with just such a method.

Before seeing how trigonometry can be used for vector addition, let's review some of the basic ideas. Suppose we have a right triangle with hypotenuse H, angle θ, side opposite the angle O, and side adjacent to the angle A, as shown in Figure 1.22.

The sine, cosine, and tangent (which we write as "sin," "cos," and "tan") of angle θ are defined as ratios of the sides of the triangle:

$$\sin\theta = \frac{O}{H} \qquad \cos\theta = \frac{A}{H} \qquad \tan\theta = \frac{O}{A} \tag{1.3}$$

If we know two of the sides of the triangle, we can find the angle θ.

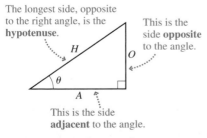

The longest side, opposite to the right angle, is the **hypotenuse**.

This is the side **opposite** to the angle.

This is the side **adjacent** to the angle.

FIGURE 1.22 A right triangle.

Conversely, if we know the angle θ and the length of one side, we can use the sine, cosine, or tangent to find the lengths of the other sides. For example, if you know θ and the length A of the adjacent side, and you need to find the hypotenuse H, you can rearrange the middle Equation 1.3 to give $H = A/\cos\theta$.

We will make regular use of these relationships in the following chapters.

EXAMPLE 1.5 Determining the sides of a triangle

A right triangle has an angle of 30° and a hypotenuse of length 10.0 m, as shown in Figure 1.23. What are the lengths of the other two sides of the triangle?

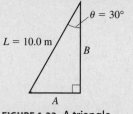

FIGURE 1.23 A triangle with two unknown sides.

SOLVE This is a problem that uses the trigonometric relationships of Equations (1.3). In Figure 1.23, the hypotenuse is labeled L, the adjacent side B, and the opposite side A. For each problem, the hypotenuse and adjacent and opposite sides will

need to be determined; they will not in general be labeled H, O, and A as in Equations (1.3).

Because we know the hypotenuse and an angle, we can compute

$$A = L\sin\theta = (10.0 \text{ m})\sin(30°) = 5.00 \text{ m}$$

$$B = L\cos\theta = (10.0 \text{ m})\cos(30°) = 8.66 \text{ m}$$

ASSESS Since we have found all three sides of a right triangle, we can check our math by seeing if the Pythagorean theorem $L^2 = A^2 + B^2$ holds for our values. We have

$$L^2 = (10.0 \text{ m})^2 = 100 \text{ m}^2$$

and

$$A^2 + B^2 = (5.00 \text{ m})^2 + (8.66 \text{ m})^2 = 100 \text{ m}^2$$

The values agree, giving us confidence that our answer is correct.

As the next example shows, we can use the rules of trigonometry to add vectors together by considering each vector to be the hypotenuse of a right triangle. If A and B are two vertices (corners) of a triangle, then we denote the length of the side between A and B as $\overline{AB}$.

EXAMPLE 1.6 Finding Sam's displacement using trigonometry

Find Sam's net displacement in Figure 1.20, this time using methods of trigonometry.

SOLVE The situation is shown once more in Figure 1.24. We will first find sides $\overline{AB}$ and $\overline{BC}$ of triangle ABC, then use these lengths to find the hypotenuse and angle of triangle DBC. (Here, $\overline{AB}$ means the side between points A and B.)

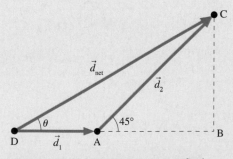

FIGURE 1.24 Using trigonometry to find Sam's displacement.

The hypotenuse AC has length 100 ft. From Equation 1.3 we have

$$\overline{AB} = \overline{AC}\cos 45° = (100 \text{ ft})\cos 45° = 71 \text{ ft}$$

$$\overline{BC} = \overline{AC}\sin 45° = (100 \text{ ft})\sin 45° = 71 \text{ ft}$$

Now consider triangle DBC. Side $\overline{BC}$, with length 71 ft, is opposite angle θ. The side adjacent to θ is $\overline{DB} = \overline{DA} + \overline{AB} = 50 \text{ ft} + 71 \text{ ft} = 121 \text{ ft}$. Then, using Equation 1.3 once again, we find

$$\tan\theta = \frac{\overline{BC}}{\overline{DB}} = \frac{71 \text{ ft}}{121 \text{ ft}} = 0.59$$

from which $\theta = \tan^{-1}(0.59) = 30°$, where $\tan^{-1}$ is the *arctangent* or *inverse tangent*. Finally, we can find the magnitude of the net displacement from the Pythagorean theorem on triangle DCB:

$$\overline{DC} = \sqrt{(\overline{DB})^2 + (\overline{BC})^2} = \sqrt{(121 \text{ ft})^2 + (71 \text{ ft})^2} = 140 \text{ ft}$$

Thus Sam's net displacement is $\vec{d}_{\text{net}} = (140 \text{ ft}, 30° \text{ north of east})$. This length, and the angle θ, agree with the values obtained using graphical methods in Example 1.4.

Velocity Vectors

We've seen that a basic quantity that describes the motion of an object is its velocity. Velocity is a vector quantity, since its specification involves not only how fast an object is moving (its speed), but also the direction in which the object is moving. We thus represent the velocity of an object by a **velocity vector** $\vec{v}$ that points in the direction of the object's motion, and whose magnitude is the object's speed. In this section, we'll learn how to draw velocity vectors on a motion diagram.

Figure 1.25a shows (using the particle model) the motion diagram of a car accelerating from rest. We've drawn vectors showing the car's displacement between successive positions of the motion diagram. To draw the car's velocity vectors, we note first that the direction of the displacement vector indicates the direction of motion between successive points in the motion diagram. But the velocity of an object also points in the direction of motion, so **an object's velocity vector points in the same direction as its displacement vector.**

Second, we've already noted that the magnitude of the velocity vector—"how fast"—is the object's speed. From Equation 1.1, the speed of an object is given by

$$\text{speed} = \frac{\text{distance traveled in a given time interval}}{\text{time interval}}$$

Now between any two successive points of the motion diagram, the distance traveled is equal to the magnitude of the displacement, that is, to the length of the displacement vector. Further, in a motion diagram the time interval between successive points is always the same. This means that the speed of an object, and thus the length of the velocity vector, is proportional to the length of the displacement vector between successive points on a motion diagram. Consequently, the vectors connecting each dot of a motion diagram to the next, which we previously labeled as displacement vectors, could equally well be identified as velocity vectors. This is shown for our car in Figure 1.25b. From now on, we'll show and label velocity vectors on motion diagrams rather than displacement vectors.

NOTE ▶ Again, the velocity vectors shown in Figure 1.25 are actually *average* velocity vectors. Because the velocity is increasing, it's actually a bit less than this average at the start of each time interval, and a bit more at the end. In Chapter 2 we'll refine these ideas and develop the idea of instantaneous velocity. ◀

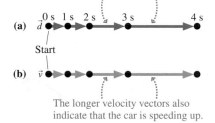

The displacement vectors are lengthening. This means the car is speeding up.

The longer velocity vectors also indicate that the car is speeding up.

FIGURE 1.25 The motion diagram for a car starting from rest.

EXAMPLE 1.7 Drawing a ball's motion diagram

Jake hits a ball at a 60° angle from the horizontal. It is caught by Jim. Draw a motion diagram of the ball that shows velocity vectors rather than displacement vectors.

SOLVE This example is typical of how many problems in science and engineering are worded. The problem does not give a clear statement of where the motion begins or ends. Are we interested in the motion of the ball just during the time it is in the air between Jake and Jim? What about the motion *as* Jake hits it (ball rapidly speeding up) or *as* Jim catches it (ball rapidly slowing down)? Should we include Jim dropping the ball after he catches it? The point is that *you* will often be called on to make a *reasonable interpretation* of a problem statement. In this problem, the details of hitting and catching the ball are complex. The motion of the ball through the air is easier to describe, and it's a motion you might expect to learn about in a physics class. So our *interpretation* is that the motion diagram should start as the ball leaves Jake's bat (ball already moving) and should end the instant it touches Jim's hand (ball still moving). We will model the ball as a particle.

With this interpretation in mind, Figure 1.26 shows the motion diagram of the ball. Notice how, in contrast to the car of Figure 1.25, the ball is already moving as the motion diagram movie begins. As before, the velocity vectors are found by connecting the dots with arrows. You can see that the velocity vectors get shorter (ball slowing down), get longer (ball speeding up), and change direction. Each $\vec{v}$ is different, so this is *not* constant velocity motion.

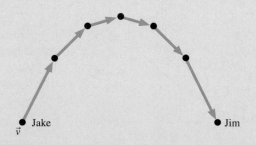

FIGURE 1.26 The motion diagram of a ball traveling from Jake to Jim.

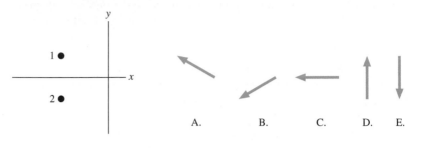

STOP TO THINK 1.5   A particle moves from position 1 to position 2 during the interval Δt. Which vector shows the particle's velocity?

1.6 Making Models: The Power of Physics

You've now seen that we often make drastic simplifications when we analyze a situation in physics. For example, we may represent a speeding car as simply a moving dot. When we analyze a situation or solve a problem, we are making a *model* of a physical situation, an idealized version of the problem that allows us to focus on its most important features.

We will introduce many different models that allow us to focus in this way. Earlier in the chapter we introduced the *particle model,* which allowed us to visualize the motion of an object as a single particle. By ignoring the details of the object, we could concentrate on the object's overall motion. This is a very useful model in many cases, as we have seen.

Another model that we will use regularly is the **atomic model.** Matter is made of atoms, and it is useful to take this into account. In Chapter 8, we will model solids as being composed of particle-like atoms connected by springs. A rubber band modeled in this way is shown in Figure 1.27. This is a simplification, but a simplification grounded in reality: For the purpose of analyzing the elastic behavior of solids, what really matters is that the atoms are connected by bonds—and that these bonds behave quite a bit like little springs.

Applying the concepts that you learn in this course to real problems is also a form of modeling: You are choosing a simplified way of looking at the problem that avoids extraneous detail. It is a skill you will acquire with practice.

Mathematical Forms

In this course, we will also make frequent use of *mathematical relationships* between physical quantities. In fact, we have already introduced some in this chapter. An important part of physics—but not the only part!—is analyzing nature using such mathematical relationships. As we do so, we find that certain ones show up in many different contexts.

As an example, consider the fairly simple mathematical equation

$$y = 4x^2 \tag{1.4}$$

This relationship between y and x is plotted in Figure 1.28a. As x increases, y increases more rapidly, so that the curve gets ever steeper.

In Figure 1.28b, we plot the important physical quantity known as the *kinetic energy K*, which we'll learn about in Chapter 10. The kinetic energy depends on the speed v of an object according to the equation

$$K = \frac{1}{2}mv^2 \tag{1.5}$$

Note how similar the shape of this curve is to that of Figure 1.28a.

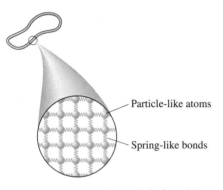

FIGURE 1.27 An atomic model of a rubber band.

Particle-like atoms

Spring-like bonds

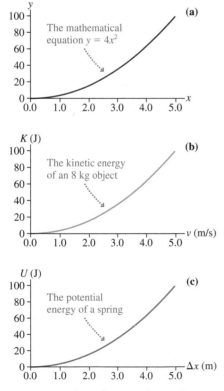

FIGURE 1.28 A mathematical equation models a physical quantity.

Finally, in Figure 1.28c we plot the *potential energy U* of a spring. In Chapter 10 we'll find that this energy depends on the displacement *x* of the end of the spring according to the equation

$$U = \frac{1}{2}kx^2 \qquad (1.6)$$

The shape of this curve is the same as the previous two.

In looking at Figure 1.28, then, it is evident that graphs of all three of these expressions have the same overall appearance. They differ in their variables; kinetic energy depends on *v* and potential energy on *x*. They also differ in some of the constants in the equations—the $\frac{1}{2}m$ for kinetic energy, the $\frac{1}{2}k$ for potential energy. But they are all versions of the same mathematical relationship. We say that all three equations have the same mathematical *form*.

The mathematical form that all three of the plots in Figure 1.28 share is that they depend on the *square* of their variable. The variable *y* in Equation 1.4 depends on the square of *x*. The kinetic energy depends on the square of *v*. And the potential energy depends on the square of *x*. We express this dependence of each equation on the square of its variable using the symbol ∝. This symbol is often read as "varies as" or "depends on." We can then write

$$y \propto x^2 \qquad K \propto v^2 \qquad U \propto x^2$$

Writing expressions in this way emphasizes how the quantity depends on its variable. In many problems the form of this variation is more important than other details in the equations, such as *m* or factors of 1/2.

Table 1.6 shows four common mathematical forms that we'll encounter in this book, with some examples of each. There will be others as well. Because certain physical principles share the same mathematical form, we can use these similarities to help us learn new concepts. Thus, much of what you learn about gravity in Chapter 6 can later be applied to electricity, because the force of gravity and the electric force share the same mathematical form, $y \propto 1/x^2$.

As we meet each mathematical model for the first time, we will insert a section in the text that gives an overview of the form. When we see the form again, we will refer back to that overview section.

TABLE 1.6 Common mathematical forms

Mathematical form	Physical example
Proportional $y \propto x$	Dependence of acceleration on force Force due to a spring
Square $y \propto x^2$	Potential energy of a spring Kinetic energy
Inverse $y \propto \dfrac{1}{x}$	Potential energy of two charges Magnetic field due to a current
Inverse square $y \propto \dfrac{1}{x^2}$	Gravitational force between two masses Electric force between two charges

1.7 Where Do We Go from Here?

This first chapter has been an introduction to some of the fundamental ideas about motion and some of the basic techniques that you will use in the rest of the course. You have seen some examples of how to make *models* of a physical situation, thereby focusing in on the essential elements of the situation. You have learned some practical ideas, such as how to convert quantities from one kind of units to another. The rest of this book—and the rest of your course—will extend these themes. You will learn how to model many kinds of physical systems, and learn the technical skills needed to set up and solve problems using these models.

As we go along, you will learn a set of very practical and useful models that will allow you to discuss and analyze a very wide range of problems. For example,

■ We will study the forces in joints, the flow of blood in arteries, how you make the sounds of speech, and how your eyes analyze light from the world around you.

■ We will look at steel beams, bones, and spiders' silk and find which is strongest. The answer may surprise you!

Radio signals travel from tower to receiver in the form of invisible *electromagnetic waves*.

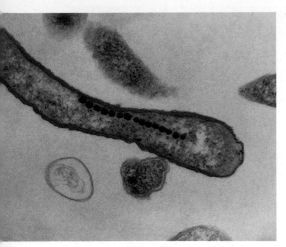

This bacterium contains microscopic magnetic particles that allow it to navigate using the earth's magnetic field.

- We will learn how radio signals are generated, how they travel, how an antenna picks them up, how the speaker in the radio turns the signals into sound, and how this sound is analyzed by your ears.
- We will look at how electricity is generated and transmitted, whether it's the electricity that comes to your house or the electrical signal that propagates through your nerve cells.

This is all very practical, and will be quite useful in your current field of study. But in the midst of solving such practical problems, you will also learn what physics teaches us about the world, and how truly amazing some of it is. Did you know that:

- All of the energy that you use on a daily basis comes from nuclear energy?
- Certain animals, such as mice, can survive falls from any height?
- There are creatures that use the earth's magnetic field to navigate?
- It is possible for something to be in two places at once, or for two things to be in the same place at the same time?
- It is possible to go on a long trip—and return years younger than your twin?

As you work through this course and learn to solve problems, don't lose sight of this big picture! The universe is a remarkable place, and physics is a wonderful tool for showing both its depth and its underlying simplicity.

SUMMARY

The goal of Chapter 1 has been to introduce the fundamental concepts of motion and to review the related basic mathematical principles.

IMPORTANT CONCEPTS

Motion Diagrams

The particle model represents a moving object as if all its mass were concentrated at a single point. Using this model, we can represent motion with a **motion diagram** where dots indicate the object's position at successive times. In a motion diagram, the time interval between successive dots is always the same.

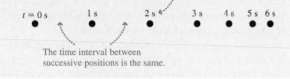

Each dot represents the position of the object. Each position is labeled with the time at which the dot was there.

The time interval between successive positions is the same.

Scalars and Vectors

Scalar quantities have only a magnitude, and can be represented by a single number. Temperature, time, and mass are scalars.

A vector is a quantity described by both a magnitude and a direction. Velocity and displacement are vectors.

Velocity vectors can be drawn on a motion diagram by connecting successive points with a vector.

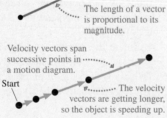

Direction

The length of a vector is proportional to its magnitude.

Velocity vectors span successive points in a motion diagram.

Start

The velocity vectors are getting longer, so the object is speeding up.

Describing Motion

Position locates an object with respect to a chosen coordinate system. It is described by a **coordinate**.

The *coordinate* is the variable used to describe the position.

This cow is at $x = -5$ miles. This car is at $x = +4$ miles.

A change in position is called a **displacement**. For motion along a line, a displacement is a signed quantity. The displacement from x_i to x_f is $\Delta x = x_f - x_i$.

Time is measured from a particular instant to which we assign $t = 0$. A **time interval** is the elapsed time between two specific instants t_i and t_f. It is given by $\Delta t = t_f - t_i$.

Velocity is the ratio of the displacement of an object to the time interval during which this displacement occurs:

$$v = \frac{\Delta x}{\Delta t}$$

Units

Every measurement of a quantity must include a unit.

The standard system of units used in science is the SI system. Common SI units include:

- Length: meters (m)
- Time: seconds (s)
- Mass: kilograms (kg)

APPLICATIONS

Working with Numbers

In scientific notation, a number is expressed as a decimal number between 1 and 10 multiplied by a power of ten. In scientific notation, the diameter of the earth is 1.27×10^7 m.

A prefix can be used before a unit to indicate a multiple of 10 or 1/10. Thus we can write the diameter of the earth as 12,700 km, where the k in km denotes 1000.

We can perform a unit conversion to convert the diameter of the earth to a different unit, such as miles. We do so by multiplying by a conversion factor equal to one, such as $1 = 1$ mi/1.61 km.

Significant figures are reliably known digits. The number of significant figures for:

- **Multiplication, division, and powers** is set by the value with the fewest significant figures.
- **Addition and subtraction** is set by the value with the smallest number of decimal places.

An order-of-magnitude-estimate is an estimate that has an accuracy of about one significant figure. Such estimates are usually made using rough numbers from everyday experience.

(MP) For instructor-assigned homework, go to www.masteringphysics.com

Problems labeled BIO are of biological or medical interest.

Problem difficulty is labeled as | (straightforward) to ||||| (challenging).

QUESTIONS

Conceptual Questions

1. a. Write a paragraph describing the *particle model*. What is it, and why is it important?
    b. Give two examples of situations, different from those described in the text, for which the particle model is appropriate.
    c. Give an example of a situation, different from those described in the text, for which it would be inappropriate.
2. A softball player slides into second base. Use the particle model to draw a motion diagram of the player from the time he begins to slide until he reaches the base. Number the dots in order, starting with zero.
3. A car travels to the left at a steady speed for a few seconds, then brakes for a stop sign. Use the particle model to draw a motion diagram of the car for the entire motion described here. Number the dots in order, starting with zero.
4. A ball is dropped from the roof of a tall building and students in a physics class are asked to sketch a motion diagram for this situation. A student submits the diagram shown in Figure Q1.4. Is the diagram correct? Explain.

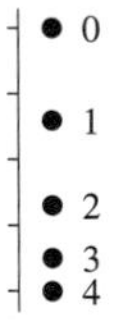

    • 0
    • 1
    • 2
    • 3
    • 4

    FIGURE Q1.4
5. Write a sentence or two describing the difference between position and displacement. Give one example of each.
6. Doug starts at position $x = 10$ m. He then undergoes a displacement $\Delta x = +15$ m. During his motion, is it possible that he passed through the origin, at $x = 0$ m?
7. Write a sentence or two describing the difference between speed and velocity. Give one example of each.
8. The motion of a skateboard along a horizontal axis is observed for 5 s. The initial position of the skateboard is negative with respect to a chosen origin, and its velocity throughout the 5 s is also negative. At the end of the observation time, is the skateboard closer to or further from the origin than initially? Explain.
9. Can the velocity of an object be positive during a time interval in which its position is always negative? Can its velocity be positive during a time interval in which its displacement is negative?
10. Two friends watch a jogger complete a 400 m lap around the track in 100 s. One of the friends states, "The jogger's velocity was 4 m/s during this lap." The second friend objects, saying, "No, the jogger's speed was 4 m/s." Who is correct? Justify your answer.
11. A softball player hits the ball and starts running toward first base. Draw a motion diagram, using the particle model, showing her position and her velocity vectors during the first few seconds of her run.
12. A child is sledding on a smooth, level patch of snow. She encounters a rocky patch and slows to a stop. Draw a motion diagram, using the particle model, showing her position and her velocity vectors.
13. A roof tile falls straight down from a two-story building. It lands in a swimming pool and settles gently to the bottom. Draw a motion diagram, using the particle model, showing the tile's position and its velocity vectors.
14. Your roommate drops a tennis ball from a third story balcony. It hits the sidewalk and bounces as high as the second story. Draw a motion diagram, using the particle model, showing the ball's velocity vectors from the time it is released until it reaches the maximum height on its bounce.
15. A car is driving north at a steady speed. It makes a gradual 90° left turn without losing speed, then continues driving to the west. Draw a motion diagram, using the particle model, showing the car's velocity vectors as seen from a helicopter hovering over the highway.
16. A toy car rolls down a ramp, then across a smooth, horizontal floor. Draw a motion diagram, using the particle model, showing the car's velocity vectors.
17. Estimate the average speed with which you go from home to campus (or another trip you commonly make) via whatever mode of transportation you use most commonly. Give your answer in both mph and m/s. Describe how you arrived at this estimate.
18. Estimate the number of times you sneezed during the past year. Describe how you arrived at this estimate.
19. Density is the ratio of an object's mass to its volume. Would you expect density to be a vector or a scalar quantity? Explain.

Multiple-Choice Questions

20. | A student walks 1.0 mi west and then 1.0 mi north. Afterward, how far is she from her starting point?
    A. 1.0 mi   B. 1.4 mi   C. 1.6 mi   D. 2.0 mi
21. | Which of the following motions is described by the motion diagram of Figure Q1.21?
    A. An ice skater gliding across the ice.
    B. An airplane braking to a stop after landing.
    C. A car pulling away from a stop sign.
    D. A pool ball bouncing off a cushion and reversing direction.

    0 1   2   3   4   5
    •• •   •   •   •   •

    FIGURE Q1.21 
22. | A bird flies 3.0 km due west and then 2.0 km due north. What is the magnitude of the bird's displacement?
    A. 2.0 km   B. 3.0 km   C. 3.6 km   D. 5.0 km

23. ⫿ A bird flies 3.0 km due west and then 2.0 km due north. Another bird flies 2.0 km due west and 3.0 km due north. What is the angle between the net displacement vectors for the two birds?
    A. 23°   B. 34°   C. 56°   D. 90°
24. ‖ A woman walks briskly at 2.00 m/s. How much time will it take her to walk one mile?
    A. 8.30 min B. 13.4 min C. 21.7 min D. 30.0 min
25. ⎮ Compute 3.24 m + 0.532 m to the correct number of significant figures.
    A. 3.7 m   B. 3.77 m   C. 3.772 m   D. 3.7720 m
26. ⎮ A rectangle has length 3.24 m and height 0.532 m. To the correct number of significant figures, what is its area?
    A. 1.72 m^2 B. 1.723 m^2
    C. 1.7236 m^2 D. 1.72368 m^2

27. ⎮ The earth formed 4.57×10^9 years ago. What is this time in seconds?
    A. 1.67×10^{12} s   B. 4.01×10^{13} s
    C. 2.40×10^{15} s   D. 1.44×10^{17} s
28. ⎮ An object's density ρ is defined as the ratio of its mass to its volume: $\rho = M/V$. The earth's mass is 5.94×10^{24} kg, and its volume is 1.08×10^{12} km^3. What is the earth's density?
    A. 5.50×10^3 kg/m^3 B. 5.50×10^6 kg/m^3
    C. 5.50×10^9 kg/m^3 D. 5.50×10^{12} kg/m^3

PROBLEMS

Section 1.1 Motion: A First Look

1. ⎮ You've made a video of a car as it skids to a halt to avoid hitting an object in the road. Use the images from the video to draw a motion diagram of the car from the time the skid begins until the car is stopped.
2. ⎮ A man rides a bike along a straight road for 5 min, then has a flat tire. He stops for 5 min to repair the flat, but then realizes he cannot fix it. He continues his journey by walking the rest of the way, which takes him another 10 min. Use the particle model to draw a motion diagram of the man for the entire motion described here. Number the dots in order, starting with zero.
3. ⎮ A jogger running east at a steady pace suddenly develops a cramp. He is lucky: A westbound bus is sitting at a bus stop just ahead. He gets on the bus and enjoys a quick ride home. Use the particle model to draw a motion diagram of the jogger for the entire motion described here. Number the dots in order, starting with zero.

Section 1.2 Position and Time: Putting Numbers on Nature

4. ⎮ Figure P1.4 shows Sue between her home and the cinema. What is Sue's position x if
    a. her home is the origin?   b. the cinema is the origin?

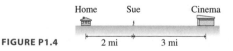

FIGURE P1.4   2 mi   3 mi

5. ⎮ Keira starts at position $x = 23$ m along a coordinate axis. She then undergoes a displacement of -45 m. What is her final position?
6. ⎮ A car travels along a straight east-west road. A coordinate system is established on the road, with x increasing to the east. The car ends up 14 mi west of the intersection with Mulberry Road. If its displacement was -23 mi, how far from and on which side of Mulberry Road did it start?

7. ⎮ Lisa is enjoying a bicycle ride on the country road shown earlier in Figure 1.7. Suppose that increasing x means moving east. At noon, she is 2 mi west of the post office. A half hour later, she is 3 mi east of the post office. What is her displacement Δx during that half hour?

Section 1.3 Velocity

8. ⎮ A security guard walks 110 m in one trip around the perimeter of the building. It takes him 240 s to make this trip. What is his speed?
9. ⎮ List the following items in order of decreasing speed, from greatest to least: (i) A wind-up toy car that moves 0.15 m in 2.5 s. (ii) A soccer ball that rolls 2.3 m in 0.55 s. (iii) A bicycle that travels 0.60 m in 0.075 s. (iv) A cat that runs 8.0 m in 2.0 s.
10. ‖ Figure P1.10 shows the motion diagram for a horse galloping in one direction along a straight path. Not every dot is labeled, but the dots are at equally spaced instants of time. What is the horse's velocity

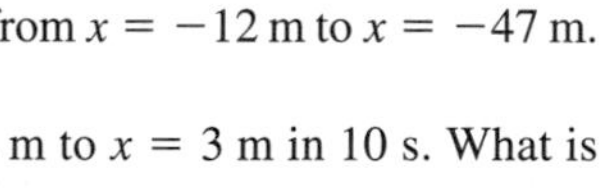

    70 s 50 s 30 s 10 s
    •   •   • • • •   •   • →x (m)
    50 150 250 350 450 550 650
    FIGURE P1.10

    a. during the first ten seconds of its gallop?
    b. during the interval from 30 s to 40 s?
    c. during the interval from 50 s to 70 s?
11. ‖ It takes Harry 35 s to walk from $x = -12$ m to $x = -47$ m. What is his velocity?
12. ‖ A dog trots from $x = -12$ m to $x = 3$ m in 10 s. What is its velocity?
13. ⎮ A ball rolling along a straight line with velocity 0.35 m/s goes from $x = 2.1$ m to $x = 7.3$ m. How much time does this take?

Section 1.4 A Sense of Scale: Significant Figures, Scientific Notation, and Units

14. ‖ Convert the following to SI units:
    a. 9.12 μs   b. 3.42 km
    c. 44 cm/ms   d. 80 km/hour

15. ‖ Convert the following to SI units:
    a. 8.0 in   b. 66 ft/s   c. 60 mph

16. ∣ Convert the following to SI units:
    a. 1.0 hour   b. 1.0 day   c. 1.0 year

17. ‖ List the following three speeds in order, from smallest to largest: 1 mm per μs, 1 km per ks, 1 cm per ms.

18. ∣ How many significant figures does each of the following numbers have?
    a. 6.21   b. 62.1   c. 0.620   d. 0.062

19. ∣ How many significant figures does each of the following numbers have?
    a. 0.621   b. 0.006200
    c. 1.0621   d. 6.21×10^3

20. ∣ Compute the following numbers to 3 significant figures.
    a. 33.3×25.4 b. $33.3 - 25.4$
    c. $\sqrt{33.3}$ d. $333.3 \div 25.4$

21. ∣ The Empire State Building has a height of 1250 ft. Express this height in meters, giving your result in scientific notation with three significant figures.

22. ‖ We interpret a measurement of $x_1 = 77$ m to mean a distance somewhere between 76.5 m and 77.5 m, thus rounding to 77 m. Similarly, $x_2 = 813.8$ m implies a distance between 813.75 m and 813.85 m. Use these values to justify the rule for determining significant figures in addition by finding the smallest and largest possible values of $x_1 + x_2$.

23. ‖ We can interpret a distance traveled of 7.43 m to mean a distance between 7.425 m and 7.435 m, thus rounding to 7.43 m. Similarly, a time interval of 0.83 s implies a time interval between 0.825 s and 0.835 s. Use these values to justify the rule for determining significant figures in division by finding the smallest and largest possible values of the speed.

24. ‖ Estimate (don't measure!) the length of a typical car. Give your answer in both feet and meters. Briefly describe how you arrived at this estimate.

25. ∣ Estimate the height of a telephone pole. Give your answer in both feet and meters. Briefly describe how you arrived at this estimate.

26. ∣ Estimate the average speed with which the hair on your
BIO head grows. Give your answer in both m/s and μm/hr. Briefly describe how you arrived at this estimate.

27. ‖ Estimate the average speed at which your fingernails grow,
BIO in both m/s and μm/hr. Briefly describe how you arrived at this estimate.

Section 1.5 Vectors and Motion: A First Look

28. ∣ Carol and Robin share a house. To get to work, Carol walks north 2.0 km while Robin drives west 7.5 km. How far apart are their workplaces?

29. ∣ Joe and Max shake hands and say goodbye. Joe walks east 0.55 km to a coffee shop, and Max flags a cab and rides north 3.25 km to a bookstore. How far apart are their destinations?

30. ‖ A city has streets laid out in a square grid, with each block 135 m long. If you drive north for three blocks, then west for two blocks, how far are you from your starting point?

31. ‖ A butterfly flies from the top of a tree in the center of a garden to rest on top of a red flower at the garden's edge. The tree is 8.0 m taller than the flower, and the garden is 12 m wide. Determine the magnitude of the butterfly's displacement.

32. ‖ A garden has a circular path of radius 50 m. John starts at the easternmost point on this path, then walks counterclockwise around the path until he is at its southernmost point. What is John's displacement? Use the (magnitude, direction) notation for your answer.

33. ‖ Anna walks 130 m due north, 50 m due east, and then 40 m due south. What is her net displacement? Use the (magnitude, direction) notation for your answer.

34. ‖‖‖ A ball on a porch rolls 60 cm to the porch's edge, drops 40 cm, continues rolling on the grass, and eventually stops 80 cm from the porch's edge. What is the magnitude of the ball's net displacement, in centimeters?

Section 1.6 Making Models: The Power of Physics

35. ‖ Sketch a graph of each function over a range of x values from 0.1 to 10. You may need to use different scales on the y axis.
    a. $y = x$ (proportional)
    b. $y = x^2$ (square)
    c. $y = \dfrac{1}{x}$ (inverse)
    d. $y = \dfrac{1}{x^2}$ (inverse square)

General Problems

Problems 36 through 42 are motion problems similar to those you will learn to solve in Chapter 2. For now, simply *interpret* the problem by drawing a motion diagram showing the object's position and its velocity vectors. **Do *not* solve these problems** or do any mathematics.

36. ‖ A Porsche accelerates from a stoplight at 5.0 m/s^2 for five seconds, then coasts for three more seconds. How far has it traveled?

37. ‖ Billy drops a watermelon from the top of a three-story building, 10 m above the sidewalk. How fast is the watermelon going when it hits?

38. ‖ Sam is recklessly driving 60 mph in a 30 mph speed zone when he suddenly sees the police. He steps on the brakes and slows to 30 mph in three seconds, looking nonchalant as he passes the officer. How far does he travel while braking?

39. ‖ A speed skater moving across frictionless ice at 8.0 m/s hits a 5.0-m-wide patch of rough ice. She slows steadily, then continues on at 6.0 m/s. What is her acceleration on the rough ice?

40. ‖ You would like to stick a wet spit wad on the ceiling, so you toss it straight up with a speed of 10 m/s. How long does it take to reach the ceiling, 3.0 m above?

41. ‖ A ball rolls along a smooth horizontal floor at 10 m/s, then starts up a 20° ramp. How high does it go before rolling back down?

42. ‖ A motorist is traveling at 20 m/s. He is 60 m from a stop light when he sees it turn yellow. His reaction time, before stepping on the brake, is 0.50 s. What steady deceleration while braking will bring him to a stop right at the light?

Problems 43 through 50 show a motion diagram. For each of these problems, write a one or two sentence "story" about a *real object* that has this motion diagram. Your stories should talk about people or objects by name and say what they are doing. Problems 36–42 are examples of motion short stories.

43. ‖

FIGURE P1.43

44. ‖

FIGURE P1.44

45. |

FIGURE P1.45

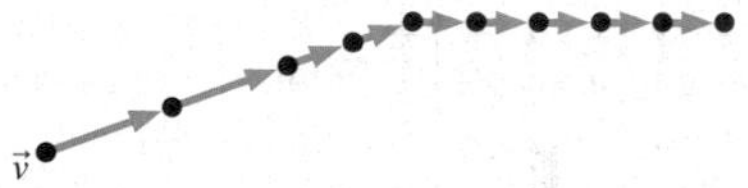

46. ‖

Start

FIGURE P1.46

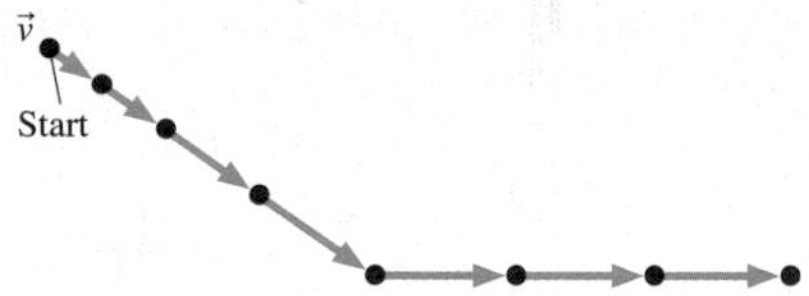

47. |

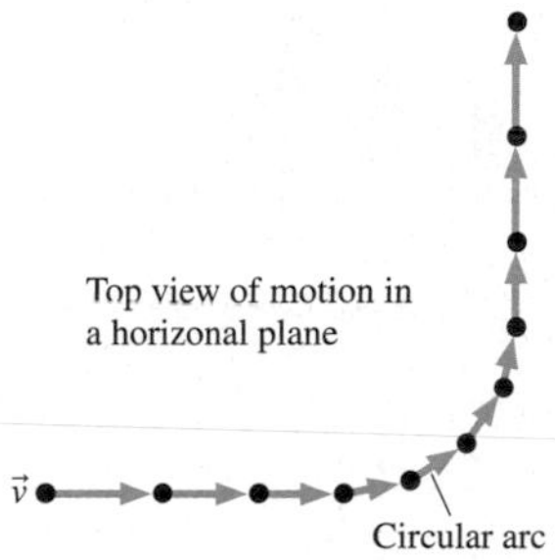

Top view of motion in
a horizonal plane

$\vec{v}$

FIGURE P1.47   Circular arc

48. ‖

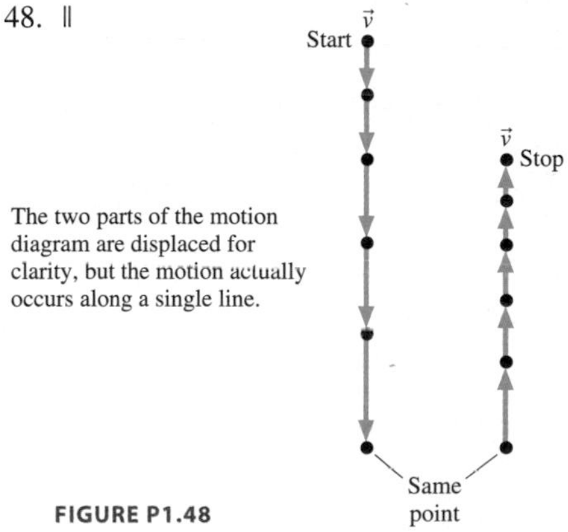

$\vec{v}$
Start

$\vec{v}$
Stop

The two parts of the motion
diagram are displaced for
clarity, but the motion actually
occurs along a single line.

Same
point

FIGURE P1.48

49. | Side view of motion
in a vertical plane
Circular arc

$\vec{v}$

FIGURE P1.49

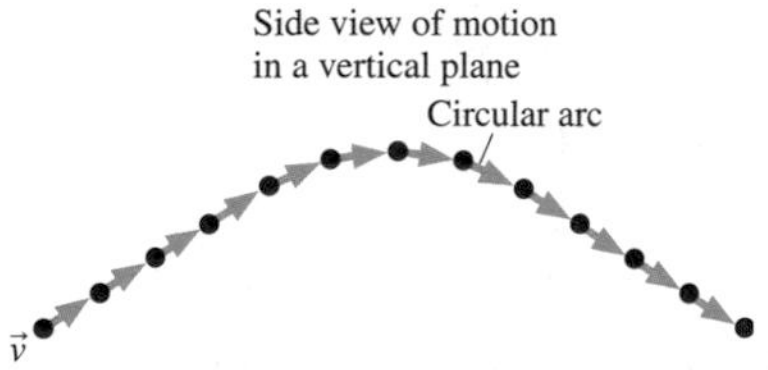

50. ‖

$\vec{v}$

FIGURE P1.50

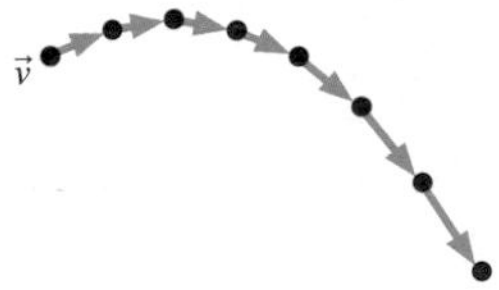

51. ‖ How many inches does light travel in one nanosecond? The speed of light is 3.0×10^8 m/s.

52. ‖ Joseph watches the roadside mile markers during a long car trip on an interstate highway. He notices that at 10:45 A.M. they are passing a marker labeled 101, and at 11:00 A.M. the car reaches marker 119. What is the car's speed, in mph?

53. ‖ Alberta is going to have dinner at her grandmother's house, but she is running a bit behind schedule. As she gets onto the highway, she knows that she must exit the highway within 45 min if she is not going to arrive late. Her exit is 32 mi away. What is the slowest speed at which she could drive and still arrive in time? Express your answer in miles per hour.

54. | Dave hits a hockey puck at 10 m/s toward the goal. If the puck travels 15 m to the goal, how long does it take to get there?

55. | In many living systems, information propagates through
BIO the body in the form of electrical signals called *nerve impulses*. Nerve impulses travel along nerve fibers consisting of many *axons* (fiber-like extensions of nerve cells) aligned in series. For example, when your foot accidentally kicks a rock the feeling of pain is transmitted from the foot to your brain along hundreds of axons that span the distance between your foot and head. Axons come in two varieties: one with a sheath called myelin around it, the other without. Myelinated (sheathed) axons conduct nerve impulses roughly 100 times faster than unmyelinated (unsheathed) axons. Nerve impulses travel at a speed of about 1.0 m/s or 2.2 mph (a walking pace) along unmyelinated axons and about 100 m/s or 220 mph (like a race car) along myelinated axons. Figure P1.55 shows three equal-length nerve fibers consisting of eight axons in a row. Nerve impulses enter at the left side simultaneously and travel to the right.
   a. Draw motion diagrams for the nerve impulses traveling along fibers A, B, and C.
   b. Which nerve impulse arrives at the right side first?
   c. Which will be last?

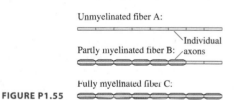

Unmyelinated fiber A:

Individual
Partly myelinated fiber B: ╱ axons

Fully myelinated fiber C:

FIGURE P1.55

56. ‖‖ The bacterium *Escherichia coli* (or *E. coli*) is a single-celled
BIO organism that lives in the gut of healthy humans and animals. Its body shape can be modeled as a 2-μm-long cylinder with a 1 μm diameter, and it has a mass of 1×10^{-12} g. Its chromosome consists of a single double-stranded chain of DNA 700 times longer than its body length. The bacterium moves at a constant speed of 20 μm/s, though not always in the same direction. Answer the following questions about *E. coli* using SI units (unless specifically requested otherwise) and correct significant figures. 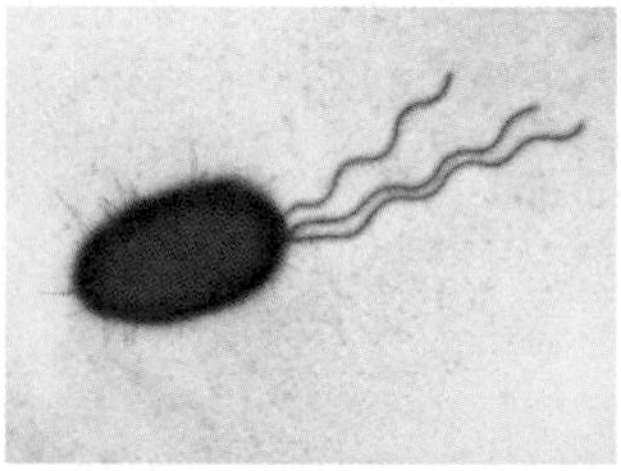
   a. What is its length?
   b. Diameter?
   c. Mass?
   d. What is the length of its DNA, in millimeters?
   e. If the organism were to move along a straight path, how many meters would it travel in one day?

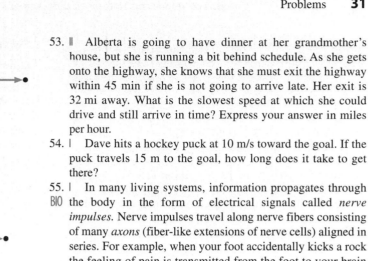

57. |||| The bacterium *Escherichia coli* (or *E. coli*) is a single-celled organism that lives in the gut of healthy humans and animals. When grown in a uniform medium rich in salts and amino acids, it swims along zig-zag paths at a constant speed. Figure P1.57 shows the positions of an *E. coli* as it moves

BIO

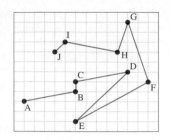

FIGURE P1.57

from point A to point J. Each segment of the motion can be identified by two letters, such as segment BC. During which segments, if any, does the bacterium have the same
   a. Displacement?   b. Speed?   c. Velocity?

58. || In 2003, the population of the United States was 291 million people. The per-capita income was $31,459. What was the total income of everyone in the United States? Express your answer in scientific notation, with the correct number of significant figures.

59. ||| The sun is 30° above the horizon. It makes a 52-m-long shadow of a tall tree. How high is the tree?

60. ||| An airplane accelerates down the runway for 500 m, then lifts into the air at an angle of 20.0 degrees above the ground. When the plane has traveled 300 m in the air, what is its distance from where it started on the runway?

61. ||| Starting from its nest, an eagle flies at constant speed for 3.0 min due east, then 4.0 min due north. From there the eagle flies directly to its nest at the same speed. How long is the eagle in the air?

62. ||| John walks 1.00 km north, then turns right and walks 1.00 km east. His speed is 1.50 m/s during the entire stroll.
   a. What is the magnitude of his displacement, from beginning to end?
   b. If Jane starts at the same time and place as John, but walks in a straight line to the endpoint of John's stroll, at what speed should she walk to arrive at the endpoint just when John does?

Passage Problems

Growth Speed

The images of trees in Figure P1.63 come from a catalog advertising fast-growing trees. If we mark the position of the top of the tree in the successive years, as shown in the graph in the figure, we obtain a motion diagram much like ones we have seen for other kinds of motion. The motion isn't steady, of course. In some months the tree grows rapidly; in other months, quite slowly. We can see, though, that the average speed of growth is fairly constant for the first few years.

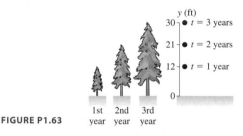

FIGURE P1.63

63. || What is the tree's speed of growth, in feet per year, from $t = 1$ yr to $t = 3$ yr?
   A. 12 ft/yr   B. 9 ft/yr   C. 6 ft/yr   D. 3 ft/yr

64. || What is this speed in m/s?
   A. 9×10^{-8} m/s   B. 3×10^{-9} m/s
   C. 5×10^{-6} m/s   D. 2×10^{-6} m/s

65. | At the end of year 3, a rope is tied to the very top of the tree to steady it. This rope is staked into the ground 15 feet away from the tree. What angle does the rope make with the ground?
   A. 63°   B. 60°   C. 30°   D. 27°

STOP TO THINK ANSWERS

Stop to Think 1.1: B. The images of B are further apart, so it travels a larger distance than does A during the same intervals of time.

Stop to Think 1.2: A. Dropped ball. **B.** Dust particle. **C.** Descending rocket.

Stop to Think 1.3: C. Depending on her initial positive position, and how far she moves in the negative direction, she could end up on either side of the origin.

Stop to Think 1.4: D > C > B = A.

Stop to Think 1.5: E. The velocity vector is found by connecting one dot on the motion diagram to the next.

2 MOTION IN ONE DIMENSION

A horse can run at 35 mph, much faster than a human. And yet, surprisingly, a man can win a race against a horse if the length of the course is right. When, and why, can a man outrun a horse?

Looking Ahead ▶▶

The goal of Chapter 2 is to describe and analyze linear motion. In this chapter you will learn to:

▶ Represent one-dimensional (straight-line) motion in different ways: using numbers, graphs, pictures, words, and equations.

▶ Use general physics problem-solving strategies and techniques.

▶ Solve problems of motion in one dimension.

Looking Back ◀◀

The material in this chapter builds on what you learned in Chapter 1 about motion and measurement. Please review:

◀ Sections 1.2–1.3 Definitions of displacement and velocity.

◀ Section 1.4 Units and significant figures.

◀ Section 1.5 Velocity vectors and motion diagrams.

A race, whether between runners, bicyclists, or drag racers, exemplifies the idea of motion. Today, we use electronic stopwatches, video recorders, and other sophisticated instruments to analyze motion, but it hasn't always been this way. Galileo, who in the early 1600s was the first scientist to study motion experimentally, used his pulse to measure time! Galileo made a useful distinction between the *cause* of motion and the *description* of motion. The modern name for the mathematical description of motion, without regard to causes, is **kinematics.** The term comes from the Greek word *kinema,* meaning "movement." You know this word through its English variation *cinema*—motion pictures!

This chapter will focus on the kinematics of motion in one dimension—that is, motion along a straight line. The motion of runners, drag racers, and skiers can be considered as motion in one dimension. But even in one dimension, there are many interesting issues to consider. For example, which is faster, a man or a horse? The answer depends on what you mean by "faster." Making such commonplace notions more precise will be one of our goals in this chapter. The other half of Galileo's distinction—the cause of motion—is a topic we will take up in Chapter 4 after first learning to describe motion.

2.1 Describing Motion

In Chapter 1 you learned about quantities that you could use to describe an object's motion. Position and velocity are measured with respect to a coordinate system, a grid or axis that *you* impose on a system. We will use an *x*-axis to analyze both horizontal motion and motion on a ramp; a *y*-axis will be used for vertical motion.

We will adopt the convention that the positive end of an *x*-axis is to the right and the positive end of a *y*-axis is up. This convention is illustrated in Figure 2.1.

Velocity is a vector; it has both a magnitude and a direction. When we draw a velocity vector on a diagram, we will use an arrow labeled with the symbol $\vec{v}$ to represent the magnitude and the direction. For motion in one dimension, vectors are restricted to point only "forward" or "backward" for horizontal motion (or "up" or "down" for vertical motion). This restriction lets us simplify our notation for vectors in one dimension. When we solve problems for motion along an *x*-axis we will represent velocity with the symbol v_x. v_x will be positive or negative, corresponding to motion to the right or the left, as shown in Figure 2.2. For motion along a *y*-axis, we will use the symbol v_y to represent the velocity; the sign conventions are also illustrated in Figure 2.2. We will use the symbol v, with no subscript, to represent the speed of an object. **Speed is the *magnitude* of the velocity vector** and is always positive.

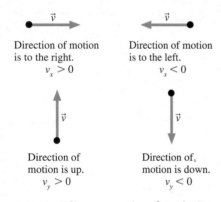

FIGURE 2.1 Sign conventions for position.

Motion Diagrams and Graphs

We saw in Chapter 1 that a motion diagram is a useful tool for analyzing the one-dimensional motion of an object. Let's continue our discussion of motion by using a motion diagram to study a straightforward situation.

Figure 2.3 is a motion diagram, made at 1 frame per minute, of a student walking to school. We have included velocity vectors connecting successive positions on the motion diagram, as we saw we could do in Chapter 1. The motion diagram shows that she leaves home at a time we choose to call $t = 0$ min, and then makes steady progress for a while. Beginning at $t = 3$ min there is a period in which the distance traveled during each time interval becomes less—perhaps she slowed down to speak with a friend. Then, at $t = 6$ min she begins walking more quickly and the distances traveled within each interval are longer.

FIGURE 2.2 Sign conventions for velocity.

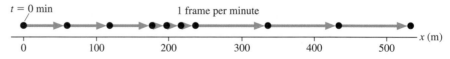

FIGURE 2.3 The motion diagram of a student walking to school and a coordinate axis for making measurements.

TABLE 2.1 Measured positions of a student walking to school

Time	Position		
t (min)	*x* (m)	*t*	*x*
0	0	5	220
1	60	6	240
2	120	7	340
3	180	8	440
4	200	9	540

Figure 2.3 includes a coordinate axis, and we can see that every dot in a motion diagram occurs at a specific position. Table 2.1 shows the student's positions at different times as measured along this axis. For example, she is at position $x = 120$ m at $t = 2$ min.

The motion diagram of Figure 2.3 is one way to represent the student's motion. Presenting the data as in Table 2.1 is a second way to represent this motion; you can easily see where the student was at any particular time. A third way to represent the motion is to make a graph of these measurements. Figure 2.4 is a graph of the position of the student at different times; we say it is a graph of x versus t for the student. We have merely taken the data from the table and plotted these particular points on the graph.

NOTE ▶ A graph of "*a* versus *b*" means that *a* is graphed on the vertical axis and *b* on the horizontal axis. ◀

We can flesh out the graph of Figure 2.4, though. Common sense tells us a few things. First, the student was *somewhere specific* at all times. That is, there was never a time when she failed to have a well-defined position, nor could she occupy two positions at one time. (As reasonable as this belief appears to be, it will be found to be not entirely accurate when we get to quantum physics!) Second, from the start to the end of her motion, the student moved *continuously* through all intervening points of space. She could not go from x = 100 m to x = 200 m without passing through every point in between. It is thus quite reasonable to believe that her motion can be shown as a continuous curve that passes through the measured points, as shown in Figure 2.5. A continuous curve that shows an object's position as a function of time is called a **position-versus-time graph** or, sometimes, just a *position graph*.

NOTE ▶ In making the graph continuous, we made some assumptions about her motion. For example, we assumed that the curve is smooth from point to point. This may not be quite true, but in most cases it is a reasonable assumption. ◀

NOTE ▶ A graph is *not* a "picture" of the motion. The student is walking along a straight line, but the graph itself is not a straight line. Further, we've graphed her position on the vertical axis even though her motion is horizontal. A graph is an *abstract representation* of motion. We will place significant emphasis on the process of interpreting graphs, and many of the exercises and problems will give you a chance to practice these skills. ◀

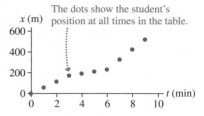

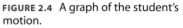

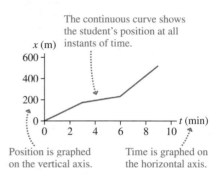

FIGURE 2.4 A graph of the student's motion.

FIGURE 2.5 Extending the graph of Figure 2.4 to a position-versus-time graph.

CONCEPTUAL EXAMPLE 2.1 Interpreting a car's position-versus-time graph

The graph in Figure 2.6 represents the motion of a car along a straight road. Describe (in words) the motion of the car.

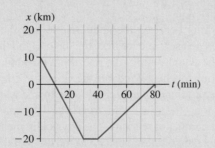

FIGURE 2.6 Position-versus-time graph for the car.

REASON The vertical axis in Figure 2.6 is labeled as "*x* (km)"; position is measured in kilometers. Our convention for motion along the *x*-axis given in Figure 2.1 tells us that *x* increases as the car moves to the right and *x* decreases as the car moves to the left. The graph thus represents the motion of a car that travels to the left for 30 minutes, stops for 10 minutes, then travels to the right for 40 minutes. It ends up 10 km to the left of where it began. Figure 2.7 gives a full explanation of the reasoning.

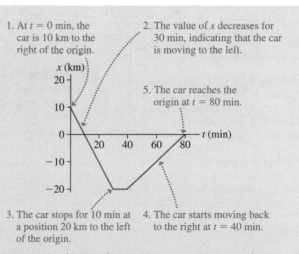

FIGURE 2.7 Looking at the position-versus-time graph in detail.

ASSESS The car travels to the left for 30 minutes and to the right for 40 minutes. Nonetheless, it ends up to the left of where it started. This means that the car was moving faster when it was moving to the left than when it was moving to the right. We can deduce this fact from the graph as well, as we will see in the next section.

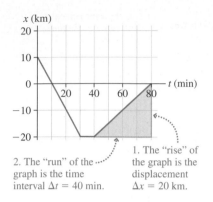

2. The "run" of the graph is the time interval $\Delta t = 40$ min.

1. The "rise" of the graph is the displacement $\Delta x = 20$ km.

3. The slope of the graph, rise over run, is the car's velocity during this phase of motion:

v_x = slope of position-versus-time graph

$= \dfrac{\Delta x}{\Delta t} = \dfrac{20 \text{ km}}{40 \text{ min}} = 0.50$ km/min

FIGURE 2.8 Calculating velocity from a position-versus-time graph.

Time lines BIO This section of the trunk of a pine tree shows the light bands of spring growth and the dark bands of summer and fall growth in successive years. If you focus on the spacing of successive dark bands, you can think of this picture as a motion diagram for the tree, representing its growth in diameter. The years of rapid growth (large distance between dark bands) during wet years and slow growth (small distance between dark bands) during years of drought are readily apparent.

Velocity Is the Slope of the Position-Versus-Time Graph

Let's look more closely at the graph of Figure 2.6 showing the motion of a car. Suppose we want to compute the velocity of the car during the 40 minutes when it moves to the right. For motion along a line, the definition of velocity from Chapter 1 can be written as

$$v_x = \frac{\Delta x}{\Delta t} \tag{2.1}$$

We can give a graphical interpretation to Equation 2.1: It is the *slope* of the section of the graph representing this motion. Recall that the slope of a straight-line graph is defined as "rise over run." Because position is graphed on the vertical axis, the "rise" of a position-versus-time graph is the object's displacement Δx. The "run" is the time interval Δt. Consequently, the slope of a straight-line segment is $\Delta x / \Delta t$. This idea is graphically illustrated in Figure 2.8. We can associate the slope of the position-versus-time graph, a *geometrical* quantity, with the *physical* quantity velocity. This will be an important aspect of interpreting position-versus-time graphs, as outlined in Tactics Box 2.1.

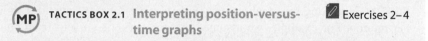

TACTICS BOX 2.1 **Interpreting position-versus-time graphs**   ✏ Exercises 2–4

Information about motion can be obtained from position-versus-time graphs as follows:

❶ Determine the *position* at time t by reading the graph at that instant of time.
❷ Determine the *speed* at time t by finding the magnitude of the slope at that point. Steeper slopes correspond to greater speeds.
❸ Determine the *direction of motion* by noting the sign of the slope. Positive slopes correspond to positive velocities and, hence, to motion to the right (or up). Negative slopes correspond to negative velocities and, hence, to motion to the left (or down).

NOTE ▶ The slope is a ratio of intervals, $\Delta x / \Delta t$, not a ratio of coordinates. That is, the slope is *not* simply x/t. ◀

NOTE ▶ We are distinguishing between the actual slope and the *physically meaningful* slope. If you were to use a ruler to measure the rise and the run of the graph, you could compute the actual slope of the line as drawn on the page. That is not the slope we are referring to when we equate the velocity with the slope of the line. Instead, we find the *physically meaningful* slope by measuring the rise and run using the scales along the axes. The "rise" Δx is some number of meters; the "run" Δt is some number of seconds. The physically meaningful rise and run include units, and the ratio of these units gives the units of the slope. ◀

EXAMPLE 2.1 Calculating a car's velocity

Looking again at Figure 2.6, what is the velocity of the car during the period from 0 to 30 min? What is the speed?

SOLVE We find the velocity by computing the slope of the graph:

$$v_x = \frac{\Delta x}{\Delta t} = \frac{-30 \text{ km}}{30 \text{ min}} = -1.0 \text{ km/min}$$

The speed is the magnitude of the velocity:

$$v = |v_x| = 1.0 \text{ km/min}$$

ASSESS The slope of the graph is negative; the value of the distance is decreasing with time, leading to a negative displacement and a negative velocity. The "−" sign in the velocity tells us that the motion is to the left. The speed for this phase of the motion is greater than that of the final phase, as we noted it should be.

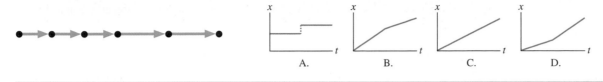

Which position-versus-time graph below best describes the motion diagram at left?

A.   B.   C.   D.

2.2 Uniform Motion

If you drive your car on a straight road at a perfectly steady 60 miles per hour (mph), you will cover 60 mi during the first hour, another 60 mi during the second hour, yet another 60 mi during the third hour, and so on. This is an example of what we call *uniform motion*. In this case, 60 mi is not your position, but rather the *change* in your position during each hour; that is, your displacement Δx. Similarly, 1 hour is a time interval Δt rather than a specific instant of time. This suggests the following definition: **Straight-line motion in which equal displacements occur during any successive equal-time intervals is called uniform motion or constant-velocity motion.**

The qualifier "any" is important. If during each hour you drive 120 mph for 30 min and stop for 30 min, you will cover 60 mi during each successive 1-hour interval. But you would *not* have equal displacements during successive 30-min intervals, so this motion is not uniform. Your constant 60 mph driving is uniform motion because you will find equal displacements no matter how you choose your successive time intervals.

Figure 2.9 shows a motion diagram and a graph for an object in uniform motion. Notice that the position-versus-time graph for uniform motion is a straight line. This follows from the requirement that all values of Δx corresponding to the same value of Δt be equal. In fact, an alternative definition of uniform motion is: **An object's motion is uniform if and only if its position-versus-time graph is a straight line.**

Equations of Uniform Motion

We have seen how to represent motion in four ways:

- Using words
- Using a motion diagram
- Using a table of data of positions at certain times
- Using a graph of position versus time

Now we will add a fifth representation: equations.

Consider an object in uniform motion along the *x*-axis with the linear position-versus-time graph shown in Figure 2.10. Recall from the previous chapter that we denote the object's initial position as x_i at time t_i. The term "initial" refers to the starting point of our analysis or the starting point in a problem. The object may or may not have been in motion prior to t_i. We use the term "final" for the ending point of our analysis or the ending point of a problem, and denote the object's final position x_f at the time t_f. The object's velocity v_x along the *x*-axis can be determined by finding the slope of the graph:

$$v_x = \frac{\text{rise}}{\text{run}} = \frac{\Delta x}{\Delta t} = \frac{x_f - x_i}{t_f - t_i} \quad (2.2)$$

Equation 2.2 can be rearranged to give

$$x_f = x_i + v_x \Delta t \quad (2.3)$$

Position equation for an object in uniform motion
(v_x is constant)

A freight train moving steadily on a straight run of track is a practical example of uniform motion.

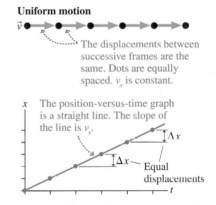

Uniform motion

The displacements between successive frames are the same. Dots are equally spaced. v_x is constant.

The position-versus-time graph is a straight line. The slope of the line is v_x.

Δx

Δx — Equal displacements

FIGURE 2.9 Motion diagram and position-versus-time graph for uniform motion.

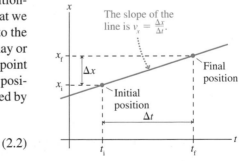

The slope of the line is $v_x = \frac{\Delta x}{\Delta t}$.

x_f

Δx

x_i

Final position

Initial position

Δt

t_i t_f

FIGURE 2.10 Position-versus-time graph for an object in uniform motion.

where $\Delta t = t_f - t_i$ is the interval of time in which the object moves from position x_i to position x_f. Equation 2.3 applies to any time interval Δt during which the velocity is constant. We can also write this in terms of the object's displacement, $\Delta x = x_f - x_i$:

$$\Delta x = v_x \Delta t \qquad (2.4)$$

The velocity of an object in uniform motion tells us the amount by which its position changes during each second. An object with a velocity of 20 m/s *changes* its position by 20 m during every second of motion: by 20 m during the first second of its motion, by another 20 m during the next second, and so on. We say that position is changing at the *rate* of 20 m/s. If the object starts at $x_i = 10$ m, it will be at $x = 30$ m after 1 s of motion and at $x = 50$ m after 2 s of motion. Thinking of velocity like this will help you develop an intuitive understanding of the connection between velocity and position—something we will explore further in the next section.

Physics may seem densely populated with equations, but most equations follow a few basic forms, as we saw in Chapter 1. The mathematical form of Equation 2.3 is a type that we will see again—a *linear relationship*.

NOTE ▶ The important features of a linear relationship are described below. In this text, the first time we use a particular mathematical form we will provide such an overview. In future chapters, when we see other examples of this type of relationship, we will refer back to this overview. ◀

Linear relationships

✏ Exercises 5, 6   (MP)

Two quantities are said to have a **linear relationship** if the graph of y versus x is a straight line. We write the mathematical relationship as

$$y = Ax + B$$

y is a linear function of x

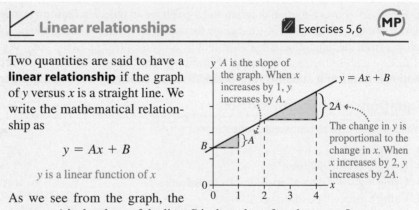

y A is the slope of the graph. When x increases by 1, y increases by A.

$y = Ax + B$

$2A$

The change in y is proportional to the change in x. When x increases by 2, y increases by $2A$.

As we see from the graph, the constant A is the slope of the line; B is the value of y when $x = 0$.

A change in x leads to a **proportional** change in y. If we have initial and final values:

$$y_f = Ax_f + B$$

$$y_i = Ax_i + B$$

We can subtract these equations to find that

$$y_f - y_i = A(x_f - x_i)$$

$$\Delta y = A\,\Delta x$$

Therefore, we can write

$$\Delta y \propto \Delta x$$

SCALING Linear scaling means, for example:

- If you double Δx, you double Δy, as you can see in the graph.
- If you decrease Δx by a factor of 3, you decrease Δy by a factor of 3.

A car takes 2.0 min to travel 1.0 mi. How far does it travel in 10 min?

SOLVE The statement of the problem assumes that the car is in uniform motion—traveling at a constant velocity. The distance will be related to the time by the *linear relationship* of Equation 2.3; this means that the change in the distance is proportional to the change in time. 10 min is a time interval that is 5 times the original 2.0 min time interval, so the car will travel 5 times as far, 5.0 mi.

A Problem-Solving Strategy

Before we introduce any more physics concepts, we'll take a break to look at a strategy for using these concepts to solve practical problems. The first step in solving a seemingly complicated problem is to break it down into a series of smaller steps. In worked examples in the text, we will use a problem-solving strategy that consists of three steps: *prepare, solve* and *assess*. Each of these steps has important elements that you should follow when you solve problems on your own.

 Problem-Solving Strategy

PREPARE The "Prepare" step of a solution is where you identify important elements of the problem and collect information you will need to solve it. It's tempting to jump right to the "solve" step, but a skilled problem solver will spend the most time on this step, the preparation. Preparation includes:

- **Drawing a picture.** In many cases, this is the most important part of a problem. The picture lets you model the problem and identify the important elements. As you add information to your picture, the outline of the solution will take shape. We will give tips for drawing effective pictures for different problems. For the problems in this chapter, a picture could be a motion diagram or a graph—or perhaps both. Later in the chapter you will learn a strategy for drawing a complete *visual overview* of a problem that incorporates these and other elements.
- **Collecting necessary information.** The problem's statement may give you some values of variables. Other important information may be implied, or must be looked up in a table. Gather everything you need to solve the problem, and include it as part of your picture or an accompanying table.
- **Doing preliminary calculations.** In some cases, there are a few calculations, such as unit conversions, that are best done in advance of the main part of the solution.

SOLVE The "Solve" step of a solution is where you actually do the mathematics or reasoning necessary to arrive at the answer needed. This is the part of the problem-solving strategy that you likely think of when you think of "solving problems." But don't make the mistake of starting here! If you just choose an equation and plug in numbers, you will likely go wrong and will waste time trying to figure out why. The "Prepare" step will help you be certain you understand the problem before you start putting numbers in equations.

ASSESS The "Assess" step of your solution is very important. When you have an answer, you should check to see if it makes sense. Ask yourself:

- **Does my solution answer the question that was asked?** Make sure you have addressed all parts of the question and clearly written down your solutions.

Building a complex structure requires careful planning. The architect's visualization and drawings have to be complete before the detailed procedures of construction get underway. The same is true for solving problems in physics.

Continued

■ **Does my answer have the correct units and number of significant figures?**

■ **Does the value I computed make physical sense?** In this book all calculations use physically reasonable numbers. You will not be given a problem to solve in which the final velocity of a bicycle is 100 miles per hour! If your final answer seems unreasonable, you should go back and check your work.

■ **Can I estimate what the answer should be to check my solution?**

■ **Does my final solution make sense in the context of the material I am learning?**

EXAMPLE 2.3 Using the problem-solving strategy to analyze the motion of a baseball

In a game of baseball, a batter hits the ball straight toward the pitcher, who stands 60 ft from the batter. The ball comes off the bat at a speed of 100 ft/s. How far is the ball from the batter 1.0 s after it passes the pitcher?

PREPARE Start by drawing a picture. In this case, we will draw a position-versus-time graph of the motion, shown in Figure 2.11. We've chosen the origin of our coordinate system to be at the position where the batter hits the ball. As the ball leaves the bat, its position increases at the rate of 100 ft/s. This is uniform motion, so the graph will be a straight line. We now add important information to our picture. One important distance to note on the graph is the position of the pitcher. When the ball is at this position, we can see that the time is a bit over 0.5 s.

We are interested in the position of the ball 1.0 s after this point. A quick look at the graph shows us that this position will be a bit over 150 ft.

SOLVE As the ball is in uniform motion, we will use Equation 2.3 for our solution. We take x_i to be 60 ft, when the ball passes the pitcher. We are interested in the position x_f 1.0 s later, so Δt is 1.0 s. Thus

$$x_f = x_i + v_x \Delta t = 60 \text{ ft} + (100 \text{ ft/s})(1.0 \text{ s}) = 160 \text{ ft}$$

ASSESS The final value is close to our estimate from the graph in the Prepare step. It also seems physically reasonable; if you have ever watched a baseball game, you know that 1.0 s after a struck ball passes the pitcher it is well on its way to the outfield.

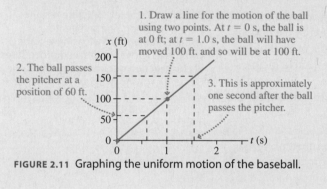

1. Draw a line for the motion of the ball using two points. At $t = 0$ s, the ball is at 0 ft; at $t = 1.0$ s, the ball will have moved 100 ft. and so will be at 100 ft.

2. The ball passes the pitcher at a position of 60 ft.

3. This is approximately one second after the ball passes the pitcher.

FIGURE 2.11 Graphing the uniform motion of the baseball.

This was a straightforward problem, but you can see that making such a complete picture in the "Prepare" step let us visualize the important points in the motion, aiding our solution. It also allowed us to estimate what the final solution should be. For more involved problems, the picture is much more important.

From Position to Velocity

We've focused thus far on graphs of position versus time. Now let's see what we can learn by graphing velocity versus time. In section 2.1, we considered the motion of a student walking to school; the motion diagram is repeated in Figure 2.12. Rather than looking at her position, let's now focus on her velocity, as represented by the green arrows. A long arrow corresponds to a large velocity, a short arrow a small velocity.

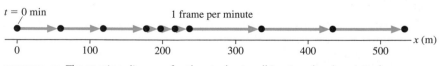

$t = 0$ min   1 frame per minute

FIGURE 2.12 The motion diagram for the student walking to school, revisited.

The green arrow between $t = 0$ min and $t = 1$ min represents her velocity during this period of time. Looking at the data in Table 2.1, we see that her velocity during this interval is

$$v_x = \frac{\Delta x}{\Delta t} = \frac{60 \text{ m}}{1 \text{ min}} \times \frac{1 \text{ min}}{60 \text{ s}} = 1.0 \text{ m/s}$$

We see from the motion diagram that she moved at this constant velocity until $t = 3$ min, then she suddenly slowed down. We can use the data in Table 2.1 to find that her velocity from $t = 3$ min to $t = 6$ min is 0.33 m/s. She then increased her velocity to 1.7 m/s. Using all this information gives us the **velocity-versus-time graph** shown in Figure 2.13.

There are three clearly defined segments to the motion: an initial segment, a middle segment with a smaller velocity, and a final segment with a larger velocity.

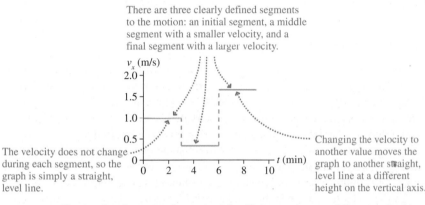

The velocity does not change during each segment, so the graph is simply a straight, level line.

Changing the velocity to another value moves the graph to another straight, level line at a different height on the vertical axis.

FIGURE 2.13 The student's motion is a series of uniform motions at different velocities.

NOTE ▶ This velocity-versus-time graph includes vertical segments in which the velocity changes instantaneously. Such rapid changes are an idealization; it actually takes a small amount of time to change velocity. ◀

Rather than begin with the motion diagram, we could deduce the velocity-versus-time graph from the position-versus-time graph as shown in Figure 2.14. Velocity is the slope of a position-versus-time graph, so we can translate from the position graph to the velocity graph in a straightforward fashion.

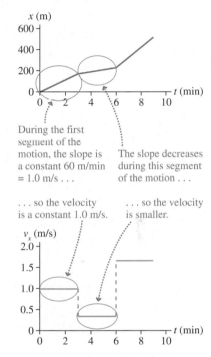

During the first segment of the motion, the slope is a constant 60 m/min = 1.0 m/s . . .

The slope decreases during this segment of the motion . . .

. . . so the velocity is a constant 1.0 m/s.

. . . so the velocity is smaller.

FIGURE 2.14 Deducing the velocity-versus-time graph from the position-versus-time graph.

EXAMPLE 2.4 Analyzing a car's position graph
Figure 2.15 gives the position-versus-time graph of a car.

a. Draw the car's velocity-versus-time graph.
b. Describe the car's motion in words.

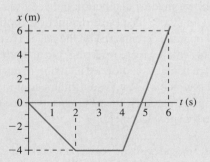

FIGURE 2.15 The position-versus-time graph of a car.

PREPARE Figure 2.15 is a graphical representation of the motion. The car's position-versus-time graph is a sequence of three straight lines. Each of these straight lines represents uniform motion at a constant velocity. We can determine the car's velocity during each interval of time by measuring the slope of the line.

SOLVE

a. From $t = 0$ s to $t = 2$ s ($\Delta t = 2$ s) the car's displacement is $\Delta x = -4$ m $- 0$ m $= -4$ m. The velocity during this interval is

$$v_x = \frac{\Delta x}{\Delta t} = \frac{-4 \text{ m}}{2 \text{ s}} = -2 \text{ m/s}$$

The car's position does not change from $t = 2$ s to $t = 4$ s ($\Delta x = 0$ m), so $v_x = 0$ m/s. Finally, the displacement between $t = 4$ s and $t = 6$ s ($\Delta t = 2$ s) is $\Delta x = 10$ m. Thus the velocity during this interval is

$$v_x = \frac{10 \text{ m}}{2 \text{ s}} = 5 \text{ m/s}$$

Continued

These velocities are represented graphically in Figure 2.16.

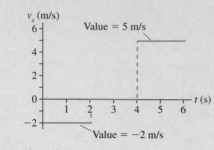

FIGURE 2.16 The velocity-versus-time graph for the car.

b. The velocity-versus-time graph of Figure 2.16 shows the motion in a way that we can describe in a straightforward manner: The car backs up for 2 s at 2 m/s, sits at rest for 2 s, then drives forward at 5 m/s for 2 s.

ASSESS Notice that the velocity graph and the position graph look completely different. The value of the velocity graph at any instant of time equals the *slope* of the position graph. Since the position graph is made up of segments of constant slope, the velocity graph will be made up of segments of constant *value*.

From Velocity to Position

We have seen how to convert from position-versus-time information to velocity-versus-time information. It is possible to go the other way as well: Given information about an object's velocity, we can determine its position.

Suppose a car moves at a constant velocity of 12 m/s for 4.0 s. How far does it travel—that is, what is its displacement during this time interval?

Equation 2.4, $\Delta x = v_x \Delta t$, describes the displacement mathematically; for a graphical interpretation, we consider the graph of velocity versus time in Figure 2.17. In the figure, we've shaded a rectangle whose height is the velocity v_x (12 m/s) and whose base is the time interval Δt (4.0 s). The area of this rectangle is $v_x \Delta t$. Looking at Equation 2.4, we see that this quantity is also equal to the displacement of the car. The area of this rectangle is the area between the axis and the line representing the velocity; we call it the "area under the graph." We see that the **displacement Δx is equal to the area under the velocity graph during interval Δt.**

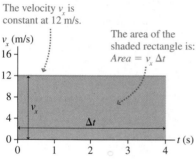

FIGURE 2.17 Displacement is the area under a velocity-versus-time graph.

Whether we use Equation 2.4 or the area under the graph to compute the displacement, we get the same result:

$$\Delta x = v_x \Delta t = (12 \text{ m/s})(4.0 \text{ s}) = 48 \text{ m}$$

Although we've shown that the displacement is the area under the graph only for uniform motion, where the velocity is constant, we'll soon see that this result applies to any one-dimensional motion.

NOTE ▶ Wait a minute! The displacement $\Delta x = x_f - x_i$ is a length. How can a length equal an area? Recall earlier, when we found that the velocity is the slope of the position graph, we made a distinction between the *actual* slope and the *physically meaningful* slope? The same distinction applies here. The velocity graph does indeed bound a certain area on the page. That is the actual area, but it is *not* the area to which we are referring. Once again, we need to measure the quantities we are using, v_x and Δt, by referring to the scales on the axes. Δt is some number of seconds while v_x is some number of meters per second. When these are multiplied together, the *physically meaningful* area has units of meters, appropriate for a displacement. ◄

STOP TO THINK 2.2   An object moves with the velocity-versus-time graph shown at left below. Which of the position-versus-time graphs matches this motion?

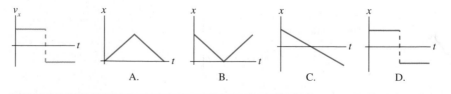

2.3 Motion with Changing Velocity

The objects we've studied so far have moved with a constant, unchanging velocity or, like the car in Example 2.4, changed abruptly from one constant velocity to another. This is not very realistic. Real moving objects speed up and slow down, *changing* their velocity. As an extreme example, think about a drag racer. In a typical race, the car begins at rest but, one second later, is moving at over 25 miles per hour!

For one-dimensional motion, an object changing its velocity is either speeding up or slowing down. When you drive your car, as you speed up or slow down—changing your velocity—a glance at your speedometer tells you how fast you're going *at that instant*. An object's velocity—a speed *and* a direction—at a specific *instant* of time *t* is called the object's **instantaneous velocity.**

But what does it mean to have a velocity "at an instant"? An instantaneous velocity of magnitude 60 mph means that the rate at which your car's position is changing—at that exact instant—is such that it would travel a distance of 60 miles in 1 hour *if* it continued at that rate without change. Said another way, if *just for an instant* your car matches the velocity of another car driving at a steady 60 mph, then your instantaneous velocity is 60 mph. Whether or not your car actually does travel at that velocity for another hour is not relevant. **From now on, the word "velocity" will always mean instantaneous velocity.**

For uniform motion, we found that an object's position-versus-time graph is a straight line and the object's velocity is the slope of that line. In contrast, Figure 2.18 shows that the position-versus-time graph for a drag racer is a *curved* line. The displacement Δx during equal intervals of time gets larger as the car speeds up. Even so, we can still use the slope of the line to measure the car's velocity.

A drag racer moves with rapidly changing velocity.

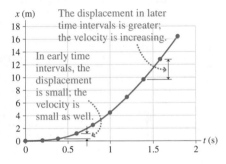

FIGURE 2.18 Position-versus-time graph for a drag racer.

Finding the instantaneous velocity

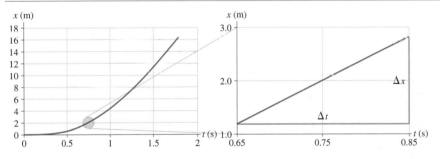

If the velocity changes, the position graph is no longer a straight line: It is clearly curved. How are we to compute the slope to find the velocity? The key is to focus in on a small segment of the motion. Let's look at the motion in a very small time interval right around $t = 0.75$ s. This is highlighted with a circle, and we show a "close-up" in the next graph, at right.

Now that we have magnified a small part of the position graph, we see that the graph in this small part appears to have a constant slope. It is always possible to make the graph appear as a straight line by choosing a small enough time interval. Now we can find the slope of the line by calculating the rise over run, just as we did before:

$$v_x = \frac{1.6 \text{ m}}{0.20 \text{ s}} = 8.0 \text{ m/s}$$

This is the slope of the graph at $t = 0.75$ s, and thus the velocity at this instant of time.

Graphically, the slope of the curve at a particular point is the same as the slope of a straight line drawn *tangent* to the curve at that point. **The slope of the tangent line is the instantaneous velocity at that instant of time.**

Calculating rise over run for the tangent line, we get

$$v_x = \frac{4.0 \text{ m}}{0.50 \text{ s}} = 8.0 \text{ m/s}$$

This is the same value we obtained from considering the close-up view.

CONCEPTUAL EXAMPLE 2.2   **Analyzing an elevator's position graph**

Figure 2.19 shows the position-versus-time graph of an elevator.

a. Sketch an approximate velocity-versus-time graph.
b. At which point or points is the elevator moving the fastest?
c. Is the elevator ever at rest? If so, at which point or points?

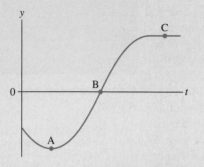

FIGURE 2.19 The position-versus-time graph for an elevator.

REASON a. Notice that the position graph shows y versus t, rather than x versus t, indicating that the motion is vertical rather than horizontal. Our analysis of one-dimensional motion has made no assumptions about the direction of motion, so it applies equally well to both horizontal and vertical motion. Let's start by sketching an approximate velocity-versus-time graph. As we just found, the velocity at a particular instant of time is the slope of a tangent line to the position-versus-time graph at that time. We can move point-by-point along the position-versus-time graph, noting the slope of the tangent at each point. This will give us the velocity at that point.

Initially, to the left of point A, the slope is negative and thus the velocity is negative (i.e., the elevator is moving downward). But the slope decreases as the curve flattens out, and by the time the graph gets to point A, the slope is zero. The slope then increases to a maximum value at point B, decreases back to zero a little before point C, and remains at zero thereafter. This reasoning process is outlined in Figure 2.20a, and Figure 2.20b shows the approximate velocity-versus-time graph that results.

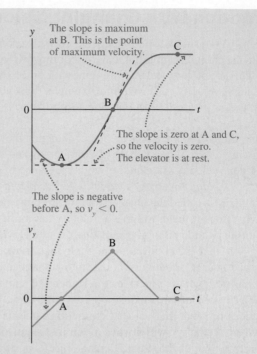

FIGURE 2.20 Finding a velocity graph from a position graph.

The other questions were really answered during the construction of the graph:

b. The elevator moves the fastest at point B where the slope of the position graph is the steepest.
c. A particle at rest has $v_y = 0$. Graphically, this occurs at points where the tangent line to the position-versus-time graph is horizontal and thus has zero slope. Figure 2.20 shows that the slope is zero at points A and C. At point A, the velocity is only instantaneously zero as the particle reverses direction from downward motion (negative velocity) to upward motion (positive velocity). At point C, the elevator has actually stopped and remains at rest.

ASSESS Once again, the shape of the velocity graph bears no resemblance to the shape of the position graph. You must translate between slope information on the position graph and value information on the velocity graph.

For uniform motion we showed that the displacement Δx is the area under the velocity-versus-time graph during time interval Δt. We can generalize this idea to the case of an object whose velocity varies. Figure 2.21a on the next page is the velocity-versus-time graph for an object whose velocity changes with time. Suppose we know the object's position to be x_i at an initial time t_i. Our goal is to find its position x_f at a later time t_f.

Because we know how to handle constant velocities, let's *approximate* the velocity function of Figure 2.21a as a series of constant-velocity steps of width Δt. The velocity during each step is constant (uniform motion), so we can calculate the displacement during each step as the area of the rectangle under the curve. The total displacement of the object between t_i and t_f can be found as the sum of all the individual displacements during each of the constant-velocity steps. We can see in Figure 2.21c that the total displacement is approximately equal to the area under the graph, even in the case where the velocity varies. Although the approximation shown in the figure is rather rough, with only nine steps, we can

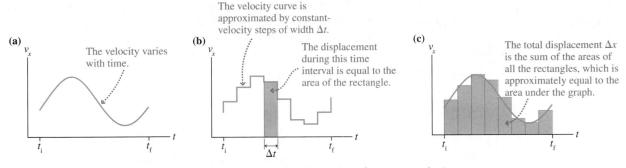

FIGURE 2.21 Approximating a velocity-versus-time graph with a series of constant-velocity steps.

imagine that it could be made as accurate as desired by having more and more ever-narrower steps.

Consequently, an object's displacement is related to its velocity by

$$x_f - x_i = \Delta x = \text{area under the velocity graph } v_x \text{ between } t_i \text{ and } t_f \quad (2.5)$$

EXAMPLE 2.5 The displacement during a rapid start

Figure 2.22 shows the velocity-versus-time graph of a car pulling away from a stop. How far does the car move during the first 3.0 s?

PREPARE Figure 2.22 is a graphical representation of the motion. The question "how far" indicates that we need to find a displacement Δx rather than a position x. According to Equation 2.5, the car's displacement $\Delta x = x_f - x_i$ between $t = 0$ s and $t = 3$ s is the area under the curve from $t = 0$ s to $t = 3$ s.

SOLVE The curve in this case is an angled line, so the area is that of a triangle:

$$\Delta x = \text{area of triangle between } t = 0 \text{ s and } t = 3 \text{ s}$$
$$= \tfrac{1}{2} \times \text{base} \times \text{height} = \tfrac{1}{2} \times 3 \text{ s} \times 12 \text{ m/s} = 18 \text{ m}$$

The car moves 18 m during the first 3 seconds as its velocity changes from 0 to 12 m/s.

ASSESS The "area" is a product of s with m/s, so Δx has the proper units of m. Let's check the numbers. The final velocity, 12 m/s, is about 25 mph. Pulling away from a stop, you might expect to reach this speed in about 3 s—at least if you have a reasonably sporty vehicle! If the car had moved at a constant 12 m/s (the final velocity) during these 3 s, the distance would be 36 m. The actual distance traveled during the 3 s is 18 m— half of 36 m. This makes sense, as the velocity was 0 m/s at the start of the problem and increased steadily to 12 m/s.

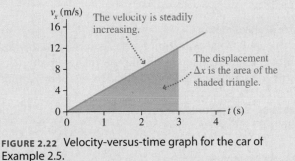

FIGURE 2.22 Velocity-versus-time graph for the car of Example 2.5.

STOP TO THINK 2.3   Which velocity-versus-time graph goes with the position-versus-time graph on the left?

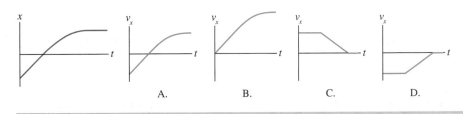

2.4 Acceleration

One of the goals of this chapter is to describe motion. Position, time, and velocity are important concepts, but we need one more motion concept, one that will describe a *change* in the velocity.

As an example, a frequently quoted measurement of car performance is the time it takes the car to change its velocity from 0 to 60 mph. Table 2.2 shows this time for two rather different cars.

TABLE 2.2 Performance data for vehicles

Vehicle	Time to go from 0 to 60 mph
1997 Porsche 911 Turbo S	3.6 s
1973 Volkswagen Super Beetle Convertible	24 s

Let's look at motion diagrams for the Porsche and the Volkswagen in Figure 2.23. We can see two important facts about the motion. First, the lengths of the velocity vectors are increasing, showing that the speeds are increasing. Second, the velocity vectors for the Porsche are increasing in length more rapidly than those of the VW. The quantity we seek is one that measures how rapidly an object's velocity vectors change in length.

Volkswagen

Porsche

FIGURE 2.23 Motion diagrams for the Porsche and Volkswagen.

When we wanted to measure changes in position, the ratio $\Delta x/\Delta t$ was useful. This ratio, which we defined as the velocity, is the *rate of change of position*. Similarly, we can measure how rapidly an object's velocity changes with the ratio $\Delta v_x/\Delta t$. Given our experience with velocity, we can say a couple of things about this new ratio:

- The ratio $\Delta v_x/\Delta t$ is the *rate of change of velocity*.
- The ratio $\Delta v_x/\Delta t$ is the *slope of a velocity-versus-time graph*.

We will define this ratio as the **acceleration,** for which we use the symbol a_x:

$$a_x = \frac{\Delta v_x}{\Delta t} \tag{2.6}$$

Definition of the acceleration as rate of change of velocity

Cushion kinematics When a car hits an obstacle head-on, the damage to the car and its occupants can be reduced by making the acceleration as small as possible. As we can see from Equation 2.6, acceleration can be reduced by making the *time* for a change in velocity as long as possible. This is the purpose of the yellow crash cushion barrels you may have seen in work zones on highways.

Similarly, $a_y = \Delta v_y/\Delta t$ for vertical motion.

As an example, let's calculate the acceleration for the Porsche and the Volkswagen. For both, the initial velocity $(v_x)_i$ is zero and the final velocity $(v_x)_f$ is 60 mph. Thus the *change* in velocity is $\Delta v_x = 60$ mph. In m/s, our SI unit of velocity, $\Delta v_x = 27$ m/s.

Now we can use Equation 2.6 to compute acceleration. Let's start with the Porsche, which speeds up to 27 m/s in $\Delta t = 3.6$ s:

$$a_{\text{Porsche }x} = \frac{\Delta v_x}{\Delta t} = \frac{27 \text{ m/s}}{3.6 \text{ s}} = 7.5 \frac{\text{m/s}}{\text{s}}$$

Notice the units; they are velocity over time. In the first second of motion, the Porsche's velocity increases by 7.5 m/s; in the next second, it increases by another 7.5 m/s, and so on. After 1 second, the velocity is 7.5 m/s; after 2 seconds, it is 15 m/s. This increase continues as long as the Porsche has an acceleration of 7.5 (m/s)/s.

The Volkswagen's acceleration is

$$a_{\text{VW }x} = \frac{\Delta v_x}{\Delta t} = \frac{27 \text{ m/s}}{24 \text{ s}} = 1.1 \frac{\text{m/s}}{\text{s}}$$

In each second, the Volkswagen changes its speed by 1.1 m/s. This is about 1/7 the acceleration of the Porsche! The reasons why the Porsche is capable of greater acceleration have to do with what *causes* the motion. We will explore the reasons for acceleration in Chapter 4. For now, we will simply note that the Porsche is capable of much larger acceleration, something you would have suspected.

NOTE ▶ It is customary to abbreviate the acceleration units (m/s)/s as m/s². For example, the Volkswagen has an acceleration of 1.1 m/s². We will use this notation, but keep in mind the *meaning* of the notation as "(meters per second) per second." ◀

Acceleration is a fairly abstract concept. Position and time are our real hands-on measurements of an object, and they are easy to understand. You can "see" where the object is located and the time on the clock. Velocity is a bit more abstract, being a relationship between the change of position and the change of time. Motion diagrams help us visualize velocity as the vector arrows connecting one position of the object to the next. Acceleration is an even more abstract idea about changes in the velocity, but we can use graphs to help us better understand it.

Acceleration Is the Slope of the Velocity-Versus-Time Graph

Let's use the values we have computed for acceleration to make a table of velocities for the Porsche and the Volkswagen. Table 2.3 uses the idea that the VW's velocity increases by 1.1 m/s every second while the Porsche's velocity increases by 7.5 m/s every second. The data in Table 2.3 are the basis for the velocity-versus-time graphs in Figure 2.24.

Look at the graph for the Porsche; it's a straight line. The slope is computed using the rise over the run:

$$\text{slope of velocity-versus-time graph} = \frac{\Delta v_x}{\Delta t}$$

Compare this with Equation 2.6; the equation for the slope is the same as that for the acceleration. **Acceleration is the slope of a velocity-versus-time graph.** The VW has a smaller acceleration, so its graph has a smaller slope. The slope could be negative; this would mean that velocity is *decreasing,* as we will see in the next example.

TABLE 2.3 Velocity data for the Volkswagen and the Porsche

Time (s)	Velocity of VW (m/s)	Velocity of Porsche (m/s)
0	0	0
1	1.1	7.5
2	2.2	15.0
3	3.3	22.5
4	4.4	30.0

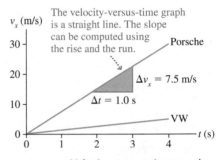

FIGURE 2.24 Velocity-versus-time graphs for the two cars.

CONCEPTUAL EXAMPLE 2.3 **Analyzing a car's velocity graph**

Figure 2.25a is a graph of velocity versus time for a car. Sketch a graph of the car's acceleration versus time.

REASON The graph can be divided into three sections:

- An initial segment, in which the velocity increases at a steady rate.
- A middle segment, in which the velocity is constant.
- A final segment, in which the velocity decreases at a steady rate.

In each section, the acceleration is the slope of the velocity-versus-time graph. Thus the initial segment has constant, posi-

tive acceleration, the middle segment has zero acceleration, and the final segment has a constant, *negative* acceleration. The acceleration graph appears in Figure 2.25b.

ASSESS This process is analogous to finding a velocity graph from the slope of a position graph. In the middle segment, notice that zero acceleration does *not* mean that the velocity is zero. The velocity is constant, which means it is *not changing* and thus the car is not accelerating. The car does accelerate during the initial and final segments. The magnitude of the acceleration is a measure of how quickly the velocity is changing. How about the sign? This is an issue we will address in the next section.

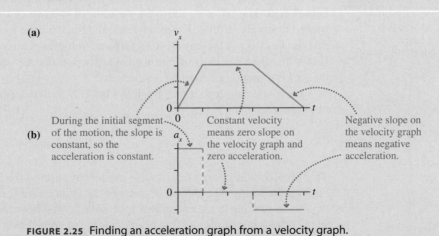

FIGURE 2.25 Finding an acceleration graph from a velocity graph.

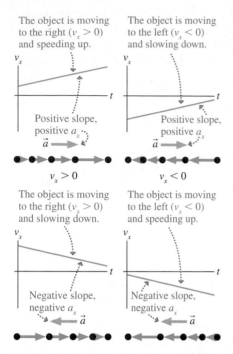

The object is moving to the right ($v_x > 0$) and speeding up.

Positive slope, positive a_x
$\vec{a}$

$v_x > 0$

The object is moving to the left ($v_x < 0$) and slowing down.

Positive slope, positive a_x
$\vec{a}$

$v_x < 0$

The object is moving to the right ($v_x > 0$) and slowing down.

Negative slope, negative a_x
$\vec{a}$

The object is moving to the left ($v_x < 0$) and speeding up.

Negative slope, negative a_x
$\vec{a}$

FIGURE 2.26 Determining the sign of the acceleration.

FIGURE 2.27 The red dots show the position of the top of the Saturn V rocket at equally spaced intervals during liftoff.

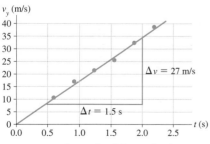

FIGURE 2.28 A graph of the rocket's velocity versus time.

The Sign of the Acceleration

A natural tendency is to think that a positive value of a_x or a_y describes an object that is speeding up while a negative value describes an object that is slowing down (decelerating). Unfortunately, this simple interpretation *does not work*.

Because an object can move right or left (or, equivalently, up and down) while either speeding up or slowing down, there are four situations to consider. Figure 2.26 shows a motion diagram and a velocity graph for each of these situations. As we've seen, an object's acceleration is the slope of its velocity graph, so a positive slope implies a positive acceleration and a negative slope implies a negative acceleration.

Acceleration, like velocity, is really a vector quantity, a concept that we will explore more fully in Chapter 3. Figure 2.26 shows the acceleration vector for the four situations. The acceleration vector points in the same direction as the velocity vector $\vec{v}$ for an object that is speeding up and opposite to $\vec{v}$ for an object that is slowing down.

An object that speeds up as it moves to the right (positive v_x) has a positive acceleration, but an object that speeds up as it moves to the left (negative v_x) has a negative acceleration. Whether or not an object that is slowing down has a negative acceleration depends on whether the object is moving to the right or to the left. This is admittedly a bit more complex than thinking that negative acceleration always means slowing down, but our definition of acceleration as the slope of the velocity graph forces us to pay careful attention to the sign of the acceleration.

STOP TO THINK 2.4 A particle moves with the following velocity-versus-time graph. At which labeled point is the magnitude of the acceleration the greatest?

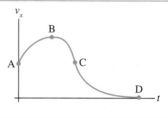

2.5 Motion with Constant Acceleration

For uniform motion—motion with constant velocity—we found in Equation 2.3 a simple relationship between position and time. It's no surprise that there are also simple relationships that connect the various kinematic variables in constant-acceleration motion. We will start with a concrete example, the launch of a Saturn V rocket like the one that carried the Apollo astronauts to the moon in the 1960s and 1970s. Figure 2.27 shows one frame from a video of a rocket lifting off the launch pad. The red dots show the position of the top of the rocket at equally spaced intervals of time. This is a motion diagram for the rocket, and we can see that the velocity is increasing. The graph of velocity versus time in Figure 2.28 shows that the velocity is increasing at a fairly constant rate. We can approximate the rocket's motion as constant acceleration.

We can use the slope of the graph in Figure 2.28 to determine the acceleration of the rocket:

$$a_y = \frac{\Delta v_y}{\Delta t} = \frac{27 \text{ m/s}}{1.5 \text{ s}} = 18 \text{ m/s}^2$$

This acceleration is more than double the acceleration of the Porsche, and it goes on for quite a long time—the first phase of the launch lasts 2 minutes! How fast is the rocket moving at the end of this acceleration, and how far has it traveled? To answer these questions, we first need to work out some basic kinematic formulas for motion with constant acceleration.

Constant Acceleration Equations

Consider an object whose acceleration a_x remains constant during the time interval $\Delta t = t_f - t_i$. At the beginning of this interval, the object has initial velocity $(v_x)_i$ and initial position x_i. Note that t_i is often zero, but it need not be. Figure 2.29a shows the acceleration-versus-time graph. It is a horizontal line between t_i and t_f, indicating a *constant* acceleration.

The object's velocity is changing because the object is accelerating. We can use the acceleration to find $(v_x)_f$ at a later time t_f. We defined acceleration as

$$a_x = \frac{\Delta v_x}{\Delta t} = \frac{(v_x)_f - (v_x)_i}{\Delta t} \qquad (2.7)$$

which is rearranged to give

$$(v_x)_f = (v_x)_i + a_x \Delta t \qquad (2.8)$$

Velocity equation for an object with constant acceleration

LINEAR
p. 38

NOTE ▶ We have expressed this equation for motion along the *x*-axis, but it is a general result that will apply to any axis. ◀

The velocity-versus-time graph for this constant-acceleration motion, shown in Figure 2.29b, is a straight line that starts at $(v_x)_i$ and has slope a_x. This is another example of a linear relationship.

We would also like to know the object's position x_f at time t_f. As you learned in the last section, the displacement Δx during the time interval Δt is the area under the velocity-versus-time graph. The shaded area in Figure 2.29b can be subdivided into a rectangle of area $(v_x)_i \Delta t$ and a triangle of area $\frac{1}{2}(a_x \Delta t)(\Delta t) = \frac{1}{2}a_x(\Delta t)^2$. Adding these gives

$$x_f = x_i + (v_x)_i \Delta t + \tfrac{1}{2}a_x(\Delta t)^2 \qquad (2.9)$$

Position equation for an object with constant acceleration

where $\Delta t = t_f - t_i$ is the elapsed time. The fact that the time interval Δt appears in the equation as $(\Delta t)^2$ causes the position-versus-time graph for constant-acceleration motion to have a parabolic shape. For the rocket launch of Figure 2.27, a graph of position of the top of the rocket versus time appears as in Figure 2.30.

Equations 2.8 and 2.9 are two of the basic kinematic equations for motion with constant acceleration. They allow us to predict an object's position and velocity at a future instant of time. We need one more equation to complete our set, a direct relationship between displacement and velocity. To derive this relationship, we first use Equation 2.8 to write $\Delta t = ((v_x)_f - (v_x)_i)/a_x$. We can substitute this into Equation 2.9 to obtain

$$(v_x)_f^2 = (v_x)_i^2 + 2a_x \Delta x \qquad (2.10)$$

Relating velocity and displacement for constant-acceleration motion

In Equation 2.10 $\Delta x = x_f - x_i$ is the *displacement* (not the distance!). Notice that Equation 2.10 does not require knowing the time interval Δt. This is an important equation in problems where you're not given information about times.

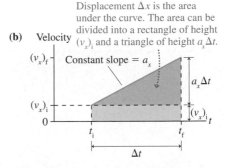

FIGURE 2.29 Acceleration and velocity graphs for motion with constant acceleration.

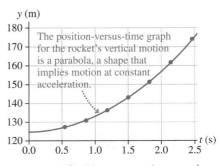

Solar sailing A rocket achieves a high speed by having a very high acceleration. A different approach is represented by a solar sail. A spacecraft with a solar sail accelerates due to the pressure of sunlight from the sun on a large, mirrored surface. The acceleration is minuscule, but it can continue for a long, long time. After an acceleration period of a few *years*, the spacecraft will reach a respectable speed!

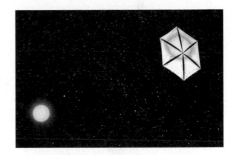

FIGURE 2.30 Position-versus-time graph for the Saturn V rocket launch.

TABLE 2.4 Kinematic equations for motion with constant acceleration

$(v_x)_f = (v_x)_i + a_x \Delta t$

$x_f = x_i + (v_x)_i \Delta t + \frac{1}{2}a_x(\Delta t)^2$

$(v_x)_f^2 = (v_x)_i^2 + 2a_x \Delta x$

1.1–1.3

Getting up to speed BIO A bird must have a minimum speed to fly. Generally, the larger the bird, the larger the takeoff speed. Small birds can get moving fast enough to fly with a vigorous jump, but larger birds may need a running start. This swan must accelerate for a long distance in order to achieve the high speed it needs to fly, so it makes a frenzied dash across the frozen surface of a pond. Swans require a long, clear stretch of water or land to become airborne. Airplanes require an even higher takeoff speed and thus an even longer runway, as we will see.

Equations 2.8, 2.9, and 2.10 are the key equations for motion with constant acceleration. These results are summarized in Table 2.4.

Figure 2.31 is a graphical comparison of motion with constant velocity (uniform motion) and motion with constant acceleration. Notice that uniform motion is really a special case of constant-acceleration motion in which the acceleration happens to be zero.

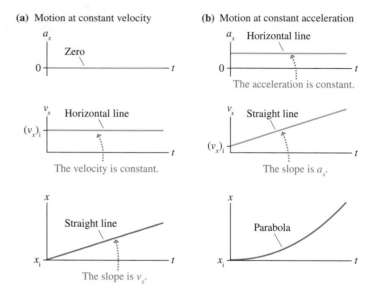

FIGURE 2.31 Motion with constant velocity and constant acceleration. These graphs assume $x_i = 0$, $(v_x)_i > 0$, and (for constant acceleration) $a_x > 0$.

For motion at constant acceleration, a graph of position versus time is a *parabola*. This is a new mathematical form, one that we will see again. If $(v_x)_i = 0$, the second equation in Table 2.4 is simply

$$\Delta x = \frac{1}{2}a_x(\Delta t)^2 \tag{2.11}$$

Δx depends on the *square* of Δt; we call this a *quadratic relationship*.

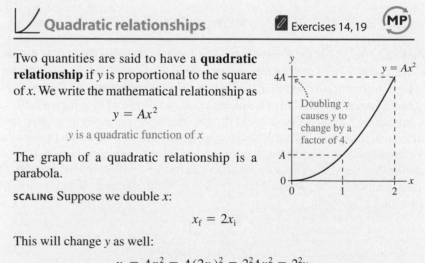

Quadratic relationships   Exercises 14, 19   (MP)

Two quantities are said to have a **quadratic relationship** if y is proportional to the square of x. We write the mathematical relationship as

$$y = Ax^2$$

y is a quadratic function of x

The graph of a quadratic relationship is a parabola.

SCALING Suppose we double x:

$$x_f = 2x_i$$

This will change y as well:

$$y_f = Ax_f^2 = A(2x_i)^2 = 2^2Ax_i^2 = 2^2y_i$$
$$= 4y_i$$

Increasing x by a factor of 2 causes y to increase by a factor of 2^2, or 4.

Generally, we can say that

Changing x by a factor C changes y by a factor C^2.

EXAMPLE 2.6 Displacement of a drag racer

A drag racer, starting from rest, travels 6.0 m in 1.0 s. Suppose the car continues this acceleration for an additional 4.0 s. How far from the starting line will the car be?

SOLVE We assume that the acceleration is constant, so the displacement will follow Equation 2.11. This is a *quadratic relationship,* so the displacement will scale as the square of the time. After 1.0 s, the car has traveled 6.0 m; after another 4.0 s, a total of 5.0 s will have elapsed. The time has increased by a factor of 5, so the displacement will increase by a factor of 5^2, or 25. The total displacement is

$$\Delta x = 25(6.0 \text{ m}) = 150 \text{ m}$$

STOP TO THINK 2.5   A car is moving toward the right. The driver puts on the brakes, reducing the speed. The car then continues to the right at a reduced speed. Which of the following velocity-versus-time graphs matches this motion?

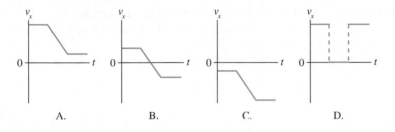

A.   B.   C.   D.

2.6 Solving One-Dimensional Motion Problems

The big challenge when solving a physics problem is to translate the words into symbols that can be manipulated, calculated, and graphed. This translation from words to symbols is the heart of problem solving in physics. Ambiguous words and phrases must be clarified, the imprecise must be made precise, and you must arrive at an understanding of exactly what the question is asking.

In this section we will explore some general problem-solving strategies that we will use throughout the text, applying them to problems of motion along a line.

> ▶ **Dinner at a distance** BIO A chameleon's tongue is a powerful tool for catching prey. Certain species can extend the tongue to a distance of over 1 ft in less than 0.1 s! A study of the kinematics of the motion of the chameleon tongue, using techniques like those in this chapter, reveals that the tongue has a period of rapid acceleration followed by a period of constant velocity. This knowledge is a very valuable clue in the analysis of the evolutionary relationships between chameleons and other animals.

The Pictorial Representation

You'll find that motion problems and other physics problems often have several variables and other pieces of information to keep track of. The best way to tackle such problems is to draw a picture, as we noted when we introduced a general problem-solving strategy. But what kind of picture should we draw?

In this section, we will begin to draw **pictorial representations** as an aid to solving problems. A pictorial representation shows all of the important details that we need to keep track of in a problem. Pictorial representations will be very important in solving motion problems.

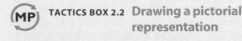

TACTICS BOX 2.2 Drawing a pictorial representation   ✎ Exercise 21

❶ **Sketch the situation.** Not just any sketch: Show the object at the *beginning* of the motion, at the *end,* and at any point where the character of the motion changes. Very simple drawings are adequate.
❷ **Establish a coordinate system.** Select your axes and origin to match the motion.
❸ **Define symbols.** Use the sketch to define symbols representing quantities such as position, velocity, acceleration, and time. *Every* variable used later in the mathematical solution should be defined on the sketch.

We will generally combine the pictorial representation with a **list of values,** which will include

■ *Known information.* Make a table of the quantities whose values you can determine from the problem statement or that can be found quickly with simple geometry or unit conversions.
■ *Desired unknowns.* What quantity or quantities will allow you to answer the question?

EXAMPLE 2.7 Drawing a pictorial representation

Complete a pictorial representation and a list of values for the following problem: A rocket sled accelerates at 50 m/s² for 5 s. What are the total distance traveled and the final velocity?

PREPARE Figure 2.32a shows a pictorial representation as drawn by an artist in the style of the figures in this book. This is cer-

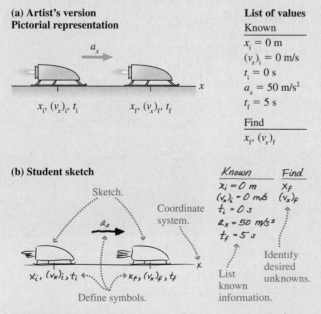

**(a) Artist's version
Pictorial representation**

List of values

Known
$x_i = 0$ m
$(v_x)_i = 0$ m/s
$t_i = 0$ s
$a_x = 50$ m/s²
$t_f = 5$ s

Find
$x_f, (v_x)_f$

(b) Student sketch

Sketch.
Coordinate system.
Define symbols.

Known
$x_i = 0$ m
$(v_x)_i = 0$ m/s
$t_i = 0$ s
$a_x = 50$ m/s²
$t_f = 5$ s

Find
x_f
$(v_x)_f$

List known information.   Identify desired unknowns.

FIGURE 2.32 Constructing a pictorial representation and a list of values.

tainly neater and more artistic than the sketches you will make when solving problems yourself! Figure 2.32b shows a sketch like one you might actually do. It's less formal, but it contains all of the important information you need to solve the problem.

NOTE ▶ Throughout this book we will illustrate select examples with actual hand-drawn figures so that you have them to refer to as you work on your own pictures for homework and practice. ◀

Let's look at how these pictures were constructed. The motion has a clear beginning and end; these are the points sketched. A coordinate system has been chosen with the origin at the starting point. The quantities x, v_x, and t are needed at both points, so these have been defined on the sketch and distinguished by subscripts. The acceleration is associated with an interval between these points. Values for two of these quantities are given in the problem statement. Others, such as $x_i = 0$ m and $t_i = 0$ s, are inferred from our choice of coordinate system. The value $(v_x)_i = 0$ m/s is part of our *interpretation* of the problem. Finally, we identify x_f and $(v_x)_f$ as the quantities that will answer the question. We now understand quite a bit about the problem and would be ready to start a quantitative analysis.

ASSESS We didn't *solve* the problem; that was not our purpose. Constructing a pictorial representation and a list of values is part of a systematic approach to interpreting a problem and getting ready for a mathematical solution.

The Visual Overview

The pictorial representation and the list of values are a very good complement to the motion diagram and other ways of looking at a problem that we have seen. As we translate a problem into a form we can solve, we will combine all of these elements into what we will term a **visual overview.** The visual overview will consist of some or all of the following elements:

- A *motion diagram.* A good strategy for solving a motion problem is to start by drawing a motion diagram.
- A *pictorial representation,* as defined above.
- A *graphical representation.* For motion problems, it is often quite useful to include a graph of position and/or velocity.
- A *list of values.* This list should sum up all of the important values in the problem.

Future chapters will add other elements (such as forces) to this visual overview of the physics.

EXAMPLE 2.8 Kinematics of a rocket launch

A Saturn V rocket is launched straight up with a constant acceleration of 18 m/s². After 150 s, how fast is the rocket moving and how far has it traveled?

PREPARE Figure 2.33 shows a visual overview of the rocket launch that includes a motion diagram, a pictorial representation and a list of values. The visual overview shows the whole problem in a nutshell. The motion diagram illustrates the motion of the rocket, the pictorial representation (produced according to Tactics Box 2.2) shows axes, identifies the important points of the motion and defines variables. Finally, we have included a list of values that gives the known and unknown quantities. In the visual overview we have taken the statement of the problem in words and made it much more precise; it contains everything you need to know about the problem.

SOLVE Our first task is to find the final velocity. We can use the first kinematic equation of Table 2.4:

$$(v_y)_f = (v_y)_i + a_y \, \Delta t = 0 \text{ m/s} + (18 \text{ m/s}^2)(150 \text{ s})$$
$$= 2700 \text{ m/s}$$

The distance traveled is found using the second equation in Table 2.4:

$$y_f = y_i + (v_y)_i \, \Delta t + \tfrac{1}{2}a_y(\Delta t)^2$$
$$= 0 \text{ m} + (0 \text{ m/s})(150 \text{ s}) + \tfrac{1}{2}(18 \text{ m/s}^2)(150 \text{ s})^2$$
$$= 2.0 \times 10^5 \text{ m} = 200 \text{ km}$$

ASSESS The acceleration is very large, and it goes on for a long time, so the large final velocity and large distance traveled seem reasonable.

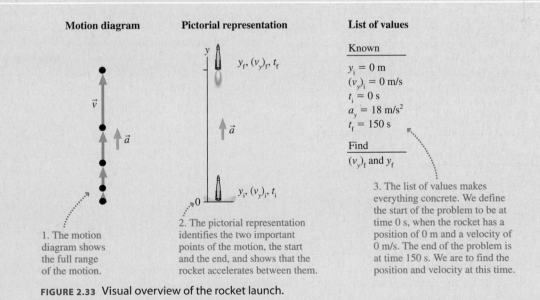

Motion diagram   **Pictorial representation**   **List of values**

Known
$y_i = 0$ m
$(v_y)_i = 0$ m/s
$t_i = 0$ s
$a_y = 18$ m/s²
$t_f = 150$ s

Find
$(v_y)_f$ and y_f

1. The motion diagram shows the full range of the motion.

2. The pictorial representation identifies the two important points of the motion, the start and the end, and shows that the rocket accelerates between them.

3. The list of values makes everything concrete. We define the start of the problem to be at time 0 s, when the rocket has a position of 0 m and a velocity of 0 m/s. The end of the problem is at time 150 s. We are to find the position and velocity at this time.

FIGURE 2.33 Visual overview of the rocket launch.

Problem-Solving Strategy for Motion with Constant Acceleration

1.4–1.6, 1.8, 1.9,
1.11–1.14

Earlier in the chapter, we introduced a general problem-solving strategy. In this section, we will adapt this general strategy to solving problems of motion with constant acceleration. We will introduce such specific problem-solving strategies in future chapters as well.

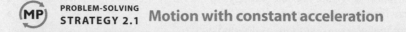

(MP) PROBLEM-SOLVING STRATEGY 2.1 Motion with constant acceleration

PREPARE Draw a visual overview of the problem. This should include a motion diagram, a pictorial representation and a list of values; a graphical representation may be useful for certain problems.

SOLVE The mathematical solution is based on the three equations in Table 2.4.

- Though the equations are phrased in terms of the variable x, it's customary to use y for motion in the vertical direction.
- Use the equation that best matches what you know and what you need to find. For example, if you know acceleration and time and are looking for a change in velocity, the first equation is the best one to use.
- Uniform motion with constant velocity has $a = 0$.

ASSESS Is your result believable? Does it have proper units? Does it make sense?

EXAMPLE 2.9 Calculating the minimum length of a runway

A fully-loaded 747 with all engines at full thrust accelerates at $2.6 \ \text{m/s}^2$. Its minimum takeoff speed is 70 m/s. How much time will the plane take to reach its takeoff speed? What minimum length of runway does the plane require for takeoff?

PREPARE The visual overview of Figure 2.34 summarizes the important details of the problem. We set x_i and t_i equal to zero at the starting point of the motion, when the plane is at rest and the acceleration begins. The final point of the motion is when the plane achieves the necessary takeoff speed of 70 m/s. The plane is accelerating to the right, so we will compute the time for the plane to reach a velocity of 70 m/s and the position of the plane at this time, giving us the minimum length of the runway.

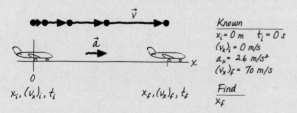

FIGURE 2.34 Visual overview for an accelerating plane.

SOLVE First we solve for the time required for the plane to reach takeoff speed. We can use the first equation in Table 2.4 to compute this time:

$$(v_x)_f = (v_x)_i + a_x \, \Delta t$$

$$70 \ \text{m/s} = 0 \ \text{m/s} + (2.6 \ \text{m/s}^2) \, \Delta t$$

$$\Delta t = \frac{70 \ \text{m/s}}{2.6 \ \text{m/s}^2} = 26.9 \ \text{s}$$

We keep an extra significant figure here as we will use this result in the next step of the calculation.

Given the time that the plane takes to reach takeoff speed, we can compute the position of the plane when it reaches this speed using the second equation in Table 2.4:

$$x_f = x_i + (v_x)_i \, \Delta t + \tfrac{1}{2} a_x (\Delta t)^2$$

$$= 0 \ \text{m} + (0 \ \text{m/s})(26.9 \ \text{s}) + \tfrac{1}{2}(2.6 \ \text{m/s}^2)(26.9 \ \text{s})^2$$

$$= 940 \ \text{m}$$

Our final answers are thus that the plane will take 27 s to reach takeoff speed, with a minimum runway length of 940 m.

ASSESS Think about the last time you flew; 27 s seems like a reasonable time for a plane to accelerate on takeoff. Actual runway lengths at major airports are 3000 m or more, a few times greater than the minimum length, because they have to allow for emergency stops during an aborted takeoff. (If we had calculated a distance far greater than 3000 m, we would know we had done something wrong!)

EXAMPLE 2.10 Finding the braking distance

A car is traveling at a speed of 30 m/s, a typical highway speed, on wet pavement. The driver sees an obstacle ahead and decides to stop. From this instant, it takes him 0.75 s to begin applying the brakes. Once the brakes are applied, the car experiences an acceleration of -6.0 m/s^2. How far does the car travel from the instant the driver notices the obstacle until stopping?

PREPARE This problem is more involved than previous problems we have solved, so we will take more care with the visual overview in Figure 2.35. In addition to a motion diagram and a pictorial representation, we include a graphical representation. Notice that there are two different phases of the motion, a constant-velocity phase before braking begins, and a steady slowing down once the brakes are applied. We will need to do two different calculations, one for each phase. Consequently, we've used numerical subscripts rather than a simple i and f.

SOLVE From t_1 to t_2 the velocity stays constant at 30 m/s. This is uniform motion, so the position at time t_2 is computed using Equation 2.3:

$$x_2 = x_1 + (v_x)_1(t_2 - t_1) = 0 \text{ m} + (30 \text{ m/s})(0.75 \text{ s})$$

$$= 22.5 \text{ m}$$

At t_2, the velocity begins to decrease at a steady -6.0 m/s^2 until the car comes to rest at t_3. This time interval can be computed using the first equation in Table 2.4, $(v_x)_3 = (v_x)_2 + a_x \Delta t$:

$$\Delta t = t_3 - t_2 = \frac{(v_x)_3 - (v_x)_2}{a_x} = \frac{0 \text{ m/s} - 30 \text{ m/s}}{-6.0 \text{ m/s}^2} = 5.0 \text{ s}$$

The position at time t_3 is computed using the second equation in Table 2.4; we take point 2 as the initial point and point 3 as the final point for this phase of the motion and use $\Delta t = t_3 - t_2$:

$$x_3 = x_2 + (v_x)_2 \Delta t + \tfrac{1}{2}a_x(\Delta t)^2$$

$$= 22.5 \text{ m} + (30 \text{ m/s})(5.0 \text{ s}) + \tfrac{1}{2}(-6.0 \text{ m/s}^2)(5.0 \text{ s})^2$$

$$= 98 \text{ m}$$

x_3 is the position of the car at the end of the problem—and so the car travels 98 m before coming to rest!

ASSESS The numbers for the reaction time and the acceleration on wet pavement are reasonable ones for an alert driver in a car with good tires. The final distance is quite large—it is more than the length of a football field.

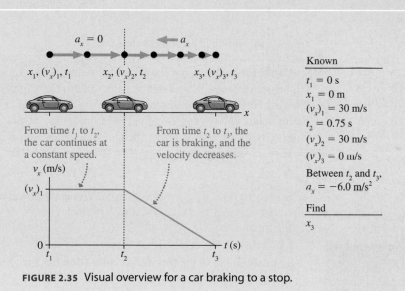

From time t_1 to t_2, the car continues at a constant speed.

From time t_2 to t_3, the car is braking, and the velocity decreases.

Known
$t_1 = 0$ s
$x_1 = 0$ m
$(v_x)_1 = 30$ m/s
$t_2 = 0.75$ s
$(v_x)_2 = 30$ m/s
$(v_x)_3 = 0$ m/s
Between t_2 and t_3,
$a_x = -6.0$ m/s^2

Find
x_3

FIGURE 2.35 Visual overview for a car braking to a stop.

2.7 Free Fall

A particularly important example of constant acceleration is the motion of an object moving under the influence of gravity only, and no other forces. This motion is called **free fall.** Strictly speaking, free fall occurs only in a vacuum, where there is no air resistance. But if you drop a hammer, air resistance is negligible, so we'll make only a very slight error in treating it *as if* it were in free fall. If you drop a feather, air resistance is *not* negligible, and we can't make this approximation. Motion with air resistance is a problem we will study in Chapter 5. Until then, we will restrict our attention to situations in which air resistance can be ignored, and will make the reasonable assumption that falling objects are in free fall.

 1.7, 1.10

"Looks like Mr. Galileo was correct . . ."
was the comment made by Apollo 15 astronaut David Scott, who dropped a hammer
and a feather on the moon. The objects were
dropped from the same height at the same
time and hit the ground simultaneously—
something that would not happen in the
atmosphere of the earth!

As part of his early studies of motion, Galileo did the first careful experiments on free fall and made the somewhat surprising observation that two objects of different weight dropped from the same height will, if air resistance can be neglected, hit the ground at the same time and with the same speed. In fact—as Galileo surmised, and as a famous demonstration on the moon showed—in a vacuum, where there is no air resistance, this holds true for *any* two objects!

Galileo's discovery about free fall means that **any two objects in free fall, regardless of their mass, have the same acceleration.** This is an especially important conclusion. Figure 2.36a shows the motion diagram of an object that was released from rest and falls freely. The motion diagram and graph would be identical for a falling baseball or a falling boulder! Figure 2.36b shows the object's velocity graph. As we can see, the velocity changes at a steady rate. The slope of the velocity-versus-time graph is the free-fall acceleration $a_{\text{free fall}}$.

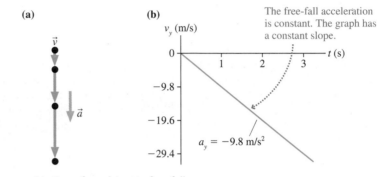

FIGURE 2.36 Motion of an object in free fall.

Careful measurements show that the value of the free-fall acceleration varies slightly at different places on the earth, due to the slightly nonspherical shape of the earth, the fact that different positions on the surface of the earth may be at different heights above sea level, and the fact that the earth is rotating. (We will consider why these factors affect the free-fall acceleration in Chapter 6.) A global average, at sea level, is

$$\vec{a}_{\text{free fall}} = (9.80 \text{ m/s}^2, \text{ vertically downward}) \qquad (2.12)$$

Standard value for the acceleration of an object in free fall

The magnitude of the **free-fall acceleration** has the special symbol g:

$$g = 9.80 \text{ m/s}^2$$

We will generally work with 2 significant figures, and so will use $g = 9.8 \text{ m/s}^2$. Several points about free fall are worthy of note:

- g, by definition, is *always* positive. **There will never be a problem that uses a negative value for g.**
- The velocity graph in Figure 2.36b has a negative slope. Even though a falling object speeds up, it has *negative* acceleration. Alternatively, notice that the acceleration vector $\vec{a}_{\text{free fall}}$ points down. Thus g is *not* the object's acceleration, simply the magnitude of the acceleration. The one-dimensional acceleration is

$$a_y = a_{\text{free fall}} = -g$$

It is a_y that is negative, not g.

Some of the children are moving up and some are moving down, but all are in free fall—and so are accelerating downward at 9.8 m/s^2.

- Because free fall is motion with constant acceleration, we can use the kinematic equations of Table 2.4 with the acceleration being due to gravity, $a_y = -g$.
- g is not called "gravity." Gravity is a force, not an acceleration. g is the *free-fall acceleration.*
- $g = 9.80 \text{ m/s}^2$ only on earth. Other planets have different values of g. You will learn in Chapter 6 how to determine g for other planets.
- We will sometimes compute acceleration in units of g. An acceleration of 9.8 m/s^2 is an acceleration of 1 g; an acceleration of 19.6 m/s^2 is 2 g. Generally, we can compute

$$\text{acceleration (in units of } g) = \frac{\text{acceleration (in units of m/s}^2)}{9.8 \text{ m/s}^2} \quad (2.13)$$

This allows us to express accelerations in units that have a definite physical reference.

NOTE ▶ Despite the name, free fall is not restricted to objects that are literally falling. Any object moving under the influence of gravity only, and no other forces, is in free fall. This includes objects falling straight down, objects that have been tossed or shot straight up, objects in projectile motion (such as a passed football) and, as we will see, satellites in orbit. In this chapter we consider only objects that move up and down along a vertical line; projectile motion will be studied in Chapter 3. ◀

EXAMPLE 2.11 Analyzing a rock's fall

A heavy rock is dropped from rest at the top of a cliff and falls 100 m before hitting the ground. How long does the rock take to fall to the ground, and what is its velocity when it hits?

PREPARE Figure 2.37 shows a visual overview with all necessary data. We have placed the origin at the ground, which makes $y_i = 100$ m.

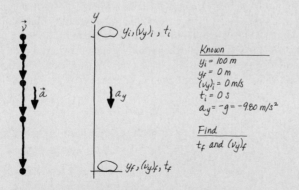

Known
$y_i = 100 \text{ m}$
$y_f = 0 \text{ m}$
$(v_y)_i = 0 \text{ m/s}$
$t_i = 0 \text{ s}$
$a_y = -g = -9.80 \text{ m/s}^2$

Find
t_f and $(v_y)_f$

FIGURE 2.37 Visual overview of a falling rock.

SOLVE Free fall is motion with the specific constant acceleration $a_y = -g$. The first question involves a relation between time and distance, a relation expressed by the second equation in Table 2.4 Using $(v_y)_i = 0 \text{ m/s}$ and $t_i = 0 \text{ s}$, we find

$$y_f = y_i + (v_y)_i \, \Delta t + \tfrac{1}{2} a_y \, \Delta t^2 = y_i - \tfrac{1}{2} g \, \Delta t^2 = y_i - \tfrac{1}{2} g t_f^2$$

We can now solve for t_f, finding:

$$t_f = \sqrt{\frac{2(y_i - y_f)}{g}} = \sqrt{\frac{2(100 \text{ m} - 0 \text{ m})}{9.80 \text{ m/s}^2}} = 4.52 \text{ s}$$

Now that we know the fall time, we can use the first kinematic equation to find $(v_y)_f$:

$$(v_y)_f = (v_y)_i - g \, \Delta t - -g t_f = -(9.80 \text{ m/s}^2)(4.52 \text{ s})$$
$$= -44.3 \text{ m/s}.$$

ASSESS Are the answers reasonable? Well, 100 m is about 300 feet, which is about the height of a 30-floor building. How long does it take something to fall 30 floors? Four or five seconds seems pretty reasonable. How fast would it be going at the bottom? Using an approximate version of our conversion factor $1 \text{ m/s} \approx 2 \text{ mph}$, we find that $44.3 \text{ m/s} \approx 90 \text{ mph}$. That also seems like a pretty reasonable speed for something that has fallen 30 floors. Suppose we had made a mistake. If we misplaced a decimal point we could have calculated a speed of 443 m/s, or about 900 mph! This is clearly *not* reasonable. If we had misplaced the decimal point in the other direction, we would have calculated a speed of $4.3 \text{ m/s} \approx 9 \text{ mph}$. This is another unreasonable result, as this is slower than a typical bicycling speed.

CONCEPTUAL EXAMPLE 2.4   Analyzing the motion of a ball tossed upward

Draw a motion diagram and a velocity-versus-time graph for a ball tossed straight up in the air from the point that it leaves the hand until just before it is caught.

REASON   You know what the motion of the ball looks like: The ball goes up, and then it comes back down again. This compli-cates the drawing of a motion diagram a bit, as the ball retraces its route as it falls. A literal motion diagram would show the upward motion and downward motion on top of each other, leading to confusion. We can avoid this difficulty by horizon-tally separating the upward motion and downward motion dia-grams. This will not affect our conclusions because it does not change any of the vectors. The motion diagram and velocity-versus-time graph appear as in Figure 2.38.

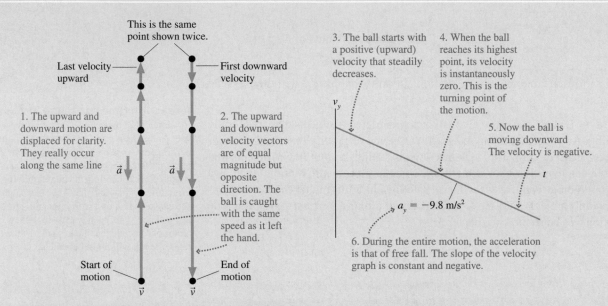

FIGURE 2.38   Motion diagram and velocity graph of a ball tossed straight up in the air.

ASSESS   The highest point in the ball's motion, where it reverses direction, is called a *turning point*. What are the velocity and the acceleration at this point? We can see from the motion diagram that the velocity vectors are pointing upward but getting shorter as the ball approaches the top. As it starts to fall, the velocity vectors are pointing downward and getting longer. There must be a moment—just an instant as $\vec{v}$ switches from pointing up to pointing down—when the velocity is zero. Indeed, the ball's velocity *is* zero for an instant at the precise top of the motion! We can also see on the velocity graph that there is one instant of time when $v_y = 0$. This is the turning point.

But what about the acceleration at the top? Many people expect the acceleration to be zero at the highest point. But recall that the velocity at the top point is changing—from up to down. If the velocity is changing, there *must* be an acceleration. The slope of the velocity graph at the instant when $v_y = 0$—that is, at the highest point—is no different than at any other point in the motion. The ball is still in free fall with acceleration $a_y = -g$!

Another way to think about this is to note that zero accelera-tion would mean no change of velocity. When the ball reached zero velocity at the top, it would hang there and not fall if the acceleration were also zero!

EXAMPLE 2.12   Finding the height of a leap

A springbok is an antelope found in southern Africa that gets its name from its remarkable jump-ing ability. When a springbok is startled, it will leap straight up into the air—a maneuver called a "pronk." If a springbok leaves the ground at 7.0 m/s, a typical value, how high will it go?

PREPARE   We will begin with the visual overview shown in Figure 2.39. Even though the springbok is moving upward, this is a free-fall problem because the springbok (after leaving the ground) is moving under the influence of gravity *only*.

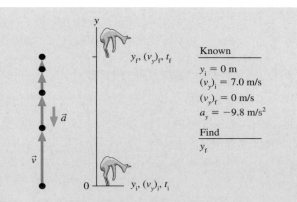

FIGURE 2.39   Visual overview of the springbok's leap.

A critical aspect of solving this sort of problem is knowing what to take as the end point of the motion. The clue is this: The very top point of the trajectory is a *turning point,* and the instantaneous velocity at this point is zero. Thus we can characterize the "top" of the trajectory—the highest point of the leap—as the point where $(v_y)_f = 0$ m/s. If we solve for y at this point, we will know the height of the leap. None of this was explicitly stated in the problem. Part of our job in solving a problem is to supply a reasonable *interpretation* of events and thus to gain the information needed to reach a solution.

SOLVE We have information about displacement and velocity, but we don't know anything about the time interval. The third kinematic equation in Table 2.4, a relationship among displacement, velocity, and acceleration, is perfect for situations like this. Because $y_i = 0$, the springbok's displacement is $\Delta y = y_f - y_i = y_f$. Its final velocity is $(v_y)_f = 0$ m/s, so we can compute

$$(v_y)_f^2 = 0 = (v_y)_i^2 - 2g \, \Delta y = (v_y)_i^2 - 2gy_f$$

which gives

$$(v_y)_i^2 = 2gy_f$$

By solving for y_f, we find that the springbok reaches a height of

$$y_f = \frac{(v_y)_i^2}{2g} = \frac{(7.0 \text{ m/s})^2}{2(9.8 \text{ m/s}^2)} = 2.5 \text{ m}$$

ASSESS 2.5 m is a remarkable leap—a bit over 8 ft—but these animals are known for their jumping ability, so this seems reasonable.

We began this chapter with another question about animals and their athletic abilities: comparing the speeds of horses and humans. Who is the winner in a race between a horse and a man? The surprising answer is "It depends." Who the winner will be depends on the length of the race! Figure 2.40 shows position and velocity graphs for an elite male sprinter and a thoroughbred racehorse. The horse's maximum velocity is about twice that of the man, but the man's acceleration from rest is about twice that of the horse. As we can see from the graphs, a man could win a *very* short race. For a longer race, the horse's higher maximum velocity will put it in the lead; the men's world record for the mile is a bit under 4 min, but a horse can easily run this distance in under 2 min.

For a race of many miles, another factor comes into play: energy. A very long race is less about velocity and acceleration than about endurance—the ability to continue expending energy for a long time. In such endurance trials, humans often win. We will explore such energy issues in Chapter 11.

In this chapter, we have explored motion along a line—that is, motion in one dimension. In the next chapter, we will explore motion in two dimensions. We will begin with a more thorough treatment of vectors, and then pick up where we left off, exploring objects moving under the influence of gravity.

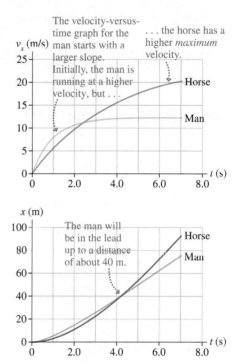

FIGURE 2.40 Velocity-versus-time and position-versus-time graphs for a sprint between a man and a horse.

STOP TO THINK 2.6 A volcano ejects a chunk of rock straight up at a velocity of $v_y = 30$ m/s. Ignoring air resistance, what will be the velocity v_y of the rock when it falls back into the volcano's crater?

A. >30 m/s   B. 30 m/s   C. 0 m/s   D. −30 m/s   E. <−30 m/s

SUMMARY

The goal of Chapter 2 has been to describe and analyze linear motion.

GENERAL STRATEGIES

Problem-Solving Strategy

Our general problem-solving strategy has three parts:

PREPARE Set up the problem:

- Draw a picture
- Collect necessary information
- Do preliminary calculations

SOLVE Do the necessary mathematics or reasoning.

ASSESS Check your answer to see if it is complete in all details and makes physical sense.

Visual Overview

A visual overview consists of several pieces that completely specify a problem. This may include any or all of the elements below:

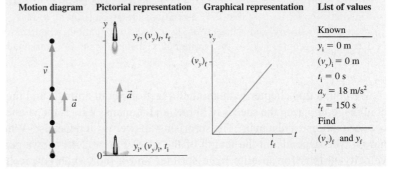

Motion diagram | Pictorial representation | Graphical representation | List of values

Known
$y_i = 0$ m
$(v_y)_i = 0$ m
$t_i = 0$ s
$a_y = 18$ m/s^2
$t_f = 150$ s

Find
$(v_y)_f$ and y_f

IMPORTANT CONCEPTS

Velocity is the rate of change of position

$$v_x = \frac{\Delta x}{\Delta t}$$

Acceleration is the rate of change of velocity.

$$a_x = \frac{\Delta v_x}{\Delta t}$$

The units of acceleration are m/s^2.

An object is speeding up if v_x and a_x have the same sign, slowing down if they have opposite signs.

A position-versus-time graph plots position on the vertical axis against time on the horizontal axis.

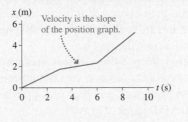

Velocity is the slope of the position graph.

A velocity-versus-time graph plots velocity on the vertical axis against time on the horizontal axis.

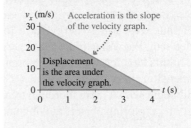

Acceleration is the slope of the velocity graph.

Displacement is the area under the velocity graph.

APPLICATIONS

Uniform motion
An object in uniform motion has a constant velocity. Its velocity graph is a horizontal line; its position graph is linear.

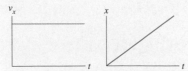

Kinematic equation for uniform motion:

$$x_f = x_i + v_x \, \Delta t$$

Uniform motion is a special case of constant-acceleration motion, with $a_x = 0$.

Motion with constant acceleration
An object with constant acceleration has a constantly changing velocity. Its velocity graph is linear; its position graph is a parabola.

Kinematic equations for motion with constant acceleration:

$$(v_x)_f = (v_x)_i + a_x \, \Delta t$$

$$x_f = x_i + (v_x)_i \, \Delta t + \tfrac{1}{2} a_x (\Delta t)^2$$

$$(v_x)_f^2 = (v_x)_i^2 + 2 a_x \, \Delta x$$

Free fall
Free fall is a special case of constant-acceleration motion; the acceleration has magnitude $g = 9.80$ m/s^2 and is always directed vertically downward whether an object is moving up or down.

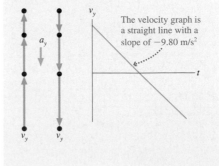

The velocity graph is a straight line with a slope of -9.80 m/s^2

(MP) For instructor-assigned homework, go to
www.masteringphysics.com

Problem difficulty is labeled as | (straightforward) to ||||| (challenging).

Problems labeled ||NT integrate significant material
from earlier chapters; BIO are of biological or medical
interest.

QUESTIONS

Conceptual Questions

1. A person gets in an elevator on the ground floor and rides it to the top floor of a building. Sketch a velocity-versus-time graph for this motion.
2. a. Give an example of a vertical motion with a positive velocity and a negative acceleration.
   b. Give an example of a vertical motion with a negative velocity and a negative acceleration.
3. Sketch a velocity-versus-time graph for a rock that is thrown straight upward, from the instant it leaves the hand until the instant it hits the ground.
4. You're driving along the highway at a steady speed of 60 mph when another driver decides to pass you. At the moment when the front of his car is exactly even with the front of your car, and you turn your head to smile at him, do the two cars have equal velocities? Explain.
5. A car is traveling north. Can its acceleration vector ever point south? Explain.
6. Certain animals are capable of running at great speeds; other
BIO animals are capable of tremendous accelerations. Speculate on which would be more beneficial to a predator—large maximum speed or large acceleration.
7. A ball is thrown straight up into the air. At each of the following instants, is the ball's acceleration a_y equal to g, $-g$, 0, $<g$, or $>g$?
   a. Just after leaving your hand?
   b. At the very top (maximum height)?
   c. Just before hitting the ground?
8. A rock is *thrown* (not dropped) straight down from a bridge into the river below.
   a. Immediately after being released, is the magnitude of the rock's acceleration greater than g, less than g, or equal to g? Explain.
   b. Immediately before hitting the water, is the magnitude of the rock's acceleration greater than g, less than g, or equal to g? Explain.
9. Figure Q2.9 at right shows an object's position-versus-time graph. The letters A to E correspond to various segments of the motion in which the graph has constant slope.
   a. Write a realistic motion short story for an object that would have this position graph.
   b. In which segment(s) is the object at rest?
   c. In which segment(s) is the object moving to the right?
   d. Is the speed of the object during segment C greater than, equal to, or less than its speed during segment E? Explain.

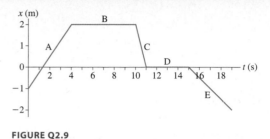

FIGURE Q2.9

10. Figure Q2.10 shows the position graph for an object moving along the horizontal axis.
    a. Write a realistic motion short story for an object that would have this position graph.
    b. Draw the corresponding velocity graph.

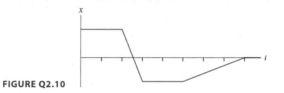

FIGURE Q2.10

11. Figure Q2.11 shows the position-versus-time graphs for two objects, A and B, that are moving along the same axis.
    a. At the instant $t = 1$ s, is the speed of A greater than, less than, or equal to the speed of B? Explain.
    b. Do objects A and B ever have the *same* speed? If so, at what time or times? Explain.

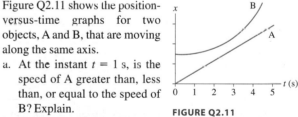

FIGURE Q2.11

12. Do the two position graphs in Figure Q2.12 represent possible motions of an object? If so, describe that motion in words. If not, why not?

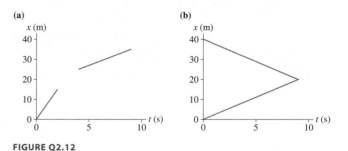

FIGURE Q2.12

13. Figure Q2.13 shows a position-versus-time graph for a moving object. At which lettered point or points
    a. Is the object moving the fastest?
    b. Is the object moving to the left?
    c. Is the object speeding up?
    d. Is the object slowing down?
    e. Is the object turning around?

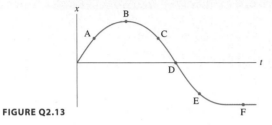

FIGURE Q2.13

14. Figure Q2.14 is the velocity-versus-time graph for an object moving along the x-axis. The graph is divided into six labeled segments.
    a. During which segment(s) is the velocity constant?
    b. During which segment(s) is the object speeding up?
    c. During which segment(s) is the object slowing down?
    d. During which segment(s) is the object standing still?
    e. During which segment(s) is the object moving to the right?

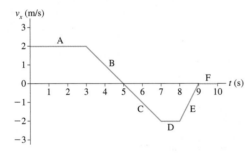

FIGURE Q2.14

15. A car traveling at velocity v takes distance d to stop after the brakes are applied. What is the stopping distance if the car is initially traveling at velocity $2v$? Assume that the acceleration due to the braking is the same in both cases.

Multiple-Choice Questions

16. | Figure Q2.16 shows the position graph of a car traveling on a straight road. At which labeled instant is the speed of the car greatest?

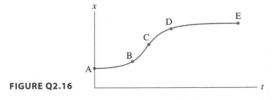

FIGURE Q2.16

17. ‖ Figure Q2.17 shows the position graph of a car traveling on a straight road. The velocity at instant 1 is _____ and the velocity at instant 2 is _____.
    A. positive, negative
    B. positive, positive
    C. negative, negative
    D. negative, zero
    E. positive, zero

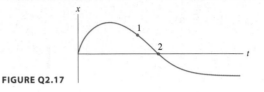

FIGURE Q2.17

18. ‖ Figure Q2.18 shows an object's position-versus-time graph. What is the velocity of the object at $t = 6$ s?
    A. 0.67 m/s   B. 0.83 m/s   C. 3.3 m/s
    D. 4.2 m/s   E. 25 m/s

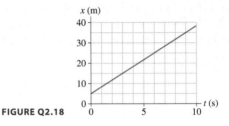

FIGURE Q2.18

19. | The following options describe the motion of four cars A–D. Which car has the largest acceleration?
    A goes from 0 m/s to 10 m/s in 5.0 s
    B goes from 0 m/s to 5.0 m/s in 2.0 s
    C goes from 0 m/s to 20 m/s in 7.0 s
    D goes from 0 m/s to 3.0 m/s in 1.0 s

20. | A car is traveling at $v_x = 20$ m/s. The driver applies the brakes, and the car slows with $a_x = -4.0$ m/s². What is the stopping distance?
    A. 5.0 m   B. 25 m   C. 40 m   D. 50 m

21. ‖ Velocity-versus-time graphs for three drag racers are shown in Figure Q2.21. At $t = 5.0$ s, which car has traveled the furthest?
    A. Andy   B. Betty   C. Carl
    D. All have traveled the same distance

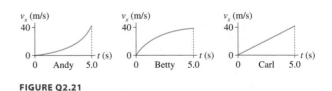

FIGURE Q2.21

22. | Which of the three drag racers in Question 21 had the greatest acceleration at $t = 0$ s?
    A. Andy   B. Betty   C. Carl
    D. All had the same acceleration

23. ‖ Ball 1 is thrown straight up in the air and, at the same instant, ball 2 is released from rest and allowed to fall. Which velocity graph in Figure Q2.23 best represents the motion of the two balls?

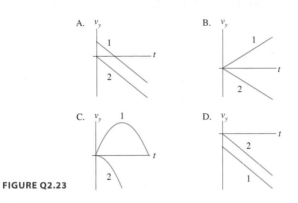

FIGURE Q2.23

Questions 24 and 25 refer to the graph in Figure Q2.24, which is the acceleration graph for an object moving along the x-axis.

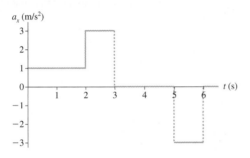

FIGURE Q2.24

24. ‖ If the object's velocity at $t = 0$ s is -1 m/s, what is its velocity at $t = 2$ s?
    A. 0 m/s   B. 0.5 m/s   C. -1 m/s
    D. 1 m/s   E. 2 m/s

25. ‖ Which graph in Figure Q2.25 could represent the velocity of the object?

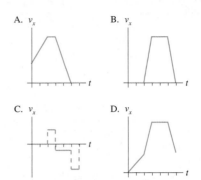

FIGURE Q2.25

26. | Figure Q2.26 shows a motion diagram with the clock reading (in seconds) shown at each position. From $t = 9$ s to $t = 15$ s the object is at the same position. After that, it returns along the same track. The positions of the dots for $t \geq 16$ s are offset for clarity. Which graph best represents the object's *velocity?*

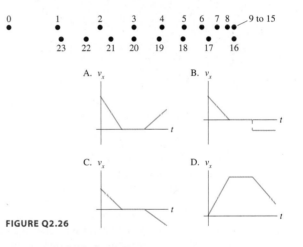

FIGURE Q2.26

PROBLEMS

Section 2.1 Describing Motion

1. | Figure P2.1 shows a motion diagram of a car traveling down a street. The camera took one frame every second. A distance scale is provided.
    a. Measure the x-value of the car at each dot. Place your data in a table, similar to Table 2.1, showing each position and the instant of time at which it occurred.
    b. Make a graph of x versus t, using the data in your table. Because you have data only at certain instants of time, your graph should consist of dots that are not connected together.

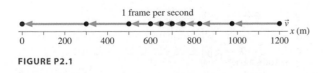

FIGURE P2.1

2. | For each motion diagram in Figure P2.2, determine the sign (positive or negative) of the position and the velocity.

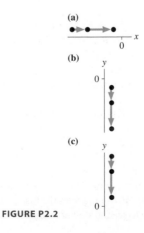

FIGURE P2.2

3. | Write a short description of the motion of a real object for which Figure P2.3 would be a realistic position-versus-time graph.

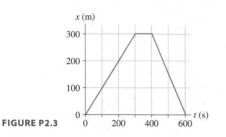

FIGURE P2.3

4. | Write a short description of the motion of a real object for which Figure P2.4 would be a realistic position-versus-time graph.

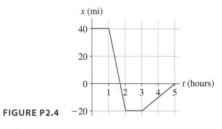

FIGURE P2.4

Section 2.2 Uniform Motion

5. ‖ Suppose a car is moving with constant velocity along a straight road. Its position was $x_1 = 15$ m at $t_1 = 0.0$ s and is $x_2 = 30$ m at $t_2 = 3.0$ s.
   a. What was its position at $t = 1.5$ s?
   b. What will its position be at $t = 5.0$ s?
6. | A car starts at the origin and moves with velocity $\vec{v} = $ (10 m/s, northeast). How far from the origin will the car be after traveling for 45 s?
7. | Larry leaves home at 9:05 A.M. and runs at constant speed to the lamppost, shown in Figure P2.7. He reaches the lamppost at 9:07 A.M., immediately turns, and runs to the tree, again at constant speed. Larry arrives at the tree at 9:10 A.M. What is Larry's velocity during each of these two intervals?

FIGURE P2.7

8. ‖ Alan leaves Los Angeles at 8:00 A.M. to drive to San Francisco, 400 mi away. He travels at a steady 50 mph. Beth leaves Los Angeles at 9:00 A.M. and drives a steady 60 mph.
   a. Who gets to San Francisco first?
   b. How long does the first to arrive have to wait for the second?
9. ‖ Richard is driving home to visit his parents. 125 mi of the trip are on the interstate highway where the speed limit is 65 mph. Normally Richard drives at the speed limit, but today he is running late and decides to take his chances by driving at 70 mph. How many minutes does he save?
10. ‖ In a 5.00 km race, one runner runs at a steady 12.0 km/h and another runs at 14.5 km/h. How long does the faster runner have to wait at the finish line to see the slower runner cross?

11. ‖‖ In an 8.00 km race, one runner runs at a steady 11.0 km/h and another runs at 14.0 km/h. How far from the finish line is the slower runner when the faster runner finishes the race?
12. ‖ A bicyclist has the position-versus-time graph shown in Figure P2.12. What is the bicyclist's velocity at $t = 10$ s, at $t = 25$ s, and at $t = 35$ s?

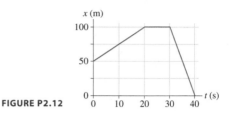

FIGURE P2.12

Section 2.3 Motion with Changing Velocity

13. ‖ Figure P2.13 shows the position graph of a particle.
   a. Draw the particle's velocity graph for the interval $0 \text{ s} \le t \le 4 \text{ s}$.
   b. Does this particle have a turning point or points? If so, at what time or times?

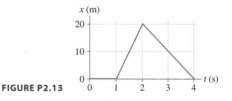

FIGURE P2.13

14. ‖‖ A car starts from $x_i = 10$ m at $t_i = 0$ s and moves with the velocity graph shown in Figure P2.14.
   a. What is the object's position at $t = 2$ s, 3 s, and 4 s?
   b. Does this car ever change direction? If so, at what time?

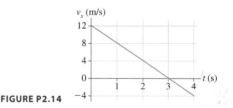

FIGURE P2.14

Section 2.4 Acceleration

15. ‖ Figure P2.15 shows the velocity graph of a bicycle. Draw the bicycle's acceleration graph for the interval $0 \text{ s} \le t \le 4 \text{ s}$. Give both axes an appropriate numerical scale.

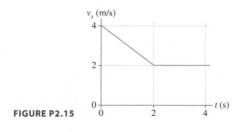

FIGURE P2.15

16. | Figure P2.16 shows the velocity graph of a train that starts from the origin at $t = 0$ s.
    a. Draw position and acceleration graphs for the train.
    b. Find the acceleration of the train at $t = 3.0$ s.

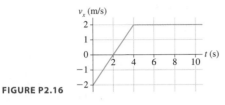

FIGURE P2.16

17. | Figure P2.17 shows the velocity graph of a bicycle moving along the x-axis. Its initial position is $x_i = 2.0$ m at $t_i = 0.0$ s. At $t = 2.0$ s, what are the bicycle's (a) position, (b) velocity, and (c) acceleration?

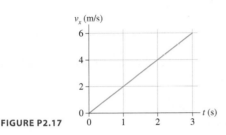

FIGURE P2.17

18. || For each motion diagram shown earlier in Figure P2.2, determine the sign (positive or negative) of the acceleration.

19. || Figure P2.19 is a somewhat simplified velocity graph for Olympic sprinter Carl Lewis starting a 100 m dash. Estimate his acceleration during each of the intervals A, B, and C.

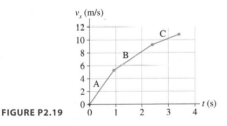

FIGURE P2.19

Section 2.5 Motion with Constant Acceleration

20. | A Thompson's gazelle can reach a speed of 13 m/s in 3.0 s.
    BIO A lion can reach a speed of 9.5 m/s in 1.0 s. A trout can reach a speed of 2.8 m/s in 0.12 s. Which animal has the greatest acceleration?

21. || When striking, the pike, a
    BIO predatory fish, can accelerate from rest to a speed of 4.0 m/s in 0.11 s.
    a. What is the acceleration of the pike during this strike?
    b. How far does the pike move during this strike?

22. ||| a. What constant acceleration, in SI units, must a car have to go from zero to 60 mph in 10 s?
    b. What fraction of g is this?
    c. How far has the car traveled when it reaches 60 mph? Give your answer both in SI units and in feet.

23. ||| A car travels with constant acceleration along a straight road. It was at the origin and at rest at $t_1 = 0$ s; its position is $x_2 = 144$ m at $t_2 = 6.00$ s.
    a. What was the car's position at $t = 3.00$ s?
    b. What will its position be at $t = 9.00$ s?

24. |||| A jet plane is cruising at 250 m/s when suddenly the pilot turns the engines up to full throttle. After traveling an additional 2.0 km, the jet is moving with a speed of 300 m/s.
    a. What is the jet's acceleration, assuming it to be a constant acceleration?
    b. Is your answer reasonable? Explain.

25. ||| A speed skater moving across frictionless ice at 8.0 m/s hits a 5.0-m-wide patch of rough ice. She slows steadily, then continues on at 6.0 m/s. What is her acceleration on the rough ice?

Section 2.6 Solving One-Dimensional Motion Problems

26. ||| A driver has a reaction time of 0.50 s, and the maximum deceleration of her car is 6.0 m/s². She is driving at 20 m/s when suddenly she sees an obstacle in the road 50 m in front of her. Can she stop the car in time to avoid a collision?

27. |||| You're driving down the highway late one night at 20 m/s when a deer steps onto the road 35 m in front of you. Your reaction time before stepping on the brakes is 0.50 s, and the maximum deceleration of your car is 10 m/s².
    a. How much distance is between you and the deer when you come to a stop?
    b. What is the maximum speed you could have and still not hit the deer?

28. | A light-rail train going from one station to the next on a straight section of track accelerates from rest at 1.4 m/s² for 15 s. It then proceeds at constant speed for 1100 m before slowing down at 2.2 m/s² until it stops at the station.
    a. What is the distance between the stations?
    b. How much time does it take the train to go between the stations?

29. || A simple model for a person running the 100 m dash is to assume the sprinter runs with constant acceleration until reaching top speed, then maintains that speed through the finish line. If a sprinter reaches his top speed of 11.2 m/s in 2.14 s, what will be his total time?

Section 2.7 Free Fall

30. ||| Ball bearings can be made by letting spherical drops of molten metal fall inside a tall tower—called a *shot tower*—and solidify as they fall.
    a. If a bearing needs 4.0 s to solidify enough for impact, how high must the tower be?
    b. What is the bearing's impact velocity?

31. | In the chapter, we saw that a person's reaction time is gen-
    BIO erally not quick enough to allow the person to catch a dollar bill dropped between the fingers. If a typical reaction time in this case is 0.25 s, how long would a bill need to be for a person to have a good chance of catching it?

32. || A ball is thrown vertically upward with a speed of 19.6 m/s.
    a. What are the ball's velocity and height after 1.00, 2.00, 3.00, and 4.00 s?
    b. Draw the ball's velocity-versus-time graph. Give both axes an appropriate numerical scale.

33. ‖ A student at the top of a building of height h throws ball A straight upward with speed v_0 and throws ball B straight downward with the same initial speed.
    a. Compare the balls' accelerations, both direction and magnitude, immediately after they leave her hand. Is one acceleration larger than the other? Or are the magnitudes equal?
    b. Compare the final speeds of the balls as they reach the ground. Is one speed larger than the other? Or are they equal?

34. ‖‖‖ In an action movie, the villain is rescued from the ocean by grabbing onto the ladder hanging from a helicopter. He is so intent on gripping the ladder that he lets go of his briefcase of counterfeit money when he is 130 m above the water. If the briefcase hits the water 6.0 s later, what was the speed at which the helicopter was ascending?

35. ‖‖‖ A rock climber stands on top of a 50-m-high cliff overhanging a pool of water. He throws two stones vertically downward 1.0 s apart and observes that they cause a single splash. The initial speed of the first stone was 2.0 m/s.
    a. How long after the release of the first stone does the second stone hit the water?
    b. What was the initial speed of the second stone?
    c. What is the speed of each stone as they hit the water?

General Problems

36. ‖ Macrophages are cells referred to as "big eaters" or "killing
BIO machines" that destroy any foreign agents (pathogens) or antibodies found in the body. When a pathogen is detected, the macrophage moves toward it and literally devours it almost instantly. Suppose a large number of pathogen molecules are lined up 1.0 μm apart on a straight fiber of tissue. A macrophage initially 0.50 μm from the nearest one moves toward it at a constant speed of 30 nm/s.
    a. How long does it take until the first pathogen molecule is devoured?
    b. How about 100 molecules?
    c. What is the total distance traveled by the macrophage by the time 100 molecules have been eaten?
    d. How many pathogen molecules will one microphage destroy in 10 minutes?

Questions 37 and 38 concern *nerve impulses,* signals that propagate along *nerve fibers* consisting of many *axons* (fiber-like extensions of nerve cells) connected end-to-end. The speed of the signal depends on whether the axon has a sheath called myelin around it or not. Nerve impulses move along axons with myelin at 100 m/s, one hundred times as quickly as on axons without it. Figure P2.37 shows small portions of three nerve fibers consisting of axons of equal size. Two-thirds of the axons in fiber B are myelinated.

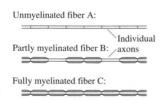

Unmyelinated fiber A:

Individual axons

Partly myelinated fiber B:

Fully myelinated fiber C:

FIGURE P2.37

37. ‖ Suppose nerve impulses simultaneously enter the left side
BIO of the nerve fibers sketched in Figure P2.37, then propagate to the right. Draw qualitatively accurate position and velocity graphs for the nerve impulses in all three cases. A nerve fiber is made up of many axons, but show the propagation of the impulses only over the six axons shown here.

38. ‖ Suppose that the nerve fibers in Figure P2.37 connect a
BIO finger to your brain, a distance of 1.2 m.
    a. What are the travel times of a nerve impulse from finger to brain along fibers A and C?
    b. For fiber B, 2/3 of the length is composed of myelinated axons, 1/3 unmyelinated axons. Compute the travel time for a nerve impulse on this fiber.
    c. When you touch a hot stove with the finger, the sensation of pain must reach your brain as a nerve signal along a nerve fiber before your muscles can react. Which of the three fibers gives you the best protection against a burn? Are any of these fibers unsuitable for transmitting urgent sensory information?

39. ‖ A driver travels along a straight east-west road. Starting from rest, he drives east for a while at highway speeds. Spotting some cows in the road, he brakes and comes to a halt. After waiting a few minutes, he backs slowly down the road until reaching the nearest crossroad.
    Draw graphs of the driver's position, velocity, and acceleration as functions of time. There are no numbers in this problem, but your graphs should have the correct shapes. Draw the three graphs one above the other so that the time axes line up.

40. ‖ A truck driver has a shipment of apples to deliver to a destination 440 miles away. The trip usually takes him 8 hours. Today he finds himself daydreaming and realizes 120 miles into his trip that he is running 15 minutes later than his usual pace at this point. At what speed must he drive for the remainder of the trip to complete the trip in the usual amount of time?

41. ‖ When you sneeze, the air in your lungs accelerates from
BIO rest to approximately 150 km/hr in about 0.50 seconds.
    a. What is the acceleration of the air in m/s^2?
    b. What is this acceleration, in units of g?

42. ‖‖‖ Two stones are released from the edge of a cliff, one a short time after the other.
    a. As they fall, the first is always going faster than the second. Does the *difference* between their speeds get larger, get smaller, or stay the same? Explain.
    b. Does their separation increase, decrease, or stay the same? Explain.
    c. Will the time interval between the instants at which they hit the ground be smaller than, equal to, or larger than the time interval between the instants of their release? Explain.

43. ‖ Figure P2.43 shows the motion diagram, made at two frames of film per second, of a ball rolling along a track. The track has a 3.0-m-long sticky section.
    a. Use the meter stick to determine the positions of the center of the ball. Place your data in a table, similar to Table 2.1, showing each position and the instant of time at which it occurred.
    b. Make a graph of x versus t for the ball. Because you have data only at certain instants of time, your graph should consist of dots that are not connected together.

c. What is the *change* in the ball's position from $t = 0$ s to $t = 1.0$ s?

d. What is the *change* in the ball's position from $t = 2.0$ s to $t = 4.0$ s?

e. What is the ball's velocity before reaching the sticky section?

f. What is the ball's velocity after passing the sticky section?

g. Determine the ball's acceleration on the sticky section of the track.

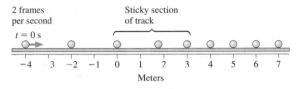

FIGURE P2.43

44. ‖ Figure P2.44 shows a velocity-versus-time graph for a particle moving along the *x*-axis. Its initial position is $x_i = 2.0$ m at $t_i = 0$ s.

a. What are the particle's position, velocity, and acceleration at $t = 1.0$ s?

b. What are the particle's position, velocity, and acceleration at $t = 3.0$ s?

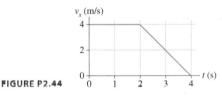

FIGURE P2.44

45. ‖ Figure P2.45 shows a velocity graph for a particle having initial position $x_i = 0$ m at $t_i = 0$ s.

a. At what time or times is the particle found at $x = 35$ m? Work with the geometry of the graph, not with kinematic equations.

b. Draw a motion diagram for the particle.

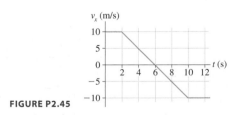

FIGURE P2.45

46. ‖‖ Julie drives 100 mi to Grandmother's house. On the way to Grandmother's, Julie drives half the *distance* at 40 mph and half the distance at 60 mph. On her return trip, she drives half the *time* at 40 mph and half the time at 60 mph.

a. How long does it take Julie to complete the trip to Grandmother's house?

b. How long does the return trip take?

47. ‖ The takeoff speed for an Airbus A320 jetliner is 80 m/s. Velocity data measured during takeoff are as follows:

t (s)	v_x (m/s)
0	0
10	23
20	46
30	69

a. What is the takeoff speed in miles per hour?

b. What is the jetliner's acceleration during takeoff?

c. At what time do the wheels leave the ground?

d. For safety reasons, in case of an aborted takeoff, the runway must be three times the takeoff distance. Can an A320 take off safely on a 2.5-mi-long runway?

48. ‖‖‖ Does a real automobile have constant acceleration? Measured data for a Porsche 944 Turbo at maximum acceleration are as follows:

t (s)	v_x (mph)
0	0
2	28
4	46
6	60
8	70
10	78

a. Convert the velocities to m/s, then make a graph of velocity versus time. Based on your graph, is the acceleration constant? Explain.

b. Draw a smooth curve through the points on your graph, then use your graph to *estimate* the car's acceleration at 2.0 s and 8.0 s. Give your answer in SI units. **Hint:** Remember that acceleration is the slope of the velocity graph.

49. ‖ a. How many days will it take a spaceship to accelerate to the speed of light (3.0×10^8 m/s) with the acceleration g?

b. How far will it travel during this interval?

c. What fraction of a light year is your answer to part b? A *light year* is the distance light travels in one year.

NOTE ▶ We know, from Einstein's theory of relativity, that no object can travel at the speed of light. So this problem, while interesting and instructive, is not realistic. ◀

50. ‖‖‖ You are driving to the grocery store at 20 m/s. You are 110 m from an intersection when the traffic light turns red. Assume that your reaction time is 0.70 s and that your car brakes with constant acceleration.

a. How far are you from the intersection when you begin to apply the brakes?

b. What acceleration will bring you to rest right at the intersection?

c. How long does it take you to stop?

51. ‖ When you blink your eye, the upper lid goes from rest with
BIO your eye open to completely covering your eye in a time of 0.024 s.

a. Estimate the distance that the top lid of your eye moves during a blink.

b. What is the acceleration of your eyelid? Assume it to be constant.

c. What is your upper eyelid's final speed as it hits the bottom eyelid?

52. ||||| A bush baby, an African
BIO primate, is capable of leaping vertically to the remarkable height of 2.26 m. To
jump this high, the bush
baby accelerates over a
distance of 0.16 m while
rapidly extending its legs.
The acceleration during the
jump is approximately constant. What is the acceleration in m/s² and in g's?

53. ||| When jumping, a flea reaches a takeoff speed of 1.0 m/s
BIO over a distance of 0.50 mm.
    a. What is the flea's acceleration during the jump phase?
    b. How long does the acceleration phase last?
    c. If the flea jumps straight up, how high will it go?
       (Ignore air resistance for this problem; in reality, air resis-
       tance plays a large role, and the flea will not reach this
       height.)

54. |||| Divers compete by diving into a 3.0-m-deep pool from a
platform 10 m above the water. What is the magnitude of the
minimum acceleration in the water needed to keep a diver
from hitting the bottom of the pool? Assume the acceleration
is constant.

55. |||| With a man on first base, Hank hits a ball to the corner
of the outfield. The outfielder has trouble retrieving the
ball, so Hank decides to try for an inside-the-park home
run. As he rounds third base at a speed of 18 ft/s, Hank sees
that the player in front of him is two-thirds of the way from
third base to home and is limping at 6.0 ft/s. It is 90 ft
between bases. If Hank wants to arrive at home plate just
behind the slower runner, what should be the magnitude of
his acceleration as he slows?

56. |||| A student standing on the ground throws a ball straight up.
The ball leaves the student's hand with a speed of 15 m/s
when the hand is 2.0 m above the ground. How long is the ball
in the air before it hits the ground? (The student moves her
hand out of the way.)

57. || A rock is tossed straight up with a speed of 20 m/s. When it
returns, it falls into a hole 10 m deep.
    a. What is the rock's velocity as it hits the bottom of the hole?
    b. How long is the rock in the air, from the instant it is
       released until it hits the bottom of the hole?

58. ||||| A 200 kg weather rocket is loaded with 100 kg of fuel and
fired straight up. It accelerates upward at 30.0 m/s² for 30.0 s,
then runs out of fuel. Ignore any air resistance effects.
    a. What is the rocket's maximum altitude?
    b. How long is the rocket in the air?
    c. Draw a velocity-versus-time graph for the rocket from
       liftoff until it hits the ground.

59. ||||| A juggler throws a ball straight up into the air with a
speed of 10 m/s. With what speed would she need to throw a
second ball half a second later, starting from the same position as the first, in order to hit the first ball at the top of its
trajectory?

60. |||| A lead ball is dropped into a lake from a diving board
5.0 m above the water. After entering the water, it sinks to the
bottom with a constant velocity equal to the velocity with
which it hit the water. The ball reaches the bottom 3.0 s after it
is released. How deep is the lake?

61. ||| A hotel elevator ascends 200 m with a maximum speed of
5.0 m/s. Its acceleration and deceleration both have a magnitude of 1.0 m/s².
    a. How far does the elevator move while accelerating to full
       speed from rest?
    b. How long does it take to make the complete trip from bot-
       tom to top?

62. ||| A car starts from rest at a stop sign. It accelerates at 2.0 m/s²
for 6.0 seconds, coasts for 2.0 s, and then slows down at a rate of
1.5 m/s² for the next stop sign. How far apart are the stop signs?

63. || A toy train is pushed forward and released at $x_i = 2.0$ m
with a speed of 2.0 m/s. It rolls at a steady speed for 2.0 s,
then one wheel begins to stick. The train comes to a stop
6.0 m from the point at which it was released. What is the
train's acceleration after its wheel begins to stick?

64. ||| Bob is driving the getaway car after the big bank robbery.
He's going 50 m/s when his headlights suddenly reveal a nail
strip that the cops have placed across the road 150 m in front
of him. If Bob can stop in time, he can throw the car into
reverse and escape. But if he crosses the nail strip, all his tires
will go flat and he will be caught. Bob's reaction time before
he can hit the brakes is 0.60 s, and his car's maximum deceleration is 10 m/s². Is Bob in jail?

65. || Heather and Jerry are standing on a bridge 50 m above a
river. Heather throws a rock straight down with a speed of
20 m/s. Jerry, at exactly the same instant of time, throws a
rock straight up with the same speed. Ignore air resistance.
    a. How much time elapses between the first splash and the
       second splash?
    b. Which rock has the faster speed as it hits the water?

66. ||| A motorist is driving at 20 m/s when she sees that a traffic
light 200 m ahead has just turned red. She knows that this
light stays red for 15 s, and she wants to reach the light just as
it turns green again. It takes her 1.0 s to step on the brakes and
begin slowing at a constant deceleration. What is her speed as
she reaches the light at the instant it turns green?

67. ||||| A "rocket car" is launched along a long straight track at
$t = 0$ s. It moves with constant acceleration $a_1 = 2.0$ m/s². At
$t = 2.0$ s, a second car is launched with constant acceleration
$a_2 = 8.0$ m/s².
    a. At what time does the second car catch up with the first
       one?
    b. How far down the track do they meet?

68. | A Porsche challenges a Honda to a 400 m race. Because the
Porsche's acceleration of 3.5 m/s² is larger than the Honda's
3.0 m/s², the Honda gets a 50-m head start. Assume, somewhat unrealistically, that both cars can maintain these accelerations the entire distance. Who wins, and by how much time?

69. |||| The minimum stopping distance for a car traveling at a
speed of 30 m/s is 60 m, including the distance traveled during the driver's reaction time of 0.50 s.
    a. What is the minimum stopping distance for the same car
       traveling at a speed of 40 m/s?
    b. Draw a position-versus-time graph for the motion of the
       car in part a. Assume the car is at $x_i = 0$ m when the driver
       first sees the emergency situation ahead that calls for a
       rapid halt.

70. |||| A rocket is launched straight up with constant acceleration.
Four seconds after liftoff, a bolt falls off the side of the rocket.
The bolt hits the ground 6.0 s later. What was the rocket's
acceleration?

In Problems 71 through 73 you are given the kinematic equation that is used to solve a problem. For each of these, you are to

a. Write a *realistic* problem for which this is the correct equation. Be sure that the answer your problem requests is consistent with the equation given.
b. Draw the motion diagram and the pictorial representation for your problem.
c. Finish the solution of the problem.

71. ‖ $64 \text{ m} = 0.0 \text{ m} + (32 \text{ m/s})(4.0 \text{ s} - 0.0 \text{ s})$
    $+ \frac{1}{2} a_x (4.0 \text{ s} - 0.0 \text{ s})^2$
72. ‖ $0.0 \text{ m/s} = 36 \text{ m/s} - (3.0 \text{ m/s}^2)t$
73. ‖ $(10 \text{ m/s})^2 = (v_y)_i^2 - 2(9.8 \text{ m/s}^2)(10 \text{ m} - 0 \text{ m})$

Passage Problems

Free Fall on Different Worlds

Objects in free fall on the earth have acceleration $a_y = -9.8 \text{ m/s}^2$. On the moon, free-fall acceleration is approximately 1/6 of the acceleration on earth. This changes the scale of problems involving free fall. For instance, suppose you jump straight upward, leaving the ground with velocity v_i and then steadily slowing until reaching zero velocity at your highest point. Because your initial velocity is determined mostly by the strength of your leg muscles, we can assume your initial velocity would be the same on the moon. But considering the final equation in Table 2.4 we can see that, with a smaller free-fall acceleration, your maximum height would be greater. The following questions ask you to think about how certain athletic feats might be performed in this reduced-gravity environment.

74. ‖   If an astronaut can jump straight up to a height of 0.50 m on earth, how high could he jump on the moon?
    A. 1.2 m   B. 3.0 m   C. 3.6 m   D. 18 m
75. ‖   On the earth, an astronaut can safely jump to the ground from a height of 1.0 m; her velocity when reaching the ground is low enough to not cause injury. From what height could the astronaut safely jump to the ground on the moon?
    A. 2.4 m   B. 6.0 m   C. 7.2 m   D. 36 m
76. ‖   On the earth, an astronaut throws a ball straight upward; it stays in the air for a total time of 3.0 s before reaching the ground again. If a ball were to be thrown upward with the same initial speed on the moon, how much time would pass before it hit the ground?
    A. 7.3 s   B. 18 s   C. 44 s   D. 108 s

STOP TO THINK ANSWERS

Stop to Think 2.1: D. The motion consists of two constant-velocity phases; the second one has a larger velocity. The correct graph has two straight-line segments, with the second one having a larger slope.

Stop to Think 2.2: A. The motion consists of two constant-velocity phases; the first one is positive velocity, the second is negative. The correct graph has two straight-line segments, the first with a positive slope and the second with a negative slope.

Stop to Think 2.3: C. Consider the slope of the position-versus-time graph; it starts out positive and constant, then decreases to zero. Thus the velocity graph must start with a constant positive value, then decrease to zero.

Stop to Think 2.4: C. Acceleration is the slope of the velocity-versus-time graph. The largest magnitude of the slope is at point C.

Stop to Think 2.5: A. The car is always moving to the right, so the velocity is positive at all times. There should be two constant-velocity segments, with the second having a smaller velocity.

Stop to Think 2.6: D. The final velocity will have the same *magnitude* as the initial velocity, but the velocity is negative because the rock will be moving downward.

3

VECTORS AND MOTION IN TWO DIMENSIONS

With astonishing precision, an archer fish secures a meal by spitting drops of water at an unsuspecting insect. What factors determine how high and how far the water drops travel?

Looking Ahead ▶▶

The goals of Chapter 3 are to learn more about vectors and to use vectors as a tool to analyze motion in two dimensions. In this chapter, you will learn to:

▶ Work with vectors, coordinate systems, and components.

▶ Find acceleration vectors on motion diagrams.

▶ Use vectors to understand motion on a ramp and relative velocity.

▶ Solve problems involving projectiles following parabolic paths.

▶ Understand the basic ideas of motion in a circle.

Looking Back ◀◀

This chapter will build naturally on the material from the previous two chapters. Please review:

◀ Sections 1.1–1.2 Basic motion concepts.

◀ Section 1.5 Vectors.

◀ Sections 2.4–2.7 Acceleration and free fall.

Once the drops of water leave the archer fish's mouth and fly through the air, their motion follows a gently curving path. This path, a parabola, is the same type of path followed by a basketball shot toward the basket, a high jumper leaping over a bar, or a skier going off a jump. These are all examples of motion in two dimensions, motion in which the objects are moving horizontally and vertically at the same time.

In order to extend our study of motion to two dimensions, we will need to develop techniques for working with vectors. We will first use vector techniques to analyze motion on a ramp and relative velocity, continue on to look at the parabolic motion of projectiles, then close the chapter by describing the motion of objects moving in circles. Along the way we'll use numerous examples drawn from sports—from skiers gliding down slopes to balls flying through the air.

3.1 Using Vectors

In the previous chapter, we solved many problems in which an object moved in a straight-line path. In this chapter, we will look at particles that take curving paths—motion in two dimensions, which we also call motion in a plane. The two most important examples are outlined in the table below. You can see that, in both cases, the changing direction of the motion will be of crucial importance.

Two-dimensional motion

Projectile motion		Circular motion	
	A ball bouncing across a table, captured here in a strobe photograph, is clearly moving in two dimensions. The ball is moving to the right while simultaneously moving up and down. Understanding the curving path of the ball will require us to analyze motion in both dimensions.	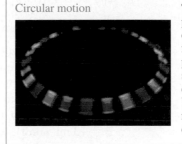	This photograph captures the path of a strobe light swung in a circle in a horizontal plane. The flashes are at equally spaced positions, so the speed of the strobe light is constant, but the direction is constantly changing. Understanding the motion will require us to understand this change in direction.

Because the direction of motion will be so important, we need to develop an appropriate mathematical language to describe it—the language of vectors. We introduced the concept of a vector in Chapter 1, and in the next few sections we will develop that concept into a useful and powerful tool. We will practice using vectors by analyzing a problem of motion in one dimension, motion on a ramp, and by studying the interesting notion of relative velocity. We will then be ready to analyze the two-dimensional motion of projectiles and of objects moving in a circle.

Recall from Chapter 1 that a vector is a quantity with both a size, or magnitude, and a direction. Figure 3.1 shows how to represent a particle's velocity as a vector $\vec{v}$. The particle's speed at this point is 5 m/s *and* it is moving in the direction indicated by the arrow. The magnitude of a vector is represented by the letter without an arrow. In this case, the particle's speed—the magnitude of the velocity vector $\vec{v}$—is $v = 5$ m/s. The magnitude of a vector, a *scalar* quantity, cannot be a negative number.

> **NOTE** ▶ Although the vector arrow is drawn across the page, from its tail to its tip, this arrow does *not* indicate that the vector "stretches" across this distance. Instead, the arrow tells us the value of the vector quantity only at the one point where the tail of the vector is placed. ◀

We found in Chapter 1 that the displacement of an object is a vector drawn from its initial position to its position at some later time. Because displacement is an easy concept to think about, we can use it to introduce some of the properties of vectors. However, **all the properties we will discuss in this chapter (addition, subtraction, multiplication, components) apply to all vectors, not just to displacement.**

Suppose that Sam, our old friend from Chapter 1, starts from his front door, walks across the street, and ends up 200 ft to the northeast of where he started. Sam's displacement, which we will label $\vec{d}_S$, is shown in Figure 3.2a on the next page. The displacement vector is a straight-line connection from his initial to his final position, not necessarily his actual path. The dotted line indicates a possible route Sam might have taken, but his displacement is the vector $\vec{d}_S$.

To describe a vector we must specify both its magnitude and its direction. We can write Sam's displacement as

$$\vec{d}_S = (200 \text{ ft, northeast})$$

Magnitude of vector   Direction of vector

$v = 5$ m/s

$\vec{v}$ ----Name of vector

The vector is drawn across the page, but it represents the particle's velocity at this one point.

FIGURE 3.1 The velocity vector $\vec{v}$ has both a magnitude and a direction.

(a)

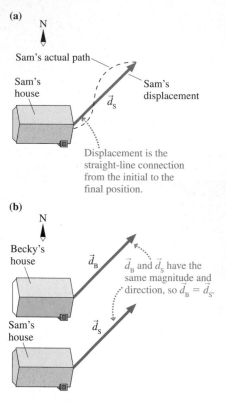

FIGURE 3.2 Displacement vectors.

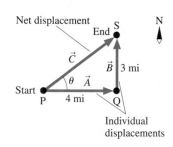

FIGURE 3.3 The net displacement $\vec{C}$ resulting from two displacements $\vec{A}$ and $\vec{B}$.

where the first number specifies the magnitude and the second item gives the direction. The magnitude of Sam's displacement is $d_S = 200$ ft, the distance between his initial and final points.

Sam's next-door neighbor Becky also walks 200 ft to the northeast, starting from her own front door. Becky's displacement $\vec{d}_B = (200$ ft, northeast) has the same magnitude and direction as Sam's displacement $\vec{d}_S$. Because vectors are defined by their magnitude and direction, **two vectors are equal if they have the same magnitude and direction.** This is true regardless of the starting points of the vectors. Thus the two displacements in Figure 3.2b are equal to each other, and we can write $\vec{d}_B = \vec{d}_S$.

Vector Addition

As we saw in Chapter 1, we can combine successive displacements by vector addition. Let's review and extend this concept. Figure 3.3 shows the displacement of a hiker who starts at point P and ends at point S. She first hikes 4 miles to the east, then 3 miles to the north. The first leg of the hike is described by the displacement vector $\vec{A} = (4$ mi, east). The second leg of the hike has displacement $\vec{B} = (3$ mi, north). Now, by definition, a vector from the initial position P to the final position S is also a displacement. This is vector $\vec{C}$ on the figure. $\vec{C}$ is the *net displacement* because it describes the net result of the hiker's first having displacement $\vec{A}$, then displacement $\vec{B}$.

The word *net* implies addition. The net displacement $\vec{C}$ is an initial displacement $\vec{A}$ *plus* a second displacement $\vec{B}$, or

$$\vec{C} = \vec{A} + \vec{B} \tag{3.1}$$

The sum of two vectors is called the **resultant vector.** It's not hard to show that vector addition is commutative: $\vec{A} + \vec{B} = \vec{B} + \vec{A}$. That is, you can add vectors in any order you wish.

Look back at Tactics Box 1.4 on page 21 to see the three-step procedure for adding two vectors. This tip-to-tail method for adding vectors, which is used to find $\vec{C} = \vec{A} + \vec{B}$ in Figure 3.3, is called *graphical addition.* Any two vectors of the same type—two velocity vectors or two force vectors—can be added in exactly the same way.

When two vectors are to be added, it is often convenient to draw them with their tails together, as shown in Figure 3.4a. To evaluate $\vec{D} + \vec{E}$, you could move vector $\vec{E}$ over to where its tail is on the tip of $\vec{D}$, then use the tip-to-tail rule of graphical addition. This gives vector $\vec{F} = \vec{D} + \vec{E}$ in Figure 3.4b. Alternatively, Figure 3.4c shows that the vector sum $\vec{D} + \vec{E}$ can be found as the diagonal of the parallelogram defined by $\vec{D}$ and $\vec{E}$. This method for vector addition, which some of you may have learned, is called the *parallelogram rule* of vector addition.

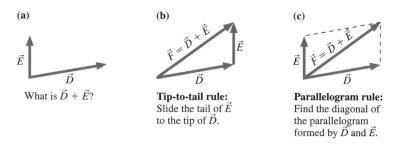

FIGURE 3.4 Two vectors can be added using the tip-to-tail rule or the parallelogram rule.

Vector addition is easily extended to more than two vectors. Figure 3.5 shows a hiker moving from initial position 0 to position 1, then position 2, then position 3, and finally arriving at position 4. These four segments are described by displacement vectors $\vec{d}_1, \vec{d}_2, \vec{d}_3$, and $\vec{d}_4$. The hiker's *net* displacement, an arrow from position 0 to position 4, is the vector $\vec{d}_{net}$. In this case,

$$\vec{d}_{net} = \vec{d}_1 + \vec{d}_2 + \vec{d}_3 + \vec{d}_4 \tag{3.2}$$

The vector sum is found by using the tip-to-tail method three times in succession.

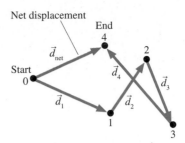
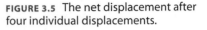

FIGURE 3.5 The net displacement after four individual displacements.

Multiplication by a Scalar

The hiker in Figure 3.3 started with displacement $\vec{A}_1 = (4\text{ mi, east})$. Suppose a second hiker walks twice as far to the east. The second hiker's displacement will then certainly be $\vec{A}_2 = (8\text{ mi, east})$. The words "twice as" indicate a multiplication, so we can say

$$\vec{A}_2 = 2\vec{A}_1$$

Multiplying a vector by a positive scalar gives another vector of *different magnitude* but pointing in the *same direction*.

Let the vector $\vec{A}$ be specified as a magnitude A and a direction θ_A. That is, $\vec{A} = (A, \theta_A)$. Now let $\vec{B} = c\vec{A}$, where c is a positive scalar constant. Then

$$\vec{B} = c\vec{A} \text{ means that } (B, \theta_B) = (cA, \theta_A) \tag{3.3}$$

In other words, the vector is stretched or compressed by the factor c (i.e., vector $\vec{B}$ has magnitude $B = cA$), but $\vec{B}$ points in the same direction as $\vec{A}$. This is illustrated in Figure 3.6.

Suppose we multiply $\vec{A}$ by zero. Using Equation 3.3,

$$0 \cdot \vec{A} = \vec{0} = (0\text{ m, direction undefined}) \tag{3.4}$$

The product is a vector having zero length or magnitude. This vector is known as the **zero vector,** denoted $\vec{0}$. The direction of the zero vector is irrelevant; you cannot describe the direction of an arrow of zero length!

What happens if we multiply a vector by a negative number? Equation 3.3 does not apply if $c < 0$ because vector $\vec{B}$ cannot have a negative magnitude. Consider the vector $-\vec{A}$, which is equivalent to multiplying $\vec{A}$ by -1. Because

$$\vec{A} + (-\vec{A}) = \vec{0} \tag{3.5}$$

the vector $-\vec{A}$ must be such that, when it is added to $\vec{A}$, the resultant is the zero vector $\vec{0}$. In other words, the *tip* of $-\vec{A}$ must return to the *tail* of $\vec{A}$, as shown in Figure 3.7. This will be true only if $-\vec{A}$ is equal in magnitude to $\vec{A}$, but opposite in direction. Thus we can conclude that

$$-\vec{A} = (A, \text{ direction opposite } \vec{A}) \tag{3.6}$$

That is, multiplying a vector by -1 reverses its direction without changing its length.

As an example, Figure 3.8 shows vectors $\vec{A}, 2\vec{A}$, and $-3\vec{A}$. Multiplication by 2 doubles the length of the vector but does not change its direction. Multiplication by -3 stretches the length by a factor of 3 *and* reverses the direction.

Vector Subtraction

How might we *subtract* vector $\vec{B}$ from vector $\vec{A}$ to form the vector $\vec{A} - \vec{B}$? With numbers, subtraction is the same as the addition of a negative number. That is, $5 - 3$ is the same as $5 + (-3)$. Similarly, $\vec{A} - \vec{B} = \vec{A} + (-\vec{B})$. We can use the rules for vector addition and the fact that $-\vec{B}$ is a vector opposite in direction to $\vec{B}$ to form rules for vector subtraction.

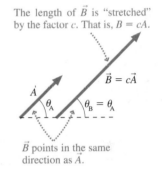

The length of $\vec{B}$ is "stretched" by the factor c. That is, $B = cA$.

$\vec{B} = c\vec{A}$

$\vec{B}$ points in the same direction as $\vec{A}$.

FIGURE 3.6 Multiplication of a vector by a scalar.

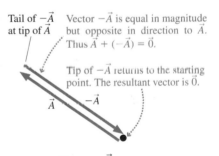

Tail of $-\vec{A}$ at tip of $\vec{A}$

Vector $-\vec{A}$ is equal in magnitude but opposite in direction to $\vec{A}$. Thus $\vec{A} + (-\vec{A}) = \vec{0}$.

Tip of $-\vec{A}$ returns to the starting point. The resultant vector is $\vec{0}$.

FIGURE 3.7 Vector $-\vec{A}$.

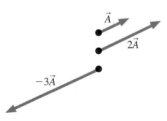

FIGURE 3.8 Vectors $\vec{A}, 2\vec{A}$, and $-3\vec{A}$.

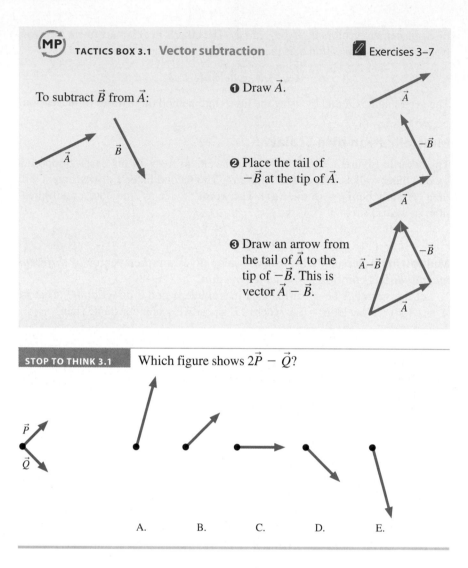

TACTICS BOX 3.1 Vector subtraction   Exercises 3–7

To subtract $\vec{B}$ from $\vec{A}$:

❶ Draw $\vec{A}$.

❷ Place the tail of $-\vec{B}$ at the tip of $\vec{A}$.

❸ Draw an arrow from the tail of $\vec{A}$ to the tip of $-\vec{B}$. This is vector $\vec{A} - \vec{B}$.

STOP TO THINK 3.1   Which figure shows $2\vec{P} - \vec{Q}$?

A.   B.   C.   D.   E.

3.2 Using Vectors on Motion Diagrams

In Chapter 2, we defined velocity for one-dimensional motion as an object's displacement—the change in position—divided by the time interval in which the change occurs:

$$v_x = \frac{\Delta x}{\Delta t} = \frac{x_f - x_i}{\Delta t}$$

In two dimensions, an object's displacement is a vector. Suppose an object undergoes displacement $\vec{d}$ during the time interval Δt. Let's define an object's velocity *vector* to be

$$\vec{v} = \frac{\vec{d}}{\Delta t} = \left(\frac{d}{\Delta t}, \text{ same direction as } \vec{d}\right) \tag{3.7}$$

Definition of velocity in two or more dimensions

Notice that we've multiplied a vector by a scalar: The velocity vector is the displacement vector multiplied by the scalar $1/\Delta t$. Consequently, as we found in Chapter 1, **the velocity vector points in the direction of displacement.** As a result, we can use the dot-to-dot vectors on a motion diagram to visualize the velocity.

NOTE ▶ Strictly speaking, the velocity defined in Equation 3.7 is the *average* velocity for the time interval Δt. This is adequate for using motion diagrams to visualize motion. As we did in Chapter 2, we'll choose Δt to be very small and thus have an *instantaneous* velocity when it's time to start doing calculations. ◀

EXAMPLE 3.1 Finding the velocity of an airplane

A small plane is 100 km due east of Denver. One hour later, it is 200 km due north of Denver. What is the plane's velocity?

PREPARE The initial and final positions of the plane are shown in Figure 3.9; the displacement $\vec{d}$ is the vector that points from the initial to the final position.

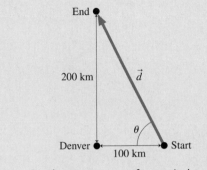

FIGURE 3.9 Displacement vector for an airplane.

SOLVE The length of the displacement vector is the hypotenuse of a right triangle:

$$d = \sqrt{(100 \text{ km})^2 + (200 \text{ km})^2} = 224 \text{ km}$$

The direction of the displacement vector is described by the angle θ in Figure 3.9. From trigonometry, this angle is

$$\theta = \tan^{-1}\left(\frac{200 \text{ km}}{100 \text{ km}}\right) = \tan^{-1}(2.00) = 63.4°$$

Thus the plane's displacement vector is

$$\vec{d} = (224 \text{ km}, 63.4° \text{ north of west})$$

Because the plane undergoes this displacement during 1 hour, its velocity is

$$\vec{v} = \left(\frac{d}{\Delta t}, \text{ same direction as } \vec{d}\right) = \left(\frac{224 \text{ km}}{1 \text{ hour}}, 63.4° \text{ north of west}\right)$$

$$= (224 \text{ km/hour}, 63.4° \text{ north of west})$$

ASSESS The plane's *speed* is $v = 224$ km/h. Speed does not have a direction.

We defined an object's acceleration in one dimension as $a_x = \Delta v_x/\Delta t$. In two dimensions, we need to use a vector to describe acceleration. The vector definition of acceleration is a straightforward extension of the one-dimensional version:

$$\vec{a} = \frac{\vec{v}_f - \vec{v}_i}{t_f - t_i} = \frac{\Delta \vec{v}}{\Delta t} \qquad (3.8)$$

Definition of acceleration in two or more dimensions

There is an acceleration whenever there is a *change* in velocity. Because velocity is a vector, it can change in two possible ways:

1. The magnitude can change, indicating a change in speed, or
2. The direction of motion can change.

In Chapter 2 we considered the first case, in which an object speeds up or slows down while moving in a straight line. Now we need to examine the second case, in which an object changes its direction of motion.

Suppose an object has an initial velocity $\vec{v}_i$ at time t_i and later, at time t_f has velocity $\vec{v}_f$. The fact that the velocity *changes* tells us the object undergoes an acceleration during the time interval $\Delta t = t_f - t_i$. We see from Equation 3.8 that the acceleration points in the same direction as the vector $\Delta \vec{v}$. This vector is the change in the velocity $\Delta \vec{v} = \vec{v}_f - \vec{v}_i$, so to know which way the acceleration vector points, we have to perform the vector subtraction $\vec{v}_f - \vec{v}_i$. Tactics Box 3.1 showed how to perform vector subtraction. Tactics Box 3.2 shows how to use vector subtraction to find the acceleration vector.

Lunging versus veering BIO The top photo shows a barracuda, a type of fish that catches prey with a rapid linear acceleration, a quick change in speed. The barracuda's body shape is optimized for such a straight-line strike. The butterfly fish in the bottom photo has a very different appearance. It can't rapidly change its speed, but its body shape lets it quickly change its direction. This is a type of acceleration the barracuda isn't good at, so the butterfly fish can avoid being caught.

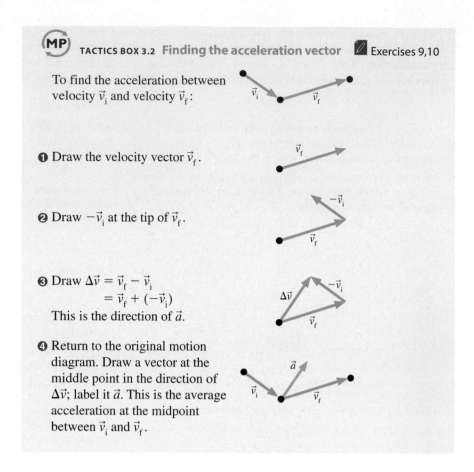

TACTICS BOX 3.2 Finding the acceleration vector   Exercises 9,10

To find the acceleration between velocity $\vec{v}_i$ and velocity $\vec{v}_f$:

❶ Draw the velocity vector $\vec{v}_f$.

❷ Draw $-\vec{v}_i$ at the tip of $\vec{v}_f$.

❸ Draw $\Delta\vec{v} = \vec{v}_f - \vec{v}_i$
$\qquad\quad = \vec{v}_f + (-\vec{v}_i)$
This is the direction of $\vec{a}$.

❹ Return to the original motion diagram. Draw a vector at the middle point in the direction of $\Delta\vec{v}$; label it $\vec{a}$. This is the average acceleration at the midpoint between $\vec{v}_i$ and $\vec{v}_f$.

Now that we know how to determine acceleration vectors, we can make a complete motion diagram with dots showing the position of the object, average velocity vectors found by connecting dots with an arrow, and acceleration vectors found using Tactics Box 3.2. Note that there is *one* acceleration vector linking each *two* velocity vectors, and $\vec{a}$ is drawn at the dot between the two velocity vectors it links.

EXAMPLE 3.2 Drawing the acceleration for a Mars descent

A spacecraft slows as it safely descends to the surface of Mars. Draw a complete motion diagram for the last few seconds of the descent.

SOLVE Figure 3.10 shows a complete motion diagram. As the spacecraft slows in its descent, the dots get closer together and the velocity vectors get shorter. The inset shows how Tactics Box 3.2 is used to determine the acceleration at one point. All the other acceleration vectors will be similar, because for each pair of velocity vectors the earlier one is longer than the later one.

ASSESS As the spacecraft in Figure 3.10 slows, the acceleration vectors and velocity vectors point in opposite directions, consistent with what we learned about the sign of the acceleration in Chapter 2.

(a) Artist version

(b) Student sketch

The acceleration vector is the same direction as $\Delta\vec{v}$

FIGURE 3.10 Motion diagram for a descending spacecraft.

EXAMPLE 3.3 Drawing the acceleration for a Ferris wheel ride

Anne rides a Ferris wheel at an amusement park. Draw Anne's motion diagram.

SOLVE Figure 3.11 shows 10 points of the motion during one complete revolution of the Ferris wheel. A person riding a Ferris wheel moves in a circle at a constant speed, so we've shown equal distances between successive dots. As before, the velocity vectors are found by connecting each dot to the next. Note that the velocity vectors are *straight lines,* not curves.

We see that all of the velocity vectors have the same length, but each has a different *direction,* and that means Anne is accelerating. This is not a "speeding up" or "slowing down" acceleration, but is, instead, a "change of direction" acceleration. The inset to Figure 3.11 shows how to find the acceleration at one particular position, at the bottom of the circle. Vector $\vec{v}_1$ is the velocity vector that leads into this dot, while $\vec{v}_2$ moves away from it. From the circular geometry of the main figure, the two angles marked α are equal. Thus we see that $\vec{v}_2$ and $-\vec{v}_1$ form an isosceles triangle and vector $\Delta\vec{v} = \vec{v}_2 - \vec{v}_1$ is exactly vertical, toward the center of the circle.

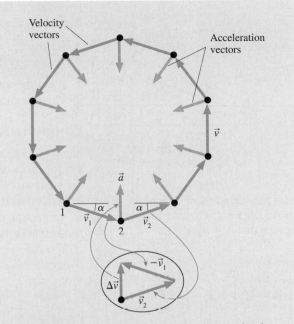

FIGURE 3.11 Motion diagram for Anne on a Ferris wheel.

Interestingly, no matter which dot you select on the motion diagram of Figure 3.11, the velocities change in such a way that the acceleration vector $\vec{a}$ points directly to the center of the circle. An acceleration vector that always points toward the center of a circle is called a *centripetal acceleration.* We will have much more to say about centripetal acceleration later in this chapter.

3.3 Coordinate Systems and Vector Components

In the past two sections, we have seen how to add and subtract vectors graphically, using these operations to deduce important details of motion. But the graphical combination of vectors is not an especially good way to find quantitative results. In this section we will introduce a *coordinate description* of vectors that will be the basis for doing vector calculations.

Coordinate Systems

We introduced the notion of a coordinate system in Chapter 1. The world does not come with a coordinate system attached to it. A coordinate system is an artificially imposed grid that you place on a problem in order to make quantitative measurements. The right choice of coordinate system will make a problem easier to solve.

We will generally use **Cartesian coordinates,** the familiar rectangular grid with perpendicular axes. By convention, the positive y-axis is located 90° *counterclockwise* from the positive x-axis, as illustrated in Figure 3.12. Figure 3.12 also identifies the four **quadrants** of the coordinate system, I through IV. Notice that, by convention, the quadrants are counted counterclockwise from the positive x-axis.

Coordinate axes have a positive end and a negative end, separated by zero at the origin where the two axes cross. When you draw a coordinate system, it is

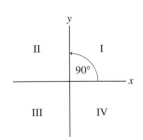

FIGURE 3.12 A Cartesian coordinate system and the quadrants of the xy-plane.

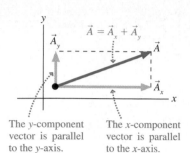

The *y*-component vector is parallel to the *y*-axis.

The *x*-component vector is parallel to the *x*-axis.

FIGURE 3.13 Component vectors $\vec{A}_x$ and $\vec{A}_y$ are drawn parallel to the coordinate axes such that $\vec{A} = \vec{A}_x + \vec{A}_y$.

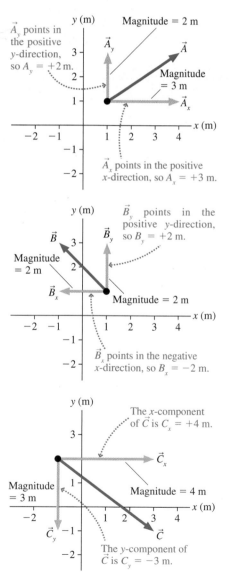

FIGURE 3.14 Determining the components of a vector.

important to label the axes. This is done by placing *x* and *y* labels at the *positive* ends of the axes, as in Figure 3.12. The purpose of the labels is twofold:

■ To identify which axis is which, and
■ To identify the positive ends of the axes.

This will be important when you need to determine whether the quantities in a problem should be assigned positive or negative values.

Component Vectors

Figure 3.13 shows a vector $\vec{A}$ and an *xy*-coordinate system that we've chosen. Once the directions of the axes are known, we can define two new vectors *parallel to the axes* that we call the **component vectors** of $\vec{A}$. Vector $\vec{A}_x$, called the *x-component vector,* is the projection of $\vec{A}$ along the *x*-axis. Vector $\vec{A}_y$, the *y-component vector,* is the projection of $\vec{A}$ along the *y*-axis. Notice that the component vectors are perpendicular to each other.

You can see, using the parallelogram rule, that $\vec{A}$ is the vector sum of the two component vectors:

$$\vec{A} = \vec{A}_x + \vec{A}_y \qquad (3.9)$$

In essence, we have broken vector $\vec{A}$ into two perpendicular vectors that are parallel to the coordinate axes. We say that we have **decomposed** or **resolved** vector $\vec{A}$ into its component vectors.

NOTE ▶ It is not necessary for the tail of $\vec{A}$ to be at the origin. All we need to know is the *orientation* of the coordinate system so that we can draw $\vec{A}_x$ and $\vec{A}_y$ parallel to the axes. ◀

Components

You learned in Chapter 2 to give the one-dimensional kinematic variable v_x a positive sign if the velocity vector $\vec{v}$ points toward the positive end of the *x*-axis, a negative sign if $\vec{v}$ points in the negative *x*-direction. The basis of this rule is that v_x is the *x-component* of $\vec{v}$. We need to extend this idea to vectors in general.

Suppose we have a vector $\vec{A}$ that has been decomposed into component vectors $\vec{A}_x$ and $\vec{A}_y$ parallel to the coordinate axes. We can describe each component vector with a single number (a scalar) called the **component.** The *x-component* and *y-component* of vector $\vec{A}$, denoted A_x and A_y, are determined as follows:

(MP) TACTICS BOX 3.3 Determining the components of a vector   ✐ Exercises 14–16

❶ The absolute value $|A_x|$ of the *x*-component A_x is the magnitude of the component vector $\vec{A}_x$.
❷ The *sign* of A_x is positive if $\vec{A}_x$ points in the positive *x*-direction, negative if $\vec{A}_x$ points in the negative *x*-direction.
❸ The *y*-component A_y is determined similarly.

In other words, the component A_x tells us two things: how big $\vec{A}_x$ is and which end of the axis $\vec{A}_x$ points toward. Figure 3.14 shows three examples of determining the components of a vector.

NOTE ▶ $\vec{A}_x$ and $\vec{A}_y$ are *component vectors;* they have a magnitude and a direction. A_x and A_y are simply *components.* The components A_x and A_y are scalars—just numbers (with units) that can be positive or negative. ◀

Much of physics is expressed in the language of vectors. We will frequently need to decompose a vector into its components or to "reassemble" a vector from its components, moving back and forth between the graphical and the component representations of a vector.

Let's start with the problem of decomposing a vector into its x- and y-components. Figure 3.15a shows a vector $\vec{A}$ at angle θ from the x-axis. It is *essential* to use a picture or diagram such as this to define the angle you are using to describe the vector's direction. $\vec{A}$ points to the right and up, so Tactics Box 3.3 tells us that the components A_x and A_y are both positive. We can use trigonometry to find

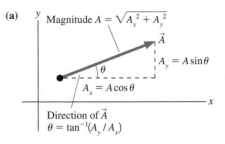

$$A_x = A\cos\theta$$
$$A_y = A\sin\theta \tag{3.10}$$

where A is the magnitude, or length, of $\vec{A}$. These equations convert the length and angle description of vector $\vec{A}$ into the vector's components, but they are correct *only* if $\vec{A}$ is in the first quadrant, pointing to the right and up.

Figure 3.15b shows vector $\vec{C}$ pointing to the right and down. In this case, where the component vector $\vec{C}_y$ is pointing *down,* in the negative y-direction, the y-component C_y is a *negative* number. The angle ϕ (the Greek letter *phi*) is measured from the y-axis, so the components of $\vec{C}$ are

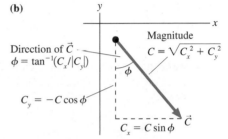

$$C_x = C\sin\phi$$
$$C_y = -C\cos\phi \tag{3.11}$$

The role of sine and cosine is reversed from that in Equations 3.10 because we are using a different angle.

FIGURE 3.15 Moving between the graphical representation and the component representation.

NOTE ▶ Each decomposition requires that you pay close attention to the direction in which the vector points and the angles that are defined. The minus sign, when needed, must be inserted manually. ◀

We can also go in the opposite direction and determine the length and angle of a vector from its x- and y-components. Because A in Figure 3.15a is the hypotenuse of a right triangle, its length is given by the Pythagorean theorem:

$$A = \sqrt{A_x^2 + A_y^2} \tag{3.12}$$

Similarly, the tangent of angle θ is the ratio of the far side to the adjacent side, so

$$\theta = \tan^{-1}\left(\frac{A_y}{A_x}\right) \tag{3.13}$$

Equations 3.12 and 3.13 can be thought of as the "reverse" of Equations 3.10.

Equation 3.12 always works for finding the length or magnitude of a vector because the squares eliminate any concerns over the signs of the components. But finding the angle requires close attention to how the angle is defined and to the signs of the components. Finding the angle of vector $\vec{C}$ in Figure 3.15b requires the length of C_y *without* the minus sign, so vector $\vec{C}$ has direction

$$\phi = \tan^{-1}\left(\frac{C_x}{|C_y|}\right) \tag{3.14}$$

Notice that the roles of x and y differ from those in Equation 3.13.

EXAMPLE 3.4   Finding the components of an acceleration vector

Find the *x*- and *y*-components of the acceleration vector $\vec{a}$ shown in Figure 3.16.

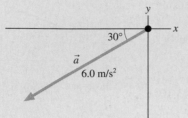

FIGURE 3.16 Acceleration vector $\vec{a}$ of Example 3.4.

PREPARE It's important to *draw* vectors. Figure 3.17 shows the original vector $\vec{a}$ decomposed into component vectors parallel to the axes.

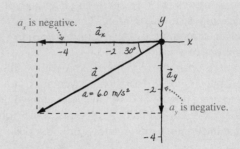

FIGURE 3.17 The components of the acceleration vector.

SOLVE The acceleration vector $\vec{a} = (6.0 \text{ m/s}^2, 30°$ below the negative *x*-axis) points to the left (negative *x*-direction) and down (negative *y*-direction), so the components a_x and a_y are both negative:

$$a_x = -a\cos 30° = -(6.0 \text{ m/s}^2)\cos 30° = -5.2 \text{ m/s}^2$$

$$a_y = -a\sin 30° = -(6.0 \text{ m/s}^2)\sin 30° = -3.0 \text{ m/s}^2$$

ASSESS The units of a_x and a_y are the same as the units of vector $\vec{a}$. Notice that we had to insert the minus signs manually by observing that the vector is in the third quadrant.

STOP TO THINK 3.2   What are the *x*- and *y*-components C_x and C_y of vector $\vec{C}$?

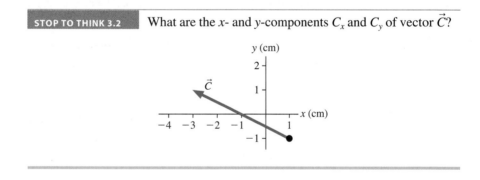

Working with Components

We've seen how to add vectors graphically, but there's an easier way: using components. To illustrate, let's look at the vector sum $\vec{C} = \vec{A} + \vec{B}$ for the vectors shown in Figure 3.18. You can see that the component vectors of $\vec{C}$ are the sum of the component vectors of $\vec{A}$ and $\vec{B}$. The same is true of the components: $C_x = A_x + B_x$ and $C_y = A_y + B_y$.

In general, if $\vec{D} = \vec{A} + \vec{B} + \vec{C} + \cdots$, then the *x*- and *y*-components of the resultant vector $\vec{D}$ are

$$D_x = A_x + B_x + C_x + \cdots$$

$$D_y = A_y + B_y + C_y + \cdots \tag{3.15}$$

This method of vector addition is called *algebraic addition*.

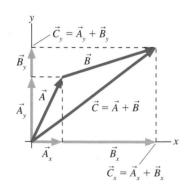

FIGURE 3.18 Using components to add vectors.

EXAMPLE 3.5 Using algebraic addition to find a bird's displacement

A bird flies 100 m due east from a tree, then 200 m northwest (that is, 45° west of north). What is the bird's net displacement?

PREPARE Figure 3.19a shows the displacement vectors $\vec{A} = (100\text{ m, east})$ and $\vec{B} = (200\text{ m, northwest})$ and also the net displacement $\vec{C}$. We draw vectors tip-to-tail if we are going to add them graphically, but it's usually easier to draw them all from the origin if we are going to use algebraic addition. Figure 3.19b redraws the vectors with their tails together.

SOLVE To add the vectors algebraically we must know their components. From the figure these are seen to be

$$A_x = 100\text{ m}$$

$$A_y = 0\text{ m}$$

$$B_x = -(200\text{ m})\cos 45° = -141\text{ m}$$

$$B_y = (200\text{ m})\sin 45° = 141\text{ m}$$

We learned *from the figure* that $\vec{B}$ has a negative x-component. Adding $\vec{A}$ and $\vec{B}$ by components gives

$$C_x = A_x + B_x = 100\text{ m} - 141\text{ m} = -41\text{ m}$$

$$C_y = A_y + B_y = 0\text{ m} + 141\text{ m} = 141\text{ m}$$

The magnitude of the net displacement $\vec{C}$ is

$$C = \sqrt{C_x^2 + C_y^2} = \sqrt{(-41\text{ m})^2 + (141\text{ m})^2} = 147\text{ m}$$

The angle θ, as defined in Figure 3.19, is

$$\theta = \tan^{-1}\left(\frac{C_y}{|C_x|}\right) = \tan^{-1}\left(\frac{141\text{ m}}{41\text{ m}}\right) = 74°$$

Thus the bird's net displacement is $\vec{C} = (147\text{ m}, 74° \text{ north of west})$.

ASSESS The final values of C_x and C_y match what we would expect from the sketch in Figure 3.19. The geometric addition was a valuable check on the answer we found by algebraic addition.

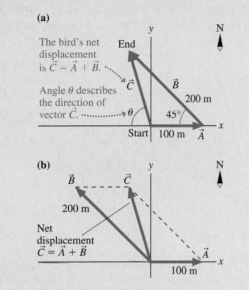

FIGURE 3.19 Finding the net displacement.

Vector subtraction and the multiplication of a vector by a scalar are also easily performed using components. To find $\vec{D} = \vec{P} - \vec{Q}$ we would compute

$$D_x = P_x - Q_x$$
$$D_y = P_y - Q_y \tag{3.16}$$

Similarly, $\vec{T} = c\vec{S}$ is

$$T_x = cS_x$$
$$T_y = cS_y \tag{3.17}$$

The next few chapters will make frequent use of *vector equations*. For example, you will learn that the equation to calculate the force on a car skidding to a stop is

$$\vec{F} = \vec{n} + \vec{w} + \vec{f} \tag{3.18}$$

Equation 3.18 is really just a shorthand way of writing the two simultaneous equations:

$$F_x = n_x + w_x + f_x$$
$$F_y = n_y + w_y + f_y \tag{3.19}$$

When you look at a trail map for a hike in a mountainous region, it will give the length of a trail *and* the elevation gain—an important variable! The elevation gain is simply d_y, the vertical component of the displacement for the hike.

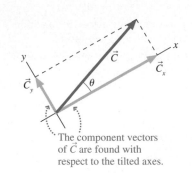

The component vectors of $\vec{C}$ are found with respect to the tilted axes.

FIGURE 3.20 A coordinate system with tilted axes.

In other words, a vector equation is interpreted as meaning: Equate the *x*-components on both sides of the equals sign, then equate the *y*-components. Vector notation allows us to write these two equations in a more compact form.

Tilted Axes

While we are used to having the *x*-axis horizontal, there is no requirement that it has to be that way. In Chapter 1, we saw that for motion on a slope we could put the *x*-axis along the slope. When we add the *y*-axis, this gives us a tilted coordinate system such as that shown in Figure 3.20.

Finding components with tilted axes is no harder than what we have done so far. Vector $\vec{C}$ in Figure 3.20 can be decomposed into component vectors $\vec{C}_x$ and $\vec{C}_y$, with $C_x = C\cos\theta$ and $C_y = C\sin\theta$.

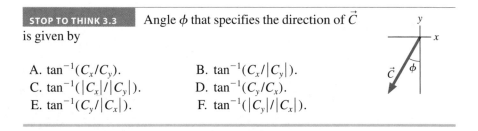

STOP TO THINK 3.3   Angle ϕ that specifies the direction of $\vec{C}$ is given by

A. $\tan^{-1}(C_x/C_y)$. B. $\tan^{-1}(C_x/|C_y|)$.
C. $\tan^{-1}(|C_x|/|C_y|)$. D. $\tan^{-1}(C_y/C_x)$.
E. $\tan^{-1}(C_y/|C_x|)$. F. $\tan^{-1}(|C_y|/|C_x|)$.

3.4 Motion on a Ramp

In this section, we will examine the problem of motion on a ramp or incline (often called an *inclined plane*). There are three reasons to look at this problem. First, it will provide good practice at using vectors to analyze motion. Second, it is a simple problem for which we can find an exact solution. Third, this seemingly abstract problem has real and important applications. As we will see, both speed skiing and roller coasters can be modeled quite well as the motion of an object sliding down a ramp.

We begin with a constant-velocity example to give us some practice with vectors and components before moving on to the more general case of accelerated motion.

EXAMPLE 3.6 Finding the height gained on a slope

A car drives up a steep 10° slope at a constant speed of 15 m/s. After 10 s, how much height has the car gained?

PREPARE Figure 3.21 is a visual overview, with *x*- and *y*-axes defined. The velocity vector $\vec{v}$ points up the slope. We are inter-

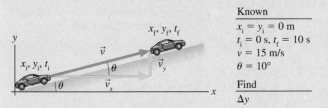

Known
$x_i = y_i = 0$ m
$t_i = 0$ s, $t_f = 10$ s
$v = 15$ m/s
$\theta = 10°$

Find
Δy

FIGURE 3.21 Visual overview of a car moving up a slope.

ested in the vertical motion of the car, so we decompose $\vec{v}$ into component vectors $\vec{v}_x$ and $\vec{v}_y$ as shown.

SOLVE The velocity component we need is v_y; this describes the vertical motion of the car. Using the rules for finding components outlined above, we find

$$v_y = v\sin\theta = (15 \text{ m/s})\sin(10°) = 2.6 \text{ m/s}$$

Because the velocity is constant, the car's vertical displacement (i.e., the height gained) during 10 s is

$$\Delta y = v_y\,\Delta t = (2.6 \text{ m/s})(10 \text{ s}) = 26 \text{ m}$$

ASSESS The car is traveling at a pretty good clip—15 m/s is a bit more than 30 mph—up a steep slope, so it should climb a respectable height in 10 s. 26 m, or about 80 ft, seems reasonable.

Accelerated Motion on a Ramp

Figure 3.22a shows a crate sliding down a frictionless (i.e., smooth) ramp tilted at angle θ. The crate accelerates due to the action of gravity, but it is *constrained* to accelerate parallel to the surface. What is the acceleration?

A motion diagram for the crate is drawn in Figure 3.22b. There is an acceleration, because the velocity is changing, with both the acceleration and velocity vectors parallel to the ramp. We can take advantage of the properties of vectors to find the crate's acceleration. To do so, Figure 3.22c sets up a coordinate system with the x-axis along the ramp and the y-axis perpendicular. This choice of coordinate system constrains all motion to be along the x-axis.

If the incline suddenly vanished, the object would have a free-fall acceleration $\vec{a}_{\text{free fall}}$ straight down. This acceleration vector can be decomposed into two component vectors: a vector $\vec{a}_x$ that is *parallel* to the incline and a vector $\vec{a}_y$ that is *perpendicular* to the incline. The vector addition rules studied earlier in this chapter tell us that $\vec{a}_{\text{free fall}} = \vec{a}_x + \vec{a}_y$.

The motion diagram shows that the object's actual acceleration $\vec{a}_x$ is parallel to the incline. The surface of the incline somehow "blocks" the other component of the acceleration $\vec{a}_y$, through a process we will examine in Chapter 5, but $\vec{a}_x$ is unhindered. It is this component of $\vec{a}_{\text{free fall}}$, parallel to the incline, that accelerates the object.

We can use trigonometry to work out the magnitude of this acceleration. Figure 3.22c shows that the three vectors $\vec{a}_{\text{free fall}}$, $\vec{a}_y$, and $\vec{a}_x$ form a right triangle with angle θ as shown. You should satisfy yourself that this is the same as the angle of the incline. By definition, the magnitude of $\vec{a}_{\text{free fall}}$ is g. This vector is the hypotenuse of the right triangle. The vector we are interested in, $\vec{a}_x$, is opposite angle θ. Thus the value of the acceleration along a frictionless slope is

$$a_x = \pm g\sin\theta \qquad (3.20)$$

NOTE ▶ The correct sign depends on the direction in which the ramp is tilted. The acceleration in Figure 3.22 is $+g\sin\theta$, but upcoming examples will show situations in which the acceleration is $-g\sin\theta$. ◀

Let's look at Equation 3.20 to see if it makes sense. A good way to do this is to consider some **limiting cases** in which the angle is at one end of its range. In these cases, the physics is clear and we can check our result. Let's look at two such possibilities:

1. Suppose the plane is perfectly horizontal, with $\theta = 0°$. If you place an object on a horizontal surface, you expect it to stay at rest with no acceleration. Equation 3.20 gives $a_x = 0$ when $\theta = 0°$, in agreement with our expectations.
2. Now suppose you tilt the plane until it becomes vertical, with $\theta = 90°$. You know what happens—the object will be in free fall, parallel to the vertical surface. Equation 3.20 gives $a_x = g$ when $\theta = 90°$, again in agreement with our expectations.

NOTE ▶ Checking your answer by looking at such limiting cases is a very good way to see if your answer makes sense. We will often do this in the "Assess" step of a solution. ◀

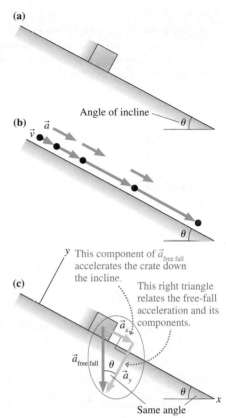

FIGURE 3.22 Acceleration on an inclined plane.

Extreme physics A speed skier, on wide skis with little friction, wearing an aerodynamic helmet and crouched low to minimize air resistance, moves in a straight line down a steep slope—pretty much like an object sliding down a frictionless ramp! There is a maximum speed that a skier could possibly achieve at the end of the slope. Course designers set the starting point to keep this maximum speed within reasonable (for this sport!) limits.

EXAMPLE 3.7 Maximum possible speed for a skier
The Willamette Pass ski area in Oregon was the site of the 1993 U.S. National Speed Skiing Competition. Each competitor's speed was measured at the end of the acceleration portion of the track, which has a reasonably constant slope. During the acceleration phase, the skiers traveled 360 m while dropping a vertical distance of 170 m. What is the fastest speed a skier could achieve at the end of this run? How much time would this fastest run take?

Continued

PREPARE We begin with the visual overview in Figure 3.23. The motion diagram shows the acceleration of the skier and the pictorial representation gives an overview of the problem including the dimensions of the slope. As above, we put the x-axis along the slope.

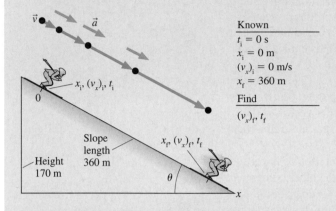

Known
$t_i = 0$ s
$x_i = 0$ m
$(v_x)_i = 0$ m/s
$x_f = 360$ m

Find
$(v_x)_f, t_f$

FIGURE 3.23 Visual overview of a skier accelerating down a slope.

SOLVE The fastest possible run would be one without any friction or air resistance, meaning the acceleration down the slope is given by Equation 3.20. The acceleration is in the positive x-direction, so we use the positive sign. What is the angle in Equation 3.20? Figure 3.23 shows that the 360-m-long slope is the hypotenuse of a triangle of height 170 m, so we use trigonometry to find

$$\sin\theta = \frac{170 \text{ m}}{360 \text{ m}} = 0.472$$

which gives $\theta = \sin^{-1}(0.472) = 28°$. Equation 3.20 then gives

$$a_x = +g\sin\theta = (9.8 \text{ m/s}^2)(\sin 28°) = 4.6 \text{ m/s}^2$$

For linear motion with constant acceleration, we can use the third of the kinematic equations in Table 2.4, $(v_x)_f^2 = (v_x)_i^2 + 2a_x \Delta x$. The initial velocity $(v_x)_i$ is zero, thus:

$$(v_x)_f = \sqrt{2a_x \Delta x} = \sqrt{2(4.6 \text{ m/s}^2)(360 \text{ m})} = 58 \text{ m/s}$$

This is the fastest that any skier could hope to be moving at the end of the run. Any friction or air resistance would decrease this speed. Because the acceleration is constant and the initial velocity $(v_x)_i$ is zero, the time of the fastest-possible run is

$$\Delta t = \frac{(v_x)_f}{a_x} = \frac{58 \text{ m/s}}{4.6 \text{ m/s}^2} = 12 \text{ s}$$

A speed skiing event is a quick affair!

ASSESS The final speed we calculated is 58 m/s, which is about 130 mph. We might expect such a high speed for this sport. In the competition noted, the actual winning speed was 111 mph, not much less than the result we calculated.

Skis on snow have very low friction, but there are other ways to reduce the friction between surfaces. For instance, a roller coaster car rolls along a track on low-friction wheels. No drive force is applied to the cars after they are released at the top of the first hill; the speed changes due to gravity alone. The cars speed up as they go down hills and slow down as they climb.

EXAMPLE 3.8 Speed of a roller coaster

A classic wooden coaster has cars that go down a big first hill, gaining speed. The cars then ascend a second hill with a slope of 30°. If the cars are going 25 m/s at the bottom and it takes them 2.0 s to climb this hill, how fast are they going at the top?

PREPARE We start with the visual overview in Figure 3.24, which includes a motion diagram, a pictorial representation, and a list of values. Notice how the motion diagram of Figure 3.24 differs from that of the previous example: The velocity decreases as the car moves up the hill, so the acceleration vector is opposite the direction of the velocity vector. The motion is along the x-axis, as before, but the acceleration vector points in the negative-x direction, so the component a_x is negative.

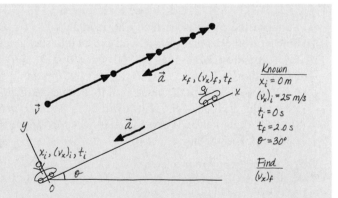

FIGURE 3.24 The coaster's speed decreases as it goes up the hill.

SOLVE To determine the final speed, we need to know the acceleration. We will assume that there is no friction or air resistance, so the magnitude of the roller coaster's acceleration is given by Equation 3.20 using the minus sign, as noted:

$$a_x = -g\sin\theta = -(9.8 \text{ m/s}^2)\sin 30° = -4.9 \text{ m/s}^2$$

The speed at the top of the hill can then be computed using our kinematic equation for velocity:

$$(v_x)_f = (v_x)_i + a_x \Delta t = 25 \text{ m/s} + (-4.9 \text{ m/s}^2)(2.0 \text{ s}) = 15 \text{ m/s}$$

ASSESS The speed is less at the top of the hill than at the bottom, as it should be, but the coaster is still moving at a pretty good clip at the top—almost 35 mph. This seems reasonable. In the motion diagram, notice that we only drew a single acceleration vector—a reasonable shortcut, as we know that the acceleration is constant. One vector can represent the acceleration for the entire motion.

STOP TO THINK 3.4 A block of ice slides down a ramp. For which height and base length is the acceleration the greatest?

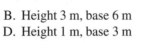

Height / Base

A. Height 4 m, base 12 m   B. Height 3 m, base 6 m
C. Height 2 m, base 5 m   D. Height 1 m, base 3 m

3.5 Relative Motion

You've now dealt many times with problems that say something like "A car travels at 30 m/s" or "A plane travels at 300 m/s." But, as we will see, we may need to be a bit more specific.

In Figure 3.25, Amy, Bill, and Carlos are watching a runner. According to Amy, the runner's velocity is $v_x = 5$ m/s. But to Bill, who's riding alongside, the runner is lifting his legs up and down but going neither forward nor backward relative to Bill. As far as Bill is concerned, the runner's velocity is $v_x = 0$ m/s. Carlos sees the runner receding in his rearview mirror, in the *negative x-direction*, getting 10 m further away from him every second. According to Carlos, the runner's velocity is $v_x = -10$ m/s. Which is the runner's *true* velocity?

Velocity is not a concept that can be true or false. The runner's velocity *relative to Amy* is 5 m/s. That is, his velocity is 5 m/s in a coordinate system attached to Amy and in which Amy is at rest. The runner's velocity relative to Bill is 0 m/s, and the velocity relative to Carlos is −10 m/s. These are all valid descriptions of the runner's motion.

Relative Velocity

Suppose we know that the runner's velocity relative to Amy is 5 m/s; we will call this velocity $(v_x)_{RA}$. The second subscript, "RA" means "**R**unner relative to **A**my." We also know that the velocity of **C**arlos relative to **A**my is 15 m/s; we write this as $(v_x)_{CA} = 15$ m/s. It is equally valid to compute Amy's velocity relative to Carlos. From Carlos's point of view, Amy is moving to the left at 15 m/s; we write Amy's velocity relative to Carlos as $(v_x)_{AC} = -15$ m/s; note that $(v_x)_{AC} = -(v_x)_{CA}$.

Given the runner's velocity relative to Amy and Amy's velocity relative to Carlos, we can compute the runner's velocity relative to Carlos by combining the two velocities we know. The subscripts as we have defined them are our guide for this combination.

$$(v_x)_{RC} = (v_x)_{RA} + (v_x)_{AC} \tag{3.21}$$

The "A" appears on the right of the first expression and on the left of the second; when we combine these velocities, we "cancel" the A to get $(v_x)_{RC}$.

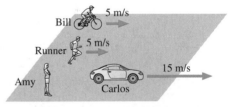

FIGURE 3.25 Amy, Bill, and Carlos each measure the velocity of the runner. The velocities are shown relative to Amy.

Throwing for the gold An athlete throwing the javelin does so while running. It's harder to throw the javelin on the run, but there's a very good reason to do so. The distance of the throw will be determined by the velocity of the javelin with respect to the ground—which is the sum of the velocity of the throw plus the velocity of the athlete. A faster run means a farther throw.

Generally, you can add two relative velocities in this manner, by "canceling" subscripts as in Equation 3.21. In Chapter 27, when we learn about relativity, we will have a more rigorous scheme for computing relative velocities, but this technique will serve our purposes at present.

EXAMPLE 3.9 How fast is a speeding bullet?

The police are chasing a bank robber. While driving at 50 m/s, they fire a bullet to shoot out a tire of his car. The police gun shoots bullets at 300 m/s. What is the bullet's speed as measured by a TV camera crew standing beside the road?

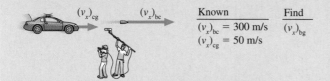

Known	Find
$(v_x)_{bc} = 300$ m/s	$(v_x)_{bg}$
$(v_x)_{cg} = 50$ m/s	

FIGURE 3.26 Relative velocities for Example 3.9.

PREPARE A visual overview of the situation is given in Figure 3.26. We assume that all motion is along the *x*-axis. The velocity of the **c**ar is given relative to the **g**round; the velocity of the **b**ullet is given relative to the **c**ar. We wish to find the velocity of the **b**ullet relative to the **g**round.

SOLVE We compute the desired velocity using the technique of Equation 3.21:

$$(v_x)_{bg} = (v_x)_{bc} + (v_x)_{cg} = 300 \text{ m/s} + 50 \text{ m/s} = 350 \text{ m/s}$$

ASSESS The bullet is fired forward from a speeding vehicle. It moves faster relative to the ground than the speed with which it was shot from the gun, so our result makes sense.

This technique for finding relative velocities also works for two-dimensional situations, as we see in the next example. Relative motion in two dimensions is another good exercise in working with vectors.

EXAMPLE 3.10 Finding the ground speed of an airplane

Cleveland is approximately 300 miles east of Chicago. A plane leaves Chicago flying due east at 500 mph. The pilot forgot to check the weather and doesn't know that the wind is blowing to the south at 100 mph. What is the plane's velocity relative to the ground?

PREPARE Figure 3.27 is a visual overview of the situation. We are given the speed of the **p**lane relative to the **a**ir ($\vec{v}_{pa}$) and the

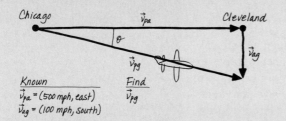

Known	Find
$\vec{v}_{pa} = (500 \text{ mph, east})$	$\vec{v}_{pg}$
$\vec{v}_{ag} = (100 \text{ mph, south})$	

FIGURE 3.27 The wind causes a plane flying due east in the air to move to the southeast relative to the ground.

speed of the **a**ir relative to the **g**round ($\vec{v}_{ag}$); the speed of the **p**lane relative to the **g**round will be the vector sum of these velocities:

$$\vec{v}_{pg} = \vec{v}_{pa} + \vec{v}_{ag}$$

This vector sum is shown in Figure 3.27.

SOLVE The plane's speed relative to the ground is the hypotenuse of the right triangle in Figure 3.27, thus

$$v_{pg} = \sqrt{v_{pa}^2 + v_{ag}^2} = \sqrt{(500 \text{ mph})^2 + (100 \text{ mph})^2} = 510 \text{ mph}$$

The plane's direction can be specified by the angle θ measured from due east:

$$\theta = \tan^{-1}\left(\frac{100 \text{ mph}}{500 \text{ mph}}\right) = \tan^{-1}(0.20) = 11°$$

The velocity of the plane relative to the ground is thus

$$\vec{v}_{pg} = (510 \text{ mph, } 11° \text{ south of east})$$

ASSESS The good news is that the wind is making the plane move a bit faster relative to the ground; the bad news is that the wind is making the plane move in the wrong direction!

3.6 Motion in Two Dimensions: Projectile Motion

Balls flying through the air, long jumpers, and cars doing stunt jumps are all examples of the two-dimensional motion that we call *projectile motion*. Projectile motion is an extension to two dimensions of the free-fall motion we studied in

Chapter 2. **A projectile is an object that moves in two dimensions under the influence of only gravity and nothing else.** Although real objects are also influenced by air resistance, the effect of air resistance is small for reasonably dense objects moving at modest speeds, so we can ignore it for the cases we consider in this chapter. As long as we can neglect air resistance, any projectile, from a football to the stream of water shot by an archer fish, will follow the same type of path, a trajectory with the mathematical form of a parabola. Because the form of the motion will always be the same, the strategies we develop to solve one projectile problem can be applied to others as well.

Figure 3.28a shows the parabolic arc of a ball tossed into the air; the camera has captured its position at equal intervals of time. In Figure 3.28b we show the motion diagram for this toss, with velocity vectors connecting the points. The acceleration vector points in the same direction as the change in velocity $\Delta\vec{v}$, which we can compute using the techniques of Tactics Box 3.2. You can see that the acceleration vector points straight down; a careful analysis would show that it has magnitude 9.80 m/s². Consequently, the acceleration of a projectile is the same as the acceleration of an object falling straight down, namely, the free-fall acceleration

$$\vec{a}_{\text{free fall}} = (9.80 \text{ m/s}^2, \text{ straight down})$$

Because the acceleration is the same for all objects, it is no wonder that the shape of the trajectory—a parabola—is the same as well.

Analyzing Projectile Motion

Suppose you toss a basketball down the court, as shown in Figure 3.29. To study this projectile motion, we've established a coordinate system with the x-axis horizontal and the y-axis vertical. The start of a projectile's motion is called the *launch,* and the angle θ of the initial velocity $\vec{v}_i$ above the horizontal (i.e., above the x-axis) is the **launch angle.** As you learned in Section 3.3, the initial velocity vector $\vec{v}_i$ can be expressed in terms of the x- and y- components $(v_x)_i$ and $(v_y)_i$. You can see from the figure that

$$(v_x)_i = v_i\cos\theta$$
$$(v_y)_i = v_i\sin\theta \qquad (3.22)$$

where v_i is the initial speed.

NOTE ▶ The components $(v_x)_i$ and $(v_y)_i$ are not always positive. A projectile launched at an angle *below* the horizontal (such as a ball thrown downward from the roof of a building) has *negative* values for θ and $(v_y)_i$. However, the speed v_i is always positive. ◀

As the projectile moves, the free-fall acceleration will change the vertical component of the velocity, but there will be no change to the horizontal component of the velocity. Therefore, the vertical and horizontal components of the acceleration are

$$a_x = 0 \text{ m/s}^2$$
$$a_y = -g = -9.80 \text{ m/s}^2 \qquad (3.23)$$

The vertical component of acceleration a_y for all projectile motion is just the familiar $-g$ of free fall while the horizontal component a_x is zero.

(a)

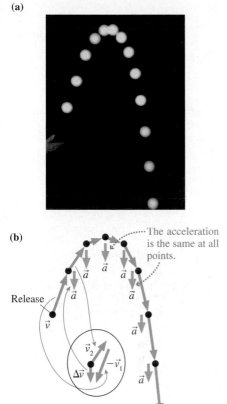

(b)

The acceleration is the same at all points.

Release

Ground

FIGURE 3.28 The motion of a tossed ball. The inset shows how to find the direction of $\Delta\vec{v}$, the change in velocity. This is the direction in which the acceleration $\vec{a}$ points.

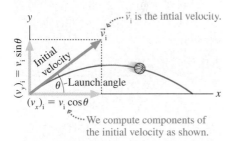

$\vec{v}_i$ is the initial velocity.

Initial velocity

θ Launch angle

$(v_x)_i = v_i\cos\theta$

We compute components of the initial velocity as shown.

FIGURE 3.29 The launch and motion of a projectile.

(a)

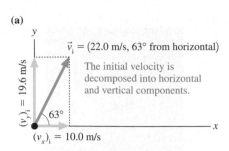

$\vec{v}_i = (22.0 \text{ m/s}, 63° \text{ from horizontal})$

The initial velocity is decomposed into horizontal and vertical components.

$(v_y)_i = 19.6 \text{ m/s}$

$63°$

$(v_x)_i = 10.0 \text{ m/s}$

(b)

The vertical component of velocity decreases by 9.8 m/s every second.

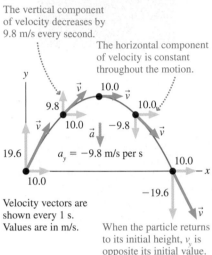

The horizontal component of velocity is constant throughout the motion.

$a_y = -9.8 \text{ m/s per s}$

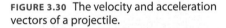

Velocity vectors are shown every 1 s. Values are in m/s.

When the particle returns to its initial height, v_y is opposite its initial value.

FIGURE 3.30 The velocity and acceleration vectors of a projectile.

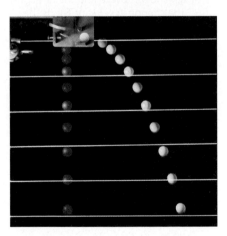

FIGURE 3.31 A projectile launched horizontally falls in the same time as a projectile that is released from rest.

To see how these accelerations determine the subsequent motion, Figure 3.30 shows a projectile launched at a speed of 22.0 m/s at an angle of 63° to the horizontal. In Figure 3.30a, the initial velocity vector is broken into its horizontal and vertical components. In Figure 3.30b, the velocity and acceleration vectors are shown every subsequent 1.0 s. Because there is no horizontal acceleration ($a_x = 0$), the value of v_x never changes. In contrast, v_y decreases by 9.8 m/s every second. This is what it *means* to accelerate at $a_y = -9.8 \text{ m/s}^2 = (-9.8 \text{ m/s})$ per second. Nothing *pushes* the projectile along the curve. Instead, the downward acceleration changes the velocity vector as shown, causing it to increase downward as the motion proceeds. At the end of the motion, when the ball is at the same height as it started, v_y is -19.6 m/s, the negative of its initial value. **The ball finishes its motion moving downward at the same speed as it started moving upward,** just as we saw in the case of one-dimensional free fall in Chapter 2.

You can see from Figure 3.30 that **projectile motion is made up of two independent motions:** uniform motion at constant velocity in the horizontal direction and free-fall motion in the vertical direction. In Chapter 2, we saw kinematic equations for constant-velocity and constant-acceleration motion. We can adapt these general equations to this current case: The horizontal motion is constant-velocity motion at $(v_x)_i$, the vertical motion is constant-acceleration motion with initial velocity $(v_y)_i$ and an acceleration of $a_y = -g$.

$$x_f = x_i + (v_x)_i \, \Delta t \qquad\qquad y_f = y_i + (v_y)_i \, \Delta t - \tfrac{1}{2}g(\Delta t)^2$$

$$(v_x)_f = (v_x)_i = \text{constant} \qquad (v_y)_f = (v_y)_i - g \, \Delta t$$

(3.24)

Equations of motion for the parabolic trajectory of a projectile

A close look at these equations reveals a surprising fact: **The horizontal and vertical components of projectile motion are independent of each other.** The initial horizontal velocity has *no* influence over the vertical motion, and vice-versa. This independence of the horizontal and vertical motions is illustrated in Figure 3.31, which shows a strobe photograph of two balls, one shot horizontally and the other released from rest at the same instant. The *vertical* motions of the two balls are identical, and they hit the floor simultaneously. Neither ball has any initial motion in the vertical direction, so both fall distance h in the same amount of time.

Let's extend these ideas to consider a "classic" problem in physics:

A hungry hunter in the jungle wants to shoot down a coconut that is hanging from the branch of a tree. He aims the gun directly at the coconut, but as luck would have it the coconut falls from the branch at the exact instant the hunter pulls the trigger. Does the bullet hit the coconut?

Figure 3.32 shows a useful way to analyze this problem. Figure 3.32a shows the trajectory of a projectile. Without gravity, a projectile would follow a

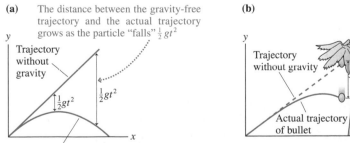

(a) The distance between the gravity-free trajectory and the actual trajectory grows as the particle "falls" $\tfrac{1}{2}gt^2$

Trajectory without gravity

$\tfrac{1}{2}gt^2$

$\tfrac{1}{2}gt^2$

Actual trajectory

(b)

Trajectory without gravity

$\tfrac{1}{2}gt^2$

Actual trajectory of bullet

FIGURE 3.32 A projectile follows a parabolic trajectory because it "falls" a distance $\tfrac{1}{2}gt^2$ below a straight-line trajectory.

straight line. Because of gravity, the particle at time t has "fallen" a distance $\frac{1}{2}gt^2$ below this line. The separation grows as $\frac{1}{2}gt^2$, giving the trajectory its parabolic shape. Figure 3.32b applies this reasoning to the bullet and coconut. Although the bullet travels very fast, it follows a slightly curved trajectory, not a straight line. Had the coconut stayed on the tree, the bullet would have curved under its target as gravity causes it to fall a distance $\frac{1}{2}gt^2$ below the straight line. But $\frac{1}{2}gt^2$ is also the distance the coconut falls while the bullet is in flight. Thus, as Figure 3.32b shows, the bullet and the coconut fall the same distance and meet at the same point!

STOP TO THINK 3.5   A 100 g ball rolls off a table and lands 2 m from the base of the table. A 200 g ball rolls off the same table with the same speed. It lands at a distance

A. <1 m.
B. 1 m.
C. Between 1 m and 2 m.
D. 2 m.
E. Between 2 m and 4 m.
F. 4 m.

3.7 Projectile Motion: Solving Problems

Now that we have a good idea of how projectile motion works, we can use that knowledge to solve some true two-dimensional motion problems.

TRY IT YOURSELF

A game of catch in a moving vehicle A great way to illustrate the independence of the horizontal and vertical components of projectile motion is to do a simple experiment. While riding in a car moving at a constant speed, toss a ball or a coin into the air. You can easily catch it! The ball and you and the car continue to move forward at a constant speed during the ball's up and down vertical motion. From the point of a view of a person watching you drive by, the ball's motion would be a parabolic arc. You can also think about this situation in terms of relative motion. How would you do this?

EXAMPLE 3.11 Planning a Hollywood stunt

To get the shots of cars flying through the air in movies, it is sometimes necessary to drive a car off a cliff and film it. Suppose a stunt man drives a car off a 10-m-high cliff at a speed of 20 m/s. How far does the car land from the base of the cliff?

PREPARE We start with a visual overview of the situation in Figure 3.33. Note that we have chosen to put the origin of the coordinate system at the base of the cliff. We assume that the car is moving horizontally as it leaves the cliff. In this case, the x- and y-components of the initial velocity are

$$(v_x)_i = v_i = 20 \text{ m/s}$$
$$(v_y)_i = 0 \text{ m/s}$$

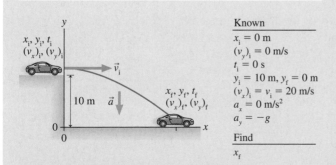

Known
$x_i = 0$ m
$(v_y)_i = 0$ m/s
$t_i = 0$ s
$y_i = 10$ m, $y_f = 0$ m
$(v_x)_i = v_i = 20$ m/s
$a_x = 0$ m/s^2
$a_y = -g$

Find
x_f

FIGURE 3.33 Visual overview for Example 3.11.

SOLVE Each point on the trajectory has x- and y-components of position, velocity, and acceleration but only *one* value of time. The time needed to move horizontally to the final position x_f is

the *same* time needed to fall 10 m vertically. **Although the horizontal and vertical motions are independent, they are connected through the time t.** This is a critical observation for solving projectile motion problems. We will call the time interval between the car leaving the cliff and landing on the ground Δt. In this problem, we'll analyze the vertical motion first. We can solve the vertical motion equations for the time interval Δt. We'll then use that value of Δt in the equation for the horizontal motion.

The vertical motion is just free fall. The initial vertical velocity is zero; the car falls from $y_i = 10$ m to $y_f = 0$ m. We can analyze this motion using the vertical-position equation from Equations 3.24:

$$y_f = y_i + (v_y)_i \Delta t - \tfrac{1}{2}g(\Delta t)^2$$
$$0 \text{ m} = 10 \text{ m} + (0 \text{ m/s})(\Delta t) - \tfrac{1}{2}(9.8 \text{ m/s}^2)(\Delta t)^2$$

Rearranging the terms and then solving for Δt gives

$$-10 \text{ m} = -\tfrac{1}{2}(9.8 \text{ m/s}^2)(\Delta t)^2$$

$$\Delta t = \sqrt{\frac{2(10 \text{ m})}{9.8 \text{ m/s}^2}} = 1.43 \text{ s}$$

Now that we have the time, we can use the horizontal-position equation from Equation 3.24 to find out where the car lands:

$$x_f = x_i + (v_x)_i \Delta t$$
$$x_f = 0 \text{ m} + (20 \text{ m/s})(1.43 \text{ s}) = 29 \text{ m}$$

ASSESS The cliff height is $h \approx 33$ ft and the initial horizontal velocity is $(v_x)_i \approx 40$ mph. At this speed, a car moves over 60 feet per second, so traveling $x_f = 29$ m ≈ 95 ft before hitting the ground seems quite reasonable.

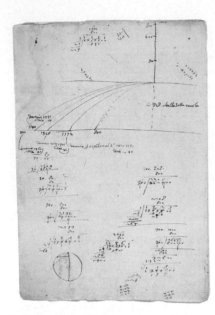

Galileo was the first person to make a serious study of projectile motion, deducing the independence of the horizontal and vertical components. This page from his notes shows his analysis of a projectile launched horizontally. In his day, this topic was cutting-edge science; now it is in Chapter 3 of a 30-chapter book!

The approach of Example 3.11 is a general one. We can condense the relevant details into a problem-solving strategy.

(MP) **PROBLEM-SOLVING STRATEGY 3.1** **Projectile motion problems**

PREPARE There are a number of steps that you should go through in setting up the solution to a projectile motion problem:

- Make simplifying assumptions. Whether the projectile is a car or a basketball, the motion will be the same.
- Draw a visual overview including a pictorial representation showing the beginning and end points of the motion.
- Establish a coordinate system with the x-axis horizontal and the y-axis vertical. In this case, you know that the horizontal acceleration will be zero and the vertical acceleration will be free-fall: $a_x = 0$ and $a_y = -g$.
- Define symbols and write down a list of known values. Identify what the problem is trying to find.

SOLVE There are two sets of kinematic equations for projectile motion, one for the horizontal component and one for the vertical:

Horizontal	Vertical
$x_f = x_i + (v_x)_i \Delta t$	$y_f = y_i + (v_y)_i \Delta t - \frac{1}{2}g(\Delta t)^2$
$(v_x)_f = (v_x)_i = \text{constant}$	$(v_y)_f = (v_y)_i - g \Delta t$

Δt **is the same for the horizontal and vertical components of the motion.** Find Δt by solving for the vertical or the horizontal component of the motion, then use that value to complete the solution for the other component.

ASSESS Check that your result has the correct units, is reasonable, and answers the question.

3.1–3.7 Activ ONLINE Physics

EXAMPLE 3.12 Checking the feasibility of a Hollywood stunt

The main characters in the movie *Speed* are on a bus that has been booby-trapped to explode if its speed drops below 50 mph. But there is a problem ahead: A 50 ft section of a freeway overpass is missing. They decide to jump the bus over the gap. The road leading up to the break has an angle of about 5°. A view of the speedometer just before the jump shows that the bus is traveling at 67 mph. The movie bus makes the jump and survives. Is this realistic, or movie fiction?

PREPARE We begin by converting speed and distance to SI units. The initial speed is $v_i = 30$ m/s and the size of the gap is $L = 15$ m. Next, following the problem-solving strategy, we make a sketch, the visual overview shown in Figure 3.34, and a list of values. In choosing our axes, we've placed the origin at the point where the bus starts its jump. The initial velocity vector is tilted 5° above horizontal, so the components of the initial velocity are

$$(v_x)_i = v_i \cos\theta = (30 \text{ m/s})(\cos 5°) = 30 \text{ m/s}$$

$$(v_y)_i = v_i \sin\theta = (30 \text{ m/s})(\sin 5°) = 2.6 \text{ m/s}$$

How do we specify the "end" of the problem? By setting $y_f = 0$ m, we'll solve for the horizontal distance x_f at which the bus returns to its initial height. If x_f exceeds 50 ft, the bus successfully clears the gap. We have optimistically drawn our diagram as if the bus makes the jump, but . . .

FIGURE 3.34 Visual overview of the bus jumping the gap.

SOLVE Problem-Solving Strategy 3.1 suggests using one component of the motion to solve for Δt. We will begin with the vertical motion. The kinematic equation for the vertical position is

$$y_f = y_i + (v_y)_i \Delta t - \tfrac{1}{2}g(\Delta t)^2$$

We know that $y_f = y_i = 0$ m. If we factor out a Δt, the position equation becomes

$$0 = \Delta t\left((v_y)_i - \tfrac{1}{2}g\,\Delta t\right)$$

One solution to this equation is $\Delta t = 0$ s. This is a legitimate solution, but it corresponds to the instant when $y = 0$ at the beginning of the trajectory. We want the second solution, for $y = 0$ at the end of the trajectory, which is when

$$0 = (v_y)_i - \tfrac{1}{2}g\,\Delta t = (2.6 \text{ m/s}) - \tfrac{1}{2}(9.8 \text{ m/s}^2)\,\Delta t$$

Which gives

$$\Delta t = \frac{2 \times (2.6 \text{ m/s})}{9.8 \text{ m/s}^2} = 0.53 \text{ s}$$

During the 0.53 s that the bus is moving vertically it is also moving horizontally. The horizontal distance it travels is $x_f = x_i + (v_x)_i \Delta t$, or

$$x_f = 0 \text{ m} + (30 \text{ m/s})(0.53 \text{ s}) = 16 \text{ m}$$

This is how far the bus has traveled horizontally when it returns to its original height. 16 m is a bit more than the width of the gap, so a bus coming off a 5° ramp at the noted speed would make it—just barely!

ASSESS We can do a quick check on our math by noting that the bus takes off and lands at the same height. This means, as we saw in Figure 3.30b, that the y-velocity at the landing should be the negative of its initial value. We can use the velocity equation for the vertical component of the motion to compute the final value and see that the final velocity value is as we predict:

$$(v_y)_f = (v_y)_i - g\,\Delta t$$

$$= (2.6 \text{ m/s}) - (9.8 \text{ m/s}^2)(0.53 \text{ s}) = -2.6 \text{ m/s}$$

During the filming of the movie, the filmmakers really did jump a bus over a gap in an overpass! The actual jump was a bit more complicated than our example because a real bus, being an extended object rather than a particle, will start rotating as the front end comes off the ramp. The actual stunt jump used an extra ramp to give a boost to the front end of the bus. Nonetheless, our example shows that the filmmakers did their homework and devised a situation in which the physics was correct.

Range and Time of Flight

There are many examples of projectiles that land at the same height as that from which they were launched, such as a football that has been kicked downfield, or a baseball hit to the outfield. In cases like this, we are often interested in how far the projectile went, or how long it was in the air.

Figure 3.35 shows a pictorial representation of a projectile that is launched with initial velocity $\vec{v}_i$ and lands at its starting height. The x- and y-axes are specified, and we have chosen our origin so that $x_i = y_i = 0$. The distance to the landing point, at which $y_f = 0$, is called the **range** of the projectile. The time that the projectile is in the air is the **time of flight.**

The x- and y-components of the projectile's initial velocity are $(v_x)_i = v_i\cos\theta$ and $(v_y)_i = v_i\sin\theta$. We can use the equation for vertical motion to find the time of flight, as we did in Example 3.12:

$$0 = 0 + (v_i\sin\theta)t_f - \tfrac{1}{2}gt_f^2 = (v_i\sin\theta - \tfrac{1}{2}gt_f)t_f$$

Here we've taken advantage of the fact that $t_i = 0$ to write $\Delta t = t_f$. We are looking for a solution for which $t_f \neq 0$, so we set the quantity in brackets equal to zero and solve for the time of flight:

$$\text{Time of flight} = t_f = \frac{2v_i\sin\theta}{g} \tag{3.25}$$

Let's look at the range now. We can use the time of flight in the equation for horizontal motion to compute the distance to the landing point:

$$x_f = x_i + (v_x)_i t_f = (v_i\cos\theta)t_f$$

$$= (v_i\cos\theta)\frac{2v_i\sin\theta}{g} = \frac{2v_i^2\sin\theta\cos\theta}{g}$$

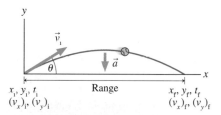

x_i, y_i, t_i Range x_f, y_f, t_f
$(v_x)_i, (v_y)_i$ $(v_x)_f, (v_y)_f$

FIGURE 3.35 The parabolic trajectory and range of a projectile.

Physics of fielding   BIO At the crack of a bat, the fielder begins running. Eyeing the ball, he makes a graceful arc to the exact spot where it will land, catching it on the run. Ball players don't estimate velocity and mentally calculate range, so how they do what they do is a fascinating—and unresolved—physics problem. Perhaps a fielder watches the relative motion of the ball as he runs and makes adjustments in his velocity to keep the ball at a constant angle, leading him to the right spot. If so, he doesn't know where the ball will land, just how to move in order to intercept it.

We can simplify this result by using the trigonometric identity $2\sin\theta\cos\theta = \sin(2\theta)$ to write the distance traveled by the projectile (i.e., its range) when launched at angle θ as

$$\text{Range} = x_f = \frac{v_i^2 \sin(2\theta)}{g} \tag{3.26}$$

NOTE ▶ Equation 3.26 is *not* a general result. It applies *only* in situations in which the projectile lands at the same elevation from which it was launched. ◀

What value of θ will give the largest distance? As you know, the sine function has a maximum value of 1 at an angle of 90°. Because the equation for x_f contains the expression $\sin(2\theta)$, it will be a maximum for $\theta_{max} = 45°$. Thus the maximum possible range, over level ground, is $x_{max} = v_i^2/g$, and this maximum is achieved only for a 45° launch.

Suppose a football player is kicking a ball down the field. If he wants the most distance for the kick, the ball should leave his foot at a 45° angle. But this angle won't give the longest "hang time," the time of flight. You can see from Equation 3.25 that the longest time of flight is achieved by kicking the ball at $\theta = 90°$—straight up! Reducing the angle from 90° will give a greater range, but a lesser hang time. The laws of physics put the range and the hang time in conflict with each other. Both cannot be maximized simultaneously, so the kicker must choose which is more important.

EXAMPLE 3.13   Finding the range of a kick

A football player kicks a ball straight up in the air, where it stays for a total time of 5.0 s. If he had kicked the ball downfield at the same speed, what is the maximum distance he could have kicked the ball?

PREPARE First, we can use the time of flight for the straight-up kick to calculate how fast the ball was kicked. We will then use that speed to determine the maximum range.

SOLVE Equation 3.25, the time-of-flight equation, with $\theta = 90°$, is

$$\text{Time of flight} = 5.0 \text{ s} = \frac{2v_i}{g}$$

Solving for v_i gives

$$v_i = \frac{g(5.0 \text{ s})}{2} = 24.5 \text{ m/s}$$

The maximum possible range is found by using this speed in Equation 3.26 with $\theta = 45°$:

$$\text{Maximum distance} = \frac{v_i^2}{g} = \frac{(24.5 \text{ m/s})^2}{9.8 \text{ m/s}^2} = 61 \text{ m}$$

ASSESS 61 m is a bit more than 61 yards—an impressive kick, but an achievable one.

3.8 Motion in Two Dimensions: Circular Motion

The 32 cars on the London Eye Ferris wheel move at a constant speed of about 0.5 m/s in a vertical circle of radius 65 m. The cars may move at a constant speed, but they do *not* move with constant velocity. Velocity is a vector that depends on both an object's speed *and* its direction of motion, and the direction of circular motion is constantly changing. This is the hallmark of uniform circular motion: Constant speed, but continuously changing direction. We will introduce some basic ideas about circular motion in this section, then return to treat it in considerably more detail in Chapter 6. For now, we will only consider objects that move around a circular trajectory at constant speed.

Period and Frequency

The time interval it takes an object to go around a circle one time, completing one revolution (abbreviated rev), is called the **period** of the motion. Period is represented by the symbol T.

The London Eye Ferris wheel.

Rather than specify the time for one revolution, we can specify circular motion by its **frequency,** the number of revolutions per second, for which we use the symbol f. An object with a period of one-half second completes 2 revolutions each second. Similarly, an object can make 10 revolutions in 1 s if its period is one-tenth of a second. This shows that frequency is the inverse of the period:

$$f = \frac{1}{T} \tag{3.27}$$

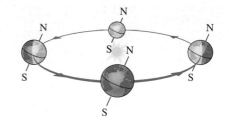

The period of the orbit of the earth around the sun is one year.

Although frequency is often expressed as "revolutions per second," *revolutions* are not true units but merely the counting of events. Thus the SI unit of frequency is simply inverse seconds, or s⁻¹. Frequency may also be given in revolutions per minute (rpm) or another time interval, but these usually need to be converted to s⁻¹ before doing calculations.

EXAMPLE 3.14 Period of an audio CD

An audio CD spins at a frequency of up to 540 rpm. At 540 rpm, how much time is required for one revolution of the CD?

SOLVE First, we convert units from rpm to rev/s:

$$f = 540 \ \frac{\text{revolutions}}{\text{minute}} \times \frac{1 \ \text{minute}}{60 \ \text{seconds}} = 9.0 \ \frac{\text{rev}}{\text{s}} = 9.0 \ \text{s}^{-1}$$

We then rearrange Equation 3.27 and compute

$$T = \frac{1}{f} = \frac{1}{9.0 \ \text{s}^{-1}} = 0.11 \ \text{s}$$

ASSESS 0.11 second is 1/9 of a second, which makes sense, as the frequency is 9 rev/s.

Angular Position

Rather than using xy-coordinates for circular motion, it will be more convenient to describe the position of the particle by its distance r from the center of the circle and its angle θ from the positive x-axis. This is shown in Figure 3.36. The angle θ is the **angular position** of the particle.

We can distinguish a position above the x-axis from a position an equal angle below the x-axis by defining θ to be positive when measured *counterclockwise* from the positive x-axis. An angle measured *clockwise* from the positive x-axis has a negative value. "Clockwise" and "counterclockwise" in circular motion are analogous, respectively, to "left of the origin" and "right of the origin" in linear motion, which we associated with negative and positive values of x.

Rather than measure angles in degrees, mathematicians and scientists usually measure the angle θ in the angular unit of *radians*. In Figure 3.36, we have noted the **arc length** s that the particle has traveled along the edge of the circle of radius r. We define the particle's angle θ in **radians** in terms of this arc length and the radius of the circle:

$$\theta \ (\text{radians}) = \frac{s}{r} \tag{3.28}$$

The radian, abbreviated rad, is the SI unit of angle. An angle of 1 rad has an arc length s exactly equal to the radius r. An important consequence of Equation 3.28 is that the arc length spanning the angle θ is

$$s = r\theta \tag{3.29}$$

NOTE ▶ Equation 3.29 is valid only if θ is measured in radians and not degrees. This very simple relationship between angle and arc length is one of the primary motivations for using radians. ◀

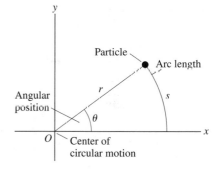

FIGURE 3.36 A particle's position is described by distance r and angle θ.

The arc length completely around a circle is the circle's circumference $2\pi r$. Thus the angle of a full circle is

$$\theta_{\text{full circle}} = \frac{s}{r} = \frac{2\pi r}{r} = 2\pi \text{ rad}$$

We can use this fact to define conversion factors among revolutions, radians, and degrees:

$$1 \text{ rev} = 360° = 2\pi \text{ rad}$$

$$1 \text{ rad} = 1 \text{ rad} \times \frac{360°}{2\pi \text{ rad}} = 57.3°$$

We will often specify angles in degrees, but keep in mind that the SI unit is the radian. You can visualize angles in radians by remembering that 1 rad is just about 60°.

Acceleration in Circular Motion

It may seem strange to think that an object moving with constant speed can be accelerating, but that's exactly what an object in uniform circular motion is doing. It is accelerating because its velocity is changing as its direction of motion changes. What is the acceleration in this case? We saw in Example 3.3, which is worth reviewing, that for circular motion at a constant speed, **the acceleration vector $\vec{a}$ points toward the center of the circle.** As you can see in Figure 3.37 the velocity is always tangent to the circle, so $\vec{v}$ and $\vec{a}$ are perpendicular to each other at all points on the circle.

An acceleration that always points directly toward the center of a circle is called a *centripetal acceleration.* The word "centripetal" comes from a Greek root meaning "center seeking." Centripetal acceleration is not a new type of acceleration; all we are doing is *naming* an acceleration that corresponds to a particular type of motion. The magnitude of the centripetal acceleration is constant because each successive $\Delta\vec{v}$ in the motion diagram has the same length.

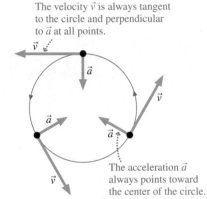

The velocity $\vec{v}$ is always tangent to the circle and perpendicular to $\vec{a}$ at all points.

The acceleration $\vec{a}$ always points toward the center of the circle.

FIGURE 3.37 The velocity and acceleration vectors for circular motion.

CONCEPTUAL EXAMPLE 3.1 Acceleration on a swing

A child is riding a playground swing. At the lowest point of her motion, is she accelerating? If so, what is the direction of her acceleration?

REASON As Figure 3.38 shows, the swing rotates in a circle around a central point where the rope or chain for the swing is attached. Even though speed isn't changing at the lowest point, the *direction* of the velocity vector is changing. Thus there is an acceleration—a centripetal acceleration directed toward the center of the circle. The velocity and acceleration vectors are perpendicular, as was shown in Figure 3.37. The velocity is directed forward, the acceleration is directed upward.

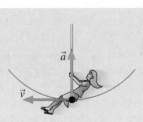

FIGURE 3.38 A child at the lowest point of motion on a swing.

ASSESS In the next chapter, we will find that there must be a force to produce an acceleration. Because the chain or rope is pulling up on the swing, it makes sense to think that the acceleration will be upward as well. But we are getting ahead of ourselves!

4.1

To complete our description of circular motion we need to find a quantitative relationship between the magnitude of the acceleration a and the speed v. Let's return to the case of the Ferris wheel, looking at the two points of the motion shown in Figure 3.39a. During a time Δt in which a car on the Ferris wheel moves from point 1 to point 2, the car moves through an angle θ and travels along an arc of length $r\theta$, as noted in the figure.

The speed of the car is constant, though the direction of motion changes. The speed is the distance divided by the time interval:

$$v = \frac{r\theta}{\Delta t}$$

Rearranging, we find the time Δt to be

$$\Delta t = \frac{r\theta}{v} \quad (3.30)$$

Figure 3.39b shows velocity vectors for the two points and the difference vector $\Delta \vec{v}$. The angle θ is the same as that in Figure 3.39a. If this angle is small, the magnitude of the difference vector is very nearly equal to the length of a circular arc between the two velocity vectors, so we can approximate

$$\Delta v = v\theta \quad (3.31)$$

The magnitude of the acceleration is computed using Δv from Equation 3.31 and Δt from Equation 3.30:

$$a = \frac{\Delta v}{\Delta t} = \frac{v\theta}{r\theta/v} = \frac{v^2}{r} \quad (3.32)$$

Combining this magnitude with the direction we noted above, we can write the centripetal acceleration as

$$\vec{a} = \left(\frac{v^2}{r}, \text{ toward center of circle}\right) \quad (3.33)$$

Centripetal acceleration of object moving in a circle of radius r at speed v

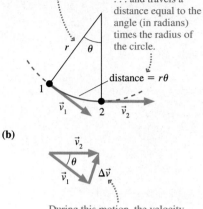

(a) As the car moves from point 1 to point 2, it goes through a circular arc of angle θ . . .

. . . and travels a distance equal to the angle (in radians) times the radius of the circle.

distance $= r\theta$

(b)

During this motion, the velocity changes direction; the difference vector points toward the center of the circle.

FIGURE 3.39 Changing velocity for an object in circular motion.

EXAMPLE 3.15 Finding the acceleration of a Ferris wheel

A typical carnival Ferris wheel has a radius of 9.0 m and rotates 6.0 times per minute. What magnitude acceleration do the riders experience?

PREPARE The cars on a Ferris wheel move in a circle at constant speed; the acceleration the riders experience is a centripetal acceleration.

SOLVE In order to use Equation 3.33 to compute an acceleration, we need to know the speed v of a rider on the Ferris wheel. The wheel rotates 6.0 times per minute; therefore, the time for one rotation (i.e., the period) is 10 s. During this time, a rider travels one circumference $2\pi r$ of the circle. Thus the speed is

$$v = \frac{\text{distance}}{\text{time interval}} = \frac{2\pi r}{\Delta t} = \frac{2\pi(9.0 \text{ m})}{10 \text{ s}} = 5.7 \text{ m/s}$$

Knowing the speed, we compute the magnitude of the acceleration as

$$a = \frac{v^2}{r} = \frac{(5.7 \text{ m/s})^2}{9.0 \text{ m}} = 3.6 \text{ m/s}^2$$

ASSESS This is about 1/3 of the free-fall acceleration; the acceleration, in units of g, is $0.37g$. This is enough to notice, but not enough to be scary!

What Comes Next: Forces

So far we have been studying motion without saying too much about what actually *causes* motion. Kinematics, the mathematical description of motion, is a good place to start because motion is very visible and very familiar. And in our study of motion we have introduced many of the basic tools, such as vectors, that we will use in the rest of the book.

But kinematics alone has some real limitations. For example, in this chapter we learned about projectile motion. But a real projectile such as a golf ball doesn't actually follow a perfectly parabolic path. Golfers know that the angle for maximum range is quite a bit less than 45°! Air resistance plays a role, and so do the dimples on a golf ball. The dimples give the ball a certain amount of "lift" as the ball spins, thus making it go further than a non-dimpled ball.

To treat more complex problems of this sort, we need to move beyond our treatment of kinematics and begin looking at the cause of motion: forces. By learning about forces, we will be able to explore a much wider range of problems in much more depth. As an example, think about the picture of a roller coaster with an inverted loop. How is it that riders can go through the loop and not fall out of their seats? This is just one of the problems that we will study once we know a bit about forces and the connection between forces and motion.

Amusement park kinematics Acceleration is fun—at least that's what the designer of this roller coaster seems to think! The coaster has ramps that give linear acceleration, parabolic segments in which the coaster follows a projectile path with a free-fall acceleration, and circular arcs in which the centripetal acceleration is larger than g. All of this acceleration means there are forces on the riders—and the coaster must be carefully designed so that these forces are well within safe limits.

STOP TO THINK 3.6   Which of the following particles has the largest centripetal acceleration?

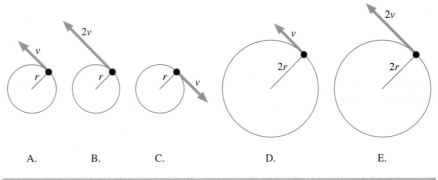

A.   B.   C.   D.   E.

SUMMARY

The goals of Chapter 3 have been to learn more about using vectors and to use vectors as a tool to analyze motion in two dimensions.

GENERAL PRINCIPLES

Projectile Motion

A projectile is an object that moves through the air under the influence of gravity and nothing else.

The path of the motion is a parabola.

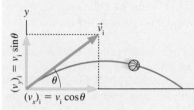

The motion consists of two pieces:

1. Vertical motion with free-fall acceleration, $a_y = -g$.

2. Horizontal motion with constant velocity.

Kinematic equations:

$$x_f = x_i + (v_x)_i \,\Delta t$$
$$(v_x)_f = (v_x)_i = \text{constant}$$
$$y_f = y_i + (v_y)_i \,\Delta t - \tfrac{1}{2}g(\Delta t)^2$$
$$(v_y)_f = (v_y)_i - g\,\Delta t$$

Circular Motion

For an object moving in a circle at a constant speed:

- The period, T, is the time for one rotation.

- The velocity is tangent to the circular path.

- The acceleration points toward the center of the circle, and has magnitude

$$a = \frac{v^2}{r}$$

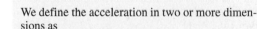
Centripetal acceleration

IMPORTANT CONCEPTS

Vectors and Components

A vector can be decomposed into **component vectors** along the x- and y-axes.

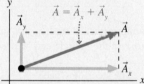

$$\vec{A} = \vec{A}_x + \vec{A}_y$$

The sign of the components depends on the **quadrant**:

$A_x < 0$	$A_x > 0$
$A_y > 0$	$A_y > 0$
$A_x < 0$	$A_x > 0$
$A_y < 0$	$A_y < 0$

Each component vector can be described with a single number specifying the component vector's length and (with a + or − sign) its direction. These two numbers are the x- and y-**components** of the vector.

The magnitude and direction of a vector can be expressed in terms of its components.

Magnitude $A = \sqrt{A_x^2 + A_y^2}$

$A_y = A\sin\theta$

$A_x = A\cos\theta$

Direction of $\vec{A}$
$\theta = \tan^{-1}(A_y / A_x)$

The Acceleration Vector

We define the acceleration in two or more dimensions as

$$\vec{a} = \frac{\vec{v}_f - \vec{v}_i}{t_f - t_i} = \frac{\Delta\vec{v}}{\Delta t}$$

We find an acceleration vector to complete the motion diagram as follows:

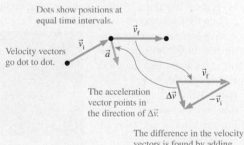

Dots show positions at equal time intervals.

Velocity vectors go dot to dot.

The acceleration vector points in the direction of $\Delta\vec{v}$.

The difference in the velocity vectors is found by adding the negative of $\vec{v}_i$ to $\vec{v}_f$.

APPLICATIONS

Relative motion

Velocities can be expressed relative to an observer. We can add relative velocities to convert to another observer's point of view.

c = car, b = bullet, g = ground

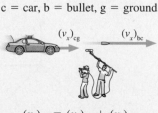

$(v_x)_{cg}$ $(v_x)_{bc}$

$$(v_x)_{bg} = (v_x)_{bc} + (v_x)_{cg}$$

Motion on a ramp

An object sliding down a ramp will accelerate parallel to the ramp:

$$a_x = \pm g\sin\theta$$

The correct sign depends on the direction in which the ramp is tilted.

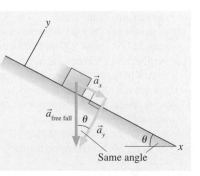

Same angle

For instructor-assigned homework, go to www.masteringphysics.com

Problem difficulty is labeled as | (straightforward) to |||| (challenging).

Problems labeled INT integrate significant material from earlier chapters; BIO are of biological or medical interest.

QUESTIONS

Conceptual Questions

1. a. Can a vector have nonzero magnitude if a component is zero? If no, why not? If yes, give an example.
    b. Can a vector have zero magnitude and a nonzero component? If no, why not? If yes, give an example.

2. The zero vector, $\vec{0}$, is defined by the property that $\vec{0} + \vec{A} = \vec{A}$ for any vector $\vec{A}$. What are the components of the zero vector? What is its magnitude?

3. Is it possible to add a scalar to a vector? If so, demonstrate. If not, explain why not.

4. Suppose two vectors have unequal magnitudes. Can their sum be $\vec{0}$? Explain.

5. Suppose $\vec{C} = \vec{A} + \vec{B}$.
    a. Under what circumstances does $C = A + B$?
    b. Could $C = A - B$? If so, how? If not, why not?

6. Suppose $\vec{C} = \vec{A} - \vec{B}$.
    a. Under what circumstances does $C = A - B$?
    b. Could $C = A + B$? If so, how? If not, why not?

7. For a projectile, which of the following quantities are constant during the flight: x, y, v_x, v_y, v, a_x, a_y? Which of the quantities are zero throughout the flight?

8. A baseball player throws a ball at a 40° angle to the ground. The ball lands on the ground some distance away.
    a. Is there any point on the trajectory where $\vec{v}$ and $\vec{a}$ are parallel to each other? If so, where?
    b. Is there any point where $\vec{v}$ and $\vec{a}$ are perpendicular to each other? If so, where?

9. A long jumper trying for the longest possible jump takes off at an angle less than 45°. Why might this be so? **Hint:** Think about the position of the center of the jumper's body on take off and on landing.

10. A person trying to throw a ball as far as possible will run forward during the throw. Explain why this increases the distance of the throw.

11. A passenger on a jet airplane claims to be able to walk at a speed in excess of 500 mph. Can this be true? Explain.

12. A roller coaster is designed with a ramp that reverses the direction of the cars. Cars start at the bottom of the ramp, roll up the ramp until stopping, then roll backward down the ramp. In what direction does the acceleration vector of the cars point
    a. as they move up the ramp?
    b. as they move down the ramp?
    c. at the highest point on the ramp?

13. During practice, two basketball players the same distance from the basket each shoot a ball at the same speed at the same time. Assuming their shots are accurate, can they avoid having the balls collide at the basket? If so, how? If not, why not? **Hint:** Are there two different launch angles that will have the same range but different times of flight?

14. A cyclist goes around a level, circular track at constant speed. Do you agree or disagree with the following statement: "Because the cyclist's speed is constant, her acceleration is zero." Explain.

15. You are driving your car in a circular path on level ground at a constant speed of 20 mph. At the instant you are driving north, and turning left, are you accelerating? If so, toward what point of the compass (N, S, E, W) does your acceleration vector point? If not, why not?

16. An airplane has been directed to fly in a clockwise circle, as seen from above, at constant speed until another plane has landed. When the plane is going north, is it accelerating? If so, in what direction does the acceleration vector point? If not, why not?

Multiple-Choice Questions

17. | The gas pedal in a car is sometimes referred to as "the accelerator." Which other controls on the vehicle can be used to produce acceleration?
    A. The brakes.   B. The steering wheel.
    C. The gear shift.   D. All of the above.

18. |||| A car travels at constant speed along the curved path shown from above in Figure Q3.18. Five possible vectors are also shown in the figure; the letter E represents the zero vector. Which vector best represents
    a. The car's *velocity* at position 1?
    b. The car's *acceleration* at point 1?
    c. The car's *velocity* at position 2?
    d. The car's *acceleration* at point 2?
    e. The car's *velocity* at position 3?
    f. The car's *acceleration* at point 3?

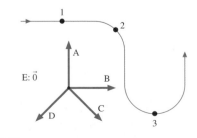

FIGURE Q3.18

19. ‖ A ball is fired from a cannon at point 1 and follows the trajectory shown in Figure Q3.19. Air resistance may be neglected. Five possible vectors are also shown in the figure; the letter E represents the zero vector. Which vector best represents
    a. The ball's *velocity* at position 2?
    b. The ball's *acceleration* at point 2?
    c. The ball's *velocity* at position 3?
    d. The ball's *acceleration* at point 3?

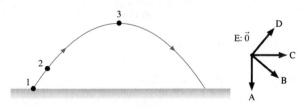

FIGURE Q3.19

20. ‖‖ A ball thrown at an initial angle of 37.0° and initial velocity of 23.0 m/s reaches a maximum height h, as shown in Figure Q3.20. With what initial speed must a ball be thrown *straight up* to reach the same maximum height h?
    A. 13.8 m/s
    B. 18.4 m/s
    C. 23.0 m/s
    D. 28.8 m/s
    E. 38.2 m/s

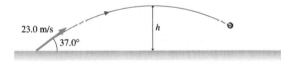

FIGURE Q3.20

21. ‖‖‖ A cannon, elevated at 40° is fired at a wall 300 m away on level ground, as shown in Figure Q3.21. The initial speed of the cannonball is 89 m/s.

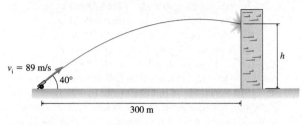

FIGURE Q3.21

   a. How long does it take for the ball to hit the wall?
      A. 1.3 s   B. 3.3 s   C. 4.4 s
      D. 6.8 s   E. 7.2 s
   b. At what height h does the ball hit the wall?
      A. 39 m   B. 47 m   C. 74 m
      D. 160 m   E. 210 m
22. ‖ A car drives horizontally off a 73-m-high cliff at a speed of 27 m/s. Ignore air resistance.
   a. How long will it take the car to hit the ground?
      A. 2.0 s   B. 3.2 s   C. 3.9 s
      D. 4.9 s   E. 5.0 s
   b. How far from the base of the cliff will the car hit?
      A. 74 m   B. 88 m   C. 100 m
      D. 170 m   E. 280 m
23. ‖‖‖ A football is kicked at an angle of 30° with a speed of 20 m/s. To the nearest second, how long will the ball stay in the air?
      A. 1 s   B. 2 s   C. 3 s   D. 4 s
24. ‖‖‖ A football is kicked at an angle of 30° with a speed of 20 m/s. To the nearest 5 m, how far will the ball travel?
      A. 15 m   B. 25 m   C. 35 m   D. 45 m
25. ‖ Riders on a Ferris wheel move in a circle with a speed of 4.0 m/s. As they go around, they experience a centripetal acceleration of 2.0 m/s². What is the diameter of this particular Ferris wheel?
      A. 4.0 m   B. 6.0 m   C. 8.0 m
      D. 16 m   E. 24 m

PROBLEMS

Section 3.1 Using Vectors

1. ‖ Trace the vectors in Figure P3.1 onto your paper. Then use graphical methods to draw the vectors (a) $\vec{A} + \vec{B}$ and (b) $\vec{A} - \vec{B}$.

FIGURE P3.1

2. ‖‖‖ Trace the vectors in Figure P3.2 onto your paper. Then use graphical methods to draw the vectors (a) $\vec{A} + \vec{B}$ and (b) $\vec{A} - \vec{B}$.

FIGURE P3.2

Section 3.2 Using Vectors on Motion Diagrams

3. ‖ a. Find the average acceleration vector at point 1 of the three-point motion diagram in Figure P3.3.
   b. Is the object's average speed between points 1 and 2 greater than, less than, or equal to its average speed between points 0 and 1? Explain how you can tell.

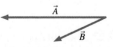

FIGURE P3.3

4. | a. Find the average acceleration vector at point 1 of the three-point motion diagram in Figure P3.4.

   b. Is the object's average speed between points 1 and 2 greater than, less than, or equal to its average speed between points 0 and 1? Explain how you can tell.

2 •

1 •

0 •

FIGURE P3.4

5. ||| Figure 3.11 showed the motion diagram for Anne as she rode a Ferris wheel that was turning at a constant speed. The inset to the figure showed how to find the acceleration vector at the lowest point in her motion. Use a similar analysis to find Anne's acceleration vector at the 12 o'clock, 4 o'clock, and 8 o'clock positions of the motion diagram. Use a ruler so that your analysis is accurate.

Section 3.3 Coordinate Systems and Vector Components

6. || A position vector in the first quadrant has an x-component of 6.0 m and a magnitude of 10 m. What is the value of its y-component?

7. ||| A velocity vector 40° above the positive x-axis has a y-component of 10 m/s. What is the value of its x-component?

8. || Jack and Jill ran up the hill at 3.0 m/s. The horizontal component of Jill's velocity vector was 2.5 m/s.

   a. What was the angle of the hill?

   b. What was the vertical component of Jill's velocity?

9. | A cannon tilted upward at 30° fires a cannonball with a speed of 100 m/s. At that instant, what is the component of the cannonball's velocity parallel to the ground?

10. || a. What are the x- and y-components of vector $\vec{E}$ in terms of the angle θ and the magnitude E shown in Figure P3.10?

    b. For the same vector, what are the x- and y-components in terms of the angle ϕ and the magnitude E?

FIGURE P3.10

11. | Draw each of the following vectors, then find its x- and y-components.

    a. $\vec{d} = (100 \text{ m}, 45° \text{ below } +x\text{-axis})$

    b. $\vec{v} = (300 \text{ m/s}, 20° \text{ above } +x\text{-axis})$

    c. $\vec{a} = (5.0 \text{ m/s}^2, -y\text{-direction})$

12. || Draw each of the following vectors, then find its x- and y-components.

    a. $\vec{d} = (2 \text{ km}, 30° \text{ left of } +y\text{-axis})$

    b. $\vec{v} = (5 \text{ cm/s}, -x\text{-direction})$

    c. $\vec{a} = (10 \text{ m/s}^2, 40° \text{ left of } -y\text{-axis})$

13. | Each of the following vectors is given in terms of its x- and y-components. Draw the vector, label an angle that specifies the vector's direction, then find the vector's magnitude and direction.

    a. $v_x = 20 \text{ m/s}, v_y = 40 \text{ m/s}$

    b. $a_x = 2.0 \text{ m/s}^2, a_y = -6.0 \text{ m/s}^2$

14. | Each of the following vectors is given in terms of its x- and y-components. Draw the vector, label an angle that specifies the vector's direction, then find the vector's magnitude and direction.

    a. $v_x = 10 \text{ m/s}, v_y = 30 \text{ m/s}$

    b. $a_x = 20 \text{ m/s}^2, a_y = 10 \text{ m/s}^2$

15. ||| While visiting England, you decide to take a jog and find yourself in the neighborhood shown on the map in

Figure P3.15. What is your displacement after running 2.0 km on Strawberry Fields, 1.0 km on Penny Lane, and 4.0 km on Abbey Road?

FIGURE P3.15

Section 3.4 Motion on a Ramp

16. ||| Suppose you begin sliding down a 15° ski slope. Ignoring friction and air resistance, how fast will you be moving after 10 s?

17. |||| A car traveling at 30 m/s runs out of gas while traveling up a 5.0° slope. How far will it coast before starting to roll back down?

18. || Santa loses his footing and slides down a frictionless, snowy roof that is tilted at an angle of 30°. If Santa slides 10 m before reaching the edge, what is his speed as he leaves the roof?

19. |||| A piano has been pushed to the top of the ramp at the back of a moving van. The workers think it is safe, but as they walk away, it begins to roll down the ramp. If the back of the truck is 1.0 m above the ground and the ramp is inclined at 20°, how much time do the workers have to get to the piano before it reaches the bottom of the ramp?

20. ||| Starting from rest, several toy cars roll down ramps of differing lengths and angles. Rank them according to their speed at the bottom of the ramp, from slowest to fastest. Car A goes down a 10 m ramp inclined at 15°, car B goes down a 10 m ramp inclined at 20°, car C goes down an 8.0 m ramp inclined at 20°, and car D goes down a 12 m ramp inclined at 12°.

Section 3.5 Relative Motion

21. | Anita is running to the right at 5 m/s, as shown in Figure P3.21. Balls 1 and 2 are thrown toward her at 10 m/s by friends standing on the ground. According to Anita, what is the speed of each ball?

FIGURE P3.21

22. | Anita is running to the right at 5 m/s, as shown in Figure P3.22. Balls 1 and 2 are thrown toward her by friends

standing on the ground. According to Anita, both balls are approaching her at 10 m/s. According to her friends, with what speeds were the balls thrown?

FIGURE P3.22

23. ⦀ A boat takes 3.0 h to travel 30 km down a river, then 5.0 h to return. How fast is the river flowing?

24. ⦀ Two children who are bored while waiting for their flight at the airport decide to race from one end of the 20-m-long moving sidewalk to the other and back. Phillippe runs on the sidewalk at 2.0 m/s (relative to the sidewalk). Renee runs on the floor at 2.0 m/s. The sidewalk moves at 1.5 m/s relative to the floor. Both make the turn instantly with no loss of speed.
    a. Who wins the race?
    b. By how much time does the winner win?

25. ⦀ An assembly line has a staple gun that rolls to the left at 1.0 m/s while parts to be stapled roll past it to the right at 3.0 m/s. The staple gun fires 10 staples per second. How far apart are the staples in the finished part?

Section 3.6 Motion in Two Dimensions: Projectile Motion

Section 3.7 Projectile Motion: Solving Problems

26. ‖ An object is launched with an initial velocity of 50.0 m/s at a launch angle of 36.9° above the horizontal.
    a. Make a table showing values of x, y, v_x, v_y, and the speed v every 1 s from $t = 0$ s to $t = 6$ s.
    b. Plot a graph of the object's trajectory during the first 6 s of motion.

27. ‖ A ball is thrown horizontally from a 20-m-high building with a speed of 5.0 m/s.
    a. Make a sketch of the ball's trajectory.
    b. Draw a graph of v_x, the horizontal velocity, as a function of time. Include units on both axes.
    c. Draw a graph of v_y, the vertical velocity, as a function of time. Include units on both axes.
    d. How far from the base of the building does the ball hit the ground?

28. | A ball with a horizontal speed of 1.25 m/s rolls off a bench 1.00 m above the floor.
    a. How long will it take the ball to hit the floor?
    b. How far from a point on the floor directly below the edge of the bench will the ball land?

29. ⦀ King Arthur's knights fire a cannon from the top of the castle wall. The cannonball is fired at a speed of 50 m/s and an angle of 30°. A cannonball that was accidentally dropped hits the moat below in 1.5 s. How far from the castle wall does the fired cannonball hit the ground?

30. | Two spheres are launched horizontally from a 1.0-m-high table. Sphere A is launched with an initial speed of 5.0 m/s. Sphere B is launched with an initial speed of 2.5 m/s.
    a. What are the times for each sphere to hit the floor?
    b. What are the distances that each travels from the edge of the table?

31. ‖ A rifle is aimed horizontally at a target 50 m away. The bullet hits the target 2.0 cm below the aim point.
    a. What was the bullet's flight time?
    b. What was the bullet's speed as it left the barrel?

32. ‖ A grey kangaroo can leap a distance of 10 m on horizontal
BIO ground. What is the minimum takeoff speed for such a leap?

33. ⦀ On the Apollo 14 mission to the moon, astronaut Alan Shepard hit a golf ball with a golf club improvised from a tool. The free-fall acceleration on the moon is 1/6 of its value on earth. Suppose he hit the ball with a speed of 25 m/s at an angle 30° above the horizontal.
    a. How long was the ball in flight?
    b. How far did it travel?
    c. How much farther would it travel on the moon than on earth?

Section 3.8 Motion in Two Dimensions: Circular Motion

34. | An old-fashioned LP record rotates at $33\frac{1}{3}$ rpm.
    a. What is its frequency in rev/s?
    b. What is its period, in seconds?

35. | An audio CD is 12.0 cm in diameter. As it rotates through an angle of 45°, how far does a point at its edge travel?

36. | To withstand "g-forces" of up to 10 g's, caused by sud-
BIO denly pulling out of a steep dive, fighter jet pilots train on a "human centrifuge." 10 g's is an acceleration of 98 m/s². If the length of the centrifuge arm is 12 m, at what speed is the rider moving when she experiences 10 g's?

37. | A particle rotates in a circle with centripetal acceleration $a = 8.0$ m/s². What is a if
    a. The radius is doubled without changing the particle's speed?
    b. The speed is doubled without changing the circle's radius?

38. ‖ A peregrine falcon in a tight, circular turn can attain a cen-
BIO tripetal acceleration 1.5 times the free-fall acceleration. If the falcon is flying at 20 m/s, what is the radius of the turn?

39. | Given the data in the chapter, what is the acceleration (magnitude and direction) of a passenger on the London Eye Ferris wheel?

General Problems

40. | Suppose $\vec{C} = \vec{A} + \vec{B}$ where vector $\vec{A}$ has components $A_x = 5$, $A_y = 2$ and vector $\vec{B}$ has components $B_x = -3$, $B_y = -5$.
    a. What are the x- and y-components of vector $\vec{C}$?
    b. Draw a coordinate system and on it show vectors $\vec{A}$, $\vec{B}$, and $\vec{C}$.
    c. What are the magnitude and direction of vector $\vec{C}$?

41. | Suppose $\vec{D} = \vec{A} - \vec{B}$ where vector $\vec{A}$ has components $A_x = 5$, $A_y = 2$ and vector $\vec{B}$ has components $B_x = -3$, $B_y = -5$.
    a. What are the x- and y-components of vector $\vec{D}$?
    b. Draw a coordinate system and on it show vectors $\vec{A}$, $\vec{B}$, and $\vec{D}$.
    c. What are the magnitude and direction of vector $\vec{D}$?

42. | Suppose $\vec{E} = 2\vec{A} + 3\vec{B}$ where vector $\vec{A}$ has components $A_x = 5$, $A_y = 2$ and vector $\vec{B}$ has components $B_x = -3$, $B_y = -5$.
    a. What are the x- and y-components of vector $\vec{E}$?
    b. Draw a coordinate system and on it show vectors $\vec{A}$, $\vec{B}$, and $\vec{E}$.
    c. What are the magnitude and direction of vector $\vec{E}$?

43. ‖ For the three vectors shown in Figure P3.43, the vector sum $\vec{D} = \vec{A} + \vec{B} + \vec{C}$ has components $D_x = 2$ and $D_y = 0$.
    a. What are the x- and y-components of vector $\vec{B}$?
    b. Write $\vec{B}$ as a magnitude and a direction.

FIGURE P3.43

44. ‖ Let $\vec{A} = (3.0 \text{ m}, 20° \text{ south of east})$, $\vec{B} = (2.0 \text{ m}, \text{north})$, and $\vec{C} = (5.0 \text{ m}, 70° \text{ south of west})$.
    a. Draw and label $\vec{A}$, $\vec{B}$, and $\vec{C}$ with their tails at the origin. Use a coordinate system with the x-axis to the east.
    b. Write the x- and y-components of vectors $\vec{A}$, $\vec{B}$, and $\vec{C}$.
    c. Find the magnitude and the direction of $\vec{D} = \vec{A} + \vec{B} + \vec{C}$.

45. ‖‖ While vacationing in the mountains you do some hiking. In the morning, your displacement is $\vec{S}_{\text{morning}} = (2000 \text{ m, east}) + (3000 \text{ m, north}) + (200 \text{ m, vertical})$. After lunch, your displacement is $\vec{S}_{\text{afternoon}} = (1500 \text{ m, west}) + (2000 \text{ m, north}) - (300 \text{ m, vertical})$.
    a. At the end of the hike, how much higher or lower are you compared to your starting point?
    b. What is your total displacement?

46. ‖‖ The minute hand on a watch is 2.0 cm long. What is the displacement vector of the tip of the minute hand
    a. From 8:00 to 8:20 A.M.?
    b. From 8:00 to 9:00 A.M.?

47. ‖‖‖ A field mouse trying to escape a hawk runs east for 5.0 m, darts southeast for 3.0 m, then drops 1.0 m down a hole into its burrow. What is the magnitude of the net displacement of the mouse?

48. ‖‖ A pilot in a small plane encounters shifting winds. He flies 26.0 km northeast, then 45.0 km due north. From this point, he flies an additional distance in an unknown direction, only to find himself at a small airstrip that his map shows to be 70.0 km directly north of his starting point. What were the length and direction of the third leg of his trip?

49. ‖‖ A contestant on "Survivor" is challenged to find buried treasure in a featureless desert. All the contestant knows is that she was driven away blindfolded from the treasure's location for 10 min with an average velocity vector having components 8.6 m/s west and 5.2 m/s north, then given a compass. She can run at a speed of 5.0 m/s. In what direction should she head and for how long should she run in order to retrieve the treasure?

50. ‖‖ The bacterium *Escherichia coli* (or *E. coli*) is a single-
BIO celled organism that lives in the gut of healthy humans and animals. When grown in a uniform medium rich in salts and amino acids, these bacteria swim along zig-zag paths at a constant speed of 20 μm/s. Figure P3.50 shows the trajectory of an *E. coli* as it moves from point A to point E. Each segment of the motion can be identified by two letters, such as segment BC.

a. For each of the four segments in the bacterium's trajectory, calculate the x and y components of its displacement and of its velocity.
b. Calculate both the total distance traveled and the magnitude of the net displacement for the entire motion.
c. What are the magnitude and the direction of the bacterium's average velocity for the entire trip?

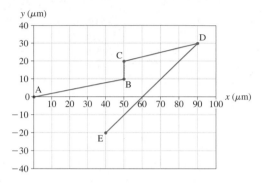

FIGURE P3.50

51. ‖‖‖ A skier is gliding along at 3.0 m/s on horizontal, frictionless snow. He suddenly starts down a 10° incline. His speed at the bottom is 15 m/s.
    a. What is the length of the incline?
    b. How long does it take him to reach the bottom?

52. ‖‖‖ A block slides along the frictionless track shown in Figure P3.52 with an initial speed of 5.0 m/s. Assume it turns all the corners smoothly, with no loss of speed.
    a. What is the block's speed as it goes over the top?
    b. What is its speed when it reaches the level track on the right side?
    c. By what percentage does the block's final speed differ from its initial speed? Is this surprising?

FIGURE P3.52

53. | One game at the amusement park has you push a puck up a long, frictionless ramp. You win a stuffed animal if the puck, at its highest point, comes to within 10 cm of the end of the ramp without going off. You give the puck a push, releasing it with a speed of 5.0 m/s when it is 8.5 m from the end of the ramp. The puck's speed after traveling 3.0 m is 4.0 m/s. Are you a winner?

54. ‖‖‖ When the moving sidewalk at the airport is broken, as it often seems to be, it takes you 50 s to walk from your gate to the baggage claim. When it is working and you stand on the moving sidewalk the entire way, without walking, it takes 75 s to travel the same distance. How long will it take you to travel from the gate to baggage claim if you walk while riding on the moving sidewalk?

55. ‖‖‖ Ships A and B leave port together. For the next two hours, ship A travels at 20 mph in a direction 30° west of north while ship B travels 20° east of north at 25 mph.
    a. What is the distance between the two ships two hours after they depart?
    b. What is the speed of ship A as seen by ship B?

56. ||| Mary needs to row her boat across a 100-m-wide river that is flowing to the east at a speed of 1.0 m/s. Mary can row the boat with a speed of 2.0 m/s relative to the water.
    a. If Mary rows straight north, how far downstream will she land?
    b. Draw a picture showing Mary's displacement due to rowing, her displacement due to the river's motion, and her net displacement.

57. ||| A flock of ducks is trying to migrate south for the winter, but they keep being blown off course by a wind blowing from the west at 12 m/s. A wise elder duck finally realizes that the solution is to fly at an angle to the wind. If the ducks can fly at 16 m/s relative to the air, in what direction should they head in order to move directly south?

58. || A kayaker needs to paddle north across a 100-m-wide harbor. The tide is going out, creating a tidal current that flows to the east at 2.0 m/s. The kayaker can paddle with a speed of 3.0 m/s.
    a. In which direction should he paddle in order to travel straight across the harbor?
    b. How long will it take him to cross?

59. ||| A sailboat is sailing due east at 8.0 mph. The wind appears to blow from the southwest at 12.0 mph.
    a. What are the true wind speed and direction?
    b. What are the true wind speed and direction if the wind appears to blow from the northeast at 12.0 mph?

60. ||||| A plane has an airspeed of 200 mph. The pilot wishes to reach a destination 600 mi due east, but a wind is blowing at 50 mph in the direction 30° north of east.
    a. In what direction must the pilot head the plane in order to reach her destination?
    b. How long will the trip take?

61. | A physics student on Planet Exidor throws a ball, and it follows the parabolic trajectory shown in Figure P3.61. The ball's position is shown at 1.0 s intervals until $t = 3.0$ s. At $t = 1.0$ s, the ball's velocity has components $v_x = 2.0$ m/s, $v_y = 2.0$ m/s.

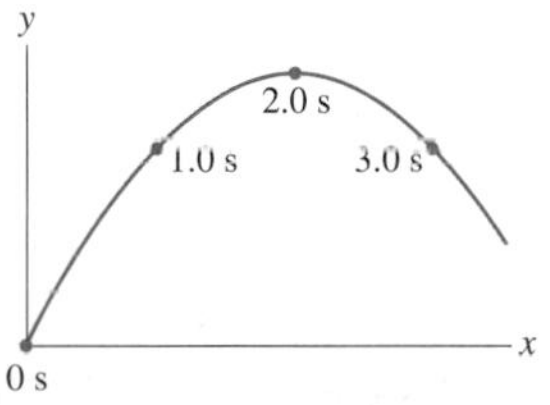

FIGURE P3.61

    a. Determine the x- and y-components of the ball's velocity at $t = 0.0$ s, 2.0 s, and 3.0 s.
    b. What is the value of g on Planet Exidor?
    c. What was the ball's launch angle?

62. || A ball thrown horizontally at 25 m/s travels a horizontal distance of 50 m before hitting the ground. From what height was the ball thrown?

63. ||| In 1780, in what is now
BIO referred to as "Brady's Leap," Captain Sam Brady of the U.S. Continental Army escaped certain death from his enemies by running over the edge of the cliff above Ohio's Cuyahoga River, which is confined at that spot to a gorge. He landed safely on the far side of the river. It was reported that he leapt 22 ft across while falling 20 ft.

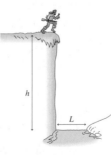

FIGURE P3.63

    a. Representing the horizontal distance jumped as L and the vertical drop as h, as shown in Figure P3.63, derive an expression for the minimum speed v he would need to make his leap if he ran straight off the cliff.
    b. Evaluate your expression for a 22 ft jump with a 20 ft drop to the other side.
    c. Is it reasonable that a person could make this leap? Use the fact that the world record for the 100 m dash is approximately 10 s to estimate the maximum speed such a runner would have.

64. ||||| A quarterback passes the football downfield at 20 m/s. It leaves his hand 1.8 m above the ground and is caught by a receiver 30 m away at the same height. What is the maximum height of the ball on its way to the receiver?

65. || A spring-loaded gun, fired vertically, shoots a marble 6.0 m straight up in the air. What is the marble's range if it is fired horizontally from 1.5 m above the ground?

66. || In a shot-put event, an athlete throws the shot with an initial speed of 12.0 m/s at a 40.0° angle from the horizontal. The shot leaves her hand at a height of 1.80 m above the ground.
    a. How far does the shot travel?
    b. Repeat the calculation of part (a) for angles 42.5°, 45.0°, and 47.5°. Put all your results, including 40.0°, in a table. At what angle of release does she throw the farthest?

67. || At what launch angle or angles will a projectile land at half of its maximum possible range on flat ground?

68. ||||| A tennis player hits a ball 2.0 m above the ground. The ball leaves his racquet with a speed of 20 m/s at an angle 5.0° above the horizontal. The horizontal distance to the net is 7.0 m, and the net is 1.0 m high. Does the ball clear the net? If so, by how much? If not, by how much does it miss?

69. | Your baseball team is up by one run in the bottom of the last inning of the game when a ground ball slips through the infield and comes straight toward you, the right fielder. As you pick up the ball 65 m from home plate, you see a runner rounding third base and heading for home. You throw the ball at an angle of 30° above the horizontal with just the right speed so that the ball is caught by the catcher, standing on home plate, at the same height as you threw it. As you release the ball, the runner is 20 m from home plate and running at 8.0 m/s. Will the ball arrive in time for the catcher to make the tag?

70. || A supply plane needs to drop a package of food to scientists working on a glacier in Greenland. The plane flies 100 m above the glacier at a speed of 150 m/s. How far short of the target should it drop the package?

71. || A child slides down a frictionless 3.0-m-long playground slide tilted upward at an angle of 40°. At the end of the slide, there is an additional section that curves so that the child is launched off the end of the slide horizontally.
    a. How fast is the child moving at the bottom of the slide?
    b. If the end of the slide is 0.40 m above the ground, how far from the end does she land?

72. || A sports car is advertised to be able to "reach 60 mph in 5 seconds flat, corner at 0.85g, and stop from 70 mph in only 168 feet."
    a. In which of those three situations is the magnitude of the car's acceleration the largest? In which is it the smallest?
    b. At 60 mph, what is the smallest turning radius that this car can navigate?

In Problems 73 and 74 you are given the equations that are used to solve a problem. For each of these, you are to

a. Write a realistic problem for which these are the correct equations. Be sure that the answer your problem requests is consistent with the equations given.

b. Finish the solution of the problem.

73. ‖‖ $100 \text{ m} = 0 \text{ m} + (50\cos\theta \text{ m/s})t_1$
    $0 \text{ m} = 0 \text{ m} + (50\sin\theta \text{ m/s})t_1 - \frac{1}{2}(9.80 \text{ m/s}^2)t_1^2$

74. ‖‖ $x_1 = 0 \text{ m} + (30 \text{ m/s})t_1$
    $0 \text{ m} = 300 \text{ m} - \frac{1}{2}(9.8 \text{ m/s}^2)t_1^2$

Passage Problems

Riding the Water Slide

A rider on a water slide goes through three different kinds of motion, as illustrated in Figure P3.75. Use the data and details from the figure to answer the following questions.

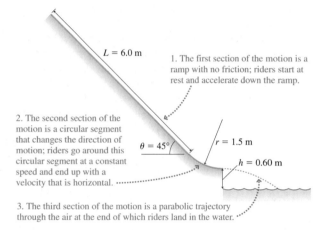

L = 6.0 m

1. The first section of the motion is a ramp with no friction; riders start at rest and accelerate down the ramp.

2. The second section of the motion is a circular segment that changes the direction of motion; riders go around this circular segment at a constant speed and end up with a velocity that is horizontal.

$\theta = 45°$ *r* = 1.5 m

h = 0.60 m

3. The third section of the motion is a parabolic trajectory through the air at the end of which riders land in the water.

FIGURE P3.75

75. | At the end of the first section of the motion, riders are moving at what approximate speed?
    A. 3 m/s   B. 6 m/s   C. 9 m/s   D. 12 m/s

76. | Suppose the acceleration during the second section of the motion is too large to be comfortable for riders. What change could be made to decrease the acceleration during this section?
    A. Reduce the radius of the circular segment.
    B. Increase the radius of the circular segment.
    C. Increase the angle of the ramp.
    D. Increase the length of the ramp.

77. ‖ What is the vertical component of the velocity of a rider as he or she hits the water?
    A. 2.4 m/s   B. 3.4 m/s   C. 5.2 m/s   D. 9.1 m/s

78. ‖ Suppose the designers of the water slide want to adjust the height *h* above the water so that riders land twice as far away from the bottom of the slide. What would be the necessary height above the water?
    A. 1.2 m   B. 1.8 m   C. 2.4 m   D. 3.0 m

79. ‖ During which section of the motion is the magnitude of the acceleration experienced by a rider the greatest?
    A. The first.   B. The second.
    C. The third.   D. It is the same in all sections.

STOP TO THINK ANSWERS

Stop to Think 3.1: A. The graphical construction of $2\vec{P} - \vec{Q}$ is shown at right.

Stop to Think 3.2: From the axes on the graph, $2\vec{P} - \vec{Q}$ we can see that the *x*- and *y*-components are -4 cm and $+2$ cm, respectively.

$-\vec{Q}$

$2\vec{P}$

Stop to Think 3.3: C. Vector $\vec{C}$ points to the left and down, so both C_x and C_y are negative. C_x is in the numerator because it is the side opposite ϕ.

Stop to Think 3.4: B. The angle of the slope is greatest in this case, leading to the greatest acceleration.

Stop to Think 3.5: D. Mass does not appear in the kinematic equations, so the mass has no effect; the balls will follow the same path.

Stop to Think 3.6: B. The magnitude of the acceleration is v^2/r; this is the case with the largest speed and smallest radius.

4 FORCES AND NEWTON'S LAWS OF MOTION

These ice boats sail across the ice at great speeds. What gets the boats moving in the first place? What keeps them from going even faster?

Looking Ahead ▶▶

The goal of Chapter 4 is to establish a connection between force and motion. In this chapter you will learn to:

▶ Recognize what a force is and is not.
▶ Identify the specific forces acting on an object.
▶ Draw free-body diagrams.
▶ Understand the connection between force and motion.

Looking Back ◀◀

To master the material introduced in this chapter, you must understand how acceleration is determined and how vectors are used. Please review:

◀ Section 1.5 Vectors and motion.
◀ Section 2.4 Acceleration.
◀ Sections 3.2 and 3.3 Vectors, coordinate systems, and vector components.

These ice sailers fly across the frozen lake at some 60 mph. Pretty amazing! We could use kinematics to describe a boat's motion with pictures, graphs, and equations. By defining position, velocity, and acceleration and dressing them in mathematical clothing, kinematics provides a language to describe *how* something moves. But kinematics would tell us nothing about *why* the boat accelerates briskly before reaching a top speed. For the more fundamental task of understanding the *cause* of motion, we turn our attention to **dynamics.** Dynamics joins with kinematics to form **mechanics,** the general science of motion. We study dynamics qualitatively in this chapter, then develop it quantitatively in the next four chapters.

The theory of mechanics originated in the mid-1600s when Sir Isaac Newton formulated his laws of motion. These fundamental principles of mechanics explain how motion occurs as a consequence of forces. Newton's laws are more than 300 years old, but they still form the basis for our contemporary understanding of motion.

A challenge in learning physics is that a textbook is not an experiment. The book can assert that an experiment will have a certain outcome, but you may not be convinced unless you see or do the experiment yourself. Newton's laws are frequently contrary to our intuition, and a lack of familiarity with the evidence for Newton's laws is a source of difficulty for many people. You should have an opportunity through lecture demonstrations and in the laboratory to see for yourself the evidence supporting Newton's laws. Physics is not an arbitrary collection of definitions and formulas, but a consistent theory as to how the universe really works. It is only with experience and evidence that we learn to separate physical fact from fantasy.

4.1 What Causes Motion?

As we remarked in Chapter 1, Aristotle (384–322 BC) and his contemporaries in the world of ancient Greece were very interested in motion. One question they asked was: What is the "natural state" of an object if left to itself? It does not take an expensive research program to see that every moving object on earth, if left to itself, eventually comes to rest. You have certainly seen countless examples of this: You must push a shopping cart to keep it rolling, but when you stop pushing, the cart soon comes to rest; a boulder bounds downhill and then tumbles to a halt. Having observed many such examples himself, Aristotle concluded that the natural state of an earthly object is to be *at rest*. An object at rest requires no explanation; it is doing precisely what comes naturally to it. We'll soon see, however, that this simple viewpoint is *incomplete*.

Aristotle further pondered moving objects. A moving object is *not* in its natural state and thus requires an explanation: Why is this object moving? What keeps it going and prevents it from being in its natural state? When a puck is sliding across the ice, what keeps it going? Why does an arrow fly through the air once it is no longer being pushed by the bowstring? Although these questions seem like reasonable ones to pose, it was Galileo who first showed that the questions being asked were, in fact, the wrong ones.

Galileo reopened the question of the "natural state" of objects. He suggested focusing on the *idealized case* in which resistance to the motion (e.g., friction or air resistance) is zero. He performed many experiments to study motion. Let's imagine a modern experiment of this kind, as shown in Figure 4.1.

The rocks in this rockslide quickly came to rest. Is this the "natural state" of objects?

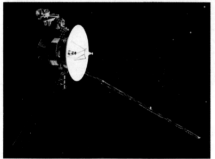

Interstellar coasting A nearly perfect example of Newton's first law is the pair of Voyager space probes launched in 1977. Both spacecraft long ago ran out of fuel and are now coasting through the frictionless vacuum of space. Although not entirely free of influence from the sun's gravity, they are now so far from the sun and other stars that gravitational influences are very nearly zero. Thus, according to the first law, they will continue at their current speed of about 40,000 miles per hour essentially forever. Billions of years from now, long after our solar system is dead, the Voyagers will still be drifting through the stars.

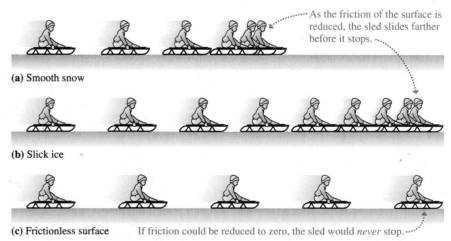

As the friction of the surface is reduced, the sled slides farther before it stops.

(a) Smooth snow

(b) Slick ice

(c) Frictionless surface   If friction could be reduced to zero, the sled would *never* stop.

FIGURE 4.1 Sleds sliding on increasingly smooth surfaces.

Tyler slides down a hill on his sled, then out onto a horizontal patch of smooth snow, which is shown in Figure 4.1a. Even if the snow is quite smooth, the friction between the sled and the snow will soon cause the sled to come to rest. What if Tyler slides down the hill onto some very slick ice, as in Figure 4.1b? This gives very low friction, and the sled could slide for quite a distance before stopping. Galileo's genius was to imagine the case where *all* sources of friction, air resistance, and other retarding influences were removed, as for the sled in Figure 4.1c sliding on imaginary *frictionless* ice. We can imagine in that case that the sled, once started in its motion, would continue in its motion *forever,* moving in a straight line with no loss of speed. In other words, **the natural state of an object—its behavior if free of external influences—is** *uniform motion* **with constant velocity!** Further, "at rest" has no special significance in Galileo's view of motion; it is simply uniform motion that happens to have a velocity of zero. This implies that an object at rest, in the absence of external influences, will remain at rest forever.

Galileo's ideas were completely counter to those of the ancient Greeks. We no longer need to explain why a sled continues to slide across the ice; that motion is its "natural" state. What needs explanation, in this new viewpoint, is why objects *don't* continue in uniform motion. Why does a sliding puck eventually slow to a stop? Why does a stone, thrown upward, slow and eventually fall back down? Galileo's new viewpoint was that the stone and the puck are *not* free of "influences": the stone is somehow pulled toward the earth, and some sort of retarding influence acted to slow the sled down. Today, we call such influences that lead to deviations from uniform motion **forces.**

Galileo's experiments were limited to motion along horizontal surfaces. It was left to Newton to generalize Galileo's conclusions, and today we call it Newton's first law of motion.

> **Newton's first law** Consider an object with no force acting on it. If it is at rest, it will remain at rest; if it is moving, it will continue to move in a straight line at a constant speed.

As an important application of Newton's first law, consider the crash test of Figure 4.2. The car contacts the wall and begins to slow. The wall is an external influence—a force—that alters the car's uniform motion. But the wall is a force on the *car,* not on the dummy. In accordance with Newton's first law, the unbelted dummy continues to move straight ahead at his original speed. Only when he collides violently with the dashboard of the stopped car is there a force acting to halt the dummy's uniform motion. If he had been wearing a seatbelt, the influence (i.e., the force) of the seatbelt would have slowed the dummy at the much lower rate with which the car slows down. We'll study the forces of collisions in detail in Chapter 10.

4.2 Force

Newton's first law tells us that an object in motion subject to no forces will continue to move in a straight line forever. But this law does not explain in any detail exactly what a force *is.* Unfortunately, there is no simple one-sentence definition of force. The concept of force is best introduced by looking at examples of some common forces and considering the basic properties shared by all forces. This will be our task in the next two sections. Let's begin by examining the properties that all forces have in common, as outlined in the table on the next page.

TRY IT YOURSELF

Getting the ketchup out The ketchup stuck at the bottom of the bottle is initially at rest. If you hit the bottom of the bottle, the bottle suddenly moves down, taking the ketchup on the bottom of the bottle with it, so that the ketchup just stays stuck to the bottom. But if instead you hit *up* on the bottle, as shown, you force the bottle rapidly upward. By the first law, the ketchup that was stuck to the bottom stays at rest, so it separates from the upward-moving bottle: the ketchup has moved forward with respect to the bottle!

At the instant of impact, the car and driver are moving at the same speed;

The car slows as it hits, but the driver continues at the same speed . . .

. . . until he hits the now-stationary dashboard. Ouch!

FIGURE 4.2 Newton's first law tells us: "Wear your seatbelts!"

What is a force?

A force is a push or a pull.

Our commonsense idea of a **force** is that it is a *push* or a *pull.* We will refine this idea as we go along, but it is an adequate starting point. Notice our careful choice of words: We refer to "*a* force," rather than simply "force." We want to think of a force as a very specific *action,* so that we can talk about a single force or perhaps about two or three individual forces that we can clearly distinguish. Hence the concrete idea of "a force" acting on an object.

A force acts on an object.

Implicit in our concept of force is that **a force acts on an object.** In other words, pushes and pulls are applied *to* something—an object. From the object's perspective, it has a force *exerted* on it. Forces do not exist in isolation from the object that experiences them.

A force requires an agent.

Every force has an **agent,** something that acts or pushes or pulls. That is, a force has a specific, identifiable *cause.* As you throw a ball, it is your hand, while in contact with the ball, that is the agent or the cause of the force exerted on the ball. *If* a force is being exerted on an object, you must be able to identify a specific cause (i.e., the agent) of that force. Conversely, a force is not exerted on an object *unless* you can identify a specific cause or agent. Note that an agent can be an inert object such as a tabletop or a wall. Such agents are the cause of many common forces.

A force is a vector.

If you push an object, you can push either gently or very hard. Similarly, you can push either left or right, up or down. To quantify a push, we need to specify both a magnitude *and* a direction. It should thus come as no surprise that a force is a vector quantity. The symbol for a force is a vector symbol such as $\vec{F}$, $\vec{w}$, or $\vec{T}$. The size or strength of such a force is its magnitude F (or w or T).

A force can be either a contact force ...

There are two basic classes of forces, depending on whether the agent touches the object or not. **Contact forces** are forces that act on an object by touching it at a point of contact. The bat must touch the ball to hit it. A string must be tied to an object to pull it. The majority of forces that we will examine are contact forces.

... or a long-range force.

Long-range forces are forces that act on an object without physical contact. Magnetism is an example of a long-range force. You have undoubtedly held a magnet over a paper clip and seen the paper clip leap up to the magnet. A coffee cup released from your hand is pulled to the earth by the long-range force of gravity.

Let's summarize these ideas as our definition of force:

- A force is a push or a pull on an object.
- A force is a vector. It has both a magnitude and a direction.
- A force requires an agent. Something does the pushing or pulling. The agent can be an inert object such as a tabletop or a wall.
- A force is either a contact force or a long-range force. Gravity is the only long-range force we will deal with until much later in the book.

There's one more important aspect of forces. If you push against a door (the object) to close it, the door pushes back against your hand (the agent). If a tow rope pulls on a car (the object), the car pulls back on the rope (the agent). In general, if an agent exerts a force on an object, the object exerts a force on the agent. We really need to think of a force as an *interaction* between two objects. Although the interaction perspective is a more exact way to view forces, it adds complications that we would like to avoid for now. Our approach will be to start by focusing on how a single object responds to forces exerted on it. Later in this chapter, we'll return to the larger issue of how two or more objects interact with each other.

Force Vectors

We can use a simple diagram to visualize how forces are exerted on objects. Because we are using the particle model, in which objects are treated as points, the process of drawing a force vector is straightforward. Here is how it goes:

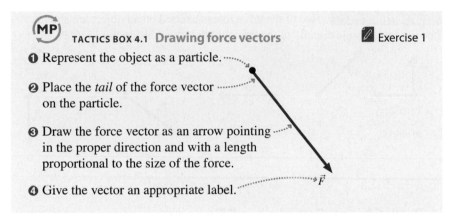

(MP)   TACTICS BOX 4.1   **Drawing force vectors**   ✏ Exercise 1

❶ Represent the object as a particle.

❷ Place the *tail* of the force vector on the particle.

❸ Draw the force vector as an arrow pointing in the proper direction and with a length proportional to the size of the force.

❹ Give the vector an appropriate label. $\vec{F}$

Step 2 may seem contrary to what a "push" should do (it may look as if the force arrow is *pulling* the object rather than *pushing* it), but recall that moving a vector does not change it as long as the length and angle do not change. The vector $\vec{F}$ is the same regardless of whether the tail or the tip is placed on the particle. Our reason for using the tail will become clear when we consider how to combine several forces.

Figure 4.3 shows three examples of force vectors. One is a pull, one a push, and one a long-range force, but in all three the *tail* of the force vector is placed on the particle representing the object.

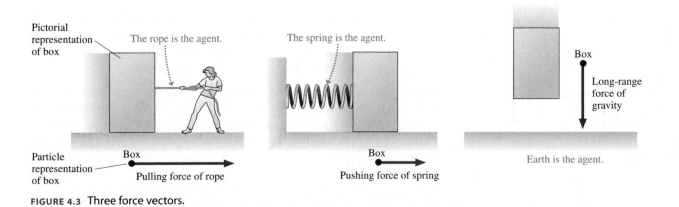

Pictorial representation of box

The rope is the agent.

The spring is the agent.

Box

Long-range force of gravity

Earth is the agent.

Particle representation of box

Box
Pulling force of rope

Box
Pushing force of spring

FIGURE 4.3 Three force vectors.

(a)

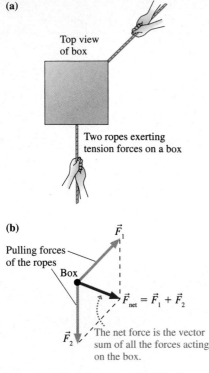

Top view
of box

Two ropes exerting
tension forces on a box

(b)

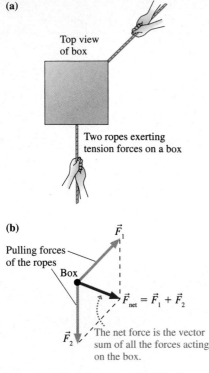

Pulling forces
of the ropes

Box

$\vec{F}_1$

$\vec{F}_{net} = \vec{F}_1 + \vec{F}_2$

$\vec{F}_2$

The net force is the vector
sum of all the forces acting
on the box.

FIGURE 4.4 Two forces applied to a box.

Combining Forces

Figure 4.4a shows a top view of a box being pulled by two ropes, each exerting a force on the box. How will the box respond? Experimentally, we find that when several forces $\vec{F}_1$, $\vec{F}_2$, $\vec{F}_3$, . . . are exerted on an object, they combine to form a **net force** that is the *vector* sum of all the forces:

$$\vec{F}_{net} = \vec{F}_1 + \vec{F}_2 + \vec{F}_3 \cdots \qquad (4.1)$$

Mathematically, this summation is called a *superposition* of forces. The net force is sometimes called the *resultant force*. Figure 4.4b shows the net force on the box.

> **NOTE** ▶ It is important to realize that the net force $\vec{F}_{net}$ is not a new force acting *in addition* to the original forces $\vec{F}_1$, $\vec{F}_2$, $\vec{F}_3$ . . . Instead, we should think of the original forces being *replaced* by $\vec{F}_{net}$. ◀

STOP TO THINK 4.1   Two of the three forces exerted on an object are shown. The net force points directly to the left. Which is the missing third force?

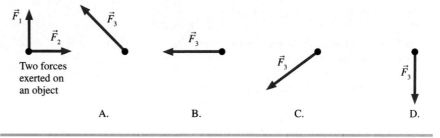

$\vec{F}_1$

$\vec{F}_2$

$\vec{F}_3$

Two forces
exerted on
an object

$\vec{F}_3$

$\vec{F}_3$

$\vec{F}_3$

A.   B.   C.   D.

4.3 A Short Catalog of Forces

There are many forces we will deal with over and over. This section will introduce you to some of them. Many of these forces have special symbols. As you learn the major forces, be sure to learn the symbol for each.

Weight

A falling rock is pulled toward the earth by the long-range force of gravity. Gravity is what keeps you in your chair, keeps the planets in their orbits around the sun, and shapes the large-scale structure of the universe. We'll have a thorough look at gravity in Chapter 6. For now we'll concentrate on objects on or near the surface of the earth (or other planet).

The gravitational pull of the earth on an object on or near the surface of the earth is called **weight**. The symbol for weight is $\vec{w}$. Weight is the only long-range force we will encounter in the next few chapters. The agent for the weight force is the *entire earth* pulling on an object. The weight force is in some ways the simplest force we'll study. As Figure 4.5 shows, **an object's weight vector always points vertically downward,** no matter how the object is moving.

> **NOTE** ▶ We often refer to "the weight" of an object. This is an informal expression for w, the magnitude of the weight force exerted on the object. Note that **weight is not the same thing as mass.** We will briefly examine mass later in the chapter and explore the connection between weight and mass in Chapter 5. ◀

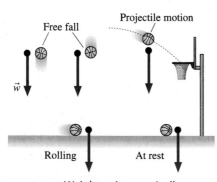

Free fall

Projectile motion

$\vec{w}$

Rolling   At rest

FIGURE 4.5 Weight points vertically downward.

Spring Force

Springs exert one of the most basic contact forces. A spring can either push (when compressed) or pull (when stretched). Figure 4.6 shows the spring force. In both cases, pushing and pulling, the tail of the force vector is placed on the particle in the force diagram. There is no special symbol for a spring force, so we simply use a subscript label: $\vec{F}_{sp}$.

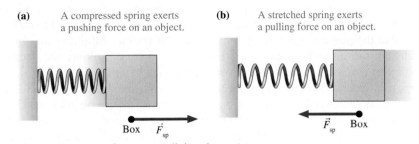

(a) A compressed spring exerts a pushing force on an object.

(b) A stretched spring exerts a pulling force on an object.

Box $\vec{F}_{sp}$ $\vec{F}_{sp}$ Box

FIGURE 4.6 The spring force is parallel to the spring.

Springs come in many forms. When deflected, they push or pull with a spring force.

Although you may think of a spring as a metal coil that can be stretched or compressed, this is only one type of spring. Hold a ruler, or any other thin piece of wood or metal, by the ends and bend it slightly. It flexes. When you let go, it "springs" back to its original shape. This is just as much a spring as is a metal coil.

Tension Force

When a string or rope or wire pulls on an object, it exerts a contact force that we call the **tension force,** represented by a capital $\vec{T}$. **The direction of the tension force is always in the direction of the string or rope,** as you can see in Figure 4.7. When we speak of "the tension" in a string, this is an informal expression for T, the size or magnitude of the tension force. Note that the tension force can only *pull* in the direction of the string; if you try to *push* with a string, it will go slack and be unable to exert a force.

We can think about the tension force using a microscopic picture. If you were to use a very powerful microscope to look inside a rope, you would "see" that it is made of *atoms* joined together by *molecular bonds*. Molecular bonds are not rigid connections between the atoms. They are more accurately thought of as tiny *springs* holding the atoms together, as in Figure 4.8. Pulling on the ends of a string or rope stretches the molecular springs ever so slightly. The tension within a rope and the tension experienced by an object at the end of the rope are really the net spring force being exerted by billions and billions of microscopic springs.

This atomic-level view of tension introduces a new idea: a microscopic **atomic model** for understanding the behavior and properties of **macroscopic** (i.e., containing many atoms) objects. We will frequently use atomic models to obtain a deeper understanding of our observations.

The atomic model of tension also helps to explain one of the basic properties of ropes and strings. When you pull on a rope tied to a heavy box, the rope in turn exerts a tension force on the box. If you pull harder, the tension force on the box becomes greater. How does the box "know" that you are pulling harder on the other end of the rope? According to our atomic model, when you pull harder on the rope its microscopic springs stretch a bit more, increasing the spring force they exert on each other—and on the box they're attached to.

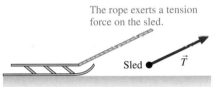

The rope exerts a tension force on the sled.

Sled $\vec{T}$

FIGURE 4.7 Tension is parallel to the rope.

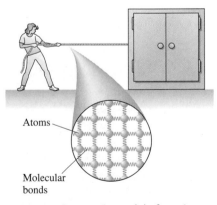

Atoms

Molecular bonds

FIGURE 4.8 An atomic model of tension.

Normal Force

If you sit on a bed, the springs in the mattress compress and, as a consequence of the compression, exert an upward force on you. Stiffer springs would show less

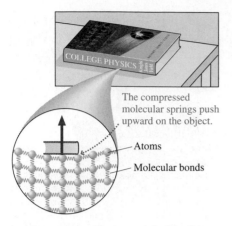

FIGURE 4.9 An atomic model of the force exerted by a table.

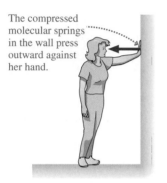

The compressed molecular springs in the wall press outward against her hand.

FIGURE 4.10 The wall pushes outward against your hand.

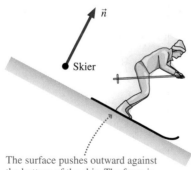

The surface pushes outward against the bottom of the skis. The force is perpendicular to the surface.

FIGURE 4.11 The normal force is perpendicular to the surface.

compression but would still exert an upward force. The compression of extremely stiff springs might be measurable only by sensitive instruments. Nonetheless, the springs would compress ever so slightly and exert an upward spring force on you.

Figure 4.9 shows a book resting on top of a sturdy table. The table may not visibly flex or sag, but—just as you do to the bed—the book compresses the molecular springs in the table. The size of the compression is very small, but it is not zero. As a consequence, the compressed molecular springs *push upward* on the book. We say that "the table" exerts the upward force, but it is important to understand that the pushing is *really* done by molecular springs. Similarly, an object resting on the ground compresses the molecular springs holding the ground together and, as a consequence, the ground pushes up on the object.

We can extend this idea. Suppose you place your hand on a wall and lean against it, as shown in Figure 4.10. Does the wall exert a force on your hand? As you lean, you compress the molecular springs in the wall and, as a consequence, they push outward *against* your hand. So the answer is "yes," the wall does exert a force on you. It's not hard to see this if you examine your hand as you lean: you can see that your hand is slightly deformed, and becomes more so the harder you lean. This deformation is direct evidence of the force that the wall exerts on your hand. Consider also what would happen if the wall suddenly vanished. Without the wall there to push against you, you would topple forward.

The force the table surface exerts is vertical, while the force the wall exerts is horizontal. In all cases, the force exerted on an object that is pressing against a surface is in a direction *perpendicular* to the surface. Mathematicians refer to a line that is perpendicular to a surface as being *normal* to the surface. In keeping with this terminology, we define the **normal force** as the force exerted by a surface (the agent) against an object that is pressing against the surface. The symbol for the normal force is $\vec{n}$.

We're not using the word *normal* to imply that the force is an "ordinary" force or to distinguish it from an "abnormal force." A surface exerts a force *perpendicular* (i.e., normal) to itself as the molecular springs press *outward*. Figure 4.11 shows an object on an inclined surface, a common situation. Notice how the normal force $\vec{n}$ is perpendicular to the surface.

We have spent a lot of time describing the normal force because many people have a difficult time understanding it. The normal force is a very real force arising from the very real compression of molecular bonds. It is in essence just a spring force, but one exerted by a vast number of microscopic springs acting at once. The normal force is responsible for the "solidness" of solids. It is what prevents you from passing right through the chair you are sitting in and what causes the pain and the lump if you bang your head into a door. Your head can then tell you that the force exerted on it by the door was very real!

Friction

You've certainly observed that a rolling or sliding object, if not pushed or propelled, slows down and eventually stops. You've probably discovered that you can slide better across a sheet of ice than across asphalt. And you also know that most objects stay in place on a table without sliding off even if the table is tilted a bit. The force responsible for these sorts of behavior is **friction.** The symbol for friction is a lower case $\vec{f}$.

Friction, like the normal force, is exerted by a surface. Unlike the normal force, however, **the frictional force is always *parallel* to the surface,** not perpendicular to it. (In many cases, a surface will exert *both* a normal and a frictional force.) On a microscopic level, friction arises as atoms from the object and atoms on the surface run into each other. The rougher the surface is, the more these atoms are forced into close proximity and, as a result, the larger the friction force. We will develop a simple model of friction in the next chapter that will be sufficient for our needs. For now, it is useful to distinguish between two kinds of friction:

- *Kinetic friction,* denoted $\vec{f}_k$, appears as an object slides across a surface. Kinetic friction is a force that always "opposes the motion," meaning that the friction force $\vec{f}_k$ on a sliding object points in a direction opposite to the direction of the object's motion.
- *Static friction,* denoted $\vec{f}_s$, is the force that keeps an object "stuck" on a surface and prevents its motion. Finding the direction of $\vec{f}_s$ is a little trickier than finding it for $\vec{f}_k$. Static friction points opposite the direction in which the object *would* move if there were no friction. That is, it points in the direction necessary to *prevent* motion.

Figure 4.12 shows examples of kinetic and static friction.

It's a drag At the high speeds attained by racing cyclists, air drag can become very significant. The world record for the longest distance traveled in one hour on an ordinary bicycle is 56.38 km, set by Chris Boardman in 1996. But a bicycle with an aerodynamic shell has a much lower drag force, allowing it to attain significantly higher speeds. The bike shown here was pedaled 84.22 km in one hour by Sam Whittingham in 2004, for an amazing average speed of 52.3 mph!

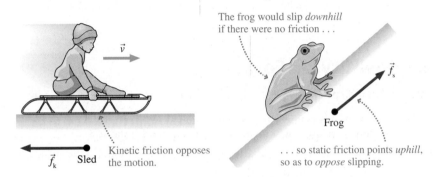

The frog would slip *downhill* if there were no friction . . .

Kinetic friction opposes the motion.

. . . so static friction points *uphill,* so as to *oppose* slipping.

FIGURE 4.12 Kinetic and static friction are parallel to the surface.

Drag

Friction at a surface is one example of a *resistive force,* a force that opposes or resists motion. Resistive forces are also experienced by objects moving through *fluids*—gases (like air) and liquids (like water). This kind of resistive force—the force of a fluid on a moving object—is called **drag** and is symbolized as $\vec{D}$. Like kinetic friction, **drag points opposite to the direction of motion.** Figure 4.13 shows an example of drag.

Drag can be a large force for objects moving at high speeds or in dense fluids. Hold your arm out the window as you ride in a car and feel how hard the air pushes against your arm; note also how the air resistance against your arm increases rapidly as the car's speed increases. Drop a lightweight bead into a beaker of water and watch how slowly it settles to the bottom. The drag force of the water on the bead is very significant.

On the other hand, for objects that are heavy and compact, that move in air, and whose speed is not too great, the drag force of air resistance is fairly small. To keep things as simple as possible, **you can neglect air resistance in all problems unless a problem explicitly asks you to include it.** The error introduced into calculations by this approximation is generally pretty small. Later, when we consider objects moving in liquids, we'll find that drag will be a very significant force that we'll *have* to include.

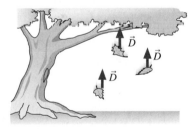

Air resistance is a significant force on falling leaves. It points opposite the direction of motion.

FIGURE 4.13 Air resistance is an example of drag.

Thrust

A jet airplane obviously has a force that propels it forward during takeoff. Likewise for the rocket being launched in Figure 4.14. This force, called **thrust,** occurs when a jet or rocket engine expels gas molecules at high speed. Thrust is a contact force, with the exhaust gas being the agent that pushes on the engine. The process by which thrust is generated is rather subtle, and requires an appreciation of Newton's third law, introduced later in this chapter. For now, we need only

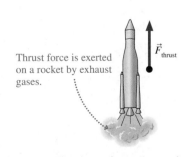

Thrust force is exerted on a rocket by exhaust gases.

FIGURE 4.14 The thrust force on a rocket is opposite to the direction of the expelled gases.

consider that **thrust is a force opposite to the direction in which the exhaust gas is expelled.** There's no special symbol for thrust, so we will call it $\vec{F}_{\text{thrust}}$.

Electric and Magnetic Forces

Electricity and magnetism, like gravity, exert long-range forces. The forces of electricity and magnetism act on charged particles. We will study electric and magnetic forces in detail in Part VI of this book. For now, it is worth noting that the forces holding molecules together—the molecular bonds—are not actually tiny springs. Atoms and molecules are made of charged particles—electrons and protons—and what we call a molecular bond is really an electric force between these particles. So when we say that the normal force and the tension force are due to "molecular springs," or that friction is due to atoms running into each other, what we're really saying is that these forces, at the most fundamental level, are actually electric forces between the charged particles in the atoms.

TABLE 4.1 Common forces and their notation

Force	Notation
General force	$\vec{F}$
Weight	$\vec{w}$
Spring force	$\vec{F}_{\text{sp}}$
Tension	$\vec{T}$
Normal force	$\vec{n}$
Static friction	$\vec{f}_{\text{s}}$
Kinetic friction	$\vec{f}_{\text{k}}$
Drag	$\vec{D}$
Thrust	$\vec{F}_{\text{thrust}}$

4.4 Identifying Forces

Force and motion problems generally have two basic steps:

1. Identify all of the forces acting on an object.
2. Use Newton's laws and kinematics to determine the motion.

Understanding the first step is the primary goal of this chapter. We'll turn our attention to step 2 in the next chapter.

A typical physics problem describes an object that is being pushed and pulled in various directions. Some forces are given explicitly, while others are only implied. In order to proceed, it is necessary to determine all the forces that act on the object. It is also necessary to avoid including forces that do not really exist. Now that you have learned the properties of forces and seen a catalog of typical forces, we can develop a step-by-step method for identifying each force in a problem. A summary of the most common forces we'll come across in the next few chapters is given in Table 4.1.

TACTICS BOX 4.2 Identifying forces   ✐ Exercises 4–8

❶ **Identify "the system" and "the environment."** The system is the object whose motion you wish to study; the environment is everything else.

❷ **Draw a picture of the situation.** Show the object—the system—and everything in the environment that touches the system. Ropes, springs, and surfaces are all parts of the environment.

❸ **Draw a closed curve around the system.** Only the object is inside the curve; everything else is outside.

❹ **Locate every point on the boundary of this curve where the environment touches the system.** These are the points where the environment exerts *contact forces* on the object.

❺ **Name and label each contact force acting on the object.** There is at least one force at each point of contact; there may be more than one. When necessary, use subscripts to distinguish forces of the same type.

❻ **Name and label each long-range force acting on the object.** For now, the only long-range force is weight.

CONCEPTUAL EXAMPLE 4.1 **Identifying forces on a bungee jumper**

A bungee jumper has leapt off a bridge and is nearing the bottom of her fall. What forces are being exerted on the bungee jumper?

REASON

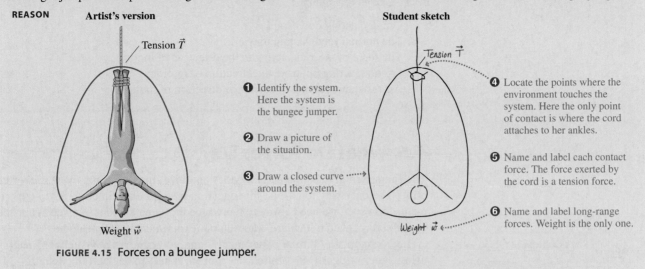

❶ Identify the system. Here the system is the bungee jumper.

❷ Draw a picture of the situation.

❸ Draw a closed curve around the system.

❹ Locate the points where the environment touches the system. Here the only point of contact is where the cord attaches to her ankles.

❺ Name and label each contact force. The force exerted by the cord is a tension force.

❻ Name and label long-range forces. Weight is the only one.

FIGURE 4.15 Forces on a bungee jumper.

CONCEPTUAL EXAMPLE 4.2 **Identifying forces on a skier**

A skier is being towed up a snow-covered hill by a tow rope. What forces are being exerted on the skier?

REASON

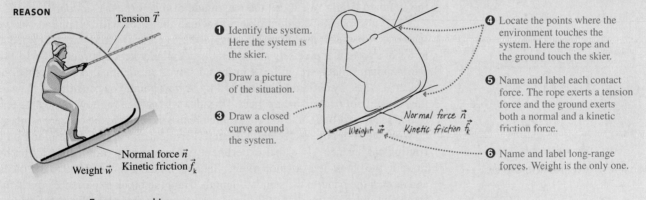

❶ Identify the system. Here the system is the skier.

❷ Draw a picture of the situation.

❸ Draw a closed curve around the system.

❹ Locate the points where the environment touches the system. Here the rope and the ground touch the skier.

❺ Name and label each contact force. The rope exerts a tension force and the ground exerts both a normal and a kinetic friction force.

❻ Name and label long-range forces. Weight is the only one.

FIGURE 4.16 Forces on a skier.

NOTE ▶ You might have expected two friction forces and two normal forces in Example 4.2, one on each ski. Keep in mind, however, that we're working within the particle model, which represents the skier by a single point. A particle has only one contact with the ground, so there is a single normal force and a single friction force. The particle model is valid if we want to analyze the motion of the skier as a whole, but we would have to go beyond the particle model to find out what happens to each ski. ◀

CONCEPTUAL EXAMPLE 4.3 **Identifying forces on a rocket**   **REASON**

A rocket is being launched to place a new satellite in orbit. Air resistance is not negligible. What forces are being exerted on the rocket?

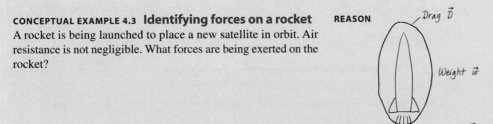

FIGURE 4.17 Forces on a rocket.

You've just kicked a rock, and it is now sliding across the ground about 2 meters in front of you. Which of these are forces acting on the rock? List all that apply.

A. Gravity, acting downward.
B. The normal force, acting upward.
C. The force of the kick, acting in the direction of motion.
D. Friction, acting opposite the direction of motion.
E. Air resistance, acting opposite the direction of motion.

4.5 What Do Forces Do?

The fundamental question is: How does an object move when a force is exerted on it? The only way to answer this question is to do experiments. To do experiments, however, we need a way to reproduce the same amount of force again and again, and we need a standard object so our experiments are repeatable.

Let's conduct a "virtual experiment," one you can easily visualize. Imagine using your fingers to stretch a rubber band to a certain length—say 10 centimeters—that you can measure with a ruler. We'll call this the *standard length*. Figure 4.18 shows the idea. You know that a stretched rubber band exerts a force because your fingers *feel* the pull. Furthermore, this is a reproducible force. The rubber band exerts the same force every time you stretch it to the standard length. We'll call the magnitude of this force the *standard force F*.

We'll also need a standard object to which the force will be applied. As we learned in Chapter 1, the SI unit of mass is the kilogram. The kilogram is defined in terms of a particular precisely machined metal cylinder kept in a vault in Paris. For our standard object, we will just make ourselves an identical copy, so that our object also has, by definition, a mass of 1 kg. At this point, you can think of mass as the "quantity of matter" in an object. This idea will suffice for now, but by the end of this section we'll be able to give a more precise meaning to the concept of mass.

Now we're ready to start the virtual experiment. First, place the block on a frictionless surface. (In a real experiment, we can nearly eliminate friction by floating the block on a cushion of air.) Second, attach a rubber band to the block and stretch the band to the standard length. Then the block experiences the same force *F* as did your finger. As the block starts to move, in order to keep the pulling force constant you must *move your hand* in just the right way to keep the length of the rubber band—and thus the force—*constant*. Figure 4.19 shows the experi-

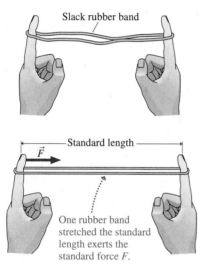

Slack rubber band

Standard length

$\vec{F}$

One rubber band stretched the standard length exerts the standard force *F*.

FIGURE 4.18 A reproducible force.

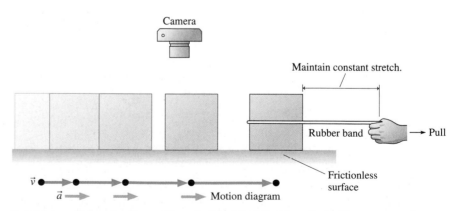

Camera

Maintain constant stretch.

Rubber band

Pull

Frictionless surface

$\vec{v}$

$\vec{a}$ Motion diagram

FIGURE 4.19 Measuring the motion of a 1 kg block that is pulled with a constant force.

ment being carried out. Once the motion is complete, you can use motion dia-grams (made from movie frames from the camera) and kinematics to analyze the block's motion.

The motion diagram in Figure 4.19 shows that the velocity vectors are getting longer, so the velocity is increasing: the block is *accelerating*. In order to study the acceleration more carefully, we make a velocity-versus-time graph, as shown in Figure 4.20. We know that the *slope* of such a graph is the acceleration. We see that the slope of the graph—and thus the acceleration—is the same at all times: it is *constant*. This is the first important finding of this experiment: **an object pulled with a constant force moves with a constant acceleration.** This finding could not have been anticipated in advance. It's conceivable that the object would speed up for a while, then move with a steady speed. Or that it would continue to speed up, but that the *rate* of increase, the acceleration, would steadily decline. These are conceivable motions, but they're not what happens. Instead, the object accel-erates *with a constant acceleration* for as long as you pull it with a constant force. Let's use the symbol a_1 for this acceleration of our standard 1 kg block when it's pulled by the standard force.

The next experiment is to see what happens if a different force is used. To get a larger force, we can use more rubber bands. If two rubber bands are each pulling equally hard, like those in Figure 4.21, the net pull is twice that of one rubber band, and the force is $2F$. Three side-by-side rubber bands, each pulled to the standard length, will exert three times the standard force, and so on. So we now have a way of applying forces of different strengths to our block. Figure 4.22 shows what happens in our experiment when we apply larger forces using more bands. When two rubber bands act instead of just one, the slope of the v-versus-t graph is twice as large, hence the acceleration is twice as large. When three bands act, the acceleration is three times greater than with just one. This is our second important finding: **The acceleration is directly proportional to the size of the force.**

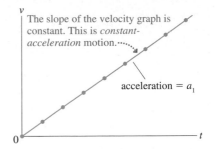

FIGURE 4.20 Velocity-versus-time graph for a 1 kg block pulled with a constant force.

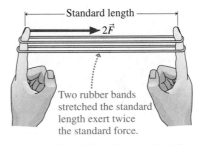

FIGURE 4.21 Doubling the standard force.

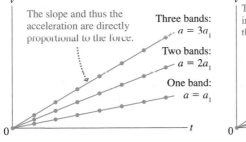

FIGURE 4.22 Velocity-versus-time graphs for a block pulled by one, two, and three rubber bands.

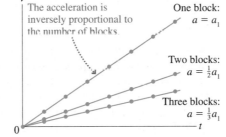

FIGURE 4.23 Velocity-versus-time graphs for one, two, and three blocks pulled with the same force.

The final question for our virtual experiment is: How does the acceleration of an object depend on the "quantity of matter" it contains? To find out, glue two identical 1 kg blocks together, so we have twice the quantity of matter. Now apply the same force—a single rubber band—as you applied to the original block. Figure 4.23 shows that the acceleration is *one-half* as great as that of the original 1 kg block. If we glue three blocks together and pull with one rubber band, we find that the acceleration is only *one-third* of the original acceleration. In general, we find that the acceleration is proportional to the *inverse* of the number of blocks. So our third important result is that **the acceleration is inversely propor-tional to the number of blocks.**

Inverse proportion

Exercise 12

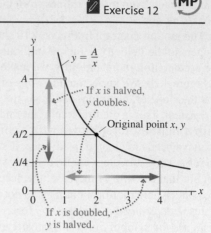

Two quantities are said to be **inversely proportional** to each other if one quantity is the *inverse* of the other. Mathematically, this means that

$$y = \frac{A}{x}$$

y is inversely proportional to *x*

Here, *A* is a constant.

If *y* is inversely proportional to *x*, *x* is also inversely proportional to *y*.

SCALING

- If you double *x*, you halve *y*.
- If you triple *x*, *y* is reduced by a factor of 3.
- If you halve *x*, *y* doubles.
- If you reduce *x* by a factor of 3, *y* becomes 3 times as large.

RATIOS For any two values of *x*, say x_1 and x_2, we have

$$y_1 = \frac{A}{x_1} \qquad \text{and} \qquad y_2 = \frac{A}{x_2}$$

so that

$$\frac{y_1}{y_2} = \frac{x_2}{x_1}$$

That is, the ratio of *y* values is the inverse of the ratio of corresponding values of *x*.

LIMITS

- As *x* gets very large, *y* approaches zero.
- As *x* approaches zero, *y* gets very large.

Mass

So far, our idea of mass is that it is a measure of the "quantity of matter" an object contains. But this definition is a little vague. Exactly what do we mean by the "quantity" of matter in an object? The object's volume? The number of atoms it contains? Or some other measure? Our experiments indicate that an unambiguous way to define the mass of an object is in terms of its *acceleration*. We've just found that the more matter an object has, the more it *resists* accelerating in response to a force. You're familiar with this idea: your car is much harder to get moving than your bicycle. The tendency of an object to resist a *change* in its velocity (i.e., to resist acceleration) is called *inertia*.

To make these ideas more quantitative, we can start with the observation made above that the acceleration is inversely proportional to the number of blocks. But the number of blocks is directly related to the mass: One block has (by definition) a mass of 1 kg, so it seems reasonable that two such blocks, with twice the "quantity of matter," should have a mass of 2 kg, and so on. So we can state more generally that, when acted upon by the standard force, **the acceleration is inversely**

proportional to the mass on which the force acts. We can write this mathematically as $a = A/m$, where A is a constant and m is the mass.

For one block we know that $m = 1$ kg and $a = a_1$. Suppose now we take an unknown object, one not made of a simple combination of 1 kg blocks. We measure its acceleration a. What is its mass m? Because $a = A/m$, we can use the discussion in the math relationship box on inverse proportion to write

$$\frac{a_1}{a} = \frac{m}{1 \text{ kg}}$$

or

$$m = 1 \text{ kg} \times \frac{a_1}{a} \tag{4.2}$$

This result tells us how to think about mass: The greater an object's mass, the greater its resistance to being accelerated. If an object is more difficult to accelerate than a 1 kg object (i.e., a is smaller than a_1), then its mass is greater than 1 kg. If the object is easier to accelerate, then its mass is less than 1 kg. Any object that undergoes the *same* acceleration as the standard 1 kg object, using the same force, must have a mass of exactly 1 kg. **Mass, then, is the property of an object that determines how it accelerates in response to an applied force.**

EXAMPLE 4.1 Finding the mass of an unknown block

When a rubber band is stretched to pull on a 1.0 kg block with a constant force, the acceleration of the block is measured to be 3.0 m/s². When a block with an unknown mass is pulled with the same rubber band, using the same force, its acceleration is 5.0 m/s². What is the mass of the unknown block?

PREPARE The acceleration a_1 of the 1 kg block is 3.0 m/s², and the acceleration a of the unknown block is 5.0 m/s².

SOLVE From the definition of mass in terms of acceleration, we find that

$$m = 1.0 \text{ kg} \times \frac{a_1}{a} = 1.0 \text{ kg} \times \frac{3.0 \text{ m/s}^2}{5.0 \text{ m/s}^2} = 0.60 \text{ kg}$$

ASSESS With the same force applied, the unknown block had a *larger* acceleration than the 1.0 kg block. It makes sense, then, that its mass—its resistance to acceleration—is *less* than 1.0 kg.

STOP TO THINK 4.3   Two rubber bands stretched to the standard length cause an object to accelerate at 2 m/s². Suppose another object with twice the mass is pulled by four rubber bands stretched to the standard length. The acceleration of this second object is

A. 1 m/s².   B. 2 m/s².   C. 4 m/s².   D. 8 m/s².   E. 16 m/s².

4.6 Newton's Second Law

We can now summarize the results of our experiments. We've seen that **a force causes an object to accelerate. The acceleration a is directly proportional to the force F and inversely proportional to the mass m.** We can express both these relationships in equation form as

$$a = \frac{F}{m} \tag{4.3}$$

Note that if we double the size of the force F, the acceleration a will double, as we found experimentally. And if we triple the mass m, the acceleration will be only one-third as great, again in accord with experiment.

Equation 4.3 tells us the magnitude of an object's acceleration in terms of its mass and the force applied. But our experiments also had another important finding: The *direction* of the acceleration was the same as the direction of the force. We can express this fact by writing Equation 4.3 in *vector* form as

$$\vec{a} = \frac{\vec{F}}{m} \tag{4.4}$$

Recall that any vector (such as $\vec{F}$), when multiplied by an ordinary number (such as $1/m$), gives a vector (such as $\vec{a}$) that points in the *same* direction as the original vector $\vec{F}$. So the relationship between force and acceleration expressed by Equation 4.4 indicates that $\vec{a}$ and $\vec{F}$ point in the same direction.

Finally, our experiment was limited to looking at an object's response to a *single* applied force. Realistically, an object is likely to be subjected to several distinct forces $\vec{F}_1, \vec{F}_2, \vec{F}_3, \ldots$ that may point in different directions. What happens then? Experiments show that the acceleration of the object is determined by the *net force* acting on it. Recall from Figure 4.4 and Equation 4.1 that the net force is the *vector sum* of all forces acting on the object. So if several forces are acting, we use the *net* force in Equation 4.4.

Newton was the first to recognize these connections between force and motion. This relationship is known today as Newton's second law.

Newton's second law An object of mass m subjected to forces $\vec{F}_1, \vec{F}_2, \vec{F}_3, \ldots$ will undergo an acceleration $\vec{a}$ given by

$$\vec{a} = \frac{\vec{F}_{\text{net}}}{m} \tag{4.5}$$

INVERSE
p. 118

where the net force $\vec{F}_{\text{net}} = \vec{F}_1 + \vec{F}_2 + \vec{F}_3 + \cdots$ is the vector sum of all forces acting on the object. **The acceleration vector $\vec{a}$ points in the same direction as the net force vector $\vec{F}_{\text{net}}$.**

We'll use Newton's second law in Chapter 5 to solve many kinds of motion problems; for the moment, however, the critical idea is that an object accelerates in the direction of the net force acting on it.

The significance of Newton's second law cannot be overstated. There was no reason to suspect that there should be any simple relationship between force and acceleration. Yet a simple but exceedingly powerful equation relates the two. Newton's work, preceded to some extent by Galileo's, marks the beginning of a highly successful period in the history of science during which it was learned that the behavior of physical objects can often be described and predicted by mathematical relationships. While some relationships are found to apply only in special circumstances, others seem to have universal applicability. Those equations that appear to apply at all times and under all conditions have come to be called "laws of nature." Newton's second law is a law of nature; you will meet others as we go through this book.

We can rewrite Newton's second law in the form

$$\vec{F}_{\text{net}} - m\vec{a} \tag{4.6}$$

which is how you'll see it presented in many textbooks and how, in practice, we'll often use the second law. Equations 4.5 and 4.6 are mathematically equivalent,

An unfair advantage? Race car driver Danica Patrick was the subject of controversial comments by other drivers who thought her small mass of 45 kg gave her an advantage over heavier drivers; the next-lightest driver's mass was 61 kg. Because every driver's car must have the same mass, Patrick's overall racing mass was lower than any other driver's. Because a car's acceleration is inversely proportional to its mass, her car could be expected to have a slightly greater acceleration.

but Equation 4.5 better describes the central idea of Newtonian mechanics: A force applied to an object causes the object to accelerate.

NOTE ▶ When several forces act on an object, be careful not to think that the strongest force "overcomes" the others to determine the motion on its own. Forces are not in competition with each other! It is $\vec{F}_{net}$, the sum of *all* the forces, that determines the acceleration $\vec{a}$. ◀

CONCEPTUAL EXAMPLE 4.4 Acceleration of a wind-blown basketball

A basketball is released from rest in a stiff breeze directed to the right. In what direction does the ball accelerate?

REASON As shown in Figure 4.24a, two forces are acting on the ball: its weight $\vec{w}$ directed downward, and a wind force $\vec{F}_{wind}$ pushing the ball to the right. Newton's second law tells us that the direction of the acceleration is the same as the direction of the net force $\vec{F}_{net}$. In Figure 4.24b we find $\vec{F}_{net}$ by graphical vector addition of $\vec{w}$ and $\vec{F}_{wind}$. We see that $\vec{F}_{net}$ and therefore $\vec{a}$ point down and to the right.

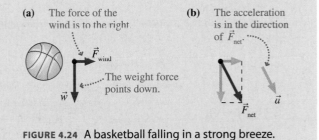

FIGURE 4.24 A basketball falling in a strong breeze.

Units of Force

Because $\vec{F}_{net} = m\vec{a}$, the unit of force must be mass units multiplied by acceleration units. We've previously specified the SI unit of mass as the kilogram. We can now define the basic unit of force as "the force that causes a 1 kg mass to accelerate at 1 m/s^2." From Newton's second law, this force is

$$1 \text{ basic unit of force} = (1 \text{ kg}) \times (1 \text{ m/s}^2) = 1 \frac{\text{kg} \cdot \text{m}}{\text{s}^2}$$

This basic unit of force is called a *newton:* One **newton** is the force that causes a 1 kg mass to accelerate at 1 m/s^2. The abbreviation for newton is N. Mathematically, $1 \text{ N} = 1 \text{ kg} \cdot \text{m/s}^2$.

The newton is a *secondary unit,* meaning that it is defined in terms of the *primary units* of kilograms, meters, and seconds. We will introduce other secondary units as needed.

It is important to develop a feeling for what the size of forces should be. Table 4.2 shows some typical forces. As you can see, "typical" forces on "typical" objects are likely to be in the range 0.01–10,000 N. Forces less than 0.01 N are too small to consider unless you are dealing with very small objects. Forces greater than 10,000 N would make sense only if applied to very massive objects.

The unit of force in the English system is the *pound* (abbreviated lb). Although the definition of the pound has varied throughout history, it is now defined in terms of the newton:

$$1 \text{ pound} = 1 \text{ lb} = 4.45 \text{ N}$$

You very likely associate pounds with kilograms rather than with newtons. Everyday language often confuses the ideas of mass and weight, but we're going to need to make a clear distinction between them. We'll have more to say about this in the next chapter.

TABLE 4.2 Approximate magnitude of some typical forces

Force	Approximate magnitude (newtons)
Weight of a U.S. nickel	0.05
Weight of a 1-pound object	5
Weight of a 110-pound person	500
Propulsion force of a car	5,000
Thrust force of a rocket motor	5,000,000

EXAMPLE 4.2 Pulling an airplane

In 2000, a team of 60 British police officers set a world record by pulling a Boeing 747, with a mass of 205,000 kg, a distance of 100 m in 53.3 s. Estimate the force with which each officer pulled on the plane.

PREPARE If we assume that the plane undergoes a constant acceleration, we can use kinematics to find the magnitude of that acceleration. Then we can use Newton's second law to find the force applied to the airplane. Figure 4.25 shows the visual overview of the airplane.

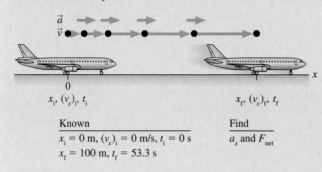

Known	Find
$x_i = 0$ m, $(v_x)_i = 0$ m/s, $t_i = 0$ s	a_x and F_{net}
$x_f = 100$ m, $t_f = 53.3$ s	

FIGURE 4.25 Visual overview of the airplane accelerating.

SOLVE Because we know the net displacement of the plane, and the time it took to move, we can use the kinematic equation

$$x_f = x_i + (v_x)_i \Delta t + \tfrac{1}{2} a_x (\Delta t)^2$$

to find the airplane's acceleration a_x. Using the known values $x_i = 0$ m and $(v_x)_i = 0$ m/s, we can solve for the acceleration to get

$$a_x = \frac{2x_f}{(\Delta t)^2} = \frac{2(100 \text{ m})}{(53.3 \text{ s})^2} = 0.0704 \text{ m/s}^2$$

Now we apply Newton's second law. The net force is

$$F_{net} = ma_x = (205,000 \text{ kg})(0.0704 \text{ m/s}^2) = 1.44 \times 10^4 \text{ N}$$

This is the force applied by all 60 men. Each man thus applies about $1/60^{th}$ of this force, or around 240 N.

ASSESS Converting this force to pounds, we have

$$F = 240 \text{ N} \times \frac{1 \text{ lb}}{4.45 \text{ N}} = 54 \text{ lb}$$

Burly policemen can certainly apply a greater force than this. We have neglected the rolling friction of the plane's tires, which apply a force opposite to the plane's motion. Friction lowers the plane's acceleration, leading to an underestimate of the force applied by the men.

STOP TO THINK 4.4   Three forces act on an object. In which direction does the object accelerate?

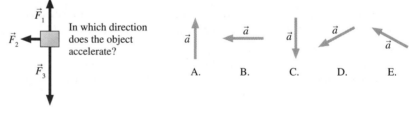

4.7 Free-Body Diagrams

Having discussed at length what is and is not a force, and what forces do to an object, we are ready to assemble our knowledge about force and motion into a single diagram called a **free-body diagram.** A free-body diagram represents the object as a particle and shows *all* of the forces acting on the object. Now that we have forces to consider, we expand our visual overview to include force identification and a free-body diagram. Learning how to draw a correct free-body diagram is a very important skill, one that in the next chapter will become a critical part of our strategy for solving motion problems. For now, let's concentrate on the basic skill of constructing a correct free-body diagram.

TACTICS BOX 4.3 Drawing a free-body diagram   📝 Exercises 17–22

❶ **Identify all forces acting on the object.** This step was described in Tactics Box 4.2.

❷ **Draw a coordinate system.** Use the axes defined in your pictorial representation (Tactics Box 2.2). If those axes are tilted, for motion along an incline, then the axes of the free-body diagram should be similarly tilted.

❸ **Represent the object as a dot at the origin of the coordinate axes.** This is the particle model.

❹ **Draw vectors representing each of the identified forces.** This was described in Tactics Box 4.1. Be sure to label each force vector.

❺ **Draw and label the *net force* vector $\vec{F}_{net}$.** Draw this vector beside the diagram, not on the particle. Or, if appropriate, write $\vec{F}_{net} = \vec{0}$. Then check that $\vec{F}_{net}$ points in the same direction as the acceleration vector $\vec{a}$ on your motion diagram.

EXAMPLE 4.3 Forces on an upward-accelerating elevator

An elevator, suspended by a cable, speeds up as it moves upward from the ground floor. Draw a free-body diagram of the elevator.

PREPARE Figure 4.26 illustrates the steps outlined in Tactics Box 4.3.

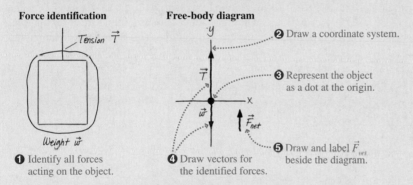

Force identification

Tension $\vec{T}$

Weight $\vec{w}$

❶ Identify all forces acting on the object.

Free-body diagram

y

$\vec{T}$

x

$\vec{w}$

$\vec{F}_{net}$

❷ Draw a coordinate system.

❸ Represent the object as a dot at the origin.

❹ Draw vectors for the identified forces.

❺ Draw and label $\vec{F}_{net}$ beside the diagram.

FIGURE 4.26 Free-body diagram of an elevator accelerating upward.

ASSESS The coordinate axes, with a vertical y-axis, are the ones we would use in a pictorial representation of the motion. The elevator is accelerating upward, so $\vec{F}_{net}$ must point upward. For this to be true, the magnitude of $\vec{T}$ must be larger than the magnitude of $\vec{w}$. The diagram has been drawn accordingly.

EXAMPLE 4.4 Forces on a rocket-propelled ice block

Bobby straps a small model rocket to a block of ice and shoots it across the smooth surface of a frozen lake. Friction is negligible. Draw a visual overview—a motion diagram, force identification diagram, and free-body diagram—of the block of ice.

PREPARE Treat the block of ice as a particle. The visual overview consists of a motion diagram to determine $\vec{a}$, a force

identification picture, and a free-body diagram. The statement of the situation tells us that friction is negligible. We can draw these three pictures using Problem Solving Strategy 1.1 for the motion diagram, Tactics Box 4.2 to identify the forces, and Tactics Box 4.3 to draw the free-body diagrams. These pictures are shown in Figure 4.27 on the next page.

Continued

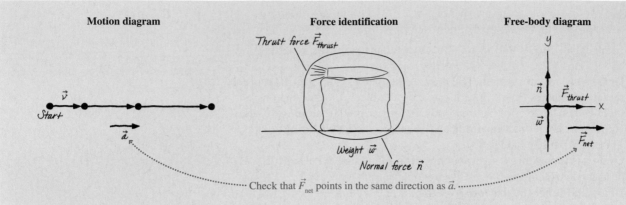

FIGURE 4.27 Visual overview for a block of ice shooting across a frictionless frozen lake.

ASSESS The motion diagram tells us that the acceleration is in the positive *x*-direction. According to the rules of vector addition, this can be true only if the upward-pointing $\vec{n}$ and the downward-pointing $\vec{w}$ are equal in magnitude and thus cancel each other. The vectors have been drawn accordingly, and this leaves the net force vector pointing toward the right, in agreement with $\vec{a}$ from the motion diagram.

EXAMPLE 4.5 Forces on a towed skier

A tow rope pulls a skier up a snow-covered hill at a constant speed. Draw a full visual overview of the skier.

PREPARE This is Example 4.2 again with the additional information that the skier is moving at constant speed. If we were doing a kinematics problem, the pictorial representation would use a tilted coordinate system with the *x*-axis parallel to the slope, so we use these same tilted coordinate axes for the free-body diagram. The motion diagram, force identification, and free-body diagram are shown in Figure 4.28.

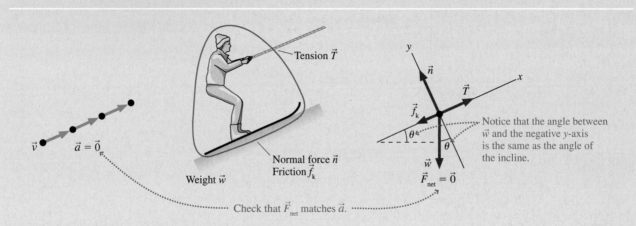

FIGURE 4.28 Visual overview for a skier being towed at a constant speed.

ASSESS We have shown $\vec{T}$ pulling parallel to the slope and $\vec{f}_k$, which opposes the direction of motion, pointing down the slope. The normal force $\vec{n}$ is perpendicular to the surface and thus along the *y*-axis. Finally, and this is important, the weight $\vec{w}$ is *vertically* downward, *not* along the negative *y*-axis.

The skier moves in a straight line with constant speed, so $\vec{a} = \vec{0}$. Newton's second law then tells us that $\vec{F}_{net} = m\vec{a} = \vec{0}$. Thus we have drawn the vectors such that the forces add to zero. We'll learn more about how to do this in Chapter 5.

Free-body diagrams will be our major tool for the next several chapters. Careful practice with the workbook exercises and homework in this chapter will pay immediate benefits in the next chapter. Indeed, it is not too much to assert that a problem is half solved, or even more, when you complete the free-body diagram.

An elevator suspended by a cable is moving upward and slowing to a stop. Which free-body diagram is correct?

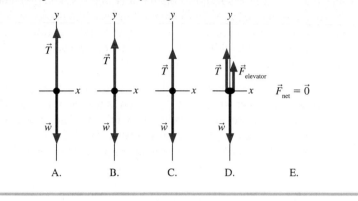

$$\vec{F}_{net} = \vec{0}$$

A.   B.   C.   D.   E.

4.8 Newton's Third Law

Thus far, we've focused on the motion of a single particle—which we called the *system*—responding to well-defined forces applied by objects outside the system—that is, from the *environment*. A skier sliding downhill, for instance, is subject to frictional and normal forces from the slope, and the pull of gravity on his body. Once we know these forces, we can use Newton's second law to calculate the acceleration, and hence the overall motion, of the skier.

But motion in the real world often involves two or more objects *interacting* with each other. Consider the two sumo wrestlers in Figure 4.29. The harder one wrestler pushes, the harder the other pushes back. A hammer and a nail, your foot and a soccer ball, and the earth-moon system are other examples of interacting objects.

Newton's second law is not sufficient to explain what happens when two or more objects interact. It does not explain how the force of the hammer on the nail is related to that of the nail on the hammer. In this section we will introduce a new law of physics, Newton's *third* law, that describes how two objects interact with each other.

FIGURE 4.29 The two sumo wrestlers are a system of interacting objects.

Interacting Objects

Think about the hammer and nail in Figure 4.30. The hammer certainly exerts a force on the nail as it drives the nail forward. At the same time, the nail exerts a force on the hammer. If you are not sure that it does, imagine hitting the nail with a glass hammer. It's the force of the nail on the hammer that would cause the glass to shatter.

Indeed, if you stop to think about it, any time that object A pushes or pulls on object B, object B pushes or pulls back on object A. As sumo wrestler A pushes on sumo wrestler B, B pushes back on A. (If A pushed forward without B pushing back, A would fall over in the same way you do if someone suddenly opens a door you're leaning against.) Your chair pushes upward on you (a normal force) while, at the same time, you push down on the chair. These are examples of what we call an *interaction*. An **interaction** is the mutual influence of two objects on each other.

These examples illustrate a key aspect of interactions: The forces involved in an interaction between two objects always occur as a *pair*. To be more specific, if object A exerts a force $\vec{F}_{A\,on\,B}$ on object B, then object B exerts a force $\vec{F}_{B\,on\,A}$ on

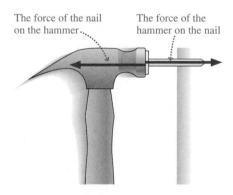

The force of the nail on the hammer   The force of the hammer on the nail

FIGURE 4.30 The hammer and nail are interacting with each other.

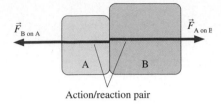

FIGURE 4.31 An action/reaction pair of forces.

object A. This pair of forces, shown in Figure 4.31, is called an **action/reaction pair.** Two objects interact by exerting an action/reaction pair of forces on each other. Notice the very explicit subscripts on the force vectors. The first letter is the *agent*—the source of the force—and the second letter is the *object* on which the force acts. $\vec{F}_{A\,on\,B}$ is thus the force exerted *by* A *on* B. The distinction is important, and this way of labeling forces will be of great help when we discuss how to identify the members of an action/reaction pair.

NOTE ▶ The name "action/reaction pair" is somewhat misleading. The forces occur simultaneously, and we cannot say which is the "action" and which the "reaction." Neither is there any implication about cause and effect; the action does not cause the reaction. **An action/reaction pair of forces exists as a pair, or not at all.** In identifying action/reaction pairs, the labels are the key. Force $\vec{F}_{A\,on\,B}$ is paired with force $\vec{F}_{B\,on\,A}$. ◀

Reasoning with Newton's Third Law

We've discovered that two objects always interact via an action/reaction pair of forces. Newton was the first to recognize how the two members of an action/reaction pair of forces are related to each other. Today we know this as Newton's third law:

> **Newton's third law** Every force occurs as one member of an action/reaction pair of forces.
>
> - The two members of an action/reaction pair act on two *different* objects.
> - The two members of an action/reaction pair point in *opposite* directions, and are *equal in magnitude.*

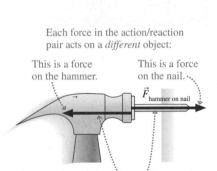

Each force in the action/reaction pair acts on a *different* object:

This is a force on the hammer.

This is a force on the nail.

$\vec{F}_{hammer\,on\,nail}$

The members of the pair point in *opposite directions*, but are of *equal magnitude.*

FIGURE 4.32 Newton's third law.

Newton's third law is often stated as "For every action there is an equal but opposite reaction." While this is a catchy phrase, it lacks the preciseness of our preferred version. In particular, it fails to capture an essential feature of action/reaction pairs—that they each act on a *different* object. This is shown in Figure 4.32, where a hammer hitting a nail exerts a force $\vec{F}_{hammer\,on\,nail}$ on the nail; by the third law, the nail must exert a force $\vec{F}_{nail\,on\,hammer}$ to complete the action/ reaction pair.

Figure 4.32 also illustrates that these two forces point in *opposite directions.* This feature of the third law is also in accord with our experience. If the hammer hits the nail with a force directed to the right, the force of the nail on the hammer is directed to the left; if the force of my chair on me pushes up, the force of me on the chair pushes down.

Finally, Figure 4.32 shows that, according to Newton's third law, the two members of an action/reaction pair have *equal* magnitudes, so that $F_{hammer\,on\,nail} = F_{nail\,on\,hammer}$. This is something new, and it is by no means obvious. Indeed, this statement causes students the most trouble when applying the third law because it seems so counter to our intuition. Consider, for instance, the collision between a bug and the windshield of a truck. The third law tells us that the magnitude of the force of the windshield on the bug is *equal* to that of the bug on the windshield! How can this be, when the bug is so small compared to the truck? The source of puzzlement in problems like this is that Newton's third law equates the size of the *forces* acting on the two objects, not their *accelerations.* The acceleration of each object depends not only on the force applied to it, but also, according to Newton's second law, its mass. The bug and the truck do in fact feel forces of equal strength from the other, but the bug, with its very small mass, undergoes an extreme acceleration from this force while the acceleration of the heavy truck is negligible. It is important to separate the *effects* of the forces (the accelerations) from the causes (the forces themselves). Because two interacting objects can have very different masses, their accelerations can be very different even though the interaction forces are of the same strength.

Identifying Forces for Interacting Objects

In order to understand the motion of a single object subject to external forces, we took the object to be the *system* and everything else to be the *environment*. Now we're interested in the motion of two or more objects that interact with each other, so we'll expand our system to include all these interacting objects. For example, for a truck pushing a car, we would take the car and truck together as the system, while the road and the earth would make up the environment, as shown in Figure 4.33.

With the system chosen in this way, we can make a distinction between two classes of forces, again shown in Figure 4.33. **External forces** are forces on objects in the system that originate from outside the system; that is, they are forces of the environment on the system. For example, the weight forces of the car and truck are external forces because they are the forces of the earth (part of the environment) on the vehicles. **Internal forces,** on the other hand, are forces *between* objects in the system. The force that the truck exerts on the car is an internal force between the two objects in this system.

Newton's third law tells us that *every* force is a member of an action/reaction pair. For an external force—a force of the environment on the system—the other member of the pair is a force of the *system* on the *environment*. But forces on the *environment* have no effect on the motion of objects in the *system*. Thus, for finding the motion of an object in the system, we don't need to identify action/reaction pairs involving external forces.

Finding both members of an action/reaction pair made up of *internal* forces is crucial, however, because both of these forces act on objects *within* the system and thus affect their motion. Only after identifying such pairs can we use Newton's third law, which relates their directions and magnitudes. The following Tactics Box 4.4 shows you how to correctly identify external forces, and the action/reaction pairs of internal forces.

Revenge of the target We normally think of the damage that the force of a bullet inflicts on its target. But according to Newton's third law, the target exerts an equal force on the bullet. The photo shows the damage sustained by bullets fired at 1600, 1800, and 2000 ft/s, after impacting a test target. The appearance of the bullet before firing is shown at the left.

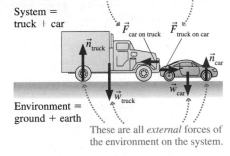

FIGURE 4.33 External forces and action/reaction pairs.

> (MP) **TACTICS BOX 4.4** Identifying forces for interacting objects   ✎ Exercises 23–27
>
> ❶ **Identify those objects whose motion you wish to study.** These objects make up the system; the environment is everything else.
> ❷ **Draw each object separately.** Place them in the correct position relative to each other.
> ❸ **Identify all forces on the system.** For each object in the system, use the techniques of Tactics Box 4.2 to find the forces acting on that object.
> ❹ ■ **Identify the action/reaction pairs.** For each force acting on an object, decide if it is of the form $\vec{F}_{A\,on\,B}$, where A and B are *both* objects in the system. If so, it is an internal force and forms an action/reaction pair with $\vec{F}_{B\,on\,A}$. Label the forces in a pair using notation like $\vec{F}_{car\,on\,truck}$ and $\vec{F}_{truck\,on\,car}$.
> ■ **Identify the external forces.** External forces are forces of the environment on an object in the system. Name each external force with its appropriate symbols such as $\vec{n}$ or $\vec{w}$. When needed, use subscripted labels such as $\vec{w}_1$ and $\vec{w}_2$ to distinguish between similar forces acting on different objects.
> ❺ **Draw separate free-body diagrams for each object.** For each object, include the forces acting on it found in Step 3. Connect the force vectors of action/reaction pairs with dotted lines.

CONCEPTUAL EXAMPLE 4.5 Action/reaction pairs in a balancing act

An acrobat balances a vase on his head. Define the system, then identify external forces acting on the system and action/reaction pairs of internal forces. Draw free-body diagrams for all objects in the system.

REASON We proceed by following the steps in Tactics Box 4.4, as shown in Figure 4.34a. We're interested in the motions of the acrobat and the vase, so they form the system. In accordance with step 2, we've drawn the objects separately with the vase above the acrobat.

Next we identify all the forces acting on both objects, using the methods of Tactics Box 4.2: We draw a closed curve around each object and decide where that object touches objects outside the curve. The vase makes contact only with the acrobat's head. Because the acrobat is part of the system, this is an internal force that we'll label $\vec{F}_{\text{A on V}}$. The only other force acting on the vase is its weight $\vec{w}_V$.

Now let's look at the forces on the acrobat. He is contacted in only two places, the floor and the top of his head where the vase touches. The floor pushes up on his feet with a normal force $\vec{n}$, an external force. The force on his head is more subtle. You might be tempted to say that this force is "the weight of the vase." But recall that *weight* is a force on an object due to the gravitational pull of the earth. Thus the "weight of the vase" is a force *on the vase* due to gravity; that is, it is the force $\vec{w}_V$ that we've already identified. So the force *on the acrobat* cannot be this weight. In fact, the force on his head is simply that due to the vase itself pushing down on him. It is thus an internal force, and we give it the label $\vec{F}_{\text{V on A}}$. Finally, we must include the long-range force of the acrobat's weight $\vec{w}_A$.

We have identified two internal forces, $\vec{F}_{\text{V on A}}$ and $\vec{F}_{\text{A on V}}$. Because every internal force must be one member of an action/reaction pair, these two must form such a pair. We can check this by noting that the subscripts are reverses of each other: A on V is just V on A flipped.

In Figure 4.34b we show the free-body diagram for the acrobat and vase. Because neither acrobat nor vase is accelerating, we know that the net force on each must be zero by Newton's second law. Thus we draw the vector $\vec{F}_{\text{A on V}}$ to have the same length as the vase's weight $\vec{w}_V$, so that these two vectors cancel. Similarly, the three forces acting on the acrobat must sum to zero. Finally, because $\vec{F}_{\text{V on A}}$ and $\vec{F}_{\text{A on V}}$ form an action/reaction pair, their *magnitudes must be the same,* as we've drawn.

Figure 4.34c shows the drawing as you should draw it. Although much simpler than the artist's version, it still contains all the steps of Tactics Box 4.4.

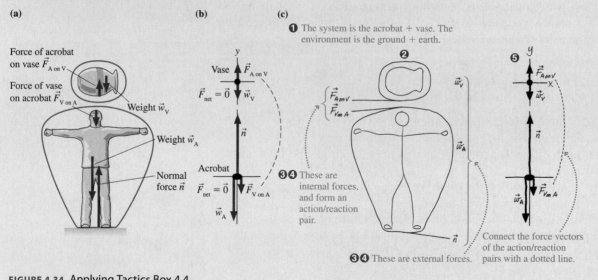

FIGURE 4.34 Applying Tactics Box 4.4.

ASSESS Using Newton's third law is the key to solving problems of interacting objects. The two forces of the action/reaction pair in Figure 4.34 really connect the two free-body diagrams together. If we know the magnitude of one of these forces in one free-body diagram, we automatically know the magnitude of its paired member in the other free-body diagram.

Propulsion

A sprinter accelerates out of the blocks. Because he's accelerating, there must be a force on him in the forward direction. For a system with an internal source of energy, a force that drives the system is a force of **propulsion.** Propulsion is an important feature not only of walking or running but also of the forward motion of cars, jets, and rockets. Propulsion is somewhat counterintuitive, so it is worth a closer look.

If you tried to walk across a frictionless floor, your foot would slip and slide *backward.* In order for you to walk, the floor needs to have friction so that your foot *sticks* to the floor as you straighten your leg, moving your body forward. The friction that prevents slipping is *static* friction. Static friction, you will recall, acts in the direction that prevents slipping, so the static friction force $\vec{f}_{\text{S on P}}$ (for *Surface* on *Person*) has to point in the *forward* direction to prevent your foot from slipping backward. As shown in Figure 4.35a, it is this forward-directed static friction force that propels you forward! The force of your foot on the floor, $\vec{f}_{\text{P on S}}$, is the other half of the action/reaction pair, and it points in the opposite direction as you push backward against the floor.

Similarly, the car in Figure 4.35b uses static friction to propel itself. The car uses its motor to turn the tires, causing the tires to push backward against the road ($\vec{f}_{\text{tire on road}}$). The road surface responds by pushing the car forward ($\vec{f}_{\text{road on tire}}$). This force of the road on the tire can be seen in photos of drag racers, where the forces are very great (Figure 4.36). Again, the forces involved are *static* friction forces. The tire is rolling, but the bottom of the tire, where it contacts the road, is instantaneously at rest. If it weren't, you would leave one giant skid mark as you drove and would burn off the tread within a few miles.

Rocket motors are somewhat different because they are not pushing *against* anything external. That's why rocket propulsion works in the vacuum of space. Instead, the rocket engine pushes hot, expanding gases out of the back of the rocket, as shown in Figure 4.37. In response, the exhaust gases push the rocket forward with the force we've called *thrust.*

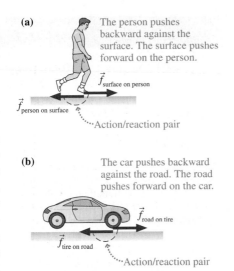

(a) The person pushes backward against the surface. The surface pushes forward on the person.

$\vec{f}_{\text{surface on person}}$

$\vec{f}_{\text{person on surface}}$

······Action/reaction pair

(b) The car pushes backward against the road. The road pushes forward on the car.

$\vec{f}_{\text{road on tire}}$

$\vec{f}_{\text{tire on road}}$

······Action/reaction pair

FIGURE 4.35 Examples of propulsion.

$\vec{F}_{\text{road on tire}}$

You can *see* that the force of the road on the tire points forward by the way it twists the rubber of the tire.

FIGURE 4.36 When the driver hits the gas, the force of the track on the tire is so great that the tire deforms.

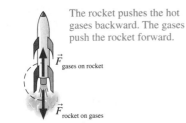

The rocket pushes the hot gases backward. The gases push the rocket forward.

$\vec{F}_{\text{gases on rocket}}$

$\vec{F}_{\text{rocket on gases}}$

FIGURE 4.37 Rocket propulsion.

STOP TO THINK 4.6 A small car is pushing a larger truck that has a dead battery. The mass of the truck is larger than the mass of the car. Which of the following statements is true?

$\vec{a}$

A. The car exerts a force on the truck, but the truck doesn't exert a force on the car.
B. The car exerts a larger force on the truck than the truck exerts on the car.
C. The car exerts the same amount of force on the truck as the truck exerts on the car.
D. The truck exerts a larger force on the car than the car exerts on the truck.
E. The truck exerts a force on the car, but the car doesn't exert a force on the truck.

Now we've assembled all the pieces we need in order to start solving problems in dynamics. We have seen what forces are and how to identify them, and we've learned how forces cause objects to accelerate according to Newton's second law. We've also found how Newton's third law governs the interaction forces between two objects. Our goal in the next several chapters is to apply Newton's laws to a variety of problems involving straight-line and circular motion.

SUMMARY

The goal of Chapter 4 has been to establish a connection between force and motion.

GENERAL PRINCIPLES

Newton's First Law

Consider an object with no net force acting on it. If it is at rest, it will remain at rest. If it is in motion, then it will continue to move in a straight line at a constant speed.

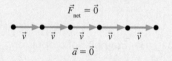

$$\vec{F}_{net} = \vec{0}$$
$$\vec{a} = \vec{0}$$

The first law tells us that no "cause" is needed for motion. Uniform motion is the "natural state" of an object.

Newton's Second Law

An object with mass m will undergo acceleration

$$\vec{a} = \frac{1}{m}\vec{F}_{net}$$

where $\vec{F}_{net} = \vec{F}_1 + \vec{F}_2 + \vec{F}_3 + \cdots$ is the vector sum of all the individual forces acting on the object.

The second law tells us that a net force causes an object to accelerate. This is the connection between force and motion. The acceleration points in the direction of $\vec{F}_{net}$.

Newton's Third Law

Every force occurs as one member of an **action/reaction** pair of forces. The two members of an action/reaction pair:

- act on two *different* objects.
- point in opposite directions and are equal in magnitude:

$$\vec{F}_{A\,on\,B} = -\vec{F}_{B\,on\,A}$$

IMPORTANT CONCEPTS

Force is a push or pull on an object.

- Force is a vector, with a magnitude and a direction.
- A force requires an agent.
- A force is either a contact force or a long-range force.

The SI unit of force is the **newton** (N). A 1 N force will cause a 1 kg mass to accelerate at 1 m/s^2.

Net force is the vector sum of all the forces acting on an object.

The net force determines the acceleration according to Newton's second law.

$$\vec{F}_{net} = \vec{F}_1 + \vec{F}_2 + \vec{F}_3$$

Mass is the property of an object that determines its resistance to acceleration.

If the same force is applied to different objects, their masses in kg are found from

$$m = \frac{a_1}{a} \times 1\ kg$$

where a is an object's acceleration, and a_1 that of a 1 kg object.

APPLICATIONS

Identifying Forces

Forces are identified by locating the points where the environment touches the system. These are points where contact forces are exerted. In addition, objects with mass feel a long-range weight force.

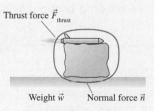

Free-Body Diagrams

A free-body diagram represents the object as a particle at the origin of a coordinate system. Force vectors are drawn with their tails on the particle. The net force vector is drawn beside the diagram.

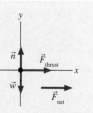

QUESTIONS

Conceptual Questions

1. A hockey puck slides along the surface of the ice. If friction and air resistance are negligible, what force is required to keep the puck moving?

2. If an object is not moving, does that mean that there are no forces acting on it? Explain.

3. An object moves in a straight line at a constant speed. Is it true that there must be no forces of any kind acting on this object? Explain.

4. Write several sentences explaining why you agree or disagree with the statement "Forces cause an object to move."

5. If you know all of the forces acting on a moving object, can you tell in which direction the object is moving? If the answer is Yes, explain how. If the answer is No, give an example.

6. Three arrows are shot horizontally. They have left the bow and are traveling parallel to the ground as shown in Figure Q4.6. Air resistance is negligible. Rank in order, from largest to smallest, the magnitudes of the *horizontal* forces F_1, F_2, and F_3 acting on the arrows. Some may be equal. State your reasoning.

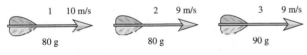

FIGURE Q4.6

7. A carpenter wishes to tighten the heavy head of his hammer onto its light handle. Which method shown in Figure Q4.7 will better tighten the head? Explain.

8. Internal injuries in vehicular
BIO accidents may be due to what is called the "third collision." The first collision is the vehicle hitting the external object. The

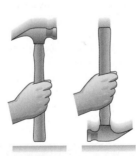

FIGURE Q4.7

second collision is the person hitting something on the inside of the car, such as the dashboard or windshield. This may cause external lacerations. The third collision, possibly the most damaging to the body, is when organs, such as the heart or brain, hit the ribcage, skull, or other confines of the body, bruising the tissues on the leading edge and tearing the organ from its supporting structures on the trailing edge.
   a. Why is there a third collision? In other words, why are the organs still moving after the second collision?
   b. If the vehicle was traveling at 60 mph before the first collision, would the organs be traveling more than, equal to, or less than 60 mph just before the third collision?

9. a. Give an example of the motion of an object in which the frictional force on the object is directed opposite to the motion.
   b. Give an example of the motion of an object in which the frictional force on the object is in the same direction as the motion.

10. Suppose you are an astronaut in deep space, far from any source of gravity. You have two objects that look identical, but one has a large mass and the other a small mass. How can you tell the difference between the two?

11. Newton's second law says that $\vec{F}_{net} = m\vec{a}$. Is $m\vec{a}$ thus a force? If so, what is its origin?

12. Superman can hover in midair if he wishes. Does this mean that he must exert a force on some other object? Explain.

13. A ball weighs 2.0 N when placed on a scale. It is then thrown straight up. What is its weight at the very top of its motion? Explain.

14. A book sits on a table. List all the forces acting on the book.

15. A person sits on a sloped hillside. Is it ever possible to have the static friction force on this person point down the hill? Explain.

16. Walking without slipping requires a static friction force
BIO between your feet (or footwear) and the floor. As described in this chapter, the force on your foot as you push off the floor is forward while the force exerted by your foot on the floor is backward. But what about your *other* foot, the one moved during a stride? What is the direction of the force on that foot as it comes in contact with the floor? Explain.

17. Figure 4.35 shows a case in which the force of the road on the car's tire points forward. In other cases, the force points backward. Give an example of such a case.

18. a. A tightrope walker at the circus steps onto the high wire, causing it to sag slightly. Is the tension in the wire less than, greater than, or equal to the performer's weight? Explain. Include a free-body diagram as part of your explanation.
   b. The leading circus magazine advertises a new wire made of a material called DreamRope. The ad says that a DreamRope wire will remain perfectly straight and horizontal, with absolutely no sag, as the performer walks across. Should you order some? Explain.

19. A very smart three-year-old child is given a wagon for her birthday. She refuses to use it. "After all," she says, "Newton's third law says that no matter how hard I pull, the wagon will exert an equal but opposite force on me. So I will never be able to get it to move forward." What would you say to her in reply?

20. Will hanging a magnet in front of an iron cart, as shown in Figure Q4.20, make it go? Explain why or why not.

FIGURE Q4.20

Multiple-Choice Questions

21. ‖ Figure Q4.21 shows the view looking down onto a frictionless sheet of ice. A puck, tied with a string to point P, slides on the surface of the ice in the circular path shown. If the string suddenly snaps when the puck is in the position shown, which path best represents the puck's subsequent motion?

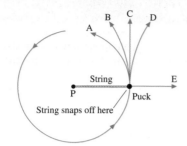

FIGURE Q4.21

22. ‖ A block has acceleration a when pulled by a string. If two identical blocks are glued together and pulled with twice the original force, their acceleration will be
    A. $(1/4)a$ B. $(1/2)a$ C. a D. $2a$ E. $4a$

23. ‖ A person gives a box a shove so that it slides up a ramp, then reverses its motion and slides down. The direction of the force of friction is
    A. Always down the ramp.
    B. Up the ramp and then down the ramp.
    C. Always down the ramp.
    D. Down the ramp and then up the ramp.

24. ‖ Tennis balls experience a large drag force. A tennis ball is hit so that it goes straight up and then comes back down. The direction of the drag force is
    A. Always up.   B. Up and then down.
    C. Always down.   D. Down and then up.

25. ‖ Rachel is pushing a box across the floor while Jon, at the same time, is hoping to stop the box by pushing in the opposite direction. There is friction between the box and floor. If the box is moving at constant speed, then the magnitude of Rachel's pushing force is
    A. Greater than the magnitude of Jon's force.
    B. Equal to the magnitude of Jon's force.
    C. Less than the magnitude of Jon's force.
    D. The problem can't be answered without knowing how large the friction force is.

26. ‖ A person is pushing horizontally on a box with a constant force, causing it to slide across the floor with a constant speed. If the person suddenly stops pushing on the box, the box will
    A. Immediately come to a stop.
    B. Continue moving at a constant speed for a while, then gradually slow down to a stop.
    C. Immediately change to a slower but constant speed.
    D. Immediately begin slowing down and eventually stop.

27. ‖‖‖ Figure Q4.27 shows block A sitting on top of block B. A constant force $\vec{F}$ is exerted on block B, causing block B to accelerate to the right. Block A rides on block B without slipping. Which statement is true?
    A. Block B exerts a friction force on block A, directed to the left.
    B. Block B exerts a friction force on block A, directed to the right.
    C. Block B does not exert a friction force on block A.

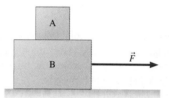

FIGURE Q4.27

28. ‖ Dave pushes his four-year-old son Thomas across the snow on a sled. As Dave pushes, Thomas speeds up. Which statement is true?
    A. The force of Dave on Thomas is larger than the force of Thomas on Dave.
    B. The force of Thomas on Dave is larger than the force of Dave on Thomas.
    C. Both forces have the same magnitude.
    D. It depends on how hard Dave pushes on Thomas.

29. ‖ A truck hits a small car. During this collision,
    A. The truck exerts a larger force on the car than the car on the truck.
    B. The car exerts a larger force on the truck than the truck on the car.
    C. The force of the truck on the car and of the car on the truck have the same magnitude.
    D. The net force is zero.

PROBLEMS

Section 4.1 What Causes Motion?

1. ‖ Whiplash injuries during an automobile accident are
   BIO caused by the inertia of the head. If someone is wearing a seatbelt, her body will tend to move with the car seat. However, her head is free to move until the neck restrains it, causing damage to the neck. Brain damage can also occur.

   Figure P4.1 shows two sequences of head and neck motion for a passenger in an auto accident. One corresponds to a head-on collision, the other to a rear-end collision. Which is which? Explain.

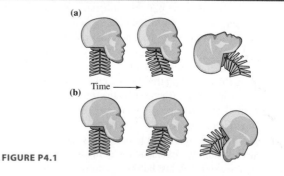

FIGURE P4.1

2. | An automobile has a head-on collision. A passenger in the
BIO car experiences a compression injury to the brain. Is this
injury most likely to be in the front or rear portion of the
brain? Explain.

3. | Passengers in the back seat of an automobile should wear
seatbelts, not only for their own protection, but also for the
protection of the people in the front seats of the car. Explain.

Section 4.2 Force

Problems 4 through 6 show two forces acting on an object at rest.
Redraw the diagram, then add a third force that will allow the
object to remain at rest. Label the new force $\vec{F}_3$.

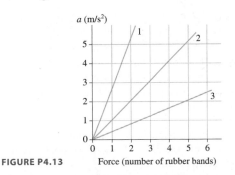

FIGURE P4.4   FIGURE P4.5   FIGURE P4.6

Section 4.3 A Short Catalog of Forces

Section 4.4 Identifying Forces

7. ‖ A mountain climber is hanging from a rope in the middle
of a crevasse. The rope is vertical. Identify the forces on the
mountain climber.

8. | A circus clown hangs from one end of a large spring. The
other end is anchored to the ceiling. Identify the forces on the
clown.

9. | A baseball player is sliding into second base. Identify the
forces on the baseball player.

10. ‖ A jet plane is speeding down the runway during takeoff.
Air resistance is not negligible. Identify the forces on the jet.

11. | A skier is sliding down a 15° slope. Friction is not negligi-
ble. Identify the forces on the skier.

12. | A tennis ball is flying horizontally across the net. Air resis-
tance is not negligible. Identify the forces on the ball.

Section 4.5 What Do Forces Do?

13. ‖‖ Figure P4.13 shows an acceleration-versus-force graph for
three objects pulled by rubber bands. The mass of object 2 is
0.20 kg. What are the masses of objects 1 and 3? Explain your
reasoning.

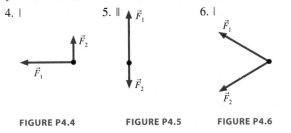

FIGURE P4.13   Force (number of rubber bands)

14. | A constant force applied to object A causes it to accelerate
at 5 m/s². The same force applied to object B causes an accel-

eration of 3 m/s². Applied to object C, it causes an acceleration of 8 m/s².
a. Which object has the largest mass?
b. Which object has the smallest mass?
c. What is the ratio of mass A to mass B (m_A/m_B)?

15. | Two rubber bands pulling on an object cause it to acceler-
ate at 1.2 m/s².
a. What will be the object's acceleration if it is pulled by four
rubber bands?
b. What will be the acceleration of two of these objects glued
together if they are pulled by two rubber bands?

16. | A constant force is applied to an object, causing the object
to accelerate at 10 m/s². What will the acceleration be if
a. The force is halved?
b. The object's mass is halved?
c. The force and the object's mass are both halved?
d. The force is halved and the object's mass is doubled?

17. | A constant force is applied to an object, causing the object
to accelerate at 8.0 m/s². What will the acceleration be if
a. The force is doubled?
b. The object's mass is doubled?
c. The force and the object's mass are both doubled?
d. The force is doubled and the object's mass is halved?

18. ‖‖ A man pulling an empty wagon causes it to accelerate at
1.4 m/s². What will the acceleration be if he pulls with the
same force when the wagon contains a child whose mass is
three times that of the wagon?

19. | A car has a maximum acceleration of 5.0 m/s². What will
the maximum acceleration be if the car is towing another car
of the same mass?

Section 4.6 Newton's Second Law

20. | Figure P4.20 shows an acceleration-versus-force graph for
a 500 g object. Redraw this graph and add appropriate acceleration values on the vertical scale.

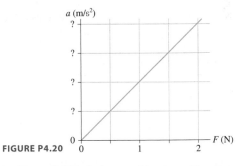

FIGURE P4.20

21. | Figure P4.21 shows an object's acceleration-versus-force
graph. What is the object's mass?

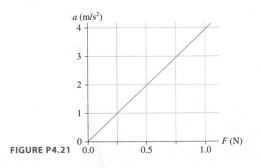

FIGURE P4.21

22. | Two children fight over a 200 g stuffed bear. The 25 kg boy pulls to the right with a 15 N force and the 20 kg girl pulls to the left with a 17 N force. Ignore all other forces on the bear (such as its weight).
    a. At this instant, can you say what the velocity of the bear is? If so, what are the magnitude and direction of the velocity?
    b. At this instant, can you say what the acceleration of the bear is? If so, what are the magnitude and direction of the acceleration?

23. | Based on the information in Table 4.2, estimate in newtons
    a. The weight of a laptop computer.
    b. The propulsion force of a bicycle.
    c. The propulsion force of a sprinter.

24. ‖ Very small forces can have tremendous effects on the motion of very small objects. Consider a single electron, with a mass of 9.1×10^{-31} kg, subject to a single force equal to the weight of a penny, 2.5×10^{-2} N. What is the acceleration of the electron?

25. ‖ The motion of a very massive object is hardly affected by what would seem to be a substantial force. Consider a super-tanker, with a mass of 3.0×10^8 kg. If it is pushed by a rocket motor (see Table 4.2) and is subject to no other forces, what will be the magnitude of its acceleration?

Section 4.7 Free-Body Diagrams

Problems 26 through 28 show a free-body diagram. For each, (a) Redraw the free-body diagram and (b) Write a short description of a real object for which this is the correct free-body diagram. Use Examples 4.3, 4.4, and 4.5 as models of what a description should be like.

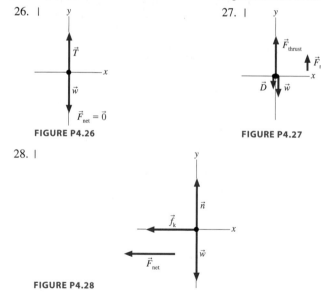

FIGURE P4.26

FIGURE P4.27

28. |

FIGURE P4.28

Problems 29 through 35 describe a situation. For each, identify all forces acting on the object and draw a free-body diagram of the object.

29. | Your car is sitting in the parking lot.
30. | Your car is accelerating from a stop.
31. ‖ Your car is slowing to a stop from a high speed.
32. | An ice hockey puck glides across frictionless ice.
33. | An elevator, hanging from a cable, descends at steady speed.
34. | Your physics textbook is sliding across the table.
35. ‖ You hold a picture motionless against a wall by pressing on it, as shown in Figure P4.35.

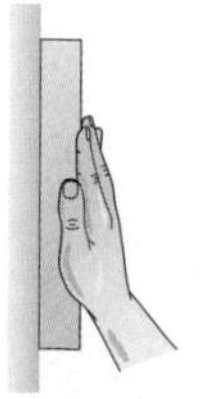

FIGURE P4.35

Section 4.8 Newton's Third Law

36. ‖ A weight lifter stands up from a squatting position while holding a heavy barbell across his shoulders. Identify all the third-law pairs of forces, then draw free-body diagrams for the weight lifter and the barbell. Use dotted lines to connect the members of all action/reaction pairs.

37. ‖ A softball player is throwing the ball. Her arm has come forward to where it is beside her head, but she hasn't yet released the ball. Identify all the third-law pairs of forces, then draw free-body diagrams for the ballplayer and the ball. Use dotted lines to connect the members of all action/reaction pairs.

38. | A soccer ball and a bowling ball roll across a hard floor and collide head on. Identify all the third-law pairs of forces at the moment of the collision, then draw free-body diagrams for each ball. Use dotted lines to connect the members of all action/reaction pairs. Friction is negligible.

General Problems

39. | Redraw the motion dia-
    INT gram shown in Figure P4.39, then draw a vector beside it to show the direction of the net force acting on the object. Explain your reasoning.

40. | Redraw the motion dia-
    INT gram shown in Figure P4.40, then draw a vector beside it to show the direction of the net force acting on the object. Explain your reasoning.

41. | Redraw the motion dia-
    INT gram shown in Figure P4.41, then draw a vector beside it to show the direction of the net force acting on the object. Explain your reasoning.

FIGURE P4.39   **FIGURE P4.40**

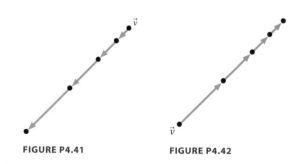

FIGURE P4.41   **FIGURE P4.42**

42. ‖ Redraw the motion diagram shown in Figure P4.42, then
    INT draw a vector beside it to show the direction of the net force acting on the object. Explain your reasoning.

Problems 43 through 49 show a free-body diagram. For each:
a. Redraw the diagram.
b. Identify the direction of the acceleration vector $\vec{a}$ and show it as a vector next to your diagram. Or, if appropriate, write $\vec{a} = \vec{0}$.
c. If possible, identify the direction of the velocity vector $\vec{v}$ and show it as a labeled vector.
d. Write a short description of a real object for which this is the correct free-body diagram. Use Examples 4.3, 4.4, and 4.5 as models of what a description should be like.

43. |

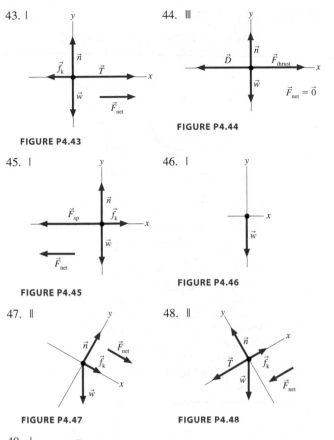

FIGURE P4.43

44. ‖

FIGURE P4.44

45. |

FIGURE P4.45

46. |

FIGURE P4.46

47. ‖

FIGURE P4.47

48. ‖

FIGURE P4.48

49. |

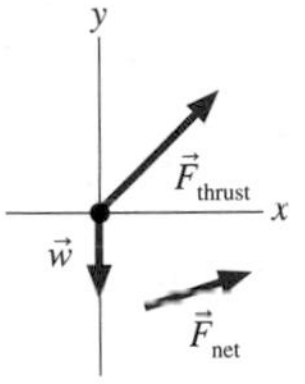

FIGURE P4.49

50. ‖ A student draws the flawed free-body diagram shown in Figure P4.50 to represent the forces acting on a car traveling at constant speed on a level road. Identify the errors in the diagram, then draw a correct free-body diagram for this situation.

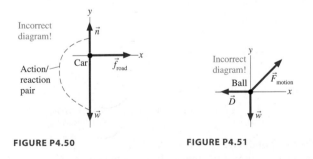

FIGURE P4.50

FIGURE P4.51

51. ‖ A student draws the flawed free-body diagram shown in Figure P4.51 to represent the forces acting on a golf ball that is traveling upward and to the right a very short time after being hit off the tee. Air resistance is assumed to be relevant. Identify the errors in the diagram, then draw a correct free-body diagram for this situation.

Problems 52 through 63 describe a situation. For each, draw a motion diagram, a force identification diagram, and a free-body diagram.

52. ‖ An elevator, suspended by a single cable, has just left the tenth floor and is speeding up as it descends toward the ground floor.

53. | A rocket is being launched straight up. Air resistance is not negligible.

54. | A jet plane is speeding down the runway during takeoff. Air resistance is not negligible.

55. | You've slammed on the brakes and your car is skidding to a stop while going down a 20° hill.

56. ‖ A skier is going down a 20° slope. A *horizontal* headwind is blowing in the skier's face. Friction is small, but not zero.

57. | You've just kicked a rock and it is now sliding down the sidewalk.

58. | A Styrofoam ball has just been shot straight up. Air resistance is not negligible.

59. ‖ A spring-loaded gun shoots a plastic ball. The trigger has just been pulled and the ball is starting to move down the barrel. The barrel is horizontal.

60. ‖ A person on a bridge throws a rock straight down toward the water. The rock has just been released.

61. ‖ A gymnast has just landed on a trampoline. She's still moving downward as the trampoline stretches.

62. ‖ A heavy box is in the back of a truck. The truck is accelerating to the right. Apply your analysis to the box.

63. ‖ A bag of groceries is on the back seat of your car as you stop for a stop light. The bag does not slide. Apply your analysis to the bag.

64. ‖ A rubber ball bounces. We'd like to understand *how* the ball bounces.
   a. A rubber ball has been dropped and is bouncing off the floor. Draw a motion diagram of the ball during the brief time interval that it is in contact with the floor. Show 4 or 5 frames as the ball compresses, then another 4 or 5 frames as it expands. What is the direction of $\vec{a}$ during each of these parts of the motion?
   b. Draw a picture of the ball in contact with the floor and identify all forces acting on the ball.
   c. Draw a free-body diagram of the ball during its contact with the ground. Is there a net force acting on the ball? If so, in which direction?
   d. During contact, is the force of the ground on the ball larger, smaller, or equal to the weight of the ball? Use your answers to parts a–c to explain your reasoning.

65. ‖ If a car stops suddenly, you feel "thrown forward." We'd like to understand what happens to the passengers as a car stops. Imagine yourself sitting on a *very* slippery bench inside a car. This bench has no friction, no seat back, and there's nothing for you to hold on to.
   a. Draw a picture and identify all of the forces acting on you as the car travels in a straight line at a perfectly steady speed on level ground.
   b. Draw your free-body diagram. Is there a net force on you? If so, in which direction?
   c. Repeat parts a and b with the car slowing down.
   d. Describe what happens to you as the car slows down.
   e. Use Newton's laws to explain why you seem to be "thrown forward" as the car stops. Is there really a force pushing you forward?

66. ▥ The fastest pitched baseball was clocked at 46 m/s. If the
BIO pitcher exerted his force (assumed to be horizontal and constant)
over a distance of 1.0 m, and a baseball has a mass of 145 g,
   a. Draw a free-body diagram of the ball during the pitch.
   b. What force did the pitcher exert on the ball during this
      record-setting pitch?
   c. Estimate the force in (b) as a fraction of the pitcher's
      weight.

67. ▎ The froghopper, champion leaper of the insect world, can
BIO jump straight up at 4.0 m/s. The jump itself lasts a mere
1.0 ms before the insect is clear of the ground.
   a. Draw a free-body diagram of this mighty leaper while the
      jump is taking place.
   b. While the jump is taking place, is the force that the ground
      exerts on the froghopper greater than, less than, or equal to
      the insect's weight? Explain.

68. ▮▮ A beach ball is thrown straight up, and some time later it
lands on the sand. Is the magnitude of the net force on the ball
greatest when it is going up or when it is on the way down? Or
is it the same in both cases? Explain. Air resistance should not
be neglected for a large, light object.

Passage Problems

A Simple Solution for a Stuck Car

If your car is stuck in the mud and you don't have a winch to pull
it out, you can use a piece of rope and a tree to do the trick. First,
you tie one end of the rope to your car and the other to a tree, then
pull as hard as you can on the middle of the rope, as shown in
Figure P4.69a. This technique applies a force to the car much larger
than the force that you can apply directly. To see why the car experiences such a large force, look at the forces acting on the center
point of the rope, as shown in Figure 4.69b. The sum of the forces
is zero, thus the tension is much greater than the force you apply. It
is this tension force that acts on the car and, with luck, pulls it free.

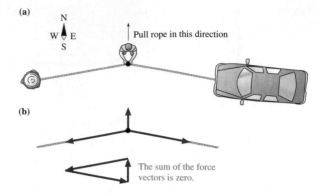

FIGURE P4.69

69. ▎ The sum of the three forces acting on the center point of
the rope is assumed to be zero because
   A. This point has a very small mass.
   B. Tension forces in a rope always cancel.
   C. This point is not accelerating.
   D. The angle of deflection is very small.

70. ▎ When you are pulling on the rope as shown, what is the
approximate direction of the tension force on the tree?
   A. North   B. South   C. East   D. West

71. ▎ Assume that you are pulling on the rope but the car is not
moving. What is the approximate direction of the force of the
mud on the car?
   A. North   B. South   C. East   D. West

72. ▎ Suppose your efforts work, and the car begins to move for-
ward out of the mud. As it does so, the force of the car on the
rope is
   A. Zero.
   B. Less than the force of the rope on the car.
   C. Equal to the force of the rope on the car.
   D. Greater than the force of the rope on the car.

STOP TO THINK ANSWERS

Stop to Think 4.1: C.

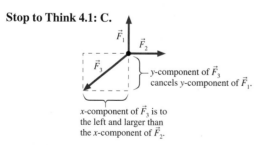

Stop to Think 4.2: A, B, and D. Friction and the normal force are
the only contact forces. Nothing is touching the rock to provide a
"force of the kick." We've agreed to ignore air resistance unless a
problem specifically calls for it.

Stop to Think 4.3: B. Acceleration is proportional to force, so
doubling the number of rubber bands doubles the acceleration of
the original object from 2 m/s² to 4 m/s². But acceleration is also
inversely proportional to mass. Doubling the mass cuts the acceleration in half, back to 2 m/s².

Stop to Think 4.4: D.

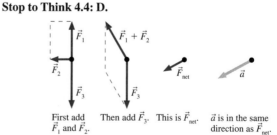

Stop to Think 4.5: C. The acceleration vector points downward as
the elevator slows. $\vec{F}_{net}$ points in the same direction as $\vec{a}$, so $\vec{F}_{net}$
also points down. This will be true if the tension is less than the
weight: $T < w$.

Stop to Think 4.6: C. Newton's third law says that the force of A
on B is *equal* and opposite to the force of B on A. This is always
true. The mass of the objects isn't relevant.

5 APPLYING NEWTON'S LAWS

Before his parachute opens, why does this skydiver fall at a constant speed? And why does he suddenly slow down when his parachute opens?

Looking Ahead ▶▶

The goal of Chapter 5 is to learn how to solve problems about motion in a straight line. In this chapter you will learn to:

▶ Use Newton's first law to solve problems of static and dynamic equilibrium.

▶ Use Newton's second law to solve dynamics problems.

▶ Understand how mass, weight, and apparent weight differ.

▶ Use simple models of friction and drag.

▶ Use Newton's third law to solve problems involving interacting objects.

Looking Back ◀◀

This chapter pulls together many strands of thought from Chapters 1–4. Please review:

◀ Sections 2.4–2.5 Constant acceleration kinematics, including free fall.

◀ Section 3.2–3.3 Vectors and components.

◀ Sections 4.4, 4.7, and 4.8 Identifying forces, drawing free-body diagrams, and identifying action/reaction pairs.

After jumping from the plane, a skydiver accelerates until reaching a *terminal speed* of about 140 mph. But when his parachute is opened, he then slows down to only some 10–20 mph. To understand the skydiver's motion, we need to look closely at the forces exerted on him. We also need to understand how those forces determine his motion.

Chapter 4 introduced Newton's three laws of motion. Now, in Chapter 5, we want to use these laws to solve force and motion problems. As will be the case throughout this textbook, our strategy will be to learn a set of *procedures* that can be applied to a wide variety of problems, not to memorize a set of equations.

This chapter focuses on objects that are at rest or that move in a straight line, such as runners, bicycles, cars, planes, and rockets. Weight, tension, thrust, friction, and drag forces will be essential to our understanding of this motion. Circular motion and rotational motion will be treated in Chapters 6 and 7.

5.1 Equilibrium

The simplest applications of Newton's laws are those for which the net force $\vec{F}_{net}$ on an object is *zero*. According to Newton's first law, an object on which there is no net force moves with constant velocity $(\vec{a} = \vec{0})$. One way an object can have

137

This human tower is in equilibrium because the net force on each man is zero.

$\vec{a} = \vec{0}$ is to be at rest. An object that remains at rest is said to be in **static equilibrium.** A second way for an object to have $\vec{a} = \vec{0}$ is to move in a straight line at a constant speed. Such an object is in **dynamic equilibrium.** The defining property of both these cases of **equilibrium** is that the net force acting on the object is $\vec{F}_{\text{net}} = \vec{0}$.

To use Newton's laws, we have to identify all the forces acting on an object and then evaluate $\vec{F}_{\text{net}}$. Recall that $\vec{F}_{\text{net}}$ is the vector sum

$$\vec{F}_{\text{net}} = \vec{F}_1 + \vec{F}_2 + \vec{F}_3 + \cdots$$

where $\vec{F}_1$, $\vec{F}_2$, and so on are the individual forces, such as tension or friction, acting on the object. We found in Chapter 3 that vector sums can be evaluated in terms of the x- and y-components of the vectors. That is, the x-component of the net force is $(F_{\text{net}})_x = F_{1x} + F_{2x} + F_{3x} + \cdots$ If we restrict ourselves to problems where all the forces are in the xy-plane, then the equilibrium requirement $\vec{F}_{\text{net}} = \vec{0}$ is a shorthand way of writing two simultaneous equations:

$$(F_{\text{net}})_x = F_{1x} + F_{2x} + F_{3x} + \cdots = 0$$
$$(F_{\text{net}})_y = F_{1y} + F_{2y} + F_{3y} + \cdots = 0$$

Recall from your math classes that the Greek letter Σ (sigma) stands for "the sum of." It will be convenient to abbreviate the sum of the x-components of all forces as

$$F_{1x} + F_{2x} + F_{3x} + \cdots = \sum F_x$$

With this notation, the requirement that an object be in equilibrium—Newton's first law—can be written as the two equations

$$\sum F_x = 0 \qquad \text{and} \qquad \sum F_y = 0 \qquad (5.1)$$

In equilibrium, the sums of the x- and y-components of the force are zero

Although this may look a bit forbidding, we'll soon see how to use a free-body diagram of the forces to evaluate these sums.

When an object is in equilibrium, we are usually interested in finding the forces that keep it in equilibrium. Newton's first law is the basis for a strategy for solving equilibrium problems.

(MP)   PROBLEM-SOLVING STRATEGY 5.1   Equilibrium problems

PREPARE First check that the object is in equilibrium: Does $\vec{a} = \vec{0}$?

- An object at rest is in static equilibrium.
- An object moving at a constant velocity is in dynamic equilibrium.

Then identify all forces acting on the object and show them on a free-body diagram. Determine which forces you know and which you need to solve for.

SOLVE An object in equilibrium must satisfy Newton's first law. In component form, the requirement is

$$\sum F_x = 0 \qquad \text{and} \qquad \sum F_y = 0$$

You can find the force components that go into these sums directly from your free-body diagram. From these two equations, solve for the unknown forces in the problem.

ASSESS Check that your result has the correct units, is reasonable, and answers the question.

Static Equilibrium

EXAMPLE 5.1 Forces supporting an orangutan

An orangutan weighing 500 N hangs from a vertical vine. What is the tension in the vine?

PREPARE The orangutan is in static equilibrium, so all the forces acting on it must cancel to give zero net force. Figure 5.1 first identifies the forces acting on the orangutan: the upward force of the tension in the vine and the downward, long-range force of gravity. These forces are then shown on a free-body diagram, where it's noted that equilibrium requires $\vec{F}_{net} = \vec{0}$.

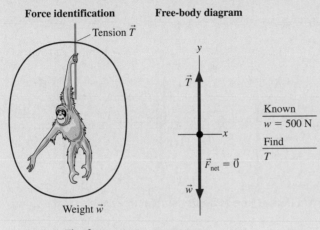

Force identification   **Free-body diagram**

Tension $\vec{T}$

Known
$w = 500$ N

Find
T

$\vec{F}_{net} = \vec{0}$

Weight $\vec{w}$

FIGURE 5.1 The forces on an orangutan.

SOLVE Neither force has an x-component, so we need to examine only the y-components of the forces. In this case, the y-component of Newton's first law is

$$\sum F_y = T_y + w_y = 0$$

You might have been tempted to write $T_y - w_y$ because the weight force points down. But remember that T_y and w_y are *components* of vectors, and can thus be positive (for a vector such as $\vec{T}$ that points up) or negative (for a vector such as $\vec{w}$ that points down). The fact that $\vec{w}$ points down is taken into account when we *evaluate* the components; that is, when we write them in terms of the *magnitudes* T and w of the vectors $\vec{T}$ and $\vec{w}$.

Because the tension vector $\vec{T}$ points straight up, in the positive y-direction, its y-component is $T_y = T$. Because the weight vector $\vec{w}$ points straight down, in the negative y-direction, its y-component is $w_y = -w$. This is where the signs enter. With these components, Newton's second law becomes

$$T - w = 0$$

This equation is easily solved for the tension in the vine:

$$T = w = 500 \text{ N}$$

ASSESS It's not surprising that the tension in the vine equals the weight of the orangutan. However, we'll soon see that this is *not* the case if the object is accelerating.

EXAMPLE 5.2 Readying a wrecking ball

A wrecking ball weighing 2500 N hangs from a cable. Prior to swinging, it is pulled back to a 20° angle by a second, horizontal cable. What is the tension in the horizontal cable?

PREPARE The ball hangs in static equilibrium until it is released. In Figure 5.2, we start by identifying all of the forces acting on the ball: a tension force from each cable and the ball's weight. We've used different symbols $\vec{T}_1$ and $\vec{T}_2$ for the two different tension forces. We then construct a free-body diagram for these three forces, noting that $\vec{F}_{net} = \vec{0}$. We're looking for the magnitude T_1 of the tension force $\vec{T}_1$ in the horizontal cable.

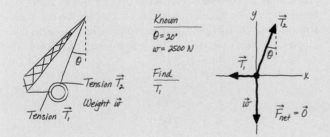

Known
$\theta = 20°$
$w = 2500$ N

Find
T_1

Tension $\vec{T}_2$

Weight $\vec{w}$

Tension $\vec{T}_1$

$\vec{F}_{net} = \vec{0}$

FIGURE 5.2 Visual overview of a wrecking ball just before release.

SOLVE The requirement of equilibrium is $\vec{F}_{net} = \vec{0}$. In component form, we have the two equations

$$\sum F_x = T_{1x} + T_{2x} + w_x = 0$$
$$\sum F_y = T_{1y} + T_{2y} + w_y = 0$$

As always, we *add* the force components together. Now we're ready to write the components of each force vector in terms of the magnitudes and directions of those vectors. We learned how to do this in Section 3.3 of Chapter 3. With practice you'll learn to read the components directly off the free-body diagram, but to begin it's worthwhile to organize the components into a table.

Force	Name of x-component	Value of x-component	Name of y-component	Value of y-component
$\vec{T}_1$	T_{1x}	$-T_1$	T_{1y}	0
$\vec{T}_2$	T_{2x}	$T_2 \sin\theta$	T_{2y}	$T_2 \cos\theta$
$\vec{w}$	w_x	0	w_y	$-w$

We see from the free-body diagram that $\vec{T}_1$ points along the negative x-axis, so $T_{1x} = -T_1$ and $T_{1y} = 0$. We need to be careful with our trigonometry as we find the components of $\vec{T}_2$.

Continued

Remembering that the side adjacent to the angle is related to the cosine, we see that the vertical (y) component of $\vec{T}_2$ is $T_2\cos\theta$. Similarly, the horizontal (x) component is $T_2\sin\theta$. The weight vector points straight down, so its y-component is $-w$. Notice that negative signs enter as we evaluate the components of the vectors, *not* when we write Newton's first law. This is a critical aspect of solving force and motion problems. With these components, Newton's first law now becomes

$$-T_1 + T_2\sin\theta + 0 = 0 \quad \text{and} \quad 0 + T_2\cos\theta - w = 0$$

We can rewrite these as

$$T_2\sin\theta = T_1 \quad \text{and} \quad T_2\cos\theta = w$$

These are two simultaneous equations with two unknowns: T_1 and T_2. To eliminate T_2 from the two equations, solve the second equation for T_2, giving $T_2 = w/\cos\theta$. Insert this expression for T_2 into the first equation to get

$$T_1 = \frac{w}{\cos\theta}\sin\theta = \frac{\sin\theta}{\cos\theta}w = w\tan\theta = (2500\text{ N})\tan 20° = 910\text{ N}$$

where we made use of the fact that $\tan\theta = \sin\theta/\cos\theta$.

ASSESS It seems reasonable that to pull the ball back to this modest angle, a force substantially less than the ball's weight will be required.

CONCEPTUAL EXAMPLE 5.1 Forces in static equilibrium

A rod is free to slide on a frictionless sheet of ice. One end of the rod is lifted by a string. Once the rod comes to rest, which diagram in Figure 5.3 shows the correct angle of the string?

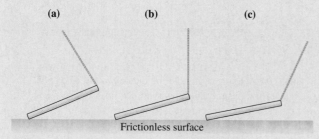

FIGURE 5.3 Which is the correct angle of the string?

REASON If the rod is to hang motionless, it must be in static equilibrium with $\sum F_x = 0$ and $\sum F_y = 0$. Figure 5.4 shows free-body diagrams for the three string orientations. Remember that tension always acts along the direction of the string and that the weight force always points straight down. The ice pushes up

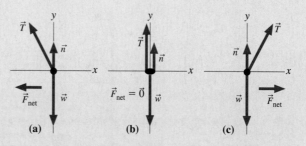

FIGURE 5.4 Free-body diagrams for three angles of the string.

with a normal force perpendicular to the surface, but frictionless ice cannot exert any horizontal force. If the string is angled, we see that its horizontal component exerts a net force on the rod. Only in case b, where the tension—and the string—are vertical, can the net force be zero.

ASSESS Frictionless surfaces are an idealization, but one that we will often use when friction is very small compared to other forces. This example illustrates that a frictionless surface has no component of force parallel to the surface.

Dynamic Equilibrium

EXAMPLE 5.3 Tension in towing a car

A car with a mass of 1500 kg is being towed at a steady speed by a rope held at a 20° angle. A friction force of 320 N opposes the car's motion. What is the tension in the rope?

PREPARE The car is moving in a straight line at a constant speed, so it is in dynamic equilibrium and must have $\vec{F}_{net} = \vec{0}$. Figure 5.5 shows three contact forces acting on the car—the tension force $\vec{T}$, friction $\vec{f}$, and the normal force $\vec{n}$—and the long-range force of gravity $\vec{w}$. These four forces are shown on the free-body diagram.

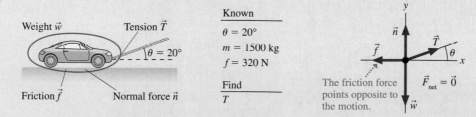

FIGURE 5.5 Visual overview of a car being towed.

SOLVE This is still an equilibrium problem, even though the car is moving, so our problem-solving procedure is unchanged. With four forces, the requirement of equilibrium is

$$\sum F_x = n_x + T_x + f_x + w_x = 0$$

$$\sum F_y = n_y + T_y + f_y + w_y = 0$$

We can again determine the horizontal and vertical components of the forces by "reading" the free-body diagram. The results are shown in the following table.

Force	Name of x-component	Value of x-component	Name of y-component	Value of y-component
$\vec{n}$	n_x	0	n_y	n
$\vec{T}$	T_x	$T\cos\theta$	T_y	$T\sin\theta$
$\vec{f}$	f_x	$-f$	f_y	0
$\vec{w}$	w_x	0	w_y	$-w$

With these components, Newton's first law reads

$$T\cos\theta - f = 0$$

$$T\sin\theta + n - w = 0$$

The first equation can be used to solve for the tension in the rope:

$$T = \frac{f}{\cos\theta} = \frac{320\text{ N}}{\cos 20°} = 340\text{ N}$$

to two significant figures. It turned out that we did not need the y-component equation in this problem. We would need it if we wanted to find the normal force $\vec{n}$.

ASSESS Had we pulled the car with a horizontal rope, the tension would need to exactly balance the friction force of 320 N. Because we are pulling at an angle, however, part of the tension in the rope pulls *up* on the car instead of in the forward direction. Thus we need a little more tension in the rope when it's at an angle.

5.2 Dynamics and Newton's Second Law

Newton's second law is the essential link between force and motion. The essence of Newtonian mechanics can be expressed in two steps:

Activ
Physics
ONLINE   2.1–2.4

■ The forces acting on an object determine its acceleration $\vec{a} = \vec{F}_{\text{net}}/m$.
■ The object's motion can be found by using $\vec{a}$ in the equations of kinematics.

We want to develop a strategy to solve a variety of problems in mechanics, but first we need to write the second law in terms of its components. To do so, let's first rewrite Newton's second law in the form

$$\vec{F}_{\text{net}} = \vec{F}_1 + \vec{F}_2 + \vec{F}_3 + \cdots = m\vec{a}$$

where $\vec{F}_1, \vec{F}_2, \vec{F}_3$, and so on are the forces acting on an object. To write the second law in component form merely requires that we use the x- and y-components of the acceleration. Thus Newton's second law, $\vec{F}_{\text{net}} = m\vec{a}$, is

$$\sum F_x = ma_x \quad \text{and} \quad \sum F_y = ma_y \qquad (5.2)$$

Newton's second law in component form

The first equation says that **the component of the acceleration in the x-direction is determined by the sum of the x-components of the forces acting on the object.** A similar statement applies to the y-direction.

There are two basic types of problems in mechanics. In the first, you use information about forces to find an object's acceleration, then use kinematics to determine the object's motion. In the second, you use information about the object's motion to determine its acceleration, then solve for unknown forces. Either way, the two equations of Equation 5.2 are the link between force and motion, and they form the basis of a problem-solving strategy. The primary goal of this chapter is to illustrate the use of this strategy. We'll then follow the strategy with some examples.

PROBLEM-SOLVING STRATEGY 5.2 Dynamics problems

PREPARE Sketch a visual overview consisting of

- A list of values that identifies known quantities and what the problem is trying to find.
- A force-identification diagram to help you identify all forces acting on the object.
- A free-body diagram that shows all the forces acting on the object.

If you'll need to use kinematics to find velocities or positions, you'll also need to sketch

- A motion diagram to determine the direction of the acceleration.
- A pictorial representation that establishes a coordinate system, shows important points in the motion, and defines symbols.

It's OK to go back and forth between these steps as you visualize the situation.

SOLVE Write Newton's second law in component form as

$$\sum F_x = ma_x \quad \text{and} \quad \sum F_y = ma_y$$

You can find the components of the forces directly from your free-body diagram. Depending on the problem, either

- Solve for the acceleration, then use kinematics to find velocities and positions, or
- Use kinematics to determine the acceleration, then solve for unknown forces.

ASSESS Check that your result has the correct units, is reasonable, and answers the question.

EXAMPLE 5.4 Putting a golf ball

A golfer putts a 46 g ball with a speed of 3.0 m/s. Friction exerts a 0.020 N retarding force on the ball, slowing it down. Will her putt reach the hole, 10 m away?

PREPARE Figure 5.6 is a visual overview of the problem. We've collected the known information, drawn a sketch, and identified what we want to find. The motion diagram shows that the ball is slowing down as it rolls to the right, so the acceleration vector points to the left. Next, we identify the forces acting on the ball and show them on a free-body diagram. Note that the net force points to the left, as it must because the acceleration points to the left.

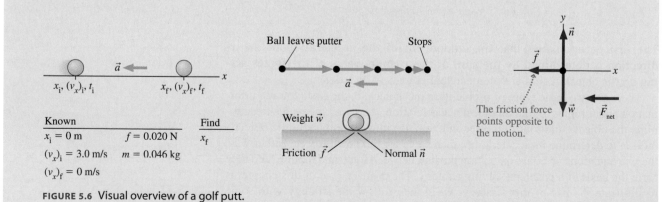

FIGURE 5.6 Visual overview of a golf putt.

Known		Find
$x_i = 0$ m	$f = 0.020$ N	x_f
$(v_x)_i = 3.0$ m/s	$m = 0.046$ kg	
$(v_x)_f = 0$ m/s		

SOLVE Newton's second law in component form is

$$\sum F_x = n_x + f_x + w_x = 0 - f + 0 = ma_x$$

$$\sum F_y = n_y + f_y + w_y = n + 0 - w = ma_y = 0$$

We've written the equations as sums, as we did with equilibrium problems, then "read" the values of the force components from the free-body diagram. The components are simple enough in this problem that we don't really need to show them in a table. It is particularly important to notice that we set $a_y = 0$ in the second equation. This is because the ball does not move in the y-direction, so it can't have any acceleration in the y-direction. This will be an important step in many problems.

The first equation is $-f = ma_x$, from which we find

$$a_x = -f/m = -(0.020 \text{ N})/(0.046 \text{ kg}) = -0.43 \text{ m/s}^2$$

(Recall from Chapter 4 that $1 \text{ N} = 1 \text{ kg} \cdot \text{m/s}^2$, so the units above work out correctly.) The negative sign shows that the acceleration is directed to the left, as expected.

Now that we know the acceleration, we can use kinematics to find how far the ball will roll before stopping. We don't have any information about the time it takes for the ball to stop, so we'll use the kinematic equation $(v_x)_f^2 = (v_x)_i^2 + 2a_x(x_f - x_i)$. This gives

$$x_f = x_i + \frac{(v_x)_f^2 - (v_x)_i^2}{2a_x} = 0 \text{ m} + \frac{(0 \text{ m/s})^2 - (3.0 \text{ m/s})^2}{2(-0.43 \text{ m/s}^2)} = 10.5 \text{ m}$$

If her aim is true, the ball will just make it into the hole.

ASSESS The key steps in any dynamics problem are to draw a correct free-body diagram and to use the free-body diagram to write Newton's second law in component form. Once you've mastered these two steps, you're well on your way to solving any dynamics problem!

EXAMPLE 5.5 Finding a rocket cruiser's acceleration

A rocket cruiser with a mass of 2200 kg and weighing 5000 N is flying horizontally over the surface of a distant planet. At its present speed, a 3000 N drag force acts on the cruiser. The cruiser's engines can be tilted so as to provide a thrust angled up or down. The pilot turns the thrust up to 14,000 N while pivoting the engines to continue flying horizontally. What is the cruiser's acceleration?

PREPARE Figure 5.7 is a visual overview in which we've listed known information, identified the forces on the cruiser, and drawn a free-body diagram. (Because kinematics is not needed to find the acceleration, we don't need a pictorial diagram.) As discussed in Chapter 4, the thrust force points *opposite* to the direction of the rocket exhaust, which we've shown at angle θ. The thrust must be directed upward to balance the weight force; otherwise the cruiser would fall. To continue flying horizontally requires the net force to be directed forward.

SOLVE Newton's second law in component form is

$$\sum F_x = (F_{\text{thrust}})_x + D_x + w_x = ma_x$$

$$\sum F_y = (F_{\text{thrust}})_y + D_y + w_y = ma_y$$

From the free-body diagram, we see that $(F_{\text{thrust}})_x = F_{\text{thrust}}\cos\theta$, $(F_{\text{thrust}})_y = F_{\text{thrust}}\sin\theta$, $D_x = -D$, $D_y = 0$, $w_x = 0$, and $w_y = -w$. We know that a_y must be zero because the cruiser is to accelerate *horizontally*. Thus the second law becomes

$$F_{\text{thrust}}\cos\theta - D = ma_x$$

$$F_{\text{thrust}}\sin\theta - w = 0$$

The first of these equations contains a_x, the quantity we want to find, but we can't solve for a_x without knowing what θ is. Fortunately, we can use the second equation to find θ, then use this value of θ in the first equation to find a_x.

The second equation gives

$$\sin\theta = w/F_{\text{thrust}} = (5000 \text{ N})/(14,000 \text{ N}) - 0.357$$

$$\theta = \sin^{-1}(0.357) = 20.9°$$

Now we can use this value in the first equation to get

$$a_x = \frac{1}{m}(F_{\text{thrust}}\cos\theta - D)$$

$$= \frac{1}{2200 \text{ kg}}[(14,000 \text{ N})\cos(20.9°) - 3000 \text{ N}] = 4.6 \text{ m/s}^2$$

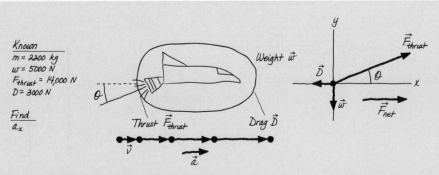

Known
$m = 2200$ kg
$w = 5000$ N
$F_{\text{thrust}} = 14,000$ N
$D = 3000$ N

Find
a_x

Weight $\vec{w}$

Thrust $\vec{F}_{\text{thrust}}$ Drag $\vec{D}$

FIGURE 5.7 Visual overview of a rocket cruiser.

Continued

ASSESS An important key to solving this problem was to use the information that the cruiser accelerates only in the horizontal direction. Mathematically, this means that $a_y = 0$. Because the thrust is much greater than the weight, we need only a mod- est downward component of the thrust to cancel the weight and let the cruiser accelerate horizontally. So our engine tilt seems reasonable.

EXAMPLE 5.6 Towing a car with acceleration

A car with a mass of 1500 kg is being towed by a rope held at a 20° angle. A friction force of 320 N opposes the car's motion. What is the tension in the rope if the car goes from rest to 12 m/s in 10 s?

PREPARE You should recognize that this problem is almost identical to Example 5.3. The difference is that the car is now accelerating, so it is no longer in equilibrium. This means, as shown in Figure 5.8, that the net force is not zero. We've already identified all the forces in Example 5.3.

SOLVE Newton's second law in component form is

$$\sum F_x = n_x + T_x + f_x + w_x = ma_x$$

$$\sum F_y = n_y + T_y + f_y + w_y = ma_y = 0$$

We've again used the fact that $a_y = 0$ for motion that is purely along the x-axis. The components of the forces were worked out in Example 5.3. Using that information, Newton's second law in component form is

$$T\cos\theta - f = ma_x$$

$$T\sin\theta + n - w = 0$$

Because the car speeds up from rest to 12 m/s in 10 s, we can use kinematics to find the acceleration:

$$a_x = \frac{\Delta v_x}{\Delta t} = \frac{(v_x)_f - (v_x)_i}{t_f - t_i} = \frac{(12\ \text{m/s}) - (0\ \text{m/s})}{(10\ \text{s}) - (0\ \text{s})} = 1.2\ \text{m/s}^2$$

We can now use the first Newton's-law equation above to solve for the tension. We have

$$T = \frac{ma_x + f}{\cos\theta} = \frac{(1500\ \text{kg})(1.2\ \text{m/s}^2) + 320\ \text{N}}{\cos 20°} = 2300\ \text{N}$$

ASSESS The tension is substantially more than the 341 N found in Example 5.3. It takes much more force to accelerate the car than to keep it rolling at constant speed.

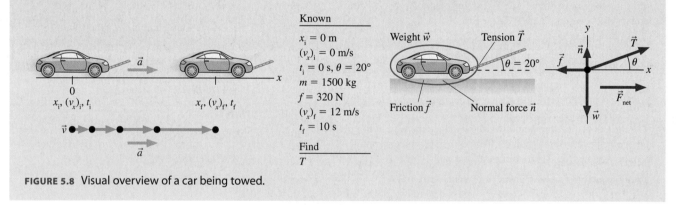

Known

$x_i = 0$ m
$(v_x)_i = 0$ m/s
$t_i = 0$ s, $\theta = 20°$
$m = 1500$ kg
$f = 320$ N
$(v_x)_f = 12$ m/s
$t_f = 10$ s

Find

T

FIGURE 5.8 Visual overview of a car being towed.

These first examples have shown all the details of our problem-solving strategy. Our purpose has been to demonstrate how the strategy is put into practice. Future examples will be briefer, but the basic *procedure* will remain the same.

STOP TO THINK 5.1 A Martian lander is approaching the surface. It is slowing its descent by firing its rocket motor. Which is the correct free-body diagram for the lander?

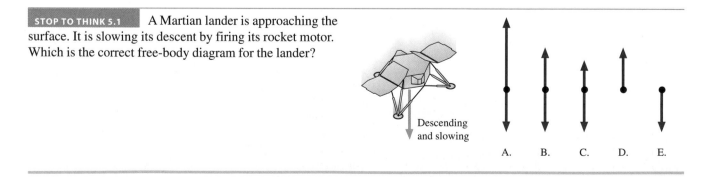

5.3 Mass and Weight

When the doctor asks what you weigh, what does she really mean? We do not make much distinction in our ordinary use of language between the terms *weight* and *mass,* but in physics their distinction is of critical importance.

Mass, you'll recall from Chapter 4, is a quantity that describes an object's inertia, its tendency to resist being accelerated. Loosely speaking, it also describes the amount of matter in an object. Mass, measured in kilograms, is an intrinsic property of an object; it has the same value wherever the object may be and whatever forces might be acting on it.

Weight, on the other hand, is a *force.* Specifically, it is the gravitational force exerted on an object by a planet. Weight is a vector, not a scalar, and the vector's direction is always straight down. Weight is measured in newtons.

Mass and weight are not the same thing, but they are related. We can use Galileo's discovery about free fall to make the connection. Figure 5.9 shows the free-body diagram of an object in free fall. The *only* force acting on this object is its weight $\vec{w}$, the downward pull of gravity. Newton's second law for this object is

$$\vec{F}_{\text{net}} = \vec{w} = m\vec{a} \tag{5.3}$$

Recall Galileo's discovery that *any* object in free fall, regardless of its mass, has the same acceleration:

$$\vec{a}_{\text{free fall}} = (g, \text{downward}) \tag{5.4}$$

where $g = 9.80 \text{ m/s}^2$ is the free-fall acceleration at the earth's surface. So a_y in Equation 5.3 is equal to $-g$, and we have $-w = -mg$, or

$$w = mg \tag{5.5}$$

The magnitude of the weight force, which we call simply "the weight," is directly proportional to the mass, with g as the constant of proportionality. Thus, for example, the weight of a 3.6 kg book is $w = (3.6 \text{ kg})(9.8 \text{ m/s}^2) = 35 \text{ N}$.

NOTE ▶ Although we derived the relationship between mass and weight for an object in free fall, the weight of an object is *independent* of its state of motion. Equation 5.5 holds for an object at rest on a table, sliding horizontally, or moving in any other way. ◀

Because an object's weight depends on g, and the value of g varies from planet to planet, weight is not a fixed, constant property of an object. The value of g at the surface of the moon is about one-sixth its earthly value, so an object on the moon would have only one-sixth its weight on earth. The object's weight on Jupiter would be larger than its weight on earth. Its mass, however, would be the same. The amount of matter has not changed, only the gravitational force exerted on that matter.

So when the doctor asks what you weigh, she really wants to know your *mass.* That's the amount of matter in your body. You can't really "lose weight" by going to the moon, even though you would weigh less there!

Measuring Mass and Weight

A *pan balance,* shown in Figure 5.10, is a device for measuring *mass.* You may have used a pan balance to "weigh" chemicals in a chemistry lab. An unknown mass is placed in one pan, then known masses are added to the other until the pans balance. Gravity pulls down on both sides, effectively *comparing* the masses, and the unknown mass equals the sum of the known masses that balance it. Although a pan balance requires gravity in order to function, it does not depend on the value of g. Consequently, the pan balance would give the same result on another planet.

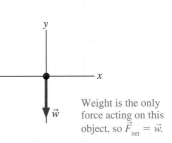

Weight is the only force acting on this object, so $\vec{F}_{\text{net}} = \vec{w}$.

FIGURE 5.9 The free-body diagram of an object in free fall.

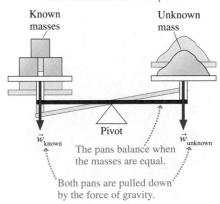

On the moon, astronaut John Young jumps 2 feet straight up, despite his spacesuit that weighs 370 pounds on earth. On the moon, where $g = 1.6 \text{ m/s}^2$, he and his suit together weighed only 90 pounds.

If the unknown mass differs from the known masses, the beam will rotate about the pivot.

Known masses   Unknown mass

Pivot

$\vec{w}_{\text{known}}$ The pans balance when the masses are equal. $\vec{w}_{\text{unknown}}$

Both pans are pulled down by the force of gravity.

FIGURE 5.10 A pan balance measures mass.

(a)

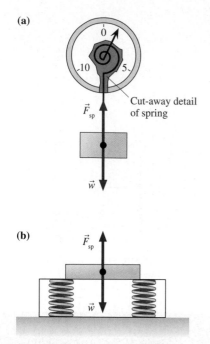

$\vec{F}_{sp}$ Cut-away detail of spring

$\vec{w}$

(b)

$\vec{F}_{sp}$

$\vec{w}$

FIGURE 5.11 A spring scale measures weight.

Spring scales, such as the two shown in Figure 5.11, measure weight, not mass. Hanging an item on the scale in Figure 5.11a, which might be used to weigh items in the grocery store, stretches the spring. The spring in the "bathroom scale" in Figure 5.11b is compressed when you stand on it.

A spring scale can be understood on the basis of Newton's first law. The object being weighed is at rest, in static equilibrium, so the net force on it must be zero. The stretched spring in Figure 5.11a *pulls* up on the object with force $\vec{F}_{sp}$; the compressed spring in Figure 5.11b *pushes* up with force $\vec{F}_{sp}$. But in both cases, in order to have $\vec{F}_{net} = \vec{0}$, the upward spring force must exactly balance the downward weight force:

$$F_{sp} = w = mg \qquad (5.6)$$

The *reading* of a spring scale is F_{sp}, the magnitude of the force that the spring is exerting. If the object is in equilibrium, then F_{sp} is exactly equal to the object's weight w. The scale does not "know" the weight of the object. All it can do is to measure how much its spring is stretched or compressed. On a different planet, with a different value for g, the expansion or compression of the spring would be different and the scale's reading would be different.

The unit of force in the English system is the *pound.* We noted in Chapter 4 that the pound is defined as 1 lb = 4.45 N. An object whose weight $w = mg$ is 4.45 N has a mass

$$m = \frac{w}{g} = \frac{4.45 \text{ N}}{9.80 \text{ m/s}^2} = 0.454 \text{ kg} = 454 \text{ g}$$

You may have learned in previous science classes that "1 pound = 454 grams" or, equivalently, that "1 kg = 2.2 lb." Strictly speaking, these well-known "conversion factors" are not true. They are comparing a weight (pounds) to a mass (kilograms). The correct statement would be, "A mass of 1 kg has a weight on *earth* of 2.2 pounds." On another planet, the weight of a 1 kg mass would be something other than 2.2 pounds.

EXAMPLE 5.7 Masses of people

What is the mass, in kilograms, of a 90 pound gymnast, a 160 pound professor, and a 240 pound football player?

SOLVE We must convert their weights into newtons; then we can find their masses from $m = w/g$. We have

$$w_{gymnast} = 90 \text{ lb} \times \frac{4.45 \text{ N}}{1 \text{ lb}} = 400 \text{ N} \qquad m_{gymnast} = \frac{w_{gymnast}}{g} = \frac{400 \text{ N}}{9.80 \text{ m/s}^2} = 41 \text{ kg}$$

$$w_{prof} = 160 \text{ lb} \times \frac{4.45 \text{ N}}{1 \text{ lb}} = 710 \text{ N} \qquad m_{prof} = \frac{w_{prof}}{g} = \frac{710 \text{ N}}{9.80 \text{ m/s}^2} = 72 \text{ kg}$$

$$w_{football} = 240 \text{ lb} \times \frac{4.45 \text{ N}}{1 \text{ lb}} = 1070 \text{ N} \qquad m_{football} = \frac{w_{football}}{g} = \frac{1070 \text{ N}}{9.80 \text{ m/s}^2} = 110 \text{ kg}$$

ASSESS It's worth remembering that a *typical* adult has a mass in the range of 60 to 80 kg.

This popular amusement park ride shoots you straight up with an acceleration of $4g$. Your apparent weight is then five times your true weight.

Apparent Weight

The weight of an object is the force of gravity on that object. You may never have thought about it, but gravity is not a force that you can feel or sense directly. Your *sensation* of weight—how heavy you feel—is due to *contact forces* pressing against you. Surfaces touch you and activate nerve endings in your skin. As you read this, your sensation of weight is due to the normal force exerted on you by the chair in which you are sitting. When you stand, you feel the contact force of the floor pushing against your feet. If you hang from a rope, your sensation of weight is due to the tension force pulling up on you.

Figure 5.12a shows how a standing man *feels* the normal force of the ground on his feet. The free-body diagram of Figure 5.12b shows that if he is at rest, with $\vec{a} = \vec{0}$, then the normal force pressing against his feet is exactly equal in magnitude to his weight. So the relationship between your weight and the forces you feel would be simple if objects were always at rest, but what happens if $\vec{a} \neq \vec{0}$?

Recall the sensations you feel while being accelerated. You feel "heavy" when an elevator suddenly accelerates upward or when an airplane accelerates for take-off. This sensation vanishes as soon as the elevator or airplane reaches a steady cruising speed. Your stomach seems to rise a little and you feel lighter than normal as the upward-moving elevator brakes to a halt or a roller coaster goes over the top. Your true weight $w = mg$ has not changed during these events, but your *apparent weight* has.

To investigate this, imagine a man weighing himself by standing on a spring scale in an elevator as it accelerates upward. What does the scale read? How does the scale reading correspond to the man's *sensation* of weight?

As Figure 5.13 shows, the only forces acting on the man are the upward spring force of the scale and the downward weight force. This seems to be the same situation as Figure 5.12, but there's one big difference. The man is accelerating; he's not in equilibrium. Thus, according to Newton's second law, there must be a net force acting on the man in the direction of $\vec{a}$.

For the net force $\vec{F}_{net}$ to point upward, the magnitude of the spring force must be *greater* than the magnitude of the weight force. That is, $F_{sp} > w$. This conclusion has major implications. Looking at the free-body diagram in Figure 5.13, we see that the y-component of Newton's second law is

$$\sum F_y = (F_{sp})_y + w_y = F_{sp} - w = ma_y \qquad (5.7)$$

where m is the man's mass.

The scale reading is the value of F_{sp}, the magnitude of the force that the scale exerts on the man. Solving Equation 5.7 for F_{sp} gives

$$F_{sp} = w + ma_y \qquad (5.8)$$

If the elevator is either at rest or moving with constant velocity, then $a_y = 0$ and the man is in equilibrium. In that case, $F_{sp} = w$ and the scale correctly reads his weight. But if $a_y \neq 0$, the scale's reading is *not* the man's true weight.

Let's define an object's **apparent weight** w_{app} as the magnitude of the contact force that supports the object. From Equation 5.8, this is

$$w_{app} = w + ma_y = mg + ma_y = m(g + a_y) \qquad (5.9)$$

If the elevator is accelerating upward, then $a_y = +a$, and Equation 5.9 reads $w_{app} = w + ma$. Thus $w_{app} > w$ and the man *feels* heavier than normal. If the elevator is accelerating downward, the acceleration vector $\vec{a}$ points downward and $a_y = -a$. Thus $w_{app} < w$ and the man feels lighter. Indeed, the scale reads less than his true weight.

An object doesn't have to be on a scale for its apparent weight to differ from its true weight. An object's apparent weight is the magnitude of the contact force supporting it. It makes no difference whether this is the spring force of the scale or simply the normal force of the floor.

The idea of apparent weight has important applications. Astronauts are nearly crushed by their apparent weight during a rocket launch when a is much greater than g. Much of the thrill of amusement park rides, such as roller coasters, comes from rapid changes in your apparent weight.

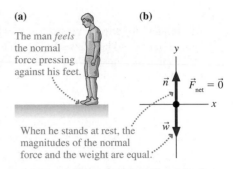

The man *feels* the normal force pressing against his feet.

When he stands at rest, the magnitudes of the normal force and the weight are equal.

FIGURE 5.12 The sensation of weight for a man at rest.

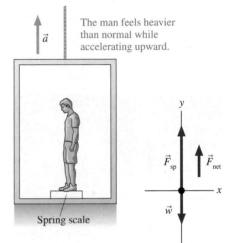

The man feels heavier than normal while accelerating upward.

Spring scale

FIGURE 5.13 A man weighing himself in an accelerating elevator.

TRY IT YOURSELF

Physics students can't jump The next time you ride up in an elevator, try jumping in the air just as the elevator starts to rise. You'll feel like you can hardly get off the ground. This is because with $a_y > 0$ your apparent weight is *greater* than your actual weight; for an elevator with a large acceleration it's like trying to jump while carrying an extra 20 pounds. What will happen if you jump as the elevator slows at the top?

EXAMPLE 5.8 Apparent weight in an elevator

Anjay's mass is 70 kg. He's standing on a scale in an elevator. As the elevator stops, the scale reads 750 N. Had the elevator been moving up or down? If the elevator had been moving at 5.0 m/s, how long does it take to stop?

PREPARE The scale reading as he stops is his apparent weight, so w_{app} = 750 N. Because we know his mass m, we can then use Equation 5.9 to find the elevator's acceleration a_y. Then we can use kinematics to find the time it takes to stop the elevator.

SOLVE From Equation 5.9 we have $w_{app} = m(g + a_y)$, so that

$$a_y = \frac{w_{app}}{m} - g = \frac{750 \text{ N}}{70 \text{ kg}} - 9.80 \text{ m/s}^2 = 0.91 \text{ m/s}^2$$

This is a *positive* acceleration. If the elevator is stopping with a positive acceleration it must have been moving *down*, with a negative velocity.

To find the stopping time, we can use the kinematic equation

$$(v_y)_f = (v_y)_i + a_y \, \Delta t$$

to get

$$\Delta t = \frac{(v_y)_f - (v_y)_i}{a_y} = \frac{(0 \text{ m/s}) - (-5.0 \text{ m/s})}{0.91 \text{ m/s}^2} = 5.5 \text{ s}$$

Notice that we used -5.0 m/s as the initial velocity because the elevator was moving down before it stopped.

ASSESS Anjay's true weight is mg = (70 kg)(9.8 m/s²) = 670 N. Thus his apparent weight is *greater* than his true weight. You have no doubt experienced this sensation in an elevator that is stopping as it reaches the ground floor. If it had stopped while going up, you'd feel *lighter* than your true weight.

Weightlessness

A weightless experience You probably wouldn't want to experience weightlessness in a falling elevator. But, as we learned in Chapter 3, objects undergoing projectile motion are in free fall as well. The special plane shown flies in the same parabolic trajectory as would a projectile with no air resistance. Objects inside, such as these passengers, are then moving along a perfect free fall trajectory. Just as for the man in the elevator, they then float with respect to the planc's interior. Such flights can last up to 30 seconds.

One last issue before leaving this topic: Suppose the elevator cable breaks and the elevator, along with the man and his scale, plunges straight down in free fall! What will the scale read? The acceleration in free fall is $a_y = -g$. When this acceleration is used in Equation 5.9, we find that w_{app} = 0! In other words, the man has *no sensation* of weight.

Think about this carefully. Suppose, as the elevator falls, the man inside releases a ball from his hand. In the absence of air resistance, as Galileo discovered, both the man and the ball would fall at the same rate. From the man's perspective, the ball would appear to "float" beside him. Similarly, the scale would float beneath him and not press against his feet. He is what we call *weightless*.

Surprisingly, "weightless" does *not* mean "no weight." An object that is **weightless** has no *apparent* weight. The distinction is significant. The man's weight is still mg, because gravity is still pulling down on him, but he has no *sensation* of weight as he free falls. The term "weightless" is a very poor one, likely to cause confusion because it implies that objects have no weight. As we see, that is not the case.

But isn't this exactly what happens to astronauts orbiting the earth? You've seen films of astronauts and various objects floating inside the Space Shuttle. If an astronaut tries to stand on a scale, it does not exert any force against her feet and reads zero. She is said to be weightless. But if the criterion to be weightless is to be in free fall, and if astronauts orbiting the earth are weightless, does this mean that they are in free fall? This is a very interesting question to which we shall return in Chapter 6.

STOP TO THINK 5.2 An elevator that has descended from the 50th floor is coming to a halt at the 1st floor. As it does, your apparent weight is

A. More than your true weight.
B. Less than your true weight.
C. Equal to your true weight.
D. Zero.

5.4 Normal Forces

In Chapter 4 we saw that an object at rest on a table is subject to an upward force due to the table. This force is called the *normal force* because it is always directed normal, or perpendicular, to the surface of contact. As we saw, the normal force has its origin in the atomic "springs" that make up the surface. The harder the object bears down on the surface, the more these springs are compressed and the harder they push back. Thus the normal force *adjusts* itself so that the object stays on the surface without penetrating it. This fact is key in solving for the normal force.

EXAMPLE 5.9 Normal force on a pressed book

A 1.2 kg book lies on a table. The book is pressed down from above with a force of 15 N. What is the normal force acting on the book from the table below?

PREPARE The book is not moving and is thus in static equilibrium. We need to identify the forces acting on the book, and prepare a free-body diagram showing these forces. These steps are illustrated in Figure 5.14.

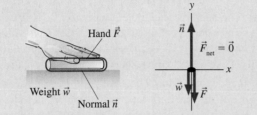

FIGURE 5.14 Finding the normal force of a book pressed from above.

SOLVE Because the book is in static equilibrium, the net force on it must be zero. The only forces acting are in the y-direction, so Newton's first law is

$$\sum F_y = n_y + w_y + F_y = n - w - F = 0$$

We learned in the last section that the weight force is $w = mg$. The weight of the book is thus

$$w = mg = (1.2 \text{ kg})(9.8 \text{ m/s}^2) = 12 \text{ N}$$

With this information, we see that the normal force exerted by the table is

$$n = F + w = 15 \text{ N} + 12 \text{ N} = 27 \text{ N}$$

ASSESS The magnitude of the normal force is *larger* than the weight of the book. From the table's perspective, the extra force from the hand pushes the book further into the atomic springs of the table. These springs then push back harder, giving a larger normal force.

A common situation is that of an object on a ramp or incline. If friction is neglected, there are only two forces acting on the object: gravity and the normal force. However, we need to carefully work out the components of these two forces. Figure 5.15a shows how. Be sure you avoid the two common errors shown in Figure 5.15b.

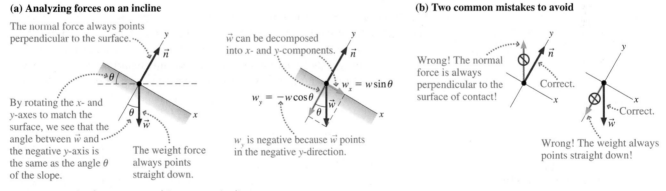

(a) Analyzing forces on an incline

The normal force always points perpendicular to the surface.

By rotating the x- and y-axes to match the surface, we see that the angle between $\vec{w}$ and the negative y-axis is the same as the angle θ of the slope.

The weight force always points straight down.

$\vec{w}$ can be decomposed into x- and y-components.

$w_x = w\sin\theta$

$w_y = -w\cos\theta$

w_y is negative because $\vec{w}$ points in the negative y-direction.

(b) Two common mistakes to avoid

Wrong! The normal force is always perpendicular to the surface of contact!

Correct.

Correct.

Wrong! The weight always points straight down!

FIGURE 5.15 The forces on an object on an incline.

EXAMPLE 5.10 Acceleration of a downhill skier

A skier slides down a steep slope of 27° on ideal, frictionless snow. What is his acceleration?

PREPARE Figure 5.16 on the next page is a visual overview. We choose a coordinate system tilted so that the x-axis points down

the slope. This greatly simplifies the analysis, because with this choice $a_y = 0$ (the skier does not move in the y-direction at all). The free-body diagram is based on the information in Figure 5.15.

Continued

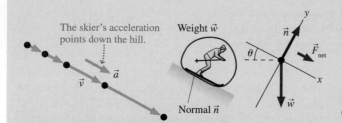

FIGURE 5.16 Visual overview of a downhill skier.

SOLVE We can now use Newton's second law in component form to find the skier's acceleration. We have

$$\sum F_x = w_x + n_x = ma_x$$

$$\sum F_y = w_y + n_y = ma_y$$

Because $\vec{n}$ points directly in the positive y-direction, $n_y = n$ and $n_x = 0$. Figure 5.15a showed the important fact that the angle between $\vec{w}$ and the negative y-axis is the *same* as the slope angle θ. With this information, the components of $\vec{w}$ are $w_x = w\sin\theta = mg\sin\theta$ and $w_y = -w\cos\theta = -mg\cos\theta$, where we used the

fact that $w = mg$. With these components in hand, Newton's second law becomes

$$\sum F_x = w_x + n_x = mg\sin\theta = ma_x$$

$$\sum F_y = w_y + n_y = -mg\cos\theta + n = ma_y = 0$$

In the second equation we used the fact that $a_y = 0$. The m cancels in the first Newton's-law equation, leaving us with

$$a_x = g\sin\theta$$

This is the expression for acceleration on a surface that we presented, without proof, in Chapter 3. Now we've justified our earlier assertion. We can use this to calculate the skier's acceleration:

$$a_x = g\sin\theta = (9.8 \text{ m/s}^2)\sin(27°) = 4.4 \text{ m/s}^2$$

ASSESS The skier's acceleration falls between $a_x = 0$, which it would be on a horizontal surface, and $a_x = 9.80 \text{ m/s}^2$, its value for a vertical drop. This seems reasonable. Notice that the mass cancelled out, so we didn't need to know the skier's mass.

5.5 Friction

In everyday life, friction is everywhere. Friction is absolutely essential for many things we do. Without friction you could not walk, drive, or even sit down (you would slide right off the chair!). It is sometimes useful to think about idealized frictionless situations, but it is equally necessary to understand a real world where friction is present. Although friction is a complicated force, many aspects of friction can be described with a simple model.

Static Friction

(a) Force identification

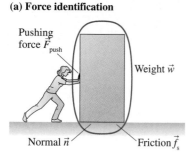

(b) Free-body diagram

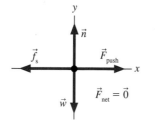

FIGURE 5.17 Static friction keeps an object from slipping.

Chapter 4 defined *static friction* $\vec{f}_s$ as the force on an object that keeps it from slipping. Figure 5.17a shows a woman pushing on a box with horizontal force $\vec{F}_{push}$. If the box remains at rest, "stuck" to the floor, it must be because of a static friction force pushing back to the left. The box is in static equilibrium, so, as shown in 5.17b, the static friction must exactly balance the pushing force:

$$f_s = F_{push}$$

To determine the *direction* of $\vec{f}_s$, decide which way the object would move if there were no friction. The static friction force $\vec{f}_s$ then points in the opposite direction, to prevent the motion. Determining the *magnitude* of $\vec{f}_s$ is a bit trickier. Unlike weight, which has the precise and unambiguous magnitude $w = mg$, the magnitude f_s of the static friction force depends on how hard you push. As shown in Figures 5.18a and 5.18b on the next page, the harder the woman pushes, the harder the friction force from the floor pushes back. If she reduces her pushing force, the friction force will automatically be reduced to match. Static friction acts in *response* to an applied force.

But there's clearly a limit to how big f_s can get. If you push hard enough, the object will slip and start to move. In other words, the static friction force has a *maximum* possible magnitude $f_{s\,max}$, as illustrated in Figure 5.18c. Experiments with friction (first done by Leonardo da Vinci) show that $f_{s\,max}$ is proportional to the magnitude of the normal force between the surface and the object. That is,

$$f_{s\,max} = \mu_s n \qquad (5.10)$$

where μ_s is called the **coefficient of static friction.** The coefficient is a number that depends on the materials of which the object and the surface are made. The higher the coefficient of static friction, the greater the "stickiness" between the object and the surface, and the harder it is to make the object slip. Table 5.1 shows some typical values of coefficients of friction. It is to be emphasized that these are only approximate. The exact value of the coefficient depends on the roughness, cleanliness, and dryness of the surfaces.

NOTE ▶ Equation 5.10 does *not* say $f_s = \mu_s n$. The value of f_s depends on the force or forces that static friction has to balance to keep the object from moving. It can have any value from zero up to, but not exceeding, $\mu_s n$. ◀

It is interesting to note that $f_{s\,max}$ does *not* depend on the contact area between the object and the surface on which it rests.

So our rules for static friction are

- The direction of static friction is such as to oppose motion.
- The magnitude f_s of static friction adjusts itself so that the net force is zero and the object doesn't move.
- The magnitude of static friction cannot exceed the maximum value $f_{s\,max}$ given by Equation 5.10. If the friction force needed to keep the object stationary is larger than $f_{s\,max}$, the object slips and starts to move.

Kinetic Friction

Once the box starts to slide, as in Figure 5.19, the static friction force is replaced by a kinetic (or sliding) friction force $\vec{f}_k$. Kinetic friction is in some ways simpler than static friction: The direction of $\vec{f}_k$ is always opposite to the direction in which an object slides across the surface, and experiments show that kinetic friction, unlike static friction, has a nearly *constant* magnitude, given by

$$f_k = \mu_k n \qquad (5.11)$$

where μ_k is called the **coefficient of kinetic friction.** Equation 5.11 also shows that kinetic friction, like static friction, is proportional to the magnitude of the

(a) Pushing gently: friction pushes back gently.

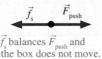

$\vec{f}_s$ balances $\vec{F}_{push}$ and the box does not move.

(b) Pushing harder: friction pushes back harder.

$\vec{f}_s$ grows as $\vec{F}_{push}$ increases, but they still cancel and the box remains at rest.

(c) Pushing harder still: $\vec{f}_s$ is now pushing back as hard as it can.

Now the magnitude of f_s has reached its maximum value $f_{s\,max}$. If $\vec{F}_{push}$ gets any bigger, the forces will *not* cancel and the box will start to move.

FIGURE 5.18 Static friction acts in *response* to an applied force.

TABLE 5.1 Coefficients of friction

Materials	Static μ_s	Kinetic μ_k	Rolling μ_r
Rubber on concrete	1.00	0.80	0.02
Steel on steel (dry)	0.80	0.60	0.002
Steel on steel (lubricated)	0.10	0.05	
Wood on wood	0.50	0.20	
Wood on snow	0.12	0.06	
Ice on ice	0.10	0.03	

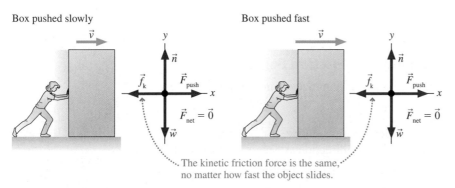

FIGURE 5.19 The kinetic friction force is *opposite* to the direction of motion.

FIGURE 5.20 The bottom of the wheel is stationary.

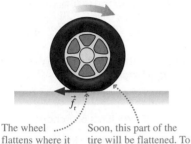

The wheel flattens where it touches the road.

Soon, this part of the tire will be flattened. To flatten it the road must push *back* on the tire.

FIGURE 5.21 Rolling friction is due to deformation of a wheel.

normal force n. Notice that **the magnitude of the kinetic friction force does not depend on how fast the object is sliding.**

Table 5.1 includes typical values of μ_k. You can see that $\mu_k < \mu_s$, which explains why it is easier to keep a box moving than it was to start it moving.

Rolling Friction

If you slam on the brakes hard enough, your car tires slide against the road surface and leave skid marks. This is kinetic friction, because the tire and the road are *sliding* against each other. A wheel *rolling* on a surface also experiences friction, but not kinetic friction: The portion of the wheel that contacts the surface is stationary with respect to the surface, not sliding. The photo in Figure 5.20 was taken with a stationary camera. Note how the part of the wheel touching the ground is not blurred, indicating that this part of the wheel is not moving with respect to the ground.

Textbooks draw wheels as circles, but no wheel is perfectly round. The weight of the wheel, and of any object supported by the wheel, causes the bottom of the wheel to flatten where it touches the surface, as Figure 5.21 shows. As a wheel rolls forward, the leading part of the tire must become deformed. This requires that the road push *backward* on the tire. In this way the road causes a backward force, even without slipping between the tire and the road.

The force of this *rolling friction* can be calculated in terms of a **coefficient of rolling friction** μ_r:

$$f_r = \mu_r n \tag{5.12}$$

with the *direction* of the force opposing the direction of motion. Thus rolling friction acts very much like kinetic friction, but values of μ_r (see Table 5.1) are much less than values of μ_k. This is why it is easier to roll an object on wheels than to slide it.

STOP TO THINK 5.3   Rank in order, from largest to smallest, the size of the friction forces $\vec{f}_A$ to $\vec{f}_E$ in the 5 different situations (one or more friction forces could be zero). The box and the floor are made of the same materials in all situations.

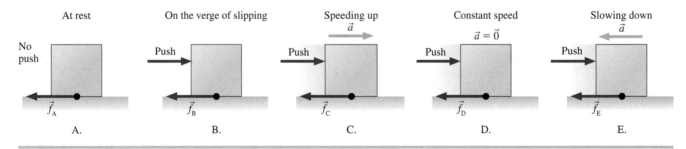

| At rest | On the verge of slipping | Speeding up $\vec{a}$ | Constant speed $\vec{a} = \vec{0}$ | Slowing down $\vec{a}$ |

No push — $\vec{f}_A$ — A.

Push — $\vec{f}_B$ — B.

Push — $\vec{f}_C$ — C.

Push — $\vec{f}_D$ — D.

Push — $\vec{f}_E$ — E.

Working with Friction Forces

2.5, 2.6   Activ Physics ONLINE

These ideas can be summarized in a *model* of friction:

> Static: $\vec{f}_s = $ (magnitude $\leq f_{s\,max} = \mu_s n$, direction as necessary to prevent motion)
>
> Kinetic: $\vec{f}_k = (\mu_k n$, direction opposite the motion) $\tag{5.13}$
>
> Rolling: $\vec{f}_r = (\mu_r n$, direction opposite the motion)

LINEAR p. 38

Here "motion" means "motion relative to the surface." The maximum value of static friction $f_{s\,max} = \mu_s n$ occurs at the point where the object slips and begins to move. Note that only one kind of friction force at a time can act on an object.

NOTE ▶ Equations 5.13 are a "model" of friction, not a "law" of friction. These equations provide a reasonably accurate, but not perfect, description of how friction forces act. For example, we've ignored the surface area of the object because surface area has little effect. Likewise, our model assumes that the kinetic friction force is independent of the object's speed. This is a fairly good, but not perfect, approximation. Equations 5.13 are a simplification of reality that works reasonably well, which is what we mean by a "model." They are not a "law of nature" on a level with Newton's laws. ◀

(MP) TACTICS BOX 5.1 Working with friction forces ✎ Exercises 20, 21

❶ If the object is *not moving* relative to the surface it's in contact with, the friction force is **static friction.** Draw a free-body diagram of the object. The *direction* of the friction force is such as to oppose sliding of the object. Then use Problem-Solving Strategy 5.1 or 5.2 to solve for f_s. If f_s is greater than $f_{s\,max} = \mu_s n$, then static friction cannot hold the object in place. The assumption that the object is at rest is not valid, and you need to redo the problem using kinetic friction.

❷ If the object is *sliding* relative to the surface, then **kinetic friction** is acting. From Newton's second law, find the normal force n. Equation 5.13 then gives the magnitude and direction of the friction force.

❸ If the object is *rolling* along the surface, then **rolling friction** is acting. From Newton's second law, find the normal force n. Equation 5.13 then gives the magnitude and direction of the friction force.

Optimized braking If you slam on your brakes, your wheels will lock up and you'll go into a skid. Then it is the *kinetic* friction force between the road and your tires that slows your car to a halt. If, however, you apply the brakes such that you don't quite skid and your tires continue to roll, the force stopping you is the *static* friction force between the road and your tires. This is a better way to brake, because the maximum static friction force is always larger than the kinetic friction force. *Antilock braking systems* (ABS) automatically do this for you when you slam on the brakes, stopping you in the shortest possible distance.

EXAMPLE 5.11 Finding the force to push a box
Carol pushes a 10.0 kg wood box across a wood floor at a steady speed of 2.0 m/s. How much force does Carol exert on the box?

PREPARE Let's assume the box slides to the right. In this case, a kinetic friction force $\vec{f}_k$ opposes the motion by pointing to the left. In Figure 5.22 we identify the forces acting on the box and construct a free-body diagram.

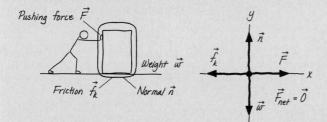

FIGURE 5.22 Forces on a box being pushed across a floor.

SOLVE The box is moving at a constant speed, so it is in dynamic equilibrium with $\vec{F}_{net} = \vec{0}$. This means that the x- and y-components of the net force must each be zero:

$$\sum F_x = n_x + w_x + F_x + (f_k)_x = 0 + 0 + F - f_k = 0$$
$$\sum F_y = n_y + w_y + F_y + (f_k)_y = n - w + 0 + 0 = 0$$

In the first equation, the x-component of $\vec{f}_k$ is equal to $-f_k$ because $\vec{f}_k$ is directed to the left. Similarly, $w_y = -w$ because the weight force points down.

From the first equation, we see that Carol's pushing force is $F = f_k$. To evaluate this, we need f_k. Here we can use our model for kinetic friction:

$$f_k = \mu_k n$$

Because the friction is wood sliding on wood, we can use Table 5.1 to find $\mu_k = 0.20$. Further, we can use the second Newton's-law equation to find that the normal force is $n = w = mg$. Thus

$$F = f_k = \mu_k n = \mu_k mg$$
$$= (0.20)(10.0\text{ kg})(9.80\text{ m/s}^2) = 20\text{ N}$$

This is the force that Carol needs to apply to the box to keep it moving at a steady speed.

ASSESS The speed of 2.0 m/s with which Carol pushes the box does not enter into the answer. This is because our model of kinetic friction does not depend on the speed of the sliding object.

CONCEPTUAL EXAMPLE 5.2 To push or pull a lawn roller?

A lawn roller is a heavy cylinder used to flatten a bumpy lawn, as shown in Figure 5.23. Is it easier to push or pull such a roller? Which is more effective for flattening the lawn, pushing or pulling?

FIGURE 5.23 Pushing and pulling a lawn roller.

REASON Figure 5.24 shows free-body diagrams for the two cases. We assume that the roller is pushed at a constant speed so that it is in dynamic equilibrium with $\vec{F}_{\text{net}} = \vec{0}$. Because the roller does not move in the y-direction, the y-component of the net force must be zero. According to our model, the magnitude f_{r} of rolling friction is proportional to the magnitude n of the normal force. If we *push* on the roller, our pushing force $\vec{F}$ will have a downward y-component. To compensate for this, the normal force must increase and, because $f_{\text{r}} = \mu_{\text{r}} n$, the rolling friction will increase as well. This makes the roller harder to move. If we *pull* on the roller, the now upward y-component of

$\vec{F}$ will lead to a *reduced* value of n and hence of f_{r}. Thus the roller is easier to pull than to push.

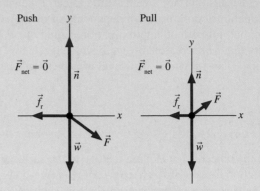

FIGURE 5.24 Free-body diagrams for the lawn roller.

However, the purpose of the roller is to flatten the soil. If the normal force $\vec{n}$ of the ground on the roller is larger, by Newton's third law the force of the roller on the ground will be larger as well. So for smoothing your lawn, it's better to push.

ASSESS You've probably experienced this effect while using an upright vacuum cleaner. The vacuum is harder to push on the forward stroke than when drawing it back.

EXAMPLE 5.12 How to dump a file cabinet

A 50.0 kg steel file cabinet is in the back of a dump truck. The truck's bed, also made of steel, is slowly tilted. What is the size of the static friction force on the cabinet when the bed is tilted 20°? At what angle will the file cabinet begin to slide?

PREPARE We'll use our model of static friction. The file cabinet will slip when the static friction force reaches its maximum possible value $f_{\text{s max}}$. Figure 5.25 shows the visual overview when the truck bed is tilted at angle θ. We can make the analysis easier if we tilt the coordinate system to match the bed of the truck. To prevent the file cabinet from slipping, the static friction force must point *up* the slope.

SOLVE Before it slips, the file cabinet is in static equilibrium. Newton's first law reads

$$\sum F_x = n_x + w_x + (f_{\text{s}})_x = 0$$
$$\sum F_y = n_y + w_y + (f_{\text{s}})_y = 0$$

From the free-body diagram we see that f_{s} has only a negative x-component and that n has only a positive y-component. We also have $w_x = w\sin\theta$ and $w_y = -w\cos\theta$. Thus the first law becomes

$$\sum F_x = w\sin\theta - f_{\text{s}} = mg\sin\theta - f_{\text{s}} = 0$$
$$\sum F_y = n - w\cos\theta = n - mg\cos\theta = 0$$

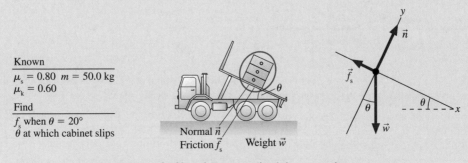

Known
$\mu_{\text{s}} = 0.80 \quad m = 50.0 \text{ kg}$
$\mu_{\text{k}} = 0.60$

Find
f_{s} when $\theta = 20°$
θ at which cabinet slips

Normal $\vec{n}$
Friction $\vec{f}_{\text{s}}$ Weight $\vec{w}$

FIGURE 5.25 Visual overview of a file cabinet in a tilted dump truck.

where we've used $w = mg$. The x-component equation allows us to determine the size of the static friction force when $\theta = 20°$:

$$f_s = mg\sin\theta = (50.0 \text{ kg})(9.80 \text{ m/s}^2)\sin 20° = 168 \text{ N}$$

This value does not require knowing μ_s. The coefficient of static friction only enters when we want to find the angle at which the file cabinet slips. Slipping occurs when the static friction reaches its maximum value

$$f_s = f_{s\,max} = \mu_s n$$

From the y-component of Newton's second law we see that $n = mg\cos\theta$. Consequently,

$$f_{s\,max} = \mu_s mg\cos\theta$$

The x-component of the second law gave

$$f_s = mg\sin\theta$$

Setting $f_s = f_{s\,max}$ then gives

$$mg\sin\theta = \mu_s mg\cos\theta$$

The mg in both terms cancels, and we find

$$\frac{\sin\theta}{\cos\theta} = \tan\theta = \mu_s$$

$$\theta = \tan^{-1}\mu_s = \tan^{-1}(0.80) = 39°$$

ASSESS Steel doesn't slide all that well on unlubricated steel, so a fairly large angle is not surprising. The answer seems reasonable. It is worth noting that $n = mg\cos\theta$ in this example. A common error is to use simply $n = mg$. Be sure to evaluate the normal force within the context of each specific problem.

Causes of Friction

It is worth a brief pause to look at the *causes* of friction. All surfaces, even those quite smooth to the touch, are very rough on a microscopic scale. When two objects are placed in contact, they do not make a smooth fit. Instead, as Figure 5.26 shows, the high points on one surface become jammed against the high points on the other surface while the low points are not in contact at all. Only a very small fraction (typically 10^{-4}) of the surface area is in actual contact. The amount of contact depends on how hard the surfaces are pushed together, which is why friction forces are proportional to n.

For an object to slip, you must push it hard enough to force these contact points over each other. Once the two surfaces are sliding against each other, their high points undergo constant collisions, deformations, and even brief bonding that leads to the resistive force of kinetic friction.

5.6 Drag

Fluids—liquids and gases—exert a drag force on objects as they move through the fluid. You experience drag forces every day as you jog, bicycle, ski, or drive your car. The drag force is especially important for the skydiver at the beginning of the chapter and for unicellular animals swimming through water.

The drag force $\vec{D}$

- Is opposite in direction to the velocity $\vec{v}$.
- Increases in magnitude as the object's speed increases.

Experimental studies have found that the drag force depends on an object's speed in a complicated way. We'll look separately at drag forces in air and in liquids because different models apply.

Drag in Air

At relatively low speeds, the drag force in air is small and can usually be neglected, but drag plays an important role as speeds increase. Fortunately, we can use a fairly simple *model* of drag if the following three conditions are met:

- The object's size (diameter) is between a few millimeters and a few meters.
- The object's speed is less than a few hundred meters per second.
- The object is moving through the air near the earth's surface.

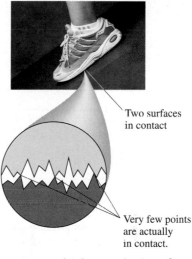

Two surfaces in contact

Very few points are actually in contact.

FIGURE 5.26 A microscopic view of friction.

These conditions are usually satisfied for balls, people, cars, and many other objects of the everyday world. Under these conditions, the drag force can be written

$$\vec{D} = \left(\tfrac{1}{2}C_D\rho A v^2, \text{ direction opposite to the motion}\right) \quad (5.14)$$

Drag force on an object of cross-section area A moving at speed v

QUADRATIC
p. 50

Here, ρ is the density of air ($\rho = 1.29$ kg/m³ at sea level), A is the cross-section area of the object (in m²), and the **drag coefficient** C_D depends on the details of the object's shape. However, the value of C_D for everyday moving objects is roughly 1/2, so a good approximation to the drag force is

$$D \approx \tfrac{1}{4}\rho A v^2 \quad (5.15)$$

This is the expression for the magnitude of the drag force that we'll use in this chapter.

The size of the drag force in air is proportional to the *square* of the object's speed: If the speed doubles, the drag increases by a factor of *four*. This model of drag fails for objects that are very small (such as dust particles) or very fast (such as jet planes) or that move in other media (such as water).

Figure 5.27 shows that the area A in Equation 5.14 is the cross section of the object as it "faces into the wind." It's interesting to note that the magnitude of the drag force depends on the object's *size and shape* but not on its *mass*. This has important consequences for the motion of falling objects.

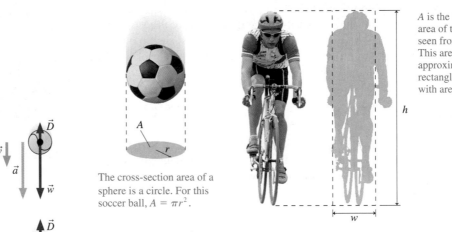

A is the cross-section area of the cyclist as seen from the front. This area is approximated by the rectangle shown, with area $A = h \times w$.

The cross-section area of a sphere is a circle. For this soccer ball, $A = \pi r^2$.

FIGURE 5.27 How to calculate the cross-section area A.

Terminal Speed

Just after an object is released from rest, its speed is low and the drag force is small (as shown in Figure 5.28a). Because the net force is nearly equal to the weight, the object will fall with an acceleration only a little less than g. As it falls further, its speed and hence the drag force increase. Now the net force is smaller, so the acceleration is smaller (as shown in Figure 5.28b). It's still speeding up, but at a lower *rate*. Eventually the speed will increase to a point such that the magnitude of the drag force *equals* the weight (as shown in Figure 5.28c). The net force—and hence the acceleration—at this speed are then *zero*, and the object falls with a *constant* speed. The speed at which the exact balance between the upward drag force and the downward weight force causes an object to fall without acceleration is called the **terminal speed** v_{term}. **Once an object has reached terminal speed, it will continue falling at that speed until it hits the ground.**

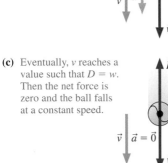

(a) At low speeds D is small and the ball falls with $a \approx g$.

(b) As v increases so does D. The net force and hence a get smaller.

(c) Eventually, v reaches a value such that $D = w$. Then the net force is zero and the ball falls at a constant speed.

FIGURE 5.28 A falling object eventually reaches terminal speed.

It's straightforward to compute the terminal speed. It is the speed, by definition, at which $D = w$ or, equivalently, $\frac{1}{4}\rho A v^2 = mg$. This speed is then

$$v_{\text{term}} \approx \sqrt{\frac{4mg}{\rho A}} \tag{5.16}$$

This equation shows that a more massive object has a larger terminal speed than a less massive object of equal size. A 10-cm-diameter lead ball, with a mass of 6 kg, has a terminal speed of 150 m/s while a 10-cm-diameter Styrofoam ball, with a mass of 50 g, has a terminal speed of only 14 m/s.

EXAMPLE 5.13   Terminal speeds of a skydiver and a mouse

A skydiver and his pet mouse jump from a plane. Estimate their terminal speeds.

PREPARE To use Equation 5.16 we need to estimate the mass m and cross-section area A of both man and mouse. Figure 5.29 shows how. A typical skydiver might be 1.8 m long and 0.40 m wide ($A = 0.72$ m^2) with a mass of 75 kg, while a mouse has a mass of perhaps 20 g (0.020 kg) and is 7 cm long and 3 cm wide ($A = 0.07$ m $\times$ 0.03 m $= 0.0021$ m^2).

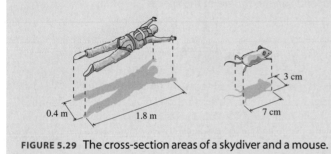

FIGURE 5.29 The cross-section areas of a skydiver and a mouse.

SOLVE We can use Equation 5.16 to find that for the skydiver

$$v_{\text{term}} \approx \sqrt{\frac{4mg}{\rho A}} = \sqrt{\frac{4(75 \text{ kg})(9.8 \text{ m/s}^2)}{(1.29 \text{ kg/m}^3)(0.72 \text{ m}^2)}} = 56 \text{ m/s}$$

This is roughly 130 mph. A higher speed can be reached by falling feet first or head first, which reduces the area A. Fortunately the skydiver can open his parachute, greatly increasing A. This brings his terminal speed down to a safe value.

For the mouse we have

$$v_{\text{term}} \approx \sqrt{\frac{4mg}{\rho A}} = \sqrt{\frac{4(0.020 \text{ kg})(9.8 \text{ m/s}^2)}{(1.29 \text{ kg/m}^2)(0.0021 \text{ m}^2)}} = 17 \text{ m/s}$$

The mouse has no parachute—nor does he need one! A mouse's terminal speed is low enough that he can fall from any height, even out of an airplane, and survive. Cats, too, have relatively low terminal speeds. In a study of cats that fell from high rises, over 90% survived—including one that fell 45 stories!

ASSESS The mouse survives the fall not only because of its lower terminal speed. The smaller an animal's body, the proportionally more robust it is. Further, a small animal's low mass and terminal speed mean that it has a very small *kinetic energy*, an idea we'll study in Chapter 10.

Although we've focused our analysis on falling objects, the same ideas apply to objects moving horizontally. If an object is thrown or shot horizontally, $\vec{D}$ causes the object to slow down. An airplane reaches its maximum speed, which is analogous to the terminal speed, when the drag is equal and opposite to the thrust: $D = F_{\text{thrust}}$. The net force is then zero and the plane cannot go any faster. The maximum speed of a passenger jet is about 550 mph.

We will continue to neglect drag unless a problem specifically calls for drag to be considered.

STOP TO THINK 5.4   The terminal speed of a Styrofoam ball is 15 m/s. Suppose a Styrofoam ball is shot straight down with an initial speed of 30 m/s. Which velocity graph is correct?

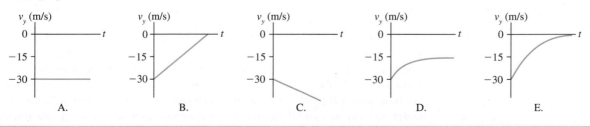

Drag in Liquids

The drag force in a liquid is quite different from that in air. Drag forces in air are largely the result of the object having to push the air out of its way as it moves. For an object moving slowly through a liquid, however, the drag force is mostly due to the *viscosity* of the liquid. We'll study viscosity in Chapter 13. For the moment, you can think of the viscosity of a fluid as a measure of how much resistance to flow the fluid has: Honey, which drizzles slowly out of its container, has a much higher viscosity than water, which flows fairly freely. Drag forces have particularly important effects on the motion of single-celled organisms in water, blood cells in the body, and many aquatic creatures.

The drag force in a liquid depends on the shape of the object, but there is a simple result called **Stokes's law** for the drag on a *sphere*. This law is useful because many biological objects, such as cells, can at least roughly be approximated as spheres. The drag on a sphere moving with speed v is

$$\vec{D} = (6\pi\eta rv, \text{ direction opposite to motion}) \qquad (5.17)$$

Stokes's law for the drag on a sphere of radius r moving at speed v

LINEAR
p.38

Here, η is the viscosity of the liquid. (Water has a viscosity $\eta = 0.0010 \text{ N} \cdot \text{s/m}^2$ at 20° C.) Notice that the drag force in a liquid depends *linearly* on the speed, whereas drag in air depends on the *square* of the speed.

EXAMPLE 5.14 Drag force on a paramecium

A paramecium is a single-celled animal able to propel itself quite rapidly through water by using its *cilia*, rapidly beating hair-like fibers that ring its body. A typical paramecium has a diameter of 50 μm. What is the drag force on a paramecium swimming at a speed of 0.25 mm/s? If its mass is 1.0×10^{-11} kg, estimate how long it takes for the paramecium to come to a stop once it stops swimming.

PREPARE Although a real paramecium has an elongated shape, we'll model the paramecium as a sphere of radius 25 μm and use Stokes's law to find the drag force on it. Once the paramecium stops swimming, the drag force will be the only force acting, and we can use Newton's second law to find its acceleration and time to come to a stop.

SOLVE The drag force is

$$D = 6\pi\eta rv = (6\pi)(0.0010 \text{ N} \cdot \text{s/m}^2)(25 \times 10^{-6} \text{ m})$$
$$\times (0.25 \times 10^{-3} \text{ m/s}) = 1.2 \times 10^{-10} \text{ N}$$

Once the paramecium stops swimming, its acceleration is $a = F_{net}/m = -D/m = -(1.2 \times 10^{-10} \text{ N})/(1.0 \times 10^{-11} \text{ kg}) = -12 \text{ m/s}^2$. This acceleration is a little greater than g. We can estimate how long it takes for the paramecium to stop by using the definition $a = \Delta v/\Delta t$. (This is only an estimate, because the drag force and hence the acceleration decrease as the paramecium slows down.) This estimated stopping time is

$$\Delta t = \Delta v/a = \frac{-0.25 \times 10^{-3} \text{ m/s}}{-12 \text{ m/s}^2} = 2 \times 10^{-5} \text{ s} = 20 \ \mu\text{s}$$

This is a *very* short time!

ASSESS In estimating the stopping time, we used the initial speed of the paramecium. Suppose its initial speed v were doubled. Then the drag force, which is proportional to the speed, would also double, and its acceleration a would double as well. But if both v and a double, $\Delta t = \Delta v/a$ would remain the same. Thus the stopping time of an object moving in a liquid is *independent of its initial speed.*

A paramecium, along with bacteria and other small organisms that live in water, lives in a very strange world where inertia—the tendency of a moving body to continue in motion—is negligible. To keep a shopping cart rolling you need to push it, but if you stop pushing it will continue to roll for some distance. A frictionless cart would roll forever at constant speed. In contrast, a paramecium comes to a halt essentially instantaneously, as soon as it stops actively swimming.

The reverse is also true: When it starts swimming again, it almost instantly reaches a constant speed, analogous to a terminal speed, at which the drag force $\vec{D}$ is equal and opposite to the paramecium's propulsion force $\vec{F}_{propulsion}$. Using Stokes's law for $\vec{D}$, we see that speed of swimming is directly proportional to the propulsion force:

$$v = \frac{F_{propulsion}}{6\pi\eta r}$$

If a small organism in water wants to go twice as fast, it has to double its propulsion force. (For objects without drag, doubling the force causes the *acceleration*, not the speed, to double.) Life in a world dominated by drag is very different from the world of accelerating objects in which we humans live.

5.7 Interacting Objects

Up to this point we have studied the dynamics of a single object subject to forces from the environment. In Example 5.11, for instance, a sliding box was acted upon by friction, normal, weight, and pushing forces that came from outside the box—from the floor, the earth, and the person pushing. As we've seen, such problems can be solved by an application of Newton's second law after all the forces have been identified.

Actlv
Physlcs 2.7–2.9

But in Chapter 4 we found that real-world motion often involves two or more objects interacting with each other. We further found that forces always come in action/reaction *pairs* that are related by Newton's third law. To remind you, Newton's third law states:

- Every force occurs as one member of an action/reaction pair of forces. The two members of the pair always act on *different* objects.
- The two members of an action/reaction pair point in *opposite* directions and are *equal* in magnitude.

Our goal in this section is to learn how to apply the second *and* third laws to systems of interacting objects.

Acceleration Constraints

Newton's third law is one relationship needed to solve problems of interacting objects. In addition, we frequently have other information about the motion in a problem. For example, consider the two boxes in Figure 5.30. As long as they're touching, box A *has* to have exactly the same acceleration as box B. If they were to accelerate differently, either box B would take off on its own, or it would suddenly slow down and box A would run over it! Our problem implicitly assumes that neither of these is happening. Thus the two accelerations are *constrained* to be equal: $\vec{a}_A = \vec{a}_B$. A well-defined relationship between the accelerations of two or more systems is called an **acceleration constraint.** It is an independent piece of information that can help solve a problem.

In practice, we'll express acceleration constraints in terms of the x- and y-components of $\vec{a}$. Consider the car being towed in Figure 5.31. As long as the cable is under tension, the accelerations are constrained to be equal: $\vec{a}_C = \vec{a}_T$. This is one-dimensional motion, so for problem solving we would use just the x-components a_{Cx} and a_{Tx}. In terms of these components, the acceleration constraint is

$$a_{Cx} = a_{Tx} = a_x$$

Because the accelerations of both systems are equal, we can drop the subscripts C and T and call both of them a_x.

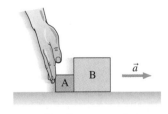

FIGURE 5.30 Two boxes moving together have the same acceleration.

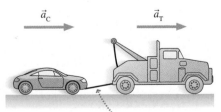

The cable is under tension.

FIGURE 5.31 The car and the truck have the same acceleration.

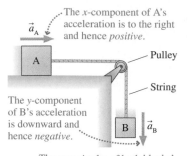

The *x*-component of A's acceleration is to the right and hence *positive*.

$\vec{a}_A$

A

Pulley

String

The *y*-component of B's acceleration is downward and hence *negative*.

B $\vec{a}_B$

The *magnitudes* of both blocks' accelerations are equal.

FIGURE 5.32 The string constrains the two systems to accelerate together.

Don't assume the accelerations of A and B will always have the same sign. Consider blocks A and B in Figure 5.32. The blocks are connected by a string, so they are constrained to move together and their accelerations have equal magnitudes. But A has a positive acceleration (to the right) in the *x*-direction while B has a negative acceleration (downward) in the *y*-direction. Thus the acceleration constraint on the components of the acceleration is

$$a_{Ax} = -a_{By}$$

This relationship does *not* say that a_{Ax} is a negative number. It is simply a relational statement, saying that a_{Ax} is (-1) times whatever a_{By} happens to be. In fact, the acceleration a_{By} in Figure 5.32 is a negative number, so a_{Ax} is actually positive. In many problems, the signs of a_{Ax} and a_{By} may not be known until the problem is solved, but the *relationship* is known from the beginning.

A Revised Strategy for Interacting-Object Problems

Problems of interacting objects can be solved with a few modifications to the basic Problem-Solving Strategy 5.2 we developed earlier in this chapter. A revised problem-solving strategy is shown below.

> **(MP)** **PROBLEM-SOLVING STRATEGY 5.3** **Interacting-object problems**
>
> **PREPARE** Identify the interacting objects that make up the system. Everything else is part of the environment. Make simplifying assumptions.
>
> Prepare a visual overview: Make a sketch of the situation. Define symbols and identify what the problem is trying to find. Identify all forces acting on each object and all action/reaction pairs. Draw a *separate* free-body diagram for each object and connect the force vectors of action/reaction pairs with dotted lines. Use subscript labels to distinguish forces, such as $\vec{n}$ and $\vec{w}$, that act independently on more than one object. Identify any acceleration constraints and add these to your list of values.
>
> **SOLVE** Use Newton's second and third laws:
>
> - Write Newton's second law in component form for each object. Find the force components from the free-body diagrams.
> - Equate the magnitudes of action/reaction pairs.
> - If relevant, include the acceleration constraints and the friction model.
> - Solve for the unknown forces or acceleration.
>
> **ASSESS** Check that your result has the correct units, is reasonable, and answers the question.

NOTE ▶ Two steps are especially important when drawing the free-body diagrams. First, draw a *separate* diagram for each object. They need not have the same coordinate system. Second, show only the forces acting *on* that object. The force $\vec{F}_{A\,on\,B}$ goes on the free-body diagram of Object B, but $\vec{F}_{B\,on\,A}$ goes on the diagram of Object A. The two members of an action/reaction pair *always* appear on two different free-body diagrams—*never* on the same diagram. ◀

You might be puzzled that the Solve step calls for the use of the third law to equate just the *magnitudes* of action/reaction forces. What about the "opposite in direction" part of the third law? You have already used it! Your free-body diagrams should show the two members of an action/reaction pair to be opposite in direction, and that information will have been utilized in writing the second-law equations. Because the directional information has already been used, all that is left is the magnitude information.

EXAMPLE 5.15 Pushing two blocks

Figure 5.33 shows a 5.0 kg block A being pushed with a 3.0 N force. In front of this block is a 10 kg block B; the two blocks move together. What force does block A exert on block B?

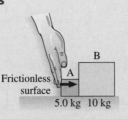

FIGURE 5.33 Two blocks are pushed by a hand.

PREPARE The visual overview of Figure 5.34 lists known information, what we want to find, and—new to problems of interacting objects—the acceleration constraint. Because the blocks move together in the positive x-direction, they both must have the same x-component of acceleration:

$$a_{Ax} = a_{Bx} = a_x$$

We've identified all the forces acting on blocks A and B, then drawn *separate* free-body diagrams. Because both blocks have a weight force and a normal force, we used subscripts A and B to distinguish between them. A key step is to include the forces of interaction between the two blocks, $\vec{F}_{A\,on\,B}$ and $\vec{F}_{B\,on\,A}$. Notice that $\vec{F}_{A\,on\,B}$ is drawn as acting on box B; it is the force *of* A *on* B. Because action/reaction pairs act in opposite directions, force $\vec{F}_{B\,on\,A}$ pushes *backward* on block A. Force vectors are always drawn on the free-body diagram of the object that *experiences* the force, not the object exerting the force.

SOLVE We begin by writing Newton's second law in component form for each block. Because the motion is only in the x-direction, we need only the x-component of the second law. For block A,

$$\sum F_x = (F_H)_x + (F_{B\,on\,A})_x = m_A a_{Ax}$$

The force components can be "read" from the free-body diagram, where we see $\vec{F}_H$ pointing to the right and $\vec{F}_{B\,on\,A}$ pointing to the left. Thus

$$F_H - F_{B\,on\,A} = m_A a_x$$

We used the acceleration constraint to write $a_{Ax} = a_x$. For B, we have

$$\sum F_x = (F_{A\,on\,B})_x = m_B a_{Bx}$$

or, again using the acceleration constraint,

$$F_{A\,on\,B} = m_B a_x$$

We have an additional piece of information: Newton's third law tells us that $F_{B\,on\,A} = F_{A\,on\,B}$. With this information, the two x-component equations become

$$F_{A\,on\,B} = m_B a_x$$
$$F_H - F_{A\,on\,B} = m_A a_x$$

Our goal is to find $F_{A\,on\,B}$, so we need to eliminate the unknown acceleration a_x. From the first equation, $a_x = F_{A\,on\,B}/m_B$. Substituting this into the second equation gives

$$F_H - F_{A\,on\,B} = \frac{m_A}{m_B} F_{A\,on\,B}$$

This can be solved for the force of block A on block B, giving

$$F_{A\,on\,B} = \frac{F_H}{1 + m_A/m_B} = \frac{3.0\ \text{N}}{1 + (5.0\ \text{kg})/(10\ \text{kg})} = \frac{3.0\ \text{N}}{1.5} = 2.0\ \text{N}$$

ASSESS The critical step in solving this problem was correctly identifying the forces and drawing the two free-body diagrams. The rest of the problem consists of writing the second-law equations directly from the free-body diagram and then using the new information of the acceleration constraint to solve for the desired quantity.

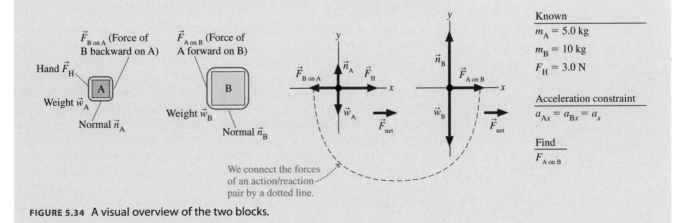

FIGURE 5.34 A visual overview of the two blocks.

STOP TO THINK 5.5   Boxes P and Q are sliding to the right across a frictionless table. The hand H is slowing them down. The mass of P is larger than the mass of Q. Rank in order, from largest to smallest, the *horizontal* forces on P, Q, and H.

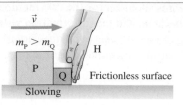

A. $F_{\text{QonH}} = F_{\text{HonQ}} = F_{\text{PonQ}} = F_{\text{QonP}}$

B. $F_{\text{QonH}} = F_{\text{HonQ}} > F_{\text{PonQ}} = F_{\text{QonP}}$

C. $F_{\text{QonH}} = F_{\text{HonQ}} < F_{\text{PonQ}} = F_{\text{QonP}}$

D. $F_{\text{HonQ}} = F_{\text{HonP}} > F_{\text{PonQ}}$

5.8 Ropes and Pulleys

2.10, 2.11   Activ Physics

Many objects are connected by strings, ropes, cables, and so on. In single-particle dynamics, we defined *tension* as the force exerted on an object by a rope or string. Now we need to think more carefully about the string itself. Just what do we mean when we talk about the tension "in" a string or rope?

Tension Revisited

Figure 5.35 shows a man pulling on a rope attached to a wall. As we saw in Chapter 4, a rope or string can be thought of as a very stiff spring. When the man pulls, the rope stretches slightly. Like a small stretched spring, each piece of the rope then pulls on its neighboring piece. By Newton's third law, the force of any segment of the rope on its neighbor has the *same* magnitude as the force of the neighbor on that segment. The magnitude of this force—the force holding the rope together—is what we call the tension *in* the rope.

The hand pulls on the rope with a force $\vec{F}_{\text{H on rope}}$. The rope pulls back with a force $\vec{T}_{\text{rope on H}}$. $\vec{F}_{\text{H on rope}}$ and $\vec{T}_{\text{rope on H}}$ form an action/reaction pair, so the tension in the rope is $\vec{T}_{\text{rope on H}} = \vec{F}_{\text{H on rope}}$.

The rope pulls on the wall and the wall pulls back on the rope. These are an action/reaction pair.

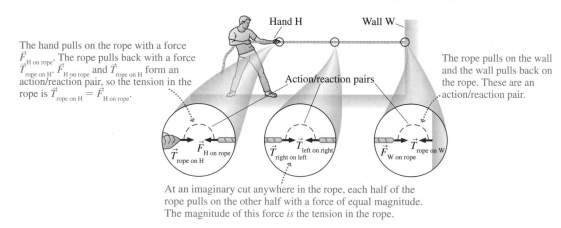

At an imaginary cut anywhere in the rope, each half of the rope pulls on the other half with a force of equal magnitude. The magnitude of this force *is* the tension in the rope.

FIGURE 5.35 Tension forces pull in both directions within a rope.

Is the tension in a rope the *same* at all points, or is there more tension at one end than the other? A rope, like any other object, must obey Newton's second law: $\vec{F}_{\text{net}} = m_{\text{rope}}\vec{a}$. The stationary rope in Figure 5.35 has no acceleration, so $\vec{F}_{\text{net}} = \vec{0}$. The external forces acting on the rope are $\vec{F}_{\text{H on rope}}$ and $\vec{F}_{\text{W on rope}}$, so $(F_{\text{net}})_x = F_{\text{W on rope}} - F_{\text{H on rope}} = T_{\text{rope on W}} - T_{\text{rope on H}}$. Here we've used the third-law result from Figure 5.35 that the tension at the end of a rope equals the strength of the force pulling on the rope. Thus $(F_{\text{net}})_x = 0$ implies $T_{\text{rope on W}} = T_{\text{rope on H}}$. In other words, **the tension in a stationary rope pulls equally hard at both ends of the rope on whatever objects the rope is attached to.**

But the tensions at the two ends of an accelerating rope are *not* the same. Because acceleration requires $(F_{\text{net}})_x \neq 0$, the force pulling on one end of the rope must be larger than the force pulling on the other end. That is, there must be *difference* between the tensions at the two ends to supply the net force needed to accelerate the rope itself. However, suppose the rope were massless. If $m_{\text{rope}} = 0$, than $(F_{\text{net}})_x = m_{\text{rope}}a_x = 0$ whether the rope is accelerating or not. It takes no effort to accelerate a massless rope, so the tension *is* the same throughout the rope. While no real rope is massless, this is a very good approximation in the common situation where the mass of a string or rope is much less than the mass of the objects it pulls on. Unless a problem explicitly mentions the mass of a rope or string, we will assume it is "massless" and thus the tension at each end is the same.

This **massless rope (or string) approximation** can greatly simplify situations where objects are connected by ropes or strings. Consider blocks A and B in Figure 5.36. If the string is assumed to be massless, the tension at both ends is the same. But the tension is also the force with which the string pulls on an object, so we see that B is pulled to the left by the string with a tension force of the *same* magnitude with which A is pulled to the right. In fact, as far as the forces on A and B are concerned, we can omit the string entirely and pretend that A and B interact directly, with forces of equal magnitude but opposite. direction. We'll often use this idea in solving problems.

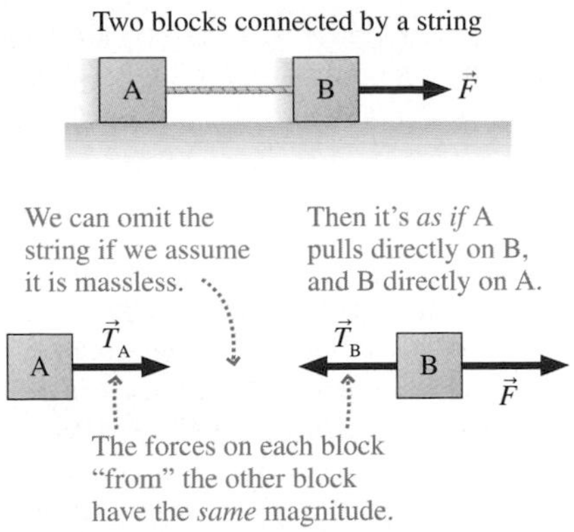

Two blocks connected by a string

We can omit the string if we assume it is massless.

Then it's *as if* A pulls directly on B, and B directly on A.

The forces on each block "from" the other block have the *same* magnitude.

FIGURE 5.36 The blocks act as if they are directly interacting.

CONCEPTUAL EXAMPLE 5.3 Pulling a rope

Figure 5.37a shows a student pulling horizontally with a 100 N force on a rope that is attached to a wall. In Figure 5.37b, two students in a tug-of-war pull on opposite ends of a rope with 100 N each. Is the tension in the second rope larger, smaller, or the same as that in the first?

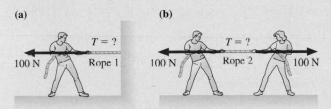

(a) $T = ?$ 100 N Rope 1

(b) $T = ?$ 100 N Rope 2 100 N

FIGURE 5.37 Pulling on a rope. Which produces a larger tension?

REASON Surely pulling on a rope from both ends causes more tension than pulling on one end. Right? Before jumping to conclusions, let's analyze the situation carefully. The situation for rope 1 is the same as that analyzed in Figure 5.35, where we learned that the tension in a rope is equal to the magnitude of the force applied to its end. Thus the tension in rope 1 is 100 N, the force with which the student pulls on the rope.

To find the tension in the second rope, suppose we make an imaginary cut through the middle of the rope, as shown in Fig-

ure 5.38. Then we can analyze the forces acting on the segment of the rope to the left of this imaginary cut. The male student pulls on the left end of this segment with a force of magnitude $F_s = 100$ N. The right end of this segment is pulled on by the rest of the rope, to its right, with a tension force of magnitude T. Because this segment of the rope is in equilibrium, we must have $T - F_s = 0$, so the tension is $T = F_s = 100$ N. The tension in the rope has not changed! It is still 100 N.

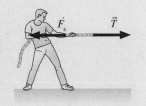

FIGURE 5.38 Analysis of the tension force in rope 2.

ASSESS We didn't need to consider the second student. Evidently it makes no difference whether the rope is tied to a wall or is being pulled on by that student. When the rope is tied to the wall, we know that it pulls on the wall with a force equal in magnitude to its tension, or 100 N. By Newton's third law, the wall must pull back on the rope with an equal force, also 100 N. But the second student pulls with a force of 100 N. The forces are the same in both situations! The rope does not care whether it is pulled by a wall or by a hand. It experiences the same forces in both cases, so the rope's tension is the same 100 N in both.

Pulleys

Strings and ropes often pass over pulleys. The application might be as simple as lifting a heavy weight or as complex as the internal cable-and-pulley arrangement that precisely moves a robot arm. Figure 5.39 shows a simple situation in which block B drags block A across a frictionless table as it falls. As the string moves, static friction between the string and pulley causes the pulley to turn. If we assume that

■ The string *and* the pulley are both massless, and
■ There is no friction where the pulley turns on its axle,

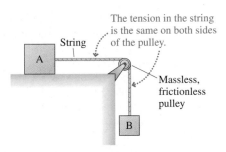

The tension in the string is the same on both sides of the pulley.

String

A

Massless, frictionless pulley

B

FIGURE 5.39 An ideal pulley changes the direction in which a tension force acts, but not its magnitude.

then no net force is needed to accelerate the string or turn the pulley. In this case, the **tension in a massless string remains constant as it passes over a massless, frictionless pulley.** We'll assume such an ideal pulley for problems in this chapter. Later, when we study rotational motion in Chapter 8, we'll consider the effect of the pulley's mass.

EXAMPLE 5.16 Lifting a stage set

A 200 kg set used in a play is stored in the loft above the stage. The rope holding the set passes up and over a pulley, then is tied backstage. The director tells a 100 kg stagehand to lower the set. When he unties the rope, the set falls and the unfortunate man is hoisted into the loft. What is the stagehand's acceleration?

PREPARE The objects of interest are the stagehand M and the set S, which we will model as particles. Assume a massless rope and a massless, frictionless pulley. Figure 5.40 shows the visual overview. The man's acceleration a_{My} is positive, while the set's acceleration a_{Sy} is negative. These two accelerations have the same magnitude because the two objects are connected by a rope, but they have opposite signs. Thus the acceleration constraint is $a_{Sy} = -a_{My}$. Tension forces $\vec{T}_S$ and $\vec{T}_M$ are due to a massless rope going over an ideal pulley, so their magnitudes are the same and we can use the same symbol T for both magnitudes.

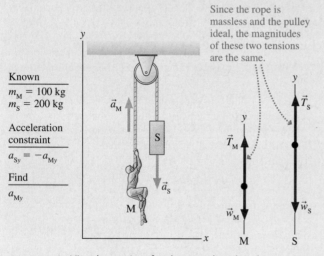

Known

$m_M = 100$ kg
$m_S = 200$ kg

Acceleration constraint

$a_{Sy} = -a_{My}$

Find

a_{My}

Since the rope is massless and the pulley ideal, the magnitudes of these two tensions are the same.

FIGURE 5.40 Visual overview for the stagehand and set.

SOLVE From the two free-body diagrams, we can write Newton's second law in component form. For the man we have

$$\sum F_{My} = T_M - w_M = T_M - m_M g = m_M a_{My}$$

while for the set we have

$$\sum F_{Sy} = T_S - w_S = T_S - m_S g = m_S a_{Sy}$$

Only the y-equations are needed. Now we can use the acceleration constraint $a_{Sy} = -a_{My}$ and the fact that the two tension forces have equal magnitude T. Inserting these into the second-law equations gives

$$T - m_M g = m_M a_{My}$$
$$T - m_S g = -m_S a_{My}$$

These are simultaneous equations in the two unknowns T and a_{My}. We can solve for T in the first equation to get

$$T = m_M a_{My} + m_M g$$

Inserting this value of T into the second equation then gives

$$m_M a_{My} + m_M g - m_S g = -m_S a_{My}$$

which we can rewrite as

$$(m_S - m_M)g = (m_S + m_M)a_{My}$$

Finally, we can solve for the hapless stagehand's acceleration:

$$a_{My} = \frac{m_S - m_M}{m_S + m_M}g = \left(\frac{100 \text{ kg}}{300 \text{ kg}}\right) \times 9.80 \text{ m/s}^2 = 3.27 \text{ m/s}^2$$

This is also the acceleration with which the set falls. If the rope's tension was needed, we could now find it from $T = m_M a_{My} + m_M g$.

ASSESS If the stagehand weren't holding on, the set would fall with free-fall acceleration g. The stagehand acts as a *counterweight* to reduce the acceleration.

EXAMPLE 5.17 A not-so-clever bank robbery

Bank robbers have pushed a 1000 kg safe to a second-story floor-to-ceiling window. They plan to break the window, then lower the safe 3.0 m to their truck. Not being too clever, they stack up 500 kg of furniture, tie a rope between the safe and the furniture, and place the rope over a pulley. Then they push the safe out the window. What is the safe's speed when it hits the truck? The coefficient of kinetic friction between the furniture and the floor is 0.50.

PREPARE This is essentially the situation that we analyzed in Figure 5.39. The systems of interest are the safe S and the furniture F, which we will model as particles. We will assume a massless rope and a massless, frictionless pulley; the tension is then the same everywhere in the rope. Thus the tension forces $\vec{T}_S$ and $\vec{T}_F$ have the same magnitude, which we'll call T.

The visual overview in Figure 5.41 establishes a coordinate system and defines the symbols that will be needed to calcu-

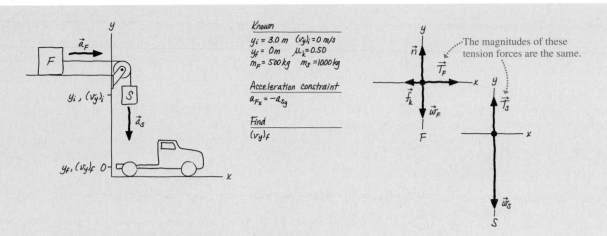

FIGURE 5.41 Visual overview of the furniture and falling safe.

late the safe's motion. The safe and the furniture are tied together, so their accelerations have the same magnitude. The safe has a y-component of acceleration a_{Sy} that is negative because the safe accelerates in the $-y$-direction. The furniture has an x-component a_F that is positive. Thus the acceleration constraint is

$$a_{Fx} = -a_{Sy}$$

SOLVE We can write Newton's second law directly from the free-body diagrams. For the furniture,

$$\sum F_{Fx} = T_F - f_k = T - f_k = m_F a_{Fx} = -m_F a_{Sy}$$

$$\sum F_{Fy} = n - w_F = n - m_F g = 0$$

And for the safe,

$$\sum F_{Sy} = T_S - w_S = T - m_S g = m_S a_{Sy}$$

Notice how we used the acceleration constraint in the first equation. We also went ahead and made use of the fact that $T_S = T_F = T$. We have one additional piece of information, the model of kinetic friction:

$$f_k = \mu_k n = \mu_k m_F g$$

where we used the y-equation of the furniture to deduce that $n = m_F g$. Substitute this result for f_k into the x-equation of the

furniture, then rewrite the furniture's x-equation and the safe's y-equation:

$$T - \mu_k m_F g = -m_F a_{Sy}$$

$$T - m_S g = m_S a_{Sy}$$

We have succeeded in reducing our knowledge to two simultaneous equations in the two unknowns a_{Sy} and T. Subtract the second equation from the first to eliminate T:

$$(m_S - \mu_k m_F) g = -(m_S + m_F) a_{Sy}$$

Finally, solve for the safe's acceleration:

$$a_{Sy} = -\left(\frac{m_S - \mu_k m_F}{m_S + m_F}\right) g$$

$$= -\frac{1000 \text{ kg} - 0.5(500 \text{ kg})}{1000 \text{ kg} + 500 \text{ kg}} \times 9.80 \text{ m/s}^2 = -4.9 \text{ m/s}^2$$

Now we need to calculate the kinematics of the falling safe. Because the time of the fall is not known or needed, we can use

$$(v_y)_f^2 = (v_y)_i^2 + 2a_{Sy} \Delta y = 0 + 2a_{Sy}(y_f - y_i) = -2a_{Sy} y_i$$

$$(v_y)_f = \sqrt{-2a_{Sy} y_i} = \sqrt{-2(-4.9 \text{ m/s}^2)(3.0 \text{ m})} = 5.4 \text{ m/s}$$

The value of $(v_y)_f$ is negative, but we only needed to find the speed, so we took the absolute value. It seems unlikely that the truck will survive the impact of the 1000 kg safe!

In many situations we are interested in the force exerted on a pulley by a rope that passes through it. Consider the pulley shown in Figure 5.42. Even though the pulley is frictionless, the tension force still pulls on the pulley. Because the rope is under tension on *both* sides of the pulley, the pulley is subject to *two* tension forces, one from each side of the rope. The net force on the pulley due to the rope is then the vector sum of these two tension forces.

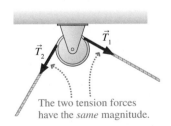

The two tension forces have the *same* magnitude.

FIGURE 5.42 Tension forces on a pulley.

EXAMPLE 5.18 Placing a leg in traction

For serious fractures of the leg, the leg may need to have a stretching force applied to it to keep contracting leg muscles from forcing the broken bones together too hard. This is often done using *traction*, an arrangement of a rope, a weight, and pulleys such as that shown in Figure 5.43. For this patient, the horizontal component of the force on the leg—the component that stretches the leg—must have a magnitude of 50 N. In order to apply this force, what must be the weight w?

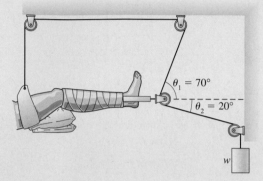

FIGURE 5.43 A leg in traction.

PREPARE To solve this problem, we need to apply three concepts that we've learned about ropes and pulleys:

1. A rope pulls on what it's attached to with a force equal to the tension in the rope. So the weight is pulled up with a force of magnitude T while being pulled down by its own weight w. But because it's in equilibrium, $w = T$.
2. The tension in a rope is unchanged when it goes around the massless, frictionless pulleys we'll assume here. So T is the same everywhere along the rope.
3. As we just learned, the rope passing through the pulley attached to the foot pulls in two directions, upwards at 70° and downwards at 20°.

We need the patient's leg to be stretched horizontally by a force of magnitude 50 N, so $(F_{PonL})_x = 50$ N, where P stands for pul-

ley and L for leg. Given that the pulley is in static equilibrium, Newton's third law gives us

$$(F_{PonL})_x = (F_{LonP})_x = (F_{RonP})_x = 50 \text{ N}$$

where R stands for rope. That is, the horizontal component of the sum of the tension forces of the rope on the pulley must also be 50 N. Figure 5.44 shows the forces on the pulley due to the tension in the rope. (This is not a true free-body diagram because we don't show the force on the pulley due to the leg).

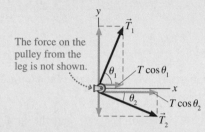

FIGURE 5.44 Forces on the pulley due to tension in the rope.

SOLVE From Figure 5.44 we can see that

$$(F_{RonP})_x = T\cos\theta_1 + T\cos\theta_2$$

From our discussion above, we know that the two tensions are equal to w, and that $(F_{RonP})_x = 50$ N. Thus we have

$$(F_{RonP})_x = w(\cos\theta_1 + \cos\theta_2) = 50 \text{ N}$$

which gives

$$w = \frac{50 \text{ N}}{(\cos\theta_1 + \cos\theta_2)} = \frac{50 \text{ N}}{(\cos 70° + \cos 20°)} = 39 \text{ N}$$

ASSESS Although the arrangement of the rope and pulleys looks complicated, the solution is straightforward using our rules for tension in ropes. Thus, for instance, the weight hanging on one end of the rope determines the tension, even though the other end of the rope helps support the unknown weight of the patient's leg.

Newton's three laws form the cornerstone of the science of mechanics. These laws allowed scientists to understand many diverse phenomena, from the motion of a raindrop to the orbits of the planets. These laws were so precise at predicting motion that they went unchallenged for well over two hundred years. At the beginning of the twentieth century, however, it began to be apparent that the laws of mechanics and the laws of electricity and magnetism were somehow inconsistent. Bringing these two apparently disconnected theories into harmony required the genius of a young patent clerk named Albert Einstein. In doing so, he shook the foundations not only of Newtonian mechanics but also of our very notions of space and time as well. We will continue to develop Newtonian mechanics for the next few chapters because of its tremendous importance to the physics of everyday life. But it's worth keeping in the back of your mind that Newton's laws aren't the ultimate statement about motion. Later in this textbook we'll reexamine motion and mechanics from the perspective of Einstein's theory of relativity.

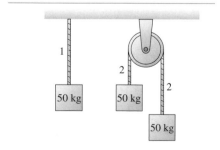

STOP TO THINK 5.6 All three 50 kg blocks are at rest. Is the tension in rope 2 greater than, less than, or equal to the tension in rope 1?

SUMMARY

The goal of Chapter 5 has been to learn how to solve problems about motion in a straight line.

GENERAL STRATEGY

All examples in this chapter follow a three-part strategy. You'll become a better problem solver if you adhere to it as you do the homework problems. The *Dynamics Worksheets* in the *Student Workbook* will help you structure your work in this way.

Equilibrium Problems

Object at rest or moving at constant velocity.

PREPARE Make simplifying assumptions.

- Check that the object is either at rest or moving with constant velocity.
- Identify forces and show them on a free-body diagram.

SOLVE Use Newton's first law in component form:

$$\sum F_x = 0$$
$$\sum F_y = 0$$

"Read" the components from the free-body diagram

ASSESS Is your result reasonable?

Dynamics Problems

Object accelerating.

PREPARE Make simplifying assumptions.
Make a **visual overview:**

- Sketch a pictorial representation
- Identify known quantities and what the problem is trying to find.
- Identify all forces and show them on a free-body diagram.

SOLVE Use Newton's second law in component form:

$$\sum F_x = ma_x \text{ and } \sum F_y = ma_y$$

"Read" the components of the vectors from the free-body diagram. If needed, use kinematics to find positions and velocities.

ASSESS Is your result reasonable?

Interacting Objects

Two or more objects interacting.

PREPARE Identify the system and the environment.
Make a **visual overview:**

- Sketch a pictorial representation.
- Identify any acceleration constraints.
- Identify all forces acting on *each* object.
- Identify action/reaction pairs between objects in the system.
- Draw separate free-body diagrams for each object. Connect action/reaction pairs with dotted lines and show them on the free-body diagram.

SOLVE Write Newton's second law for each object. Use Newton's third law to equate the magnitudes of action/reaction pairs. Include acceleration constraints.

ASSESS Is your result reasonable?

IMPORTANT CONCEPTS

Specific information about three important forces:

Weight $\vec{w} = (mg, \text{downwards})$

Friction $\vec{f_s} = (0 \text{ to } \mu_s n, \text{direction as necessary to prevent motion})$
$\vec{f_k} = (\mu_k n, \text{direction opposite the motion})$
$\vec{f_r} = (\mu_r n, \text{direction opposite the motion})$

Drag $\vec{D} \approx \left(\frac{1}{4}\rho A v^2, \text{direction opposite to the motion}\right)$ for motion in air
$\vec{D} = (6\pi\eta r v, \text{direction opposite the motion})$ for motion in liquid

Newton's laws are vector expressions. You must write them out by components:

$$(F_{\text{net}})_x = \sum F_x = ma_x \text{ or } 0$$
$$(F_{\text{net}})_y = \sum F_y = ma_y \text{ or } 0$$

APPLICATIONS

Apparent weight is the magnitude of the contact force supporting an object. It is what a scale would read, and it is your sensation of weight.

$$w_{\text{app}} = m(g + a_y)$$

Apparent weight equals your true weight $w = mg$ only when $a_y = 0$.

Acceleration constraints
Objects that are constrained to move together must have accelerations of equal magnitude: $a_A = a_B$.

This must be expressed in terms of components, such as $a_{Ax} = -a_{By}$.

Strings and pulleys

- A string or rope pulls what it's connected to with a force equal to its tension.
- The tension in a rope is equal to the force pulling on the rope.
- The tension in a massless rope is the same at all points in the rope.
- Tension does not change when a rope passes over a massless, frictionless pulley.

$F_{\text{rope on wall}} = \text{tension}$
$F_{\text{hand on rope}} = \text{tension}$

MP For instructor-assigned homework, go to www.masteringphysics.com

Problem difficulty is labeled as I (straightforward) to IIIII (challenging).

Problems labeled can be done on a Workbook Dynamics Worksheet; INT integrate significant material from earlier chapters; BIO are of biological or medical interest.

QUESTIONS

Conceptual Questions

1. An object is subject to two forces that do not point in opposite directions. Is it possible to choose their magnitudes so that the object is in equilibrium? Explain.
2. Are the objects described here in static equilibrium, dynamic equilibrium, or not in equilibrium at all?
   a. A girder is lifted at constant speed by a crane.
   b. A girder is lowered by a crane. It is slowing down.
   c. You're straining to hold a 200 lb barbell over your head.
   d. A jet plane has reached its cruising speed and altitude.
   e. A rock is falling into the Grand Canyon.
   f. A box in the back of a truck doesn't slide as the truck stops.
3. What forces are acting on you right now? What net force is acting on you right now?
4. Decide whether each of the following is true or false. Give a reason!
   a. The mass of an object depends on its location.
   b. The weight of an object depends on its location.
   c. Mass and weight describe the same thing in different units.
5. An astronaut takes his bathroom scale to the moon and then stands on it. Is the reading of the scale his true weight? Explain.
6. A light block of mass m and a heavy block of mass M are attached to the ends of a rope. A student holds the heavier block and lets the lighter block hang below it, as shown in Figure Q5.6. Then she lets go. Air resistance can be neglected.
   a. What is the tension in the rope while the blocks are falling, before either hits the ground?
   b. Would your answer be different if she had been holding the lighter block initially?

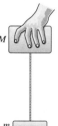

FIGURE Q5.6

7. Four balls are thrown straight up. Figure Q5.7 is a "snapshot" showing their velocities. They have the same size but different mass. Air resistance is negligible. Rank in order, from largest to smallest, the magnitudes of the net forces, F_{net1}, F_{net2}, F_{net3}, F_{net4}, acting on the balls. Some may be equal. Give your answer in the form A > B = C > D, and state your reasoning.

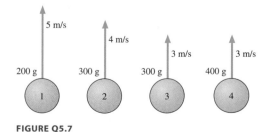

FIGURE Q5.7

8. Suppose you attempt to pour out 100 g of salt, using a pan balance for measurements, while in an elevator that is accelerating upward. Will the quantity of salt be too much, too little, or the correct amount? Explain.
9. a. Can the normal force on an object be directed horizontally? If not, why not? If so, provide an example.
   b. Can the normal force on an object be directed downward? If not, why not? If so, provide an example.
10. A ball is thrown straight up. Taking the drag force of air into account, does it take longer for the ball to travel to the top of its motion or for it to fall back down again?
11. Three objects move through the air as shown in Figure Q5.11. Rank in order, from largest to smallest, the three drag forces D_1, D_2, and D_3. Some may be equal. Give your answer in the form A > B = C and state your reasoning.

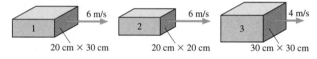

FIGURE Q5.11

12. A skydiver is falling at her terminal speed. Right after she opens her parachute, which has a very large area, what is the direction of the net force on her?
13. Raindrops can fall at different speeds; some fall quite quickly, others quite slowly. Why might this be true?
14. An airplane moves through the air at a constant speed. The jet engine's thrust applies a force in the direction of motion. Reducing thrust will cause the plane to fly at a slower—but still constant—speed. Explain why this is so.
15. Is it possible for an object to travel in air faster than its terminal speed? If not, why not? If so, explain how this might happen.

For Questions 16 to 19, determine the reading of the spring scale.

- All objects are at rest.
- The strings, scales, and pulleys are massless and the pulleys are frictionless.
- The spring scale reads in kg.

16.

FIGURE Q5.16

17.

FIGURE Q5.17

18.

FIGURE Q5.18

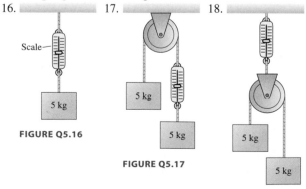

19.

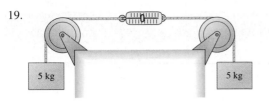

FIGURE Q5.19

For Questions 20 through 23, write the acceleration constraint in terms of *components*. For example, write $a_{1x} = a_{2y}$, if that is the appropriate answer, rather than $a_1 = a_2$.

20.

21.

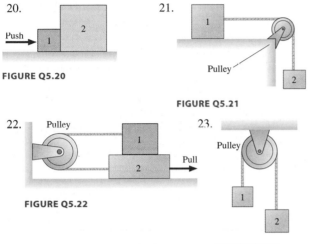

FIGURE Q5.20

FIGURE Q5.21

22.

23.

FIGURE Q5.22

FIGURE Q5.23

Multiple-Choice Questions

24. | A 2.0 kg ball is suspended by two light strings as shown in Figure Q5.24. What is the tension T in the angled string?
    A. 9.5 N   B. 15 N
    C. 20 N   D. 26 N
    E. 30 N

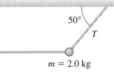

FIGURE Q5.24

25. | While standing in a low tunnel, you raise your arms and push against the ceiling with a force of 100 N. Your mass is 70 kg.
    a. What force does the ceiling exert on you?
       A. 10 N   B. 100 N   C. 690 N
       D. 790 N   E. 980 N
    b. What force does the floor exert on you?
       A. 10 N   B. 100 N   C. 690 N
       D. 790 N   E. 980 N

26. | A 5.0 kg dog sits on the floor of an elevator that is accelerating *downward* at 1.20 m/s².
    a. What is the magnitude of the normal force of the elevator floor on the dog?
       A. 34.2 N   B. 43.0 N   C. 49.0 N
       D. 55.0 N   E. 74.0 N
    b. What is the magnitude of the force of the dog on the elevator floor?
       A. 4.2 N   B. 49.0 N   C. 55.0 N
       D. 43.0 N   E. 74.0 N

27. ‖ A 3.0 kg puck slides due east on a horizontal frictionless surface at a constant speed of 4.5 m/s. Then a force of magnitude 6.0 N, directed due north, is applied for 1.5 s. Afterward,
    a. What is the northward component of the puck's velocity?
       A. 0.50 m/s   B. 2.0 m/s   C. 3.0 m/s
       D. 4.0 m/s   E. 4.5 m/s
    b. What is the speed of the puck?
       A. 4.9 m/s   B. 5.4 m/s   C. 6.2 m/s
       D. 7.5 m/s   E. 11 m/s

28. ‖ A rocket in space, initially at rest, fires its main engines at a constant thrust. As it burns fuel, the mass of the rocket decreases. Which of the graphs in Figure Q5.28 best represents the velocity of the rocket as a function of time?

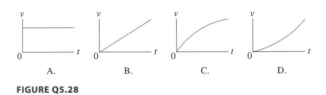

FIGURE Q5.28

29. | Eric has a mass of 60 kg. He is standing on a scale in an elevator that is accelerating downward at 1.7 m/s². What is the approximate reading on the scale?
    A. 0 N   B. 400 N   C. 500 N   D. 600 N

30. | The two blocks in Figure Q5.30 are at rest on frictionless surfaces. What must be the mass of the right block in order that the two blocks remain stationary?
    A. 4.9 kg   B. 6.1 kg   C. 7.9 kg
    D. 9.8 kg   E. 12 kg

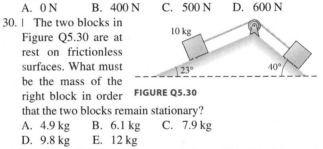

FIGURE Q5.30

31. | A football player at practice pushes a 60 kg blocking sled across the field at a constant speed. The coefficient of kinetic friction between the grass and the sled is 0.30. How much force must he apply to the sled?
    A. 18 N   B. 60 N   C. 180 N   D. 600 N

32. | Two football players are pushing a 60 kg blocking sled across the field at a constant speed of 2.0 m/s. The coefficient of kinetic friction between the grass and the sled is 0.30. Once they stop pushing, how far will the sled slide before coming to rest?
    A. 0.20 m   B. 0.68 m   C. 1.0 m   D. 6.6 m

33. ‖ Land Rover ads used to claim that their vehicles could climb a slope of 45°. For this to be possible, what must be the minimum coefficient of static friction between the vehicle's tires and the road?
    A. 0.5   B. 0.7   C. 0.9   D. 1.0

34. ‖ A truck is traveling at 30 m/s on a slippery road. The driver slams on the brakes and the truck starts to skid. If the coefficient of kinetic friction between the tires and the road is 0.20, how far will the truck skid before stopping?
    A. 230 m   B. 300 m   C. 450 m   D. 680 m

PROBLEMS

Section 5.1 Equilibrium

1. | The three ropes in Figure P5.1 are tied to a small, very light ring. Two of the ropes are anchored to walls at right angles, and the third rope pulls as shown. What are T_1 and T_2, the magnitudes of the tension forces in the first two ropes?

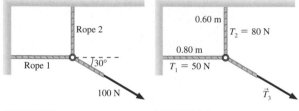

FIGURE P5.1   **FIGURE P5.2**

2. ‖ The three ropes in Figure P5.2 are tied to a small, very light ring. Two of these ropes are anchored to walls at right angles with the tensions shown in the figure. What are the magnitude and direction of the tension $\vec{T}_3$ in the third rope?

3. ‖‖ A 20 kg loudspeaker is suspended 2.0 m below the ceiling by two cables that are each 30° from vertical. What is the tension in the cables?

4. | A 1000 kg steel beam is supported by the two ropes shown in Figure P5.4. Each rope can support a maximum sustained tension of 5600 N. Do the ropes break?

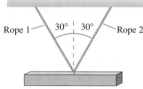

FIGURE P5.4

5. ‖‖‖ A football coach sits on a sled while two of his players build their strength by dragging the sled across the field with ropes. The friction force on the sled is 1.0 kN and the angle between the two horizontal ropes is 20°. How hard must each player pull to drag the coach at a steady 2.0 m/s?

6. ‖ When you bend your knee, the quadriceps muscle is stretched. This increases the tension in the quadriceps tendon attached to your kneecap (patella), which, in turn, increases the tension in the patella tendon that attaches your kneecap to your lower leg bone (tibia). Simultaneously, the end of your upper leg bone (femur) pushes outward on the patella. Figure P5.6 shows how these parts of a knee joint are arranged. What size force does the femur exert on the kneecap if the tendons are oriented as in the figure and the tension in each tendon is 60 N?

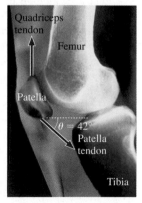

FIGURE P5.6

BIO

Section 5.2 Dynamics and Newton's Second Law

7. | A force with x-component F_x acts on a 2.0 kg object as it moves along the x-axis. The object's acceleration graph (a_x versus t) is shown in Figure P5.7. Draw a graph of F_x versus t.

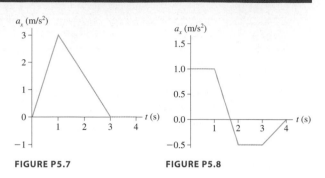

FIGURE P5.7   **FIGURE P5.8**

8. | A force with x-component F_x acts on a 500 g object as it moves along the x-axis. The object's acceleration graph (a_x versus t) is shown in Figure P5.8. Draw a graph of F_x versus t.

9. | A force with x-component F_x acts on a 2.0 kg object as it moves along the x-axis. A graph of F_x versus t is shown in Figure P5.9. Draw an acceleration graph (a_x versus t) for this object.

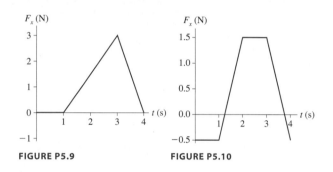

FIGURE P5.9   **FIGURE P5.10**

10. | A force with x-component F_x acts on a 500 g object as it moves along the x-axis. A graph of F_x versus t is shown in Figure P5.10. Draw an acceleration graph (a_x versus t) for this object.

11. ‖ The forces in Figure P5.11 are acting on a 2.0 kg object. Find the values of a_x and a_y, the x- and y-components of the object's acceleration.

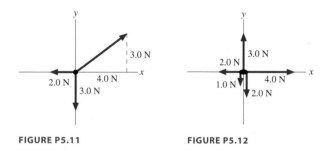

FIGURE P5.11   **FIGURE P5.12**

12. | The forces in Figure P5.12 are acting on a 2.0 kg object. Find the values of a_x and a_y, the x- and y-components of the object's acceleration.

13. | A horizontal rope is tied to a 50 kg box on frictionless ice. What is the tension in the rope if
    a. The box is at rest?
    b. The box moves at a steady 5.0 m/s?
    c. The box has $v_x = 5.0$ m/s and $a_x = 5.0$ m/s²?

14. ||| A crate pushed along the floor with velocity $\vec{v}_i$ slides a distance d after the pushing force is removed.
    a. If the mass of the crate is doubled but the initial velocity is not changed, what distance does the crate slide before stopping? Explain.
    b. If the initial velocity of the crate is doubled to $2\vec{v}_i$ but the mass is not changed, what distance does the crate slide before stopping? Explain.

15. | In a head-on collision, a car stops in 0.10 s from a speed of 14 m/s. The driver has a mass of 70 kg, and is, fortunately, tightly strapped into his seat. What force is applied to the driver by his seat belt during that fraction of a second?

Section 5.3 Mass and Weight

16. | An astronaut's weight on earth is 800 N. What is his weight on Mars, where $g = 3.76$ m/s²?

17. | A woman has a mass of 55.0 kg.
    a. What is her weight on earth?
    b. What are her mass and her weight on the moon, where $g = 1.62$ m/s²?

18. | A box with a 75 kg passenger inside is launched straight up into the air by a giant rubber band. After the box has left the rubber band but is still moving *upward,*
    a. What is the passenger's true weight?
    b. What is the passenger's apparent weight?

19. | a. How much force does an 80 kg astronaut exert on his chair while sitting at rest on the launch pad?
    b. How much force does the astronaut exert on his chair while accelerating straight up at 10 m/s²?

20. | It takes the elevator in a skyscraper 4.0 s to reach its cruising speed of 10 m/s. A 60 kg passenger gets aboard on the ground floor. What is the passenger's apparent weight
    a. Before the elevator starts moving?
    b. While the elevator is speeding up?
    c. After the elevator reaches its cruising speed?

21. | Zach, whose mass is 80 kg, is in an elevator descending at 10 m/s. The elevator takes 3.0 s to brake to a stop at the first floor.
    a. What is Zach's apparent weight before the elevator starts braking?
    b. What is Zach's apparent weight while the elevator is braking?

22. | Figure P5.22 shows the velocity graph of a 75 kg passenger in an elevator. What is the passenger's apparent weight at $t = 1.0$ s? At 5.0 s? At 9.0 s?

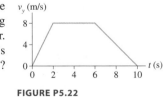

FIGURE P5.22

Section 5.4 Normal Forces

23. | a. A 0.60 kg bullfrog is sitting at rest on a level log. How large is the normal force of the log on the bullfrog?
    b. A second 0.60 kg bullfrog is on a log tilted 30° above horizontal. How large is the normal force of the log on this bullfrog?

24. ||| A 23 kg child goes down a straight slide inclined 38° above horizontal. The child is acted on by his weight, the normal force from the slide, and kinetic friction.
    a. Draw a free-body diagram of the child.
    b. How large is the normal force of the slide on the child?

Section 5.5 Friction

25. || Bonnie and Clyde are sliding a 300 kg bank safe across the floor to their getaway car. The safe slides with a constant speed if Clyde pushes from behind with 385 N of force while Bonnie pulls forward on a rope with 350 N of force. What is the safe's coefficient of kinetic friction on the bank floor?

26. || A 4000 kg truck is parked on a 15° slope. How big is the friction force on the truck?

27. || A 1000 kg car traveling at a speed of 40 m/s skids to a halt on wet concrete where $\mu_k = 0.60$. How long are the skid marks?

28. | A stubborn 120 kg mule sits down and refuses to move. To drag the mule to the barn, the exasperated farmer ties a rope around the mule and pulls with his maximum force of 800 N. The coefficients of friction between the mule and the ground are $\mu_s = 0.80$ and $\mu_k = 0.50$. Is the farmer able to move the mule?

29. || A 10 kg crate is placed on a horizontal conveyor belt. The materials are such that $\mu_s = 0.50$ and $\mu_k = 0.30$.
    a. Draw a free-body diagram showing all the forces on the crate if the conveyer belt runs at constant speed.
    b. Draw a free-body diagram showing all the forces on the crate if the conveyer belt is speeding up.
    c. What is the maximum acceleration the belt can have without the crate slipping?
    d. If acceleration of the belt exceeds the value determined in part c, what is the acceleration of the crate?

Section 5.6 Drag

30. | What is the drag force on a 1.6 m wide, 1.4 m high car traveling at
    a. 10 m/s ($\approx$22 mph)?   b. 30 m/s ($\approx$65 mph)?

31. || A 75 kg skydiver can be modeled as a rectangular "box" with dimensions 20 cm $\times$ 40 cm $\times$ 1.8 m. What is his terminal speed if he falls feet first?

32. | A 22-cm-diameter bowling ball has a terminal speed of 85 m/s. What is the ball's mass?

33. || In Example 5.14, the stopping of a paramecium due to the Stokes's law drag force was treated in an approximate fashion, taking the drag force to be independent of time.
    a. Use the same approximation to determine how *far* the paramecium drifts after it stops actively swimming at 0.25 mm/s.
    b. Is your answer to part a much less than, comparable to, or much greater than the size of the paramecium?

Section 5.7 Interacting Objects

34. || A 1000 kg car pushes a 2000 kg truck that has a dead battery. When the driver steps on the accelerator, the drive wheels of the car push against the ground with a force of 4500 N.
    a. What is the magnitude of the force of the car on the truck?
    b. What is the magnitude of the force of the truck on the car?

35. |||| Blocks with masses of 1.0 kg, 2.0 kg, and 3.0 kg are lined up in a row on a frictionless table. All three are pushed forward by a 12 N force applied to the 1.0 kg block. How much force does the 2.0 kg block exert on (a) the 3.0 kg block and (b) the 1.0 kg block?

Section 5.8 Ropes and Pulleys

36. ‖ What is the tension in the rope of Figure P5.36?

37. ‖ A 2.0-m-long, 500 g rope pulls a 10 kg block of ice across a horizontal, frictionless surface. The block accelerates at 2.0 m/s². How much force pulls forward on (a) the block of ice, (b) the rope?

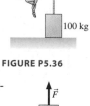

60 kg

100 kg

FIGURE P5.36

38. ‖ Figure P5.38 shows two 1.00 kg blocks connected by a rope. A second rope hangs beneath the lower block. Both ropes have a mass of 250 g. The entire assembly is accelerated upward at 3.00 m/s² by force $\vec{F}$.
   a. What is F?
   b. What is the tension at the top end of rope 1?
   c. What is the tension at the bottom end of rope 1?
   d. What is the tension at the top end of rope 2?

$\vec{F}$

A

Rope 1

B

Rope 2

FIGURE P5.38

General Problems

39. ‖‖ A 500 kg piano is being lowered into position by a crane while two people steady it with ropes pulling to the sides. Bob's rope pulls to the left, 15° below horizontal, with 500 N of tension. Ellen's rope pulls toward the right, 25° below horizontal.
   a. What tension must Ellen maintain in her rope to keep the piano descending vertically?
   b. What is the tension in the vertical main cable supporting the piano?

40. ‖‖ In an electricity experiment, a 1.00 g plastic ball is suspended on a 60.0-cm-long string and given an electric charge. A charged rod brought near the ball exerts a horizontal electrical force $\vec{F}_{\text{elec}}$ on it, causing the ball to swing out to a 20.0° angle and remain there.
   a. What is the magnitude of $\vec{F}_{\text{elec}}$?
   b. What is the tension in the string?

41. ‖ Figure P5.41 shows the velocity graph of a 2.0 kg object as it moves along the x-axis. What is the net force acting on this object at $t = 1$ s? At 4 s? At 7 s?

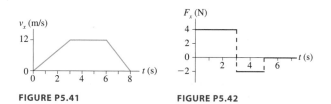

FIGURE P5.41   **FIGURE P5.42**

42. ‖ Figure P5.42 shows the force acting on a 2.0 kg object as it moves along the x-axis. The object is at rest at the origin at $t = 0$ s. What are its acceleration and velocity at $t = 6.0$ s?

43. ‖ A 50 kg box hangs from a rope. What is the tension in the rope if
   a. The box is at rest?
   b. The box has $v_y = 5.0$ m/s and is speeding up at 5.0 m/s²?

44. ‖ A 50 kg box hangs from a rope. What is the tension in the rope if
   a. The box moves up at a steady 5.0 m/s?
   b. The box has $v_y = 5.0$ m/s and is slowing down at 5.0 m/s²?

45. ‖ Your forehead can withstand a force of about 6.0 kN force before fracturing, while your cheekbone can only withstand about 1.3 kN.
   a. If a 140 g baseball strikes your head at 30 m/s and stops in 0.0015 s, what is the magnitude of the ball's acceleration?
   b. What is the magnitude of the force that stops the baseball?
   c. What force does the baseball apply to your head? Explain.
   d. Are you in danger of a fracture if the ball hits you in the forehead? In the cheek?

46. ‖ Seat belts and air bags save lives by reducing the forces exerted on the driver and passengers in an automobile collision. Cars are designed with a "crumple zone" in the front of the car. In the event of an impact, the passenger compartment decelerates over a distance of about 1 m as the front of the car crumples. An occupant restrained by seat belts and air bags decelerates with the car. By contrast, an unrestrained occupant keeps moving forward with no loss of speed (Newton's first law!) until hitting the dashboard or windshield, as we saw in Figure 4.2. These are unyielding surfaces, and the unfortunate occupant then decelerates over a distance of only about 5 mm.
   a. A 60 kg person is in a head-on collision. The car's speed at impact is 15 m/s. Estimate the net force on the person if he or she is wearing a seat belt and if the air bag deploys.
   b. Estimate the net force that ultimately stops the person if he or she is not restrained by a seat belt or air bag.
   c. How do these two forces compare to the person's weight?

47. ‖‖ Bob, who has a mass of 75 kg, can throw a 500 g rock with a speed of 30 m/s. The distance through which his hand moves as he accelerates the rock forward from rest until he releases it is 1.0 m.
   a. What constant force must Bob exert on the rock to throw it with this speed?
   b. If Bob is standing on frictionless ice, what is his recoil speed after releasing the rock?

48. ‖‖‖ An 80 kg spacewalking astronaut pushes off a 640 kg satellite, exerting a 100 N force for the 0.50 s it takes him to straighten his arms. How far apart are the astronaut and the satellite after 1.0 min?

49. ‖ What thrust does a 200 g model rocket need in order to have a vertical acceleration of 10.0 m/s²
   a. On earth?
   b. On the moon, where $g = 1.62$ m/s²?

50. ‖‖ A 20,000 kg rocket has a rocket motor that generates 3.0×10^5 N of thrust.
   a. What is the rocket's initial upward acceleration?
   b. At an altitude of 5.0 km the rocket's acceleration has increased to 6.0 m/s². What mass of fuel has it burned?

51. ‖‖‖ You've always wondered about the acceleration of the elevators in the 101-story-tall Empire State Building. One day, while visiting New York, you take your bathroom scales into the elevator and stand on them. The scales read 150 lb as the door closes. The reading varies between 120 lb and 170 lb as the elevator travels 101 floors.
   a. What is the magnitude of the acceleration as the elevator starts upward?
   b. What is the magnitude of the acceleration as the elevator brakes to a stop?

52. ⦀ A 23 kg child goes down a straight slide inclined 38° above horizontal. The child is acted on by his weight, the normal force from the slide, kinetic friction, and a horizontal rope exerting a 30 N force as shown in Figure P5.52. How large is the normal force of the slide on the child?

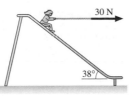

FIGURE P5.52

53. ⦀ Researchers often use *force plates* to measure the forces that
BIO people exert against the floor during movement. A force plate
INT works like a bathroom scale, but it keeps a record of how the reading changes with time. Figure P5.53 shows the data from a force plate as a woman jumps straight up and then lands.
    a. What was the vertical component of her acceleration during push-off?
    b. What was the vertical component of her acceleration while in the air?
    c. What was the vertical component of her acceleration during the landing?
    d. What was her speed as her feet left the force plate?
    e. How high did she jump?

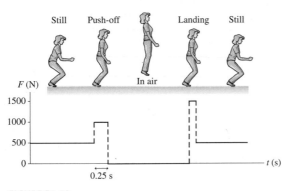

FIGURE P5.53

54. ⦀ A 77 kg sprinter is running the 100 m dash. At one instant,
BIO early in the race, his acceleration is 4.7 m/s^2.
    a. What *total* force does the track surface exert on the sprinter? Assume his acceleration is parallel to the ground. Give your answer as a magnitude and an angle with respect to the horizontal.
    b. This force is applied to one foot (the other foot is in the air), which for a fraction of a second is stationary with respect to the track surface. Because the foot is stationary, the net force on it must be zero. Thus the force of the lower leg bone on the foot is equal but opposite to the force of the track on the foot. If the lower leg bone is 60° from horizontal, what are the components of the leg's force on the foot in the directions parallel and perpendicular to the leg? (Force components perpendicular to the leg can cause dislocation of the ankle joint.)

55. ⦀ Sam, whose mass is 75 kg, takes off across level snow on his jet-powered skis. The skis have a thrust of 200 N and a coefficient of kinetic friction on snow of 0.10. Unfortunately, the skis run out of fuel after only 10 s.
    a. What is Sam's top speed?
    b. How far has Sam traveled when he finally coasts to a stop?

56. ⦀ A person with compromised pinch
strength in his fingers can only exert a nor-
BIO mal force of 6.0 N to either side of a pinch-held object, such as the book shown in Figure P5.56. What is the heaviest book he can hold onto vertically before it slips out of his fingers? The coefficient of static friction of the surface between the fingers and the book cover is 0.80.

FIGURE P5.56

57. ⦀ A 1.0 kg wood block is pressed against a vertical wood wall by a 12 N force as shown in Figure P5.57. If the block is initially at rest, will it move upward, move downward, or stay at rest?

FIGURE P5.57

58. ⦀ A 50,000 kg locomotive, with steel wheels, is traveling at 10 m/s on steel rails when its engine and brakes both fail. How far will the locomotive roll before it comes to a stop?

59. ⦀ An Airbus A320 jetliner has a takeoff mass of 75,000 kg. It reaches its takeoff speed of 82 m/s (180 mph) in 35 s. What is the thrust of the engines? You can neglect air resistance but not rolling friction.

60. ⦀ A 2.0 kg wood block is launched up a wooden ramp that is inclined at a 35° angle. The block's initial speed is 10 m/s.
    a. What vertical height does the block reach above its starting point?
    b. What speed does it have when it slides back down to its starting point?

61. ⦀ A 2.7 g Ping-Pong ball has a diameter of 4.0 cm.
    a. The ball is shot straight up at twice its terminal speed. What is its initial acceleration?
    b. The ball is shot straight down at twice its terminal speed. What is its initial acceleration?

62. ⦀ The fastest recorded skydive was by an Air Force officer who jumped from a helium balloon at an elevation of 103,000 ft, three times higher than airliners fly. Because the density of air is so low at these altitudes, he reached a speed of 614 mph at an elevation of 90,000 ft, then gradually slowed as the air became more dense. Assume that he fell in the spread-eagle position and that his low-altitude terminal speed is 125 mph. Use this information to determine the density of air at 90,000 ft.

63. ⦀ The cable cars in San Francisco are pulled along their tracks by an underground steel cable that moves along at 4.0 m/s. The cable is driven by large motors at a central power station and extends, via an intricate pulley arrangement, for several miles beneath the city streets. The length of a cable stretches by up to 30 m during its lifetime. To keep the tension constant, the cable passes around a 1.5-m-diameter "tensioning pulley" that rolls back and forth on rails, as shown in Figure P5.63. A 2000 kg block is attached to the tensioning pulley's cart, via a rope and pulley, and is suspended in a deep hole. What is the tension in the cable car's cable?

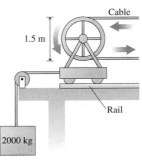

FIGURE P5.63

64. ||| The 100 kg block in Figure P5.64 takes 6.0 s to reach the floor after being released from rest. What is the mass of the block on the left?

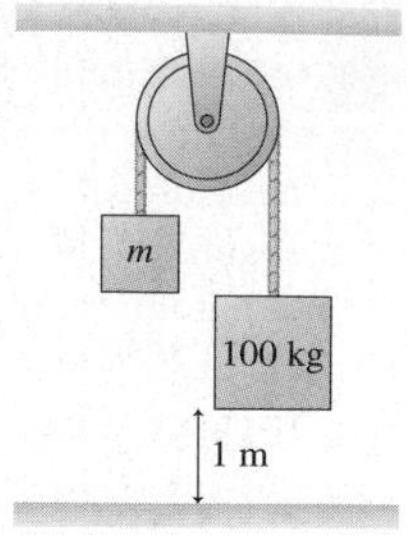

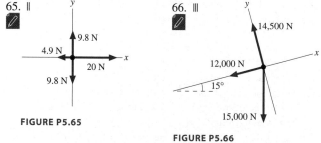

FIGURE P5.64

Problems 65 and 66 show free-body diagrams. For each,

a. Write a realistic dynamics problem for which this is the correct free-body diagram. Your problem should ask a question that can be answered with a value of position or velocity (such as "How far?" or "How fast?"), and should give sufficient information to allow a solution.

b. Solve your problem!

65. ||

FIGURE P5.65

66. |||

FIGURE P5.66

In Problems 67 through 69 you are given the dynamics equations that are used to solve a problem. For each of these, you are to

a. Write a realistic problem for which these are the correct equations.

b. Draw the free-body diagram and the pictorial representation for your problem.

c. Finish the solution of the problem.

67. ||| $-0.80n = (1500 \text{ kg})a_x$
$n - (1500 \text{ kg})(9.8 \text{ m/s}^2) = 0$

68. ||| $T - 0.2n - (20 \text{ kg})(9.8 \text{ m/s}^2)\sin 20° = (20 \text{ kg})(2.0 \text{ m/s}^2)$
$n - (20 \text{ kg})(9.8 \text{ m/s}^2)\cos 20° = 0$

69. ||| $(100 \text{ N})\cos 30° - f_k = (20 \text{ kg})a_x$
$n + (100 \text{ N})\sin 30° - (20 \text{ kg})(9.8 \text{ m/s}^2) = 0$
$f_k = 0.20n$

Passage Problems

Sliding on the Ice

In the winter sport of curling, players give a 20 kg stone a push across a sheet of ice. The stone moves approximately 40 m before coming to rest. The final position of the stone, in principle, only depends on the initial speed at which it is launched and the force of friction between the ice and the stone, but team members can use brooms to sweep the ice in front of the stone to adjust its speed and trajectory a bit; they must do this without touching the stone. Judicious sweeping can lengthen the travel of the stone by 3 m.

70. | A curler pushes a stone to a speed of 3.0 m/s over a time of 2.0 s. Ignoring the force of friction, how much force must the curler apply to the stone to bring it up to speed?
A. 3.0 N   B. 15 N   C. 30 N   D. 150 N

71. | The sweepers in a curling competition adjust the trajectory of the stone by
A. Decreasing the coefficient of friction between the stone and the ice.
B. Increasing the coefficient of friction between the stone and the ice.
C. Changing friction from kinetic to static.
D. Changing friction from static to kinetic.

72. | Suppose the stone is launched with a speed of 3 m/s and travels 40 m before coming to rest. What is the *approximate* magnitude of the friction force on the stone?
A. 0 N   B. 2 N   C. 20 N   D. 200 N

73. | Suppose the stone's mass is increased to 40 kg, but it is launched at the same 3 m/s. Which one of the following is true?
A. The stone would now travel a longer distance before coming to rest.
B. The stone would now travel a shorter distance before coming to rest.
C. The coefficient of friction would now be greater.
D. The force of friction would now be greater.

STOP TO THINK ANSWERS

Stop to Think 5.1: A. The lander is descending and slowing. The acceleration vector points upward, and so $\vec{F}_{net}$ points upward. This can be true only if the thrust has a larger magnitude than the weight.

Stop to Think 5.2: A. You are descending and slowing, so your acceleration vector points upward and there is a net upward force on you. The floor pushes up against your feet harder than gravity pulls down.

Stop to Think 5.3: $f_B > f_C = f_D = f_E > f_A$. Situations C, D, and E are all kinetic friction, which does not depend on either velocity or acceleration. Kinetic friction is smaller than the maximum static friction that is exerted in B. $f_A = 0$ because no friction is needed to keep the object at rest.

Stop to Think 5.4: D. The ball is shot *down* at 30 m/s, so $v_{0y} = -30$ m/s. This exceeds the terminal speed, so the upward drag force is *larger* than the downward weight force. Thus the ball *slows down* even though it is "falling." It will slow until $v_y = -15$ m/s, the terminal velocity, then maintain that velocity.

Stop to Think 5.5: B. $F_{QonH} = F_{HonQ}$ and $F_{PonQ} = F_{QonP}$ because these are action/reaction pairs. Box Q is slowing down and therefore must have a net force to the left. So from Newton's second law we also know that $F_{HonQ} > F_{PonQ}$.

Stop to Think 5.6: Equal to. Each block is hanging in equilibrium, with no net force, so the upward tension force is mg.

6

CIRCULAR MOTION, ORBITS, AND GRAVITY

Motorcyclists in the "Globe of Death" ride their bikes on the inside of a spherical steel frame, seeming to defy gravity as they ride up the sides and upside down over the top. What prevents them from falling?

Looking Ahead ▶▶

The goal of Chapter 6 is to learn about motion in a circle, including orbital motion under the influence of a gravitational force. In this chapter, you will learn to:

▶ Understand the kinematics of circular motion.

▶ Use Newton's laws to analyze the dynamics of circular motion.

▶ Understand the circular orbits of satellites and planets.

▶ Further your understanding of the force of gravity as a long-range force.

Looking Back ◀◀

This chapter uses what you have learned about one-dimensional motion, circular motion, Newton's laws, and the concepts of mass and weight. Please review:

◀ Section 2.2 Uniform motion.

◀ Section 3.8 Circular motion.

◀ Section 5.2 Using Newton's second law.

◀ Section 5.3 Mass, weight, and apparent weight.

Engines revving, the motorcyclists build up speed until they are riding in excess of 30 mph in small circles inside the spherical enclosure. As their speeds increase, they begin to ride up the sides, culminating in vertical loops that turn them completely upside down. In this chapter we will examine circular motion problems that range from such extreme examples to more everyday situations such as rounding a corner in a car.

You learned in Chapter 3 that circular motion involves an acceleration because the velocity is always changing direction. In this chapter, we will develop a new set of kinematic variables to describe circular motion, then look at the dynamics of uniform circular motion: the forces that produce the motion. One especially important kind of circular motion—the orbital motion of the moon about the earth and the planets about the sun—will lead us to consider the nature of gravity, the force that keeps things in orbit.

This is a very practical chapter, one in which we will learn why you need to turn corners more slowly on slippery roads, why highway curves are banked, and why water stays in a bucket when you swing it over your head.

6.1 Uniform Circular Motion

We began our study of circular motion in Section 3.8. Figure 6.1 is a review of the variables we used to specify circular motion and the relationships among them. We will start this chapter by extending and clarifying these concepts.

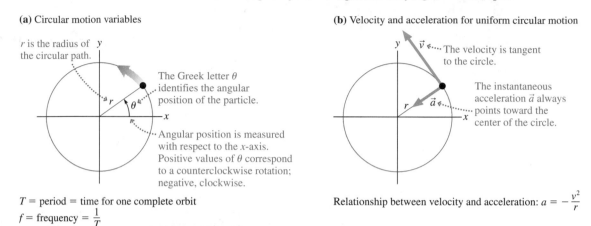

(a) Circular motion variables

r is the radius of the circular path.

The Greek letter θ identifies the angular position of the particle.

Angular position is measured with respect to the *x*-axis. Positive values of θ correspond to a counterclockwise rotation; negative, clockwise.

T = period = time for one complete orbit

f = frequency = $\frac{1}{T}$

(b) Velocity and acceleration for uniform circular motion

The velocity is tangent to the circle.

The instantaneous acceleration $\vec{a}$ always points toward the center of the circle.

Relationship between velocity and acceleration: $a = -\frac{v^2}{r}$

FIGURE 6.1 Reviewing circular motion.

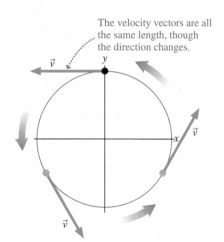

The velocity vectors are all the same length, though the direction changes.

FIGURE 6.2 A particle in uniform circular motion.

If the particle in circular motion is moving at *constant speed* we say that it is in **uniform circular motion.** Figure 6.2 shows a particle in uniform circular motion. The particle might be a satellite moving in an orbit, a ball on the end of a string, or even just a dot painted on the side of a wheel. Regardless of what the particle represents, its velocity vector is always tangent to the circular path. The particle's speed *v* is constant, so the vector's length stays constant as the particle moves around the circle.

Angular Displacement and Angular Velocity

To fully describe uniform circular motion, we will define some new variables. Figure 6.3 shows a particle in uniform circular motion that moves from an initial angular position θ_i at time t_i to a final angular position θ_f at a later time t_f. The change $\Delta\theta = \theta_f - \theta_i$ is called the **angular displacement.** We can measure the particle's circular motion in terms of the rate of change of θ just as we measured a particle's linear motion in terms of the rate of change of its position *x*.

In analogy with linear motion, where $v_x = \Delta x/\Delta t$, we define the **angular velocity** to be

$$\omega = \frac{\text{angular displacement}}{\text{time interval}} = \frac{\Delta\theta}{\Delta t} \tag{6.1}$$

Angular velocity of a particle in uniform circular motion

The symbol ω is a lowercase Greek omega, *not* an ordinary *w*. The SI unit of angular velocity is rad/s.

NOTE ▶ Recall our discussion of average and instantaneous velocity in Chapter 2. The definition of angular velocity in Equation 6.1 is really the *average* angular velocity. For a very small time interval Δt, the calculation will approximate the *instantaneous* angular velocity. If the angular velocity is constant, as it will be for the problems we consider in this chapter, the average angular velocity and instantaneous angular velocity are the same. ◄

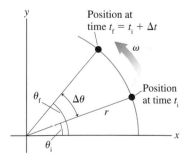

Position at time $t_f = t_i + \Delta t$

ω

Position at time t_i

FIGURE 6.3 A particle moves from θ_i to θ_f with angular velocity ω.

Angular velocity is the *rate* at which a particle's angular position is changing as it rotates around the circle. A particle that starts from $\theta = 0$ rad with an angular velocity of 0.5 rad/s will be at angle $\theta = 0.5$ rad after 1 s, at $\theta = 1.0$ rad after 2 s, at $\theta = 1.5$ rad after 3 s, and so on. Its angular position is increasing at the *rate* of 0.5 radians per second. In analogy with uniform linear motion, which you studied in Chapter 2, uniform circular motion is motion in which the angle increases at a *constant* rate: **A particle moves with uniform circular motion if and only if its angular velocity ω is constant.**

NOTE ▶ It's important to keep in mind which quantities are changing and which are constant. For a particle in uniform circular motion, the angular velocity and the speed are constant. This is how we define uniform circular motion. But the velocity is constantly changing because the direction of motion changes. ◀

Angular velocity, like the velocity v_x of one-dimensional motion, can be positive or negative. The signs noted in Figure 6.4 are based on the fact that angles are positive when measured counterclockwise from the positive x-axis.

Circular motion is analogous to linear motion, with angular variables replacing linear variables. Much of what you learned about linear kinematics and dynamics carries over to circular motion. For example, Equation 2.6 gave us a formula for computing a linear displacement during a time interval:

$$x_f - x_i = \Delta x = v_x \, \Delta t$$

You can see from Equation 6.1 that we can write a similar equation for the angular displacement:

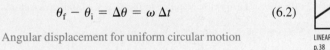

$$\theta_f - \theta_i = \Delta\theta = \omega \, \Delta t \tag{6.2}$$

Angular displacement for uniform circular motion

LINEAR
p. 38

For linear motion, we use the term *speed v* when we are not concerned with the direction of motion, *velocity v_x* when we are. For circular motion, we define the **angular speed** to be the absolute value of the angular velocity, irrespective of its direction of rotation. Although potentially confusing, it is customary to use the symbol ω for angular speed *and* for angular velocity. If the direction of rotation is not important, we will interpret ω to mean angular speed, a positive quantity. In kinematic equations, such as Equation 6.2, ω is the angular velocity, and you need to use a negative value for a clockwise rotation.

The angular speed ω is closely related to the period T and the frequency f of the motion. If a particle in uniform circular motion moves around a circle once, which by definition takes time T, its angular displacement is $\Delta\theta = 360° = 2\pi$ rad. The angular speed is thus

$$\omega = \frac{2\pi \text{ rad}}{T} \tag{6.3}$$

We can also write the angular speed in terms of the frequency:

$$\omega = (2\pi \text{ rad})f \tag{6.4}$$

where f must be in rev/s. For example, a particle in circular motion with frequency 10 rev/s would have angular speed $\omega = 20\pi$ rad/s $= 62.8$ rad/s.

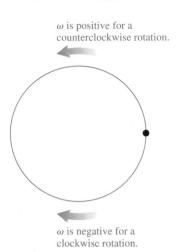

ω is positive for a counterclockwise rotation.

ω is negative for a clockwise rotation.

FIGURE 6.4 Positive and negative angular velocities.

Why do clocks go clockwise? In the northern hemisphere, the rotation of the earth causes the sun to follow a circular arc through the southern sky, rising in the east and setting in the west. For millennia, humans have marked passing time by noting the position of shadows cast by the sun, which sweep in an arc from west to east—eventually leading to the development of the sundial, the first practical timekeeping device. In the northern hemisphere, sundials point north, and the shadow sweeps around the dial from left to right. Early clockmakers used the same convention, which is how it came to be clockwise.

EXAMPLE 6.1 Kinematics at the roulette wheel

A small, steel ball rolls counterclockwise around the inside of a 30.0-cm-diameter roulette wheel. The ball completes exactly 2 rev in 1.20 s.

a. What is the ball's angular velocity?
b. What is the ball's angular position at $t = 2.00$ s? Assume $\theta_i = 0$.

PREPARE Treat the ball as a particle in uniform circular motion.

SOLVE

a. The period of the ball's motion, the time for 1 rev, is $T = 0.600$ s, so the angular speed is

$$\omega = \frac{2\pi \text{ rad}}{T} = \frac{2\pi \text{ rad}}{0.600 \text{ s}} = 10.5 \text{ rad/s}$$

Because the motion is counterclockwise, the angular velocity is positive: $\omega = +10.5$ rad/s.

b. The ball moves with constant angular velocity, so its position is given by Equation 6.2. When we solved problems of this sort for linear motion, we found that a graphical representation was helpful. The same is true here. A graph of θ vs. t is shown in Figure 6.5. The angular position increases steadily with time, starting at $\theta = 0$ rad at $t = 0$ s, and increasing to $\theta = 4\pi$ rad (two full revolutions) at $t = 1.20$ s, and a bit more than three revolutions at $t = 2.00$ s.

The ball's angular position at $t = 2.00$ s is

$$\theta_f = \theta_i + \omega \, \Delta t = 0 \text{ rad} + (10.5 \text{ rad/s})(2.00 \text{ s}) = 21.0 \text{ rad}$$

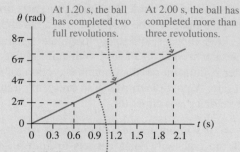

At 1.20 s, the ball has completed two full revolutions. At 2.00 s, the ball has completed more than three revolutions.

Because the ball moves at constant angular velocity, a graph of the angular position against time will be a straight line.

FIGURE 6.5 Angular position for the ball on the roulette wheel.

But the ball is back to where it started after each revolution. If what we want is the ball's final angular position, complete revolutions need not be counted. It is common practice to subtract out integer multiples of 2π rad, representing the completed revolutions. Because $21.0/2\pi = 3.34$, we can write

$$\theta_f = 21.0 \text{ rad} = 3.34 \times 2\pi \text{ rad}$$
$$= 3 \times 2\pi \text{ rad} + 0.34 \times 2\pi \text{ rad}$$
$$= 3 \times 2\pi \text{ rad} + 2.1 \text{ rad}$$

In other words, at $t = 2.00$ s, the ball has completed 3 rev and is 2.1 rad = 120° into its fourth revolution. An observer would say that the ball's angular position is $\theta = 120°$.

ASSESS Figure 6.5 shows that the kinematics of uniform circular motion is quite similar to that of uniform motion.

6.2 Speed, Velocity, and Acceleration in Uniform Circular Motion

The previous section described uniform circular motion in terms of angular variables. In Chapter 3, we introduced a description of uniform circular motion in terms of velocity and acceleration vectors. We will now unite these two different descriptions, which will let us consider a much wider range of problems.

Speed

For a particle in uniform circular motion, it's easy to relate the particle's period T to its speed v, as we saw at the end of Chapter 3. For a particle moving with constant speed, speed is simply distance/time. In one period, the particle moves once around a circle of radius r and travels the circumference $2\pi r$. Thus

$$v = \frac{1 \text{ circumference}}{1 \text{ period}} = \frac{2\pi r}{T} \tag{6.5}$$

If we combine this result with Equation 6.3 for the angular speed, we find that speed v and angular speed ω are related by

$$v = \omega r \tag{6.6}$$

Relationship between speed and angular speed

NOTE ▶ In Equation 6.6, ω **must be in units of rad/s.** If you are given a frequency in rev/s or rpm, you should convert it to an angular speed in rad/s. ◀

EXAMPLE 6.2 Finding the speed at two points on a CD

The diameter of an audio compact disc is 12.0 cm. When the disc is spinning at its maximum rate of 540 rpm, what is the speed of a point (a) at the outside edge of the disc, 6.0 cm from the center, and (b) at a distance 3.0 cm from the center?

PREPARE Consider two points A and B on a rotating compact disc in Figure 6.6. During one period, the disc rotates one time, and both points rotate through the same angle, 2π rad. Thus the angular speed, $\omega = 2\pi/T$, is the same for these two points; in fact, it is the same for all points on the disc. But as they go around one time, the two points move different *distances;* the outer point B goes around a larger circle. The two points thus have different *speeds.* We will solve this problem by first finding the angular speed of the disc, then computing the speeds at the two points.

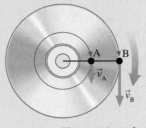

FIGURE 6.6 The rotation of an audio compact disc.

SOLVE We first convert the frequency of the disc from rpm to rev/s:

$$f = \left(540 \frac{\text{rev}}{\text{min}}\right) \times \left(\frac{1 \text{ min}}{60 \text{ s}}\right) = 9.00 \text{ rev/s}$$

We can compute the angular speed using Equation 6.4:

$$\omega = (2\pi \text{ rad})(9.00 \text{ rev/s}) = 56.5 \text{ rad/s}$$

The angular speed has the appropriate units. We can now use Equation 6.6 to compute the speeds of points on the disc. At point B, $r = 6.0$ cm $= 0.060$ m so the speed at the outside edge is

$$v_B = \omega r = (56.5 \text{ rad/s})(0.060 \text{ m}) = 3.4 \text{ m/s}$$

At point A, $r = 3.0$ cm $= 0.030$ m, so the speed is

$$v_B = \omega r = (56.5 \text{ rad/s})(0.030 \text{ m}) = 1.7 \text{ m/s}$$

ASSESS The speeds are a few m/s, which seems reasonable. The point farther from the center is moving at a higher speed, as we expected.

CONCEPTUAL EXAMPLE 6.1 Varying the angular speed of a CD

As we saw in Example 6.2, different points on a rotating CD move at different speeds. This creates a technical problem. The data on a CD are read by a laser beam that slowly scans from the inner part of the disc to the outside as the disc is played. The laser beam reflects from "pits" as they go by on the moving disc, so the data *rate* is proportional to the speed at that point on the disc. To have a constant data rate, so the music plays at a constant rate, the position where the data are being read must move at a constant speed. To keep the *speed* constant at the point of the laser beam, the *angular speed* of the disc must change as different parts of the disc are read. As a disc is played, and the laser scans from the inside to the outside, does the angular speed increase or decrease?

REASON We can rewrite Equation 6.6 as follows:

$$\omega = \frac{v}{r}$$

The speed of the disc at the position being read is constant. Thus, as the laser reads points farther from the center, at larger values of r, the angular speed must decrease.

ASSESS The angular speed of an audio compact disc does change as it is read. At the very start, when the laser is near the center of the disc, the frequency is about 540 rpm; by the time the laser is reading data near the outside of the disc, the frequency has dropped to just over 200 rpm.

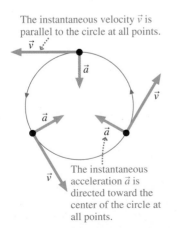

The instantaneous velocity $\vec{v}$ is parallel to the circle at all points.

The instantaneous acceleration $\vec{a}$ is directed toward the center of the circle at all points.

FIGURE 6.7 Velocity and acceleration for uniform circular motion.

Acceleration

Although the speed of a particle in uniform circular motion is constant, its velocity is *not* constant because the direction of the motion is always changing. As you learned in Chapter 3, and as Figure 6.7 reminds you, there is an acceleration at every point in the motion, with the acceleration vector $\vec{a}$ pointing toward the center of the circle. We called this the *centripetal acceleration,* and we showed that for uniform circular motion the acceleration was given by $a = v^2/r$. Because $v = \omega r$, we can also write this relationship in terms of the angular speed:

$$a = \frac{v^2}{r} = \omega^2 r \tag{6.7}$$

QUADRATIC
p. 50

Centripetal acceleration for uniform circular motion

Acceleration depends on speed, but also distance from the center of the circle.

CONCEPTUAL EXAMPLE 6.2 Who has the larger acceleration?

Two children are riding in circles on a merry-go-round, as shown in Figure 6.8. Which child experiences the larger acceleration?

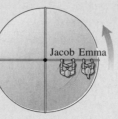

FIGURE 6.8 Top view of a merry-go-round.

REASON All points on the merry-go-round move at the same angular speed. The second expression for the acceleration in Equation 6.7 tells us that $a = \omega^2 r$. As the two children are moving with the same angular speed, Emma, with a larger value of r, experiences a larger acceleration.

ASSESS In the previous example, we saw that points farther from the center move at a higher speed. This would imply a higher acceleration as well, so our answer makes sense.

EXAMPLE 6.3 Finding the period of a carnival ride

In the Quasar carnival ride passengers travel in a horizontal 5.0-m-radius circle. For safe operation, the maximum sustained acceleration that riders may experience is 20 m/s², approximately twice the acceleration of gravity. What is the period of the ride when it is being operated at the maximum acceleration?

PREPARE We will assume that the cars on the ride are in uniform circular motion. The visual overview of Figure 6.9 shows a top view of the motion of the ride.

SOLVE The angular speed can be computed from the acceleration by rearranging Equation 6.7:

$$\omega = \sqrt{\frac{a}{r}} = \sqrt{\frac{20 \text{ m/s}^2}{5.0 \text{ m}}} = 2.0 \text{ rad/s}$$

At this angular speed, the period is $T = \dfrac{2\pi}{\omega} = 3.1$ s.

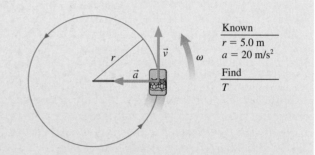

Known
$r = 5.0$ m
$a = 20$ m/s²
Find
T

FIGURE 6.9 Visual overview for the Quasar carnival ride.

ASSESS One rotation in just over three seconds seems reasonable for a pretty zippy carnival ride. The period for this particular ride is actually 3.7 s, so it runs a bit slower than the maximum safe speed.

Hurling the heavy hammer This man is throwing a hammer that weighs over 30 pounds as far as he can by spinning the hammer around in a circle and then letting go. While he holds the handle, the hammer follows a circular path. He must provide a very large force directed toward the center of the circle to produce the centripetal acceleration, as you can see by how he is leaning away from the hammer. When he lets go, there is no longer a force directed toward the center, and the hammer will stop going in a circle and fly across the field.

STOP TO THINK 6.1   Rank in order, from largest to smallest, the centripetal accelerations of particles A to D.

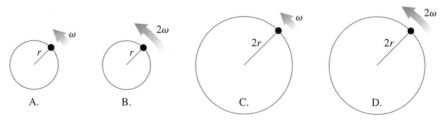

6.3 Dynamics of Uniform Circular Motion

Riders spinning around on a circular carnival ride are accelerating—they are *not* traveling at constant velocity in a straight line, as we have seen. Consequently, according to Newton's first law, the riders *must* have a net force acting on them. In this section, we'll look at the forces necessary for uniform circular motion.

We've already determined the acceleration of a particle in uniform circular motion—the centripetal acceleration of Equation 3.36 and Equation 6.7. Newton's second law tells us exactly what the net force must be to cause this acceleration:

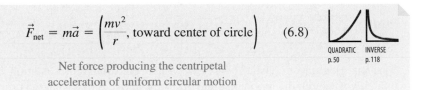

$$\vec{F}_{net} = m\vec{a} = \left(\frac{mv^2}{r}, \text{ toward center of circle}\right) \quad (6.8)$$

Net force producing the centripetal
acceleration of uniform circular motion

QUADRATIC   INVERSE
p. 50   p. 118

In other words, **a particle of mass m moving at constant speed v around a circle of radius r must always have a net force of magnitude mv^2/r pointing toward the center of the circle,** as in Figure 6.10. Without such a force, the particle would move off in a straight line tangent to the circle.

The force described by Equation 6.8 is not a *new* force. Our rules for identifying forces have not changed. What we are saying is that a particle moves with uniform circular motion *if and only if* a net force always points toward the center of the circle. The force itself must have identifiable agents and must be a combination of our familiar forces, such as tension, friction, or the normal force. Equation 6.8 simply tells us how the net force needs to act—how strong and in which direction—to cause the particle to move with speed v in a circle of radius r.

In each example of circular motion that we will consider in this chapter a physical force or a combination of forces directed toward the center produces the necessary acceleration. In some cases, the circular motion and the force are quite obvious, as in the hammer throw. Other cases are a bit more subtle. For instance, a car going around a corner is following a circular path on a level road; the necessary force is provided by the friction force between the tires and the road.

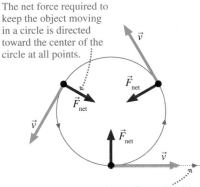

The net force required to keep the object moving in a circle is directed toward the center of the circle at all points.

In the absence of a net force, the object would continue in a straight line.

FIGURE 6.10 Net force for circular motion.

CONCEPTUAL EXAMPLE 6.3 **Will the rope break?**

A man has used a rope to hang a swing from a tree branch. The rope is just strong enough to support his weight, but no more. A friend pulls him back on the swing and lets go. Will the rope break?

REASON The swing moves along a circular arc. Although this is not uniform circular motion—it speeds up and then slows down—the speed is nearly constant for a short interval at the bottom of the swing. Thus the motion at the bottom is well approximated as uniform circular motion. There must be a net force toward the center of the circle to keep the swing moving in a circle. You can see from the free-body diagram of Figure 6.11 that the forces acting on it are the man's weight $\vec{w}$, downward, and the tension $\vec{T}$ in the rope, upward. A net force toward the center of the circle requires $T > w$. But the rope can barely support his weight, so $T_{max} = w$. The rope will break!

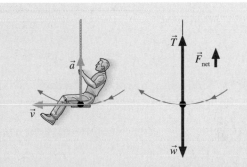

FIGURE 6.11 Visual overview for the motion of the swing.

ASSESS Note that the swing is in circular motion even though it does not travel through a complete circle.

Solving Circular Dynamics Problems

We have one basic equation for circular dynamics problems: Equation 6.8. This is just a version of Newton's second law; the right side is a net force, the left a mass times an acceleration, the centripetal acceleration. The techniques for solving circular dynamics problems are thus quite similar to those we have seen for solving other Newton's second law problems.

4.2–4.5

(MP) **PROBLEM-SOLVING STRATEGY 6.1**   **Circular dynamics problems**

PREPARE Begin your visual overview with a pictorial representation in which you sketch the motion, define symbols, and identify what the problem is trying to find. It is best to draw the free-body diagram with the circle viewed edge-on, the x-axis pointing toward the center of the circle, and the y-axis perpendicular to the plane of the circle.

SOLVE Newton's second law for uniform circular motion, $\vec{F}_{net} = (mv^2/r$, toward center of circle), is a vector equation. Some forces act in the plane of the circle, some act perpendicular to the circle, and some may have components in both directions. In the coordinate system described above, Newton's second law is

$$\sum F_x = \frac{mv^2}{r} \quad \text{and} \quad \sum F_y = 0$$

That is, the net force toward the center of the circle is mv^2/r, while the net force perpendicular to the circle is zero. The components of the forces are found directly from the free-body diagram. Depending on the problem, either

- Use the net force to determine the speed v, then use circular kinematics to find frequencies or angular velocities.
- Use circular kinematics to determine the speed v, then solve for unknown forces.

ASSESS Make sure your net force points toward the center of the circle. Check that your result has the correct units, is reasonable, and answers the question.

EXAMPLE 6.4 Analyzing the motion of a cart

An energetic father places his 20 kg child on a 5.0 kg cart to which a 2.0-m-long rope is attached. He then holds the end of the rope and spins the cart and child around in a circle, keeping the rope parallel to the ground. If the tension in the rope is 100 N, how much time does it take for the cart to make one rotation?

PREPARE We proceed according to the steps of Problem-Solving Strategy 6.1. Figure 6.12 shows a visual overview of the problem. The main idea of the pictorial representation on the left is to illustrate the relevant geometry and to define the symbols that will be used. A circular dynamics problem usually does not have starting and ending points like a projectile problem, so numerical subscripts such as x_1 or y_2 are usually not needed. Here we need to define the cart's speed v and the radius r of the circle.

The object moving in the circle is the cart plus the child, a total mass of 25 kg; the free-body diagram below shows the forces. **For uniform circular motion, we'll draw the free-body diagram looking at the edge of the circle because this is the plane of the forces.** Three forces are acting on the cart: the

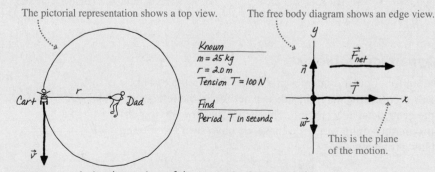

The pictorial representation shows a top view.   The free body diagram shows an edge view.

FIGURE 6.12 A visual overview of the cart spinning in a circle.

weight force $\vec{w}$, the normal force of the ground $\vec{n}$, and the tension force of the rope $\vec{T}$. As the problem-solving strategy suggests, we've shown these forces on a free-body diagram with the x-axis pointing toward the center of the circle and the y-axis perpendicular to the plane of motion.

Notice that there are two quantities for which we use the symbol T: the tension and the period. We will include additional information when necessary to distinguish the two.

SOLVE There is no net force in the y-direction, perpendicular to the circle, so $\vec{w}$ and $\vec{n}$ must be equal and opposite. There is a net force in the x-direction, toward the center of the circle, as there must be to cause the centripetal acceleration of circular motion. Only the tension force has an x-component, so Newton's second law is

$$\sum F_x = T = \frac{mv^2}{r}$$

We know the mass, the radius of the circle, and the tension, so we can solve for v:

$$v = \sqrt{\frac{Tr}{m}} = \sqrt{\frac{(100 \text{ N})(2.0 \text{ m})}{25 \text{ kg}}} = 2.83 \text{ m/s}$$

From this, we can compute the period with a slight rearrangement of Equation 6.5:

$$T = \frac{2\pi r}{v} = \frac{(2\pi)(2.0 \text{ m})}{2.83 \text{ m/s}} = 4.4 \text{ s}$$

ASSESS The speed is about 3 m/s. Because 1 m/s $\approx$ 2mph, this implies that the child is going about 6 mph. A trip around the circle in just over 4 s at a speed of about 6 mph sounds reasonable; it's a fast ride, but not so fast as to be scary!

EXAMPLE 6.5 Finding the maximum speed to turn a corner

What is the maximum speed with which a 1500 kg car can make a turn around a curve of radius 20 m on a level (unbanked) road without sliding? (This radius turn is about what you might expect at a major intersection in a city.)

PREPARE We start with the visual overview in Figure 6.13. The car moves along a circular arc at constant speed—uniform circular motion—for the quarter-circle necessary to complete the turn. The motion before and after the turn is not relevant to the problem. The more interesting issue is *how* a car turns a corner. What force or forces can we identify that cause the direction of the velocity vector to change? Imagine you are driving a car on a completely frictionless road, such as a very icy road. You would not be able to turn a corner. Turning the steering wheel would be of no use; the car would slide straight ahead, in

accordance with both Newton's first law and the experience of anyone who has ever driven on ice! So it must be *friction* that somehow allows the car to turn.

The top view of the tire in Figure 6.13 shows the force on the tire as it turns a corner. If the road surface were frictionless, the tire would slide straight ahead. The force that prevents an object from sliding across a surface is *static friction*. Static friction $\vec{f_s}$ pushes *sideways* on the tire, toward the center of the circle. How do we know the direction is sideways? If $\vec{f_s}$ had a component either parallel to $\vec{v}$ or opposite to $\vec{v}$, it would cause the car to speed up or slow down. Because the car changes direction but not speed, static friction must be perpendicular to $\vec{v}$. Thus $\vec{f_s}$ causes the centripetal acceleration of circular motion around the curve. With this in mind, the free-body diagram, drawn from behind the car, shows the static friction force pointing toward the center of the circle.

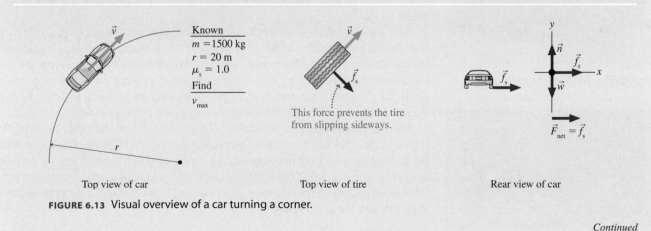

Known
$m = 1500 \text{ kg}$
$r = 20 \text{ m}$
$\mu_s = 1.0$

Find

v_{max}

This force prevents the tire from slipping sideways.

$\vec{F}_{net} = \vec{f_s}$

Top view of car   Top view of tire   Rear view of car

FIGURE 6.13 Visual overview of a car turning a corner.

Continued

SOLVE The only force in the x-direction, toward the center of the circle, is static friction. Newton's second law along the x-axis is

$$\sum F_x = f_s = \frac{mv^2}{r}$$

The only difference between this example and the previous one is that the tension force toward the center has been replaced by a static friction force toward the center.

Three forces are acting on the car: the weight force $\vec{w}$, the normal force $\vec{n}$, and the force of static friction $\vec{f_s}$. Because there is no vertical motion, the weight force and the normal force must be equal and opposite:

$$n = w = mg$$

The net force toward the center of the circle is the force of static friction. Recall from Equation 5.10 in Chapter 5 that static friction has a maximum possible value:

$$f_{s\,max} = \mu_s n = \mu_s mg$$

Because the static friction force has a maximum value, there will be a maximum speed with which a car can turn without sliding. The maximum speed is reached when the static friction force reaches its maximum $f_{s\,max} = \mu_s mg$. If the car enters the curve at a speed higher than the maximum, static friction will not be large enough to provide the necessary centripetal acceleration and the car will slide.

The maximum speed occurs at the maximum value of the force of static friction:

$$f_{s\,max} = \frac{mv_{max}^2}{r}$$

Using the known value of $f_{s\,max}$, we find

$$\frac{mv_{max}^2}{r} = f_{s\,max} = \mu_s mg$$

Rearrangement of this gives

$$v_{max}^2 = \mu_s gr$$

For rubber tires on pavement, we can find the relevant value of μ_s in Table 5.2; it is 1.0. We can then complete our calculation:

$$v_{max} = \sqrt{\mu_s gr} = \sqrt{(1.0)(9.8 \text{ m/s}^2)(20 \text{ m})} = 14 \text{ m/s}$$

ASSESS 14 m/s $\approx$ 30 mph, which seems like a reasonable upper limit for the speed at which a car can turn a corner without sliding. There are a few other things to note about the solution:

- The car's mass canceled out. The maximum speed *does not* depend on the mass of the vehicle, though this may seem surprising.
- The final expression for v_{max} *does* depend on the coefficient of friction and the radius of the turn. v_{max} decreases if μ is less (a slipperier road) or if r is smaller (a tighter turn). Both of these make sense.

FIGURE 6.14 Wings on an Indy racer.

A banked turn on a racetrack.

Because v_{max} depends on μ_s and because μ_s depends on road conditions, the maximum safe speed through turns can vary dramatically. Wet or icy roads lower the value of μ_s and thus lower the maximum speed of turns. A car that handles normally while driving straight ahead on a wet road can suddenly slide out of control when turning a corner. Icy conditions are even worse. If you lower the value of the coefficient of friction in Example 6.5 from 1.0 (dry pavement) to 0.1 (icy pavement), the maximum speed for the turn goes down to 4.4 m/s—about 10 mph!

Race cars turn corners at much higher speeds than normal passenger vehicles. One design modification of the *cars* to allow this is the addition of wings, as on the car in Figure 6.14. The wings provide an additional force pushing the car *down* onto the pavement by deflecting air upward, causing the air to push down on the car. The extra downward force increases the normal force, thus increasing the maximum static friction force and making faster turns possible.

There are also design modifications of the *track* to allow race cars to corner at high speeds. If the track is banked by raising the outside edge of curved sections, the normal force can provide some of the necessary centripetal force, as we will see in the next example. The curves on race tracks may be quite sharply banked. Curves on ordinary highways are often banked as well, though at more modest angles suiting the lower speeds.

EXAMPLE 6.6 Finding speed on a banked turn

A curve on a racetrack of radius 70 m is banked at a 15° angle. At what speed can a car take this curve without assistance from friction?

PREPARE The purpose of banking becomes clear if you look at the free-body diagram in Figure 6.15. The normal force $\vec{n}$ is perpendicular to the road, so tilting the road causes $\vec{n}$ to have a component toward the center of the circle. The horizontal component n_x is the inward force that causes the centripetal acceleration needed to turn the car. Notice that no friction force is shown because the car is assumed to make the turn without assistance from friction.

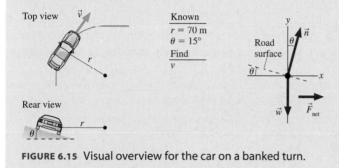

FIGURE 6.15 Visual overview for the car on a banked turn.

SOLVE Without friction, $n_x = n\sin\theta$ is the only component of force toward the center of the circle. It is this inward component of the normal force on the car that causes it to turn the corner. Newton's second law is

$$\sum F_x = n\sin\theta = \frac{mv^2}{r}$$

$$\sum F_y = n\cos\theta - w = 0$$

where θ is the angle at which the road is banked and we've assumed that the car is traveling at the correct speed v. From the y-equation,

$$n = \frac{w}{\cos\theta} = \frac{mg}{\cos\theta}$$

Substituting this into the x-equation and solving for v gives

$$\left(\frac{mg}{\cos\theta}\right)\sin\theta = mg\tan\theta = \frac{mv^2}{r}$$

$$v = \sqrt{rg\tan\theta} = 14 \text{ m/s}$$

ASSESS This is $\approx$30 mph, a reasonable speed. Only at this very specific speed can the turn be negotiated without reliance on friction forces.

Maximum Walking Speed

Humans and other two-legged animals have two basic gaits: walking and running. At slow speeds, you walk. When you need to go faster, you run. Why don't you just walk faster? There is an upper limit to the speed of walking, and this limit is set by the physics of circular motion.

Think about the motion of your body as you take a walking stride. You put one foot forward, then push off with your rear foot. Your body pivots over your front foot, and you bring your rear foot forward to take the next stride. As we can see in Figure 6.16a, the pivoting of your body over the forward foot is the arc of a circle. In a walking gait, your body is in circular motion as you pivot on your forward foot.

A force toward the center of the circle is required for this circular motion, as shown in Figure 6.16. Figure 6.16b shows the forces acting on the person's body during the midpoint of the stride: weight, directed down, and the normal force of the ground, directed up. Newton's second law for the x-axis is

$$\sum F_x = w - n = \frac{mv^2}{r}$$

Notice that $n < w$. Your body tries to "lift off" as it pivots over your foot, decreasing the normal force exerted on you by the ground. The normal force becomes smaller as you walk faster, but n cannot be less than zero. Thus the maximum possible walking speed v_{max} occurs when $n = 0$. Setting $n = 0$,

$$w = mg = \frac{mv_{max}^2}{r}$$

Thus

$$v_{max} = \sqrt{gr} \qquad (6.9)$$

The maximum possible walking speed is limited by r, the length of the leg, and g, the free-fall acceleration. This formula is a good approximation to the maximum

(a) Walking stride The speed of the circular motion is the walking speed.

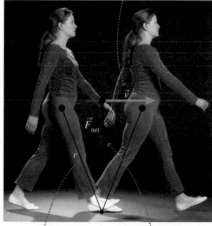

The radius of the circular motion is the length of the leg from the foot to the hip.

The circular motion requires a force directed toward the center of the circle.

(b) Forces in the stride Side view (same as photo)

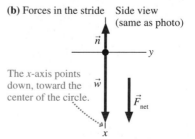

The x-axis points down, toward the center of the circle.

FIGURE 6.16 Analysis of a walking stride.

walking speed for humans and other animals. The maximum walking speed is higher for animals with longer legs. Giraffes, with their very long legs, can walk at high speeds. Animals such as mice with very short legs have such a low maximum walking speed that they rarely use this gait. Mice will generally run to get from one place to another.

For humans, the length of the leg is approximately 0.7 m, so we calculate a maximum speed of

$$v_{max} \approx 2.6 \text{ m/s} \approx 6 \text{ mph}$$

You *can* walk this fast, though it becomes energetically unfavorable to do so at speeds above 4 mph. Most people will make a transition to a running gait at about this speed. Children, with their shorter legs, must make a transition to a running gait at a much lower speed.

STOP TO THINK 6.2 A block on a string spins in a horizontal circle on a frictionless table. Rank in order, from largest to smallest, the tensions T_A to T_E acting on the blocks A to E.

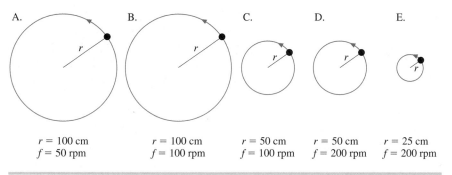

A.	B.	C.	D.	E.
$r = 100$ cm	$r = 100$ cm	$r = 50$ cm	$r = 50$ cm	$r = 25$ cm
$f = 50$ rpm	$f = 100$ rpm	$f = 100$ rpm	$f = 200$ rpm	$f = 200$ rpm

---

FIGURE 6.17 Inside the Gravitron, a rotating circular room.

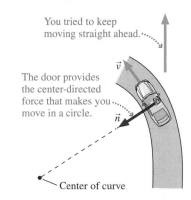

You tried to keep moving straight ahead.

The door provides the center-directed force that makes you move in a circle.

$\vec{v}$

$\vec{n}$

Center of curve

FIGURE 6.18 Bird's-eye view of a passenger in a car turning a corner.

6.4 Apparent Forces in Circular Motion

Figure 6.17 shows a carnival ride that spins the riders around inside a large cylinder. The people are "stuck" to the inside wall of the cylinder! As you probably know from experience, the riders *feel* that they are being pushed outward, into the wall. But our analysis has found that an object in circular motion must have an *inward* force to create the centripetal acceleration. How can we explain this apparent difference?

Centrifugal Force?

If you are a passenger in a car that turns a corner quickly, you may feel "thrown" against the door. But is there really such a force? Figure 6.18 shows a bird's-eye view of you riding in a car as it makes a left turn. You try to continue moving in a straight line, obeying Newton's first law, when—without having been provoked—the door suddenly turns in front of you and runs into you! You do, indeed, then feel the force of the door because it is now the force of the door, pushing *inward* toward the center of the curve, that is causing you to turn the corner. But you were not "thrown" into the door; the door ran into you.

A "force" that *seems* to push an object to the outside of a circle is called a *centrifugal force*. Despite having a name, there really is no such force. What you feel is your body trying to move ahead in a straight line (which would take you away from the center of the circle) as outside forces act to turn you in a circle. The only real forces, those that appear on free-body diagrams, are the ones pushing inward toward the center. **A centrifugal force will never appear on a free-body diagram and never be included in Newton's laws.**

With this in mind, let's revisit the rotating carnival ride. A person watching from above would see the riders in the cylinder moving in a circle with the walls providing the inward force that causes their centripetal acceleration. The riders *feel* as if they're being pushed outward because their attempts to move in a straight line are being resisted by the wall of the cylinder, which keeps getting in the way. But feelings aren't forces. The only actual force is the contact force of the cylinder wall pushing *inward*.

Apparent Weight in Circular Motion

Imagine swinging a bucket of water over your head. If you swing the bucket quickly, the water stays in. But you'll get a shower if you swing too slowly. Why does the water stay in the bucket? Or think about a roller coaster that does a loop-the-loop. How does the car stay on the track when it's upside down? You might have said that there was a centrifugal force holding the water in the bucket and the car on the track, but we have seen that there really isn't a centrifugal force. Analyzing these questions will tell us a lot about forces in general and circular motion in particular.

Figure 6.19a shows a roller coaster car going around a vertical loop-the-loop of radius r. If you've ever ridden a roller coaster, you know that your sensation of weight changes as you go over the crests and through the dips. To understand why, let's look at the forces on passengers going through the loop. To simplify our analysis, we will assume that the speed of the car stays constant as it moves through the loop.

Figure 6.19b shows a passenger's free-body diagram at the *bottom* of the loop. The only forces acting on the passenger are her weight $\vec{w}$ and the normal force $\vec{n}$ of the seat pushing up on her. Recall, from Chapter 5, that the passenger's apparent weight, her sensation of weight, is the magnitude of the force supporting her. Here the seat is supporting her with the normal force $\vec{n}$, so her apparent weight is $w_{app} = n$. Based on our understanding of circular motion, we can say that

- She's moving in a circle, so there *must* be a net force toward the center of the circle—above her head—to provide the centripetal acceleration.
- The net force points *upward*, so it must be the case that $n > w$.
- Her apparent weight is $w_{app} = n$, so her apparent weight is larger than her true weight ($w_{app} > w$). Thus she "feels heavy" at the bottom of the circle.

In short, the normal force has to *exceed* the weight force to provide the net force she needs to "turn the corner" at the bottom of the circle. To analyze the situation quantitatively, we note from the force vectors in Figure 6.19b that the magnitude of the net force at the bottom of the loop is

$$(F_{net})_{bottom} = n - w$$

The net force provides the centripetal acceleration, so we can say that

$$(F_{net})_{bottom} = n - w = \frac{mv^2}{r} \tag{6.10}$$

From Equation 6.10 we find her apparent weight to be:

$$(w_{app})_{bottom} = n = w + \frac{mv^2}{r} \tag{6.11}$$

The passenger's apparent weight at the bottom is *larger* than her true weight w, which agrees with your experience when you go through a dip or a valley.

Now let's look at the roller coaster car as it crosses the top of the loop. Things are a little trickier here. Whereas the normal force of the track pushes up when the car is at the bottom of the circle, it pushes *down* when the car is at the top and the

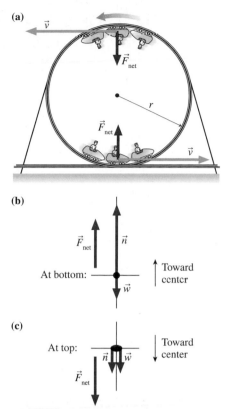

FIGURE 6.19 A roller coaster car going around a loop-the-loop.

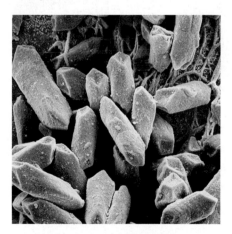

When "down" is up BIO You can tell, even with your eyes closed, what direction is down. This sense is due to small crystals of calcium carbonate, called *otoliths,* in your inner ears. Gravity pulls the otoliths down, so a normal force from a sensitive supporting membrane must push them up. Your brain interprets "down" as the opposite of this normal force. Normally, what your ears tell you is "down" is really down. But at the top of a loop in a roller coaster, the normal force is directed down, so your inner ear tells you that "down" is up! If your ears tell you one thing and your eyes another, it can be quite disorienting.

track is above the car. Figure 6.19c shows the car's free-body diagram at the top of the loop. It's worth thinking carefully about this diagram to make sure you understand what it is showing.

The car is still moving in a circle, so there must be a net force *downward,* toward the center of the circle, to provide the centripetal acceleration. We see from force vectors in the free-body diagram of Figure 6.19c that the net force is now

$$(F_{net})_{top} = n + w$$

Thus we can compute

$$(F_{net})_{top} = n + w = \frac{mv^2}{r} \qquad (6.12)$$

The normal force that the track exerts on the car at the top of the circle is

$$n = \frac{mv^2}{r} - w \qquad (6.13)$$

The apparent weight of the car (and the passengers in the car) is

$$(w_{app})_{top} = n = \frac{mv^2}{r} - w \qquad (6.14)$$

If v is sufficiently large, the apparent weight can exceed the true weight.

Let's look at what happens if the car goes slower. Notice from Equation 6.14 that, as v decreases, there comes a point when n reaches zero. At that point, the track is *not* pushing against the car. Instead, the car is able to complete the circle because the weight force alone provides sufficient centripetal acceleration.

The speed at which $n = 0$ is called the *critical speed* v_c:

$$v_c = \sqrt{\frac{rw}{m}} = \sqrt{\frac{rmg}{m}} = \sqrt{rg} \qquad (6.15)$$

The critical speed is the slowest speed at which the car can complete the circle. To understand why, notice that Equation 6.14 gives a negative value for n if $v < v_c$. But that is physically impossible. The track can push against the wheels of the car ($n > 0$), but it can't pull on them, so the slowest possible speed is the speed for which $n = 0$ at the top. If $v < v_c$, the car cannot turn the full loop but, instead, comes off the track and becomes a projectile!

Water stays in a bucket swung over your head for the same reason. The bottom of the bucket pushes against the water to provide the inward force that causes circular motion. If you swing the bucket too slowly, the force of the bucket on the water drops to zero. At that point, the water leaves the bucket and becomes a projectile following a parabolic trajectory onto your head!

A fast-spinning world Saturn, a gas giant planet composed largely of fluid matter, is quite a bit larger than the earth. It also rotates much more quickly, completing one rotation in just under 11 hours. The rapid rotation decreases the apparent weight at the equator enough to distort the fluid surface; the planet is noticeably out of round, as the red circle shows. The diameter at the equator is 11% greater than the diameter at the poles.

EXAMPLE 6.7 How slow can you go?

A motorcyclist in the Globe of Death, pictured at the start of the chapter, rides in a 2.2-m-radius vertical loop. To keep control of the bike, the rider wants the normal force on his tires at the top of the loop to equal or exceed his and the bike's combined weight. What is the minimum speed at which the rider can take the loop?

PREPARE The visual overview for this problem is shown in Figure 6.20. At the top of the loop, the normal force of the cage on the tires is a *downward* force. In accordance with Problem-Solving Strategy 6.1, we've chosen the x-axis to point toward the center of the circle.

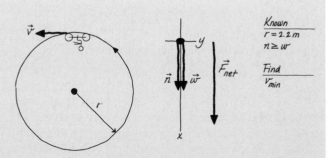

FIGURE 6.20 Riding in a vertical loop around the Globe of Death.

SOLVE We will consider the forces at the top point of the loop. Because the x-axis points downward, Newton's second law is

$$\sum F_x = w + n = \frac{mv^2}{r}$$

The minimum acceptable speed occurs when $n = w$, thus

$$2w = 2mg = \frac{mv_{min}^2}{r}$$

Solving for the speed, we find

$$v_{min} = \sqrt{2gr} = \sqrt{2(9.8 \text{ m/s}^2)(2.2 \text{ m})} = 6.6 \text{ m/s}$$

ASSESS The minimum speed is ≈ 15 mph, which isn't all that fast; the bikes can easily reach this speed. But normally several bikes are in the globe at one time. The big challenge is to keep all of the riders in the cage moving at this speed in synchrony. The period for the circular motion at this speed is $T = 2\pi r/v \approx 2$ s, leaving little room for error!

Centrifuges

The *centrifuge,* an important biological application of circular motion, is used to separate the components of a liquid with different densities. Typically these are different types of cells, or the components of cells, suspended in water. By using very high angular velocities, the centrifuge produces centripetal accelerations that are thousands of times greater than the acceleration due to gravity.

You probably know that a less-dense liquid, such as oil, floats on top of a more-dense liquid, such as water. That is, gravity naturally separates the components of a liquid so that the most dense components are on the bottom and the least dense on top. But suppose the liquid consists of many small objects, such as cells, suspended in water. Although gravity causes the cells to slowly drift toward the bottom of the container, the terminal velocity for very small objects can be minuscule. It could take days or even months for the cells to settle out and separate by density. And the slightest agitation would destroy the fragile separation. It's not practical to wait for biological samples to separate due to gravity alone.

The separation would go faster if the force of gravity could be increased. Although we can't change gravity, we can increase the apparent weight of objects in the sample by spinning them very fast, and that is what the centrifuge in Figure 6.21 does. As the centrifuge effectively increases gravity to thousands of times its normal value, the cells or cell components settle out and separate by density in a matter of minutes or hours.

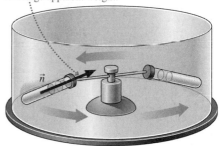

A centrifuge.

The high angular velocity requires a large normal force, which leads to a large apparent weight.

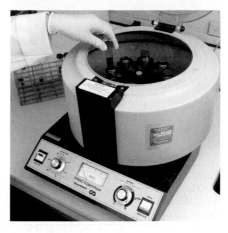

FIGURE 6.21 The operation of a centrifuge.

EXAMPLE 6.8 Analyzing the ultracentrifuge

An 18-cm-diameter ultracentrifuge produces an extraordinarily large centripetal acceleration of 250,000g. What is the frequency in rpm? What is the apparent weight of a sample with a mass of 0.0030 kg?

PREPARE The acceleration in SI units is

$$a = 250{,}000(9.80 \text{ m/s}^2) = 2.45 \times 10^6 \text{ m/s}^2$$

The radius is half the diameter, or $r = 9.0$ cm $= 0.090$ m.

SOLVE The centripetal acceleration is related to the angular speed by $a = \omega^2 r$. Thus

$$\omega = \sqrt{\frac{a}{r}} = \sqrt{\frac{2.45 \times 10^6 \text{ m/s}^2}{0.090 \text{ m}}} = 5.22 \times 10^3 \text{ rad/s}$$

This corresponds to a frequency

$$f = \frac{\omega}{2\pi} = \frac{5.22 \times 10^3 \text{ rad/s}}{2\pi} = 830 \text{ rev/s}$$

Converting to rpm, we find

$$830 \frac{\text{rev}}{\text{s}} \times \frac{60 \text{ s}}{1 \text{ min}} = 50{,}000 \text{ rpm}$$

Continued

Human centrifuge BIO If you spin your arm rapidly in a vertical circle, the motion will produce an effect like that in a centrifuge. The motion will assist outbound blood flow in your arteries and retard inbound blood flow in your veins. There will be a buildup of fluid in your hand that you will be able to see (and feel!) quite easily.

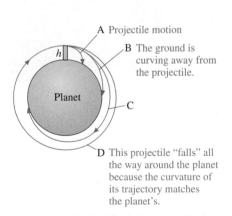

FIGURE 6.22 Projectiles being launched at increasing speeds from height *h* on a smooth, airless planet.

At this rotation rate, the 0.0030 kg mass has an apparent weight

$$w_{app} = ma = (3.0 \times 10^{-3}\,\text{kg})(2.45 \times 10^6\,\text{m/s}^2) = 7.4 \times 10^3\,\text{N}$$

ASSESS Because the acceleration is 250,000*g*, the apparent weight is 250,000 times the actual weight. The forces in the ultracentrifuge are very large and can destroy the machine if it is not carefully balanced.

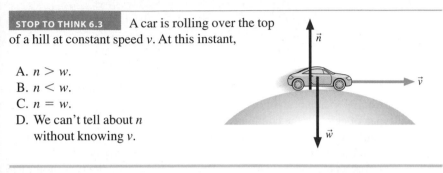

STOP TO THINK 6.3   A car is rolling over the top of a hill at constant speed *v*. At this instant,

A. $n > w$.
B. $n < w$.
C. $n = w$.
D. We can't tell about *n* without knowing *v*.

6.5 Circular Orbits and Weightlessness

The Space Shuttle orbits the earth in a circular path at a speed of over 15,000 miles per hour. What forces act on it? Why does it move in a circle? Before we start considering the physics of orbital motion, let's return, for a moment, to projectile motion. Projectile motion occurs when the only force on an object is gravity. Our analysis of projectiles made an implicit assumption that the earth is flat and that the free-fall acceleration, due to gravity, is everywhere straight down. This is an acceptable approximation for projectiles of limited range, such as baseballs or cannon balls, but there comes a point where we can no longer ignore the curvature of the earth.

Orbital Motion

Figure 6.22 shows a perfectly smooth, spherical, airless planet with one vertical tower of height *h*. A projectile is launched from this tower with initial speed v_i parallel to the ground. If v_i is very small, as in trajectory A, the "flat-earth approximation" is valid and the problem is identical to Example 3.11 in which a car drove off a cliff. The projectile simply falls to the ground along a parabolic trajectory.

As the initial speed v_i is increased, it seems to the projectile that the ground is curving out from beneath it. It is still falling the entire time, always getting closer to the ground, but the distance that the projectile travels before finally reaching the ground—that is, its range—increases because the projectile must "catch up" with the ground that is curving away from it. Trajectories B and C are of this type.

If the launch speed v_i is sufficiently large, there comes a point at which the curve of the trajectory and the curve of the earth are parallel. In this case, the projectile "falls" but it never gets any closer to the ground! This is the situation for trajectory D. The projectile returns to the point from which it was launched, at the same speed at which it was launched, making a closed trajectory. Such a closed trajectory around a planet or star is called an **orbit.**

The most important point of this qualitative analysis is that, in the absence of air resistance, **an orbiting projectile is in free fall.** This is, admittedly, a strange idea, but one worth careful thought. An orbiting projectile is really no different from a thrown baseball or a car driving off a cliff. The only force acting on it is gravity, but its tangential velocity is so large that the curvature of its trajectory matches the curvature of the earth. When this happens, the projectile "falls" under the influence of gravity but never gets any closer to the surface, which curves away beneath it.

When we first studied free fall in Chapter 2, we said that the free-fall acceleration is always directed vertically downward. As we see in Figure 6.23, this is due to our perspective; "downward" really means "toward the center of the earth." For a projectile in orbit, the direction of the force of gravity changes, always pointing toward the center of the earth.

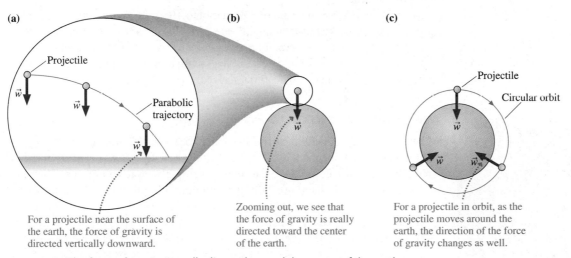

(a)   (b)   (c)

For a projectile near the surface of the earth, the force of gravity is directed vertically downward.

Zooming out, we see that the force of gravity is really directed toward the center of the earth.

For a projectile in orbit, as the projectile moves around the earth, the direction of the force of gravity changes as well.

FIGURE 6.23 The force of gravity is really directed toward the center of the earth.

As you have learned, a force of constant magnitude that always points toward the center of a circle causes the centripetal acceleration of uniform circular motion. Because the only force acting on the orbiting projectile in Figure 6.23 is gravity, and we're assuming the projectile is very near the surface of the earth, we can write

$$a = \frac{F_{net}}{m} = \frac{w}{m} = \frac{mg}{m} = g \qquad (6.16)$$

An object moving in a circle of radius r at speed v_{orbit} will have this centripetal acceleration if

$$a = \frac{(v_{orbit})^2}{r} = g \qquad (6.17)$$

That is, if an object moves parallel to the surface with the speed

$$v_{orbit} = \sqrt{rg} \qquad (6.18)$$

then the free-fall acceleration provides exactly the centripetal acceleration needed for a circular orbit of radius r. An object with any other speed will not follow a circular orbit.

The earth's radius is $r = R_e = 6.37 \times 10^6$ m. The orbital speed of a projectile just skimming the surface of an airless, bald earth is

$$v_{orbit} = \sqrt{R_e g} = \sqrt{(6.37 \times 10^6 \text{ m})(9.80 \text{ m/s}^2)} = 7900 \text{ m/s} \approx 18,000 \text{ mph}$$

We can use v_{orbit} to calculate the period of the satellite's orbit:

$$T = \frac{2\pi r}{v_{orbit}} = 2\pi\sqrt{\frac{r}{g}} \qquad (6.19)$$

For this earth-skimming orbit, $T = 5065$ s $= 84.4$ min.

Of course, this orbit is unrealistic; even if there were no trees and mountains, a real projectile moving at this speed would burn up from the friction of air resistance. Suppose, however, that we launched the projectile from a tower of height $h = 200$ mi $\approx 3.2 \times 10^5$ m, above most of the earth's atmosphere. This is approximately the height of low-earth-orbit satellites, such as the Space Shuttle. Note that $h \ll R_e$, so the radius of the orbit $r = R_e + h = 6.69 \times 10^6$ m is only 5% greater than the earth's radius. Many people have a mental image that satellites orbit far above the earth, but in fact most satellites come pretty close to skimming the surface.

At this slightly larger value of r, Equation 6.19 gives $T = 87$ min. The actual period of the Space Shuttle at an altitude of 200 mi is about 91 minutes, so our calculation is very good—but not perfect. As we'll see in the next section, a correct calculation must take into account the fact that gravity gradually gets weaker at higher elevations above the earth's surface.

Weightlessness in Orbit

When we discussed *weightlessness* in Chapter 5, we saw that it occurs during free fall. We asked the question, at the end of Section 5.4, whether astronauts and their spacecraft are in free fall. We can now give an affirmative answer: They are, indeed, in free fall. They are falling continuously around the earth, under the influence of only the gravitational force, but never getting any closer to the ground because the earth's surface curves beneath them. Weightlessness in space is no different from the weightlessness in a free-falling elevator. **Weightlessness does *not* occur from an absence of weight or an absence of gravity.** Instead, the astronaut, the spacecraft, and everything in it are "weightless" (i.e., their *apparent* weight is zero) because they are all falling together.

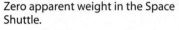

Zero apparent weight in the Space Shuttle.

Rotating space stations BIO The weightlessness astronauts experience in orbit has serious physiological consequences. Astronauts who spend time in weightless environments lose bone and muscle mass and suffer other adverse effects. One solution is to introduce "artificial gravity." On a space station, the easiest way to do this would be to make the station rotate, producing an apparent weight. The designers of this space station model for the movie *2001: A Space Odyssey* made it rotate for just that reason.

The Orbit of the Moon

If a satellite is simply "falling" around the earth, with the gravitational force causing a centripetal acceleration, then what about the moon? Is it obeying the same laws of physics? Or do celestial objects obey laws that we cannot discover by experiments here on earth?

The radius of the moon's orbit around the earth is 3.84×10^8 m. If we use Equation 6.19 to calculate the period of the moon's orbit, the time the moon takes to circle the earth once, we get

$$T = 2\pi\sqrt{\frac{r}{g}} = 2\pi\sqrt{\frac{3.84 \times 10^8 \text{ m}}{9.80 \text{ m/s}^2}} = 655 \text{ min} \approx 11 \text{ hours} \qquad (6.20)$$

This is clearly wrong; the period of the moon's orbit is approximately one month.

Newton believed that the laws of motion he had discovered were *universal* and so should apply to the motion of the moon as well as to the motion of objects in the laboratory. But why should we assume that the free-fall acceleration g is the same at the distance of the moon as it is on or near the earth's surface? If gravity is the force of the earth pulling on an object, it seems plausible that the size of that force, and thus the size of g, should diminish with increasing distance from the earth.

Newton proposed the idea that the earth's force of gravity decreases with the square of the distance from the earth. This is the basis of *Newton's law of gravity,* a topic we will study in the next section. Gravity is less at the distance of the moon—exactly the strength needed to make the moon orbit at the observed rate. The moon, just like the space shuttle, is simply "falling" around the earth!

6.6 Newton's Law of Gravity

A popular image has Newton thinking of the idea of gravity after an apple fell on his head. This amusing story is at least close to the truth. Newton himself said that the "notion of gravitation" came to him as he "sat in a contemplative mood" and "was occasioned by the fall of an apple." It occurred to him that, perhaps, the apple was attracted to the center of the earth but was prevented from getting there by the earth's surface. And if the apple was so attracted, why not the moon? Newton's genius was his sudden realization that **the force of the earth on the moon (and of the sun on the planets) was identical to the force of the earth on the apple.** In other words, gravitation is a *universal* force between all objects in the universe! This is not shocking today, but no one before Newton had ever thought that the mundane motion of objects on earth had any connection at all with the stately motion of the planets around the sun.

Isaac Newton was born to a poor farming family in 1642, the year of Galileo's death. He entered Trinity College at Cambridge University at age 19 as a "subsizar," a poor student who had to work his way through school. Newton graduated in 1665, at age 23, just as an outbreak of the plague in England forced the universities to close for two years. He returned to his family farm for that period, during which he made important experimental discoveries in optics, laid the foundations for his theories of mechanics and gravitation, and made major progress toward his invention of calculus as a whole new branch of mathematics.

Gravity Obeys an Inverse-Square Law

Newton also recognized that the strength of gravity must decrease with distance, as we saw at the end of the previous section. These two notions about gravity—that it is universal, and that it decreases with distance—form the basis for Newton's law of gravity.

Newton proposed that *every* object in the universe attracts *every other* object with a force that has the following properties:

1. The force is inversely proportional to the square of the distance between the objects.
2. The force is directly proportional to the product of the masses of the two objects.

Figure 6.24 shows two spherical masses m_1 and m_2 separated by distance r. Each mass exerts an attractive force on the other, a force that we call the **gravitational force.** These two forces form an action/reaction pair, so $\vec{F}_{1\,on\,2}$ is equal in magnitude and opposite in direction to $\vec{F}_{2\,on\,1}$. The magnitude of the forces is given by Newton's law of gravity.

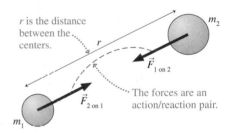

FIGURE 6.24 The gravitational forces on masses m_1 and m_2.

Newton's law of gravity If two objects with masses m_1 and m_2 are a distance r apart, the objects exert attractive forces on each other of magnitude

$$F_{1\,on\,2} = F_{2\,on\,1} = \frac{Gm_1m_2}{r^2} \qquad (6.21)$$

The forces are directed along the line joining the two objects.

The constant G is called the **gravitational constant.** In the SI system of units, G has the value

$$G = 6.67 \times 10^{-11} \text{ N} \cdot \text{m}^2/\text{kg}^2$$

NOTE ▶ Strictly speaking, Newton's law of gravity applies to *particles* with masses m_1 and m_2. Fortunately, it can be shown that the law also applies to the force between two spherical objects if r is the distance between their centers. ◀

As the distance r between two objects increases, the gravitational force between them decreases. Because the distance appears squared in the denominator, Newton's law of gravity is what we call an **inverse-square** law. Doubling the distance between two masses causes the force between them to decrease by a factor of 4. This mathematical form is one we will see again, so it is worth our while to explore it in more detail.

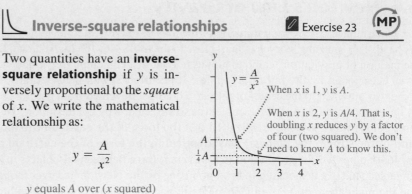

Inverse-square relationships ✏ Exercise 23 (MP)

Two quantities have an **inverse-square relationship** if y is inversely proportional to the *square* of x. We write the mathematical relationship as:

$$y = \frac{A}{x^2}$$

y equals *A* over (*x* squared)

$y = \dfrac{A}{x^2}$

When x is 1, y is A.

When x is 2, y is $A/4$. That is, doubling x reduces y by a factor of four (two squared). We don't need to know A to know this.

SCALING Inverse-square scaling means, for example:

- If you double x, you decrease y by a factor of 4, as you can see in the graph.
- If you increase x by a factor of 3, you decrease y by a factor of 9.
- If you decrease x by a factor of 3, you increase y by a factor of 9.

Generally: **An *increase* in x by a factor C results in a *decrease* of y by a factor C^2.**

LIMITS As x becomes large, y becomes very small; as x becomes small, y becomes very large.

CONCEPTUAL EXAMPLE 6.4 Varying gravitational force

The gravitational force between two giant lead spheres is 0.010 N when the spheres are 20 m apart, measured between their centers. What is the distance between their centers when the gravitational force between them is 0.160 N?

REASON We can solve this problem without knowing the masses of the two spheres. The key is to consider the ratios of forces and distances. Gravity is an inverse-square relationship; the force is related to the inverse square of the distance. The force *increases* by a factor $(0.160\ \text{N})/(0.010\ \text{N}) = 16$, so the distance must *decrease* by a factor $\sqrt{16} = 4$. The distance is thus $(20\ \text{m})/4 = 5.0\ \text{m}$.

ASSESS This type of ratio reasoning is a very good way to get a quick handle on the solution to a problem.

EXAMPLE 6.9 Gravitational force between two people

You are seated in your physics class next to another student 0.60 m away. Estimate the magnitude of the gravitational force between you. Assume that you each have a mass of 65 kg.

PREPARE We will model each of you as a sphere; this is not a particularly good model, but it will do for making an estimate. We will take the 0.60 m as the distance between your centers.

SOLVE The gravitational force is given by Equation 6.21:

$$F_{(\text{you}) \text{on} (\text{other student})} = \frac{Gm_{\text{you}} m_{\text{other student}}}{r^2}$$

$$= \frac{(6.67 \times 10^{-11}\ \text{N} \cdot \text{m}^2/\text{kg}^2)(65\ \text{kg})(65\ \text{kg})}{(0.60\ \text{m})^2}$$

$$= 7.8 \times 10^{-7}\ \text{N}$$

ASSESS The force is quite small, roughly the weight of one hair on your head.

There is a gravitational force between all objects in the universe, but the gravitational force between two ordinary-sized objects is very small, and is not something you have ever noticed. Only when one (or both) of the masses is exceptionally large does the force of gravity become important. The downward force of the earth on you—your weight—is large, because the earth has an enormous mass. And the attraction is mutual; by Newton's third law, you exert an upward force on the earth that is equal to your weight. The large mass of the earth makes the effect of this force on the earth negligible, though.

EXAMPLE 6.10 Gravitational force of the earth on a person

What is the magnitude of the gravitational force of the earth on a 60 kg person? The earth has mass 5.98×10^{24} kg and radius 6.37×10^6 m.

PREPARE We'll again model the person as a sphere. The distance r in Newton's law of gravity is the distance between the centers of the two spheres. The size of the person is negligible compared to the size of the earth, so we can use the earth's radius as r.

SOLVE The force of gravity on the person due to the earth can be computed using Equation 6.21:

$$F_{\text{earth on } m} = \frac{GM_e m}{R_e^2} = \frac{(6.67 \times 10^{-11} \text{ N} \cdot \text{kg}^2/\text{m}^2)(5.98 \times 10^{24} \text{ kg})(60 \text{ kg})}{(6.37 \times 10^6 \text{ m})^2}$$

$$= 590 \text{ N}$$

ASSESS This force is exactly the same as we would calculate using the formula for the weight force, $w = mg$. This isn't surprising, though. Chapter 5 introduced the weight of an object as simply the "force of gravity" acting on it. Newton's law of gravity is a more fundamental law for calculating the force of gravity, but it's still the same force that we earlier called "weight."

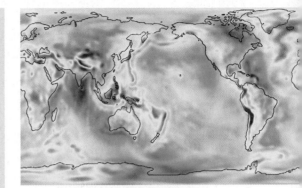

Variable gravity When we calculated the force of the earth's gravity we assumed that the earth's shape and composition are uniform. Neither is quite true, so there is a very small variation in gravity at the surface of the earth, as shown in this image. Red means slightly greater surface gravity; blue, slightly weaker. These variations are caused by differing distances from the earth's center and by unevenness in density of the earth's crust. Though these variations are important for scientists studying the earth, they are small enough that we can ignore them for the computations we'll do in this textbook.

NOTE ▶ We will use uppercase R and M to represent the large mass and radius of a star or planet, as we did in Example 6.10. ◀

The force of gravitational attraction between the earth and you is responsible for your weight. If you were to venture to another planet, your *mass* would be the same but your *weight* would vary, as we discussed in Chapter 5. We will now explore this concept in more detail.

Gravity on Other Worlds

When astronauts ventured to the moon, television images showed them walking—and even jumping and skipping—with some ease, even though they were wearing life support systems with a mass of over 80 kg. This was a visible reminder that the weight of objects is less on the moon. Let's consider why this is so.

Figure 6.25 shows an astronaut on the moon weighing a rock of mass m. When we compute the weight of an object on the surface of the earth, we use the formula $w = mg$. We can do the same calculation for a mass on the moon, as long as we use the value of g on the moon.

$$w = mg_{\text{moon}} \tag{6.22}$$

This is the "little g" perspective. Falling body experiments on the moon would give the value of g_{moon} as 1.62 m/s^2.

But we can also take a "big G" perspective. The weight of the rock comes from the gravitational attraction of the moon, and we can compute this weight using Equation 6.21. The distance r is the radius of the moon, which we'll call R_{moon}. Thus

$$F_{\text{moon on } m} = \frac{GM_{\text{moon}} m}{R_{\text{moon}}^2} \tag{6.23}$$

Because Equations 6.22 and 6.23 are two names and two expressions for the same force, we can equate the right-hand sides to find that

$$g_{\text{moon}} = \frac{GM_{\text{moon}}}{R_{\text{moon}}^2}$$

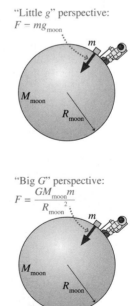

"Little g" perspective:
$F = mg_{\text{moon}}$

"Big G" perspective:
$F = \frac{GM_{\text{moon}} m}{R_{\text{moon}}^2}$

FIGURE 6.25 An astronaut weighing a mass on the moon.

We have done this calculation for an object on the moon, but the result is completely general. At the surface of a planet (or a star), the free-fall acceleration g, a consequence of gravity, can be computed as

$$g_{planet} = \frac{GM_{planet}}{R_{planet}^2} \qquad (6.24)$$

Free-fall acceleration on the surface of a planet

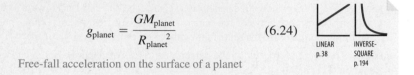

LINEAR p.38   INVERSE-SQUARE p.194

If we use values for the mass and the radius of the moon from the table inside the cover of the book, we can compute $g_{moon} = 1.62$ m/s^2. This means that an object will weigh less on the moon than it would on the earth, where g is 9.80 m/s^2. A 70 kg astronaut wearing an 80 kg spacesuit will weigh over 330 lb on the earth but only 54 lb on the moon.

The low lunar gravity makes walking very easy, but a walking pace on the moon will be very slow. Earlier in the chapter we found that the maximum walking speed was $v_{max} = \sqrt{gr}$ where r is the length of the leg. For a typical leg length of 0.7 m and the gravity of the moon, the *maximum* walking speed is about 1 m/s, just over 2 mph—a very gentle stroll!

Equation 6.24 gives g at the surface of a planet. More generally, imagine an object at distance $r > R$ from the center of a planet. Its free-fall acceleration at this distance is

$$g = \frac{GM}{r^2} \qquad (6.25)$$

This more general result agrees with Equation 6.24 if $r = R$, but it allows us to determine the "local" free-fall acceleration at distances $r > R$. Equation 6.25 expresses Newton's idea that the size of g should decrease as you get farther from the earth.

As you're flying in a jet airplane at a height of about 10 km, the free-fall acceleration is about 0.3% less than on the ground. At the height of the space shuttle, about 300 km, Equation 6.25 gives $g = 8.9$ m/s^2, about 10% less than the free-fall acceleration on the earth's surface. If you use this slightly smaller value of g in Equation 6.19 for the period of a satellite's orbit, you'll get the correct period of about 90 minutes. This value of g, only slightly less than the ground-level value, emphasizes the point that an object in orbit is not "weightless" because there is no gravity in space, but because it is in free fall.

Walking on the moon BIO The low lunar gravity made walking at a reasonable pace difficult for the Apollo astronauts, but the reduced weight made jumping quite easy. Videos from the surface of the moon often show the astronauts getting from place to place by hopping or skipping—not for fun, but for speed and efficiency.

EXAMPLE 6.11 Gravity on Saturn

Saturn, at 5.68×10^{26} kg, has nearly 100 times the mass of the earth. It is also much larger, with a radius of 5.85×10^7 m. What is the value of g on the surface of Saturn?

SOLVE We can use Equation 6.24 to compute the value of g_{Saturn}:

$$g_{Saturn} = \frac{GM_{Saturn}}{R_{Saturn}^2} = \frac{(6.67 \times 10^{-11} \text{ N} \cdot \text{m}^2/\text{kg}^2)(5.68 \times 10^{26} \text{ kg})}{(5.85 \times 10^7 \text{ m})^2} = 11.1 \text{ m/s}^2$$

$$= 11.1 \text{ m/s}^2$$

ASSESS Even though Saturn is much more massive than the earth, its larger radius gives it a surface gravity that is not markedly different from that of the earth. If Saturn had a solid surface, you could walk and move around quite normally.

EXAMPLE 6.12 Finding the speed to orbit Deimos

Mars has two moons, each much smaller than the earth's moon. The smaller of these two bodies, Deimos, has an average radius of only 6.3 km and a mass of 1.8×10^{15} kg. At what speed would a projectile move in a very low orbit around Deimos?

SOLVE The free-fall acceleration at the surface of Deimos is quite small:

$$g_{\text{Deimos}} = \frac{GM_{\text{Deimos}}}{R_{\text{Deimos}}^2} = \frac{(6.67 \times 10^{-11} \text{ N} \cdot \text{m}^2/\text{kg}^2)(1.8 \times 10^{15} \text{ kg})}{(6.3 \times 10^3 \text{ m})^2}$$

$$= 0.0030 \text{ m/s}^2$$

Given this, we can use Equation 6.18 to calculate the orbital speed:

$$v_{\text{orbit}} = \sqrt{gr} = \sqrt{(0.0030 \text{ m/s}^2)(6.3 \times 10^3 \text{ m})} = 4.3 \text{ m/s} \approx 10 \text{ mph}$$

ASSESS This is a quite slow speed. With a good jump, you could easily launch yourself into an orbit around Deimos!

STOP TO THINK 6.4   Rank in order, from largest to smallest, the free-fall accelerations on the surfaces of the following planets.

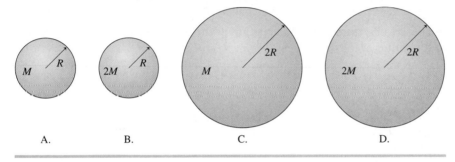

A.   B.   C.   D.

6.7 Gravity and Orbits

The planets of the solar system orbit the sun because the sun's gravitational pull, a force that points toward the center, causes the centripetal acceleration of circular motion. Mercury, the closest planet, experiences the largest acceleration, while Pluto, the most distant, has the smallest.

Figure 6.26 shows a massive body of mass M, such as the earth or the sun, with a lighter body of mass m orbiting it. The lighter body is called a **satellite,** even though it may be a planet orbiting the sun. Newton's second law for the satellite is

$$F_{M \text{ on } m} = \frac{GMm}{r^2} = ma = \frac{mv^2}{r} \tag{6.26}$$

Solving for v, we find that the speed of a satellite in a circular orbit is

$$v = \sqrt{\frac{GM}{r}} \tag{6.27}$$

Speed of a satellite in a circular orbit of radius r
about a star or planet of mass M

Actlv
ONLINE
Physics   4.6

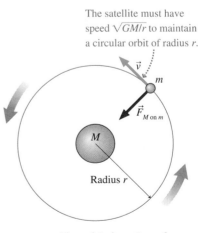

The satellite must have speed $\sqrt{GM/r}$ to maintain a circular orbit of radius r.

FIGURE 6.26 The orbital motion of a satellite is due to the force of gravity.

A satellite must have this specific speed in order to maintain a circular orbit of radius r about the larger mass M. If the velocity differs from this value, the orbit will become elliptical rather than circular. Notice that the orbital speed does not depend on the satellite's mass m. This is consistent with our previous discoveries that free-fall motion and projectile motion due to gravity are independent of the mass.

For a planet orbiting the sun, the period T is the time to complete one full orbit. The relationship among speed, radius, and period is the same as for any circular motion, $v = 2\pi r/T$. Combining this with the value of v for a circular orbit from Equation 6.27 gives

$$\sqrt{\frac{GM}{r}} = \frac{2\pi r}{T}$$

If we square both sides and rearrange, we find that the period of a satellite is given by

$$T^2 = \left(\frac{4\pi^2}{GM}\right)r^3 \tag{6.28}$$

Relationship between the orbital period T and radius r for a satellite in a circular orbit around an object of mass M

In other words, **the square of the period of the orbit is proportional to the cube of the radius of the orbit.**

NOTE ▶ The mass M in Equation 6.28 is the mass of the object at the center of the orbit. ◀

This relationship between radius and period had been deduced from naked-eye observations of planetary motions by the 17th century astronomer Johannes Kepler. One of Newton's major scientific accomplishments was to use his law of gravity and his laws of motion to prove what Kepler had deduced from observations. Even today, Newton's law of gravity and equations such as Equation 6.28 are essential tools for the NASA engineers who launch probes to other planets in the solar system.

The table inside the back cover of this book contains astronomical information about the sun and the planets that will be useful for many of the end-of-chapter problems. Note that planets farther from the sun have longer periods, in agreement with Equation 6.28.

EXAMPLE 6.13 Locating a geostationary satellite

Communication satellites appear to "hover" over one point on the earth's equator. A satellite that appears to remain stationary as the earth rotates is said to be in a *geostationary orbit*. What is the radius of the orbit of such a satellite?

PREPARE For the satellite to remain stationary with respect to the earth, the satellite's orbital period must be 24 hours; in seconds this is $T = 8.64 \times 10^4$ s.

SOLVE We solve for the radius of the orbit by rearranging Equation 6.28. The mass at the center of the orbit is the earth.

$$r = \left(\frac{GM_eT^2}{4\pi^2}\right)^{\frac{1}{3}} = \left(\frac{(6.67 \times 10^{-11} \text{ N} \cdot \text{m}^2/\text{kg}^2)(5.98 \times 10^{24} \text{ kg})(8.64 \times 10^4 \text{ s})^2}{4\pi^2}\right)^{\frac{1}{3}}$$

$$= 4.22 \times 10^7 \text{ m}$$

ASSESS This is quite a high orbit; the radius is about 7 times the radius of the earth.

Gravity on a Grand Scale

Although relatively weak, gravity is a long-range force. No matter how far apart two objects may be, there is a gravitational attraction between them. Consequently, gravity is the most ubiquitous force in the universe. It not only keeps your feet on the ground, but is also at work at a much larger scale. The Milky Way galaxy, the collection of stars of which our sun is a part, is held together by gravity. But why doesn't the attractive force of gravity simply pull all of the stars together?

The reason is that all of the stars in the galaxy are in orbit around the center of the galaxy. The gravitational attraction keeps the stars moving in orbits around the center of the galaxy rather than falling inward, much as the planets orbit the sun rather than falling into the sun. In the nearly 5 billion years that our solar system has existed, it has orbited the center of the galaxy approximately 20 times.

The galaxy as a whole doesn't rotate at a fixed angular speed, though. All of the stars in the galaxy are different distances from the galaxy's center, and so orbit with different periods. Stars closer to the center complete their orbits in less time, as we would expect from Equation 6.28. As the stars orbit, their relative positions shift. Stars that are relatively near neighbors now could be on opposite sides of the galaxy at some later time.

The rotation of a *rigid body* like a wheel is much simpler. As a wheel rotates, all of the points keep the same relationship to each other; every point on the wheel moves with the same angular velocity. The rotational dynamics of such rigid bodies is a topic we will take up in the next chapter.

A spiral galaxy, similar to our Milky Way galaxy.

STOP TO THINK 6.5 If the mass of the moon were doubled but it stayed in its present orbit, how would its orbital period change?

A. The period would increase.
B. The period would decrease.
C. The period would stay the same.

SUMMARY

The goal of Chapter 6 has been to learn about motion in a circle, including orbital motion under the influence of a graviational force.

GENERAL PRINCIPLES

Uniform Circular Motion

An object moving in a circular path is in uniform circular motion if v is constant.

- The speed is constant, but the direction is constantly changing.

- The **centripetal acceleration** is directed toward the center of the circle.

$$a = \frac{v^2}{r}$$

- The acceleration requires a force directed toward the center of the circle. Newton's second law for circular motion is

$$\vec{F}_{net} = m\vec{a} = \left(\frac{mv^2}{r}, \text{ toward center of circle}\right)$$

Universal Gravitation

Two objects with masses m_1 and m_2 at a distance r apart exert attractive gravitational forces on each other of magnitude:

$$F_{1\,on\,2} = F_{2\,on\,1} = \frac{Gm_1m_2}{r^2}$$

where the gravitational constant is

$$G = 6.67 \times 10^{-11} \text{ N} \cdot \text{m}^2/\text{kg}^2$$

This is **Newton's Law of Gravity.** Gravity is an inverse square law.

IMPORTANT CONCEPTS

Describing circular motion

Circular motion repeats:

Period: T = time for one complete circle.

Frequency: $f = \dfrac{1}{T}$

We define new variables for circular motion. By convention, counterclockwise is positive.

Angular displacement: $\Delta\theta = \theta_f - \theta_i$

Angular velocity: $\omega = \dfrac{\Delta\theta}{\Delta t}$

Uniform circular motion kinematics

For uniform circular motion:

$$\omega = 2\pi f \qquad \theta_f - \theta_i = \Delta\theta = \omega\,\Delta t$$

The velocity, acceleration and circular motion variables are related as follows:

$$v = \frac{2\pi r}{T}$$

$$v = \omega r$$

$$a = \frac{v^2}{r} = \omega^2 r$$

APPLICATIONS

Apparent weight and weightlessness

Circular motion requires a net force pointing to the center. The apparent weight $w_{app} = n$ is usually not the same as the true weight w. n must be > 0 for the object to be in contact with a surface.

In orbital motion, the net force is provided by gravity. An astronaut and his spacecraft are both in free fall, so he feels weightless.

Planetary gravity and orbital motion

For a planet of mass M and radius R, the free-fall acceleration on the surface is

$$g = \frac{GM}{R^2}$$

The speed of a satellite in a low orbit is

$$v = \sqrt{gR}$$

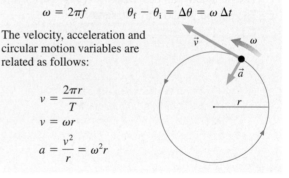

A **satellite** in a circular orbit of radius r around an object of mass M moves at a speed v given by

$$v = \sqrt{\frac{GM}{r}}$$

The period and radius are related as follows:

$$T^2 = \left(\frac{4\pi^2}{GM}\right)r^3$$

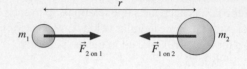

For instructor-assigned homework, go to
www.masteringphysics.com

Problem difficulty is labeled as I (straightforward) to IIIII (challenging).

Problems labeled [pencil icon] can be done on a Workbook Dynamics Worksheet; INT integrate significant material from earlier chapters; BIO are of biological or medical interest.

QUESTIONS

Conceptual Questions

1. The batter in a baseball game hits a home run. As he circles the bases, is his angular velocity positive or negative?

2. Viewed from somewhere in space above the north pole, would a point on the earth's equator have a positive or negative angular velocity due to the earth's rotation?

3. A cyclist goes around a level, circular track at constant speed. Do you agree or disagree with the following statement? "Since the cyclist's speed is constant, her acceleration is zero." Explain.

4. In uniform circular motion, which of the following quantities are constant: speed, instantaneous velocity, angular velocity, centripetal acceleration, the magnitude of the net force?

5. A particle moving along a straight line can have nonzero acceleration even when its speed is zero (for instance, a ball in free fall at the top of its path). Can a particle moving in a circle have nonzero *centripetal* acceleration when its speed is zero? If so, give an example. If not, why not?

6. Would having four-wheel drive on a car make it possible to drive faster around corners on an icy road, without slipping, than the same car with two-wheel drive? Explain.

7. BIO Large birds like pheasants often walk short distances. Small birds like chickadees never walk. They either hop or fly. Why might this be?

8. When you drive fast on the highway with muddy tires, you can hear the mud flying off the tires into your wheel wells. Why does the mud fly off?

9. A ball on a string moves in a vertical circle as in Figure Q6.9. When the ball is at its lowest point, is the tension in the string greater than, less than, or equal to the ball's weight? Explain. (You may want to include a free-body diagram as part of your explanation.)

FIGURE Q6.9

10. Give an everyday example of circular motion for which the centripetal acceleration is mostly or completely due to a force of the type specified: (a) Static friction. (b) Tension.

11. Give an everyday example of circular motion for which the centripetal acceleration is mostly or completely due to a force of the type specified: (a) Gravity. (b) Normal force.

12. It's been proposed that future space stations create "artificial gravity" by rotating around an axis. (The space station would have to be much larger than the present space station for this to be feasible.)
    a. How would this work? Explain.
    b. Would the artificial gravity be equally effective throughout the space station? If not, where in the space station would the residents want to live and work?

13. Variation in your apparent weight is desirable when you ride a roller coaster; it makes the ride fun. However, too much variation over a short period of time can be painful. For this reason, the loops of real roller coasters are not simply circles like Figure 6.19a. A typical loop is shown in Figure Q6.13. The radius of the circle that matches the track at the top of the loop is much smaller than that of a matching circle at other places on the track. Explain why this shape gives a more comfortable ride than a circular loop.

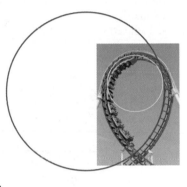

FIGURE Q6.13

14. Riding in the back of a pickup truck can be very dangerous. If the truck turns suddenly, the riders can be thrown from the truck bed. Why are the riders ejected from the bed?

15. How far away from the earth does an orbiting spacecraft have to be in order for the astronauts inside to be "weightless"?

16. A small projectile is launched parallel to the ground at height $h = 1$ m with sufficient speed to orbit a completely smooth, airless planet. A bug rides in a small hole inside the projectile. Is the bug weightless? Explain.

17. Why is it impossible for an astronaut inside an orbiting space shuttle to go from one end to the other by walking normally?

18. If every object in the universe feels an attractive gravitational force due to every other object, why don't you feel a pull from someone seated next to you?

19. A mountain climber's weight is less on the top of a tall mountain than at the base, though his mass is the same. Why?

20. Is the earth's gravitational force on the sun larger, smaller, or equal to the sun's gravitational force on the earth? Explain.

21. The mass of Jupiter is $M_{\text{Jupiter}} = 300 M_{\text{earth}}$. Jupiter orbits around the sun with $T_{\text{Jupiter}} = 11.9$ yr in an orbit with $r_{\text{Jupiter}} = 5.2 r_{\text{earth}}$. Suppose the earth could be moved to the distance of Jupiter and placed in a circular orbit around the sun. Would the new period of the earth's orbit be less than, equal to, or greater than T_{Jupiter}? Or is the value impossible to determine, because it would depend on the speed the earth is given? Or is the question meaningless because it is not possible for a planet of earth's mass to orbit at the distance of Jupiter? Explain.

Multiple-Choice Questions

22. | As seen from above, a car rounds the curved path shown in Figure Q6.22 at a constant speed. Which vector best represents the net force acting on the car?

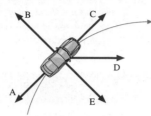

FIGURE Q6.22

23. | Suppose you and a friend, each of mass 60 kg, go to the park and get on a 4.0-m-diameter merry-go-round. You stand on the outside edge of the merry-go-round, while your friend pushes so that it rotates once every 6.0 s. What is the magnitude of the (apparent) outward force that you feel?
    A. 7 N   B. 63 N   C. 130 N   D. 260 N

24. | The cylindrical space station in Figure Q6.24, 200 m in diameter, rotates in order to provide artificial gravity of 1 g for the occupants. How much time does the station take to complete one rotation?
    A. 3 s   B. 20 s   C. 28 s   D. 32 s

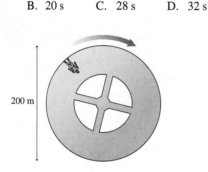

200 m

FIGURE Q6.24

25. || Two cylindrical space stations, the second four times the diameter of the first, rotate so as to provide the same amount of artificial gravity. If the first station makes one rotation in the time T, then the second station makes one rotation in time
    A. $T/4$ B. $2T$ C. $4T$ D. $16T$

26. ||| A newly discovered planet has twice the mass and three times the radius of the earth. What is the free-fall acceleration at its surface, in terms of the free-fall acceleration g at the surface of the earth?
    A. $\frac{2}{9}g$ B. $\frac{2}{3}g$ C. $\frac{3}{4}g$ D. $\frac{4}{3}g$

27. || Suppose one night the radius of the earth doubled but its mass stayed the same. What would be an approximate new value for the free-fall acceleration at the surface of the earth?
    A. 2.5 m/s²   B. 5.0 m/s²   C. 10 m/s²   D. 20 m/s²

28. | Currently, the moon goes around the earth once every 27.3 days. If the moon could be brought into a new orbit with a smaller radius, its orbital period would be
    A. More than 27.3 days.
    B. 27.3 days.
    C. Less than 27.3 days.

29. || Two planets orbit a star. Planet 1 has orbital radius r_1 and planet 2 has $r_2 = 4r_1$. Planet 1 orbits with period T_1. Planet 2 orbits with period
    A. $T_2 = \frac{1}{2}T_1$ B. $T_2 = 2T_1$ C. $T_2 = 4T_1$ D. $T_2 = 8T_1$

Questions 30 through 32 concern a classic figure-skating jump called the axel. A skater starts the jump moving forward as shown in Figure Q6.30, leaps into the air, and turns one-and-a-half revolutions before landing. The typical skater is in the air for about 0.5 s, and the skater's hands are located about 0.8 m from the rotation axis.

FIGURE Q6.30

30. | What is the approximate angular speed of the skater during the leap?
    A. 2 rad/s   B. 6 rad/s   C. 9 rad/s   D. 20 rad/s

31. | The skater's arms are fully extended during the jump. What is the approximate centripetal acceleration of the skater's hand?
    A. 10 m/s²   B. 30 m/s²   C. 300 m/s²   D. 450 m/s²

32. | What is the approximate speed of the skater's hand?
    A. 1 m/s   B. 3 m/s   C. 9 m/s   D. 15 m/s

PROBLEMS

Section 6.1 Uniform Circular Motion

1. || What is the angular speed of the tip of the minute hand on a clock, in rad/s?

2. | An old-fashioned vinyl record rotates on a turntable at 45 rpm. What are (a) the angular speed in rad/s and (b) the period of the motion?

3. || The earth's radius is about 4000 miles. Kampala, the capital of Uganda, and Singapore are both nearly on the equator. The distance between them is 5000 miles.
    a. Through what angle do you turn, relative to the earth, if you fly from Kampala to Singapore? Give your answer in both radians and degrees.
    b. The flight from Kampala to Singapore takes 9 hours. What is the plane's angular speed relative to the earth?

4. || A Ferris wheel rotates at an angular velocity of 0.036 rad/s. At $t = 0$ min, your friend Seth is at the very top of the ride. What is Seth's angular position at $t = 3.0$ min, measured counterclockwise from the top? Give your answer as an angle in degrees between 0° and 360°.

5. |||| A turntable rotates counterclockwise at 78 rpm. A speck of dust on the turntable is at $\theta = 0.45$ rad at $t = 0$ s. What is the angle of the speck at $t = 8.0$ s? Your answer should be between 0 and 2π rad.

6. || A fast-moving superhero in a comic book runs around a circular, 70-m-diameter track five and a half times (ending up directly opposite her starting point) in 3.0 s. What is her angular speed, in rad/s?

Section 6.2 Speed, Velocity, and Acceleration in Uniform Circular Motion

7. || A 5.0-m-diameter merry-go-round is turning with a 4.0 s period. What is the speed of a child on the rim?

8. ‖ A skater holds her arms outstretched as she spins at 180 rpm. What is the speed of her hands if they are 140 cm apart?

9. │ The horse on a carousel is 4.0 m from the central axis.
   a. If the carousel rotates at 0.10 rev/s, how long does it take the horse to go around twice?
   b. How fast is a child on the horse going (in m/s)?

10. ‖ The radius of the earth's very nearly circular orbit around the sun is 1.50×10^{11} m. Find the magnitude of the earth's (a) velocity, (b) angular velocity, and (c) centripetal acceleration as it travels around the sun. Assume a year of 365 days.

11. ‖ Your roommate is working on his bicycle and has the bike upside down. He spins the 60-cm-diameter wheel, and you notice that a pebble stuck in the tread goes by three times every second. What are the pebble's speed and acceleration?

12. │ To withstand "g-forces" of up to 10 g's, caused by suddenly pulling out of a steep dive, fighter jet pilots train on a "human centrifuge." 10 g's is an acceleration of 98 m/s². If the length of the centrifuge arm is 12 m, at what speed is the rider moving when she experiences 10 g's?

Section 6.3 Dynamics of Uniform Circular Motion

13. ‖‖‖ Figure P6.13 is a bird's-eye view of particles on a string moving in horizontal circles on a table top. All are moving at the same speed. Rank in order, from largest to smallest, the tensions T_1 to T_4.

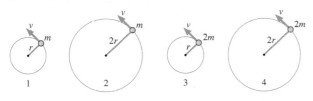

FIGURE P6.13

14. ‖ A 200 g block on a 50-cm-long string swings in a circle on a horizontal, frictionless table at 75 rpm.
   a. What is the speed of the block?
   b. What is the tension in the string?

15. │ A 1500 kg car drives around a flat 200-m-diameter circular track at 25 m/s. What are the magnitude and direction of the net force on the car? What causes this force?

16. │ A fast pitch softball player does a "windmill" pitch, illustrated in Figure P6.16, moving her hand through a circular arc to pitch a ball at 70 mph. The 0.19 kg ball is 50 cm from the pivot point at her shoulder. At the lowest point of the circle, the ball has reached its maximum speed.
   a. At the bottom of the circle, just before the ball leaves her hand, what is its centripetal acceleration?
   b. What are the magnitude and direction of the force her hand exerts on the ball at this point?

FIGURE P6.16

17. │ A baseball pitching machine works by rotating a light and stiff rigid rod about a horizontal axis until the ball is moving toward the target. Suppose a 144 g baseball is held 85 cm from the axis of rotation and released at the major league pitching speed of 85 mph.
   a. What is the ball's centripetal acceleration just before it is released?
   b. What is the magnitude of the net force that is acting on the ball just before it is released?

Section 6.4 Apparent Forces in Circular Motion

18. │ You hold a bucket in one hand. In the bucket is a 500 g rock. You swing the bucket so the rock moves in a vertical circle 2.2 m in diameter. What is the minimum speed the rock must have at the top of the circle if it is to always stay in contact with the bottom of the bucket?

19. ‖ The passengers in a roller coaster car feel 50% heavier than their true weight as the car goes through a dip with a 30 m radius of curvature. What is the car's speed at the bottom of the dip?

20. ‖‖ A roller coaster car crosses the top of a circular loop-the-loop at twice the critical speed. What is the ratio of the car's apparent weight to its true weight?

21. ‖ As a roller coaster car crosses the top of a 40-m-diameter loop-the-loop, its apparent weight is the same as its true weight. What is the car's speed at the top?

22. ‖ A typical laboratory centrifuge rotates at 4000 rpm. Test
   BIO tubes have to be placed into a centrifuge very carefully
   INT because of the very large accelerations.
   a. What is the acceleration at the end of a test tube that is 10 cm from the axis of rotation?
   b. For comparison, what is the magnitude of the acceleration a test tube would experience if dropped from a height of 1.0 m and stopped in a 1.0-ms-long encounter with a hard floor?

Section 6.5 Circular Orbits and Weightlessness

23. ‖‖ A satellite orbiting the moon very near the surface has a period of 110 min. Use this information, together with the radius of the moon from the table on the inside of the back cover, to calculate the free-fall acceleration on the moon's surface.

Section 6.6 Newton's Law of Gravity

24. ‖ The centers of a 10 kg lead ball and a 100 g lead ball are separated by 10 cm.
   a. What gravitational force does each exert on the other?
   b. What is the ratio of this gravitational force to the weight of the 100 g ball?

25. ‖ The gravitational force of a star on an orbiting planet 1 is F_1. Planet 2, which is twice as massive as planet 1 and orbits at twice the distance from the star, experiences gravitational force F_2. What is the ratio F_2/F_1?

26. │ The free-fall acceleration at the surface of planet 1 is 20 m/s². The radius and the mass of planet 2 are twice those of planet 1. What is the free-fall acceleration on planet 2?

27. ‖‖‖ What is the ratio of the sun's gravitational force on you to the earth's gravitational force on you?

28. | a. What is the free-fall acceleration at the surface of the sun?

b. Billions of years from now, the sun will become a red giant star and its radius will swell to the size of the earth's orbit. Assuming its mass doesn't change, what will be the free-fall acceleration on the surface at that time?

29. ‖ a. What is the gravitational force of the sun on the earth?

b. What is the gravitational force of the moon on the earth?

c. The moon's force is what percent of the sun's force?

30. | What is the free-fall acceleration at the surface of (a) Mars and (b) Jupiter?

Section 6.7 Gravity and Orbits

31. | Planet X orbits the star Omega with a "year" that is 200 earth days long. Planet Y circles Omega at four times the distance of Planet X. How long is a year on Planet Y?

32. ‖‖ Satellite A orbits a planet with a speed of 10,000 m/s. Satellite B is twice as massive as satellite A and orbits at twice the distance from the center of the planet. What is the speed of satellite B?

33. ‖ The space shuttle is in a 250-mile-high orbit. What are the shuttle's orbital period, in minutes, and its speed?

34. ‖ The *asteroid belt* circles the sun between the orbits of Mars and Jupiter. One asteroid has a period of 5.0 earth years. What are the asteroid's orbital radius and speed?

35. ‖‖ An earth satellite moves in a circular orbit at a speed of 5500 m/s. What is its orbital period?

General Problems

36. | How fast must a plane fly along the earth's equator so that the sun stands still relative to the passengers? In which direction must the plane fly, east to west or west to east? Give your answer in both km/h and mph. The radius of the earth is 6400 km.

37. ‖ The car in Figure P6.37 travels at a constant speed along the road shown. Draw vectors showing its acceleration at the three points A, B, and C, or write $\vec{a} = \vec{0}$. The lengths of your vectors should correspond to the magnitudes of the accelerations.

FIGURE P6.37

38. ‖‖ In the Bohr model of the hydrogen atom, an electron (mass $m = 9.1 \times 10^{-31}$ kg) orbits a proton at a distance of 5.3×10^{-11} m. The proton pulls on the electron with an electric force of 8.2×10^{-8} N. How many revolutions per second does the electron make?

39. ‖‖ A 75 kg man weighs himself at the north pole and at the equator. Which scale reading is higher? By how much? Assume the earth is a perfect sphere. Explain why the readings differ.

40. | A 1500 kg car takes a 50-m-radius unbanked curve at 15 m/s. What is the size of the friction force on the car?

41. ‖ A 500 g ball swings in a vertical circle at the end of a 1.5-m-long string. When the ball is at the bottom of the circle, the tension in the string is 15 N. What is the speed of the ball at that point?

42. ‖ Suppose the moon were held in its orbit not by gravity but by a massless cable attached to the center of the earth. What would be the tension in the cable? See the inside of the back cover for astronomical data.

43. ‖‖‖ A 30 g ball rolls around a 40-cm-diameter L-shaped track, shown in Figure P6.43, at 60 rpm. Rolling friction can be neglected.

FIGURE P6.43

a. How many different contact forces does the track exert on the ball? Name them.

b. What is the magnitude of the net force of the track on the ball?

44. ‖‖ A 5.0 g coin is placed 15 cm from the center of a turntable. The coin has static and kinetic coefficients of friction with the turntable surface of $\mu_s = 0.80$ and $\mu_k = 0.50$. The turntable very slowly speeds up to 60 rpm. Does the coin slide off?

45. ‖‖‖ A *conical pendulum* is formed by attaching a 500 g ball to a 1.0-m-long string, then allowing the mass to move in a horizontal circle of radius 20 cm. Figure P6.45 shows that the string traces out the surface of a cone, hence the name.

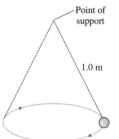

a. What is the tension in the string?

b. What is the ball's angular velocity, in rpm?

Hint: Determine the horizontal and vertical components of the forces

FIGURE P6.45

acting on the ball, and use the fact that the vertical component of acceleration is zero since there is no vertical motion.

46. ‖‖‖‖ In an old-fashioned amusement park ride, passengers stand inside a 3.0-m-tall, 5.0-m-diameter hollow steel cylinder with their backs against the wall. The cylinder begins to rotate about a vertical axis. Then the floor on which the passengers are standing suddenly drops away! If all goes well, the passengers will "stick" to the wall and not slide. Clothing has a static coefficient of friction against steel in the range 0.60 to 1.0 and a kinetic coefficient in the range 0.40 to 0.70. What is the minimum rotational frequency, in rpm, for which the ride is safe?

47. | A car drives over the top of a hill that has a radius of 50 m. What maximum speed can the car have without flying off the road at the top of the hill?

48. | While at the county fair, you decide to ride the Ferris wheel. Having eaten too many candy apples and elephant ears, you find the motion somewhat unpleasant. To take your mind off your stomach, you wonder about the motion of the ride. You estimate the radius of the big wheel to be 15 m, and you use your watch to find that each loop around takes 25 s.

a. What are your speed and magnitude of your acceleration?

b. What is the ratio of your apparent weight to your true weight at the top of the ride?

c. What is the ratio of your apparent weight to your true weight at the bottom?

49. ‖ While a person is walking, his arms (each with typical length 70 cm measured from the shoulder joint) swing through approximately a 45° angle in 0.5 s. As a reasonable approximation, we can assume that the arm moves with constant speed during each swing.
    a. What is the acceleration of a 1.0 g drop of blood in the fingertips at the bottom of the swing?
    b. Draw a free-body diagram for the drop of blood in part a.
    c. Find the magnitude and direction of the force that the blood vessel must exert on the drop of blood.
    d. What force would the blood vessel exert if the arm were not swinging?

50. ‖ A 100 g ball on a 60-cm-long string is swung in a vertical circle about a point 200 cm above the floor. The string suddenly breaks when it is parallel to the ground and the ball is moving upward. The ball reaches a height 600 cm above the floor. What was the tension in the string an instant before it broke?

51. ‖‖ The ultracentrifuge is an important tool for separating and analyzing proteins in biological research. Because of the enormous centripetal accelerations that can be achieved, the apparatus (see Figure 6.21) must be carefully balanced so that each sample is matched by another on the opposite side of the rotor shaft. Failure to do so is a costly mistake, as seen in Figure P6.51. Any difference in mass of the opposing samples will cause a net force in the horizontal plane on the shaft of the rotor. Suppose that a scientist makes a slight error in sample preparation, and one sample has a mass 10 mg greater than the opposing sample. If the samples are 10 cm from the axis of the rotor and the ultracentrifuge spins at 70,000 rpm, what is the magnitude of the net force on the rotor due to the unbalanced samples?

FIGURE P6.51

52. ‖‖‖ A 1.0-m-diameter lead sphere has a mass of 5900 kg. A dust particle rests on the surface. What is the ratio of the gravitational force of the sphere on the dust particle to the weight of the dust particle?

53. ‖‖‖ The space shuttle orbits 300 km above the surface of the earth.
    a. What is the force of gravity on a 1.0 kg sphere inside the space shuttle?
    b. The sphere floats around inside the space shuttle, apparently "weightless." How is this possible?

54. ‖‖‖ A starship is circling a distant planet of radius R. The astronauts find that the free-fall acceleration at their altitude is half the value at the planet's surface. How far above the surface are they orbiting? Your answer will be a multiple of R.

55. ‖‖‖ A sensitive gravimeter at a mountain observatory finds that the free-fall acceleration is 0.0075 m/s² less than that at sea level. What is the observatory's altitude?

56. ‖ Suppose we could shrink the earth without changing its mass. At what fraction of its current radius would the free-fall acceleration at the surface be three times its present value?

57. ‖ Planet Z is 10,000 km in diameter. The free-fall acceleration on Planet Z is 8.0 m/s².
    a. What is the mass of Planet Z?
    b. What is the free-fall acceleration 10,000 km above Planet Z's north pole?

58. ‖‖‖ What are the speed and altitude of a geostationary satellite (see Example 6.13) orbiting Mars? Mars rotates on its axis once every 24.8 hours.

59. ‖ a. What is the free-fall acceleration on Mars?
    b. Estimate the maximum speed at which an astronaut can walk on the surface of Mars.

60. ‖‖‖ How long will it take a rock dropped from 2.0 m above the surface of Mars to reach the ground?

61. ‖‖‖‖ A 20 kg sphere is at the origin and a 10 kg sphere is at $(x, y) = (20\text{ cm}, 0\text{ cm})$. At what point or points could you place a small mass such that the net gravitational force on it due to the spheres is zero?

62. ‖ a. At what height above the earth is the free-fall acceleration 10% of its value at the surface?
    b. What is the speed of a satellite orbiting at that height?

63. ‖ Mars has a small moon, Phobos, that orbits with a period of 7 h 39 min. The radius of Phobos' orbit is 9.4×10^6 m. Use only this information (and the value of G) to calculate the mass of Mars.

64. ‖ You are the science officer on a visit to a distant solar system. Prior to landing on a planet you measure its diameter to be 1.80×10^7 m and its rotation period to be 22.3 h. You have previously determined that the planet orbits 2.20×10^{11} m from its star with a period of 402 earth days. Once on the surface you find that the free-fall acceleration is 12.2 m/s². What are the mass of (a) the planet and (b) the star?

65. ‖ Europa, a satellite of Jupiter, is believed to have a liquid ocean of water (with a possibility of life) beneath its icy surface. In planning a future mission to Europa, what is the fastest that an astronaut with legs of length 0.70 m could walk on the surface of Europa? Europa is 3100 km in diameter and has a mass of 4.8×10^{22} kg.

In Problems 66 through 69 you are given the equation (or equations) used to solve a problem. For each of these, you are to

a. Write a realistic problem for which this is the correct equation. The last two questions should involve real planets. Be sure that the answer your problem requests is consistent with the equation given.

b. Finish the solution of the problem.

66. ‖ $60\text{ N} = (0.30\text{ kg})\omega^2(0.50\text{ m})$

67. ‖‖ $(1500\text{ kg})(9.80\text{ m/s}^2) - 11{,}760\text{ N} = (1500\text{ kg})v^2/(200\text{ m})$

68. ‖ $\dfrac{(6.67 \times 10^{-11}\text{ N} \cdot \text{m}^2/\text{kg}^2)(1.90 \times 10^{27}\text{ kg})}{r^2}$
$= \dfrac{(6.67 \times 10^{-11}\text{ N} \cdot \text{m}^2/\text{kg}^2)(5.98 \times 10^{24}\text{ kg})}{(6.37 \times 10^6\text{ m})^2}$

69. ‖ $\dfrac{(6.67 \times 10^{-11}\text{ N} \cdot \text{m}^2/\text{kg}^2)(5.98 \times 10^{24}\text{ kg})(1000\text{ kg})}{r^2}$
$= \dfrac{(1000\text{ kg})(1997\text{ m/s})^2}{r}$

Passage Problems

Orbiting the Moon

Suppose a spacecraft orbits the moon in a very low, circular orbit, just a few hundred meters above the lunar surface. The moon has a diameter of 3500 km, and the free-fall acceleration at the surface is 1.6 m/s^2.

70. | The direction of the net force on the craft is
    A. Away from the surface of the moon.
    B. In the direction of motion.
    C. Toward the center of the moon.
    D. Nonexistent, because the net force is zero.
71. | How fast is this spacecraft moving?
    A. 53 m/s   B. 75 m/s   C. 1700 m/s   D. 2400 m/s
72. | How much time does it take for the spacecraft to complete one orbit?
    A. 38 min   B. 76 min   C. 110 min   D. 220 min

73. | The material that comprises the side of the moon facing the earth is actually slightly more dense than the material on the far side. When the spacecraft is above a more dense area of the surface, the moon's gravitational force on the craft is a bit stronger. In order to stay in a circular orbit of constant height and speed, the spacecraft could fire its rockets while passing over the denser area. The rockets should be fired so as to generate a force on the craft
    A. Away from the surface of the moon.
    B. In the direction of motion.
    C. Toward the center of the moon.
    D. Opposite the direction of motion.

STOP TO THINK ANSWERS

Stop to Think 6.1: D > B > C > A. The centripetal acceleration is $\omega^2 r$. Changing r by a factor of 2 changes the centripetal acceleration by a factor of 2, but changing ω by a factor of 2 changes the centripetal acceleration by a factor of 4.

Stop to Think 6.2: $T_D > T_B = T_E > T_C > T_A$. The center-directed force is $m\omega^2 r$. Changing r by a factor of 2 changes the tension by a factor of 2, but changing f (and thus ω) by a factor of 2 changes the tension by a factor of 4.

Stop to Think 6.3: B. The car is moving in a circle, so there must be a net force toward the center of the circle. The circle is below the car, so the net force must point downward. This can be true only if $w > n$.

Stop to Think 6.4: B > A > D > C. The free-fall acceleration is proportional to the mass, but inversely proportional to the square of the radius.

Stop to Think 6.5: C. The period of the orbit does not depend on the mass of the orbiting object.

7

ROTATIONAL MOTION

As the earth rotates on its axis, the distant stars appear to move in eternal circles in the sky overhead. In reality, however, the angular velocity of the earth is very slowly decreasing, leading to an increase in the length of the day of $18 \ \mu s$ each year. What causes the angular velocity of a rotating object to change?

Looking Ahead ▶▶

The goal of Chapter 7 is to understand the physics of rotating objects. In this chapter you will learn to:

▶ Understand the angular and tangential acceleration of a rotating object.

▶ Calculate the torque exerted on an object.

▶ Determine an object's center of gravity and its moment of inertia.

▶ Apply the concepts of torque and moment of inertia to the rotation of an object about a fixed axis.

▶ Understand the motion of a rolling object.

Looking Back ◀◀

Rotational motion will revisit many of the major themes introduced in previous chapters, especially the properties of circular motion. Please review:

◀ Section 4.6 Newton's second law.

◀ Section 6.1 Uniform circular motion.

◀ Section 6.2 Speed, velocity, and acceleration in circular motion.

◀ Section 6.3 Dynamics of circular motion.

In Chapter 6 we studied circular motion, the motion of a single particle around a circular path. Our goal in this chapter is to extend this treatment to focus on extended rotating objects such as wheels, axles, and spinning tops. Objects such as these, whose shape does not change as they rotate, are called *rigid bodies*. Our own earth is an important example of a rotating rigid body. As we ride along on its surface, we see the stars making circular paths about a pole.

How do forces act to increase the angular velocity of a rotating object? What makes an object easy or hard to get rotating, and what is needed to keep it rotating? How do we describe a rolling object—one that is both rotating and moving as a whole? These are some of the questions we'll address in this chapter.

You will quickly discover that the physics of rotational motion is analogous to the physics of linear motion that you studied in earlier chapters. For example, the new concepts of torque and angular acceleration are the rotational analogs of force and acceleration, and we'll find a new version of Newton's second law that is the rotational equivalent of $\vec{F}_{net} = m\vec{a}$.

7.1 The Rotation of a Rigid Body

Thus far, our study of physics has focused almost exclusively on the *particle model* in which an entire object is represented as a single point in space. The particle model is entirely adequate for understanding motion in a wide variety of situations, but there are also cases for which we need to consider the motion of an *extended object*—a system of particles for which the size and shape *do* make a difference and cannot be neglected.

A **rigid body** is an extended object whose size and shape do not change as it moves. For example, a bicycle wheel can be thought of as a rigid body. Figure 7.1 shows a rigid body as a collection of atoms held together by the rigid "massless rods" of molecular bonds.

Real molecular bonds are, of course, not perfectly rigid. That's why an object seemingly as rigid as a bicycle wheel can flex and bend. Thus Figure 7.1 is really a simplified *model* of an extended object, the **rigid-body model.** The rigid-body model is a very good approximation of many real objects of practical interest, such as wheels and axles. Even nonrigid objects can often be modeled as a rigid body during parts of their motion. For example, a diver is well described as a rotating rigid body while she's in the tuck position.

Figure 7.2 illustrates the three basic types of motion of a rigid body: **translational motion, rotational motion,** and **combination motion.** We've already studied translational motion of a rigid body using the particle model. If a rigid body doesn't rotate, this model is often adequate for describing its motion. The rotational motion of a rigid body will be the main focus of this chapter. We'll also discuss an important case of combination motion—that of a *rolling* object—later in this chapter.

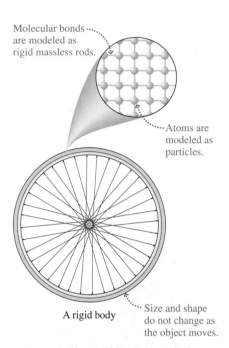

Molecular bonds are modeled as rigid massless rods.

Atoms are modeled as particles.

A rigid body

Size and shape do not change as the object moves.

FIGURE 7.1 The rigid-body model of an extended object.

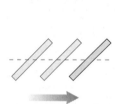

Parabolic trajectory

Translational motion:
The object as a whole moves along a trajectory but does not rotate.

Rotational motion:
The object rotates about a fixed point. Every point on the object moves in a circle.

Combination motion:
An object rotates as it moves along a trajectory.

FIGURE 7.2 Three basic types of motion of a rigid body.

Uniform Rotational Motion of a Rigid Body

Figure 7.3 shows a wheel rotating on an axle. Notice that as the wheel rotates for a time interval Δt, two points on the wheel, marked with dots, turn through the *same angle,* even through their distances r from the axis of rotation may be different. That is, $\Delta\theta_1 = \Delta\theta_2$ during the time interval Δt. As a consequence, the two points have equal angular velocities: $\omega_1 = \omega_2$. In general, **every point on a rotating rigid body has the same angular velocity.** Because of this, we can refer to the angular velocity ω *of the wheel.*

Recall from Chapter 6 that the speed of a particle moving in a circle is $v = \omega r$, so two points of a rotating object will have different *speeds* if they have different distances from the axis of rotation, but *all* points have the *same* angular velocity ω. Thus angular velocity is one of the most important parameters of a rotating object.

Because every point on a rotating object moves in a circle, we can carry forward all the results for circular motion from Chapter 6. Thus the angular displacement of any point on the wheel shown in Figure 7.3 is found from Equation 6.1 as $\Delta\theta = \omega\,\Delta t$; the speed of any particle in the wheel is $v = \omega r$, where r is the particle's distance from the axis; and the particle's centripetal acceleration is $a = \omega^2 r$.

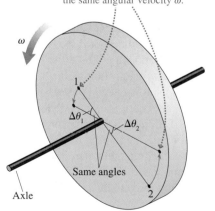

Every point on the wheel undergoes circular motion with the same angular velocity ω.

ω

Same angles

Axle

FIGURE 7.3 All points on a wheel rotate with the same angular velocity.

Angular Acceleration

If you push on the edge of a bicycle wheel, it begins to rotate. If you continue to push, it rotates ever faster. Its angular velocity is *changing*. To understand the dynamics of rotating objects, we'll need to be able to describe this case of changing angular velocity, that is, the case of *nonuniform* circular motion.

Figure 7.4 shows a bicycle wheel whose angular velocity is changing. The dot represents a particular point on the wheel at successive times. At time t_i the angular velocity is ω_i; at a later time $t_f = t_i + \Delta t$ the angular velocity has changed to ω_f. The change in angular velocity during this time interval is

$$\Delta \omega = \omega_f - \omega_i$$

Recall that in Chapter 2 we defined the *linear* acceleration as

$$a_x = \frac{\Delta v_x}{\Delta t} = \frac{(v_x)_f - (v_x)_i}{\Delta t}$$

By analogy, we now define the **angular acceleration** as

$$\alpha = \frac{\text{change in angular velocity}}{\text{time interval}} = \frac{\Delta \omega}{\Delta t} \qquad (7.1)$$

Angular acceleration for a particle in nonuniform circular motion

We use the symbol α (Greek *alpha*) for angular acceleration. Because the units of ω are rad/s, the units of angular acceleration are (rad/s)/s, or rad/s^2. The sign of α depends on the sign of the change in angular velocity—again by analogy with one-dimensional motion. Like ω, the angular acceleration α is the same for every point on a rotating rigid body.

In Chapter 6 we found analogies between linear and angular positions and velocities. Here we've extended those analogies to include linear and angular accelerations. Table 7.1 summarizes all of these analogies between linear and circular motion.

NOTE ▶ Don't confuse the angular acceleration with the centripetal acceleration introduced in Chapter 6. The angular acceleration indicates how rapidly the *angular* velocity is changing. The centripetal acceleration is a vector quantity that points toward the center of a particle's circular path; it is nonzero even if the angular velocity is constant. ◀

In addition, the various equations of one-dimensional kinematics have analogs for rotational or circular motion. Table 7.2 lists the equations for one-dimensional motion and the analogous equations for the kinematics of circular motion.

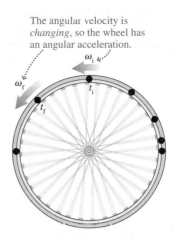

The angular velocity is *changing*, so the wheel has an angular acceleration.

FIGURE 7.4 A rotating wheel with a changing angular velocity.

TABLE 7.1 Linear and circular motion variables

Linear motion	Circular motion
Position x	Angular position θ
Velocity $v_x = \Delta x/\Delta t$	Angular velocity $\omega = \Delta \theta/\Delta t$
Acceleration $a_x = \Delta v_x/\Delta t$	Angular acceleration $\alpha = \Delta \omega/\Delta t$

Activ
Physics 7.7

TABLE 7.2 Linear and circular motion equations

Linear motion	Circular motion
Displacement at constant speed: $\Delta x = v \, \Delta t$	Angular displacement at constant angular speed: $\Delta \theta = \omega \, \Delta t$
Displacement at constant acceleration: $\Delta x = v_i \, \Delta t + \frac{1}{2} a \, \Delta t^2$	Angular displacement at constant angular acceleration: $\Delta \theta = \omega_i \, \Delta t + \frac{1}{2} \alpha \, \Delta t^2$
Change in velocity at constant acceleration: $\Delta v = a \, \Delta t$	Change in angular velocity at constant angular acceleration: $\Delta \omega = \alpha \, \Delta t$

EXAMPLE 7.1 Spinning up a computer disk

The disk in a computer disk drive spins up to 5400 rpm in 2.00 s. What is the angular acceleration of the disk? At the end of 2.00 s, how many revolutions has the disk made?

PREPARE The initial angular velocity is $\omega_i = 0$ rad/s. The final angular velocity is 5400 rpm. However, this value is not in the correct SI units of rad/s. The conversion is

$$\omega_f = \frac{5400 \text{ rev}}{\text{min}} \times \frac{1 \text{ min}}{60 \text{ s}} \times \frac{2\pi \text{ rad}}{1 \text{ rev}} = 565 \text{ rad/s}$$

SOLVE From the definition of angular acceleration,

$$\alpha = \frac{\Delta\omega}{\Delta t} = \frac{565 \text{ rad/s} - 0 \text{ rad/s}}{2.00 \text{ s}} = 283 \text{ rad/s}^2$$

We can compute the angular displacement during this acceleration by using the angular displacement equation from Table 7.2:

$$\Delta\theta = \omega_i t + \tfrac{1}{2}\alpha \, \Delta t^2$$

$$= (0 \text{ rad/s})(2.00 \text{ s}) + \tfrac{1}{2}(283 \text{ rad/s}^2)(2.00 \text{ s})^2$$

$$= 566 \text{ rad}$$

Each revolution corresponds to an angular displacement of 2π, so we have

$$\text{Number of revolutions} = \frac{566 \text{ rad}}{2\pi \text{ rad/revolution}}$$

$$= 90.1 \text{ revolutions}$$

The disk completes 90 revolutions during the first two seconds.

ASSESS We solved this problem using the same methods you learned for one-dimensional motion, replacing linear variables and equations with their rotational equivalents.

Tangential Acceleration

(a) Uniform circular motion

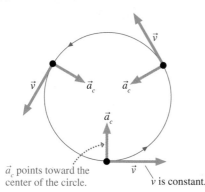

$\vec{a}_c$ points toward the center of the circle.

v is constant.

(b) Nonuniform circular motion

a_t is causing the particle to speed up.

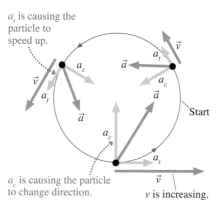

a_c is causing the particle to change direction.

v is increasing.

FIGURE 7.5 Uniform and nonuniform circular motion.

As you learned in Chapter 6, and as Figure 7.5a reminds you, a particle undergoing uniform circular motion has an acceleration directed inward toward the center of the circle. This centripetal acceleration $\vec{a}_c$ is due to the change in the *direction* of the particle's velocity. Recall that the magnitude of the centripetal acceleration is $a_c = v^2/r = \omega^2 r$.

NOTE ▶ Centripetal acceleration will now be denoted a_c to distinguish it from the tangential acceleration a_t, discussed below. ◀

If the particle's circular motion is *nonuniform,* so that the particle's speed is changing, then the particle will have another component to its acceleration. Figure 7.5b shows a particle whose speed is increasing as it moves around its circular path. Because the *magnitude* of the velocity is increasing, this second component of the acceleration is directed *tangentially* to the circle, in the same direction as the velocity. This component of acceleration is called the **tangential acceleration**. As shown in Figure 7.5b, **the full acceleration $\vec{a}$ is then the vector sum of these two components,** the centripetal acceleration $\vec{a}_c$ and the tangential acceleration $\vec{a}_t$.

The tangential acceleration measures the rate at which the particle's speed increases. Thus its magnitude is

$$a_t = \frac{\Delta v}{\Delta t}$$

We can relate the tangential acceleration to the *angular* acceleration by using the relation $v = \omega r$ between the speed of a particle moving in a circle of radius r and its angular velocity ω. We have

$$a_t = \frac{\Delta v}{\Delta t} = \frac{\Delta(\omega r)}{\Delta t} = \frac{\Delta\omega}{\Delta t}r$$

or, because $\alpha = \Delta\omega/\Delta t$ from Equation 7.1,

$$a_t = \alpha r \qquad (7.2)$$

Relationship between tangential and angular acceleration

LINEAR
p.38

We've seen that all points on a rotating rigid body have the same angular acceleration. From Equation 7.2, however, the tangential acceleration of a point on a rotating object depends on the point's distance r from the axis, so that the tangential acceleration is *not* the same for all points.

STOP TO THINK 7.1   A ball on the end of a string swings in a horizontal circle once every second. State whether the magnitudes of each of the following quantities is zero, constant (but not zero), or changing.

a. Velocity
b. Angular velocity
c. Centripetal acceleration

d. Angular acceleration
e. Tangential acceleration

7.2 Torque

Newton's genius, summarized in his second law of motion, was to recognize force as the cause of acceleration. But what about *angular* acceleration? What do Newton's laws have to tell us about rotational motion? To begin our study of rotational motion, we'll need to find a rotational equivalent of force.

Consider the common experience of pushing open a door. Figure 7.6 is a top view of a door that is hinged on the left. Four pushing forces are shown, all of equal strength. Which of these will be most effective at opening the door?

Force $\vec{F}_1$ will open the door, but force $\vec{F}_2$, which pushes straight at the hinge, will not. Force $\vec{F}_3$ will open the door, but not as easily as $\vec{F}_1$. What about $\vec{F}_4$? It is perpendicular to the door, it has the same magnitude as $\vec{F}_1$, but you know from experience that pushing close to the hinge is not as effective as pushing at the outer edge of the door.

The ability of a force to cause a rotation or a twisting motion thus depends on three factors:

1. The magnitude F of the force.
2. The distance r from the pivot to the point at which the force is applied.
3. The angle at which the force is applied.

We can incorporate these three observations into a single quantity called the **torque** τ (Greek tau). Loosely speaking, τ measures the "effectiveness" of a force at causing an object to rotate about a pivot. **Torque is the rotational equivalent of force.** In Figure 7.6, for instance, the torque τ_1 due to $\vec{F}_1$ is greater than τ_4 due to $\vec{F}_4$.

To make these ideas specific, Figure 7.7 shows a force $\vec{F}$ applied at one point of a wrench that can rotate about the nut that it's turning. Figure 7.7 defines the distance r from the pivot to the point at which the force is applied; the **radial line,** the line starting at the pivot and extending through the point at which the force is applied; and the angle ϕ (Greek phi) measured from the radial line to the direction of the force.

We saw in Figure 7.6 that force $\vec{F}_1$, which was directed perpendicular to the door, was quite effective in opening it, but force $\vec{F}_2$, directed toward the hinges, had no effect on its rotation. As shown in Figure 7.8, this suggests breaking the force $\vec{F}$ applied to the wrench into two component vectors: $\vec{F}_\perp$ directed perpendicular to the radial line, and $\vec{F}_\parallel$ directed parallel to it. Because $\vec{F}_\parallel$ points either directly toward or away from the pivot, it has no effect on the wrench's rotation, and thus contributes nothing to the torque. Only $\vec{F}_\perp$ tends to cause rotation of the wrench, so it is this component of the force that determines the torque.

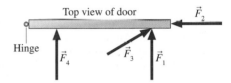

FIGURE 7.6 The four forces are the same strength, but they have different effects on the swinging door.

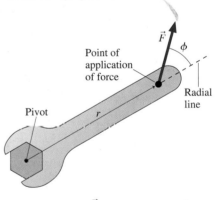

FIGURE 7.7 Force $\vec{F}$ exerts a torque about the pivot point.

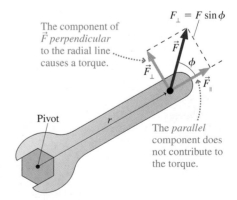

FIGURE 7.8 Torque is due to the component of the force perpendicular to the radial line.

NOTE ▶ The perpendicular component $\vec{F}_\perp$ is pronounced "F perpendicular" and the parallel component $\vec{F}_\parallel$ is pronounced "F parallel." ◀

We've seen that a force applied at a larger distance r from the pivot has a greater effect on rotation, so we expect a larger value of r to give a greater torque. We also saw that only $\vec{F}_\perp$ contributes to the torque. Both these observations are contained in our first expression for torque:

$$\tau = rF_\perp \tag{7.3}$$

Torque due to a force with perpendicular component $F_\perp$
acting at a distance r from the pivot

7.1   Activ
      Physics

From this equation, we see that the SI units of torque are newton-meters, abbreviated N · m.

EXAMPLE 7.2   Torque in opening a door

In trying to open a stuck door, Ryan pushes perpendicular to the door's surface with a force of 240 N at a distance of 0.75 m from the hinges. What torque does Ryan exert on the door?

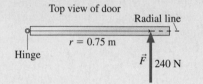

Top view of door

Radial line

$r = 0.75$ m

Hinge

$\vec{F}$ 240 N

FIGURE 7.9 Ryan's force exerts a torque on the door.

PREPARE In Figure 7.9 the radial line is shown drawn from the pivot—the hinge—through the point at which the force $\vec{F}$ is applied. We see that $\vec{F}$ is perpendicular to the radial line, so $F_\perp = F = 240$ N and $F_\parallel = 0$. The distance from the hinge to the point at which the force is applied is $r = 0.75$ m.

SOLVE We can find the torque on the door from Equation 7.3. We have

$$\tau = rF_\perp = (0.75 \text{ m})(240 \text{ N}) = 180 \text{ N} \cdot \text{m}$$

ASSESS Ryan could slightly increase the torque he exerts by pushing at the very edge of the door in Figure 7.9.

We can see from Figure 7.8 that the component of force $\vec{F}$ perpendicular to the radial line is $F_\perp = F \sin\phi$. Thus a more general expression for the torque exerted by force $\vec{F}$ is

$$\tau = rF \sin\phi \tag{7.4}$$

Torque due to a force F applied at a distance r from the pivot,
at an angle ϕ to the radial line

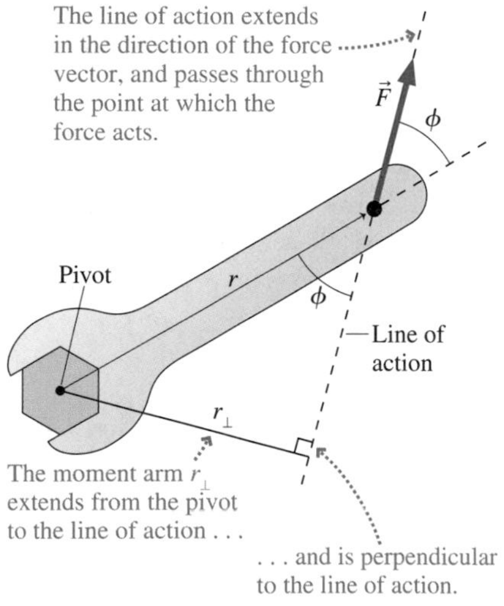

NOTE ▶ Torque differs from force in a very important way. Torque is calculated or measured *about a particular point*. To say that a torque is 20 N · m is meaningless without specifying the point about which the torque is calculated. Torque can be calculated about any point, but its value depends on the point chosen because this choice determines r and ϕ. In practice, we usually calculate torques about a hinge, pivot, or axle. ◀

Returning to the door of Figure 7.6, you can see that $\vec{F}_1$ is most effective at opening the door because $\vec{F}_1$ exerts the *largest torque* about the pivot point: It has the largest distance r from the pivot *and* acts at $\phi = 90°$, so that $\sin\phi = 1$. $\vec{F}_3$ has equal magnitude, but it is applied at an angle less than 90° and thus exerts less torque. $\vec{F}_2$, pushing straight at the hinge with $\phi = 0°$ and $\sin\phi = 0$, exerts no torque at all. And $\vec{F}_4$, with a smaller value for r, exerts less torque than $\vec{F}_1$.

Figure 7.10 shows an alternative way to calculate torque. The line that is in the direction of the force, and passes through the point at which the force acts,

The line of action extends in the direction of the force vector, and passes through the point at which the force acts.

$\vec{F}$

ϕ

Pivot r ϕ

—Line of action

$r_\perp$

The moment arm $r_\perp$ extends from the pivot to the line of action . . .

. . . and is perpendicular to the line of action.

FIGURE 7.10 You can also calculate torque in terms of the moment arm between the pivot and the line of action.

is called the *line of action*. The perpendicular distance from this line to the pivot is the **moment arm** (or *lever arm*) $r_\perp$. You can see from the figure that $r_\perp = r\sin\phi$. We can then write Equation 7.4 as $\tau = rF\sin\phi = F(r\sin\phi) = Fr_\perp$. Thus an equivalent expression for the torque is

$$\tau = r_\perp F \qquad (7.5)$$

Torque due to a force F with moment arm $r_\perp$

Equations 7.3–7.5 are three different ways of thinking about—and calculating—the torque due to a force. Depending on the problem at hand, one might be easier to use than the others. But they are all calculating the *same* torque, and will give the same value for the torque.

CONCEPTUAL EXAMPLE 7.1 Starting a bike

It is hard to get going if you try to start your bike with the pedal at the highest point. Why is this?

REASON Aided by the weight of the body, the greatest force can be applied to the pedal straight down. But with the pedal at the top, this force is exerted almost directly toward the pivot, causing only a small torque. We could say either that the perpendicular component of the force is small or that the moment arm is small.

ASSESS If you've ever climbed a steep hill while standing on the pedals, you know that you get the greatest forward motion when one pedal is completely forward with the crank parallel to the ground. This gives the maximum possible torque because the force you apply is entirely perpendicular to the radial line, and the moment arm is as long as it can be.

These equations give only the magnitude of the torque. But torque, like force, has a sign. **A torque that tends to rotate the object in a counterclockwise direction is positive, while a negative torque gives a clockwise rotation.** Figure 7.11 summarizes the signs. Notice that a force pushing straight toward the pivot or pulling straight out from the pivot exerts *no* torque.

NOTE ▶ When calculating a torque, you must supply the appropriate sign by observing the direction in which the torque acts. ◀

Torque versus speed To start and stop quickly, the basketball player needs to apply a large torque to her wheel. To make the torque as large as possible, the handrim— the outside wheel that she actually grabs— is almost as big as the wheel itself. The racer needs to move continuously at high speed, so his wheel spins much faster. To allow his hands to keep up, his handrim is much smaller than his chair's wheel, making its linear velocity correspondingly lower. The smaller radius means, however, that the torque he can apply is lower as well.

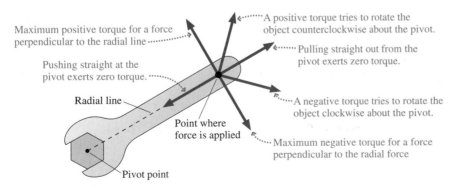

Maximum positive torque for a force perpendicular to the radial line

A positive torque tries to rotate the object counterclockwise about the pivot.

Pushing straight at the pivot exerts zero torque.

Pulling straight out from the pivot exerts zero torque.

Radial line

Point where force is applied

A negative torque tries to rotate the object clockwise about the pivot.

Maximum negative torque for a force perpendicular to the radial force

Pivot point

FIGURE 7.11 Signs and strengths of the torque.

EXAMPLE 7.3 Calculating the torque on a nut

Luis uses a 20-cm-long wrench to turn a nut. The wrench handle is tilted 30° above the horizontal, and Luis pulls straight down on the end with a force of 100 N. How much torque does Luis exert on the nut?

PREPARE Figure 7.12 shows the situation. The two illustrations correspond to two methods of calculating torque, corresponding to Equations 7.3 and 7.5.

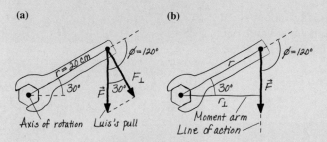

FIGURE 7.12 A wrench being used to turn a nut.

SOLVE According to Equations 7.3 and 7.4, the torque can be calculated as $\tau = rF_\perp = rF\sin\phi$. From Figure 7.12a we see that the angle between the force and the radial line is $\phi = 30° + 90° = 120°$.

The torque is then

$$\tau = -rF\sin\phi = -(0.20\ \text{m})(100\ \text{N})(\sin 120°) = -17\ \text{N·m}$$

We put in the minus sign because the torque is negative—it tries to rotate the nut in a *clockwise* direction.

Alternatively, we can use Equation 7.5 to find the torque. Figure 7.12b shows the moment arm $r_\perp$, the shortest distance from the pivot to the line of action. From the figure we see that

$$r_\perp = r\cos 30° = (0.20\ \text{m})(\cos 30°) = 0.17\ \text{m}$$

Then the torque is

$$\tau = -r_\perp F = -(0.17\ \text{m})(100\ \text{N}) = -17\ \text{N·m}$$

Again, we insert the minus sign because the torque acts to give a clockwise rotation.

ASSESS Both methods give the same answer for the torque, as expected. In general, however, you need use only one of Equations 7.3–7.5 to find the torque in any given situation. In using any of these methods to find the torque, remember to include the minus sign if the torque acts to rotate the object in a clockwise direction.

STOP TO THINK 7.2 Rank in order, from largest to smallest, the five torques τ_A to τ_E. The rods all have the same length and are pivoted at the dot.

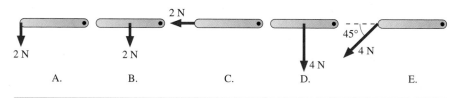

Net Torque

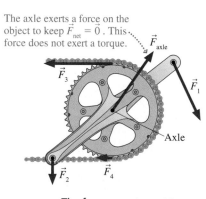

The axle exerts a force on the object to keep $\vec{F}_{net} = \vec{0}$. This force does not exert a torque.

FIGURE 7.13 The forces exert a net torque about the pivot point.

Figure 7.13 shows the forces acting on the crankset of a bicycle. Forces $\vec{F}_1$ and $\vec{F}_2$ are due to the rider pushing on the pedals, and $\vec{F}_3$ and $\vec{F}_4$ are tension forces from the chain. The crankset is free to rotate about a fixed axle, but the axle prevents it from having any translational motion with respect to the bike frame. It does so by exerting force $\vec{F}_{axle}$ on the object to balance the other forces and keep $\vec{F}_{net} = \vec{0}$.

Forces $\vec{F}_1$, $\vec{F}_2$, $\vec{F}_3$, and $\vec{F}_4$ exert torques τ_1, τ_2, τ_3, τ_4 on the crank (measured about the axle), but $\vec{F}_{axle}$ does *not* exert a torque because it is applied at the pivot point—the axle—and so has zero moment arm. Thus the *net* torque about the axle is the sum of the torques due to the *applied* forces:

$$\tau_{net} = \tau_1 + \tau_2 + \tau_3 + \cdots = \sum \tau \tag{7.6}$$

In practice, usually only a small number of forces exert torques. For example, the only torque we considered in Example 7.3 was due to Luis's pull.

EXAMPLE 7.4 Force in turning a capstan

A capstan is a device used on old sailing ships to raise the anchor. A sailor pushes the long lever, turning the capstan and winding up the anchor rope. If the sailor lifts the anchor at a constant speed, the net torque, as we'll learn later in the chapter, is zero.

Suppose the rope tension due to the weight of the anchor is 1500 N. If the distance from the axis to the point on the lever where the sailor pushes is exactly seven times the radius of the capstan around which the rope is wound, with what force must the sailor push if the net torque on the capstan is to be zero?

PREPARE Shown in Figure 7.14 is a view looking down from above the capstan. The rope pulls with a tension force $\vec{T}$ at distance R from the axis of rotation. The sailor pushes with a force $\vec{F}$ at distance $7R$ from the axis. Both forces are perpendicular to their radial lines, so ϕ in Equation 7.4 is $90°$.

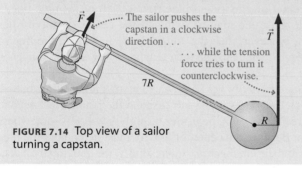

The sailor pushes the capstan in a clockwise direction . . .

. . . while the tension force tries to turn it counterclockwise.

$7R$

R

FIGURE 7.14 Top view of a sailor turning a capstan.

SOLVE The torque due to the tension in the rope is

$$\tau_T = RT\sin 90° = RT$$

We don't know the capstan radius, so we'll just leave it as R for now. This torque is positive because it tries to turn the capstan counterclockwise. The torque due to the sailor is

$$\tau_S = -(7R)F\sin 90° = -7RF$$

We put the minus sign in because this torque acts in the clockwise (negative) direction. The net torque is zero, so we have $\tau_T + \tau_S = 0$, or

$$RT - 7RF = 0$$

Note that the radius R cancels, leaving

$$F = \frac{T}{7} = \frac{1500 \text{ N}}{7} = 210 \text{ N}$$

ASSESS The force the sailor must exert is one-seventh the force the rope exerts: the long lever helps him lift the heavy anchor. In the HMS *Warrior*, built in 1860, it took 200 men turning the capstan to lift the huge anchor that weighed close to 55 000 N!

Note that forces $\vec{F}$ and $\vec{T}$ point in very different directions. Their torques only depend on their directions with respect to their own radial lines, not on the directions of the forces with respect to each other. The force the sailor needs to apply remains unchanged as he circles the capstan.

STOP TO THINK 7.3   Two forces act on the wheel shown. What third force, acting at point P, will make the net torque on the wheel zero?

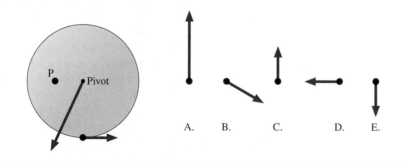

P   •Pivot

A.   B.   C.   D.   E.

7.3 Gravitational Torque and the Center of Gravity

As the gymnast in Figure 7.15 pivots around the bar, a torque due to the force of gravity causes her to rotate toward a vertical position. A falling tree or a car hood slamming shut are other examples where gravity exerts a torque on an object. Stationary objects can also experience a torque due to gravity. A diving board experiences a gravitational torque about its fixed end. It doesn't rotate because of a counteracting torque provided by forces from the base at its fixed end.

(a) Gravity exerts a force and a torque on each particle that makes up the gymnast.   Rotation axis

(b) The weight force provides a torque about the rotation axis.

Center of gravity

$\vec{w}$

The gymnast responds *as if* her entire weight acts at her center of gravity.

FIGURE 7.15 The center of gravity is the point where the weight appears to act.

We've learned how to calculate the torque due to a single force acting on an object. But gravity doesn't act at a single point on an object. It pulls downward on *every particle* that makes up the object, as shown for the gymnast in Figure 7.15a, and so each particle experiences a small torque due to the force of gravity that acts upon it. The gravitational torque on the object as a whole is then the *net* torque exerted on all the particles. We won't prove it, but the gravitational torque can be calculated by assuming that the net force of gravity—that is, the object's weight $\vec{w}$—acts at single special point on the object called its **center of gravity** (symbol ☺). Then we can calculate the torque due to gravity by our methods learned earlier for a single force ($\vec{w}$) acting at a single point (the center of gravity). Figure 7.15b shows how we can consider the gymnast's weight as acting at her center of gravity.

Finding the Center of Gravity

To calculate the gravitational torque, we need to know where the object's center of gravity is located. There is a simple experimental method for finding the center of gravity of any object, based on the observation that **any object free to rotate about a pivot will come to rest with its center of gravity directly below the pivot.** To see this, consider the cutout map of the continental United States in Figure 7.16a. If the center of gravity is to the right or left of the blue line, a gravitational torque will cause the map to swing. If the center of gravity lies directly *below* the pivot, however, then the weight force lies along the line of action, and the torque is zero. The map can then remain at rest with no tendency to swing.

We know that the center of gravity lies somewhere along the blue line, but we don't yet know where. To find out, we need to suspend the map from a second pivot, as shown in Figure 7.16b. Then the center of gravity will fall somewhere along the red line shown. Because the center of gravity must lie on both the blue and red lines, it must be at their *intersection*. Interestingly, the geographical center of the continental United States is defined in just this way, as the center of gravity of a map of the contiguous United States. This point is one mile northwest of Lebanon, Kansas.

For a simple symmetrical object, such as a rod, sphere, or cube made from a uniform material, Figure 7.17 shows that **the center of gravity of a symmetrical object lies at its geometrical center.** A particularly simple case of this is a point particle, whose center of gravity lies at the position of the particle.

The center of gravity of an unsymmetrical object is harder to locate and, in some cases, may not even lie inside the object. Consider the high-wire act of Figure 7.18a on the next page, which appears to be a masterful feat of skill and balance. But because the man hangs so far below the motorcycle, the center of gravity of the acrobats and motorcycle actually lies in midair as shown, *below* the wire. This means that the act hangs stably, with no effort on the part of the acrobats. It is actually quite common for an object's center of gravity to lie outside the object. Because of the position of her arms and legs, the center of gravity of the high jumper in Figure 7.18b actually passes *under* the bar as she clears it.

The Center of Mass

In the particle model, all the mass of an object is considered to be concentrated at a single point, and all forces act at this point. The acceleration of this point particle is determined from Newton's second law as $\vec{a} = \vec{F}_{net}/m$. The motion of an extended object is generally more complex. Forces can act at different points on the object, leading to a net torque. And the object can undergo both translational

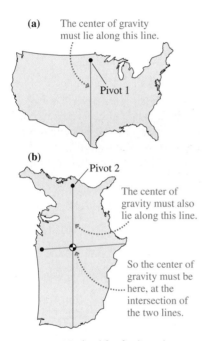

(a) The center of gravity must lie along this line.

Pivot 1

(b) Pivot 2

The center of gravity must also lie along this line.

So the center of gravity must be here, at the intersection of the two lines.

FIGURE 7.16 Method for finding the center of gravity of an object.

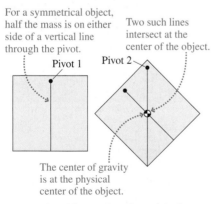

For a symmetrical object, half the mass is on either side of a vertical line through the pivot.

Two such lines intersect at the center of the object.

Pivot 1   Pivot 2

The center of gravity is at the physical center of the object.

FIGURE 7.17 The center of gravity of a symmetrical object lies at its center.

and rotational motion at the same time. This means that different points on the object can have *different* accelerations. To which point on the object does the $\vec{a}$ found from Newton's second law refer?

There is such a special point of an object called its **center of mass.** The concepts of center of mass and center of gravity are closely related. In fact, for everyday objects near the surface of the earth the positions of the center of mass and center of gravity are essentially identical. Their positions differ appreciably only when the size of the object is large enough that $\vec{g}$, the acceleration due to gravity, is different at different points on the object. The distinction between the positions of the center of gravity and center of mass is important only for huge objects such as planets or galaxies, and in this book we'll consider the positions to be the same.

For an extended object, then, Newton's second law gives the acceleration of the center of mass. We can thus write

$$\vec{a}_{\rm cm} = \frac{\vec{F}_{\rm net}}{m} \tag{7.7}$$

where $\vec{a}_{\rm cm}$ is the acceleration of the center of mass.

An important application of this principle is to the case of objects thrown into the air. The net force in this case is just the object's weight $\vec{w} = m\vec{g}$, so that from Equation 7.7 we have $\vec{a}_{\rm cm} = \vec{g}$. This is just the acceleration of a *point particle* during projectile motion, so that **the motion of the center of mass of an extended object undergoing projectile motion is the same as that of a point object undergoing projectile motion.** Figure 7.19 shows a hammer that has been tossed in the air. The hammer is undergoing both translational and rotational motion. Although the overall motion of the hammer is quite complex, the motion of the center of mass is simple: It has the same parabolic path as the ball moving through the air in Figure 3.28.

This is the advantage of the high jump technique illustrated in Figure 7.18b. Once the high jumper leaves the ground, her center of mass (which is the same point as her center of gravity) moves along a fixed parabolic path, no matter how she flexes her body. For a jump just at the limit of her ability, her center of mass will pass *below* the bar. But if she can flex her body as shown, she can just make it over the bar.

FIGURE 7.18 An object's center of gravity can lie outside the object.

Calculating the Position of the Center of Gravity

It is nice to know you can locate an object's center of gravity by suspending it from a pivot, but it is rarely a practical technique. More often, we would like to calculate the center of gravity of an object made up of a combination of particles and other objects whose center of gravity positions are known.

Recall that for the purpose of calculating the torque due to gravity, the center of gravity is the point at which the total weight of the object can be considered as acting. This suggests a method for finding the center of gravity of an object made up of several point particles:

1. Calculate the gravitational torque on the object as the sum of the gravitational torques due to the weight of each of its particles, including the appropriate signs. We can use our already-developed methods for finding torques due to forces acting at known positions.
2. Calculate the torque due to the entire weight $\vec{w}$ acting at the center of gravity, whose position is still unknown. The torques calculated in these two ways must be the same; setting them equal will allow us to find the center-of-gravity position.

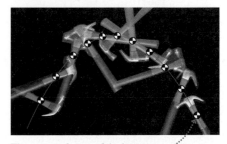

The center of mass of the hammer follows the same trajectory as a point particle undergoing projectile motion.

FIGURE 7.19 A hammer undergoing rotational and translational motion.

(a) Calculating the torque due to the weight of each mass

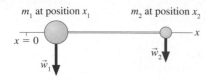

m_1 at position x_1 m_2 at position x_2

(b) Calculating the torque due to the total weight acting at the center of gravity

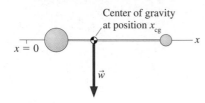

Center of gravity at position x_{cg}

FIGURE 7.20 Finding the center of gravity of a dumbbell.

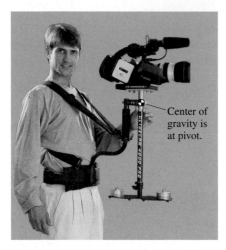

Holding steady Many moviemaking scenes require handheld shots for which the cameraman must walk along, following the action. The stabilizer shown reduces unwanted camera motion as the cameraman walks. The center of gravity of the camera and its hanging weight arm is located exactly at a pivot that can swing freely in any direction. The frictionless pivot exerts no torque on the camera and arm. Neither does the weight, because it acts at the pivot. With no torque acting on it, the camera has no tendency to rotate. The long arm also increases the system's *moment of inertia,* further decreasing unwanted rotations. More about this later!

To see how this works, consider the unbalanced dumbbell shown in Figure 7.20. Where along the bar should you lift it so that it remains balanced horizontally? From our discussion above, we know that it will remain balanced if the center of gravity is at the pivot—so you need to grasp it at its center of gravity. Let's find the position of the center of gravity by using our method outlined above.

First, we'll calculate the torque due to the weight of each particle; the net torque is the sum of these. Remember that the torque must be calculated about a specific point, so let's calculate the torque around the origin at $x = 0$.

Referring to Figure 7.20a, the torque τ_1 due to the weight of the left particle is

$$\tau_1 = -x_1 w_1 = -x_1 m_1 g$$

We put in the minus sign because the torque is negative—it tries to rotate the mass in a *clockwise* direction around the origin. Similarly, the torque τ_2 due to the right particle is

$$\tau_2 = -x_2 w_2 = -x_2 m_2 g$$

The net torque due to gravity is then

$$\tau_{net} = \tau_1 + \tau_2 = -(x_1 m_1 + x_2 m_2)g$$

In Figure 7.20b, we calculate the torque by considering the entire weight $\vec{w}$ as acting at the center of gravity. The (as yet unknown) position of the center of gravity is x_{cg}, so we have

$$\tau = -x_{cg} w = -x_{cg} m_{total} g = -x_{cg}(m_1 + m_2)g$$

This torque is also negative because it tries to rotate the entire dumbbell clockwise.

The torques calculated in these two ways must be the same, so we have

$$-x_{cg}(m_1 + m_2)g = -(x_1 m_1 + x_2 m_2)g$$

which we can solve to find that the center-of-gravity position is

$$x_{cg} = \frac{x_1 m_1 + x_2 m_2}{m_1 + m_2} \tag{7.8}$$

The following Tactics Box shows how Equation 7.8 can be generalized to find the center of gravity of *any* number of particles. If the particles don't all lie along the x-axis, then we'll also need to find the y-coordinate of the center of gravity.

(MP)   **TACTICS BOX 7.1 Finding the center of gravity**   ✎ Exercise 12

❶ Choose an origin for your coordinate system. You can choose any convenient point as the origin.
❷ Determine the coordinates $(x_1, y_1), (x_2, y_2), (x_3, y_3), \ldots$ for the particles of mass $m_1, m_2, m_3, \ldots$ respectively.
❸ The x-coordinate of the center of gravity is

$$x_{cg} = \frac{x_1 m_1 + x_2 m_2 + x_3 m_3 + \cdots}{m_1 + m_2 + m_3 + \cdots} \tag{7.9}$$

❹ Similarly, the y-coordinate of the center of gravity is

$$y_{cg} = \frac{y_1 m_1 + y_2 m_2 + y_3 m_3 + \cdots}{m_1 + m_2 + m_3 + \cdots} \tag{7.10}$$

Because the center of gravity depends on products such as $x_1 m_1$, objects with large masses count more heavily than objects with small masses. Consequently, **the center of gravity tends to lie closer to the heavier objects or particles** that make up the entire object.

EXAMPLE 7.5 Where should the dumbbell be lifted?

A 1.0-m-long dumbbell has a 10 kg mass on the left and a 5.0 kg mass on the right. Find the position of the center of gravity, the point where the dumbbell should be lifted in order to remain balanced.

PREPARE We'll treat the two masses as point particles separated by a massless rod. Then we can use the steps from Tactics Box 7.1 to find the center of gravity. Let's choose the origin to be at the position of the 10 kg mass on the left, making $x_1 = 0$ m and $x_2 = 1.0$ m. Because the dumbbell masses lie on the x-axis, the y-coordinate of the center of gravity must also lie on the x-axis. Thus we only need to solve for the x-coordinate of the center of gravity.

SOLVE The x-coordinate of the center of gravity is found from Equation 7.9:

$$x_{cg} = \frac{x_1 m_1 + x_2 m_2}{m_1 + m_2} = \frac{(0\text{ m})(10\text{ kg}) + (1.0\text{ m})(5.0\text{ kg})}{(10\text{ kg}) + (5.0\text{ kg})}$$

$$= 0.33\text{ m}$$

The center of gravity is 0.33 m from the 10 kg mass, or, equivalently, 0.17 m left of the center of the bar.

ASSESS The position of the center of gravity is closer to the greater mass. This is in agreement with our general statement above that the center of gravity tends to lie closer to the heavier particles.

The center of gravity of an extended object can often be found by considering the object as made up of pieces, each of whose mass and center of gravity is known or can be found. Then the coordinates of the entire object's center of gravity are given by Equations 7.9 and 7.10, with (x_1, y_1), (x_2, y_2), (x_3, y_3), . . . the coordinates of the centers of gravity of each piece and m_1, m_2, m_3, . . . their masses.

This method is widely used in biomechanics and kinesiology to calculate the center of gravity of the human body. Figure 7.21 shows how the body can be considered to be made of several segments, each of whose mass and center of gravity has been measured. The numbers shown are appropriate for a man with a total mass of 80 kg. For a given posture the positions of the segments and their centers of gravity can be found, and thus the whole-body center of gravity from Equations 7.9 and 7.10 (and a third equation for the z-coordinate). Example 7.6 explores a simplified version of this method.

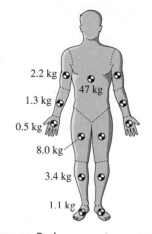

FIGURE 7.21 Body segment masses and centers of gravity.

EXAMPLE 7.6 Finding the center of gravity of a gymnast

A gymnast performing on the rings holds himself in the pike position. Figure 7.22 shows how we can consider his body to be made up of two segments whose masses and center-of-gravity positions are shown. The upper segment includes his head, trunk, and arms, while the lower segment consists of his legs. Locate the overall center of gravity of the gymnast.

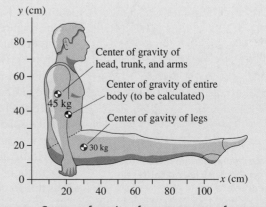

FIGURE 7.22 Centers of gravity of two segments of a gymnast.

Continued

PREPARE From Figure 7.22 we can find the x- and y-coordinates of the segment centers of gravity. We have

$$x_{\text{trunk}} = 15 \text{ cm}; \quad y_{\text{trunk}} = 50 \text{ cm};$$
$$x_{\text{legs}} = 30 \text{ cm}; \quad y_{\text{legs}} = 20 \text{ cm}$$

SOLVE The x- and y-coordinates of the center of gravity are given by Equations 7.9 and 7.10:

$$x_{\text{cg}} = \frac{x_{\text{trunk}}m_{\text{trunk}} + x_{\text{legs}}m_{\text{legs}}}{m_{\text{trunk}} + m_{\text{legs}}}$$
$$= \frac{(15 \text{ cm})(45 \text{ kg}) + (30 \text{ cm})(30 \text{ kg})}{45 \text{ kg} + 30 \text{ kg}} = 21 \text{ cm}$$

and

$$y_{\text{cg}} = \frac{y_{\text{trunk}}m_{\text{trunk}} + y_{\text{legs}}m_{\text{legs}}}{m_{\text{trunk}} + m_{\text{legs}}}$$
$$= \frac{(50 \text{ cm})(45 \text{ kg}) + (20 \text{ cm})(30 \text{ kg})}{45 \text{ kg} + 30 \text{ kg}} = 38 \text{ cm}$$

ASSESS The center of gravity position of the entire body, shown in Figure 7.22, is closer to that of the heavier trunk segment than to that of the lighter legs. It also lies along a line connecting the two segment centers of gravity, just as it would for the center of gravity of two point particles. Note also that the gymnast's hands—the pivot point—must lie directly below his center of gravity. Otherwise he would rotate forward or backward.

STOP TO THINK 7.4 The balls shown below are connected by very lightweight rods pivoted at the point indicated by a dot. The rod lengths are all equal except for A, which is twice as long. Rank in order, from least to greatest, the magnitudes of the gravitational torques about the pivot for arrangements A to E.

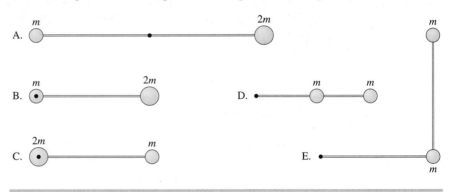

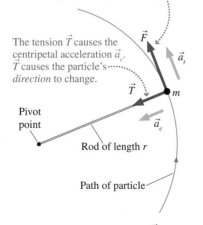

The tangential force $\vec{F}$ causes the tangential acceleration $\vec{a}_t$. $\vec{F}$ causes the particle's *speed* to change.

The tension $\vec{T}$ causes the centripetal acceleration $\vec{a}_c$. $\vec{T}$ causes the particle's *direction* to change.

Pivot point

Rod of length r

Path of particle

FIGURE 7.23 A tangential force $\vec{F}$ exerts a torque on the particle and causes an angular acceleration.

7.4 Rotational Dynamics and Moment of Inertia

We've found that torque is the rotational equivalent of force. Now we need to learn what torque *does*. Let's start by examining a *single particle* subject to a torque. Figure 7.23 shows a particle of mass m attached to a lightweight, rigid rod of length r that constrains the particle to move in a circle. The particle is subject to two forces. Because it's moving in a circle there must be a centripetal force— here, the tension $\vec{T}$ from the rod—directed toward the center of the circle. As we learned in Chapter 6, this is the force responsible for changing the *direction* of the particle's velocity. The acceleration associated with this change in the particle's velocity is the centripetal acceleration $\vec{a}_c$.

But the particle in Figure 7.23 is also subject to a second force $\vec{F}$, one directed *tangentially* to the circle. Because this force points in the direction of the particle's motion, it is the force responsible for changing the *speed* of the particle. Thus this force causes a tangential acceleration $\vec{a}_t$.

To make the connection between torque and angular motion, let's begin by applying Newton's second law in the direction tangent to the circle. We have

$$a_t = \frac{F}{m} \qquad (7.11)$$

Now the tangential and angular accelerations are related by $a_t = \alpha r$, so we can write Equation 7.11 as $\alpha r = F/m$, or

$$\alpha = \frac{F}{mr} \qquad (7.12)$$

We can write the angular acceleration in terms of the *torque* exerted by F (the tension force exerts no torque because it is directed along the radial line). Because force $\vec{F}$ is perpendicular to the radial line, it exerts torque

$$\tau - rF$$

With this relation between F and τ, we can write Equation 7.12 as

$$\alpha = \frac{\tau}{mr^2} \qquad (7.13)$$

In Section 7.2 we asked the question, "What do Newton's laws have to tell us about rotational motion?" We now have the first part of an answer: **A torque causes an angular acceleration.** This is the rotational equivalent of our prior discovery, for motion along a line, that a force causes an acceleration. Now all that remains is to expand this idea from a single particle to an extended object.

Newton's Second Law for Rotational Motion

Figure 7.24 shows a rigid body that undergoes rotation about a fixed and unmoving axis. According to the rigid-body model, we can think of the object as consisting of particles with masses $m_1, m_2, m_3, \ldots$ at fixed distances $r_1, r_2, r_3, \ldots$ from the axis. Suppose forces $\vec{F}_1, \vec{F}_2, \vec{F}_3, \ldots$ act on these particles. These forces exert torques around the rotation axis, so the object will undergo an angular acceleration α. Because all the particles that make up the object rotate together, each particle has this *same* angular acceleration α. Rearranging Equation 7.13 slightly, we can write the torques on the particles as

$$\tau_1 = m_1 r_1^2 \alpha \qquad \tau_2 = m_2 r_2^2 \alpha \qquad \tau_3 = m_3 r_3^2 \alpha$$

and so on for every particle in the object. If we add up all these torques, the *net* torque on the object is

$$\tau_{\text{net}} = \tau_1 + \tau_2 + \tau_3 + \cdots = m_1 r_1^2 \alpha + m_2 r_2^2 \alpha + m_3 r_3^2 \alpha + \cdots$$

$$= \alpha(m_1 r_1^2 + m_2 r_2^2 + m_3 r_3^2 + \cdots) = \alpha \sum m_i r_i^2 \qquad (7.14)$$

By factoring α out of the sum, we're making explicit use of the fact that every particle in a rotating rigid body has the *same* angular acceleration α.

The quantity $\sum mr^2$ in Equation 7.14, which is the proportionality constant between angular acceleration and net torque, is called the object's **moment of inertia** I:

$$I = m_1 r_1^2 + m_2 r_2^2 + m_3 r_3^2 + \cdots = \sum m_i r_i^2 \qquad (7.15)$$

Moment of inertia of a collection of particles

This large granite ball, with a mass of 26,400 kg, floats with nearly zero friction on a thin layer of pressurized water. As the girl pushes on it, she exerts a torque on the ball. Because of the ball's large moment of inertia, the torque causes only a small angular acceleration, so the ball's angular velocity increases only slowly as she pushes.

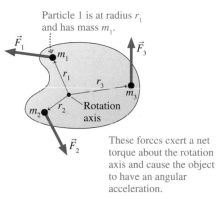

Particle 1 is at radius r_1 and has mass m_1.

These forces exert a net torque about the rotation axis and cause the object to have an angular acceleration.

FIGURE 7.24 The forces on a rigid body exert a torque about the rotation axis.

Activ Physics 7.6

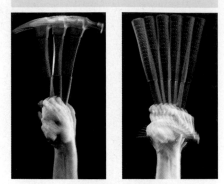

Hammering home inertia Most of the mass of a hammer is in its head, so the hammer's moment of inertia is large when calculated about an axis passing through the end of the handle (far from the head), but small when calculated about an axis passing through the head itself. You can *feel* this difference by attempting to wave a hammer back and forth about the handle end and the head end. It's much harder to do about the handle end because the large moment of inertia keeps the angular acceleration small.

The units of moment of inertia are those of mass times distance squared, or $kg \cdot m^2$. An object's moment of inertia, like torque, *depends on the axis of rotation.* Once the axis is specified, allowing the values of r_1, r_2, r_3, ... to be determined, the moment of inertia *about that axis* can be calculated from Equation 7.15.

NOTE ▶ The word *moment* in *moment of inertia* and *moment arm* has nothing to do with time. These terms stem from the Latin *momentum,* meaning "motion." ◀

Substituting the moment of inertia into Equation 7.14 puts the final piece of the puzzle into place, giving us the fundamental equation for rigid-body dynamics:

Newton's second law for rotation An object that experiences a net torque τ_{net} about the axis of rotation undergoes an angular acceleration

$$\alpha = \frac{\tau_{net}}{I} \qquad (7.16)$$

LINEAR p. 38   INVERSE p. 118

where I is the moment of inertia of the object *about the rotation axis.*

In practice we often write $\tau_{net} = I\alpha$, but Equation 7.16 better conveys the idea that **a net torque is the cause of angular acceleration.** In the absence of a net torque ($\tau_{net} = 0$), the object either does not rotate ($\omega = 0$) or rotates with *constant angular velocity* ($\omega = $ constant).

Interpreting the Moment of Inertia

Before rushing to calculate moments of inertia, let's get a better understanding of its meaning. First, notice that **moment of inertia is the rotational equivalent of mass.** It plays the same role in Equation 7.16 as does mass m in the now-familiar $\vec{a} = \vec{F}_{net}/m$. Recall that objects with larger mass have a larger *inertia,* meaning that they're harder to accelerate. Similarly, an object with a larger moment of inertia is harder to get rotating: It takes a larger torque to spin up an object with a larger moment of inertia than an object with a smaller moment of inertia. The fact that *moment of inertia* retains the word *inertia* reminds us of this.

But why does the moment of inertia depend on the distances r from the rotation axis? Think about trying to start a merry-go-round from rest, as shown in Figure 7.25. By pushing on the rim of the merry-go-round, you exert a torque on it and its angular velocity begins to increase. If your friends sit at the rim of the merry-go-round, as in Figure 7.25a, their distances r from the axle are large. The moment of inertia is large, according to Equation 7.15, and it will be difficult to get the merry-go-round rotating. If, however, your friends sit near the axle, as in Figure 7.25b, then r and the moment of inertia are small. You'll find it's much easier to get the merry-go-round going.

Thus an object's moment of inertia depends not only on the object's mass but on *how the mass is distributed* around the rotation axis. This is well known to bicycle racers. Every time a cyclist accelerates, she has to "spin up" the wheels and tires. The larger the moment of inertia, the more effort it takes and the slower her acceleration. For this reason, racers use the lightest possible tires, and they put those tires on wheels that have been designed to keep the mass as close as possible to the center without sacrificing the necessary strength and rigidity.

Table 7.3 summarizes the analogies between linear and rotational dynamics.

(a) Mass concentrated around the rim.

(b) Mass concentrated at the center.

Larger moment of inertia, harder to get rotating.

Smaller moment of inertia, easier to get rotating.

FIGURE 7.25 Moment of inertia depends both on the mass and how the mass is distributed.

TABLE 7.3 Rotational and linear dynamics

Rotational dynamics		Linear dynamics	
Torque	τ_{net}	Force	$\vec{F}_{net}$
Moment of inertia	I	Mass	m
Angular acceleration	α	Acceleration	$\vec{a}$
Second law	$\alpha = \tau_{net}/I$	Second law	$\vec{a} = \vec{F}_{net}/m$

EXAMPLE 7.7 **Calculating the moment of inertia**

Your friend is creating an abstract sculpture that consists of three small, heavy spheres attached by very lightweight 10-cm-long rods as shown in Figure 7.26. The spheres have masses $m_1 = 1.0$ kg, $m_2 = 1.5$ kg, and $m_3 = 1.0$ kg. What is the object's moment of inertia if it is rotated about axis a? About axis b?

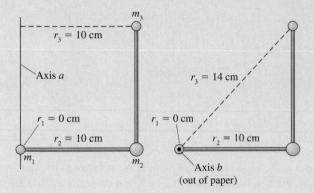

FIGURE 7.26 Three point particles separated by lightweight rods.

PREPARE We'll use Equation 7.15 for the moment of inertia:

$$I = m_1 r_1^2 + m_2 r_2^2 + m_3 r_3^2$$

In this expression, r_1, r_2, and r_3 are the distances of each particle from the axis of rotation, so they depend on the axis chosen. Particle 1 lies directly on both axes, so $r_1 = 0$ cm in both cases.

Particle 2 lies 10 cm (0.10 m) from both axes. Particle 3 is 10 cm from axis a, but farther from axis b. We can find r_3 for axis b by using the Pythagorean theorem, which gives $r_3 = 14$ cm. These distances are indicated in the figure.

SOLVE For each axis, we can prepare a table of the values of r, m, and mr^2 for each particle, then add the values of mr^2. For axis a we have

Particle	r	m	mr^2
1	0 m	1.0 kg	0 kg · m²
2	0.10 m	1.5 kg	0.015 kg · m²
3	0.10 m	1.0 kg	0.010 kg · m²
			$I_a = 0.025$ kg · m²

For axis b we have

Particle	r	m	mr^2
1	0 m	1.0 kg	0 kg · m²
2	0.10 m	1.5 kg	0.015 kg · m²
3	0.14 m	1.0 kg	0.020 kg · m²
			$I_b = 0.035$ kg · m²

ASSESS We've already noted that the moment of inertia of an object is higher when its mass is distributed farther from the axis of rotation. Here, m_3 is farther from axis b than from axis a, leading to a higher moment of inertia about that axis.

The Moments of Inertia of Common Shapes

Newton's second law for rotational motion is easy to write, but we can't make use of it without knowing an object's moment of inertia. Unlike mass, we can't measure moment of inertia by putting an object on a scale. And while we can guess that the center of gravity of a symmetrical object is at the physical center of the object, we can *not* guess the moment of inertia of even a simple object.

For an object consisting of only a few point particles connected by massless rods, we can use Equation 7.15 to directly calculate I. But such an object is pretty unrealistic. All real objects are made up of solid material that is itself comprised of countless atoms. To calculate the moment of inertia of even a simple object requires integral calculus and is beyond the scope of this text. A short list of common moments of inertia is given in Table 7.4 on the next page. We use a capital M for the total mass of an extended object.

TABLE 7.4 Moments of inertia of objects with uniform density and total mass M

Object and axis	Picture	I	Object and axis	Picture	I
Thin rod, about center		$\frac{1}{12}ML^2$	Cylinder or disk, about center		$\frac{1}{2}MR^2$
Thin rod (of any cross section), about end		$\frac{1}{3}ML^2$	Cylindrical hoop, about center		MR^2
Plane or slab, about center		$\frac{1}{12}Ma^2$	Solid sphere, about diameter		$\frac{2}{5}MR^2$
Plane or slab, about edge		$\frac{1}{3}Ma^2$	Spherical shell, about diameter		$\frac{2}{3}MR^2$

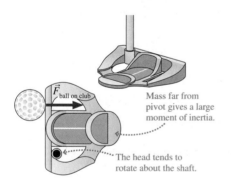

Novel golf clubs The latest craze in golf putters is heads with high moment of inertia. When the putter hits the ball, the ball—by Newton's third law—exerts a force on the putter and thus exerts a torque that causes the head of the putter to rotate around the shaft. Any rotation while the putter is still in contact with the ball will affect the ball's direction. If the putter's mass is largely placed rather far from the shaft (the rotation axis), the moment of inertia about the shaft can be greatly increased. The large moment of inertia of the head will keep its angular acceleration small—reducing unwanted rotation and allowing a truer putt.

We can make some general observations about the moments of inertia in Table 7.4. For instance, the cylindrical hoop is composed of particles that are all the same distance R from the axis. Thus each particle of mass m makes the *same* contribution mR^2 to the hoop's moment of inertia. Adding up all these contributions gives

$$I = m_1R^2 + m_2R^2 + m_3R^2 + \cdots = (m_1 + m_2 + m_3 + \cdots)R^2 = MR^2$$

as given in the table. The solid cylinder of the same mass and radius has a *lower* moment of inertia than the hoop because much of the cylinder's mass is nearer its center. In the same way we can see why a slab rotated about its center has a lower moment of inertia than the same slab rotated about its edge: in the latter case, some of the mass is twice as far from the axis as the farthest mass in the former case. Those particles contribute *four times* as much to the moment of inertia, leading to an overall larger moment of inertia for the slab rotated about its edge.

STOP TO THINK 7.5 Four very lightweight disks of equal radii each have three identical heavy marbles glued to them as shown. Rank in order, from largest to smallest, the moments of inertia of the disks about the indicated axis.

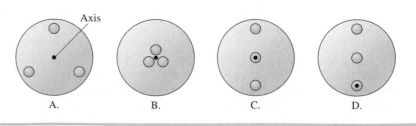

7.5 Using Newton's Second Law for Rotation

In this section we'll look at several examples of rotational dynamics for rigid bodies that rotate about a *fixed axis*. The restriction to a fixed axis avoids complications that arise for an object undergoing a combination of rotational and translational motion. The problem-solving strategy for rotational dynamics is very similar to that of linear dynamics in Chapter 5.

Activ
Physics   7.8–7.10

(MP) **PROBLEM-SOLVING STRATEGY 7.1** **Rotational dynamics problems**

PREPARE Model the object as a simple shape. Draw a pictorial representation to clarify the situation, define coordinates and symbols, and list known information.

■ Identify the axis about which the object rotates.
■ Identify forces and determine their distance from the axis.
■ Identify the torques caused by the forces and the signs of the torques.

SOLVE The mathematical representation is based on Newton's second law for rotational motion

$$\tau_{net} = I\alpha \quad \text{or} \quad \alpha = \frac{\tau_{net}}{I}$$

■ Find the moment of inertia either by direct calculation using Equation 7.15 or from Table 7.4 for common shapes of objects.
■ Use rotational kinematics to find angular positions and velocities.

ASSESS Check that your result has the correct units, is reasonable, and answers the question.

EXAMPLE 7.8 Starting an airplane engine

The engine in a small airplane is specified to have a torque of 500 N · m. This engine drives a 2.0-m-long, 40 kg single-blade propeller. On start-up, how long does it take the propeller to reach 2000 rpm?

PREPARE The propeller can be modeled as a rod that rotates about its center. The engine exerts a torque on the propeller. Figure 7.27 shows the propeller and the rotation axis.

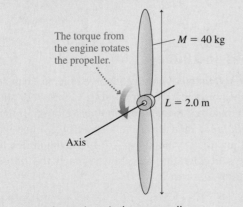

The torque from the engine rotates the propeller.

$M = 40$ kg

$L = 2.0$ m

Axis

FIGURE 7.27 A rotating airplane propeller.

SOLVE The moment of inertia of a rod rotating about its center is found from Table 7.4:

$$I = \tfrac{1}{12}ML^2 = \tfrac{1}{12}(40 \text{ kg})(2.0 \text{ m})^2 = 13.3 \text{ kg} \cdot \text{m}^2$$

The 500 N · m torque of the engine causes an angular acceleration

$$\alpha = \frac{\tau}{I} = \frac{500 \text{ N} \cdot \text{m}}{13.3 \text{ kg} \cdot \text{m}^2} = 37.5 \text{ rad/s}^2$$

The time needed to reach $\omega_f = 2000$ rpm $= 33.3$ rev/s $= 209$ rad/s is

$$\Delta t = \frac{\Delta\omega}{\alpha} = \frac{\omega_f - \omega_i}{\alpha} = \frac{209 \text{ rad/s} - 0 \text{ rad/s}}{37.5 \text{ rad/s}^2} = 5.6 \text{ s}$$

ASSESS We've assumed a constant angular acceleration, which is reasonable for the first few seconds while the propeller is still turning slowly. Eventually, air resistance and friction will cause opposing torques and the angular acceleration will decrease. At full speed, the negative torque due to air resistance and friction cancels the torque of the engine. Then $\tau_{net} = 0$ and the propeller turns at *constant* angular velocity with no angular acceleration.

EXAMPLE 7.9 Angular acceleration of a falling pole

A 7.0-m-tall telephone pole with a mass of 260 kg has just been placed in the ground. Before the wires can be connected, the pole is hit by lightning at its base, nearly severing the pole from its base. The pole begins to fall, rotating about the part still connected to the base. Estimate the pole's angular acceleration when it has fallen by 25° from the vertical.

PREPARE The situation is shown in Figure 7.28, where we define our symbols and list the known information. Two forces are acting on the pole: the pole's weight $\vec{w}$, which acts at the center of gravity, and the force of the base on the pole

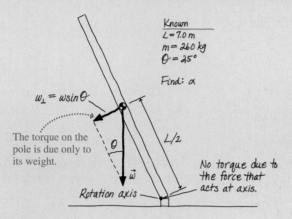

The torque on the pole is due only to its weight.

Known
$L = 7.0\,m$
$m = 260\,kg$
$\theta = 25°$

Find: α

No torque due to the force that acts at axis.

FIGURE 7.28 A falling telephone pole undergoes an angular acceleration due to a gravitational torque.

(not shown). This second force exerts no torque because it acts at the axis of rotation. The torque on the pole is thus due only to gravity. From the figure we see that this torque tends to rotate the pole in a counterclockwise direction, so the torque is positive.

SOLVE We'll model the pole as a uniform thin rod rotating about one end. Its center of gravity is at its center, a distance $L/2$ from the axis. You can see from the figure that the perpendicular component of $\vec{w}$ is $w_\perp = w\sin\theta$. Thus the torque due to gravity is

$$\tau_{net} = \left(\frac{L}{2}\right)w_\perp = \left(\frac{L}{2}\right)w\sin\theta = \frac{mgL}{2}\sin\theta$$

From Table 7.4, the moment of inertia of a thin rod rotated about its end is $I = \frac{1}{3}mL^2$. Thus from Newton's second law for rotational motion the angular acceleration is

$$\alpha = \frac{\tau_{net}}{I} = \frac{\frac{1}{2}mgL\sin\theta}{\frac{1}{3}mL^2} = \frac{3g\sin\theta}{2L}$$

$$= \frac{3(9.8\ \text{m/s}^2)\sin 25°}{2(7.0\ \text{m})} = 0.9\ \text{rad/s}^2$$

ASSESS The answer is given to only one significant figure because the problem asked for an *estimate* of the angular acceleration. This is usually a hint that you should make some simplifying assumptions, as we did here in modeling the pole as a thin rod.

CONCEPTUAL EXAMPLE 7.2 Balancing a meter stick

You've probably tried balancing a rod-shaped object vertically on your fingertip. If the object is very long, like a meter stick or a baseball bat, it's not too hard. But if it's short, like a pencil, it's almost impossible. Why is this?

REASON Suppose you've managed to balance a vertical stick on your fingertip, but then it starts to fall. You'll need to quickly adjust your finger to bring the stick back into balance. As Example 7.9 showed, the angular acceleration α of a thin rod is *inversely proportional* to L. Thus a long object like a

meter stick topples much more slowly than a short one like a pencil. Your reaction time is fast enough to correct for a slowly falling meter stick but not for a rapidly falling pencil.

ASSESS If we double the length of a rod, its mass doubles and its center of gravity is twice as high, so the gravitational torque τ on it is four times as much. But because a rod's moment of inertia is $I = \frac{1}{3}ML^2$, the longer rod's moment of inertia will be *eight* times greater so the angular acceleration will be only half as large.

Constraints Due to Ropes and Pulleys

Many important applications of rotational dynamics involve objects, such as pulleys, that are connected via ropes or belts to other objects. Figure 7.29 shows a rope passing over a pulley and connected to an object in linear motion. If the rope turns on the pulley *without slipping,* then the rope's speed v_{rope} must exactly match the speed of the rim of the pulley, which is $v_{rim} = \omega R$. If the pulley has an angular acceleration, the rope's acceleration a_{rope} must match the *tangential* acceleration of the rim of the pulley, $a_t = \alpha R$.

The object attached to the other end of the rope has the same speed and acceleration as the rope. Consequently, the object must obey the constraints

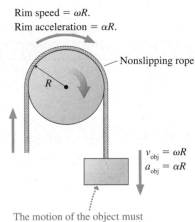

Rim speed = ωR.
Rim acceleration = αR.

Nonslipping rope

$$v_{obj} = \omega R$$
$$a_{obj} = \alpha R \tag{7.17}$$

Motion constraints for an object connected to a pulley of radius R by a nonslipping rope

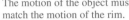

$v_{obj} = \omega R$
$a_{obj} = \alpha R$

These constraints are very similar to the acceleration constraints introduced in Chapter 5 for two objects connected by a string or rope.

NOTE ▶ The constraints are given as magnitudes. Specific problems will require you to specify signs that depend on the direction of motion and on the choice of coordinate system. ◀

The motion of the object must match the motion of the rim.

FIGURE 7.29 The rope's motion must match the motion of the rim of the pulley.

EXAMPLE 7.10 Time for a bucket to fall

Josh has just raised a 2.5 kg bucket of water using a well's winch when he accidentally lets go of the handle. The rope is wrapped around a 3.0 kg, 4.0-cm-diameter cylinder, which rotates on an axle through the center. The bucket is released from rest 4.0 m above the water level of the well. How long does it take to reach the water?

PREPARE Assume the rope is massless and does not slip. Figure 7.30a gives a visual overview of the falling bucket. Figure 7.30b shows the free-body diagrams for the cylinder and the bucket. The rope tension exerts an upward force on the bucket and a downward force on the outer edge of the cylinder. The rope is massless, so these two tension forces act as if they are an action/reaction pair with $T_b = T_c$. Because the tensions have equal magnitudes, we'll call both their magnitudes T.

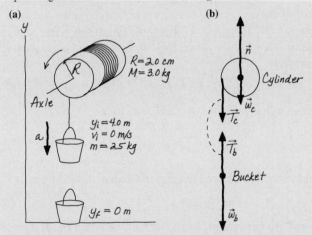

(a)

$R = 2.0$ cm
$M = 3.0$ kg

Axle

$y_i = 4.0$ m
$v_i = 0$ m/s
$m = 2.5$ kg

$y_f = 0$ m

(b)

$\vec{n}$
Cylinder
$\vec{w_c}$
$\vec{T_c}$
$\vec{T_b}$
Bucket
$\vec{w_b}$

FIGURE 7.30 Visual overview of a falling bucket.

SOLVE Newton's second law applied to the linear motion of the bucket is

$$ma_y = T - mg$$

where, as usual, the y-axis points upward. What about the cylinder? There is a normal force $\vec{n}$ on the cylinder due to the axle and the weight of the cylinder $\vec{w_c}$. However, neither of these forces exerts a torque because each passes through the rotation

axis. The only torque comes from the rope tension. The moment arm for the tension is $r_\perp = R$, and the torque is positive because the rope turns the cylinder counterclockwise. Thus $\tau_{rope} = TR$ and Newton's second law for the rotational motion is

$$\alpha = \frac{\tau_{net}}{I} = \frac{TR}{\frac{1}{2}MR^2} = \frac{2T}{MR}$$

The moment of inertia of a cylinder rotating about a center axis was taken from Table 7.4.

The last piece of information we need is the constraint due to the fact that the rope doesn't slip. Equation 7.17 relates only the magnitudes of the linear and angular accelerations, but in this problem α is positive (counterclockwise acceleration), while a_y is negative (downward acceleration). Hence

$$a_y = -\alpha R$$

Using α from the cylinder's equation in the constraint, we find

$$a_y = -\alpha R = -\frac{2T}{MR}R = -\frac{2T}{M}$$

Thus the tension is $T = -\frac{1}{2}Ma_y$. If we use this value of the tension in the bucket's equation, we can solve for the acceleration:

$$ma_y = -\frac{1}{2}Ma_y - mg$$

$$a_y = -\frac{g}{(1 + M/2m)} = -6.1 \text{ m/s}^2$$

The time to fall through $\Delta y = y_f - y_i = -4.0$ m is found from kinematics:

$$\Delta y = \frac{1}{2}a_y(\Delta t)^2$$

$$\Delta t = \sqrt{\frac{2\Delta y}{a_y}} = \sqrt{\frac{2(-4.0 \text{ m})}{-6.1 \text{ m/s}^2}} = 1.1 \text{ s}$$

ASSESS The expression for the acceleration gives $a_y = -g$ if $M = 0$. This makes sense because the bucket would be in free fall if there were no cylinder. When the cylinder has mass, the downward force of gravity on the bucket has to accelerate the bucket *and* spin the cylinder. Consequently, the acceleration is reduced and the bucket takes longer to fall.

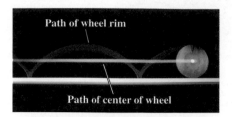

FIGURE 7.31 The trajectories of the center of a wheel and of a point on the rim are seen in a time-exposure photograph.

7.6 Rolling Motion

Rolling is a *combination motion* in which an object rotates about an axis that is moving along a straight-line trajectory. For example, Figure 7.31 is a time-exposure photo of a rolling wheel with one light bulb on the axis and a second light bulb at the edge. The axis light moves straight ahead, but the edge light follows a curve called a *cycloid*. Let's see if we can understand this interesting motion. We'll consider only objects that roll without slipping.

Figure 7.32 shows a round object—a wheel or a sphere—that rolls forward exactly one revolution. The point that had been on the bottom follows the cycloid, the curve you saw in Figure 7.31, to the top and back to the bottom. The overall position of the object is measured by the position x_{cm} of its center of mass, located at the object's center. *Because the object doesn't slip,* the center of mass moves forward exactly one circumference: $\Delta x_{cm} = 2\pi R$.

NOTE ▶ We'll use center of mass rather than center of gravity in this section because we're not concerned with gravitational torque. ◀

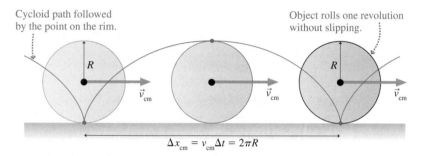

FIGURE 7.32 An object rolling through one revolution.

We can also write the distance traveled in terms of the velocity of the center of mass: $\Delta x_{cm} = v_{cm} \Delta t$. But Δt, the time it takes the object to make one complete revolution, is nothing other than the rotation period T. In other words, $\Delta x_{cm} = v_{cm} T$.

These two expressions for Δx_{cm} come from two perspectives on the motion, one looking at the rotation and the other at the translation of the center of mass. But it's the same distance no matter how you look at it, so these two expressions must be equal. Consequently,

$$\Delta x_{cm} = 2\pi R = v_{cm} T \tag{7.18}$$

If we divide by T, we can write the center-of-mass velocity as

$$v_{cm} = \frac{2\pi}{T} R \tag{7.19}$$

But $2\pi/T$ is the angular velocity ω, as you learned in Chapter 7, leading to

$$v_{cm} = \omega R \tag{7.20}$$

Equation 7.20 is the **rolling constraint,** the basic link between translation and rotation for objects that roll without slipping.

NOTE ▶ The rolling constraint is equivalent to Equation 7.17 for the speed of a rope that doesn't slip as it passes over a pulley. ◀

We can find the velocity for any point on a rolling object by adding the velocity of that point when the object is in pure translation, without rolling, to the velocity of the point when the object is in pure rotation, without translating. Figure 7.33 shows how the velocity vectors at the top, center, and bottom of a rotating wheel are found in this way.

Ancient movers The great stone *moai* of Easter Island were moved as much as 16 km from a quarry to their final positions. Archeologists believe that one possible method of moving these 14 ton statues was to place them on rollers. One disadvantage of this method is that the statues, placed on top of the rollers, move twice as fast as the rollers themselves. Thus rollers are continuously left behind, and have to be carried back to the front and reinserted. Sadly, the indiscriminate cutting of trees for moving *moai* may have hastened the demise of this island civilization.

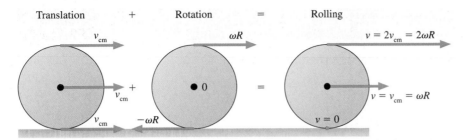

FIGURE 7.33 Rolling is a combination of translation and rotation.

Thus the point at the top of the wheel has a forward speed of v_{cm} due to its translational motion plus a forward speed of $\omega R = v_{cm}$ due to its rotational motion. The speed of a point at the top of a wheel is then $2v_{cm} = 2\omega R$, or *twice* the speed of its center of mass. On the other hand, the point at the bottom of the wheel, where it touches the ground, still has a forward speed of v_{cm} due to its translational motion. But its velocity due to rotation points *backward* with a magnitude of $\omega R = v_{cm}$. Adding these, we find that the velocity of this lowest point is *zero*. In other words, **the point on the bottom of a rolling object is instantaneously at rest.**

Although this seems surprising, it is really what we mean by "rolling without slipping." If the bottom point had a velocity, it would be moving horizontally relative to the surface. In other words, it would be slipping or sliding across the surface. To roll without slipping, the bottom point, the point touching the surface, must be at rest.

EXAMPLE 7.11 Rotating your tires

The diameter of your tires is 0.60 m. You take a 60 mile trip at a speed of 45 mph.

a. During this trip, what was your tires' angular speed?
b. How many times did they revolve?

PREPARE The angular speed is related to the speed of a wheel's center by Equation 7.20, $v_{cm} = \omega R$. Because the center of the wheel turns on an axle fixed to the car, the speed v_{cm} of the wheel's center is the same as that of the car. We prepare by converting the car's speed to SI units to get

$$v_{cm} = (45 \text{ mph}) \times \left(0.447 \frac{\text{m/s}}{\text{mph}}\right) = 20 \text{ m/s}$$

Once we know the angular speed, we can find the number of times the tires turned from the rotational-kinematic equation $\Delta\theta = \omega \Delta t$. We'll need to find the time traveled Δt from $v_{cm} = \Delta x/\Delta t$.

SOLVE a. From Equation 7.20 we have

$$\omega = \frac{v_{cm}}{R} = \frac{20 \text{ m/s}}{0.30 \text{ m}} = 67 \text{ rad/s}$$

b. The time of the trip is

$$\Delta t = \frac{\Delta x}{v_{cm}} = \frac{60 \text{ mi}}{45 \text{ mi/h}} = 1.33 \text{ h} \times \frac{3600 \text{ s}}{1 \text{ h}} = 4800 \text{ s}$$

Thus the total angle through which the tires turn is

$$\Delta\theta = \omega \Delta t = (67 \text{ rad/s})(4800 \text{ s}) = 3.2 \times 10^5 \text{ rad}$$

Because each turn of the wheel is 2π rad, the number of turns is

$$\frac{3.2 \times 10^5 \text{ rad}}{2\pi \text{ rad}} = 51,000 \text{ turns}$$

ASSESS You can see that your tires rotate roughly a thousand times per mile. During the lifetime of a tire, about 50,000 miles, it will rotate about 50 million times!

STOP TO THINK 7.6   A wheel rolls without slipping. Which is the correct velocity vector for point P on the wheel?

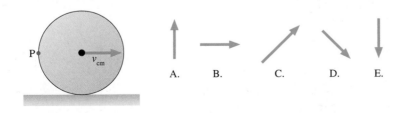

A.   B.   C.   D.   E.

SUMMARY

The goal of Chapter 7 has been to understand the physics of rotating objects.

GENERAL PRINCIPLES

Newton's Second Law for Rotational Motion

If a net torque τ_{net} acts on an object, the object will experience an angular acceleration given by $\alpha = \tau_{net}/I$, where I is the object's moment of inertia about the pivot point.

This law is analogous to Newton's second law for linear motion, $\vec{a} = \vec{F}_{net}/m$.

IMPORTANT CONCEPTS

Torque is the rotational analog of force. Just as a force causes an object to undergo a linear acceleration, torque causes an object to undergo an angular acceleration.

There are two interpretations of torque:

$$1.\ \tau = rF_\perp \qquad 2.\ \tau = r_\perp F$$

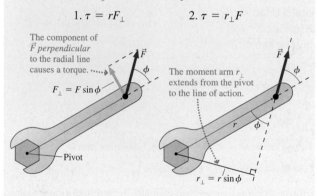

The component of $\vec{F}$ *perpendicular* to the radial line causes a torque.

$F_\perp = F\sin\phi$

Pivot

The moment arm $r_\perp$ extends from the pivot to the line of action.

$r_\perp = r\sin\phi$

Both interpretations give the same expression for the magnitude of the torque: $\tau = rF\sin\phi$

The moment of inertia is the rotational equivalent of mass. The larger the moment of inertia, the more difficult it is to get the object rotating. For an object made up of particles of mass $m_1, m_2, \ldots$ at distances $r_1, r_2, \ldots$ from the axis, the moment of inertia is

$$I = m_1 r_1^2 + m_2 r_2^2 + m_3 r_3^2 + \cdots = \sum mr^2$$

Angular and tangential acceleration

A particle moving in a circle has

- A velocity tangent to the circle
- A centripetal acceleration $\vec{a}_c$ directed towards the center of the circle.

If the particle's speed is increasing, it will also have

- A tangential acceleration $\vec{a}_t$ directed tangent to the circle.
- An angular acceleration α.

The angular and tangential accelerations are related by $a_t = \alpha r$.

For a **rigid body,** the angular velocity and angular acceleration is the same for every point on the object.

Centripetal acceleration

Angular velocity

Tangential acceleration

Velocity

Center of gravity

The **center of gravity** of an object is the point at which gravity can be considered as acting.

Gravity acts on each particle that makes up the object.

The object responds *as if* its entire weight acts at the center of gravity.

$\vec{w}$

The **position of the center of gravity** depends on the distances $x_1, x_2, \ldots$ of each particle of mass $m_1, m_2, \ldots$ from the origin:

$$x_{cg} = \frac{x_1 m_1 + x_2 m_2 + x_3 m_3 + \cdots}{m_1 + m_2 + m_3 + \cdots}$$

APPLICATIONS

Moments of inertia of common shapes

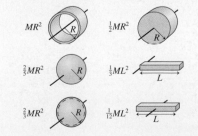

MR^2 $\frac{1}{2}MR^2$

$\frac{2}{5}MR^2$ $\frac{1}{3}ML^2$

$\frac{2}{3}MR^2$ $\frac{1}{12}ML^2$

Rotation about a fixed axis

When a net torque is applied to an object that rotates about a fixed axis, the object will undergo an **angular acceleration** given by

$$\alpha = \frac{\tau_{net}}{I}$$

If a rope unwinds from a pulley of radius R, the angular quantities of the pulley are related to the linear quantities of the rope as

$$a_{obj} = \alpha R \qquad v_{obj} = \omega R$$

Rolling motion

For an object that rolls without slipping

$$v_{cm} = \omega R$$

The velocity of a point at the top of the object is twice that of the center.

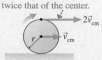

$2\vec{v}_{cm}$

$\vec{v}_{cm}$

For instructor-assigned homework, go to www.masteringphysics.com

Problems labeled INT integrate significant material from earlier chapters; BIO are of biological or medical interest.

Problem difficulty is labeled as I (straightforward) to IIIII (challenging).

QUESTIONS

Conceptual Questions

1. Figure Q7.1 shows four rotating wheels. What are the signs (+ or −) of the angular velocity and angular acceleration in each case?

Speeding up   Slowing down   Slowing down   Speeding up

FIGURE Q7.1

2. If you are using a wrench to loosen a very stubborn nut, you can make the job easier by using a "cheater pipe." This is a piece of pipe that slides over the handle of the wrench, as shown in Figure Q7.2, making it effectively much longer. Explain why this would help you loosen the nut.

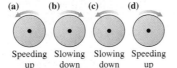

FIGURE Q7.2

3. Five forces are applied to a door, as seen from above in Figure Q7.3. For each force, is the torque about the hinge positive, negative, or zero?

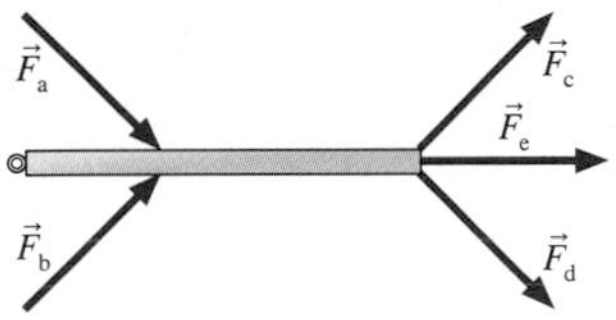

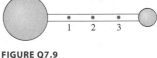

FIGURE Q7.3

4. A screwdriver with a very thick handle requires less force to operate than one with a very skinny handle. Explain why this is so.

5. If you have ever driven a truck, you likely found that it had a steering wheel with a larger diameter than that of a passenger car. Why is this?

6. A common type of door stop is a wedge made of rubber. Is such a stop more effective when jammed under the door near or far from the hinges? Why?

7. Suppose you are hanging from a tree branch. If you move out along the branch, farther away from the trunk, the branch will be more likely to break. Explain why this is so.

8. A student gives a quick push to a ball at the end of a massless, rigid rod, causing the ball to rotate clockwise in a *horizontal* circle as shown in Figure Q7.8. The rod's pivot is frictionless.
   a. As the student is pushing, is the torque about the pivot positive, negative, or zero?
   b. After the push has ended, what does the ball's angular velocity do? Steadily increase? Increase for a while, then hold steady? Hold steady? Decrease for a while, then hold steady? Steadily decrease? Explain.
   c. Right after the push has ended, is the torque positive, negative, or zero?

Pivot

Push

Top view

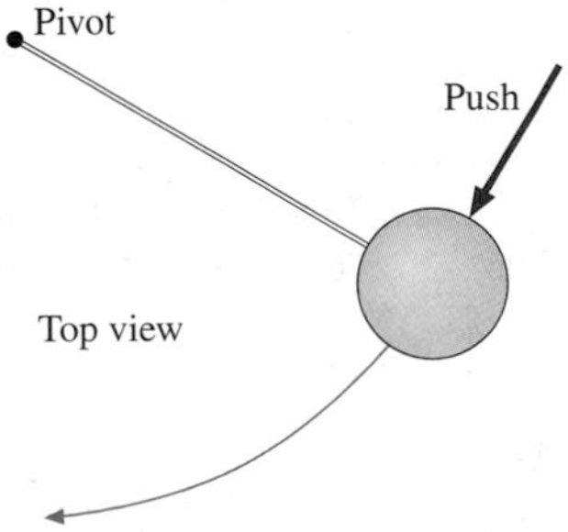

FIGURE Q7.8

9. Is the center of gravity of the dumbbell shown in Figure Q7.9 at point 1, 2, or 3? Explain.

1   2   3

FIGURE Q7.9

10. When you rise from a chair, you have to lean quite far forward (try it!). Why is this?

11. Suppose you have two identical-looking metal spheres of the same size and the same mass. One of them is solid, the other is hollow. How can you tell which is which?

12. The moment of inertia of a uniform rod about an axis through its center is $ML^2/12$. The moment of inertia about an axis at one end is $ML^2/3$. Explain *why* the moment of inertia is larger about the end than about the center.

13. A heavy steel rod, 1.0 m long, and a light pencil, 0.15 m long, are held 15° from the vertical with one end on a table, then released simultaneously. Which will hit the table first? Or will it be a tie? Explain.

14. When they ski, small children fall down more than adults do. This is partially a matter of coordination, but it is partially a matter of physics. Why might you expect shorter people to fall more easily on the ski slope?

15. A car traveling at 60 mph has a pebble stuck in one of its tires. Eventually the pebble works loose, and at the instant of release it is at the top of the tire. Explain why the pebble then slams *hard* into the *front* of the wheel well.

Multiple-Choice Questions

16. I A nut needs to be tightened with a wrench. Which force shown in Figure Q7.16 will apply the greatest torque to the nut?

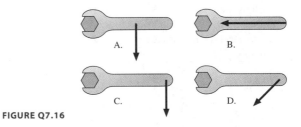

A.   B.   C.   D.

FIGURE Q7.16

17. I Suppose a bolt on your car engine needs to be tightened to a torque of 20 N · m. You are using a 15-cm-long wrench, and you apply a force at the very end in the direction that produces maximum torque. What force should you apply?
   A. 1300 N   B. 260 N   C. 130 N   D. 26 N

18. ‖ A machine part is made up of two pieces, with centers of gravity shown in Figure Q7.18. Which point could be the center of gravity of the entire part?

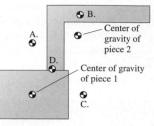

A.
B.
Center of gravity of piece 2
D.
Center of gravity of piece 1
C.

FIGURE Q7.18

19. ‖ A typical compact disk has a mass of 15 g and a diameter of 120 mm. What is its moment of inertia about an axis through its center, perpendicular to the disk?
    A. 2.7×10^{-5} kg · m^2 B. 5.4×10^{-5} kg · m^2
    C. 1.1×10^{-4} kg · m^2 D. 2.2×10^{-4} kg · m^2

20. ‖ Suppose you make a new kind of compact disk that is the same thickness as a current disk but twice the diameter. By what factor will the moment of inertia increase?
    A. 2   B. 4   C. 8   D. 16

21. ‖ Doors 1 and 2 have the same mass, height, and thickness. Door 2 is twice as wide as door 1. Bob pushes straight against the outer edge of door 2 with force F, and Barb pushes straight against the outer edge of door 1 with force $2F$. How do the angular accelerations α_1 and α_2 of the two doors compare?
    A. $\alpha_1 > \alpha_2$ B. $\alpha_1 = \alpha_2$ C. $\alpha_1 < \alpha_2$

22. ‖‖ A baseball bat has a heavy barrel and a thin handle. If you want to hold a baseball bat on your palm so that it balances vertically, you should
    A. Put the end of the handle in your palm, with the barrel up.
    B. Put the end of the barrel in your palm, with the handle up.
    C. The bat will be equally easy to balance in either configuration.

23. ‖ A car traveling at a steady 30 m/s has 74-cm-diameter tires. What is the approximate acceleration of a piece of the tread on any of the tires?
    A. 24 m/s^2 B. 48 m/s^2 C. 2400 m/s^2 D. 4800 m/s^2

PROBLEMS

Section 7.1 The Rotation of a Rigid Body

1. ‖‖ To throw the discus, the thrower holds the discus with a fully outstretched arm and makes one revolution as rapidly as possible to give maximum speed to the discus at release. The diameter of the circle in which the discus moves is about 1.8 m. If the thrower takes 1.0 s to complete one revolution, starting from rest, what will be the speed of the discus at release?

2. ‖‖ A computer hard disk starts from rest, then speeds up with an angular acceleration of 190 rad/s^2 until it reaches its final angular speed of 7200 rpm. How many revolutions has the disk made 10.0 s after it starts up?

3. ‖ The crankshaft in a race car goes from rest to 3000 rpm in 2.0 s.
   a. What is the crankshaft's angular acceleration?
   b. How many revolutions does it make while reaching 3000 rpm?

Section 7.2 Torque

4. ‖ Reconsider the situation in Example 7.3. If Luis pulls straight down on the end of a wrench that is in the same orientation but is 35 cm long, rather than 20 cm, what force must he apply to exert the same torque?

5. ‖ Balls are attached to light rods and can move in horizontal circles as shown in Figure P7.5. Rank in order, from smallest to largest, the torques τ_1 to τ_4 about the centers of the circles.

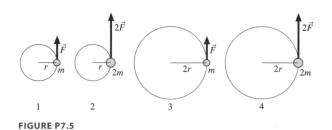

FIGURE P7.5

6. ‖‖ Six forces, each of magnitude either F or $2F$, are applied to a door as seen from above in Figure P7.6. Rank in order, from smallest to largest, the six torques τ_1 to τ_6 about the hinge.

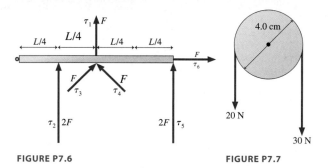

FIGURE P7.6   **FIGURE P7.7**

7. ‖ What is the net torque on the pulley about the axle in Figure P7.7?

8. ‖ The tune-up specifications of a car call for the spark plugs to be tightened to a torque of 38 N · m. You plan to tighten the plugs by pulling on the end of a 25-cm-long wrench. Because of the cramped space under the hood, you'll need to pull at an angle of 120° with respect to the wrench shaft. With what force must you pull?

9. ‖‖ A professor's office door is 0.91 m wide, 2.0 m high, and 4.0 cm thick; has a mass of 25 kg; and pivots on frictionless hinges. A "door closer" is attached to the door and the top of the door frame. When the door is open and at rest, the door closer exerts a torque of 5.2 N · m. What is the least force that you need to apply to the door to hold it open?

10. ‖ In Figure P7.10, force $\vec{F}_2$ acts half as far from the pivot as $\vec{F}_1$. What magnitude of $\vec{F}_2$ causes the net torque on the rod to be zero?

FIGURE P7.10

11. ‖ Tom and Jerry both push on the 3.0-m-diameter merry-go-round shown in Figure P7.11.

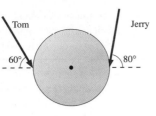

FIGURE P7.11

   a. If Tom pushes with a force of 40.0 N and Jerry pushes with a force of 35.2 N, what is the net torque on the merry-go-round?

   b. What is the net torque if Jerry reverses the direction he pushes by 180° without changing the magnitude of his force?

12. ‖ What is the net torque of the bar shown in Figure P7.12, about the axis indicated by the dot?

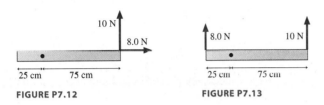

FIGURE P7.12   **FIGURE P7.13**

13. ‖‖ What is the net torque of the bar shown in Figure P7.13, about the axis indicated by the dot?

14. ‖‖ What is the net torque of the bar shown in Figure P7.14, about the axis indicated by the dot?

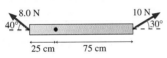

FIGURE P7.14

Section 7.3 Gravitational Torque and the Center of Gravity

15. ‖ A 1.7-m-long barbell has a 20 kg weight on its left end and a 35 kg weight on its right end.

   a. If you ignore the weight of the bar itself, how far from the left end of the barbell is the center of gravity?

   b. Where is the center of gravity if the 8.0 kg mass of the barbell itself is taken into account?

16. ‖ Three identical coins lie on three corners of a square 10.0 cm on a side, as shown in Figure P7.16. Determine the x and y coordinates of the center of gravity of the three coins.

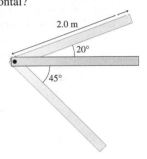

FIGURE P7.16

17. ‖ Hold your arm out-
BIO stretched so that it is horizontal. Estimate the mass of your arm and the position of its center of gravity. What is the gravitational torque on your arm in this position, computed around the shoulder joint?

18. ‖ While sitting in a chair, extend your lower leg so that it is
BIO horizontal. Estimate the mass of your leg and the position of its center of gravity. What is the gravitational torque on your leg in this position, computed around the thigh joint?

19. ‖‖ A flag pole consists of a 2.0-m-long, 5.0 kg pole with a 3.0 kg decorative ball on its end, as shown in Figure P7.19. What is the gravitational torque on the flagpole, calculated about an axis at the fixed end of the pole?

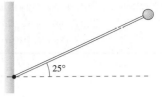

FIGURE P7.19

20. ‖‖ A 4.00-m-long, 500 kg steel beam extends horizontally from the point where it has been bolted to the framework of a new building under construction. A 70.0 kg construction worker stands at the far end of the beam. What is the magnitude of the torque about the point where the beam is bolted into place?

21. ‖ An athlete at the gym holds a 3.0 kg steel ball in his hand.
BIO His arm is 70 cm long and has a mass of 4.0 kg. What is the magnitude of the torque about his shoulder if he holds his arm

   a. Straight out to his side, parallel to the floor?

   b. Straight, but 45° below horizontal?

22. ‖ The 2.0-m-long, 15 kg beam in Figure P7.22 is hinged at its left end. It is "falling" (rotating clockwise, under the influence of gravity), and the figure shows its position at three different times. What is the gravitational torque on the beam about an axis through the hinged end when the beam is at the

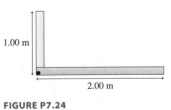

FIGURE P7.22

   a. Upper position?

   b. Middle position?

   c. Lower position?

23. ‖‖‖ Two thin beams are joined end-to-end as shown in Figure P7.23 to make a single object. The left beam is 10.0 kg and 1.00 m long and the right one is 40.0 kg and 2.00 m long.

   a. How far from the left end of the left beam is the center of gravity of the object?

   b. What is the gravitational torque on the object about an axis through its left end?

FIGURE P7.23

24. ‖‖ Figure P7.24 shows two thin beams joined at right angles. The vertical beam is 15.0 kg and 1.00 m long and the horizontal beam is 25.0 kg and 2.00 m long.

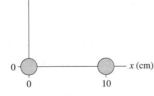

FIGURE P7.24

   a. Find the center of gravity of the two joined beams. Express your answer in the form (x, y), taking the origin at the corner where the beams join.

   b. Calculate the gravitational torque on the joined beams about an axis through the corner.

Section 7.4 Rotational Dynamics and Moment of Inertia

25. ‖‖ A regulation table tennis ball has a mass of 2.7 g and is 40 mm in diameter. What is its moment of inertia about an axis that passes through its center?

26. ‖ Three pairs of balls are connected by very light rods as shown in Figure P7.26. Rank in order, from smallest to largest, the moments of inertia I_1, I_2, and I_3 about axes through the centers of the rods.

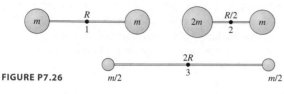

FIGURE P7.26

27. ‖ A playground toy has four seats, each 5.0 kg, attached to very light, 1.5-m-long rods, as seen from above in Figure P7.27. If two children, with masses of 15 kg and 20 kg, sit in seats opposite one another, what is the moment of inertia about the rotation axis?

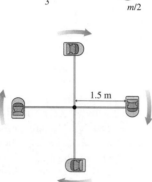

FIGURE P7.27

28. ‖ What is the moment of inertia in Problem 27 if the children sit in adjacent seats, instead of opposite ones?

29. ‖‖ You have two steel spheres. Sphere 2 has twice the radius of sphere 1. By what *factor* does the moment of inertia I_2 of sphere 2 exceed the moment of inertia I_1 of sphere 1?

Section 7.5 Using Newton's Second Law for Rotation

30. ‖ The left part of Figure P7.30 shows a bird's-eye view of two identical balls connected by a light rod that rotates about a vertical axis through its center. The right part shows a ball of twice the mass connected to a light rod of half the length, that rotates about its left end. If equal forces are applied as shown in the figure, which of the two rods will have the greater angular acceleration?

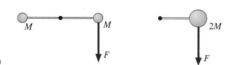

FIGURE P7.30

31. ‖ a. What is the moment of inertia of the door in Problem 9?
    b. If you let go of the open door, what is its angular acceleration immediately afterward?

32. ‖ A small grinding wheel has a moment of inertia of 4.0×10^{-5} kg · m². What net torque must be applied to the wheel for its angular acceleration to be 150 rad/s²?

33. ‖ While sitting in a swivel chair, you push against the floor with your heel to make the chair spin. The 7.0 N frictional force is applied at a point 40 cm from the chair's rotation axis, in the direction that causes the greatest angular acceleration. If that angular acceleration is 1.8 rad/s², what is the total moment of inertia about the axis of you and the chair?

34. ‖ An object's moment of inertia is 2.0 kg · m². Its angular velocity is increasing at the rate of 4.0 rad/s per second. What is the net torque on the object?

35. ‖‖‖ A 200 g, 20-cm-diameter plastic disk is spun on an axle through its center by an electric motor. What torque must the motor supply to take the disk from 0 to 1800 rpm in 4.0 s?

36. ‖ The 2.5 kg object shown in Figure P7.36 has a moment of inertia about the rotation axis of 0.085 kg · m². The rotation axis is horizontal. When released, what will be the object's initial angular acceleration?

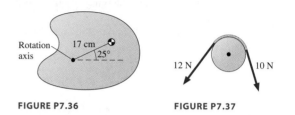

FIGURE P7.36   **FIGURE P7.37**

37. ‖ A frictionless pulley, which can be modeled as a 0.80 kg solid cylinder with a 0.30 m radius, has a rope going over it, as shown in Figure P7.37. The tension in the rope is 10 N on one side and 12 N on the other. What is the angular acceleration of the pulley?

38. ‖ If you lift the front wheel of a poorly maintained bicycle off the ground and then start it spinning at 0.72 rev/s, friction in the bearings causes the wheel to stop in just 12 s. If the moment of inertia of the wheel about its axle is 0.30 kg · m², what is the magnitude of the frictional torque?

39. ‖‖‖ A 5.0-cm-diameter toy top has a moment of inertia of 3.0×10^{-5} kg · m² about its rotation axis. To get the top spinning, its string is pulled with a tension of 0.30 N. How long does it take for the top to complete the first five revolutions? The string is long enough that it is wrapped around the top more than five turns.

40. ‖ Reconsider Example 7.9, but imagine that the pole is an electrical pole with a a 42 kg electrical transformer attached to its very top. What is the angular acceleration in this case?

41. ‖‖‖ A 1.5 kg block and a 2.5 kg block are attached to opposite ends of a light rope. The rope hangs over a solid, frictionless pulley that is 30 cm in diameter and has a mass of 0.75 kg. What is the acceleration of the lighter block?

Section 7.6 Rolling Motion

42. ‖‖‖ A bicycle with 0.80-m-diameter tires is coasting on a level road at 5.6 m/s. A small blue dot has been painted on the tread of the rear tire.
    a. What is the angular speed of the tires?
    b. What is the speed of the blue dot when it is 0.80 m above the road?
    c. What is the speed of the blue dot when it is 0.40 m above the road?

43. ‖‖‖‖ A 1.2 g pebble is stuck in a tread of a 0.75-m-diameter automobile tire, held in place by static friction that can be at most 3.6 N. The car starts from rest and gradually accelerates on a straight road. How fast is the car moving when the pebble flies out of the tire tread?

General Problems

44. ‖ Figure P7.44 shows the angular-position-versus-time graph
INT   for a particle moving in a circle.
   a. Write a description of the particle's motion.
   b. Draw the angular-velocity-versus-time graph.

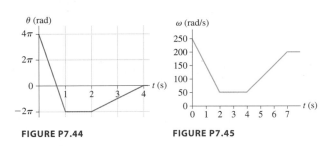

FIGURE P7.44   FIGURE P7.45

45. | The graph in Figure P7.45 shows the angular velocity of the
INT   crankshaft in a car. Draw a graph of the angular acceleration
   versus time. Include appropriate numerical scales on both
   axes.
46. ‖‖ A computer disk is 8.0 cm in diameter. A reference dot on
   the edge of the disk is initially located at $\theta = 45°$. The disk
   accelerates steadily for 0.50 s, reaching 2000 rpm, then coasts
   at steady angular velocity for another 0.50 s.
   a. What is the tangential acceleration of the reference dot at
      $t = 0.25$ s?
   b. What is the centripetal acceleration of the reference dot at
      $t = 0.25$ s?
   c. What is the angular position of the reference dot at
      $t = 1.0$ s?
   d. What is the speed of the reference dot at $t = 1.0$ s?
47. ‖‖ The 20-cm-diameter disk in Figure P7.47 can rotate on an
   axle through its center. What is the net torque about the
   axle?

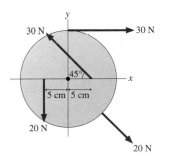

FIGURE P7.47

48. ‖‖‖ A combination lock has a 1.0-cm-diameter knob that is
INT   part of the dial you turn to unlock the lock. To turn that knob,
   you grip it between your thumb and forefinger with a force
   of 0.60 N as you twist your wrist. Suppose the coefficient of
   static friction between the knob and your fingers is only 0.12
   because some oil accidentally got onto the knob. What is the
   most torque you can exert on the knob without having it slip
   between your fingers?
49. ‖‖‖ A 70 kg man's arm, including the hand, can be modeled as a
BIO   75-cm-long uniform cylinder with a mass of 3.5 kg. In raising
   both his arms, from hanging down to straight up, by how
   much does he raise his center of gravity?

50. ‖‖‖ A penny has a mass of 2.5 g and is 1.5 mm thick; a nickel
   has a mass of 5.7 g and is 1.9 mm thick. If you make a stack of
   coins on a table, starting with five nickels and finishing with
   four pennies, how far above the tabletop is the center of gravity of the stack?
51. ‖‖ The machinist's square shown in Figure P7.51 consists of a
   thin, rectangular blade connected to a rectangular handle.
   a. Determine the x and y coordinates of the center of gravity.
      Let the lower-left corner be $x = 0, y = 0$.
   b. Sketch how the tool would hang if it were allowed to
      freely pivot about the point $x = 0, y = 0$.
   c. When hanging from that point, what angle would the long
      side of the blade make with the vertical?

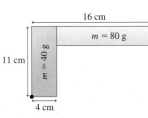

FIGURE P7.51   4 cm

52. ‖‖‖‖ The four masses shown in
   Figure P7.52 are connected
   by massless, rigid rods.
   a. Find the coordinates of
      the center of gravity.
   b. Find the moment of
      inertia about an axis that
      passes through mass A
      and is perpendicular to
      the page.
   c. Find the moment of
      inertia about a diagonal   FIGURE P7.52
      axis that passes through masses B and D.

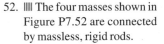

53. ‖ Three 0.10 kg balls are connected by light rods to form an
   equilateral triangle with a side length of 0.30 m. What is the
   moment of inertia of this triangle about an axis perpendicular
   to its plane and passing through one of the balls?
54. ‖‖‖‖ The three masses shown in
   Figure P7.54 are connected by
   massless, rigid rods.
   a. Find the coordinates of the
      center of gravity.
   b. Find the moment of inertia
      about an axis that passes
      through mass A and is per-
      pendicular to the page.
   c. Find the moment of inertia   FIGURE P7.54
      about an axis that passes through masses B and C.

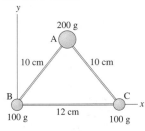

55. ‖‖‖‖ A closed 55 gallon drum is a 90-cm-long, 58-cm-diameter
   cylinder with flat ends that is made with 15 kg of sheet metal.
   If the thickness of the ends is the same as that of the curved
   sides, what is the moment of inertia about an axis that runs
   down the center of the drum?
   Hint: The uniform thickness of the sheet metal will allow you
   to determine the mass of the ends and the mass of the cylinder.
   What fraction of the surface area of the drum is associated
   with its ends?

56. ‖ A reasonable estimate of the moment of inertia of an ice
BIO skater spinning with her arms at her sides can be made by
modeling most of her body as a uniform cylinder. Suppose the
skater has a mass of 64 kg. One eighth of that mass is in her
arms, which are 60 cm long and 20 cm from the vertical axis
about which she rotates. The rest of her mass is approximately
in the form of a 20-cm-radius cylinder.
   a. Estimate the skater's moment of inertia to two significant
      figures.
   b. If she were to hold her arms outward, rather than at her
      sides, would her moment of inertia increase, decrease, or
      remain unchanged? Explain.
57. ‖ What is the moment of inertia of a 6.0-cm-diameter, 1.2 kg
cylinder that has a 2.0-cm-diameter cylindrical hole drilled
along its axis?
   Hint: Mathematically, removing mass from an object is
   equivalent to adding "negative mass" at the same location.
58. ‖ Starting from rest, a 12-cm-diameter compact disk takes
3.0 s to reach its operating angular velocity of 2000 rpm.
Assume that the angular acceleration is constant. The disk's
moment of inertia is 2.5×10^{-5} kg·m².
   a. How much torque is applied to the disk?
   b. How many revolutions does it make before reaching full
      speed?
59. ‖ A 60-cm-long, 500 g bar rotates in a horizontal plane on
an axle that passes through the center of the bar, perpendicular to its long axis. Compressed air is fed in through the
axle, passes through a small hole down the length of the bar,
and escapes as air jets from holes at the ends of the bar. The
jets are perpendicular to the bar's axis. Starting from rest,
the bar spins up to an angular velocity of 150 rpm at the end
of 10 s.
   a. How much force does each jet of escaping air exert on the
      bar?
   b. If the axle is moved to one end of the bar while the air jets
      are unchanged, what will be the bar's angular velocity at
      the end of 10 s?
60. ‖ Flywheels are large, massive wheels used to store energy.
They can be spun up slowly, then the wheel's energy can be
released quickly to accomplish a task that demands high
power. An industrial flywheel has a 1.5 m diameter and a
mass of 250 kg. A motor spins up the flywheel with a constant torque of 50 N·m. How long does it take the flywheel
to reach top angular speed of 1200 rpm?
61. ‖ A 3.0 kg block is attached to a string that is wrapped around
a 2.0 kg, 4.0-cm-diameter *hollow* cylinder that is free to
rotate. (Use Figure 7.30 but treat the cylinder as hollow.) The
block is released 1.0 m above the ground. What is the speed of
the block as it hits the ground?
62. ‖ A 1.5 kg block is connected by a rope
across a 50-cm-diameter, 2.0 kg, frictionless
pulley, as shown in Figure P7.62. A constant
10 N tension is applied to the other end of
the rope. Starting from rest, how long does
it take the block to move 30 cm?

FIGURE P7.62

63. ‖ A 1.0 kg ball and a 2.0 kg ball are connected by a 1.0-m-long rigid, massless rod.
The rod is rotating clockwise about its center of gravity at 20 rpm. What torque will
bring the balls to a halt in 5.0 s?

64. ‖ The 2.0 kg, 30-cm-diameter disk in Figure P7.64 is spin-
INT ning at 300 rpm. How much friction force must the brake
apply to the rim to bring the disk to a halt in 3.0 s?

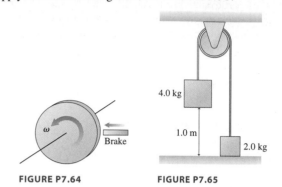

FIGURE P7.64   **FIGURE P7.65**

65. ‖ The two blocks in Figure P7.65 are connected by a massless
rope that passes over a pulley. The pulley is 12 cm in diameter
and has a mass of 2.0 kg. As the pulley turns, friction at the axle
exerts a torque of magnitude 0.50 N·m. If the blocks are
released from rest, how long does it take the 4.0 kg block to
reach the floor?
66. ‖ The bunchberry flower has the fastest-moving parts ever
BIO seen in a plant. Initially, the stamens are held by the petals in a
bent position, storing elastic energy like a coiled spring. As
the petals release, the tips of the stamens act like medieval
catapults, flipping through a 60° angle in just 0.30 ms to
launch pollen from the anther sacs at their ends. The human
eye just sees a burst of pollen; careful photography (see Figure P7.66a) reveals the details. As shown in Figure P7.66b, we
can model a stamen tip as a 1.0-mm-long, 10 μg rigid rod with
a 10 μg anther sac at one end and a pivot point at the opposite
end. Although oversimplifying, we will assume that the angular acceleration is constant throughout the motion.

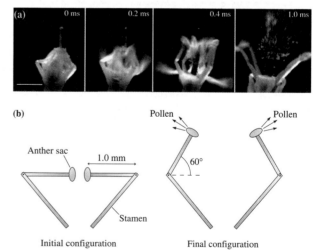

FIGURE P7.66

   a. What is the tangential acceleration of the anther sac during
      the motion?
   b. What is the speed of the anther sac as it releases its pollen?
   c. How large is the "straightening torque"? Neglect gravita-
      tional forces in your calculation.
   d. Compute the gravitational torque on the stamen tip
      (including the anther sac) in its initial orientation. Was it
      reasonable to neglect the gravitational torque in part c?

Passage Problems

The Illusion of Flight

The grand jeté is a classic ballet maneuver in which a dancer executes a horizontal leap while moving her arms and legs up and then down. At the center of the leap, the arms and legs are gracefully extended, as we see in Figure P7.67a. The goal of the leap is to create the illusion of flight. As discussed in Section 7.3, the center of mass—and hence the center of gravity—of an extended object follows a parabolic trajectory when undergoing projectile motion. But when you watch a dancer leap through the air, you don't watch her center of gravity, you watch her head. If the translational motion of her head is horizontal—not parabolic—this creates the illusion that she is flying through the air, held up by unseen forces.

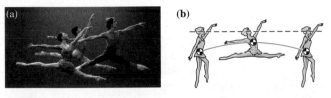

FIGURE P7.67

Figure P7.67b illustrates how the dancer creates this illusion. While in the air, she changes the position of her center of gravity relative to her body by moving her arms and legs up, then down. Her center of gravity moves in a parabolic path, but her head moves in a straight line. It's not flight, but it will appear that way, at least for a moment.

67. | To perform this maneuver, the dancer relies on the fact that the position of her center of gravity
   A. Is near the center of the torso.
   B. Is determined by the positions of her arms and legs.
   C. Moves in a horizontal path.
   D. Is outside of her body.

68. | Suppose you wish to make a vertical leap with the goal of getting your head as high as possible above the ground. At the top of your leap, your arms should be
   A. Held at your sides.
   B. Raised above your head.
   C. Outstretched, away from your body.

69. | When the dancer is in the air, is there a gravitational torque on her? Take the dancer's rotation axis to be through her center of gravity.
   A. Yes, there is a gravitational torque.
   B. No, there is not a gravitational torque.
   C. It depends on the positions of her arms and legs.

70. | In addition to changing her center of gravity, a dancer may change her moment of inertia. Consider her moment of inertia about a vertical axis through the center of her body. When she raises her arms and legs, this
   A. Increases her moment of inertia.
   B. Decreases her moment of inertia.
   C. Does not change her moment of inertia.

STOP TO THINK ANSWERS

Stop to Think 7.1: The angular velocity ω is constant. Thus the magnitude of the velocity $v = \omega r$ and the centripetal acceleration $a_c = \omega^2 r$ is constant. This also means that the ball's angular acceleration α and tangential acceleration $a_t = \alpha r$ are both zero.

Stop to Think 7.2: $\tau_E > \tau_A = \tau_D > \tau_B > \tau_C$. The perpendicular component in E is larger than 2 N.

Stop to Think 7.3: A. The force acting at the axis exerts no torque. Thus the third force needs to exert an equal but opposite torque to that exerted by the force acting at the rim. Force A, which has twice the magnitude but acts at half the distance from the axis, does so.

Stop to Think 7.4: $\tau_E = \tau_B > \tau_D > \tau_A = \tau_C$. The torques are $\tau_B = \tau_E = 2mgL$, $\tau_D = \frac{3}{2}mgL$, $\tau_A = \tau_C = mgL$, where L is the length of the rod in B.

Stop to Think 7.5: $I_D > I_A > I_C > I_B$. The moments of inertia are $I_B \approx 0$, $I_C = 2mr^2$, $I_A = 3mr^2$, $I_D = mr^2 + m(2r)^2 = 5mr^2$.

Stop to Think 7.6: C. The velocity of P is the vector sum of v_{cm} directed to the right, and an upward velocity of the same magnitude due to the rotation of the wheel.

8

EQUILIBRIUM AND ELASTICITY

How does a dancer balance so gracefully *en pointe*? And how does her foot withstand the great stresses concentrated on her toes? In this chapter we'll find answers to both these questions.

Looking Ahead ▶▶

The goal of Chapter 8 is to learn about the static equilibrium of extended objects and to understand the basic properties of springs and elastic materials. In this chapter you will learn to:

▶ Understand how torque determines the equilibrium of an extended object.

▶ Use force and torque to analyze extended objects in static equilibrium.

▶ Understand what determines the stability of an object.

▶ Use Hooke's law to calculate the force due to a spring.

▶ Understand the elastic properties of materials and use the concepts of stress, strain, and Young's modulus.

Looking Back ◀◀

This chapter uses what you have learned about equilibrium, torque, and the spring force. Please review:

◀ Section 4.3 Spring forces.

◀ Section 5.1 Equilibrium.

◀ Sections 7.2–7.3 Torque, center of gravity, and gravitational torque.

When you stand on tiptoe, you without thinking about it. with balance, unconsciously adjust the position of your body You have years of experience adapting to the forces and torques that act on your body as you stand upright in different postures. How any object—a dancer, a chair, or a suspension bridge—remains upright and stable will be a key question we answer in this chapter. We will also look at why some objects are more stable than others, and why it is harder to balance on the toes of one foot than to stand with both feet flat on the floor.

To balance on her toes, the dancer in the photo must adjust the positions of her arms and legs by applying forces with her muscles and tendons. As we will see, these forces may be quite large. How do your bones react to such stresses? We will consider this question by looking at how elastic materials, ranging from spider silk to bones, respond to applied forces.

8.1 Torque and Static Equilibrium

We have now spent several chapters studying motion and its causes. In many disciplines, it is just as important to understand the conditions under which objects do *not* move. In structural engineering, buildings and dams must be designed such that they remain motionless, even when huge forces act on them. In sports science, a correct stationary position is often the starting point for a successful athletic event. And joints in the body must sustain large forces when the body is supporting heavy stationary loads, as in lifting or carrying heavy objects.

Recall from Section 5.1 that an object at rest is in *static equilibrium*. As long as an object can be modeled as a *particle*, the condition necessary for static equilibrium is that the net force $\vec{F}_{net}$ on the particle is zero. Such a situation is shown in Figure 8.1a, where the two forces applied to the particle balance and the particle can remain at rest.

But in Chapter 7 we moved beyond the particle model to study extended objects that can rotate. Consider, for example, the block in Figure 8.1b. In this case the two forces act along the same line, the net force is zero, and the block is in equilibrium. But what about the block in Figure 8.1c? The net force is still zero, but this time the block begins to rotate because the two forces exert a net *torque*. For an extended object, $\vec{F}_{net} = \vec{0}$ is not by itself enough to ensure static equilibrium. There is a second condition for static equilibrium of an extended object: The net torque τ_{net} on the object must also be zero.

If we write the net force in component form, the conditions for static equilibrium of an extended object are

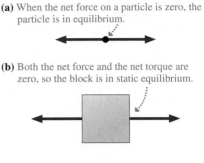

(a) When the net force on a particle is zero, the particle is in equilibrium.

(b) Both the net force and the net torque are zero, so the block is in static equilibrium.

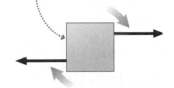

(c) The net force is still zero, but the net torque is *not* zero. The block is not in equilibrium.

FIGURE 8.1 A block with no net force acting on it may still be out of equilibrium.

$$\left.\begin{array}{l} \sum F_x = 0 \\ \sum F_y = 0 \end{array}\right\} \text{ No net force}$$

$$\sum \tau = 0 \quad \} \text{ No net torque}$$

(8.1)

Conditions for static equilibrium of an extended object

If motion is possible in the z-direction, we'd also require that $\sum F_z = 0$. In this chapter, however, we'll only consider motion restricted to the xy-plane.

EXAMPLE 8.1 Finding the force from the biceps tendon
Weightlifting can exert extremely large forces on the body's joints and tendons. In the *strict curl* event, a standing athlete lifts a barbell by moving only his forearms, which pivot at the elbow. The record weight lifted in the strict curl is over 200 pounds (about 900 N). Figure 8.2 shows the arm bones and the main lifting muscle when the forearm is horizontal. The distance from the tendon to the elbow joint is 4.0 cm, and from the barbell to the elbow 35 cm.

a. What is the tension in the tendon connecting the biceps muscle to the bone while holding a 900 N barbell stationary in this position?
b. What is the force exerted by the elbow on the forearm bones?

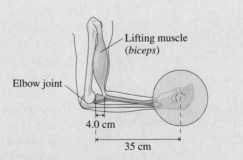

FIGURE 8.2 An arm holding a barbell.

Continued

PREPARE Figure 8.3 shows a simplified model of the arm and the forces acting on the forearm. $\vec{F}_{tendon}$ is the tension due to the muscle, $\vec{F}_{barbell}$ is the downward force of the barbell, and $\vec{F}_{elbow}$ is the force of the elbow joint on the forearm. As a simplification, we've neglected the weight of the arm itself because it is so much less than the weight of the barbell.

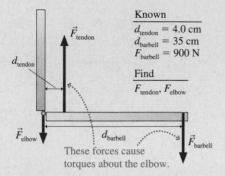

Known
$d_{tendon} = 4.0$ cm
$d_{barbell} = 35$ cm
$F_{barbell} = 900$ N

Find
F_{tendon}, F_{elbow}

These forces cause torques about the elbow.

FIGURE 8.3 Visual overview of holding a barbell.

SOLVE a. For the forearm to be in static equilibrium, the net force and net torque on it must both be zero. Setting the net force to zero gives

$$\sum F_y = F_{tendon} - F_{elbow} - F_{barbell} = 0$$

Because each arm supports half the weight of the barbell, the magnitude of the weight force is $F_{barbell} = 450$ N. We don't know either of the forces F_{tendon} and F_{elbow}, nor does the force equation give us enough information to find them. But the fact that the torque must be zero gives us that extra information. Choosing the elbow joint to be the axis of rotation, the net torque about this point is

$$\tau_{net} = F_{elbow} \times 0 + F_{tendon}d_{tendon} - F_{barbell}d_{barbell} = 0$$

Note that $\vec{F}_{elbow}$ makes no contribution to the torque because it acts directly at the pivot, so its moment arm is zero. The tension in the tendon tries to rotate the arm counterclockwise, so it produces a positive torque. Similarly, the torque due to the barbell is negative. We can solve the torque equation for F_{tendon} to find

$$F_{tendon} = F_{barbell}\frac{d_{barbell}}{d_{tendon}} = (450 \text{ N})\frac{35 \text{ cm}}{4.0 \text{ cm}} = 3900 \text{ N}$$

b. We now need to make use of the force equation. We have

$$F_{elbow} = F_{tendon} - F_{barbell} = 3900 \text{ N} - 450 \text{ N} = 3450 \text{ N}$$

ASSESS The short distance d_{tendon} from the tendon to the elbow joint means that the force supplied by the biceps has to be very large to counter the torque generated by a force applied at the opposite end of the forearm.

STOP TO THINK 8.1   Which of these objects is in static equilibrium?

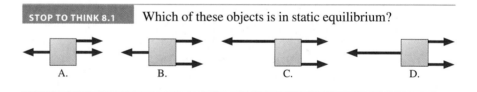

A.   B.   C.   D.

Choosing the Pivot Point

In Example 8.1, we calculated the net torque using the elbow joint as the axis of rotation or pivot point. But we learned in Chapter 7 that the torque depends on which point is chosen as the pivot point. Was there something special about our choice of the elbow joint?

Consider the hammer shown in Figure 8.4, supported on a pegboard by two pegs A and B. Because the hammer is in static equilibrium, the net torque around the pivot at A must be zero: the clockwise torque due to the weight $\vec{w}$ is exactly balanced by the counterclockwise torque due to the force $\vec{n}_B$ of peg B. (Recall that the torque due to $\vec{n}_A$ is zero, because here $\vec{n}_A$ acts at the pivot A.) But if instead we take B as the pivot, the net torque is still zero. The counterclockwise torque due to $\vec{w}$ (with a large force but small moment arm) balances the clockwise torque due to $\vec{n}_A$ (with a small force but large moment arm). Indeed, **for an object in static equilibrium, the net torque about *every* point must be zero.** This means you can pick *any* point you wish as a pivot point for calculating the torque.

Although any choice of a pivot point will work, some choices are better because they simplify the calculations. Often, there is a "natural" axis of rotation in the problem, an axis about which rotation *would* occur if the object were not in static equilibrium. Example 8.1 is of this type, with the elbow joint as a natural axis of rotation.

If no point naturally suggests itself as an axis, look for a point on the object at which several forces act, or at which a force acts whose magnitude you don't know. Such points are good choices because any force acting at that point does not contribute to the torque. For instance, the woman in Figure 8.5 is in equilib-

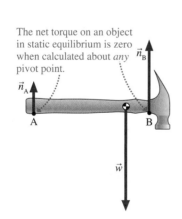

The net torque on an object in static equilibrium is zero when calculated about *any* pivot point.

FIGURE 8.4 A hammer resting on two pegs.

rium as she rests on the rock wall. A good choice of pivot point would be where her foot contacts the wall because this choice eliminates the torque due to the force $\vec{F}$ of the wall on her foot. But don't agonize over the choice of a pivot point! You can still solve the problem no matter which point you choose.

 PROBLEM-SOLVING STRATEGY 8.1 Static equilibrium problems

PREPARE Model the object as a simple shape. Draw a visual overview that shows all forces and distances. List known information.

- Pick an axis or pivot about which the torques will be calculated.
- Determine the torque about this pivot point due to each force acting on the object.
- Determine the sign of each torque about this pivot point.

SOLVE The mathematical steps are based on the fact that an object in static equilibrium has no net force and no net torque.

$$\vec{F}_{net} = \vec{0} \quad \text{and} \quad \tau_{net} = 0$$

- Write equations for $\sum F_x = 0$, $\sum F_y = 0$, $\sum \tau = 0$.
- Solve the resulting equations.

ASSESS Check that your result is reasonable and answers the question.

The torque due to $\vec{F}$ about this point is zero. This makes this point a good choice as the pivot.

FIGURE 8.5 Choosing the pivot for a woman rappelling down a rock wall.

Activ **Physics** 7.2–7.5

EXAMPLE 8.2 Forces on a board on sawhorses

A board weighing 100 N sits across two sawhorses, as shown in Figure 8.6. What are the magnitudes of the normal forces of the sawhorses acting on the board?

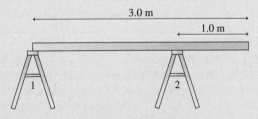

FIGURE 8.6 A board sitting on two sawhorses.

PREPARE The board and the forces acting on it are shown in Figure 8.7. $\vec{n}_1$ and $\vec{n}_2$ are the normal forces on the board due to the sawhorses, and $\vec{w}$ is the weight of the board acting at the center of gravity. The distance d_1 to the center of the board is half the board's length, or 1.5 m. Then d_2 is $d_1 - 1.0$ m, or 0.5 m.

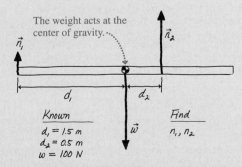

The weight acts at the center of gravity.

Known
$d_1 = 1.5$ m
$d_2 = 0.5$ m
$w = 100$ N

Find
n_1, n_2

FIGURE 8.7 Visual overview of a board on two sawhorses.

As discussed above, a good choice for the pivot is a point at which an unknown force acts, because that force contributes nothing to the torque. Either the point where $\vec{n}_1$ acts or the point where $\vec{n}_2$ acts will work; let's choose the left end of the board, where $\vec{n}_1$ acts, for this example. With this choice of pivot point, the moment arm for $\vec{w}$, which acts halfway down the board, is $d_1 = 1.5$ m. Because $\vec{w}$ tends to rotate the board clockwise, its torque is negative. The moment arm for $\vec{n}_2$ is the distance $d_1 + d_2 = 2.0$ m, and its torque is positive.

SOLVE The board is in static equilibrium, so the net force $\vec{F}_{net}$ and the net torque τ_{net} must both be zero. The forces have only y-components, so the force equation is

$$F_{net} = n_1 - w + n_2 = 0$$

The torque equation, computed around the left end of the board, is

$$\tau_{net} = -d_1 w + (d_1 + d_2)n_2 = 0$$

We now have two simultaneous equations with the two unknowns n_1 and n_2. To solve these, let's solve for n_2 in the torque equation and then substitute that result into the force equation. From the torque equation,

$$n_2 = \frac{d_1 w}{d_1 + d_2} = \frac{(1.5 \text{ m})(100 \text{ N})}{2.0 \text{ m}} = 75 \text{ N}$$

The force equation is then $n_1 - 100 \text{ N} + 75 \text{ N} = 0$, which we can solve for n_1:

$$n_1 = w - n_2 = 100 \text{ N} - 75 \text{ N} = 25 \text{ N}$$

ASSESS It seems reasonable that $n_2 > n_1$ because more of the board sits over the right sawhorse.

EXAMPLE 8.3 Choosing a different axis

Repeat Example 8.2, but put the pivot point at the center of gravity.

PREPARE The forces causing torques about the center of gravity are now the normal forces $\vec{n}_1$ and $\vec{n}_2$, with moment arms d_1 and d_2. Force $\vec{n}_2$ creates a positive (counterclockwise) torque and force $\vec{n}_1$ creates a negative (clockwise) torque. The weight $\vec{w}$, acting at the pivot, does not contribute to the torque.

SOLVE The torque about the center of gravity is

$$\tau_{\text{net}} = -d_1 n_1 + d_2 n_2 = 0$$

The force equation still gives $n_1 = w - n_2$. Substituting this into the torque equation we get

$$-d_1(w - n_2) + d_2 n_2 = 0$$

We can solve this for n_2 to get

$$n_2 = \frac{d_1 w}{d_1 + d_2} = \frac{(1.5\text{ m})(100\text{ N})}{2.0\text{ m}} = 75\text{ N}$$

as before. We can then use the force equation to find that $n_1 = 25$ N.

ASSESS Even with a very different choice of pivot point from Example 8.2, we get the same result for the forces.

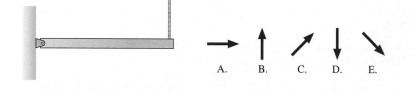

STOP TO THINK 8.2   A beam with a pivot on its left end is suspended from a rope. In which direction is the force of the pivot on the beam?

A. B. C. D. E.

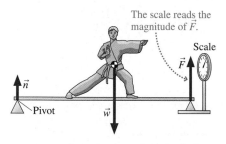

The scale reads the magnitude of $\vec{F}$.

FIGURE 8.8 Reaction board method of finding a person's center of gravity.

An interesting application of static equilibrium is to find the center of gravity of the human body. Because the human body is highly flexible, the position of the center of gravity is quite variable and depends on just how the body is posed. The horizontal position of the body's center of gravity can be located accurately from simple measurements with a *reaction board* and a scale, as shown in Figure 8.8. The subject lies or stands on the board in the desired posture, and the scale reading F is recorded. The following example shows how a person's center of gravity can be found using this information.

EXAMPLE 8.4 Finding the center of gravity of the human body

A woman weighing 600 N lies on a 2.5-m-long, 60 N reaction board with her feet over the pivot. The scale on the right reads 250 N. What is the distance d from the woman's feet to her center of gravity?

PREPARE The forces and distances in the problem are shown in Figure 8.9. We'll consider the board and woman as a single object. We've assumed that the board is uniform, so its center of

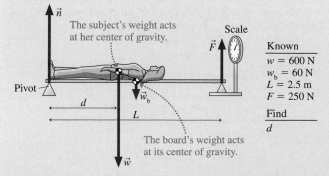

The subject's weight acts at her center of gravity.

The board's weight acts at its center of gravity.

Known
$w = 600$ N
$w_b = 60$ N
$L = 2.5$ m
$F = 250$ N

Find
d

FIGURE 8.9 Visual overview of the reaction board and woman.

gravity is at its midpoint. We'll calculate all torques around the pivot at the left end of the board. The torque due to $\vec{F}$ is positive, and those due to $\vec{w}$ and $\vec{w}_b$ are negative. The torque due to $\vec{n}$, which acts at the pivot, is zero.

SOLVE Because the board and woman are in static equilibrium, the net force and net torque on them must be zero. The force equation reads

$$\sum F_y = n - w_b - w + F = 0$$

and the torque equation gives

$$\sum \tau = -\frac{L}{2}w_b - dw + LF = 0$$

In this case, the force equation isn't needed because we can solve the torque equation for d:

$$d = \frac{LF - \frac{1}{2}Lw_b}{w} = \frac{(2.5\text{ m})(250\text{ N}) - \frac{1}{2}(2.5\text{ m})(60\text{ N})}{600\text{ N}}$$
$$= 0.92\text{ m}$$

ASSESS If the woman is 5′ 6″ (1.68 m) tall, her center of gravity is $(0.92\text{ m})/(1.68\text{ m}) = 55\%$ of her height, or a little more than halfway up her body. This seems reasonable.

EXAMPLE 8.5 Will the ladder slip?

A 3.0-m-long ladder leans against a frictionless wall at an angle of 60°. What is the minimum value of μ_s, the coefficient of static friction with the ground, that will prevent the ladder from slipping?

PREPARE The ladder is a rigid rod of length L. To not slip, both the net force and net torque on the ladder must be zero. Figure 8.10 shows the ladder and the forces acting on it. The bottom corner of the ladder is a good choice of a pivot point because two of the forces pass through this point and thus produce no torque about it. With this choice, the weight of the ladder, acting at the center of gravity, exerts torque $d_1 w$ and the force of the wall exerts torque $-d_2 n_2$. The signs are based on the observation that $\vec{w}$ would cause the ladder to rotate counterclockwise, while $\vec{n}_2$ would cause it to rotate clockwise.

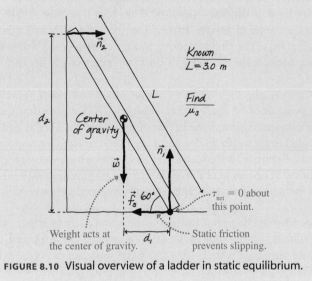

Known
$L = 3.0$ m

Find
μ_s

d_2

Center of gravity

$\vec{n}_2$

$\vec{n}_1$

$\vec{w}$

$\vec{f}_s$ 60°

$\tau_{net} = 0$ about this point.

Weight acts at the center of gravity.

d_1

Static friction prevents slipping.

FIGURE 8.10 Visual overview of a ladder in static equilibrium.

SOLVE The x- and y-components of $\vec{F}_{net} = \vec{0}$ are

$$\sum F_x = n_2 - f_s = 0$$
$$\sum F_y = n_1 - w = n_1 - Mg = 0$$

The torque about the bottom corner is

$$\tau_{net} = d_1 w - d_2 n_2 = \frac{1}{2}(L\cos 60°)Mg - (L\sin 60°)n_2 = 0$$

Altogether, we have three equations in the three unknowns n_1, n_2, and f_s. If we solve the third for n_2,

$$n_2 = \frac{\frac{1}{2}(L\cos 60°)Mg}{L\sin 60°} = \frac{Mg}{2\tan 60°}$$

we can then substitute this into the first to find

$$f_s = \frac{Mg}{2\tan 60°}$$

Our model of static friction is $f_s \leq f_{s\,max} = \mu_s n_1$. We can find n_1 from the second equation: $n_1 = Mg$. Using this, the model of friction tells us that

$$f_s \leq \mu_s Mg$$

Comparing these two expressions for f_s, we see that μ_s must obey

$$\mu_s \geq \frac{1}{2\tan 60°} = 0.29$$

Thus the minimum value of the coefficient of static friction is 0.29.

ASSESS You know from experience that you can lean a ladder or other object against a wall if the ground is "rough," but it slips if the surface is too smooth. 0.29 is a "medium" value for the coefficient of static friction, which is reasonable.

8.2 Stability and Balance

Figure 8.11 shows several common items on a table. As you know from experience, the tall candlestick can be easily toppled by a bump against the table, while the squat bowl would take a great disturbance to knock it over. To see why this is so, we'll need to understand how the gravitational torque on an object that has been tilted from its equilibrium position acts either to bring the object back upright or to pull it completely over.

Figure 8.12a on the next page shows a soda can and its **base of support,** the area between the points it would pivot on if tilted slightly in either direction. If the can is tilted slightly, as in Figure 8.12b, the gravitational torque tends to rotate it back to its equilibrium position. If the can is tilted further it reaches a special angle, shown in Figure 8.12c, at which the center of gravity is just above the pivot. Then the weight force is directed toward the pivot and so there is no torque on the can; it could fall in either direction if released. Finally, if the can is tilted even further, as in Figure 8.12d, the gravitational torque will rotate the can *away* from its equilibrium position, and the can will topple.

FIGURE 8.11 The tall candlestick is easy to knock over, but the squat bowl is very stable. Why are they different?

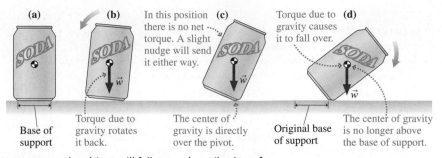

FIGURE 8.12 An object will fall over when tilted too far.

(a) Base of support

(b) Torque due to gravity rotates it back.

In this position there is no net torque. A slight nudge will send it either way.

(c) The center of gravity is directly over the pivot.

Torque due to (d) gravity causes it to fall over.

Original base of support

The center of gravity is no longer above the base of support.

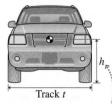

Track *t*

For the SUV, the center of gravity height *h* is 47% of *t*.

Track *t*

For the car the center of gravity height *h* is 33% of *t*.

FIGURE 8.13 Compared to a passenger car, an SUV has a high center of gravity relative to its width.

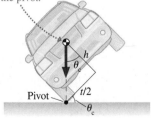

The vehicle is at the critical angle θ_c when its center of gravity is exactly over the pivot.

Pivot

FIGURE 8.14 The critical angle determines how far an object can tilt before rolling over.

From this discussion, we can see that if a vertical line through an object's center of gravity passes through its base of support, the object will tend to rotate back to its equilibrium position. In this case, we say that the object is **stable.** If such a line lies outside the base of support, the object will fall over; the object is **unstable.** As an example, the photo that opens this chapter shows a dancer balanced on her toes. Her base of support—the tips of her toes—is quite small. To balance, she must get her center of gravity directly over her toes. She does this by positioning her arms and legs. The slightest movement of an arm or leg will cause her to lose her balance because the motion will shift her center of gravity outside the base of support.

One area where stability plays a vital role is in the design of motor vehicles. With the increasing popularity of sport utility vehicles (SUVs), this topic has become one of major concern. Because the center of gravity of an SUV is generally quite high compared to its width, SUVs are more prone to *rollover* accidents than are passenger cars. Figure 8.13 shows an SUV and a passenger car with their centers of gravity marked. The height of the SUV's center of gravity is 47% of its track width whereas the passenger car's center-of-gravity height is only 33% of its track.

Figure 8.14 shows why the height of the center of gravity is important. As a result of an accident, running off the edge of the road, or taking a curve too fast, the vehicle may tilt up onto one set of wheels. Just as for the soda can in Figure 8.12, there is a special *critical angle* θ_c at which the center of gravity is directly over the pivot point. If the vehicle's maximum tilt is less than θ_c it will fall back upright, but if its tilt is greater than θ_c it will continue to tilt further, leading to rollover. The angle θ_c is thus the maximum safe angle to which the vehicle can tilt.

We can see from Figure 8.14 that the critical angle is given by $\tan\theta_c = (t/2)/h$, or

$$\theta_c = \tan^{-1}\left(\frac{t/2}{h}\right) = \tan^{-1}\left(\frac{t}{2h}\right) \tag{8.2}$$

That is, the maximum possible angle of tilt is determined by the *ratio* of track width to center-of-gravity height. Stability does not depend on the absolute height of the center of gravity but, instead, on the center-of-gravity height as a fraction of the track width.

For a passenger car with $h = 0.33t$, the critical angle is $\theta_c = 57°$. But for an SUV with $h = 0.47t$, the critical angle is only $\theta_c = 47°$. Loading an SUV with cargo further raises the center of gravity, especially if the roof rack is used, thus reducing θ_c even more. Various automobile safety groups have determined that a vehicle with $\theta_c > 50°$ is unlikely to roll over in an accident. A rollover becomes increasingly likely when θ_c is reduced below 50°. The same argument that leads to Equation 8.2 for tilted vehicles can be made for any object, leading to the general rule that **a wider base of support and/or a lower center of gravity improve stability.**

CONCEPTUAL EXAMPLE 8.1 How far to walk the plank?

A cat walks along a plank that extends out from a table. If the cat walks too far out on the plank, the plank will begin to tilt. What determines when this happens?

REASON An object is stable if its center of gravity lies over its base of support, and unstable otherwise. Let's take the cat and the plank to be one combined object whose center of gravity lies along a line between the cat's center of gravity and that of the plank.

In Figure 8.15a, when the cat is near the left end of the plank, the combined center of gravity is over the base of support and the plank is stable. As the cat moves to the right, he reaches a point where the combined center of gravity is directly over the edge of the table, as shown in Figure 8.15b. If the cat takes one more step, the cat and plank will become unstable and the plank will begin to tilt.

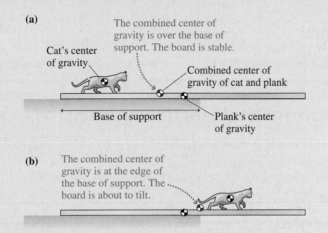

(a)

Cat's center of gravity

The combined center of gravity is over the base of support. The board is stable.

Combined center of gravity of cat and plank

Base of support

Plank's center of gravity

(b)

The combined center of gravity is at the edge of the base of support. The board is about to tilt.

FIGURE 8.15 Changing stability as a cat walks on a plank.

ASSESS Because the plank's center of gravity must be to the left of the edge for it to be stable by itself, the cat can actually walk a short distance out onto the unsupported part of the plank before it starts to tilt. The heavier the plank is, the further the cat can walk.

Stability and Balance of the Human Body

The human body is remarkable for its ability to constantly adjust its stance to remain stable on just two points of support. In walking, running, or even in the simple act of rising from a chair, the position of the body's center of gravity is constantly changing. To maintain stability, we unconsciously adjust the positions of our arms and legs to keep our center of gravity over our base of support.

A simple example of how the body naturally realigns its center of gravity is found in the act of standing up on tiptoes. Figure 8.16a shows the body in its normal standing position. Notice that the center of gravity is well-centered over the base of support (the feet), ensuring stability. If the subject were now to stand on tiptoes *without* otherwise adjusting the body position, her center of gravity would fall behind the base of support, which is now the balls of the feet, and she would fall backwards. To prevent this, as shown in Figure 8.16b, the body naturally leans forward, regaining stability by moving the center of gravity over the balls of the feet.

The simple act of walking can be broken down into a sequence of stable and unstable positions. As you extend your leg forward to initiate a step, a point is reached where your body is no longer stable—your center of gravity has moved forward of the base of support, the rear foot. At this point, your body begins to topple forward—onto the leading foot. You then lean slightly forward, placing your center of gravity over your front foot so that your rear foot can be lifted

Balancing soda can Try to balance a soda can—full or empty—on the narrow bevel at the bottom. It can't be done because, either full or empty, the center of gravity is near the center of the can. If the can is tilted enough to sit on the bevel, the center of gravity lies far outside this small base of support. But if you put about 2 ounces (60 ml) of water in an empty can, the center of gravity will be right over the bevel and the can will balance.

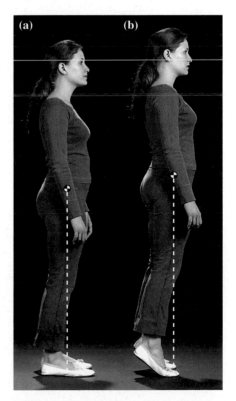

(a)   **(b)**

FIGURE 8.16 When you stand on your tiptoes, you must lean forward in order to remain balanced.

Impossible balance Stand facing a wall with your toes touching the base of the wall. Now rise onto your tiptoes. You will not be able to stand without falling backwards. As we see from Figure 8.16b, your body has to lean forward to stand on tiptoes. With the wall in your way, you cannot lean enough to maintain your balance, and you will begin to topple backwards.

without falling. As your rear foot swings forward, the process begins again. This description is a bit oversimplified because it ignores the side-to-side loss of stability as each foot is raised, but it illustrates how physics concepts can be used to understand human motion.

STOP TO THINK 8.3   Rank in order, from least stable to most stable, the three objects shown in the figure. The positions of their centers of gravity are marked. (For the centers of gravity to be positioned like this, the objects must have a nonuniform composition.)

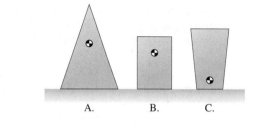

8.3 Springs and Elastic Materials

We have assumed that objects in equilibrium maintain their shape as forces and torques are applied to them. In reality this is an oversimplification. Every solid object stretches, compresses, or deforms when a force acts upon it. This change is easy to see when you press on a green twig on a tree, but even the largest branch on the tree will bend slightly under your weight.

If you stretch a rubber band, there is a force that tries to pull the rubber band back to its equilibrium, or unstretched, length. A force that restores a system to an equilibrium position is called a **restoring force.** Systems that exhibit such restoring forces are called **elastic.** The most basic examples of **elasticity** are things like springs and rubber bands. If you stretch a spring, a tension-like force pulls back. Similarly, a compressed spring tries to re-expand to its equilibrium length. Elasticity and restoring forces are properties of much stiffer systems as well. The steel beams of a bridge bend slightly as you drive your car over it, but they are restored to equilibrium after your car passes by. Your leg bones flex a bit during each step you take. Nearly everything that stretches, compresses, bends, or twists exhibits a restoring force and can be called elastic.

Springs

The behavior of a simple spring will serve to illustrate the basic ideas of elasticity. When no forces act on a spring to compress or extend it, it will relax to its **equilibrium length.** If we now compress the spring by a displacement Δx, how hard does it push back? Figure 8.17 shows what happens: The further we push the spring, the harder the restoring force of the spring pushes back.

In Figure 8.18, data for the magnitude of the restoring force of a real spring show that **the force of the spring is** *proportional* **to the displacement of the end of the spring.** That is, compressing or stretching the spring twice as far results in a restoring force that is twice as large. This is a *linear relationship,* and the slope k of the line is the proportionality constant:

$$F_{sp} = k \, \Delta x \qquad (8.3)$$

A second important fact about spring forces is illustrated in Figure 8.19. If the spring is compressed, as in Figure 8.19a, Δx is positive and, because $\vec{F}_{sp}$ points to the left, its component $(F_{sp})_x$ is negative. If, however, the spring is stretched, as in Figure 8.19b, Δx is negative and, because $\vec{F}_{sp}$ points to the right, its component

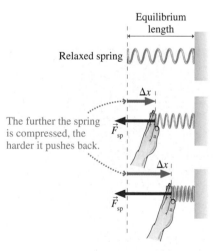

FIGURE 8.17 The spring force depends on how far the spring is compressed.

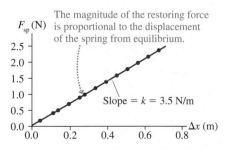

FIGURE 8.18 Measured data for the restoring force of a real spring.

$(F_{sp})_x$ is positive. In general, **the spring force always points in the opposite direction to the displacement.** We can express this fact, along with what we've learned about the magnitude of the spring force, by rewriting Equation 8.3 in terms of the *component* of the spring force:

$$(F_{sp})_x = -k\,\Delta x \qquad (8.4)$$

Hooke's law for the force due to a spring

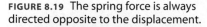

LINEAR
p. 38

The minus sign in Equation 8.4 reflects the fact that $(F_{sp})_x$ and Δx are always of opposite sign.

The proportionality constant k is called the **spring constant.** The units of the spring constant are N/m. The spring constant k is a property that characterizes a spring, just as mass m characterizes a particle. If k is large, it takes a large pull to cause a significant stretch, and we call the spring a "stiff" spring. If k is small, we can stretch the spring with very little force, and we call it a "soft" spring. Every spring has its own, unique value of k. The spring constant for the spring in Figure 8.18 can be determined from the slope of the straight line to be $k = 3.5$ N/m.

Equation 8.4 for the restoring force of a spring was first suggested by Robert Hooke, a contemporary (and sometimes bitter rival) of Newton. Hooke's law is not a true "law of nature," in the sense that Newton's laws are, but is actually just a *model* of a restoring force. It works extremely well for some springs, as in Figure 8.18, but less well for others. Hooke's law will fail for any spring if it is compressed or stretched too far.

NOTE ▶ Just as we used massless strings, we will adopt the idealization of a *massless spring*. While not a perfect description, it is a good approximation if the mass attached to a spring is much larger than the mass of the spring itself. ◀

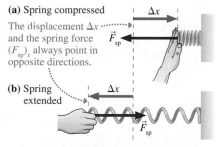

(a) Spring compressed

The displacement Δx and the spring force $(F_{sp})_x$ always point in opposite directions.

(b) Spring extended

FIGURE 8.19 The spring force is always directed opposite to the displacement.

Elasticity in action When a golf ball is struck by a club, it compresses quite a bit during the short time of the collision. The restoring force that pushes the ball back into its original shape helps launch the ball off the face of the club, making for a longer drive. Other deformations happen as well. The club face compresses a small amount, and the flexible shaft of the club may bend by as much as 15°.

EXAMPLE 8.6 Weighing a fish

A scale used to weigh fish consists of a spring connected to the ceiling. The spring's equilibrium length is 30 cm. When a 4.0 kg fish is suspended from the end of the spring, it stretches to a length of 42 cm.

a. What is the spring constant k for this spring?
b. If an 8.0 kg fish is suspended from the spring, what will be the length of the spring?

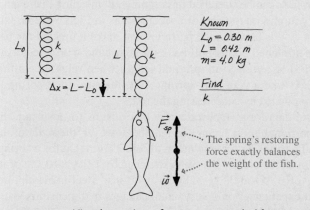

FIGURE 8.20 Visual overview of a mass suspended from a spring.

PREPARE The visual overview in Figure 8.20 shows the details for the first part of the problem. Because the fish is in static equilibrium, the upward restoring force of the spring must exactly balance the downward weight of the fish.

SOLVE a. The spring stretches by $\Delta x = L - L_0 = 0.42$ m $-$ 0.30 m $= 0.12$ m. The magnitude of the restoring force is equal to the weight of the fish, $w = mg = (4.0\text{ kg})(9.8\text{ m/s}^2) = 39.2$ N. To find k we need only use magnitudes, so we use Equation 8.3 to solve for the spring constant:

$$k = \frac{F_{sp}}{\Delta x} = \frac{39.2\text{ N}}{0.12\text{ m}} = 330\text{ N/m}$$

b. The restoring force (equal to the weight of the fish) is proportional to the stretch of the spring. If we double the mass (and thus the weight) of the fish, the displacement of the end of the spring will double as well, to $\Delta x = 0.24$ m. The new length of the spring will be

$$L = L_0 + \Delta x = 0.30\text{ m} + 0.24\text{ m} = 0.54\text{ m}$$

ASSESS The greater the weight, the more the spring stretches. We could use the displacement as a measure of the weight of an object hung from the end of the spring. This is the principle behind a spring scale.

EXAMPLE 8.7 When does the block slip?

Figure 8.21 shows a spring attached to a 2.0 kg block. The other end of the spring is pulled by a motorized toy train that moves forward at 5.0 cm/s. The spring constant is 50 N/m and the coefficient of static friction between the block and the surface is 0.60. The spring is at its equilibrium length at $t = 0$ s when the train starts to move. When does the block slip?

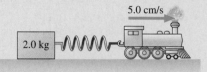

FIGURE 8.21 A toy train stretches the spring until the block slips.

PREPARE We model the block as a particle and the spring as a massless spring. Figure 8.22 is a free-body diagram for the block. We convert the speed of the train into m/s, $v = 0.050$ m/s.

When the spring force exceeds the maximum force of static friction, the block will slip.

FIGURE 8.22 Free-body diagram for the block.

SOLVE Recall that the tension in a massless string pulls equally at *both* ends of the string. The same is true for the spring force: It pulls (or pushes) equally at *both* ends. Imagine holding a rubber band with your left hand and stretching it with your right

hand. Your left hand feels the pulling force, even though it was the right end of the rubber band that moved.

This is the key to solving the problem. As the right end of the spring moves, stretching the spring, the spring pulls backward on the train *and* forward on the block with equal strength. The train is moving to the right, and so the spring force pulls to the left on the train—as we would expect. But the block is at the other end of the spring; the spring force pulls to the right on the block, as shown in Figure 8.22. As the spring stretches, the static friction force on the block increases in magnitude to keep the block at rest. The block is in static equilibrium, so

$$\sum (F_{net})_x = (F_{sp})_x + (f_s)_x = F_{sp} - f_s = 0$$

where F_{sp} is the magnitude of the spring force. This magnitude is $F_{sp} = k\,\Delta x$, where $\Delta x = vt$ is the distance the train has moved. Thus

$$f_s = F_{sp} = k\,\Delta x$$

The block slips when the static friction force reaches its maximum value $f_{s\,max} = \mu_s n = \mu_s mg$. This occurs when the train has moved

$$\Delta x = \frac{f_{s\,max}}{k} = \frac{\mu_s mg}{k} = \frac{(0.60)(2.0 \text{ kg})(9.8 \text{ m/s}^2)}{50 \text{ N/m}} = 0.235 \text{ m}$$

The time at which the block slips is

$$t = \frac{\Delta x}{v} = \frac{0.235 \text{ m}}{0.050 \text{ m/s}} = 4.7 \text{ s}$$

ASSESS Note that the distance Δx that the spring can be stretched before the block slips is *inversely proportional* to k. If the spring has a large k, so that it is quite stiff, it need be stretched only slightly before it exerts a large force on the block.

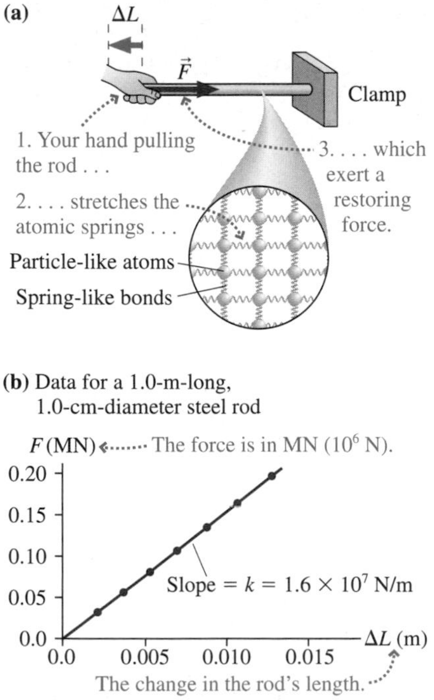

(a)

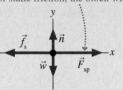

1. Your hand pulling the rod . . .

2. . . . stretches the atomic springs . . .

3. . . . which exert a restoring force.

Particle-like atoms

Spring-like bonds

(b) Data for a 1.0-m-long, 1.0-cm-diameter steel rod

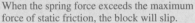

F (MN) ······· The force is in MN (10^6 N).

Slope = $k = 1.6 \times 10^7$ N/m

ΔL (m)

The change in the rod's length.

FIGURE 8.23 Stretching a steel rod.

Elasticity: The Springiness of Matter

In Chapter 4 we noted that we could model most solid materials as being made of particle-like atoms connected by spring-like bonds. We can model a steel rod this way, as illustrated in Figure 8.23a. The spring-like bonds between the atoms in steel are quite stiff, but they can be stretched or compressed, meaning that even a steel rod is elastic. If you pull on the end of a steel rod, as in Figure 8.23a, you will slightly stretch the bonds between the particles that make it up, and the rod itself will stretch. The stretched bonds pull back on your hand with a restoring force that causes the rod to return to its original length when released. In this sense, the entire rod acts like a very stiff spring. As is the case for a spring, a restoring force is also produced by compressing the rod.

In Figure 8.23b, real data for a steel rod show that, just as for a spring, the restoring force is proportional to the change in length. However, the *scale* of the stretch of the rod and the restoring force is much different from that for a spring. If a 1.0 m steel rod with a diameter of 1.0 cm is stretched by 1.0 mm (0.1% of its length), the restoring force is 16,000 N—almost two tons! Steel is elastic, but under normal forces, it experiences only very small changes in dimension. Materials of this sort are called **rigid.**

The behavior of other materials, such as the rubber in a rubber band, is quite different. A rubber band can be stretched quite far—several times its equilibrium length—with a very small force, snapping back to its original shape when released. Materials that show large deformations with small forces are called **pliant.**

The molecular structure of rubber is illustrated in Figure 8.24. The structure is quite different from that of steel. Steel and other rigid materials have atoms in a lattice with strong bonds among them; these spring-like bonds are responsible for the elasticity of the materials. In rubber, the atoms are not in a lattice, but are linked to form long-chain molecules. When the rubber is relaxed, the molecules have no overall ordering, as shown in Figure 8.24a. In Figure 8.24b, we see that as the rubber band is stretched, these long-chain molecules are pulled into alignment with each other. This ordering of the system induces a restoring force, because the molecules "want" to return to their usual random state.

This is a more complicated situation than the spring-like bonds between the atoms of steel, and you can see in Figure 8.24c that the restoring force of a rubber band is not simply proportional to the displacement of the end. Hooke's law typically does not apply to rubber and other pliant materials.

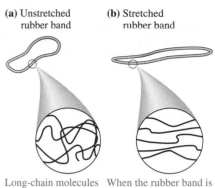

(a) Unstretched rubber band

(b) Stretched rubber band

Long-chain molecules in the rubber are randomly oriented.

When the rubber band is stretched, the molecules straighten out along the stretch direction.

(c) Data for the stretch of a rubber band

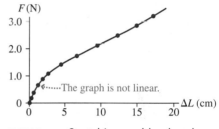

The graph is not linear.

FIGURE 8.24 Stretching a rubber band.

CONCEPTUAL EXAMPLE 8.2 Determining the stretch of a rubber band

A rubber band connected to the ceiling has an equilibrium length of 5.0 cm. A 100 g weight attached to the end stretches the rubber band to a length of 7.0 cm. Suppose an additional 100 g is attached to the end. Will the new length be less than 9.0 cm, exactly 9.0 cm, or more than 9.0 cm? Assume that the rubber band is similar to that of Figure 8.24.

REASON Consider any point on the graph of Figure 8.24c. Because the graph curves downward, doubling the force F (going upward by a factor of two) requires increasing ΔL by more than a factor of 2. If the rubber band stretches by 2.0 cm when supporting 100 g, it will stretch by more than 4.0 cm when supporting 200 g, so the length will be more than 9.0 cm.

▶ **Stretching a molecule** BIO DNA is a long-chain molecule that is normally tightly coiled as in the top figure. Amazingly, it is possible to "grab" the two ends of a DNA molecule and gently stretch it while measuring the restoring force. The computer-generated images show the unstretched molecule (top) beginning to unwind as the force on its ends is increased. The bottom frame shows the molecule just before it ruptures. By measuring the restoring force as the DNA molecule is stretched, scientists can learn, for instance, how various enzymes act to cut and then reseal coils in the DNA structure.

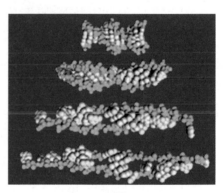

Although Hooke's law does not apply for pliant materials, we can still define an average "spring constant" by finding the average slope of the graph of restoring force versus change in length. Later in the chapter, you will see tables of values for the stiffness and the strength of pliant materials. These numbers are average values defined in just this way.

Elastic behavior occurs in other ways than just stretching or compressing. Consider again a steel rod. If you bend the rod gently, as shown in Figure 8.25, the top surface of the rod stretches slightly while the lower edge compresses. The stretching and compressing of atomic springs in this rigid material produce a restoring force that is trying to return the rod to its original shape. The rod exhibits elasticity in bending as well as in stretching and compressing. Pliant materials exhibit similar restoring forces, but the associated deflections will be much greater.

All elastic systems thus follow the same general rule: if you compress, bend, twist, or otherwise deform them, they develop a restoring force that acts to restore them to their original shape. For rigid materials like steel, these restoring forces all have their origin in the spring-like bonds between atoms. Because this

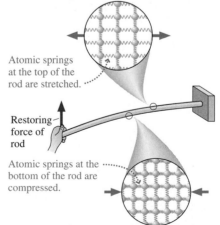

Atomic springs at the top of the rod are stretched.

Restoring force of rod

Atomic springs at the bottom of the rod are compressed.

FIGURE 8.25 When an object is stretched or bent, the spring-like molecular bonds stretch and compress.

restoring force is due to the atomic springs, it is no surprise that (for small deformations) the restoring force is proportional to the deformation; that is, the restoring force obeys Hooke's law. For pliant materials like rubber, the restoring force is typically not linear, but we may use Hooke's law to approximate the behavior of such materials.

STOP TO THINK 8.4   A 1.0 kg weight is suspended from a spring, stretching it by 5.0 cm. How much does the spring stretch if the 1.0 kg weight is replaced by a 3.0 kg weight?

A. 5.0 cm   B. 10.0 cm   C. 15.0 cm   D. 20.0 cm

8.4 Stretching, Compressing, and Bending Materials

An atomic model explains why there are restoring forces accompanying a deformation of rigid and pliant materials. In this section, we will make our models quantitative. We will also take a look at *structures*. Bone is a material with certain properties, but to consider the strength of particular bones in the body, we must consider their geometry. How thick is the bone? Where is the material dense, and where is it porous?

In Figure 8.23a, we saw that a steel rod behaves like a spring with a very large spring constant k. A rod's spring constant depends on several factors, as shown in Figure 8.26. First, we expect that a thick rod, with a large cross-section area A, will be more difficult to stretch than a thinner rod. Second, a rod with a long length L will be easier to stretch by a given amount than a short rod (think of trying to stretch a rope by 1 cm—this would be easy to do for a 10-m-long rope, but it would be pretty hard for a piece of rope only 10 cm long). Finally, the stiffness of the rod will depend on the material that it's made of. Experiments bear out these observations, and it is found that the spring constant of the rod can be written as

$$k = \frac{YA}{L} \tag{8.5}$$

where the constant Y is called **Young's modulus.** Young's modulus characterizes the elasticity of the material from which an object is made, but it does not depend on the object's shape or size. All rods made from steel have the same Young's modulus, regardless of their length or area. Aluminum rods have a different Young's modulus. In this regard, Young's modulus is somewhat analogous to density; all pieces of steel have the same density, regardless of their size and shape, while pieces of aluminum have a different density.

From Equation 8.3, the magnitude of the restoring force for a spring is related to the change in its length as $F_{sp} = k\,\Delta x$. Writing the change in the length of a rod as ΔL, as shown in Figure 8.26, we can use Equation 8.5 to write the restoring force F of a rod as

$$F = \frac{YA}{L}\,\Delta L \tag{8.6}$$

Equation 8.6 applies both to elongation and to compression.

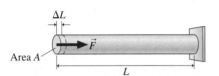

FIGURE 8.26 A stretched rod.

It's useful to write Equation 8.6 in terms of two new ratios, the *stress* and the *strain,* as follows:

The ratio of force to cross-section area is ·········➤ $\frac{F}{A} = Y\left(\frac{\Delta L}{L}\right)$ ◀········ The ratio of the change in length to the original length is called **strain**.   (8.7)

called **stress**.

The unit of stress is N/m^2. If the stress is due to stretching, we call it a **tensile stress.** The strain is the fractional change in the rod's length. If the rod's length changes by 1%, the strain is 0.01. Because strain is dimensionless, Young's modulus Y has the same units as stress. Table 8.1 gives values of Young's modulus for several rigid materials. Large values of Y characterize materials that are stiff. "Softer" materials have smaller values of Y. Because the values of Young's modulus for rigid materials are very large, it takes a significant stress to produce even a small strain.

TABLE 8.1 Young's modulus for rigid materials

Material	Young's modulus $(10^{10}\ N/m^2)$
Cast iron	20
Steel	20
Silicon	13
Copper	11
Aluminum	7
Glass	7
Concrete	3
Wood (Douglas Fir)	1

EXAMPLE 8.8 Finding the stretch of a wire

A *Foucault pendulum* in a physics department (used to prove that the earth rotates) consists of a 120 kg steel ball that swings at the end of a 6.0-m-long steel cable. The cable has a diameter of 2.5 mm. When the ball was first hung from the cable, by how much did the cable stretch?

PREPARE The amount by which the cable stretches depends on the elasticity of the steel cable. Young's modulus for steel is given in Table 8.1 as $Y = 20 \times 10^{10}\ N/m^2$.

SOLVE Equation 8.7 relates the stretch of the cable ΔL to the restoring force F and to the properties of the cable. Rearranging terms, we find that the cable stretches by

$$\Delta L = L\left(\frac{F}{AY}\right)$$

The cross-section area of the cable is

$$A = \pi r^2 = \pi(0.00125\ m)^2 = 4.91 \times 10^{-6}\ m^2$$

The restoring force of the cable is equal to the ball's weight:

$$F = w = mg = (120\ kg)(9.8\ m/s^2) = 1180\ N$$

The change in length is thus

$$\Delta L = \frac{(6.0\ m)(1180\ N)}{(4.9 \times 10^{-6}\ m^2)(20 \times 10^{10}\ N/m^2)}$$

$$= 0.0072\ m = 7.2\ mm$$

ASSESS Though the ball is quite heavy—about 270 pounds—the cable stretches just over 0.25 in, 0.12% of the total length of the cable. The strain is only 0.0012. An aluminum or copper cable of similar dimensions would stretch more because they have smaller values of Young's modulus.

Bending Beams

We saw, at the end of the previous section, how bending a rod compresses parts of the rod and stretches others. We can extend our treatment of restoring forces to treat this type of deformation. Figure 8.27 shows a beam of length L, width w, and thickness t fixed at one end and free to move at the other. Deflecting the end of the beam causes a restoring force F at the end of the beam.

The magnitude of the restoring force F depends on the dimensions of the beam, Young's modulus for the material, and the deflection d. For small values of the deflection, it can be shown that the restoring force is given by

$$F = \left(\frac{Ywt^3}{4L^3}\right)d \qquad (8.8)$$

Again, this expression has the same form as Equation 8.3 for a spring, with the quantity in brackets playing the role of the spring constant k. A higher value means a stiffer beam.

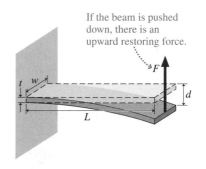

If the beam is pushed down, there is an upward restoring force.

FIGURE 8.27 A deflected beam exerts a restoring force.

EXAMPLE 8.9 Imaging biological samples using atomic force microscopy

In atomic force microscopy (AFM), the sharp tip at the end of a flexible microscopic beam is dragged very lightly over a sample. As the tip rides over high spots on the sample, such as the DNA strands shown in the AFM image to the right, the beam flexes very slightly. This bending can be detected using a laser.

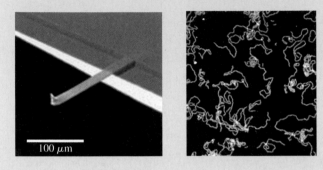

100 μm

The beam shown here is 120 μm long, 35 μm wide, and 1.0 μm thick and is made of silicon. The smallest deflection of the beam that can be detected is 5.0×10^{-11} m. At this deflection, what force is applied to the sample by the AFM tip?

PREPARE We can use Equation 8.8 to find the restoring force of the beam. From Table 8.1, Young's modulus for silicon is 1.3×10^{11} N/m².

SOLVE We have

$$F = \left[\frac{(1.3 \times 10^{11} \text{ N/m}^2)(35 \times 10^{-6} \text{ m})(1.0 \times 10^{-6} \text{ m})^3}{4(120 \times 10^{-6} \text{ m})^3} \right] \times$$
$$(5.0 \times 10^{-11} \text{ m}) = 3.3 \times 10^{-11} \text{ N} = 33 \text{ pN}$$

ASSESS In order to image soft biological materials without damage, the force applied must be extremely small. Interestingly, the spring constant of the AFM beam is $k = F/d = 0.7$ N/m, roughly that of an ordinary Slinky. The force is so small because the *deflection* is so small.

When the beam is bent, the restoring force comes mostly from the stretching of the top and the compression of the bottom.

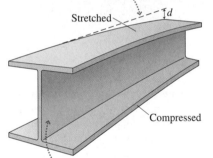

Stretched

↕d

Compressed

Thus the stiffness of the beam is only slightly lessened by removing much of the steel in the middle of the beam.

FIGURE 8.28 The structure of an I-beam.

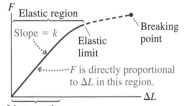

FIGURE 8.29 Stretch data for a steel rod.

TABLE 8.2 Tensile strengths of rigid materials

Material	Tensile strength (N/m²)
Glass	60×10^6
Cast iron	150×10^6
Aluminum	400×10^6
Steel	1000×10^6

Beams used in buildings and bridges usually don't have the simple rectangular cross section shown in Figure 8.27. A common form for buildings and bridges is the "I-beam" with a cross section similar to a capital "I". As we see in Figure 8.28, this form eliminates most of the weight of the beam while preserving most of the restoring force when the beam is bent. This beam will be much stiffer than a square beam of the same mass.

Another shape that is lightweight and resistant to bending is a tube; again, the material is concentrated where the strain is the greatest. Tubes with thin walls are used in many applications where rigidity combined with light weight is of importance. Backpacking tents, hang gliders, and bicycle frames all take advantage of the stiffness of tubes. The long bones in your body also take the form of hollow tubes, as we'll see in the next section.

Beyond the Elastic Limit

In the previous section, we found that if we stretch a rod by a small amount ΔL, it will pull back with a restoring force F, according to Equation 8.6. But if we continue to stretch the rod, this simple linear relationship between ΔL and F will eventually break down. Figure 8.29 is a graph of the rod's restoring force from the start of the stretch until the rod finally breaks.

As you can see, the graph has a *linear region,* the region where F and ΔL are proportional to each other, obeying $F = k \Delta L$—Hooke's law. **As long as the stretch stays within the linear region, a solid rod acts like a spring and obeys Hooke's law.**

How far can you stretch the rod before damaging it? As long as the stretch is less than the **elastic limit,** the rod will return to its initial length L when the force is removed. The elastic limit is the end of the **elastic region.** Stretching the rod beyond the elastic limit will permanently deform it, and the rod won't return to its original length. Finally, at a certain point the rod will reach a breaking point, where it will snap in two. The maximum stress that a material can be subjected to before failing is called the **tensile strength.** Table 8.2 lists values of tensile strength for rigid materials. When we speak of the strength of a material, we are referring to its tensile strength.

EXAMPLE 8.10 Breaking a pendulum cable

After a late night of studying physics, several 80 kg students decide it would be fun to swing on the Foucault pendulum of Example 8.8. What's the maximum number of students that the pendulum cable could support?

SOLVE The tensile strength of steel is given in Table 8.2 as 1000×10^6 N/m², or 1.0×10^9 N/m². This is the largest stress steel can sustain. The stress of the cable is F/A, so we can write

$$F_{max} = A(1.0 \times 10^9 \text{ N/m}^2)$$

From Example 8.8, the diameter of the cable is 2.5 mm, so its radius is 0.00125 m. Thus

$$F_{max} = (\pi(0.00125 \text{ m})^2)(1.0 \times 10^9 \text{ N/m}^2) = 4.9 \times 10^3 \text{ N}$$

This force is the weight of the heaviest mass the cable can support, $w = m_{max}g$. The maximum mass that can be supported is

$$m_{max} = F_{max}/g = 500 \text{ kg}$$

The ball has a mass of 120 kg, leaving 380 kg for the students. Four students have a mass of 320 kg, which is less than this value. But five students, totaling 400 kg, would cause the cable to break.

ASSESS Steel has a very large tensile strength. This very narrow wire can still support 4900 N ≈ 1100 lb.

Spider silk BIO The glands on the abdomen of a spider produce different kinds of silk. The silk that is used in webs can be quite stretchy; that used to subdue prey is generally not. An individual strand of silk may be a mix of fibers of different types, allowing spiders great flexibility in their material.

Biological Materials

Suppose we take equal lengths of spider silk and steel wire, stretch each, and measure the restoring force of each until they break. The graph of stress versus strain might appear as in Figure 8.30.

The spider silk is certainly less stiff: For a given stress, the silk will stretch about 100 times farther than steel. Interestingly, though, spider silk and steel eventually fail at approximately the same stress. In this sense, spider silk is "as strong as steel." Many pliant biological materials share this combination of low stiffness and large tensile strength. These materials can undergo significant deformations without failing. Tendons, the walls of arteries, and the web of a spider are all quite strong but nonetheless capable of significant stretch.

Bone is an interesting example of a rigid biological material. Most bones in your body are made of two different kinds of bony material: dense and rigid cortical (or compact) bone on the outside, and porous, flexible cancellous (or spongy) bone on the inside. Figure 8.31 shows a cross section of a typical bone. Cortical and cancellous bone have very different values of Young's modulus. Young's modulus for cortical bone approaches that of concrete, so it is very rigid with little ability to stretch or compress. In contrast, cancellous bone has a much smaller Young's modulus. Consequently, the elastic properties of bones can be well modeled as those of a hollow cylinder.

The structure of bones in birds actually approximates a hollow cylinder quite well. Figure 8.32 shows that a typical bone is a thin-walled tube of cortical bone with a tenuous structure of cancellous bone inside. Much like an I-beam, most of a cylinder's rigidity comes from the material near its surface. A hollow cylinder retains most of the rigidity of a solid one, but it is much lighter. Bird bones carry this idea to its extreme.

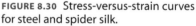
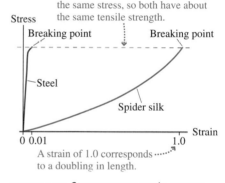

Both materials fail at approximately the same stress, so both have about the same tensile strength.

A strain of 1.0 corresponds to a doubling in length.

FIGURE 8.30 Stress-versus-strain curves for steel and spider silk.

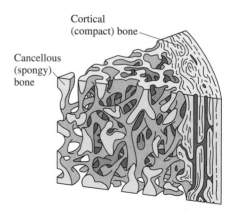

Cortical (compact) bone

Cancellous (spongy) bone

FIGURE 8.31 Cross section of a long bone.

FIGURE 8.32 Cross section of a bone from a bird.

TABLE 8.3 Young's modulus for biological materials

Material	Young's modulus (10^{10} N/m^2)
Tooth enamel	6
Cortical bone	1.6
Cancellous bone	0.02–0.3
Spider silk	0.2
Tendon	0.15
Blood vessel (aorta)	0.00005

TABLE 8.4 Tensile strength of biological materials

Material	Tensile strength (N/m^2)
Cancellous bone	5×10^6
Cortical bone	100×10^6
Spider silk	1000×10^6

Table 8.3 gives values of Young's modulus for biological materials. Note the large difference between pliant and rigid materials. Table 8.4 shows the tensile strength for biological materials. Interestingly, spider silk, a pliant material, has a greater tensile strength than bone!

The values in Table 8.4 are for static forces—forces applied for a long time in a testing machine. Bone can withstand significantly higher stresses if the forces are applied for only a very short period of time.

EXAMPLE 8.11 Finding the compression of a bone

The femur, the long bone in the thigh, can be modeled as a tube of cortical bone for most of its length. A 70 kg person has a femur with a cross-section area (of the cortical bone) of 4.8×10^{-4} m^2, a typical value.

a. If this person supports his entire weight on one leg, what fraction of the tensile strength of the bone does this stress represent?
b. By what fraction of its length does the femur shorten?

SOLVE

a. The force compressing the femur is the person's weight, $F = mg = (70 \text{ kg})(9.8 \text{ m/s}^2) = 690$ N. The resulting stress on the femur is

$$\frac{F}{A} = \frac{690 \text{ N}}{4.8 \times 10^{-4} \text{ m}^2} = 1.4 \times 10^6 \text{ N/m}^2$$

A stress of 1.4×10^6 N/m^2 is 1.4% of the tensile strength of cortical bone given in Table 8.4.

b. We can use Equation 8.7 to compute the strain $\Delta L/L$, with the value of Young's modulus for cortical bone from Table 8.3:

$$\frac{\Delta L}{L} = \left(\frac{1}{Y}\right)\frac{F}{A} = \left(\frac{1}{1.6 \times 10^{10} \text{ N/m}^2}\right)(1.4 \times 10^6 \text{ N/m}^2) = 8.8 \times 10^{-5} \approx 0.0001$$

The femur compression is $\Delta L \approx 0.0001L$, or $\approx 0.01\%$ of its length.

ASSESS Under normal forces, the bones of the body do not compress significantly, and they experience stresses much less than the tensile strength of bone.

The dancer in the chapter opening photo stands *en pointe,* balanced delicately on the tip of her shoe with her entire weight supported on a very small area. The stress on the bones in her toes is very large, but it is still well below the tensile strength of bone.

STOP TO THINK 8.5 A 10 kg mass is hung from a 1-m-long cable, causing the cable to stretch by 2 mm. Suppose a 10 kg mass is hung from a 2 m length of the same cable. By how much does the cable stretch?

A. 0.5 mm   B. 1 mm   C. 2 mm   D. 3 mm   E. 4 mm

The goal of Chapter 8 has been to learn about the static equilibrium of extended objects and to understand the basic properties of springs and elastic materials.

GENERAL PRINCIPLES

Static Equilibrium

An object in **static equilibrium** must have no net force on it and no net torque. Mathematically, we express this as

$$\sum F_x = 0$$
$$\sum F_y = 0$$
$$\sum \tau = 0$$

Since the net torque is zero about *any* point, the pivot point for calculating the torque can be chosen at any convenient location.

Springs and Hooke's Law

When a spring is stretched or compressed, it exerts a force proportional to the change Δx in the spring's length but in the opposite direction. This is known as **Hooke's Law:**

$$(F_{sp})_x = -k\,\Delta x$$

The constant of proportionality k is called the **spring constant.** It is larger for a "stiff" spring.

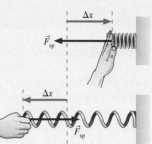

IMPORTANT CONCEPTS

Stability

An object is **stable** if its center of gravity is over its base of support, and **unstable** if it is not.

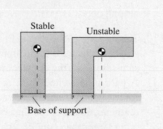

If an object is tipped, it will reach the limit of its stability when its center of gravity is over the edge of the base. This defines the **critical angle** θ_c.

Greater stability is possible with a lower center of gravity or a broader base of support.

This object is at its critical angle.

This object has a wider base of support and hence a larger critical angle.

This object has a lower center of gravity, so its critical angle is larger too.

Elastic materials and Young's modulus

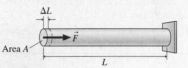

Area A L

A solid rod illustrates how materials respond when stretched or compressed.

Stress is the restoring force of the rod divided by its cross-section area. $\left(\dfrac{F}{A}\right) = Y\left(\dfrac{\Delta L}{L}\right)$ **Strain** is the fractional change in the rod's length.

Young's modulus.

This can be written as

This is the "spring constant" k for the rod. $F = \left(\dfrac{YA}{L}\right)\Delta L$

showing that a rod obeys Hooke's law, and acts like a very stiff spring.

APPLICATIONS

Forces in the body

Muscles and tendons apply the forces and torques needed to maintain static equilibrium. These forces may be quite large.

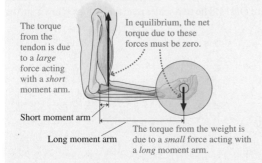

The torque from the tendon is due to a *large* force acting with a *short* moment arm.

In equilibrium, the net torque due to these forces must be zero.

Short moment arm

Long moment arm

The torque from the weight is due to a *small* force acting with a *long* moment arm.

The elastic limit and beyond

If a rod or other shape is not stretched too far, when released it will return to its original shape.

If stretched too far, it will permanently deform, and finally break. The stress at which an object breaks is its **tensile stress.**

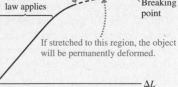

If not stretched beyond here, the object will return to its original length.

Hooke's law applies

Breaking point

If stretched to this region, the object will be permanently deformed.

(MP) For instructor-assigned homework, go to www.masteringphysics.com

Problem difficulty is labeled as | (straightforward) to ||||| (challenging).

Problems labeled |N| integrate significant material from earlier chapters; B|O are of biological or medical interest.

QUESTIONS

Conceptual Questions

1. An object is acted upon by two (and only two) forces that are of equal magnitude and oppositely directed. Is the object necessarily in static equilibrium?

2. Sketch a force acting at point P in Figure Q8.2 that would make the rod be in static equilibrium. Is there only one such force?

3. Could a ladder on a level floor lean against a wall in static equilibrium if there were no friction forces? Explain.

4. Suppose you are hanging from a tree branch. If you move out the branch, farther away from the trunk, the branch will be more likely to break. Explain why this is so.

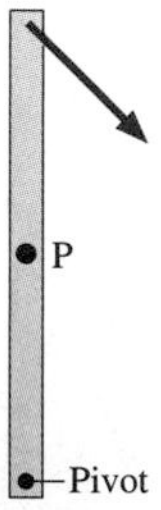

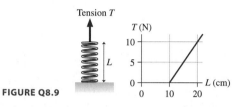

FIGURE Q8.2

5. As divers stand on tiptoes on the edge of a diving platform, in preparation for a high dive, as shown in Figure Q8.5, they usually extend their arms in front of them. Why do they do this?

6. Where are the centers of gravity of the two people doing the classic yoga poses shown in Figure Q8.6?

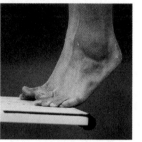

FIGURE Q8.5

(a)  (b)

FIGURE Q8.6

7. A spring exerts a 10 N force after being stretched by 1 cm from its equilibrium length. By how much will the spring force *increase* if the spring is stretched from 4 cm away from equilibrium to 5 cm from equilibrium?

8. The left end of a spring is attached to a wall. When Bob pulls on the right end with a 200 N force, he stretches the spring by 20 cm. The same spring is then used for a tug-of-war between Bob and Carlos. Each pulls on his end of the spring with a 200 N force.
   a. How far does Bob's end of the spring move? Explain.
   b. How far does Carlos's end of the spring move? Explain.

9. A spring is attached to the floor and pulled straight up by a string. The string's tension is measured. The graph in Figure Q8.9 shows the tension in the spring as a function of the spring's length L.
   a. Does this spring obey Hooke's Law? Explain why or why not.
   b. If it does, what is the spring constant?

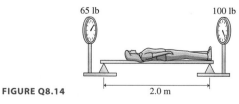

FIGURE Q8.9

10. Take a spring and cut it in half to make two springs. Is the spring constant of these smaller springs larger, smaller, or the same as the spring constant of the original spring? Explain.

11. A wire is stretched right to its breaking point by a 5000 N force. A longer wire made of the same material has the same diameter. Is the force that will stretch it right to its breaking point larger than, smaller than, or equal to 5000 N? Explain.

12. The beam in Figure Q8.12 is being pressed down by a large force on its middle. If the beam is turned 90° about its long axis and then subjected to the same downward force, will its deflection be larger, smaller, or unchanged? Explain.

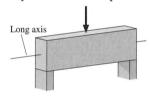

FIGURE Q8.12

13. Steel nails are rigid and unbending. Steel wool is soft and squishy. How would you account for this difference?

Multiple-Choice Questions

Questions 14 and 15 use the information in the following paragraph and figure.

A student lies on a very light, rigid board with a scale under each end. Her feet are directly over one scale, and her body is positioned as shown in Figure Q8.14. The two scales read the values shown in the figure.

65 lb   100 lb

FIGURE Q8.14   2.0 m

14. | What is the student's weight?
    A. 65 lb   B. 75 lb   C. 100 lb   D. 165 lb
15. ||| Approximately how far from her feet is the student's center of gravity?
    A. 0.6 m   B. 0.8 m   C. 1.0 m   D. 1.2 m

Questions 16 through 18 use the information in the following paragraph and figure.

Suppose you stand on one foot while holding your other leg up behind you. Your muscles will have to apply a force to hold your leg in this raised position. We can model this situation as in Figure Q8.16. The leg pivots at the knee joint, and the force to hold the leg up is provided by a tendon attached to the lower leg as shown. Assume that the lower leg and the foot together have a combined mass of 4.0 kg, and that their combined center of gravity is at the center of the lower leg.

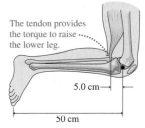

The tendon provides the torque to raise the lower leg.

5.0 cm
50 cm

FIGURE Q8.16

16. || How much force must the tendon exert to keep the leg in
BIO this position?
    A. 40 N   B. 200 N   C. 400 N   D. 1000 N
17. || As you hold your leg in this position, the upper leg exerts a
BIO force on the lower leg at the knee joint. What is the direction of this force?
    A. Up   B. Down   C. Right   D. Left
18. || What is the magnitude of the force of the upper leg on the
BIO lower leg at the knee joint?
    A. 40 N   B. 160 N   C. 200 N   D. 240 N

19. ||| A tall ladder is leaning against a wall as shown in Figure Q8.19. There is no friction between the top of the ladder and the wall. The coefficient of static friction between the bottom of the ladder and the ground is small but not zero. A painter climbs up the ladder to reach a high spot on the wall. At which location should the painter be most worried about the ladder slipping?
    A. Near the bottom.
    B. At the middle of the ladder.
    C. Near the top.
    D. The risk of the ladder slipping is the same at all locations.

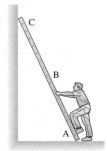

FIGURE Q8.19

20. ||| A 30.0-cm-long board is placed on a table such that its right end hangs over the edge by 8.0 cm. A second identical board is stacked on top of the first such that its right edge is 8.0 cm past the edge of the first board, as shown in Figure Q8.20. If additional boards are stacked in the same manner, each shifted 8.0 cm to the right, what is the largest number of boards (including the first) that can lie in static equilibrium?

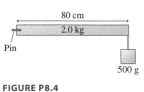

30.0 cm
8.0 cm
8.0 cm

FIGURE Q8.20

    A. 1   B. 2   C. 3   D. 4
    E. As many boards as one likes can be stacked this way without falling over.

PROBLEMS

Section 8.1 Torque and Static Equilibrium

1. | A 64 kg woman stands on a very light, rigid board that rests on a bathroom scale at each end, as shown in Figure P8.1. What is the reading on each of the scales?

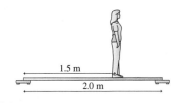

1.5 m
2.0 m

FIGURE P8.1

2. || Suppose the woman in Figure P8.1 is 54 kg, and the board she is standing on has a 10 kg mass. What is the reading on each of the scales?
3. || You're carrying a 3.6-m-long, 25 kg pole to a construction site when you decide to stop for a rest. You place one end of the pole on a fence post and hold the other end of the pole 35 cm from its tip. How much force must you exert to keep the pole motionless in a horizontal position?

4. || How much torque must the pin exert to keep the rod in Figure P8.4 from rotating? Calculate this torque about an axis that passes through the point where the pin enters the rod and is perpendicular to the plane of the figure.

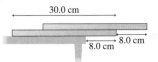

80 cm
2.0 kg
Pin
500 g

FIGURE P8.4

5. | Is the object in Figure P8.5 in equilibrium? Explain.

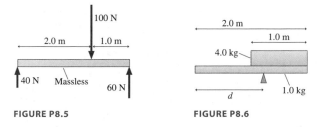

100 N
2.0 m   1.0 m
40 N   Massless   60 N

FIGURE P8.5

2.0 m
1.0 m
4.0 kg
d
1.0 kg

FIGURE P8.6

6. ||| The two objects in Figure P8.6 are balanced on the pivot. What is distance d?

7. | A 60 kg diver stands at the end of a 30 kg spring-board, as shown in Figure P8.7. The board is attached to a hinge at the left end but simply rests on the right support. What is the magnitude of the vertical force exerted by the hinge on the board?

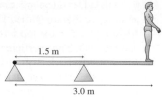

FIGURE P8.7

8. ‖ A uniform beam of length 1.0 m and mass 10 kg is attached to a wall by a cable, as shown in Figure P8.8. The beam is free to pivot at the point where it attaches to the wall. What is the tension in the cable?

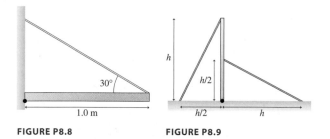

FIGURE P8.8   **FIGURE P8.9**

9. ‖‖ Figure P8.9 shows a vertical pole of height h that can rotate about a hinge at the bottom. The pole is held in position by two wires under tension. What is the ratio of the tension in the left wire to the tension in the right wire?

Section 8.2 Stability and Balance

10. | You want to slowly push a stiff board across a 20 cm gap between two tabletops that are at the same height. If you apply only a horizontal force, how long must the board be so that it doesn't tilt down into the gap before reaching the other side?

11. | A magazine rack has a center of gravity 16 cm above the floor, as shown in Figure P8.11. Through what maximum angle, in degrees, can the rack be tilted without falling over?

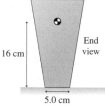

FIGURE P8.11

12. | A car manufacturer claims that you can drive its new vehicle across a hill with a 47° slope before the vehicle starts to tip. If the vehicle is 2.0 m wide, how high is its center of gravity?

13. ‖ A thin 2.00 kg box rests on a 6.00 kg board that hangs over the end of a table, as shown in Figure P8.13. How far can the center of the box be from the end of the table before the board begins to tilt?

FIGURE P8.13

14. ‖‖‖ The object shown in Figure P8.14 is made of a uniform material. What is the greatest that x can be without the object tipping over?

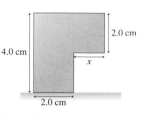

FIGURE P8.14

Section 8.3 Springs and Elastic Materials

15. | One end of a spring is attached to a wall. A 25 N pull on the other end causes the spring to stretch by 3.0 cm. What is the spring constant?

16. | Experiments using "optical tweezers" measure the elastic-
BIO ity of individual DNA molecules. For small enough changes in length, the elasticity has the same form as that of a spring. A DNA molecule is anchored at one end, then a force of 1.5 nN (1.5×10^{-9} N) pulls on the other end, causing the molecule to stretch by 5.0 nm (5.0×10^{-9} m). What is the spring constant of that DNA molecule?

17. ‖‖ A spring has an unstretched length of 10 cm. It exerts a restoring force F when stretched to a length of 11 cm.
   a. For what total stretched length of the spring is its restoring force $3F$?
   b. At what compressed length is the restoring force $2F$?

18. ‖ A 10-cm-long spring is attached to the ceiling. When a 2.0 kg mass is hung from it, the spring stretches to a length of 15 cm.
   a. What is the spring constant?
   b. How long is the spring when a 3.0 kg mass is suspended from it?

19. | A spring stretches 5.0 cm when a 0.20 kg block is hung from it. If a 0.70 kg block replaces the 0.20 kg block, how far does the spring stretch?

20. | A 1.2 kg block is hung from a vertical spring, causing the spring to stretch by 2.4 cm. How much further will the spring stretch if a 0.60 kg block is added to the 1.2 kg block?

21. | A runner wearing spiked shoes pulls a 20 kg sled across frictionless ice using a horizontal spring with spring constant 1.5×10^2 N · m. The spring is stretched 20 cm from its equilibrium length. What is the acceleration of the sled?

22. | You need to make a spring scale to measure the mass of objects hung from it. You want each 1.0 cm length along the scale to correspond to a mass difference of 0.10 kg. What should be the value of the spring constant?

Section 8.4 Stretching, Compressing, and Bending Materials

23. ‖ A force stretches a wire by 1.0 mm.
   a. A second wire of the same material has the same cross section and twice the length. How far will it be stretched by the same force?
   b. A third wire of the same material has the same length and twice the diameter as the first. How far will it be stretched by the same force?

24. ‖‖ What hanging mass will stretch a 2.0-m-long, 0.50-mm-diameter steel wire by 1.0 mm?

25. | How much force does it take to stretch a 10-m-long, 1.0-cm-diameter steel cable by 5.0 mm?

26. | An 80-cm-long, 1.0-mm-diameter steel guitar string must be tightened to a tension of 2.0 kN by turning the tuning screws. By how much is the string stretched?

27. ‖‖ A 2000 N force stretches a wire by 1.0 mm.
   a. A second wire of the same material is twice as long and has twice the diameter. How much force is needed to stretch it by 1.0 mm? Explain.
   b. A third wire of the same material is twice as long as the first and has the same diameter. How far is it stretched by a 4000 N force?

28. | A 1.2-m-long steel rod with a diameter of 0.50 cm hangs vertically from the ceiling. An auto engine weighing 4.7 kN is hung from the rod. By how much does the rod stretch?

29. | A mine shaft has an elevator hung from a single steel-wire cable of diameter 2.5 cm. When the cable is fully extended, the end of the cable is 500 m below the support. How much does the fully extended cable stretch when 3000 kg of ore is loaded into the elevator?

30. || A 3.0-m-tall, 50-cm-diameter concrete column supports a 200,000 kg load. By how much is the column compressed?

31. | A three-legged wooden bar stool made out of solid Douglas fir has legs that are 2.0 cm in diameter. When a 75 kg man sits on the stool, by what percent does the length of the legs decrease? Assume, for simplicity, that the stool's legs are vertical and that each bears the same load.

32. || A force applied to the end of a beam like the one shown in Figure 8.27 deflects the beam by 1.0 cm.
    a. If the same force is applied to the end of a beam that is twice as thick but otherwise identical, by how much will that beam deflect?
    b. If the same force is applied to the end of a beam that is twice as thick, wide, and long, by how much will that beam deflect?

33. | An aluminum beam, like the one shown in Figure 8.27, is deflected by 4.0 mm under a load. If that beam is replaced by another of the same shape and dimensions, but made of cast iron, what will be the deflection?

34. | A long steel bar with a 3.0 cm × 3.0 cm cross-section is imbedded in a concrete wall. 2.0 m of the bar extends horizontally from the wall. If a 0.75 kg brick is placed at the end of the bar, how far is the end of the bar deflected downward?

General Problems

35. ||| A 3.0-m-long rigid beam with a mass of 100 kg is supported at each end, as shown in Figure P8.35. An 80 kg student stands 2.0 m from support 1. How much upward force does each support exert on the beam?

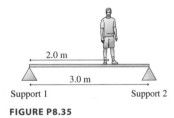

FIGURE P8.35

36. || An 80 kg construction worker sits down 2.0 m from the end of a 1450 kg steel beam to eat his lunch, as shown in Figure P8.36. The cable supporting the beam is rated at 15,000 N. Should the worker be worried?

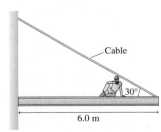

FIGURE P8.36

37. ||| A forearm can be modeled as a 1.2 kg, 35-cm-long "beam" that pivots at the elbow and is supported by the biceps, as shown in Figure P8.37. How much force must the biceps exert to hold a 500 g ball with the forearm parallel to the floor?

FIGURE P8.37

38. ||| If the person in Figure P8.37 lowers his forearm to be 15° below horizontal, how much force must the biceps exert to hold the 500 g ball? Note that the "insertion point" where the biceps attaches to the forearm is always 4.0 cm from the elbow joint, and assume that the person's wrist remains unbent as he lowers his forearm.

39. ||| A man is attempting to raise a 7.5-m-long, 28 kg flagpole that has a hinge at the base by pulling on a rope attached to the top of the pole, as shown in Figure P8.39. With what force does the man have to pull on the rope to hold the pole motionless in this position?

FIGURE P8.39

40. |||| Children with masses m_1 and m_2 sit at opposite ends of a seesaw with length L and mass M. What distance D should the pivot point be from the first child if the seesaw is to balance? Your answer will be an expression written in terms of the three masses and the length L.

41. || A 70 kg skier is on flat snow waiting in line for a ski lift. While goofing around, he bends forward stiffly, as shown in Figure P8.41. To hold this position, tension in his back muscles must support the weight of his upper body. His upper body has 55% of his mass, and the center of gravity of his upper body is 43 cm from his hips, in the middle of his trunk. What torque do his back muscles exert about an axis through his hips?

FIGURE P8.41   **FIGURE P8.42**

42. || The object shown in Figure P8.42 is made of three identical cubes, each 2.0 cm on a side, glued together.
    a. Taking the pivot to be the origin, what are the x- and y-coordinates of the object's center of gravity?
    b. What is the maximum angle, in degrees, it can be rotated around the pivot and still fall back when released, rather than falling over?

43. ‖ A 40 kg, 5.0-m-long beam is supported by, but not attached to, the two posts in Figure P8.43. A 20 kg boy starts walking along the beam. How close can he get to the right end of the beam without it tipping?

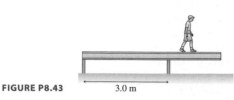

FIGURE P8.43   3.0 m

44. ‖‖ A very light tripod has 70-cm-long legs that can be spread out so that each makes a 20° angle with the vertical. The points where the three legs rest on the ground form an equilateral triangle. If you attach an object to the tripod, such as a camera with a long lens, what is the maximum possible horizontal distance between the center of gravity of the object and the center of the tripod, if the tripod is to remain stable?

45. ‖ A 5.0 kg mass hanging from a spring scale is slowly lowered onto a vertical spring, as shown in Figure P8.45. The scale reads in newtons.

Scale

FIGURE P8.45

   a. What does the spring scale read just before the mass touches the lower spring?
   b. The scale reads 20 N when the lower spring has been compressed by 2.0 cm. What is the value of the spring constant for the lower spring?
   c. At what compression distance will the scale read zero?

46. ‖‖ Two identical, side-by-side springs with spring constant 240 N/m support a 2.00 kg hanging box. By how much is each spring stretched?

47. ‖ Two springs have the same equilibrium length but different spring constants. They are arranged as shown in Figure P8.47, then a block is pushed against them, compressing both by 1.00 cm. With what net force do they push back on the block?

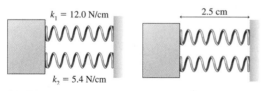

$k_1 = 12.0$ N/cm   2.5 cm

$k_2 = 5.4$ N/cm

FIGURE P8.47   **FIGURE P8.48**

48. ‖ Two springs have the same spring constant $k = 130$ N/m but different equilibrium lengths, 3.0 cm and 5.0 cm. They are arranged as shown in Figure P8.48, then a block is pushed against them, compressing both to a length of 2.5 cm. With what net force do they push back on the block?

49. ‖‖ Figure P8.49 shows two springs attached to a block that can slide on a frictionless surface. In the block's equilibrium position, the left spring is compressed by 2.0 cm.

$k_1 = 10$ N/m $k_2 = 20$ N/m

FIGURE P8.49

   a. By how much is the right spring compressed?
   b. What is the net force on the block if it is moved 15 cm to the right of its equilibrium position?

50. ‖‖‖ Figure P8.50 shows two springs attached to a box that can slide on a frictionless surface. In the block's equilibrium position, neither spring is stretched. What is the net force on the block if it is moved 15 cm to the right of its equilibrium position? **Hint:** There is zero net force on the point where the two springs meet. This implies a relationship between the amounts the two springs stretch.

$k_1 = 10$ N/m $k_2 = 20$ N/m

FIGURE P8.50

51. ‖ A 60 kg student is standing atop a spring in an elevator that
INT is accelerating upward at 3.0 m/s². The spring constant is 2.5×10^3 N/m. By how much is the spring compressed?

52. ‖ A 25 kg child bounces on a pogo stick. The pogo stick has a
INT spring with spring constant 2.0×10^4 N/m. When the child makes a nice big bounce, she finds that at the bottom of the bounce she is accelerating *upwards* at 9.8 m/s². How much is the spring compressed?

53. ‖‖‖ Two 3.0 kg blocks on a level, frictionless surface are con-
INT nected by a spring with spring constant 1000 N/m, as shown in Figure P8.53. The left block is pushed by a horizontal force $\vec{F}$. At $t = 0$ s, both blocks have velocity 3.2 m/s. For the next second, the spring's compression is a constant 1.5 cm.
   a. What is the velocity of the right block at $t = 1.0$ s?
   b. What is the magnitude of $\vec{F}$ during that 1.0 s interval?

$\vec{F}$

FIGURE P8.53

54. ‖ What is the effective spring constant (that is, the ratio of force to change in length) of a copper cable that is 5.0 mm in diameter and 5.0 m long?

55. ‖ A 3.5 kg potted plant hangs from the end of a 35-cm-long aluminum wall bracket. The bracket is 2.5 cm wide and 5.0 mm thick. Ignoring the weight of the aluminum, how much does the end of the bracket bend downward?

56. ‖ When you walk, your Achilles tendon, which connects your
BIO heel to your calf muscles, repeatedly stretches and contracts, much like a spring. This helps make walking more efficient. Suppose your Achilles tendon is 15 cm long and has a cross-section area of 110 mm², typical values. If you model the Achilles tendon as a spring, what is its spring constant?

57. ‖ There is a disk of cartilage between each pair of vertebrae
BIO in your spine. Healthy cartilage has a Young's modulus of approximately 1.0×10^6 N/m². Suppose a disk is 0.50 cm thick and 4.0 cm in diameter. If this disk supports half the weight of a 65 kg person, by what fraction of its thickness does the disk compress?

58. ‖‖‖ When you stand on your tiptoes, as the woman in Fig-
BIO ure 8.16 is doing, your feet pivot about your ankle. The forces on your foot are an upward force on your toes from the floor, a downward force on your ankle from the lower leg bone, and an upward force on the heel of your foot from your Achilles tendon. Suppose a 60 kg woman stands on tiptoes with the sole of her foot making a 25° angle with the floor. The distance from her toes to her ankle is 12 cm, the distance from her ankle to the point on the heel where the tendon attaches is

5.0 cm, and each foot supports half her weight. Assume that all forces act vertically.

a. What is the upward force on the toes of one foot?

b. What upward force does the Achilles tendon exert on the heel of that foot?

c. The tension in the Achilles tendon will cause it to stretch. If the Achilles tendon is 15 cm long and has a cross-section area of 110 mm^2, by how much will it stretch under this force?

59. ‖ In Example 8.2, the tension in the biceps tendon for a per-
BIO son doing a strict curl of a 900 N barbell was found to be 3900 N. What fraction does this represent of the maximum possible tension the biceps tendon can support? The tensile strength of tendon is approximately 100×10^6 N/m^2, and you can assume a typical cross-section area of 130 mm^2.

60. ‖ Orb spiders make silk with a typical diameter of 0.15 mm.
BIO a. A typical large orb spider has a mass of 0.50 g. If this spider suspends itself from a single 12-cm-long strand of silk, by how much will the silk stretch?

b. What is the maximum weight that a single thread of this silk could support?

61. ‖‖ Larger animals have sturdier bones than smaller animals. A
BIO mouse's skeleton is only a few percent of its body weight, compared to 16% for an elephant. To see why this must be so, recall, from Example 8.11, that the stress on the femur for a man standing on one leg is 1.4% of the bone's tensile strength. Suppose we scale this man up by a factor of 10 in all dimensions, keeping the same body proportions. Use the data for Example 8.11 to compute the following.

a. Both the inside and outside diameter of the femur, the region of cortical bone, will increase by a factor of 10. What will be the new cross-section area?

b. The man's body will increase by a factor of 10 in each dimension. What will be his new mass?

c. If the scaled up man now stands on one leg, what fraction of the tensile strength is the stress on the femur?

Passage Problems

The Diving Board

A diving board is a long, flexible board that is mounted as shown in Figure P8.62. The board's left end is fastened to a fixed support by a connection that is free to pivot. The board is supported toward its middle on another fixed support point, as shown. A diver of mass 60 kg stands at the end of the board. In the following problems we will ignore the mass of the board itself.

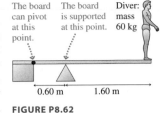

FIGURE P8.62

62. ‖ There is a force on the board from the pivot point. What is the direction of this force?

A. Up   B. Down

C. To the right, toward the diver

D. To the left, away from the diver

63. ‖ The support point exerts an upward force on the board. What is the magnitude of this force?

A. 590 N   B. 1000 N   C. 1600 N   D. 2200 N

64. ‖ Suppose the board deflects by 3.0 cm when the 60 kg diver stands on it. Now he jumps up and comes down on the board; at the lowest point of his motion, the deflection of the board is 6.0 cm. What is the net force on the diver at this point?

A. 590 N, up   B. 590 N, down

C. 1180 N, up   D. zero

65. ‖ A new board is installed that is twice as thick as the old board but is made of the same material and is the same length and width. Approximately how much does this board deflect when the 60 kg diver stands on it?

A. 0.4 cm   B. 0.8 cm   C. 1.5 cm   D. 3.0 cm

STOP TO THINK ANSWERS

Stop to Think 8.1: D. Only object D has both zero net force and zero net torque.

Stop to Think 8.2: B. The tension in the rope and the weight have no horizontal component. To make the net force zero, the force due to the pivot must also have no horizontal component, so we know it points either up or down. Now consider the torque about the point where the rope is attached. The tension provides no torque. The weight exerts a counterclockwise torque. To make the net torque zero, the pivot force must exert a *clockwise* torque, which it can only do if it points *up*.

Stop to Think 8.3: B, A, C. The critical angle θ_c, shown in the figure, measures how far the object can be tipped before falling. B has the smallest critical angle, followed by A, then C.

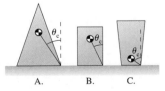

Stop to Think 8.4: C. The restoring force of the spring is proportional to the stretch. Increasing the restoring force by a factor of 3 requires increasing the stretch by a factor of 3.

Stop to Think 8.5: E. The cables have the same diameter, and the force is the same, so the stress is the same in both cases. This means that the strain, $\Delta L/L$, is the same. The 2 m cable will experience twice the change in length of the 1 m cable.

Force and Motion

The goal of Part I has been to discover the connection between force and motion. We started with kinematics, the mathematical description of motion, then we proceeded to dynamics, the explanation of motion in terms of forces. We then used these descriptions to analyze and explain motions ranging from the motion of the moon about the earth to the forces in your elbow when you lift a weight. Newton's three laws of motion formed the basis of all of our explanations.

The table below is a *knowledge structure* for force and motion. The knowledge structure does not represent everything you have learned over the past eight chapters. It's a summary of the "big picture," outlining the basic goals, the general principles, and the primary applications of the part of the book we have just finished. When you are immersed in a chapter, it may be hard to see the

connections among all of the different topics. Before we move on to new topics, we will finish each part of the book with a knowledge structure to make these connections clear.

Work through the knowledge structure from top to bottom. First are the goals and general principles. There aren't that many general principles, but we can use them along with the general problem-solving strategy to solve a wide range of problems. Once you recognize a problem as a dynamics problem, you immediately know to start with Newton's laws. You can then determine the category of motion and apply Newton's second law in the appropriate form. The kinematic equations for that category of motion then allow you to reach the solution you seek. These equations and other detailed information from the chapters are summarized in the bottom section.

KNOWLEDGE STRUCTURE I Force and Motion

BASIC GOALS	How can we describe motion? How does an object respond to a force? How do systems interact? What is the nature of the force of gravity? How can we analyze the motion and deformation of extended objects?

GENERAL PRINCIPLES	**Newton's first law**	An object will remain at rest or will continue to move with constant velocity (i.e., be in equilibrium), if and only if $\vec{F}_{net} = \vec{0}$.
	Newton's second law	$\vec{F}_{net} = m\vec{a}$
	Newton's third law	$\vec{F}_{A \, on \, B} = -\vec{F}_{B \, on \, A}$
	Newton's law of gravity	$F_{1 \, on \, 2} = F_{2 \, on \, 1} = \dfrac{Gm_1 m_2}{r^2}$

BASIC PROBLEM-SOLVING STRATEGY
Use Newton's second law for each particle or system. Use Newton's third law to equate the magnitudes of the two members of an action/reaction pair.

Types of forces:
$\vec{w} = (mg, \text{ downwards})$
$\vec{f}_k = (\mu_k n, \text{ opposite motion})$
$(F_{sp})_x = -k\Delta x$

Linear and projectile motion:

$$\sum F_x = ma_x \qquad \sum F_x = 0$$
$$\text{or}$$
$$\sum F_y = 0 \qquad \sum F_y = ma_y$$

Circular motion:
The force is directed to the center:
$$\vec{F}_{net} = \left(\frac{mv^2}{r}, \text{ toward center of circle}\right)$$

Rigid body motion:
When a torque is exerted on an object with moment of inertia I:
$$\tau_{net} = I\alpha$$

Equilibrium:
For an object at rest:
$$\sum F_x = 0$$
$$\sum F_y = 0 \qquad \sum \tau = 0$$

Linear and projectile kinematics

Uniform motion: $x_f = x_i + v_x \Delta t$
($a_x = 0, v_x = $ constant)

Constant acceleration: $(v_x)_f = (v_x)_i + a_x \Delta t$
($a_x = $ constant)
$$x_f = x_i + (v_x)_i \Delta t + \tfrac{1}{2}a_x(\Delta t)^2$$
$$(v_x)_f^2 = (v_x)_i^2 + 2a_x \Delta x$$

Projectile motion:
Projectile motion is uniform horizontal motion and constant acceleration vertical motion with $a_y = -g$.

Circular kinematics

Uniform circular motion:
$$f = \frac{1}{T} \qquad \omega = 2\pi f$$
$$v = \frac{2\pi r}{T} = \omega r \qquad a = \frac{v^2}{r} = \omega^2 r$$

Velocity is the slope of the position-versus-time graph.
Acceleration is the slope of the velocity-versus-time graph.

Rigid bodies

Torque $\tau = rF_\perp = r_\perp F$
Center of gravity $x_{cm} = \dfrac{x_1 m_1 + x_2 m_2 + \cdots}{m_1 + m_2 + \cdots}$
Moment of inertia $I = \sum mr^2$

Dark Matter and the Structure of the Universe

The idea that the earth exerts a gravitational force on us is something we now accept without questioning. But when Isaac Newton developed this idea to show that the gravitational force also holds the moon in its orbit, it was a remarkable, ground-breaking insight. It changed the way that we look at the universe we live in.

Newton's laws of motion and gravity are tools that allow us to continue Newton's quest to better understand our place in the cosmos. But it sometimes seems that the more we learn, the more we realize how little we actually know and understand.

Here's an example. Advances in astronomy over the past 100 years have given us great insight into the structure of the universe. But everything our telescopes can see appears to be only a small fraction of what is out there. As much as 90% of the mass in the universe is *dark matter*— matter that gives off no light or other radiation that we can detect. Everything that we have ever seen through a telescope is merely the tip of the cosmic iceberg.

What is this dark matter? Black holes? Neutrinos? Some form of exotic particle? No one knows. It could be any of these, or all of them—or something entirely different that no one has yet dreamed of. You might wonder how we know that such matter exists if no one has seen it. Even though we can't directly observe dark matter, we see its effects. And you now know enough physics to understand why.

Whatever dark matter is, it has mass, and so it has gravity. This picture of the Andromeda galaxy shows a typical spiral galaxy structure: a dense collection of stars in the center surrounded by a disk of stars and other matter. This is the shape of our own Milky Way galaxy.

The spiral Andromeda galaxy.

This structure is reminiscent of the structure of the solar system, a dense mass (the sun) in the center surrounded by a disk of other matter (the planets, asteroids, and comets.) The sun's gravity keeps the planets in their orbits, but the planets would fall into the sun unless they were in constant motion around it. The same is true of a spiral galaxy; everything in the galaxy orbits its center. Our solar system orbits the center of our galaxy with a period of about 200 million years.

The orbital speed of an object depends on the mass that pulls on it. If you analyze our sun's motion about the center of the Milky Way, or the motion of stars in the Andromeda galaxy about its center, you find that the orbits are much faster than they should be, based on how many stars we see. There must be some other mass present.

There's another problem with the orbital motion of stars around the center of their galaxies. We know that the orbital speeds of planets decrease with distance from the sun; Pluto orbits at a much slower speed than the earth. We might expect something similar for galaxies: Stars farther from the center should orbit at reduced speeds. But they don't. As we measure outward from the center of the galaxy, the orbital speed stays about the same—even as we get to the edge of the visible disk. There must be some other mass—the invisible dark matter—exerting a gravitational force on the stars. This dark matter, which far outweighs the matter we can see, seems to form a halo around the centers of galaxies, providing the gravitational force necessary to produce the observed rotation. Other observations of the motions of galaxies with respect to each other verify this basic idea.

Ultimately, the fate of the universe depends on its mass. The universe is currently expanding. If the universe contains enough mass, the mutual attraction will cause a final collapse—what you might term a "Big Crunch," the antithesis of the Big Bang. If there is not enough mass, the universe will expand forever. Dark matter may well be 90% of the universe. It may hold the key to the universe's future. And we don't know what it is.

This sort of mystery is what drives scientific investigation. It's what drove Newton to wonder about the connection between the fall of an apple and the motion of the moon, and what drove investigators to develop all of the techniques and theories you will learn about in the coming chapters.

PART II

CONSERVATION LAWS

Why Some Things Stay the Same

Part I of this textbook was about *change*. Simple observations show us that most things in the world around us are changing. Even so, there are some things that *don't* change even as everything else is changing around them. Our emphasis in Part II will be on things that stay the same.

Consider, for example, a strong, sealed box in which you have replaced all the air with a mixture of hydrogen and oxygen. The mass of the box plus the gases inside is 600.0 g. Now, suppose you use a spark to ignite the hydrogen and oxygen. As you know, this is an explosive reaction, with the hydrogen and oxygen combining to create water—and quite a bang. But the strong box contains the explosion and all of its products.

What is the mass of the box after the reaction? The gas inside the box is different now, but a careful measurement would reveal that the mass hasn't changed—it's still 600.0 g! We say that the mass is *conserved*. Of course, this is only true if the box has stayed sealed. For conservation of mass to apply, the system must be *closed*.

Conservation Laws

A closed system of interacting particles has another remarkable property. Each system is characterized by a certain number, and no matter how complex the interactions, the value of this number never changes. This number is called the *energy* of the system, and the fact that it never changes is called the *law of conservation of energy*. It is, perhaps, the single most important physical law ever discovered.

The law of conservation of energy is much more general than Newton's laws. Energy can be converted to many different forms, and, in all cases, the total energy stays the same:

- Gasoline, diesel, and jet engines convert the energy of a fuel into the mechanical energy of moving pistons, wheels, and gears.
- A solar cell converts the electromagnetic energy of light into electrical energy.
- An organism converts the chemical energy of food into a variety of other forms of energy, including kinetic energy, sound energy, and thermal energy.

Energy will be *the* most important concept throughout the remainder of this textbook, and much of Part II will focus on understanding what energy is and how it is used.

But energy is not the only conserved quantity. We will begin Part II with the study of two other quantities that are conserved in a closed system: *momentum* and *angular momentum*. Their conservation will help us understand a wide range of physical processes, from the forces when two rams butt heads to the graceful spins of ice skaters.

Conservation laws will give us a new and different *perspective* on motion. Some situations are most easily analyzed from the perspective of Newton's laws, but others make much more sense when analyzed from a conservation-law perspective. An important goal of Part II is to learn which perspective is best for a given problem.

The kestrel is pulling in its wings to begin a steep dive, in which it can achieve a speed of 60 mph. How does the bird achieve such a speed, and why does this speed help the kestrel catch its prey? Such questions are best answered by considering the conservation of energy and momentum.

9

MOMENTUM

Male rams butt heads at high speeds in a ritual to assert their dominance. How can the force of this collision be minimized so as to avoid damage to their brains?

Looking Ahead ▶▶

The goals of Chapter 9 are to introduce the ideas of impulse, momentum, and angular momentum and to learn a new problem-solving strategy based on conservation laws. In this chapter you will learn to:

▶ Understand and use the concepts of impulse and momentum.

▶ Draw a new "before-and-after" visual overview.

▶ Solve problems using the law of conservation of momentum.

▶ Apply these ideas to collisions and explosions.

▶ Understand the concept of angular momentum and solve problems using its conservation law.

Looking Back ◀◀

The law of conservation of momentum is based on Newton's second and third laws, and angular momentum builds on concepts of torque and moment of inertia. Please review:

◀ Section 4.8 Action/reaction force pairs and Newton's third law.

◀ Section 5.2 Newton's second law.

◀ Section 7.2, 7.4 Torque and moment of inertia.

Two rams crashing together is a vivid example of what we'll call a *collision*. During a collision, two objects exert forces on each other that vary in a complex way during the time of the collision. Using Newton's second law to predict the outcome of a collision would thus be a daunting challenge. Nevertheless, some collisions have very simple outcomes. For example, consider a train car rolling along the tracks toward an identical car at rest. The two cars couple together upon impact and then roll down the tracks together. The forces between the train cars during the collision are exceedingly complex. Yet if you were to measure their speeds just before and after the collision, you would find that the two coupled cars have exactly half the speed of the single car before impact. How can such a complex interaction give rise to such a simple outcome?

The opposite of a collision is an interaction that forces two objects apart. These interactions are called *explosions,* even though they may lack a flash or a pop. As an example, imagine you're standing on ice skates, and you throw a ball whose mass is 1% of your mass directly forward. The interaction between the ball and your hand as you propel it forward is quite complex, yet you'll find that no matter how you throw it, your recoil speed is always 1% of the ball's final speed. Another simple outcome.

Our goal in this chapter is to learn how to predict these simple outcomes without having to know all the details of the interaction forces. Two new ideas will make this possible. The first is *momentum,* a concept that unites the masses and velocities of the particles in a system. The second is the idea of a *conservation law,* a statement that will relate the situation "before" an interaction to the situation "after" the interaction. This before-and-after perspective will be a powerful new problem-solving tool.

9.1 Impulse

Suppose that two or more objects have an intense and perhaps complex interaction, such as a collision or an explosion. Our goal is to find a relationship between the velocities of the objects before the interaction and their velocities after the interaction. We'll start by looking at collisions.

A **collision** is a short-duration interaction between two objects. The collision between a tennis ball and a racket, or a baseball and a bat, may seem instantaneous to your eye, but that is a limitation of your perception. A careful look at the tennis ball/racket collision in Figure 9.1 reveals that the right side of the ball is flattened and pressed up against the strings of the racket. It takes time to compress the ball, and more time for the ball to re-expand as it leaves the racket.

The duration of a collision depends on the materials from which the objects are made, but 1 to 10 ms (0.001 to 0.010 s) is typical. This is the time during which the two objects are in contact with each other. The harder the objects, the shorter the contact time. A collision between two steel balls lasts less than 1 ms, while that between a tennis ball and racket might last 10 ms.

Let's begin our discussion by considering a collision that most of us have experienced: kicking a soccer ball. A sequence of high-speed photos of a soccer kick is shown in Figure 9.2. As the foot and the ball just come into contact, as shown in the top frame, the ball is just beginning to compress. By the middle frame of Figure 9.2, the ball has sped up and become greatly compressed. Finally, as shown in the bottom frame, the ball, now moving very fast, is again only slightly compressed.

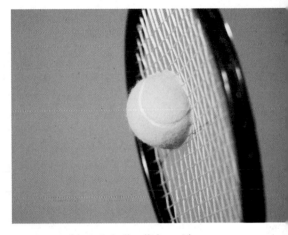

FIGURE 9.1 A tennis ball collides with a racket. Notice that the right side of the ball is flattened.

FIGURE 9.2 A sequence of high-speed photos of a soccer ball being kicked.

The amount by which the ball is compressed is a measure of the magnitude of the force the foot exerts on the ball; more compression indicates a greater force. If we were to graph this force versus time, it would look something like Figure 9.3, although the detailed shaped of the curve might be different. The force is zero until the foot first contacts the ball, rises quickly to a maximum value, and then falls back to zero as the ball leaves the foot. Thus there is a well-defined duration Δt of the force. A large force like this exerted during a small interval of time is called an **impulsive force.** The force of a hammer on a nail, or of a bat on a baseball, are other examples of impulsive forces.

The *effect* of an impulsive force depends on the area between the axis and the force-versus-time curve (i.e., the area "under" the force curve). This area, shown in Figure 9.4a on the next page, is called the **impulse** J of the force. A harder kick or a kick of longer duration delivers a larger impulse and thus causes the ball to leave the kicker's foot with a higher speed (i.e., it has a larger effect).

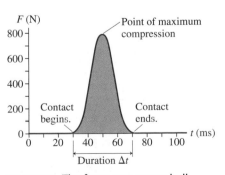

FIGURE 9.3 The force on a soccer ball changes rapidly.

(a) F

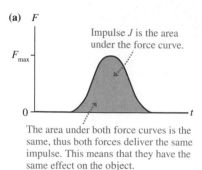

Impulse J is the area under the force curve.

F_{max}

0 ————————— t

The area under both force curves is the same, thus both forces deliver the same impulse. This means that they have the same effect on the object.

(b) F

The area under the rectangle is $F_{avg} \Delta t$, its width Δt multiplied by its height F_{avg}.

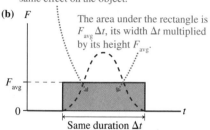

F_{avg}

0 ————————— t

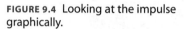

Same duration Δt

FIGURE 9.4 Looking at the impulse graphically.

Impulsive forces can be complex, and the shape of the force-versus-time graph often changes in a complicated way. Consequently, it is often useful to think of the collision in terms of an *average* force F_{avg}. As Figure 9.4b shows, F_{avg} is defined to be the constant force that has the same duration Δt and the same area under the force curve as the real force. You can see from the figure that the area under the force curve can be written simply as $F_{avg} \Delta t$. Thus

$$\text{impulse } J = \text{area under the force curve} = F_{avg} \Delta t \qquad (9.1)$$

Impulse due to a force acting for a duration Δt

From Equation 9.1 we see that impulse has units of N · s, but you should be able to show that N · s are equivalent to kg · m/s. We'll see shortly why the latter are the preferred units for impulse.

So far, we've been assuming the force is directed along a coordinate axis, such as the x-axis. In this case impulse is a *signed* quantity—it can be positive or negative. A positive impulse results from an average force directed in the positive x-direction (that is, F_{avg} is positive), while a negative impulse is due to a force directed in the negative x-direction (F_{avg} is negative). More generally, the impulse is a *vector* quantity pointing in the direction of the average force vector:

$$\vec{J} = \vec{F}_{avg} \Delta t \qquad (9.2)$$

EXAMPLE 9.1 Finding the impulse on a bouncing ball

A rubber ball experiences the force shown in Figure 9.5 as it bounces off the floor.

a. What is the impulse on the ball?
b. What is the average force on the ball?

PREPARE The impulse due to an impulsive force is the area under the force curve. Here the shape of the graph is triangular, so we'll need to use

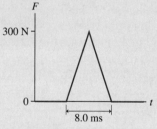

FIGURE 9.5 The force of the floor on a bouncing ball.

the fact that the area of a triangle is $\frac{1}{2} \times$ height $\times$ base.

SOLVE a. The impulse is

$$J = \tfrac{1}{2}(300 \text{ N})(0.0080 \text{ s}) = 1.2 \text{ N} \cdot \text{s} = 1.2 \text{ kg} \cdot \text{m/s}$$

b. From Equation 9.1, $J = F_{avg} \Delta t$, we can find the average force that would give this same impulse:

$$F_{avg} = \frac{J}{\Delta t} = \frac{1.2 \text{ N} \cdot \text{s}}{0.0080 \text{ s}} = 150 \text{ N}$$

ASSESS In this particular example, the average value of the force is half the maximum value. This is not surprising for a triangular force because the area of a triangle is *half* the base times the height.

9.2 Momentum and the Impulse-Momentum Theorem

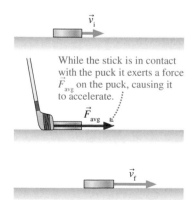

While the stick is in contact with the puck it exerts a force $\vec{F}_{avg}$ on the puck, causing it to accelerate.

FIGURE 9.6 The stick exerts an impulse on the puck, changing its speed.

We've noted that the effect of an impulsive force depends on the impulse delivered to the object. The effect also depends on the object's mass. Our experience tells us that giving a kick to a heavy object will change its velocity much less than giving the same kick to a light object. We want now to find a quantitative relationship between impulse, mass, and velocity change.

Consider the puck of mass m in Figure 9.6, sliding with an initial velocity $\vec{v}_i$. It is struck by a hockey stick that delivers an impulse $\vec{J} = \vec{F}_{avg} \Delta t$ to the puck. After the impulse, the puck leaves the stick with a final velocity $\vec{v}_f$. How is this final velocity related to the initial velocity?

From Newton's second law, the average acceleration of the puck during the time the stick is in contact with it is

$$\vec{a}_{avg} = \frac{\vec{F}_{avg}}{m} \qquad (9.3)$$

The average acceleration is related to the change in the velocity by

$$\vec{a}_{avg} = \frac{\Delta \vec{v}}{\Delta t} = \frac{\vec{v}_f - \vec{v}_i}{\Delta t} \qquad (9.4)$$

Combining Equations 9.3 and 9.4, we have

$$\frac{\vec{F}_{avg}}{m} = \vec{a}_{avg} = \frac{\vec{v}_f - \vec{v}_i}{\Delta t}$$

or, rearranging,

$$\vec{F}_{avg} \Delta t = m\vec{v}_f - m\vec{v}_i \qquad (9.5)$$

We recognize the left side of this equation as the impulse $\vec{J}$. The right side is the *change* in the quantity $m\vec{v}$. This quantity, the product of the object's mass and velocity, is called the **momentum** of the object. The symbol for momentum is $\vec{p}$:

$$\vec{p} = m\vec{v} \qquad (9.6)$$

Momentum of an object of mass m and velocity $\vec{v}$

From Equation 9.6, the units of momentum are those of mass times velocity, or kg · m/s. We noted above that kg · m/s are the preferred units of impulse. Now we see that the reason for that preference is to match the units of momentum.

Figure 9.7 shows that the momentum $\vec{p}$ is a *vector* quantity that points in the same direction as the velocity vector $\vec{v}$. Like any vector, $\vec{p}$ can be decomposed into x- and y-components. Equation 9.6, which is a vector equation, is a shorthand way to write the two equations

$$\begin{aligned} p_x &= mv_x \\ p_y &= mv_y \end{aligned} \qquad (9.7)$$

NOTE ▶ One of the most common errors in momentum problems is a failure to use the appropriate signs. The momentum component p_x has the same sign as v_x. Just like velocity, momentum is positive for a particle moving to the right (on the x-axis) or up (on the y-axis), but *negative* for a particle moving to the left or down. ◀

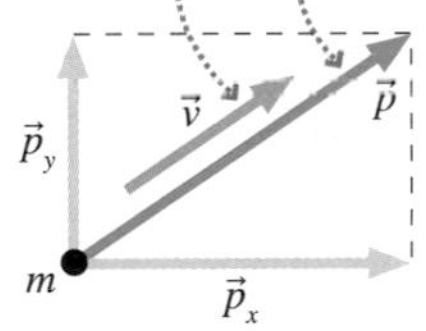

The *magnitude* of an object's momentum is simply the product of the object's mass and speed, or $p = mv$. A heavy, fast-moving object will have a great deal of momentum, while a light, slow-moving object will have very little. Two objects with very different masses can have similar momenta if their speeds are very different as well. Table 9.1 gives some typical values of the momenta (the plural of *momentum*) of various moving objects. You can see that the momenta of a bullet and a fastball are similar. The momentum of a moving car is almost a billion times greater than that of a falling raindrop.

The Impulse-Momentum Theorem

Returning to Equation 9.5, we can now write this equation in terms of impulse and momentum as

$$\vec{J} = \vec{p}_f - \vec{p}_i = \Delta \vec{p} \qquad (9.8)$$

Impulse-momentum theorem

where $\vec{p}_i = m\vec{v}_i$ is the object's initial momentum, $\vec{p}_f = m\vec{v}_f$ is its final momentum after the impulse, and $\Delta \vec{p} = \vec{p}_f - \vec{p}_i$ is the *change* in its momentum. This expression is known as the **impulse-momentum** theorem. It states that **an impulse**

Momentum is a vector that points in the same direction as the object's velocity.

FIGURE 9.7 A particle's momentum vector $\vec{p}$ can be decomposed into x- and y-components.

TABLE 9.1 Some typical momenta (approximate)

Object	Mass (kg)	Speed (m/s)	Momentum (kg · m/s)
Falling raindrop	2×10^{-5}	5	10^{-4}
Bullet	0.004	500	2
Pitched baseball	0.15	40	6
Running person	70	3.0	200
Car on highway	1000	30	3×10^4

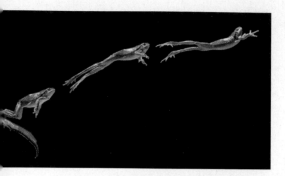

Legging it BIO A frog making a jump wants to gain as much momentum as possible before leaving the ground. This means that he wants the greatest impulse $J = F_{avg} \Delta t$ delivered to him by the ground. There is a maximum force that muscles can exert, limiting F_{avg}. But the time interval Δt over which the force is exerted can be greatly increased by having long legs. Many animals that are good jumpers have particularly long legs.

delivered to an object causes the object's momentum to change. That is, the *effect* of an impulsive force is to change the object's momentum from $\vec{p}_i$ to

$$\vec{p}_f = \vec{p}_i + \vec{J} \tag{9.9}$$

Equation 9.8 can also be written in terms of its *x*- and *y*-components as

$$J_x = \Delta p_x = (p_x)_f - (p_x)_i = m(v_x)_f - m(v_x)_i$$
$$J_y = \Delta p_y = (p_y)_f - (p_y)_i = m(v_y)_f - m(v_y)_i \tag{9.10}$$

The impulse-momentum theorem is illustrated by two examples in Figure 9.8. In the first, the putter strikes the ball, exerting a force on it and delivering an impulse $\vec{J} = \vec{F}_{avg} \Delta t$. Notice that the direction of the impulse is the same as that of the force. Because $\vec{p}_i = \vec{0}$ in this situation, we can use the impulse-momentum theorem to find that the ball leaves the putter with momentum $\vec{p}_f = \vec{p}_i + \vec{J} = \vec{J}$.

NOTE ▶ You can think of the putter as changing the ball's momentum by transferring momentum to it as an impulse. Thus we say the putter *delivers* an impulse to the ball, and the ball *receives* an impulse from the putter. ◀

The soccer player in Figure 9.8b presents a more complicated case. Here, the initial momentum of the ball is directed downward to the left. The impulse exerted on it by the player's head, upward to the right, is strong enough to reverse the ball's motion and send it off in a new direction. The graphical addition of vectors in Figure 9.8b again shows that $\vec{p}_f = \vec{p}_i + \vec{J}$.

(a)

(b)

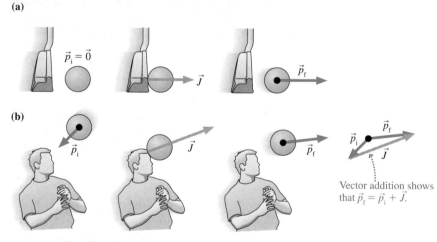

FIGURE 9.8 Impulse causes a *change* in momentum.

EXAMPLE 9.2 Calculating the change in momentum

A ball of mass $m = 0.25$ kg rolling to the right at 1.3 m/s strikes a wall and rebounds to the left at 1.1 m/s. What is the change in the ball's momentum? What is the impulse delivered to it by the wall?

PREPARE A visual overview of the ball bouncing is shown in Figure 9.9. This is a new kind of visual overview, one in which we show the situation "before" and "after" the interaction. We'll have more to say about before-and-after pictures in the next section. The ball is moving along the *x*-axis, so we'll write the momentum in component form, as in Equation 9.7. The change in momentum is then the difference between the final and initial values of the momentum. By the impulse-momentum theorem, the impulse is equal to this change in momentum.

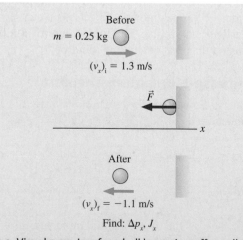

FIGURE 9.9 Visual overview for a ball bouncing off a wall.

SOLVE The x-component of the initial momentum is

$$(p_x)_i = m(v_x)_i = (0.25 \text{ kg})(1.3 \text{ m/s}) = 0.33 \text{ kg} \cdot \text{m/s}$$

The y-component of the momentum is zero both before and after the bounce. After the ball rebounds, the x-component of its momentum is

$$(p_x)_f = m(v_x)_f = (0.25 \text{ kg})(-1.1 \text{ m/s}) = -0.28 \text{ kg} \cdot \text{m/s}$$

It is particularly important to notice that the x-component of the momentum, like that of the velocity, is negative. This indicates that the ball is moving to the *left*. The change in momentum is

$$\Delta p_x = (p_x)_f - (p_x)_i = (-0.28 \text{ kg} \cdot \text{m/s}) - (0.33 \text{ kg} \cdot \text{m/s})$$
$$= -0.61 \text{ kg} \cdot \text{m/s}$$

The change in the momentum is negative. By the impulse-momentum theorem, the impulse delivered to the ball by the wall is equal to this change, so

$$J_x = \Delta p_x = -0.61 \text{ kg} \cdot \text{m/s}$$

ASSESS The impulse is negative, indicating that the force causing the impulse is pointing to the left, which makes sense.

An interesting application of the impulse-momentum theorem is to the question of how to slow down a fast-moving object in the gentlest possible way. For instance, a car is headed for a collision with a bridge abutment. How can this crash be made survivable? How do the rams in the chapter-opening photo avoid injury when they collide?

In these examples, the object has momentum $\vec{p}_i$ just before impact and zero momentum after (i.e., $\vec{p}_f = \vec{0}$). The impulse-momentum theorem tells us that

$$\vec{J} = \vec{F}_{avg} \Delta t = \Delta \vec{p} = \vec{p}_f - \vec{p}_i = -\vec{p}_i$$

or

$$\vec{F}_{avg} = -\frac{\vec{p}_i}{\Delta t} \qquad (9.11)$$

INVERSE
p.118

That is, the average force needed to stop an object is *inversely proportional* to the duration Δt of the collision. **If the duration of the collision can be increased, the force of the impact will be decreased.** This is the principle used in most impact-lessening techniques.

For example, obstacles such as bridge abutments are made safer by placing a line of water-filled barrels in front of them. Water is heavy but deformable. In case of a collision, the time it takes for the car to plow through these barrels is much longer than the time it would take it to stop if it hit the abutment head on. The force on the car (and on the driver from his or her seat belt) is greatly reduced by the longer-duration collision with the barrels.

The spines of a hedgehog obviously help protect it from predators. But they serve another function as well. If a hedgehog falls from a tree—a not uncommon occurrence—it simply rolls itself into a ball before it lands. Indeed, hedgehogs have been observed to purposely descend to the ground by simply dropping from the tree. Its thick spines then cushion the blow by increasing the time it takes for the animal to come to rest. Along with its small size, this adaptation allows for the hedgehog to easily survive long falls unhurt.

The butting rams shown in the photo at the beginning of this chapter also have adaptations that allow them to collide at high speeds without injury to their brains. The cranium has a double wall to prevent skull injuries, and there is a thick spongy mass that increases the time it takes for the brain to come to rest upon impact, again reducing the magnitude of the force on the brain.

TRY IT YOURSELF

Water balloon catch If you've ever tried to catch a water balloon, you may have learned the hard way not to catch it with your arms rigidly extended. The brief collision time implies a large, balloon-bursting force. A better way to catch a water balloon is to pull your arms in toward your body as you catch it, lengthening the collision time and hence reducing the force on the balloon.

A hedgehog is its own crash cushion!

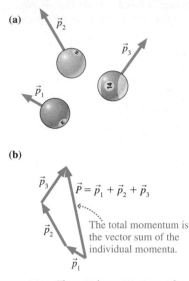

FIGURE 9.10 The total momentum of three pool balls.

Total Momentum

If we have more than one object moving—a *system* of particles—then the system as a whole has an overall momentum. The **total momentum** $\vec{P}$ (note the capital P) of a system of particles is the vector sum of the momenta of each of the individual particles:

$$\vec{P} = \vec{p}_1 + \vec{p}_2 + \vec{p}_3 + \cdots$$

Figure 9.10 shows how the momentum vectors of three moving pool balls are graphically added to find the total momentum. The concept of total momentum will be of key importance when we discuss the conservation law for momentum in Section 9.4.

STOP TO THINK 9.1 The cart's change of momentum is

A. $-30 \text{ kg} \cdot \text{m/s}$.
B. $-20 \text{ kg} \cdot \text{m/s}$.
C. $-10 \text{ kg} \cdot \text{m/s}$.
D. $10 \text{ kg} \cdot \text{m/s}$.
E. $20 \text{ kg} \cdot \text{m/s}$.
F. $30 \text{ kg} \cdot \text{m/s}$.

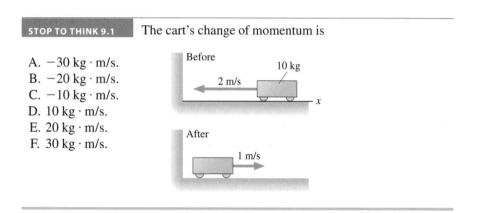

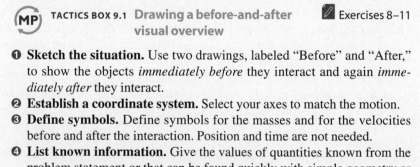

9.3 Solving Impulse and Momentum Problems

Visual overviews have become an important problem-solving tool. The visual overviews and free-body diagrams that you learned to draw in Chapters 1–8 were oriented toward the use of Newton's laws and a subsequent kinematical analysis. Now we are interested in making a connection between "before" and "after."

(MP) **TACTICS BOX 9.1** Drawing a before-and-after visual overview   ✏ Exercises 8–11

❶ **Sketch the situation.** Use two drawings, labeled "Before" and "After," to show the objects *immediately before* they interact and again *immediately after* they interact.

❷ **Establish a coordinate system.** Select your axes to match the motion.

❸ **Define symbols.** Define symbols for the masses and for the velocities before and after the interaction. Position and time are not needed.

❹ **List known information.** Give the values of quantities known from the problem statement or that can be found quickly with simple geometry or unit conversions. Before-and-after pictures are usually simpler than the pictures you used for dynamics problems, so listing known information on the sketch is often adequate.

❺ **Identify the desired unknowns.** What quantity or quantities will allow you to answer the question? These should have been defined as symbols in step 3.

EXAMPLE 9.3 Force in hitting a baseball

A 150 g baseball is thrown with a speed of 20 m/s. It is hit straight back toward the pitcher at a speed of 40 m/s. The impulsive force of the bat on the ball has the shape shown in Figure 9.11. What is the *maximum* force F_{max} that the bat exerts on the ball? What is the *average* force that the bat exerts on the ball?

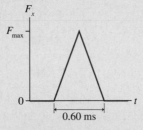

FIGURE 9.11 The interaction force between the baseball and the bat.

PREPARE Model the interaction as a collision. Figure 9.12 is a before-and-after visual overview in which the steps from Tactics Box 9.1 are explicitly noted. Because F_x is positive (a force to the right), we know the ball was initially moving toward the left and is hit back toward the right. Thus we converted the statements about *speeds* into information about *velocities*, with $(v_x)_i$ negative.

SOLVE In the last several chapters we've started the mathematical solution with Newton's second law. Now we want to use the impulse-momentum theorem:

$$\Delta p_x = J_x = \text{area under the force curve}$$

We know the velocities before and after the collision, so we can find the change in the ball's momentum:

$$\Delta p_x = m(v_x)_f - m(v_x)_i = (0.15 \text{ kg})(40 \text{ m/s} - (-20 \text{ m/s}))$$
$$= 9.0 \text{ kg} \cdot \text{m/s}$$

The force curve is a triangle with height F_{max} and width 0.60 ms. As in Example 9.1, the area under the curve is

$$J_x = \text{area} = \tfrac{1}{2} \times F_{max} \times (6.0 \times 10^{-4} \text{ s})$$
$$= (F_{max})(3.0 \times 10^{-4})$$

According to the impulse-momentum theorem, $\Delta p_x = J_x$, so we have

$$9.0 \text{ kg} \cdot \text{m/s} = (F_{max})(3.0 \times 10^{-4} \text{ s})$$

Thus the *maximum* force is

$$F_{max} = \frac{9.0 \text{ kg} \cdot \text{m/s}}{3.0 \times 10^{-4} \text{ s}} = 30{,}000 \text{ N}$$

The *average* force, which depends on the collision duration $\Delta t = 6.0 \times 10^{-4}$ s, has the smaller value

$$F_{avg} = \frac{J_x}{\Delta t} = \frac{\Delta p_x}{\Delta t} = \frac{9.0 \text{ kg} \cdot \text{m/s}}{6.0 \times 10^{-4} \text{ s}} = 15{,}000 \text{ N}$$

ASSESS F_{max} is a large force, but quite typical of the impulsive forces during collisions. The main thing to focus on is our new perspective: An impulse changes the momentum of an object.

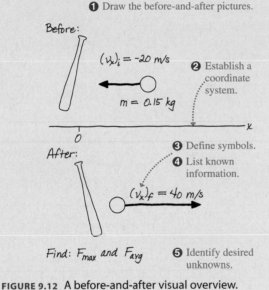

1 Draw the before-and-after pictures.

Before:

$(v_x)_i = -20$ m/s

$m = 0.15$ kg

2 Establish a coordinate system.

3 Define symbols.

4 List known information.

After:

$(v_x)_f = 40$ m/s

Find: F_{max} and F_{avg}

5 Identify desired unknowns.

FIGURE 9.12 A before-and-after visual overview.

The Impulse Approximation

Other forces often act on an object during a collision or other brief interaction. In Example 9.3, for instance, the baseball also has a weight force acting on it. Usually these other forces are *much* smaller than the interaction forces. The 1.5 N weight of the ball is vastly less than the 30,000 N force of the bat on the ball. We can reasonably neglect these small forces *during* the brief time of the impulsive force. Doing so is called the **impulse approximation.**

When we use the impulse approximation, $(p_x)_i$ and $(p_x)_f$—and $(v_x)_i$ and $(v_x)_f$—are then the momenta (and velocities) *immediately* before and *immediately* after the collision. For example, the velocities in Example 9.3 are those of the ball just before and after it collides with the bat. We could then do a follow-up problem, including weight and drag, to find the ball's speed a second later as the second baseman catches it.

EXAMPLE 9.4   Height of a bouncing ball

A 100 g rubber ball is thrown straight down onto a hard floor so that it strikes the floor with a speed of 11 m/s. Figure 9.13 shows the force that the floor exerts on the ball. Estimate the height of the ball's bounce.

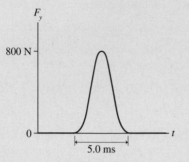

FIGURE 9.13 The force of the floor on a bouncing rubber ball.

PREPARE The ball experiences an impulsive force while in contact with the floor. Using the impulse approximation, we'll neglect the ball's weight during these 5.0 ms. The ball's rise after the bounce is free-fall motion; that is, motion subject only to the force of gravity. We'll use free-fall kinematics to describe the motion after the bounce.

Figure 9.14 is a visual overview. Here we have a two-part problem, an impulsive collision followed by upward free fall.

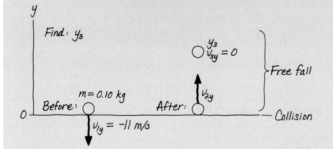

FIGURE 9.14 Before-and-after visual overview for a bouncing ball.

The overview thus shows the ball just before the collision, where we label its velocity as v_{1y}; just after the collision, where its velocity is v_{2y}; and at the highest point of its rising free fall, where its velocity is $v_{3y} = 0$.

SOLVE The impulse-momentum theorem tells us that $J_y = \Delta p_y = p_{2y} - p_{1y}$, so that $p_{2y} = p_{1y} + J_y$. The initial momentum, just before the collision, is $p_{1y} = mv_{1y} = (0.10 \text{ kg})(-11 \text{ m/s}) = -1.1 \text{ kg} \cdot \text{m/s}$.

Next, we need to find the impulse J_y, which is the area under the curve in Figure 9.13. Because the force is just given as a smooth curve, we'll have to *estimate* this area. Recall that the area can be written as $F_{avg} \Delta t$. From the curve, we might estimate F_{avg} to be about 400 N, or half the maximum value of the force. With this estimate we have

$$J_y = \text{area under the force curve} \approx (400 \text{ N}) \times (0.0050 \text{ s})$$
$$= 2.0 \text{ N} \cdot \text{s} = 2.0 \text{ kg} \cdot \text{m/s}$$

Thus

$$p_{2y} = p_{1y} + J_y = (-1.1 \text{ kg} \cdot \text{m/s}) + 2.0 \text{ kg} \cdot \text{m/s}$$
$$= 0.9 \text{ kg} \cdot \text{m/s}$$

and the post-collision velocity is

$$v_{2y} = \frac{p_{2y}}{m} = \frac{0.9 \text{ kg} \cdot \text{m/s}}{0.10 \text{ kg}} = 9 \text{ m/s}$$

The rebound speed is less than the impact speed, as expected. Finally, we can use free-fall kinematics to find

$$v_{3y}^2 = 0 = v_{2y}^2 - 2g \Delta y = v_{2y}^2 - 2gy_3$$
$$y_3 = \frac{v_{2y}^2}{2g} = \frac{(9 \text{ m/s})^2}{2(9.8 \text{ m/s}^2)} = 4 \text{ m}$$

We estimate that the ball bounces to a height of 4 m.

ASSESS This is a reasonable height for a rubber ball thrown down quite hard.

STOP TO THINK 9.2   A 10 g rubber ball and a 10 g clay ball are each thrown at a wall with equal speeds. The rubber ball bounces, the clay ball sticks. Which ball receives the greater impulse from the wall?

A. The clay ball receives a larger impulse because it sticks.
B. The rubber ball receives a larger impulse because it bounces.
C. They receive equal impulses because they have equal momenta.
D. Neither receives an impulse because the wall doesn't move.

9.4 Conservation of Momentum

The impulse-momentum theorem was derived from Newton's second law and is really just an alternative way of looking at that law. It is used in the context of single-particle dynamics, much as we used Newton's law in Chapters 4–7.

This chapter opened by noting that very complex interactions, such as two train cars coupling together, sometimes have very simple outcomes. To predict the outcomes, we need to see how Newton's *third* law looks in the language of impulse and momentum. Newton's third law will lead us to one of the most important conservation laws in physics.

Figure 9.15 shows two balls with initial velocities $(v_{1x})_i$ and $(v_{2x})_i$. The balls collide, then bounce apart with final velocities $(v_{1x})_f$ and $(v_{2x})_f$. The forces during the collision, as the balls are interacting, are the action/reaction pair $\vec{F}_{1\,on\,2}$ and $\vec{F}_{2\,on\,1}$. For now, we'll continue to assume that the motion is one dimensional along the x-axis.

> **NOTE ▶** The notation, with all the subscripts, may seem excessive. But there are two balls, and each has an initial and a final velocity, so we need to distinguish between four different velocities. ◀

Figure 9.16 shows the x-components of the two forces $\vec{F}_{1\,on\,2}$ and $\vec{F}_{2\,on\,1}$. Because they make up an action/reaction pair, the two forces always are of equal magnitude but of opposite sign. The impulse J_{2x} delivered to ball 2 is the area under the $\vec{F}_{1\,on\,2}$ curve; this area and hence this impulse are positive, indicating that the impulse on ball 2 is directed to the right. Likewise, the impulse J_{1x} delivered to ball 1 is the area "under" the $\vec{F}_{2\,on\,1}$ curve—but this area is considered to be *negative* because it is below the axis. As a result, the two impulses J_{1x} and J_{2x} are also equal in magnitude but opposite in sign, so that $J_{1x} = -J_{2x}$. For the soccer player in Figure 9.8, for example, this means that the impulse his head delivers to the ball is equal in magnitude but oppositely directed to the impulse the ball delivers to his head.

According to the impulse-momentum theorem, the changes in momenta of the two balls are

$$\Delta p_{1x} = J_{1x} \qquad \text{and} \qquad \Delta p_{2x} = J_{2x}$$

But because $J_{1x} = -J_{2x}$, we have

$$\Delta p_{1x} = -\Delta p_{2x}$$

or

$$\Delta p_{1x} + \Delta p_{2x} = \Delta(p_{1x} + p_{2x}) = 0$$

If the change in the quantity $p_{1x} + p_{2x}$ is zero, then it must be the case that

$$p_{1x} + p_{2x} = \text{constant} \qquad (9.12)$$

If $p_{1x} + p_{2x}$ is a constant, then the sum of the momenta *after* the collision equals the sum of the momenta *before* the collision. That is,

$$(p_{1x})_f + (p_{2x})_f = (p_{1x})_i + (p_{2x})_i \qquad (9.13)$$

Furthermore, this equality is independent of the interaction force. We don't need to know anything about $\vec{F}_{1\,on\,2}$ and $\vec{F}_{2\,on\,1}$ to make use of Equation 9.13.

Section 9.2 defined the total momentum $\vec{P}$ of a *system* of particles to be the sum of the momenta of each particle. Because Equation 9.13 is a sum of momenta, we can write it as

$$(P_x)_f = (P_x)_i \qquad (9.14)$$

Because it doesn't change during the collision, we say that the x-component of total momentum is *conserved*. Equations 9.13 and 9.14 are our first example of a *conservation law*.

Law of Conservation of Momentum

The same arguments just presented for the two colliding balls can be extended to systems containing any number of objects. Figure 9.17 on the next page

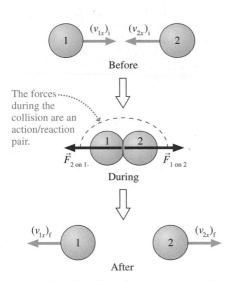

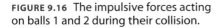

FIGURE 9.15 A collision between two balls.

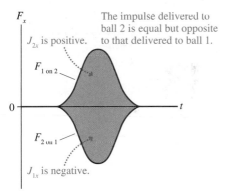

FIGURE 9.16 The impulsive forces acting on balls 1 and 2 during their collision.

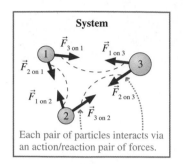

FIGURE 9.17 An isolated system of three particles.

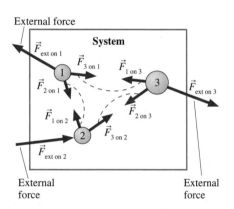

FIGURE 9.18 A system of particles subject to external forces.

shows the idea. Each pair of particles in the system interacts via forces that are an action/reaction pair. Exactly as for the two-particle collision, the change in momentum of particle 2 due to the force from particle 3 is exactly equal in magnitude, but opposite in direction, to the change in particle 3's momentum due to particle 2. The *net* change in the momentum of these two particles due to their interaction forces is thus zero. The same argument holds for every pair, with the net result that the change in the *total* momentum $\vec{P}$ of the system is zero. The total momentum of the system remains constant; that is, it is *conserved*.

Note in Figure 9.17 that the particles interact only with other particles that are part of the system. There are no forces on any of the particles from agents *outside* of the system. Forces that act only between particles within the system are called **internal forces**. A system subject only to internal forces is called an **isolated system** and, as we have just seen, **the total momentum of an isolated system is conserved.**

Most systems are not isolated but are subject to forces from agents outside the system. These forces are called **external forces.** For example, if you toss a ball (the system) into the air, an external force—its weight, the force of the earth's gravity on the ball—acts on the system. How do external forces affect the momentum of a system of particles?

In Figure 9.18 we show the same three-particle system of Figure 9.17, but now with *external* forces acting on the three particles. These external forces *can* change the momentum of the system. During a time interval Δt, for instance, the external force $\vec{F}_{\text{ext on 1}}$ acting on particle 1 changes its momentum, according to the impulse-momentum theorem, by $\Delta \vec{p}_1 = (\vec{F}_{\text{ext on 1}}) \Delta t$. The momenta of the other two particles change similarly. Thus the change in the total momentum is

$$
\begin{aligned}
\Delta \vec{P} &= \Delta \vec{p}_1 + \Delta \vec{p}_2 + \Delta \vec{p}_3 \\
&= (\vec{F}_{\text{ext on 1}} \Delta t) + (\vec{F}_{\text{ext on 2}} \Delta t) + (\vec{F}_{\text{ext on 3}} \Delta t) \\
&= (\vec{F}_{\text{ext on 1}} + \vec{F}_{\text{ext on 2}} + \vec{F}_{\text{ext on 3}}) \Delta t \\
&= \vec{F}_{\text{net}} \Delta t
\end{aligned}
\tag{9.15}
$$

where $\vec{F}_{\text{net}}$ is the net force due to *external forces*.

Equation 9.15 has two very important implications. First, it tells us that we can analyze the motion of the system as a whole without needing to consider the internal forces between the particles that make up the system. In fact, we have been using this idea all along as an *assumption* of the particle model. When we treat cars and rocks and baseballs as particles, we assume that the internal forces between the atoms—the forces that hold the object together—do not affect the motion of the object as a whole. Now we have *justified* that assumption.

The second implication of Equation (9.15), and the key one from the perspective of this chapter, applies to systems for which $\vec{F}_{\text{net}} = \vec{0}$. According to Equation 9.15, **if $\vec{F}_{\text{net}} = \vec{0}$, the *total* momentum $\vec{P}$ of the system does not change.** The total momentum remains constant, *regardless* of whatever interactions are going on *inside* the system.

Earlier, we found that the total momentum is also conserved when the system is isolated, with no external forces acting on it. With no external forces, it is certainly the case that $\vec{F}_{\text{net}} = \vec{0}$. Thus we can extend our definition of an isolated system: **An isolated system is one for which $\vec{F}_{\text{net}} = \vec{0}$.**

The importance of these results is sufficient to elevate them to a law of nature, alongside Newton's laws.

Law of conservation of momentum The total momentum $\vec{P}$ of an isolated system is a constant. Interactions within the system do not change the system's total momentum.

NOTE ▶ It is worth emphasizing the critical role of Newton's third law in the derivation of Equation 9.15. The law of conservation of momentum is a direct consequence of the fact that interactions within an isolated system are action/reaction pairs. ◀

Mathematically, the law of conservation of momentum for an isolated system is

$$\vec{P}_f = \vec{P}_i \qquad (9.16)$$

Law of conservation of momentum for an isolated system

The total momentum after an interaction is equal to the total momentum before the interaction. Because Equation 9.16 is a vector equation, the equality is true for each of the components of the momentum vector. That is,

$$\underbrace{(p_{1x})_f + (p_{2x})_f + (p_{3x})_f + \cdots}_{\text{Final momentum}} = \underbrace{(p_{1x})_i + (p_{2x})_i + (p_{3x})_i + \cdots}_{\text{Initial momentum}}$$

x-component

Particle 1 Particle 2 Particle 3

$$(9.17)$$

y-component $\cdots\blacktriangleright (p_{1y})_f + (p_{2y})_f + (p_{3y})_f + \cdots = (p_{1y})_i + (p_{2y})_i + (p_{3y})_i + \cdots$

EXAMPLE 9.5 Speed of ice skaters pushing off

Two ice skaters, Sandra and David, stand facing each other on frictionless ice. Sandra has a mass of 45 kg, David a mass of 80 kg. They then push off from each other. After the push, Sandra moves off at a speed of 2.2 m/s. What is David's speed?

PREPARE The two skaters interact with each other, but they form an isolated system because, for each skater, the upward normal force of the ice balances their downward weight force to make $\vec{F}_{net} = \vec{0}$. Thus the total momentum of the system of the two skaters is conserved.

Figure 9.19 shows a before-and-after visual overview for the two skaters. The total momentum before they push off is $\vec{P}_i = \vec{0}$

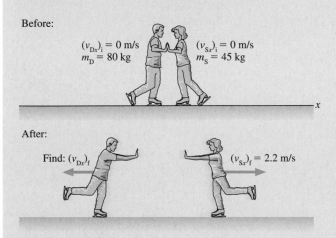

Before:

$(v_{Dx})_i = 0$ m/s
$m_D = 80$ kg

$(v_{Sx})_i = 0$ m/s
$m_S = 45$ kg

After:

Find: $(v_{Dx})_f$

$(v_{Sx})_f = 2.2$ m/s

FIGURE 9.19 Before-and-after visual overview for two skaters pushing off from each other.

because both skaters are at rest. Consequently, the total momentum will still be $\vec{0}$ *after* they push off.

SOLVE Since the motion is only in the x-direction, we'll only need to consider x-components of momentum. We write Sandra's initial momentum as $(p_{Sx})_i = m_S(v_{Sx})_i$, where m_S is her mass and $(v_{Sx})_i$ her initial velocity. Similarly, we write David's initial momentum as $(p_{Dx})_i = m_D(v_{Dx})_i$. Both these momenta are zero because both skaters are initially at rest.

We can now apply the mathematical statement of momentum conservation, Equation 9.17. Writing the final momentum of Sandra as $m_S(v_{Sx})_f$, and that of David as $m_D(v_{Dx})_f$, we have

$$\underbrace{m_S(v_{Sx})_f + m_D(v_{Dx})_f}_{\substack{\text{The skaters' final} \\ \text{momentum} \ldots}} = \underbrace{m_S(v_{Sx})_i + m_D(v_{Dx})_i}_{\substack{\ldots \text{equals their initial} \\ \text{momentum} \ldots}} = \underbrace{0}_{\substack{\ldots \text{which} \\ \text{was zero.}}}$$

Solving for $(v_{Dx})_f$, we find

$$(v_{Dx})_f = -\frac{m_S}{m_D}(v_{Sx})_f = -\frac{45 \text{ kg}}{80 \text{ kg}} \times 2.2 \text{ m/s} = -1.2 \text{ m/s}$$

David moves backward with a *speed* of 1.2 m/s.

ASSESS The *total* momentum of the system is zero both before and after they push off, but the individual momenta are not. Because $(p_{Sx})_f$ is positive (Sandra moves to the right), $(p_{Dx})_f$ must have the same magnitude but the opposite sign (David moves to the left).

Notice that we didn't need to know any details about the force between David and Sandra in order to find David's final speed. Conservation of momentum *mandates* this result.

A Strategy for Conservation of Momentum Problems

6.3, 6.4, 6.6, 6.7, 6.10 Actiy ONLINE Physics

Our derivation of the law of conservation of momentum, and the conditions under which it holds, suggests a problem-solving strategy.

(MP) **PROBLEM-SOLVING STRATEGY 9.1** **Conservation of momentum problems**

PREPARE Clearly define *the system.*

- If possible, choose a system that is isolated ($\vec{F}_{net} = \vec{0}$) or within which the interactions are sufficiently short and intense that you can ignore external forces for the duration of the interaction (the impulse approximation). Momentum is then conserved.
- If it's not possible to choose an isolated system, try to divide the problem into parts such that momentum is conserved during one segment of the motion. Other segments of the motion can be analyzed using Newton's laws or, as you'll learn in Chapter 10, conservation of energy.

Following Tactics Box 9.1, draw a before-and-after visual overview. Define symbols that will be used in the problem, list known values, and identify what you're trying to find.

SOLVE The mathematical representation is based on the law of conservation of momentum: $\vec{P}_f = \vec{P}_i$. In component form, this is

$$(p_{1x})_f + (p_{2x})_f + (p_{3x})_f + \cdots = (p_{1x})_i + (p_{2x})_i + (p_{3x})_i + \cdots$$
$$(p_{1y})_f + (p_{2y})_f + (p_{3y})_f + \cdots = (p_{1y})_i + (p_{2y})_i + (p_{3y})_i + \cdots$$

ASSESS Check that your result has the correct units, is reasonable, and answers the question.

EXAMPLE 9.6 Getaway speed of a cart

Bob is running from the police and thinks he can make a faster getaway by jumping on a stationary cart in front of him. He runs toward the cart, jumps on, and rolls along the horizontal street. Bob has a mass of 75 kg and the cart's mass is 25 kg. If Bob's speed is 4.0 m/s when he jumps onto the cart, what is the cart's speed after Bob jumps on?

PREPARE When Bob lands on and sticks to the cart, a "collision" occurs between Bob and the cart. If we take Bob and the cart together to be the system, the forces involved in this collision—friction forces between Bob's feet and the cart—are internal forces. The net external force on the system is zero, so the total momentum of Bob + cart is conserved: It is the same before and after the collision.

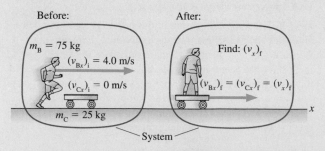

FIGURE 9.20 Before-and-after visual overview of Bob and the cart.

The visual overview in Figure 9.20 shows the important point that Bob and the cart move together after he lands on the cart, so $(v_x)_f$ is their common final velocity.

SOLVE To solve for the final velocity of Bob and the cart, we'll use conservation of momentum: $(P_x)_f = (P_x)_i$. Written in terms of the individual momenta, we have

$$(P_x)_i = m_B(v_{Bx})_i + m_C\underbrace{(v_{Cx})_i}_{0\ m/s} = m_B(v_{Bx})_i$$
$$(P_x)_f = m_B(v_x)_f + m_C(v_x)_f = (m_B + m_C)(v_x)_f$$

In the second equation, we've used the fact that both Bob and the cart travel at the common velocity of $(v_x)_f$. Equating the final and initial total momenta gives

$$(m_B + m_C)(v_x)_f = m_B(v_{Bx})_i$$

Solving this for $(v_x)_f$, we find

$$(v_x)_f = \frac{m_B}{m_B + m_C}(v_{Bx})_i = \frac{75\ \text{kg}}{100\ \text{kg}} \times 4.0\ \text{m/s} = 3.0\ \text{m/s}$$

The cart's speed is 3.0 m/s immediately after Bob jumps on.

ASSESS Bob has *lost* speed because he had to share his initial momentum with the cart. Not a good way to make a getaway!

Notice how easy this was! No forces, no acceleration constraints, no simultaneous equations. Why didn't we think of this before? While conservation laws are indeed powerful, they can only answer certain questions. Had we wanted to know how far Bob slid across the cart before sticking to it, how long the slide took, or what the cart's acceleration was during the collision, we would not have been able to answer such questions on the basis of the conservation law. There is a price to pay for finding a simple connection between before and after, and that price is the loss of information about the details of the interaction. If we are satisfied with knowing only about before and after, then conservation laws are a simple and straightforward way to proceed. But many problems *do* require us to understand the interaction, and for these there is no avoiding Newton's laws and all they entail.

It Depends on the System

The first step in the problem-solving strategy asks you to clearly define *the system*. This is worth emphasizing, because many problem-solving errors arise from trying to apply momentum conservation to an inappropriate system. **The goal is to choose a system whose momentum will be conserved.** Even then, it is the *total* momentum of the system that is conserved, not the momenta of the individual particles within the system.

As an example, consider what happens if you drop a rubber ball and let it bounce off a hard floor. Is momentum conserved during the collision of the ball with the floor? You might be tempted to answer yes because the ball's rebound speed is very nearly equal to its impact speed. But there are two errors in this reasoning.

First, momentum depends on *velocity,* not speed. The ball's velocity and momentum just before the collision are negative. They are positive after the collision. Even if their magnitudes are equal, the ball's momentum after the collision is *not* equal to its momentum before the collision.

But more importantly, we haven't defined the system. The momentum of what? Whether or not momentum is conserved depends on the system. Figure 9.21 shows two different choices of systems. In Figure 9.21a, where the ball itself is chosen as the system, the gravitational force of the earth on the ball is an external force. This force causes the ball to accelerate toward the earth, changing the ball's momentum. The force of the floor on the ball is also an external force. The impulse of $\vec{F}_{\text{floor on ball}}$ changes the ball's momentum from "down" to "up" as the ball bounces. The momentum of this system is most definitely *not* conserved.

Figure 9.21b shows a different choice. Here the system is ball + earth. Now the gravitational forces and the impulsive forces of the collision are interactions *within* the system. This is an isolated system, so the *total* momentum $\vec{P} = \vec{p}_{\text{ball}} + \vec{p}_{\text{earth}}$ is conserved.

In fact, the total momentum is $\vec{P} = \vec{0}$. Before you release the ball, both the ball and the earth are at rest (in the earth's reference frame). The total momentum is zero before you release the ball, so it will *always* be zero. Consider the situation just before the ball hits the floor. If the ball's velocity is $v_{\text{B}y}$, it must be the case that

$$m_{\text{B}} v_{\text{B}y} + m_{\text{E}} v_{\text{E}y} = 0$$

and thus

$$v_{\text{E}y} = -\frac{m_{\text{B}}}{m_{\text{E}}} v_{\text{B}y}$$

In other words, as the ball is pulled down toward the earth, the ball pulls up on the earth (action/reaction pair of forces) until the entire earth reaches velocity $v_{\text{E}y}$. The earth's momentum is equal and opposite to the ball's momentum.

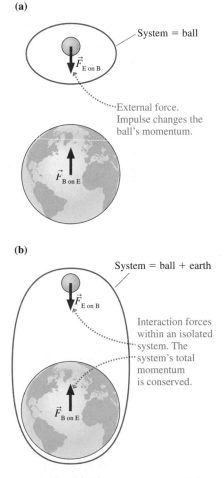

(a)

System = ball

$\vec{F}_{\text{E on B}}$

External force. Impulse changes the ball's momentum.

$\vec{F}_{\text{B on E}}$

(b)

System = ball + earth

$\vec{F}_{\text{E on B}}$

Interaction forces within an isolated system. The system's total momentum is conserved.

$\vec{F}_{\text{B on E}}$

FIGURE 9.21 Whether or not momentum is conserved as a ball falls to earth depends on your choice of the system.

Why don't we notice the earth "leaping up" toward us each time we drop something? Because of the earth's enormous mass relative to everyday objects. A typical rubber ball has a mass of 60 g and hits the ground with a velocity of about -5 m/s. The earth's upward velocity is thus

$$v_{Ey} \approx -\frac{6 \times 10^{-2} \text{ kg}}{6 \times 10^{24} \text{ kg}}(-5 \text{ m/s}) = 5 \times 10^{-26} \text{ m/s}$$

The earth does, indeed, have a momentum equal and opposite to that of the ball, but the earth is so massive that it needs only an infinitesimal velocity to match the ball's momentum. At this speed, it would take the earth 300 million years to move the diameter of an atom!

Explosions

An **explosion,** where the particles of the system move apart after a brief, intense interaction, is the opposite of a collision. The explosive forces, which could be from an expanding spring or from expanding hot gases, are *internal* forces. If the system is isolated, its total momentum during the explosion will be conserved.

EXAMPLE 9.7 Recoil speed of a rifle

A 30 g ball is fired from a 1.2 kg spring-loaded toy rifle with a speed of 15 m/s. What is the recoil speed of the rifle?

PREPARE As the ball moves down the barrel, there are complicated forces exerted on the ball and on the rifle. However, if we take the system to be the ball + rifle, these are *internal* forces that do not change the total momentum.

The *external* forces of the rifle's and ball's weights are balanced by the external force exerted by the person holding the

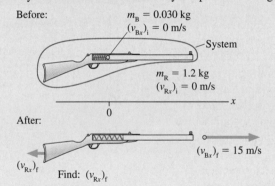

FIGURE 9.22 Before-and-after visual overview for a toy rifle.

rifle, so $\vec{F}_{net} = \vec{0}$. This is an isolated system and the law of conservation of momentum applies.

Figure 9.22 shows a visual overview before and after the ball is fired. We'll assume the ball is fired in the $+x$-direction.

SOLVE The x-component of the total momentum is $P_x = p_{Bx} + p_{Rx}$. Everything is at rest before the trigger is pulled, so the initial momentum is zero. After the trigger is pulled, the internal force of the spring pushes the ball down the barrel *and* pushes the rifle backwards. Conservation of momentum gives

$$(P_x)_f = m_B(v_{Bx})_f + m_R(v_{Rx})_f = (P_x)_i = 0$$

Solving for the rifle's velocity, we find

$$(v_{Rx})_f = -\frac{m_B}{m_R}(v_{Bx})_f = -\frac{0.030 \text{ kg}}{1.2 \text{ kg}} \times 15 \text{ m/s} = -0.38 \text{ m/s}$$

The minus sign indicates that the rifle's recoil is to the left. The recoil *speed* is 0.38 m/s.

ASSESS Real rifles fire their bullets at much higher velocities, and their recoil is correspondingly higher. Shooters need to brace themselves against the "kick" of the rifle back against their shoulder.

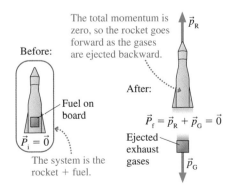

FIGURE 9.23 Rocket propulsion is an example of conservation of momentum.

We would not know where to begin to solve a problem such as this using Newton's laws. But Example 9.7 is a simple problem when approached from the before-and-after perspective of a conservation law. The selection of ball + rifle as "the system" was the critical step. For momentum conservation to be a useful principle, we had to select a system in which the complicated forces due to the spring and to friction were all internal forces. The rifle by itself is *not* an isolated system, so its momentum is *not* conserved.

Much the same reasoning explains how a rocket or jet aircraft accelerates. Figure 9.23 shows a rocket with a parcel of fuel on board. Burning converts the fuel to hot gases that are expelled from the rocket motor. If we choose rocket + gases to be the system, the burning and expulsion are all internal forces. In deep space there are no other forces, so the total momentum of the rocket + gases sys-

▶ **Squid propulsion** BIO Squids use a form of "rocket propulsion" to make quick movements to escape enemies or catch prey. The squid draws in water through a pair of valves in its outer sheath, or mantle, and then quickly expels the water through a funnel, propelling the squid backward. The funnel's direction is adjustable, allowing the squid to move in any backward direction.

tem must be conserved. The rocket gains forward velocity and momentum as the exhaust gases are shot out the back, but the *total* momentum of the system remains zero.

Many people find it hard to understand how a rocket can accelerate in the vacuum of space because there is nothing to "push against." Thinking in terms of momentum, you can see that the rocket does not push against anything *external,* but only against the gases that it pushes out the back. In return, in accordance with Newton's third law, the gases push forward on the rocket. The details of rocket propulsion are more complex than we want to handle, because the mass of the rocket is changing, but you should be able to use the law of conservation of momentum to understand the basic principle by which rocket propulsion occurs.

STOP TO THINK 9.3 An explosion in a rigid pipe shoots out three pieces. A 6 g piece comes out the right end. A 4 g piece comes out the left end with twice the speed of the 6 g piece. From which end, left or right, does the third piece emerge?

9.5 Inelastic Collisions

Collisions can have different possible outcomes. A rubber ball dropped on the floor bounces—it's *elastic*—but a ball of clay sticks to the floor without bouncing; we call such a collision *inelastic.* A golf club hitting a golf ball causes the ball to rebound away from the club (elastic), but a bullet striking a block of wood embeds itself in the block (inelastic).

A collision in which the two objects stick together and move with a common final velocity is called a **perfectly inelastic collision.** The clay hitting the floor and the bullet embedding itself in the wood are examples of perfectly inelastic collisions. Other examples include railroad cars coupling together upon impact and darts hitting a dart board. Figure 9.24 emphasizes the fact that the two objects have a common final velocity after they collide. (We have drawn the combined object moving to the right, but it could have ended up moving to the left, depending on the objects' masses and initial velocities.)

In an *elastic collision,* by contrast, the two objects bounce apart. We've looked at some examples of elastic collisions, but a full analysis requires some ideas about energy. We will return to elastic collisions in Chapter 10.

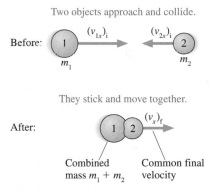

Two objects approach and collide.

Before:

They stick and move together.

After:

Combined mass $m_1 + m_2$

Common final velocity

FIGURE 9.24 A perfectly inelastic collision.

EXAMPLE 9.8 Speeds in a perfectly inelastic glider collision

In a laboratory experiment, a 200 g air-track glider and a 400 g air-track glider are pushed toward each other from opposite ends of the track. The gliders have Velcro tabs on the front so that they will stick together when they collide. The 200 g glider is pushed with an initial speed of 3.0 m/s. The collision causes it to reverse direction at 0.50 m/s. What was the initial speed of the 400 g glider?

Continued

PREPARE Model the gliders as particles. Define the two gliders as the system. This is an isolated system, so its total momentum is conserved in the collision. The gliders stick together, so this is a perfectly inelastic collision.

Figure 9.25 shows a visual overview. We've chosen to let the 200 g glider (glider 1) start out moving to the right, so $(v_{1x})_i$ is a positive 3.0 m/s. The gliders move to the left after the colli-

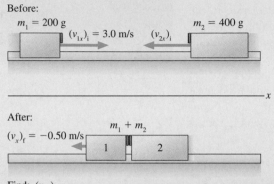

Before:

$m_1 = 200$ g $m_2 = 400$ g

$(v_{1x})_i = 3.0$ m/s $(v_{2x})_i$

—————————————————————— x

After:

$(v_x)_f = -0.50$ m/s $m_1 + m_2$

           1   2

Find: $(v_{2x})_i$

FIGURE 9.25 Before-and-after visual overview for two gliders colliding on an air track.

sion, so their common final velocity is $(v_x)_f = -0.50$ m/s. You can see that velocity $(v_{2x})_i$ must be negative in order to "turn around" both gliders.

SOLVE The law of conservation of momentum, $(P_x)_f = (P_x)_i$, is

$$(m_1 + m_2)(v_x)_f = m_1(v_{1x})_i + m_2(v_{2x})_i$$

where we made use of the fact that the combined mass $m_1 + m_2$ moves together after the collision. We can easily solve for the initial velocity of the 400 g glider:

$$(v_{2x})_i = \frac{(m_1 + m_2)(v_x)_f - m_1(v_{1x})_i}{m_2}$$

$$= \frac{(0.60 \text{ kg})(-0.50 \text{ m/s}) - (0.20 \text{ kg})(3.0 \text{ m/s})}{0.40 \text{ kg}}$$

$$= -2.3 \text{ m/s}$$

The negative sign, which we anticipated, indicates that the 400 g glider started out moving to the left. The initial *speed* of the glider, which we were asked to find, is 2.3 m/s.

ASSESS The key step in solving inelastic collision problems is that both objects move after the collision with the same velocity. You should thus choose a single symbol (here, $(v_x)_f$) for this common velocity.

STOP TO THINK 9.4 The two particles shown collide and stick together. After the collision, the combined particles

A. Move to the right as shown.   B. Move to the left.   C. Are at rest.

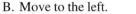

2.5 m/s   1.5 m/s

① ⟶ ⟵ ②   ①②⟶

0.40 kg   Before   0.80 kg   After

9.6 Momentum and Collisions in Two Dimensions

Our examples thus far have been confined to motion along a one-dimensional axis. Many practical examples of momentum conservation involve motion in a plane. The total momentum $\vec{P}$ is a *vector* sum of the momenta $\vec{p} = m\vec{v}$ of the individual particles. Consequently, as we found in Section 9.3, momentum is conserved only if each component of $\vec{P}$ is conserved:

$$(p_{1x})_f + (p_{2x})_f + (p_{3x})_f + \cdots = (p_{1x})_i + (p_{2x})_i + (p_{3x})_i + \cdots$$
$$(p_{1y})_f + (p_{2y})_f + (p_{3y})_f + \cdots = (p_{1y})_i + (p_{2y})_i + (p_{3y})_i + \cdots \quad (9.18)$$

In this section we'll apply momentum conservation to motion in two dimensions.

Collisions and explosions often involve motion in two dimensions.

EXAMPLE 9.9 Analyzing a peregrine falcon strike

Peregrine falcons often grab their prey from above while both falcon and prey are in flight. A falcon, flying at 18 m/s, swoops down at a 45° angle from behind a pigeon flying horizontally at 9.0 m/s. The falcon has a mass of 0.80 kg and the pigeon a mass of 0.36 kg. What are the speed and direction of the falcon (now holding the pigeon) immediately after impact?

PREPARE This is a perfectly inelastic collision because after the collision the falcon and pigeon move at a common velocity. The total momentum of the falcon + pigeon system is conserved. For a two-dimensional collision, this means that the x-component of the total momentum before the collision must equal the x-component of the total momentum after the collision, and similarly for the y-components. Figure 9.26 is a before-and-after visual overview.

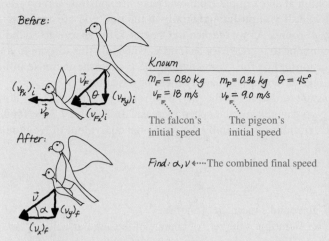

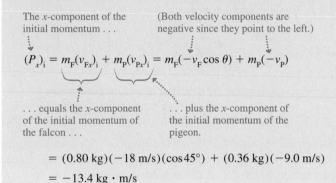

FIGURE 9.26 Before-and-after visual overview for a falcon catching a pigeon.

SOLVE We'll start by finding the x- and y-components of the momentum before the collision. For the x-component we have

The x-component of the initial momentum . . .   (Both velocity components are negative since they point to the left.)

$$(P_x)_i = m_F(v_{Fx})_i + m_P(v_{Px})_i = m_F(-v_F \cos\theta) + m_P(-v_P)$$

. . . equals the x-component of the initial momentum of the falcon . . .   . . . plus the x-component of the initial momentum of the pigeon.

$$= (0.80 \text{ kg})(-18 \text{ m/s})(\cos 45°) + (0.36 \text{ kg})(-9.0 \text{ m/s})$$

$$= -13.4 \text{ kg} \cdot \text{m/s}$$

Similarly, for the y-component of the initial momentum we have

$$(P_y)_i = m_F(v_{Fy})_i + m_P(v_{Py})_i = m_F(-v_F \sin\theta) + 0$$

$$= (0.80 \text{ kg})(-18.0 \text{ m/s})(\sin 45°) = -10.2 \text{ kg} \cdot \text{m/s}$$

After the collision, the two birds move with a common velocity $\vec{v}$ that is directed at an angle α from the horizontal. The x-component of the final momentum is then

$$(P_x)_f = (m_F + m_P)(v_x)_f$$

Momentum conservation requires $(P_x)_f = (P_x)_i$, so

$$(v_x)_f = \frac{(P_x)_i}{(m_F + m_P)} = \frac{-13.4 \text{ kg} \cdot \text{m/s}}{(0.80 \text{ kg}) + (0.36 \text{ kg})} = -11.6 \text{ m/s}$$

Similarly, $(P_y)_f = (P_y)_i$ gives

$$(v_y)_f = \frac{(P_y)_i}{(m_F + m_P)} = \frac{-10.2 \text{ kg} \cdot \text{m/s}}{(0.80 \text{ kg}) + (0.36 \text{ kg})} = -8.79 \text{ m/s}$$

From the figure we see that $\tan\alpha = (v_y)_f/(v_x)_f$, so that

$$\alpha = \tan^{-1}\left(\frac{(v_y)_f}{(v_x)_f}\right) = \tan^{-1}\left(\frac{-8.79 \text{ m/s}}{-11.6 \text{ m/s}}\right) = 37°$$

The magnitude of the final velocity (i.e., the speed) can be found from the Pythagorean theorem as

$$v = \sqrt{(v_x)_f^2 + (v_y)_f^2}$$

$$= \sqrt{(-11.6 \text{ m/s})^2 + (-8.79 \text{ m/s})^2} = 15 \text{ m/s}$$

Thus immediately after impact the falcon, with its meal, is moving 37° below horizontal at a speed of 15 m/s.

It's instructive to examine this collision with a picture of the momentum vectors. The vectors $\vec{p}_F$ and $\vec{p}_P$ before the collision, and their sum $\vec{P} = \vec{p}_F + \vec{p}_P$, are shown in Figure 9.27. You can see that the total momentum vector makes a 53° angle with the negative y-axis. The individual momenta change in the collision, *but the total momentum does not.*

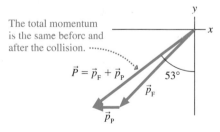

FIGURE 9.27 The momentum vectors of the falcon strike.

9.7 Angular Momentum

For a single particle, we can think of the law of conservation of momentum as an alternative way of stating Newton's first law. Rather than saying that a particle will continue to move in a straight line at constant velocity unless acted on by a net force, we can say that the momentum of an isolated particle is conserved. Both express the idea that a particle moving in a straight line tends to "keep going" unless something acts on it to change its motion.

Another important motion you've studied is motion in a circle. The momentum $\vec{p}$ is *not* conserved for a particle undergoing circular motion. Momentum is a vector, and the momentum of a particle in circular motion changes as the direction of motion changes.

Nonetheless, a spinning bicycle wheel would keep turning if it were not for friction, and a ball moving in a circle at the end of a string tends to "keep going" in a circular path. The quantity that expresses this idea for circular motion is called *angular momentum*.

FIGURE 9.28 By applying a torque to the merry-go-round, the girl is increasing its angular momentum.

Let's start by looking at an example from everyday life: pushing a merry-go-round, as in Figure 9.28. If you push tangentially to the rim, you are applying a *torque* to the merry-go-round. As we learned in Chapter 7, the merry-go-round's angular speed will continue to increase for as long as you apply this torque. If you push *harder* (greater torque) or for a *longer time,* the greater the increase in its angular velocity will be. How can we quantify these observations?

Let's apply a constant torque τ_{net} to the merry-go-round for a time Δt. By how much will the merry-go-round's angular speed increase? In Section 7.3 we found that the angular acceleration α is given by the rotational equivalent of Newton's second law, or

$$\alpha = \frac{\tau_{net}}{I} \tag{9.19}$$

where I is the merry-go-round's moment of inertia.

Now the angular acceleration is the rate of change of the angular velocity, so

$$\alpha = \frac{\Delta \omega}{\Delta t} \tag{9.20}$$

Setting Equations 9.19 and 9.20 equal to each other gives

$$\frac{\Delta \omega}{\Delta t} = \frac{\tau_{net}}{I}$$

or, rearranging,

$$\tau_{net}\, \Delta t = I\, \Delta \omega \tag{9.21}$$

If you recall the impulse-momentum theorem for *linear* motion, which is

$$\vec{F}_{net}\, \Delta t = m\, \Delta \vec{v} = \Delta \vec{p} \tag{9.22}$$

you can see that Equation 9.21 is an analogous statement about rotational motion. Because the quantity $I\omega$ is evidently the rotational equivalent of $m\vec{v}$, the linear momentum $\vec{p}$, it seems reasonable to define the **angular momentum** L to be

$$L = I\omega \tag{9.23}$$

Angular momentum of an object with moment
of inertia I rotating at angular velocity ω

The SI units of angular momentum are those of moment of inertia times those of angular velocity, or $kg \cdot m^2/s$.

Just as an object in linear motion can have a large momentum either by having a large mass or a large speed, a rotating object can have a large angular momentum by having a large moment of inertia or a large angular velocity. The merry-go-round in Figure 9.28 has a larger angular momentum if it's spinning fast than if it's spinning slowly. Also, the merry-go-round (large I) has a much larger angular momentum than a toy top (small I) spinning with the same angular velocity.

Table 9.2 summarizes the analogies between linear and rotational quantities that you learned in Chapter 7 and adds the analogy between linear momentum and angular momentum.

TABLE 9.2 Rotational and linear dynamics

Rotational dynamics	Linear dynamics
Torque τ_{net}	Force $\vec{F}_{net}$
Moment of inertia I	Mass m
Angular velocity ω	Velocity $\vec{v}$
Angular momentum $L = I\omega$	Linear momentum $\vec{p} = m\vec{v}$

Conservation of Angular Momentum

Having now defined angular momentum, Equation 9.21 reads

$$\tau_{net}\, \Delta t = \Delta L \tag{9.24}$$

in exact analogy with its linear dynamics equivalent, Equation 9.22. This equation states that the change in the angular momentum of an object is proportional to the net torque applied to the object. If the net torque on an object is *zero,* the rotational analog of an isolated system, then the change in the angular momentum is zero as well. That is, a rotating object will continue to rotate with *constant* angular velocity—to "keep going"—unless acted upon by an external torque. We can state this conclusion as the *law of conservation of angular momentum:*

Actlv
Physlcs   7.14

> **Law of conservation of angular momentum** The angular momentum of a rotating object subject to no net torque ($\tau_{net} = 0$) is a constant. The final angular momentum L_f is equal to the initial angular momentum L_i.

This law is analogous to that for the conservation of linear momentum; there, linear momentum is conserved if the net *force* is zero. Because the angular momentum is $L = I\omega$, the mathematical statement of the law of conservation of angular momentum is

$$I_i\omega_i = I_f\omega_f \qquad (9.25)$$

EXAMPLE 9.10 Period of a merry-go-round

Joey, whose mass is 36 kg, stands at the center of a 200 kg merry-go-round that is rotating once every 2.5 s. While it is rotating, Joey walks out to the edge of the merry-go-round, 2.0 m from its center. What is the rotational period of the merry-go-round when Joey gets to the edge?

PREPARE Take the system to be Joey + merry-go-round and assume frictionless bearings. There is no external torque on this system, so the angular momentum of the system will be conserved. As shown in the visual overview of Figure 9.29, we model the merry-go-round as a uniform disk of radius $R = 2.0$ m. From Table 7.2, the moment of inertia of a disk is $I_{disk} = \frac{1}{2}MR^2$. Modeling Joey as a particle of mass m, his moment of inertia is zero when he is at the center, but it increases to mR^2 when he reaches the edge.

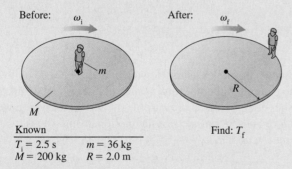

Known		Find: T_f
$T_i = 2.5$ s	$m = 36$ kg	
$M = 200$ kg	$R = 2.0$ m	

FIGURE 9.29 Visual overview of the merry-go-round.

SOLVE The mathematical statement of the law of conservation of momentum is $L_i = L_f$, or, from Equation 9.25, $I_i\omega_i = I_f\omega_f$ which we can rewrite as

$$\omega_f = \frac{I_i}{I_f}\omega_i$$

As Joey moves out to the edge, the moment of inertia of the system increases and, as a result, the angular velocity decreases. Initially, the moment of inertia of the system is just that of the merry-go-round because Joey's contribution is zero. Thus

$$I_i = I_{disk} = \frac{1}{2}MR^2 = \frac{1}{2}(200 \text{ kg})(2.0 \text{ m})^2 = 400 \text{ kg} \cdot \text{m}^2$$

When Joey reaches the edge, the total moment of inertia becomes

$$I_f = I_{disk} + mR^2 = 400 \text{ kg} \cdot \text{m}^2 + (36 \text{ kg})(2.0 \text{ m})^2$$
$$= 540 \text{ kg} \cdot \text{m}^2$$

The initial angular speed is related to the initial period of rotation T_i by

$$\omega_i = \frac{2\pi}{T_i} = \frac{2\pi}{2.5 \text{ s}} = 2.5 \text{ rad/s}$$

Thus the final angular speed is

$$\omega_f = \frac{I_i}{I_f}\omega_i = \frac{400 \text{ kg} \cdot \text{m}^2}{540 \text{ kg} \cdot \text{m}^2}(2.5 \text{ rad/s}) = 1.9 \text{ rad/s}$$

When Joey reaches the edge, the period of the merry-go-round has increased to

$$T_f = \frac{2\pi}{\omega_f} = \frac{2\pi}{1.9 \text{ rad/s}} = 3.3 \text{ s}$$

ASSESS The merry-go-round rotates *more slowly* after Joey moves out to the edge. If the system's moment of inertia increases, as it does when Joey moves out, the angular velocity must decrease to keep the angular momentum constant.

Large moment of inertia; slow spin.

Small moment of inertia; fast spin.

Conservation of momentum enters into many aspects of sports. Because no external torques act, the angular momentum of a platform diver is conserved while she's in the air. Just as for Joey and the merry-go-round of Example 9.10, she spins slowly when her moment of inertia is large; by decreasing her moment of inertia, she increases her rate of spin. Divers can thus markedly increase their spin rate by changing their body from an extended posture to a tuck position. Figure skaters also increase their spin rate by decreasing their moment of inertia, as shown in Figure 9.30. The following example gives a simplified treatment of this process.

FIGURE 9.30 A spinning figure skater.

EXAMPLE 9.11 **Analyzing a spinning ice skater**

An ice skater spins around on the tips of his blades while holding a 5.0 kg weight in each hand. He begins with his arms straight out from his body and his hands 140 cm apart. While spinning at 2.0 rev/s, he pulls the weights in and holds them 50 cm apart against his shoulders. If we neglect the mass of the skater, how fast is he spinning after pulling the weights in?

PREPARE Although the mass of the skater is larger than the mass of the weights, neglecting the skater's mass is not a bad approximation. Angular momentum depends on the radius of the circle. The skater's mass is concentrated in his torso, which has an effective radius (i.e., where most of the mass is concentrated) of only 9 or 10 cm. The weights move in much larger circles and have a disproportionate influence on his motion. The skater's arms exert radial forces on the weights just to keep them moving in circles, and even larger radial forces as he pulls them in. But there is no torque on the weights, so their total angular momentum is conserved. Figure 9.31 shows a before-and-after visual overview, as seen from above.

SOLVE The two weights have the same mass, move in circles with the same radius, and have the same angular velocity. Thus the total angular momentum is twice that of one weight. The mathematical statement of angular momentum conservation, $I_f \omega_f = I_i \omega_i$, is

$$(2mr_f^2)\omega_f = (2mr_i^2)\omega_i$$

Because the angular velocity is related to the rotation frequency f by $\omega = 2\pi f$, this equation simplifies to

$$f_f = \left(\frac{r_i}{r_f}\right)^2 f_i$$

When he pulls the weights in, his rotation frequency increases to

$$f_f = \left(\frac{0.70 \text{ m}}{0.25 \text{ m}}\right)^2 \times 2.0 \text{ rev/s} = 16 \text{ rev/s}$$

ASSESS Pulling in the weights increases the skater's spin from 2 rev/s to 16 rev/s. This is somewhat high, because we neglected the mass of the skater, but it illustrates how skaters do "spin up" by pulling their mass in toward the rotation axis.

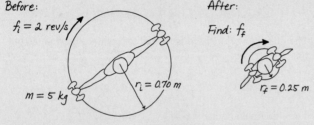

Before:

$f_i = 2$ rev/s

$m = 5$ kg $r_i = 0.70$ m

After:

Find: f_f

$r_f = 0.25$ m

FIGURE 9.31 Top view visual overview of the spinning ice skater.

The eye of a hurricane As air masses from the slowing rotating outer zones are drawn toward the low-pressure center, their moment of inertia decreases. Because the angular momentum of these air masses is conserved, their speed must *increase* as they approach the center, leading to the high wind speeds near the center of the storm.

Solving either of these two examples using Newton's laws would be quite difficult. We would have to deal with internal forces, such as Joey's feet against the merry-go-round, and other complications. For problems like these, where we're interested only in the before-and-after aspects of the motion, using a conservation law makes the solution much simpler.

STOP TO THINK 9.5 The left figure shows two boys of equal mass standing halfway to the edge on a turntable that is freely rotating at angular speed ω_i. They then walk to the positions shown in the right figure. The final angular speed ω_f is

ω_i $\omega_f = ?$

A. Greater than ω_i.   B. Less than ω_i.   C. Equal to ω_i.

SUMMARY

The goals of Chapter 9 have been to introduce the ideas of impulse, momentum, and angular momentum and to learn a new problem-solving strategy based on conservation laws.

GENERAL PRINCIPLES

Law of Conservation of Momentum

The total momentum $\vec{P} = \vec{p}_1 + \vec{p}_2 + \cdots$ of an isolated system is a constant. Thus

$$\vec{P}_f = \vec{P}_i$$

Conservation of Angular Momentum

The angular momentum L of a rotating object subject to zero external torque does not change. Thus

$$L_f = L_i$$

This can be written in terms of the moment of inertia and angular velocity as

$$I_f \omega_f = I_i \omega_i$$

Solving Momentum Conservation Problems

PREPARE Choose an isolated system or a system that is isolated during at least part of the problem. Draw a visual overview of the system before and after the interaction.

SOLVE Write the law of conservation of momentum in terms of vector components

$$(p_{1x})_f + (p_{2x})_f + \cdots = (p_{1x})_i + (p_{2x})_i + \cdots$$
$$(p_{1y})_f + (p_{2y})_f + \cdots = (p_{1y})_i + (p_{2y})_i + \cdots$$

In terms of masses and velocities, this is

$$m_1(v_{1x})_f + m_2(v_{2x})_f + \cdots = m_1(v_{1x})_i + m_2(v_{2x})_i + \cdots$$
$$m_1(v_{1y})_f + m_2(v_{2y})_f + \cdots = m_1(v_{1y})_i + m_2(v_{2y})_i + \cdots$$

ASSESS Is the result reasonable?

IMPORTANT CONCEPTS

Momentum $\vec{p} = m\vec{v}$

Impulse J_x = area under force curve

Impulse and momentum are related by the impulse-momentum theorem

$$\Delta p_x = J_x$$

This is an alternative statement of Newton's second law.

Angular momentum $L = I\omega$ is the rotational analog of linear momentum $\vec{p} = m\vec{v}$.

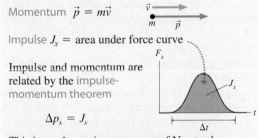

System A group of interacting particles.

Isolated system A system on which the net external force is zero.

Before-and-after visual overview

- Define the system.

- Use two drawings to show the system *before* and *after* the interaction.

- List known information and identify what you are trying to find.

APPLICATIONS

Collisions Two or more particles come together. In a perfectly inelastic collision, they stick together and move with a common final velocity.

Explosions Two or more particles move away from each other.

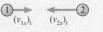

Two dimensions Both the x- and y-components of the total momentum P must be conserved, giving two simultaneous equations.

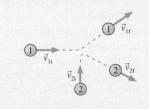

For instructor-assigned homework, go to
www.masteringphysics.com

Problems labeled 🖉 can be done on a Workbook
Momentum Worksheet; INT integrate significant material from earlier chapters; BIO are of biological or medical interest.

Problem difficulty is labeled as I (straightforward) to IIIII (challenging).

QUESTIONS

Conceptual Questions

1. Rank in order, from largest to smallest, the momenta p_{1x} through p_{5x} of the objects presented in Figure Q9.1. Explain.

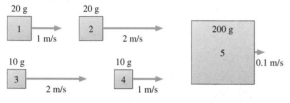

FIGURE Q9.1

2. Explain the concept of *impulse* in nonmathematical language. That is, don't simply put the equation in words to say that "impulse is the area under the force-versus-time curve." Explain it in terms that would make sense to an educated person who had never heard of it.

3. A 0.2 kg plastic cart and a 20 kg lead cart can roll without friction on a horizontal surface. Equal forces are used to push both carts forward for a time of 1 s, starting from rest. After the force is removed at $t = 1$ s, is the momentum of the plastic cart greater than, less than, or equal to the momentum of the lead cart? Explain.

4. Two pucks, of mass m and $4m$, lie on a frictionless table. Equal forces are used to push both pucks forward a distance of 1 m.
   a. Which puck takes longer to travel the distance? Explain.
   b. Which puck has the greater momentum upon completing the distance? Explain.

5. Explain the concept of "isolated system" in nonmathematical language.

6. A golf club continues forward after hitting the golf ball. Is momentum conserved in the collision? Explain, making sure you are careful to identify the "system."

7. Two particles collide, one of which was initially moving and the other initially at rest.
   a. Is it possible for *both* particles to be at rest after the collision? Give an example in which this happens, or explain why it can't happen.
   b. Is it possible for *one* particle to be at rest after the collision? Give an example in which this happens, or explain why it can't happen.

8. Automobiles are designed with "crumple zones" intended to collapse in a collision. Why would a manufacturer design part of a car so that it collapses in a collision?

9. You probably know that it feels better to catch a baseball if you are wearing a padded glove. Explain why this is so, using the ideas of momentum and impulse.

10. In the early days of rocketry, some people claimed that rockets couldn't fly in outer space as there was no air for the rockets to push against. Suppose you were an early investigator in the field of rocketry and met someone who made this argument. How would you convince the person that rockets could travel in space?

11. Two ice skaters, Megan and Jason, push off from each other on frictionless ice. Jason's mass is twice that of Megan.
    a. Which skater, if either, experiences the greater impulse during the push? Explain.
    b. Which skater, if either, has the greater speed after the push-off? Explain.

12. Suppose a rubber ball and a steel ball collide. Which, if either, receives the larger impulse? Explain.

13. While standing still on a basketball court, you throw the ball to a teammate. Why do you not move backward as a result? Is the law of conservation of momentum violated?

14. As you release a ball, it falls, gaining speed and momentum. Is momentum conserved?
    a. Answer this question from the perspective of choosing the ball alone as the system.
    b. Answer this question from the perspective of choosing the ball and the earth together as the system.

15. Rank in order, from largest to smallest, the angular momenta L_1 through L_5 of the particles shown in Figure Q9.15. Explain.

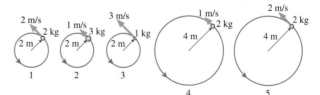

FIGURE Q9.15

16. Figure Q9.16 shows two masses held together by a thread on a rod that is rotating about its center with angular velocity ω. If the thread breaks, the masses will slide out to the ends of the rod. If that happens, will the rod's angular velocity increase, decrease, or remain unchanged? Explain.

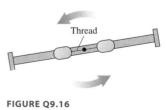

FIGURE Q9.16

17. The disks shown in Figure Q9.17 have equal mass. Is the angular momentum of disk 2, on the right, larger than, smaller than, or equal to the angular momentum of disk 1? Explain.

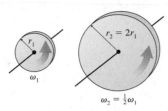

FIGURE Q9.17

18. If the earth warms significantly, the polar ice caps will melt. Water will move from the poles, near the earth's rotation axis, and will spread out around the globe. In principle, this will change the length of the day. Why? Will the length of the day increase or decrease?

Multiple-Choice Questions

19. | Curling is a sport played with 20 kg stones that slide across an ice surface. Suppose a curling stone sliding at 1 m/s strikes another stone and comes to rest in 2 ms. Approximately how much force is there on the stone during the impact?
   A. 200 N   B. 1000 N   C. 2000 N   D. 10,000 N

20. | Two balls are hung from cords. The first ball, of mass 1.0 kg, is pulled to the side and released, reaching a speed of 2.0 m/s at the bottom of its arc. Then, as shown in Figure Q9.20, it hits and sticks to another ball. The speed of the pair just after the collision is 1.2 m/s. What is the mass of the second ball?
   A. 0.67 kg   B. 2.0 kg   C. 1.7 kg   D. 1.0 kg

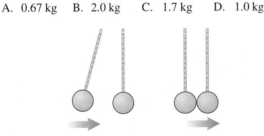

FIGURE Q9.20

21. ‖ Figure Q9.21 shows two blocks sliding on a frictionless surface. Eventually the smaller block overtakes the larger one, collides with it, and sticks. What is the speed of the two blocks after the collision?
   A. $v_i/2$ B. $4v_i/5$ C. v_i D. $5v_i/4$ E. $2v_i$

FIGURE Q9.21

22. ‖ Two friends are sitting in a stationary canoe. At $t = 3.0$ s the person at the front tosses a sack to the person in the rear, who catches the sack 0.2 s later. Which plot in Figure Q9.22 shows the velocity of the boat as a function of time? Positive velocity is forward, negative velocity is backward. Neglect any drag force on the canoe from the water.

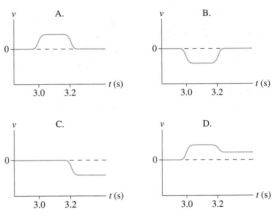

FIGURE Q9.22

23. ‖ A small puck is sliding to the right with momentum $\vec{p}_i$ on a horizontal, frictionless surface, as shown in Figure Q9.23. A force is applied to the puck for a short time and its momentum afterward is $\vec{p}_f$. Which lettered arrow shows the direction of the impulse that was delivered to the puck?

FIGURE Q9.23

24. ‖‖ A red ball, initially at rest, is simultaneously hit by a blue ball traveling from west to east at 3 m/s and a green ball traveling east to west at 3 m/s. All three balls have equal mass. Afterward, the red ball is traveling south and the green ball is moving to the east. In which direction is the blue ball traveling?
   A. West.
   B. North.
   C. Between north and west.
   D. Between north and east.
   E. Between south and west.

25. ‖‖ A 24 g, 3-cm-diameter thin, hollow sphere rotates at 30 rpm about a vertical, frictionless axis through its center. A 4 g bug stands at the top of the sphere. He then walks along the surface of the sphere until he reaches its "equator." When he reaches the equator, the sphere is rotating at
   A. 15 rpm   B. 24 rpm   C. 30 rpm
   D. 37 rpm   E. 45 rpm

26. ‖ A 5.0 kg solid cylinder of radius 12 cm rotates with $\omega_i = 3.7$ rad/s about an axis through its center. A torque of 0.040 N · m is applied to the cylinder for 5.0 s. By how much does the cylinder's angular momentum change?
   A. 0.12 kg · m²/s   B. 0.20 kg · m²/s   C. 0.38 kg · m²/s
   D. 0.52 kg · m²/s   E. 0.88 kg · m²/s

PROBLEMS

Section 9.1 Impulse

Section 9.2 Momentum and the Impulse-Momentum Theorem

1. | At what speed do a bicycle and its rider, with a combined mass of 100 kg, have the same momentum as a 1500 kg car traveling at 5.0 m/s?

2. ‖ A 57 g tennis ball is served at 45 m/s. If the ball started from rest, what impulse was applied to the ball by the racket?

3. | A student throws a 120 g snowball at 7.5 m/s at the side of the schoolhouse, where it hits and sticks. What is the magnitude of the average force on the wall if the duration of the collision is 0.15 s?

4. ‖‖ In Figure P9.4, what value of F_{max} gives an impulse of 6.0 N · s?

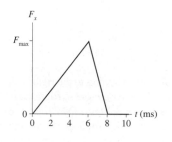

FIGURE P9.4

5. | What is the magnitude of the momentum of
   a. A 1500 kg car traveling at 10 m/s?
   b. A 200 g baseball thrown at 40 m/s?

Section 9.3 Solving Impulse and Momentum Problems

6. | Use the impulse-momentum theorem to find how long a stone falling straight down takes to increase its speed from 5.5 m/s to 10.4 m/s.

7. ‖‖ a. A 2.0 kg object is moving to the right with a speed of 1.0 m/s when it experiences the force shown in Figure P9.7a. What are the object's speed and direction after the force ends?
   b. Answer this question for the force shown in Figure P9.7b.

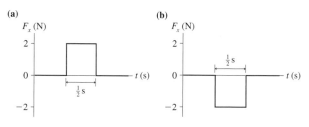

FIGURE P9.7

8. ‖‖ A 60 g tennis ball with an initial speed of 32 m/s hits a wall and rebounds with the same speed. Figure P9.8 shows the force of the wall on the ball during the collision. What is the value of F_{max}, the maximum value of the contact force during the collision?

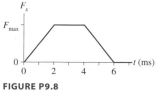

FIGURE P9.8

9. | A child is sliding on a sled at 1.5 m/s to the right. You stop the sled by pushing on it for 0.50 s in a direction opposite to its motion. If the mass of the child and sled is 35 kg, what average force do you need to apply to stop the sled? Use the concepts of impulse and momentum.

10. ‖ An ice hockey puck slides along the ice at 12 m/s. A hockey stick delivers an impulse of 4.0 kg · m/s, causing the puck to move off in the opposite direction with the same speed. What is the mass of the puck?

11. | As part of a safety investigation, two 1400 kg cars traveling at 20 m/s are crashed into different barriers. Find the average forces exerted on (a) the car that hits a line of water barrels and takes 1.5 s to stop, and (b) the car that hits a concrete barrier and takes 0.1 s to stop.

12. | In a Little League baseball game, the 145 g ball enters the strike zone with a speed of 15.0 m/s. The batter hits the ball, and it leaves his bat with a speed of 20.0 m/s in exactly the opposite direction.
    a. What is the magnitude of the impulse delivered by the bat to the ball?
    b. If the bat is in contact with the ball for 1.5 ms, what is the magnitude of the average force exerted by the bat on the ball?

Section 9.4 Conservation of Momentum

13. ‖‖ A small, 100 g cart is moving at 1.20 m/s on an air track when it collides with a larger, 1.00 kg cart at rest. After the collision, the small cart recoils at 0.850 m/s. What is the speed of the large cart after the collision?

14. ‖ A man standing on very slick ice fires a rifle horizontally. The mass of the man together with the rifle is 70 kg, and the mass of the bullet is 10 g. If the bullet leaves the muzzle at a speed of 500 m/s, what is the final speed of the man?

15. | An 80 kg quarterback jumps straight up in the air right before throwing a 0.43 kg football horizontally at 15 m/s. How fast will he be moving backward just after releasing the ball? Do you think that this would be noticeable?

16. | A strong man is compressing a lightweight spring between two weights. One weight has a mass of 2.3 kg, the other a mass of 5.3 kg. He is holding the weights stationary, but then he loses his grip and the weights fly off in opposite directions. The lighter of the two is shot out at a speed of 6.0 m/s. What is the speed of the heavier weight?

17. | A 10,000 kg railroad car is rolling at 2.00 m/s when a 4000 kg load of gravel is suddenly dropped in. What is the car's speed just after the gravel is loaded?

18. | A 5000 kg open train car is rolling on frictionless rails at 22.0 m/s when it starts pouring rain. A few minutes later, the car's speed is 20.0 m/s. What mass of water has collected in the car?

19. | A 50.0 kg archer, standing on frictionless ice, shoots a 100 g arrow at a speed of 100 m/s. What is the recoil speed of the archer?

20. | A 9.5 kg dog takes a nap in a canoe and wakes up to find the canoe has drifted out onto the lake but now is stationary. He walks along the length of the canoe at 0.50 m/s, relative to the water, and the canoe simultaneously moves in the opposite direction at 0.15 m/s. What is the mass of the canoe?

Section 9.5 Inelastic Collisions

21. ‖ A 300 g bird flying along at 6.0 m/s sees a 10 g insect heading straight toward it with a speed of 30 m/s. The bird opens its mouth wide and enjoys a nice lunch. What is the bird's speed immediately after swallowing?

22. │ A 71 kg baseball player jumps straight up to catch a line drive. If the 140 g ball is moving horizontally at 28 m/s, and the catch is made when the ballplayer is at the highest point of his leap, what is his speed immediately after stopping the ball?

23. ‖ A kid at the junior high cafeteria wants to propel an empty milk carton along a lunch table by hitting it with a 3.0 g spit ball. If he wants the speed of the 20 g carton just after the spit ball hits it to be 0.30 m/s, at what speed should his spit ball hit the carton?

24. │ The parking brake on a 2000 kg Cadillac has failed, and it is rolling slowly, at 1 mph, toward a group of small children. Seeing the situation, you realize you have just enough time to drive your 1000 kg Volkswagen head-on into the Cadillac and save the children. With what speed should you impact the Cadillac to bring it to a halt?

25. │ A 1500 kg car is rolling at 2.0 m/s. You would like to stop the car by firing a 10 kg blob of sticky clay at it. How fast should you fire the clay?

Section 9.6 Momentum and Collisions in Two Dimensions

26. ‖ A 20 g ball of clay traveling east at 3.0 m/s collides with a 30 g ball of clay traveling north at 2.0 m/s. What are the speed and the direction of the resulting 50 g ball of clay?

27. ‖ Two particles collide and bounce apart. Figure P9.27 shows the initial momenta of both and the final momentum of particle 2. What is the final momentum of particle 1? Show your answer by copying the figure and drawing the final momentum vector on the figure.

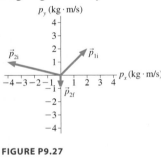

FIGURE P9.27

28. ‖‖ A 20 g ball of clay traveling east at 2.0 m/s collides with a 30 g ball of clay traveling 30° south of west at 1.0 m/s. What are the speed and direction of the resulting 50 g blob of clay?

29. ‖ A firecracker in a coconut blows the coconut into three pieces. Two pieces of equal mass fly off south and west, perpendicular to each other, at 20 m/s. The third piece has twice the mass as the other two. What are the speed and direction of the third piece?

Section 9.7 Angular Momentum

30. ‖‖ What is the angular momentum of the moon around the earth? The moon's mass is 7.4×10^{22} kg and it orbits 3.8×10^8 m from the earth.

31. │ A little girl is going on the merry-go-round for the first time, and wants her 47 kg mother to stand next to her on the ride, 2.6 m from the merry-go-round's center. If her mother's speed is 4.2 m/s when the ride is in motion, what is her angular momentum around the center of the merry-go-round?

32. ‖ What is the angular momentum of the 500 g rotating bar in Figure P9.32?

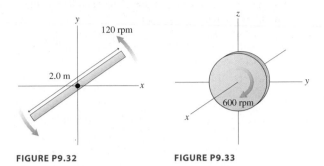

FIGURE P9.32　　　　　　　**FIGURE P9.33**

33. ‖ What is the angular momentum of the 2.0 kg, 4.0-cm-diameter rotating disk in Figure P9.33?

34. │ Divers change their body position in midair while rotating about their center of mass. In one dive, the diver leaves the board with her body nearly straight, then tucks into a somersault position. If the moment of inertia of the diver in a straight position is 14 kg · m² and in a tucked position is 4.0 kg · m², by what factor is her angular velocity when tucked greater than when straight?

35. │ Ice skaters often end their performances with spin turns, where they spin very fast about their center of mass with their arms folded in and legs together. Upon ending, their arms extend outward, proclaiming their finish. Not quite as noticeably, one leg goes out as well. Suppose that the moment of inertia of a skater with arms out and one leg extended is 3.2 kg · m² and for arms and legs in is 0.80 kg · m². If she starts out spinning at 5.0 rev/s, what is her angular speed (in rev/s) when her arms and one leg open outward?

General Problems

36. ‖ What is the impulse on a 3.0 kg particle that experiences the force described by the graph in Figure P9.36?

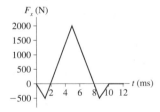

FIGURE P9.36

37. ‖‖ A 600 g air-track glider collides with a spring at one end of the track. Figure P9.37 shows the glider's velocity and the force exerted on the glider by the spring. How long is the glider in contact with the spring?

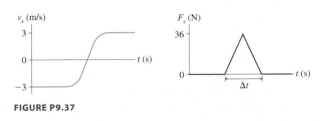

FIGURE P9.37

38. ‖ Far in space, where gravity is negligible, a 425 kg rocket traveling at 75.0 m/s fires its engines. Figure P9.38 shows the thrust force as a function of time. The mass lost by the rocket during these 30.0 s is negligible.
    a. What impulse does the engine impart to the rocket?
    b. At what time does the rocket reach its maximum speed? What is the maximum speed?

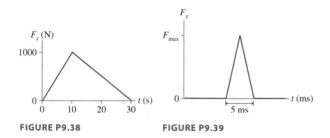

FIGURE P9.38   **FIGURE P9.39**

39. ‖ A 200 g ball is dropped from a height of 2.0 m, bounces on a hard floor, and rebounds to a height of 1.5 m. Figure P9.39 shows the impulse received from the floor. What maximum force does the floor exert on the ball?
40. ‖‖ A 200 g ball is dropped from a height of 2.0 m and bounces on a hard floor. The force on the ball from the floor is shown in Figure P9.40. How high does the ball rebound?

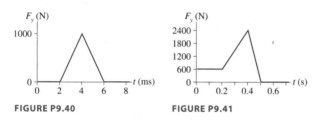

FIGURE P9.40   **FIGURE P9.41**

41. ‖‖‖ Figure P9.41 is a graph of the force exerted by the floor on a woman making a vertical jump. At what speed does she leave the ground?
    Hint: The force of the floor is not the only force acting on the woman.
42. ‖ A sled slides along a horizontal surface for which the coefficient of kinetic friction is 0.25. Its velocity at point A is 8.0 m/s and at point B is 5.0 m/s. Use the impulse-momentum theorem to find how long the sled takes to travel from A to B.
43. ‖ A 50.0 g ball is launched from ground level at an angle 30.0° above the horizon. Its initial speed is 25.0 m/s.
    a. What are the values of p_x and p_y an instant after the ball is launched, at the point of maximum altitude, and an instant before the ball hits the ground?
    b. Why is one component of $\vec{p}$ constant? Explain.
    c. For the component of $\vec{p}$ that changes, show that the change in momentum is equal to the weight of the ball multiplied by the time of flight. Explain why this is so.
44. ‖ Squids rely on jet propulsion, a versatile technique to move around in water. A 1.5 kg squid at rest suddenly expels 0.10 kg of water backward to quickly get itself moving forward at 3.0 m/s. If other forces (such as the drag force on the squid) are ignored, what is the speed with which the squid expels the water?
45. ‖‖‖ The flowers of the bunchberry plant open with astonishing force and speed, causing the pollen grains to be ejected out of the flower in a mere 0.30 ms at an acceleration of

2.5×10^4 m/s². If the acceleration is constant, what impulse is delivered to a pollen grain with a mass of 1.0×10^{-7} g?
46. ‖ a. With what speed are pollen grains ejected from a bunchberry flower? See Problem 45 for information.
    b. Suppose that 1000 ejected pollen grains slam into the abdomen of a 5.0 g bee that is hovering just above the flower. If the collision is inelastic, what is the bee's speed immediately afterward? Is the bee likely to notice?
47. ‖‖‖ A tennis player swings her 1000 g racket with a speed of 10 m/s. She hits a 60 g tennis ball that was approaching her at a speed of 20 m/s. The ball rebounds at 40 m/s.
    a. How fast is her racket moving immediately after the impact? You can ignore the interaction of the racket with her hand for the brief duration of the collision.
    b. If the tennis ball and racket are in contact for 10 ms, what is the average force that the racket exerts on the ball?
48. ‖‖‖ A 20 g ball of clay is thrown horizontally at 30 m/s toward a 1.0 kg block sitting at rest on a frictionless surface. The clay hits and sticks to the block.
    a. What is the speed of the block and clay right after the collision?
    b. Use the block's initial and final speeds to calculate the impulse the clay exerts on the block.
    c. Use the clay's initial and final speeds to calculate the impulse the block exerts on the clay.
    d. Does $\vec{J}_{\text{block on clay}} = -\vec{J}_{\text{clay on block}}$?
49. ‖ Dan is gliding on his skateboard at 4.0 m/s. He suddenly jumps backward off the skateboard, kicking the skateboard forward at 8.0 m/s. How fast is Dan going as his feet hit the ground? Dan's mass is 50 kg and the skateboard's mass is 5.0 kg.
50. ‖ James and Sarah stand on a stationary cart with frictionless wheels. The total mass of the cart and riders is 130 kg. At the same instant, James throws a 1.0 kg ball to Sarah at 4.5 m/s, while Sarah throws a 0.50 kg ball to James at 1.0 m/s. James's throw is to the right and Sarah's is to the left.
    a. While the two balls are in the air, what are the speed and direction of the cart and its riders?
    b. After the balls are caught, what are the speed and direction of the cart and riders?
51. ‖ A 10-m-long glider with a mass of 680 kg (including the passengers) is gliding horizontally through the air at 30 m/s when a 60 kg skydiver drops out by releasing his grip on the glider. What is the glider's velocity just after the skydiver lets go?
52. ‖ The cars of a long coal train are filled by pulling them under a hopper, from which coal falls into the cars at a rate of 10,000 kg/s. Ignoring friction due to the rails, what is the average force that the engine must exert on the coal train to keep it moving under the hopper at a speed of 0.50 m/s?
53. ‖ Three identical train cars, coupled together, are rolling east at 2.0 m/s. A fourth car traveling east at 4.0 m/s catches up with the three and couples to make a four-car train. A moment later, the train cars hit a fifth car that was at rest on the tracks, and it couples to make a five-car train. What is the speed of the five-car train?
54. ‖ A 110 kg linebacker running at 2.0 m/s and an 82 kg quarterback running at 3.0 m/s have a head-on collision in midair. The linebacker grabs and holds onto the quarterback. Who ends up moving forward after they hit?

55. | Most geologists believe that the dinosaurs became extinct 65 million years ago when a large comet or asteroid struck the earth, throwing up so much dust that the sun was blocked out for a period of many months. Suppose an asteroid with a diameter of 2.0 km and a mass of 1.0×10^{13} kg hits the earth with an impact speed of 4.0×10^4 m/s.
   a. What is the earth's recoil speed after such a collision? (Use a reference frame in which the earth was initially at rest.)
   b. What percentage is this of the earth's speed around the sun? (Use the astronomical data inside the cover.)

56. | At the center of a 50-m-diameter circular ice rink, a 75 kg skater traveling north at 2.5 m/s collides with and holds onto a 60 kg skater who had been heading west at 3.5 m/s.
   a. How long will it take them to glide to the edge of the rink?
   b. Where will they reach it? Give your answer as an angle north of west.

57. ‖ Two ice skaters, with masses of 50 kg and 75 kg, are at the center of a 60-m-diameter circular rink. The skaters push off against each other and glide to opposite edges of the rink. If the heavier skater reaches the edge in 20 s, how long does the lighter skater take to reach the edge?

58. ‖ One billiard ball is shot east at 2.00 m/s. A second, identical billiard ball is shot west at 1.00 m/s. The balls have a glancing collision, not a head-on collision, deflecting the second ball by 90° and sending it north at 1.41 m/s. What are the speed and direction of the first ball after the collision?

59. ‖ A 10 g bullet is fired into a 10 kg wood block that is at rest on a wood table. The block, with the bullet embedded, slides 5.0 cm across the table. What was the speed of the bullet?

60. ‖ You are part of a search-and-rescue mission that has been called out to look for a lost explorer. You've found the missing explorer, but you're separated from him by a 200-m-high cliff and a 30-m-wide raging river, as shown in Figure P9.60. To save his life, you need to get a 5.0 kg package of emergency supplies across the river. Unfortunately, you can't throw the package

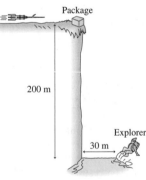

FIGURE P9.60

hard enough to make it across. Fortunately, you happen to have a 1.0 kg rocket intended for launching flares. Improvising quickly, you attach a sharpened stick to the front of the rocket, so that it will impale itself into the package of supplies, then fire the rocket at ground level toward the supplies. What minimum speed must the rocket have just before impact in order to save the explorer's life?

61. ‖ A 1500 kg weather rocket accelerates upward at 10.0 m/s². It explodes 2.00 s after liftoff and breaks into two fragments, one twice as massive as the other. Photos reveal that the lighter fragment traveled straight up and reached a maximum height of 530 m. What were the speed and direction of the heavier fragment just after the explosion?

62. | Two 500 g blocks of wood are 2.0 m apart on a frictionless table. A 10 g bullet is fired at 400 m/s toward the blocks. It passes all the way through the first block, then embeds itself in the second block. The speed of the first block immediately afterward is 6.0 m/s. What is the speed of the second block after the bullet stops?

63. | A 500 kg cannon fires a 10 kg cannonball with a speed of 200 m/s relative to the muzzle. The cannon is on wheels that roll without friction. When the cannon fires, what is the speed of the cannonball relative to the earth?

64. | Laura, whose mass is 35 kg, jumps horizontally off a 55 kg canoe at 1.5 m/s relative to the canoe. What is the canoe's speed just after she jumps?

65. ‖‖ A spaceship of mass 2.0×10^6 kg is cruising at a speed of 5.0×10^6 m/s when the antimatter reactor fails, blowing the ship into three pieces. One section, having a mass of 5.0×10^5 kg, is blown straight backward with a speed of 2.0×10^6 m/s. A second piece, with mass 8.0×10^5 kg, continues forward at 1.0×10^6 m/s. What are the direction and speed of the third piece?

66. ‖‖ A proton is shot at 5.0×10^7 m/s toward a gold target. The nucleus of a gold atom, with a mass 197 times that of the proton, repels the proton and deflects it straight back with 90% of its initial speed. What is the recoil speed of the gold nucleus?

67. ‖ Figure P9.67 shows a collision between three balls of clay. The three hit simultaneously and stick together. What are the speed and direction of the resulting blob of clay?

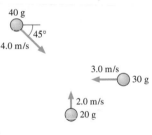

FIGURE P9.67

68. ‖ The carbon isotope ^{14}C is used for carbon dating of archeological artifacts. ^{14}C (mass 2.34×10^{-26} kg) decays by the process known as *beta decay* in which the nucleus emits an electron (the beta particle) and a subatomic particle called a neutrino. In one such decay, the electron and the neutrino are emitted at right angles to each other. The electron (mass 9.11×10^{-31} kg) has a speed of 5.00×10^7 m/s and the neutrino has a momentum of 8.00×10^{-24} kg · m/s. What is the recoil speed of the nucleus?

69. ‖‖ A 1.0-m-long massless rod is pivoted at one end and swings around in a circle on a frictionless table. A block with a hole through the center can slide in and out along the rod. Initially, a small piece of wax holds the block 30 cm from the pivot. The block is spun at 50 rpm, then the temperature of the rod is slowly increased. When the wax melts, the block slides out to the end of the rod. What is the final angular speed? Give your answer in rpm.

70. ‖ A 200 g puck revolves in a circle on a frictionless table at the end of a 50.0-cm-long string. The puck's angular momentum is 3.00 kg · m²/s. What is the tension in the string?

71. ▌▌▌ Figure P9.71 shows a 100 g puck revolving in a 20-cm-radius circle on a frictionless table. The string passes through a hole in the center of the table and is tied to two 200 g weights.

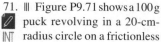
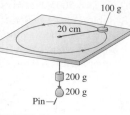

FIGURE P9.71

   a. What speed does the puck need to support the two weights?
   b. The lower weight is a light bag filled with sand. Suppose a hole is poked in the bag and the sand slowly leaks out while the puck is revolving. What will be the puck's speed and the radius of its trajectory after all of the sand is gone?

72. ▌▌ A 2.0 kg, 20-cm-diameter turntable rotates at 100 rpm on frictionless bearings. Two 500 g blocks fall from above, hit the turntable simultaneously at opposite ends of a diagonal, and stick. What is the turntable's angular speed, in rpm, just after this event?

73. ▌▌▌ A 200 g, 40.0-cm-diameter turntable rotates on frictionless bearings at 60.0 rpm. A 20.0 g block sits at the center of the turntable. A compressed spring shoots the block radially outward along a frictionless groove in the surface of the turntable. What is the turntable's angular speed when the block reaches the outer edge?

74. ▌▌ A 3.0-m-diameter merry-go-round with a mass of 250 kg is spinning at 20 rpm. John runs around the merry-go-round at 5.0 m/s, in the same direction that it is turning, and jumps onto the outer edge. John's mass is 30 kg. What is the merry-go-round's angular speed, in rpm, after John jumps on?

75. ▌▌ Disk A, with a mass of 2.0 kg and a radius of 40 cm, rotates clockwise about a frictionless vertical axle at 30 rev/s. Disk B, also 2.0 kg but with a radius of 20 cm, rotates counterclockwise about that same axle, but at a greater height than disk A,

at 30 rev/s. Disk B slides down the axle until it lands on top of disk A, after which they rotate together. After the collision, what is their common angular speed (in rev/s) and in which direction do they rotate?

Passage Problems

Hitting a Golf Ball

Consider a golf club hitting a golf ball. To a good approximation, we can model this as a collision between the rapidly moving head of the golf club and the stationary golf ball, ignoring the shaft of the club and the golfer.

A golf ball has a mass of 46 g. Suppose a 200 g club head is moving at a speed of 40 m/s just before striking the golf ball. After the collision, the golf ball's speed is 60 m/s.

76. ▎ What is the momentum of the club-ball system right before the collision?
   A. 1.8 kg · m/s   B. 8.0 kg · m/s
   C. 3220 kg · m/s   D. 8000 kg · m/s

77. ▎ Immediately after the collision, the momentum of the club-ball system will be
   A. Less than before the collision.
   B. The same as before the collision.
   C. More than before the collision.

78. ▎ A manufacturer makes a golf ball that compresses more than a traditional golf ball when struck by a club. How will this affect the average force during the collision?
   A. The force will decrease.
   B. The force will not be affected.
   C. The force will increase.

79. ▎ By approximately how much does the club head slow down as a result of hitting the ball?
   A. 4 m/s   B. 6 m/s   C. 14 m/s   D. 26 m/s

STOP TO THINK ANSWERS

Stop to Think 9.1: F. The cart is initially moving in the negative x-direction, so $(p_x)_i = -20$ kg · m/s. After it bounces, $(p_x)_f = 10$ kg · m/s. Thus $\Delta p = (10 \text{ kg} \cdot \text{m/s}) - (-20 \text{ kg} \cdot \text{m/s}) = 30$ kg · m/s.

Stop to Think 9.2: B. The clay ball goes from $(v_x)_i = v$ to $(v_x)_f = 0$, so $J_{clay} = \Delta p_x = -mv$. The rubber ball rebounds, going from $(v_x)_i = v$ to $(v_x)_f = -v$ (same speed, opposite direction). Thus $J_{rubber} = \Delta p_x = -2mv$. The rubber ball has a larger momentum change, and this requires a larger impulse.

Stop to Think 9.3: Right end. The pieces started at rest, so the total momentum of the system is zero. It's an isolated system, so the total momentum after the explosion is still zero. The 6 g piece has momentum $6v$. The 4 g piece, with velocity $-2v$, has momentum $-8v$. The combined momentum of these two pieces is $-2v$. In order for P to be zero, the third piece must have a *positive* momentum ($+2v$) and thus a positive velocity.

Stop to Think 9.4: B. The momentum of ball 1 is $(0.40 \text{ kg})(2.5 \text{ m/s}) = 1.0$ kg · m/s, while that of ball 2 is $(0.80 \text{ kg})(-1.5 \text{ m/s}) = -1.2$ kg · m/s. The total momentum is then 1.0 kg · m/s − 1.2 kg · m/s = −0.2 kg · m/s. Because it's negative, the total momentum, and hence the final velocity of the balls, is directed to the left.

Stop to Think 9.5: B. Angular momentum $L = I\omega$ is conserved. Both boys have mass m and initially stand distance $R/2$ from the axis. Thus the initial moment of inertia is $I_i = I_{disk} + 2 \times m(R/2)^2 = I_{disk} + \frac{1}{2}mR^2$. The final moment of inertia is $I_f = I_{disk} + 0 + mR^2$, because the boy standing at the axis contributes nothing to the moment of inertia. Because $I_f > I_i$ we must have $\omega_f < \omega_i$.

10

ENERGY AND WORK

Using just a fast run-up and flexible pole, how can a pole vaulter reach an astonishing 6 m (20 ft) off the ground?

Looking Ahead ▶▶

The goal of Chapter 10 is to introduce the concept of energy and learn a new problem-solving strategy based on conservation of energy. In this chapter you will learn to:

▶ Understand some of the important forms of energy, and how energy can be transformed and transferred.

▶ Understand what work is, and how to calculate it.

▶ Understand and use the concepts of kinetic, potential, and thermal energy.

▶ Solve problems using the law of conservation of energy.

▶ Apply these ideas to elastic collisions.

Looking Back ◀◀

Part of our introduction to energy will be based on the kinematics of constant acceleration. In addition, we will need ideas from rotational motion. We will also use the before-and-after pictorial representation developed for impulse and momentum problems. Please review

◀ Section 2.4 Constant-acceleration kinematics.

◀ Section 7.5 Moment of inertia.

◀ Sections 9.2–9.3 Before-and-after visual overviews and conservation of momentum.

Energy. It's a word you hear all the time. We use chemical energy to heat our homes and bodies, electrical energy to run our lights and computers, and solar energy to grow our crops and forests. We're told to use energy wisely and not to waste it. Athletes and weary students consume "energy bars" and "energy drinks."

But just what is energy? The concept of energy has grown and changed with time, and it is not easy to define in a general way just what energy is. Rather than starting with a formal definition, we'll let the concept of energy expand slowly over the course of several chapters. In this chapter we introduce several fundamental forms of energy, including kinetic energy, potential energy, and thermal energy. Our goal is to understand the characteristics of energy, how energy is used, and, especially important, how energy is transformed from one form to another. For example, this pole vaulter, after years of training, has become extraordinarily proficient at transforming his energy of motion into energy associated with height from the ground.

We'll also discover a very powerful conservation law for energy. Some scientists consider the law of conservation of energy to be the most important of all the laws of nature. But all that in due time. First we have to start with the basic ideas.

10.1 A "Natural Money" Called Energy

We will start by discussing what seems to be a completely unrelated topic: money. As you will discover, monetary systems have much in common with energy. Let's begin with a short story.

The Parable of the Lost Penny

John was a hard worker. His only source of income was the paycheck he received each month. Even though most of each paycheck had to be spent on basic necessities, John managed to keep a respectable balance in his checking account. He even saved enough to occasionally buy a few savings bonds, his investment in the future.

John never cared much for pennies, so he kept a jar by the door and dropped all his pennies into it at the end of each day. Eventually, he reasoned, his saved pennies would be worth taking to the bank and converting into crisp new dollar bills.

John found it fascinating to keep track of these various forms of money. He noticed, to his dismay, that the amount of money in his checking account did not spontaneously increase overnight. Furthermore, there seemed to be a definite correlation between the size of his paycheck and the amount of money he had in the bank. So John decided to embark on a systematic study of money.

He began, as would any good scientist, by using his initial observations to formulate a hypothesis, which he called a *model* of the monetary system. He found that he could represent his monetary model with the flowchart in Figure 10.1.

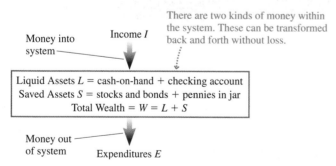

FIGURE 10.1 John's model of the monetary system.

As the chart shows, John divided his money into two basic types, liquid assets and saved assets. The *liquid assets* L, which included his checking account and the cash in his pockets, were moneys available for immediate use. His *saved assets* S, which included his savings bonds as well as the jar of pennies, had the potential to be converted into liquid assets, but they were not available for immediate use.

John decided to call the sum total of assets his *wealth:* $W = L + S$.

John's assets were, more or less, simply definitions. The more interesting question, he thought, was how his wealth depended on his *income I* and *expenditures E*. These represented money transferred *to* him by his employer and money transferred *by* him to stores and bill collectors. After painstakingly collecting and analyzing his data, John finally determined that the relationship between monetary transfers and wealth is

$$\Delta W = I - E$$

John interpreted this equation to mean that the *change* in his wealth, ΔW, was numerically equal to the *net* monetary transfer $I - E$.

During a week-long period when John stayed home sick, isolated from the rest of the world, he had neither income nor expenses. In grand confirmation of his hypothesis, he found that his wealth W_f at the end of the week was identical to his wealth W_i at the week's beginning. That is, $W_f = W_i$. This occurred despite the fact that he had moved pennies from his pocket to the jar and also, by telephone, had sold some bonds and transferred the money to his checking account. In other words, John found that he could make all of the *internal* conversions of assets from one form to another that he wanted, but his total wealth remained constant ($W = $ constant) as long as he was isolated from the world. This seemed such a remarkable rule that John named it the *law of conservation of wealth.*

One day, however, John added up his income and expenditures for the week, and the changes in his various assets, and he was 1¢ off! Inexplicably, some money seemed to have vanished. He was devastated. All those years of careful research, and now it seemed that his monetary hypothesis might not be true. Under some circumstances, yet to be discovered, it looked like $\Delta W \neq I - E$. Off by a measly penny. A wasted scientific life. . . .

But wait! In a flash of inspiration, John realized that perhaps there were other types of assets, yet to be discovered, and that his monetary hypothesis would still be valid if *all* assets were included. Weeks went by as John, in frantic activity, searched fruitlessly for previously *hidden* assets. Then one day, as John lifted the cushion off the sofa to vacuum out the potato chip crumbs—lo and behold, there it was!—the missing penny!

John raced to complete his theory, now including money in the sofa, the washing machine, and behind the radiator as previously unknown forms of assets that were easy to convert from other forms, but often rather difficult to recover.

Other researchers soon discovered other types of assets, such as the remarkable find of the "cash in the mattress." To this day, when *all* known assets are included, monetary scientists have never found a violation of John's simple hypothesis that $\Delta W = I - E$. John was last seen sailing for Stockholm to collect the Nobel Prize for his Theory of Wealth.

10.2 The Basic Energy Model

John, despite his diligent efforts, did not discover a law of nature. The monetary system is a human construction that, by design, obeys John's "laws." Monetary system laws, such as that you cannot print money in your basement, are enforced by society, not by nature. But suppose that physical objects possessed a "natural money" that was governed by a theory, or model, similar to John's. An object might have several forms of natural money that could be converted back and forth, but the total amount of an object's natural money would *change* only if natural money were *transferred* to or from the object. Two key words here, as in John's model, are *transfer* and *change*.

One of the greatest and most significant discoveries of science is that there is such a "natural money" called **energy.** You have heard of some of the many forms of energy, such as solar energy or nuclear energy, but others may be new to you. These forms of energy can differ as much as a checking account differs from loose change in the sofa. Much of our study is going to be focused on the *transformation* of energy from one form to another. Much of modern technology is concerned with transforming energy, such as changing the chemical energy of oil molecules to electrical energy or to the kinetic energy of your car.

As we use energy concepts, we will be "accounting" for energy that is transferred in or out of a system or that is transformed from one form to another within a system. Figure 10.2 shows a simple model of energy that is based on John's model of the monetary system. Many details must be added to this model, but it's a good starting point. The fact that nature "balances the books" for energy is one of the most profound discoveries of science.

A major goal of ours is to discover the conditions under which energy is conserved. Surprisingly, the *law of conservation of energy* was not recognized until the mid-nineteenth century, long after Newton. The reason, similar to John's lost penny, was that it took scientists a long time to realize how many types of energy there are and the various ways that energy can be converted from one form to another. As you'll soon learn, energy ideas go well beyond Newtonian mechanics to include new concepts about heat, about chemical energy, and about the energy of the individual atoms and molecules that comprise an object. All of these forms of energy will ultimately have to be included in our accounting scheme for energy.

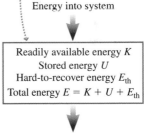

There are several kinds of energy within the system. These can be transformed back and forth without loss.

Energy into system

Readily available energy K
Stored energy U
Hard-to-recover energy E_{th}
Total energy $E = K + U + E_{th}$

Energy out of system

FIGURE 10.2 An initial model of energy.

Systems and Energy

In Chapter 9 we introduced the idea of a *system* of interacting objects. A system can be quite simple, such as a saltshaker sliding across the table, or much more complex, such as a city or a human body. But whether simple or complex, every system in nature has associated with it a quantity we call its **total energy** E. Like John's total wealth, which was made up of assets of many kinds, the total energy of a system is made up of many kinds of energies. In the table below, we give a brief overview of some of the more important forms of energy; in the rest of the chapter we'll look at several of these forms of energy in much greater detail.

Some important forms of energy

Kinetic energy K

Kinetic energy is the energy of *motion*. All moving objects have kinetic energy. The heavier an object, and the faster it moves, the more kinetic energy it has. The wrecking ball in this picture is effective in part because of its large kinetic energy.

Gravitational potential energy U_g

Gravitational potential energy is *stored* energy associated with an object's *height above the ground*. As this roller coaster ascends the track, energy is stored as increased gravitational potential energy. As it descends, this stored energy is converted into kinetic energy.

Elastic or spring potential energy U_s

Elastic potential energy is energy stored when a spring or other elastic object, such as this archer's bow, is *stretched*. This energy can later be transformed into the kinetic energy of the arrow. We'll sometimes use the symbol U to represent potential energy when it is not important to distinguish between U_g and U_s.

Thermal energy E_{th}

Hot objects have more *thermal energy* than cold ones because the molecules in a hot object jiggle around more than those in a cold object. Thermal energy is really just the sum of the microscopic kinetic and potential energies of all the molecules in an object. In boiling water, some molecules have enough energy to escape the water as steam.

Chemical energy E_{chem}

Electric forces cause atoms to bind together to make molecules. Energy can be stored in these bonds, energy that can later be released as the bonds are rearranged during chemical reactions. When we burn fuel to run our car, or eat food to power our bodies, we are using *chemical energy*.

Nuclear energy $E_{nuclear}$

An enormous amount of energy is stored in the *nucleus,* the tiny core of an atom. Certain nuclei can be made to break apart, releasing some of this *nuclear energy,* which is transformed into the kinetic energy of the fragments and then into thermal energy. This is the source of energy of nuclear power plants and nuclear weapons.

A system may have many of these kinds of energy present in it at once. For instance, a moving car has kinetic energy of motion, chemical energy stored in its gasoline, thermal energy in its hot engine, and other forms of energy in its many other parts. The total energy of the system, E, is just the *sum* of the different energies present in the system, so that we have

$$E = K + U_g + U_s + E_{th} + E_{chem} + \cdots \qquad (10.1)$$

The energies shown in this sum are the forms of energy in which we'll be most interested in this and the next chapter. The ellipses (. . .) represent other forms of energy, such as nuclear or electric, that also might be present. We'll treat these and others in later chapters.

Energy Transformations

We've seen that all systems contain energy in many different forms. But if the amounts of each form of energy never changed, the world would be a very dull place. What makes the world interesting is that **energy of one kind can** *trans-*

form **into energy of another kind.** The gravitational potential energy of the roller coaster at the top of the track is rapidly converted into kinetic energy as the coaster descends; the chemical energy of gasoline is converted into the kinetic energy of your moving car. The following table illustrates a few common energy transformations. In this table, we'll use an arrow → as a shorthand way of representing an energy transformation.

Some energy transformations

A weightlifter lifts a barbell over her head
The barbell has much more gravitational potential energy when high above her head than when on the floor. To lift the barbell, she is transforming chemical energy in her body into gravitational potential energy of the barbell.

$$E_{chem} \rightarrow U_g$$

A base runner slides into the base
When running, he has lots of kinetic energy. After sliding, he has none. His kinetic energy is transformed mainly into thermal energy: the ground and his legs are slightly warmer.

$$K \rightarrow E_{th}$$

A burning campfire
The wood contains considerable chemical energy. When the carbon in the wood combines chemically with oxygen in the air, this chemical energy is transformed largely into thermal energy of the hot gases and embers.

$$E_{chem} \rightarrow E_{th}$$

A springboard diver
Here's a two-step energy transformation. The picture shows the diver after his first jump onto the board itself. At the instant shown, the board is flexed to its maximum extent. There is a large amount of elastic potential energy stored in the board. Soon this energy will begin to be transformed into kinetic energy; as he rises into the air and slows, this kinetic energy will be transformed into gravitational potential energy.

$$U_s \rightarrow K \rightarrow U_g$$

Figure 10.3 reinforces the idea that **energy transformations are changes of energy *within* the system from one form to another.** Note that it is easy to convert kinetic, potential, or chemical energies into thermal energy. But converting thermal energy back into these other forms is not so easy. How it can be done, and what possible limitations there might be in doing so, will form a large part of the next chapter.

Energy Transfers: Work and Heat

We've just seen that energy *transformations* occur between forms of energy *within* a system. In our monetary model, these transformations are like John's shifting of money between his own various assets, such as from his savings

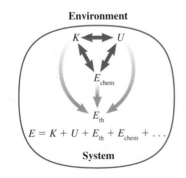

FIGURE 10.3 Energy transformations occur within the system.

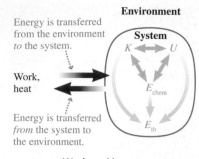

Environment

Energy is transferred from the environment *to* the system.

Work, heat

Energy is transferred *from* the system to the environment.

System

$K \leftrightarrow U$

E_{chem}

E_{th}

FIGURE 10.4 Work and heat are energy transfers into and out of the system.

One dictionary defines *work* as:

1. Physical or mental effort; labor.
2. The activity by which one makes a living.
3. A task or duty.
4. Something produced as a result of effort, such as a *work of art.*
5. Plural *works:* The essential or operating parts of a mechanism.
6. The transfer of energy to a body by application of a force.

---

account to stocks. But John also interacted with the greater world around him, receiving money as income and outlaying it as expenditures. Every physical system also interacts with the world around it, that is, with its *environment.* In the course of these interactions, the system can exchange energy with the environment. **An exchange of energy between system and environment is called an energy *transfer.*** There are two primary energy transfer processes: **work,** the *mechanical* transfer of energy to or from a system by pushing or pulling on it, and **heat,** the *nonmechanical* transfer of energy from the environment to the system (or vice versa) *because of a temperature difference between the two.* Figure 10.4 shows how our energy model is modified to include energy transfers. In this chapter we'll focus mainly on work; the concept of heat will be developed much further in Chapters 11 and 12.

Work is a common word in the English language, with many meanings. When you first think of work, you probably think of the first two definitions in this list. After all, we talk about "working out," or we say, "I just got home from work." But that is *not* what work means in physics.

In physics we use *work* in the sense of definition 6: Work is the process of *transferring* energy from the environment to a system, or from a system to the environment, by the application of mechanical forces—pushes and pulls—to the system. Once the energy has been transferred to the system, it can appear in many forms. Exactly what form it takes depends on the details of the system and how the forces are applied. The table below gives a few examples of energy transfers due to work. We use W as the symbol for work.

Energy transfers: work

Putting a shot

The system: The shot.

The environment: The athlete.

As the athlete pushes on the shot to get it moving, he is doing work on the system. That is, he is transferring energy from himself to the ball. The energy transferred to the system appears as kinetic energy.

The transfer: $W \to K$

Striking a match

The system: The match and matchbox.

The environment: The hand.

As the hand quickly pulls the match across the box, the hand does work on the system, increasing its thermal energy. The match-head becomes hot enough to ignite.

The transfer: $W \to E_{th}$

Firing a slingshot

The system: The slingshot.

The environment: The boy.

As the boy pulls back on the elastic bands, he does work on the system, increasing its elastic potential energy.

The transfer: $W \to U_s$

Notice that in each example above, the environment applies a force while the system undergoes a *displacement.* Energy is transferred as work only when the system *moves* while the force acts. A force applied to a stationary object, such as when you push against a wall, transfers no energy to the object and thus does no work.

NOTE ▶ In the table above, energy is being transferred *from* the athlete *to* the shot by the force of his hand. We say he "does work" on the shot, or "work is done" by the force of his hand. ◀

It is also possible to convert work into gravitational potential, electric, or even chemical energy. We'll have much more to say about work in the next section. But the key points to remember are that **work is the transfer of energy to or from a system by the application of forces,** and that **the system must undergo a displacement for this energy to be transferred.**

There is a second, nonmechanical means of transferring energy between a system and its environment, which we discuss here only briefly. As mentioned before, we'll have much more to say about heat in the next two chapters. When a hot object is placed in contact with a cooler one, energy flows naturally from the hot object to the cool one. **The transfer of energy from a hot to a cold object is called** *heat,* and it is given the symbol Q. It is important to note that heat is not an energy *of* a system, as are kinetic energy and chemical energy. Rather, heat is energy *transferred* between two systems.

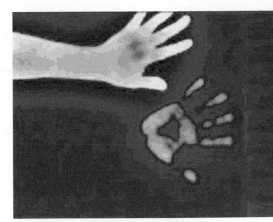

As the hand in the photo was held against the wall, heat was transferred from the warm hand to the cool wall, warming up the wall. The warm "handprint" can be imaged using a special camera sensitive to the temperature of objects.

STOP TO THINK 10.1 A child slides down a playground slide at constant speed. The energy transformation is

A. $U_g \rightarrow K$ B. $K \rightarrow U_g$ C. $W \rightarrow K$ D. $U_g \rightarrow E_{th}$ E. $K \rightarrow E_{th}$

10.3 The Law of Conservation of Energy

Remember that when John was *isolated* from the rest of the world—having neither income nor expenses—his internal wealth could be converted between its many forms, but his *total* wealth remained constant. A similar but much more fundamental law is found for the "natural money" of energy.

Let's start our study of this law by considering an **isolated system** that is separated from its surrounding environment in such a way that no energy can flow into or out of the system. This means that no work is done on the system, nor is any energy transferred as heat. We've already seen that the total energy of a system is made up of many forms of energy that are continually transforming from one kind to another. It is a deep and remarkable fact of nature that during these transformations, the total energy of an isolated system—the *sum* of all of the individual kinds of energy—remains *constant.* Any increase in, say, the system's kinetic energy must be accompanied by a decrease in its potential or thermal energies so that the total energy remains unchanged, as shown in Figure 10.5. We say that **the total energy of an isolated system is** *conserved,* giving us the following *law of conservation of energy.*

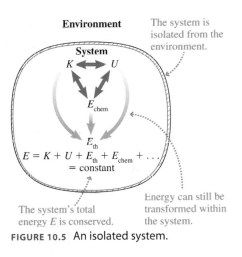

FIGURE 10.5 An isolated system.

Law of conservation of energy for an isolated system The total energy of an isolated system remains constant:

The energies in the system are constantly transforming from one kind to another . . .   . . . but their *sum* is a constant: it doesn't change.

$$K + U_g + U_s + E_{th} + E_{chem} + \ldots = E = \text{constant}$$

(10.2)

Another way to think of this conservation law is in terms of energy *changes.* Recall that we denote the change in a quantity by the symbol Δ, so we write the change in a system's kinetic energy, for instance, as ΔK. Now suppose that an isolated system has its kinetic energy change by ΔK, its gravitational potential energy by ΔU_g, and so on. Then the sum of these changes is the change in the total energy. But since the total energy is constant, its change is *zero.* We can thus write the law of conservation of energy in an alternate form as

Any increase in one form of energy must be accompanied by a decrease in other forms, so that the total change is zero.

The law of conservation of energy sets a fundamental constraint on those processes that can occur in nature. In any process that occurs within an isolated system, the changes in each form of energy must add up to zero, as required by Equation 10.3.

CONCEPTUAL EXAMPLE 10.1 **Energy changes in a bungee ride**

A popular fair attraction is the trampoline bungee ride. The rider bounces up and down on large bungee cords. During part of her motion she is found to be moving upward with the cords becoming more stretched. Is she speeding up or slowing down during this interval?

REASON We'll take our system to include the rider, the bungee cords, and the *earth*. We'll see later how gravitational potential energy is stored in the *system* consisting of the earth and an object such as the rider. With this choice of system,

to a good approximation the system is isolated, with no energy being transferred into or out of the system. Thus the total energy of the system is constant: $\Delta E = 0$.

Because she's moving upward, her height is increasing—and thus so is her gravitational potential energy. Thus $\Delta U_g > 0$. We also know that the cords are getting more stretched, hence more elastic potential energy is being stored. Thus $\Delta U_s > 0$ as well. Now the law of conservation of energy, Equation 10.3, states that $\Delta E = \Delta K + \Delta U_g + \Delta U_s = 0$, so that $\Delta K = -(\Delta U_g + \Delta U_s)$. Both ΔU_g and ΔU_s are positive, so ΔK must be *negative*. This means that her kinetic energy is *decreasing*. Since kinetic energy is energy of motion, this means that she's slowing down.

ASSESS In that part of her motion where she's moving upward and the cords are stretching, she's approaching the highest point of her motion. It makes sense that she's slowing down here, since at the high point her speed is instantaneously zero.

Systems That Aren't Isolated

When John had income and expenses, his total wealth could change. Indeed, he found that his wealth increased by exactly the amount of his income, and decreased by exactly the amount of his expenditures. Similarly, if a system is *not* isolated, so that it can exchange energy with its environment, the system's energy can change. We have seen that the two primary means of energy exchange are work and heat. If an amount of work W is done on the system, this means that an amount of energy W is transferred from the environment to the system, increasing the system's energy by exactly W. Similarly, if a certain amount of energy is transferred from a hot environment to a cooler system as heat Q, the system's energy will increase by exactly the amount Q. As illustrated in Figure 10.6, **the change in the system's energy is simply the sum of the work done on the system and the heat transferred to the system:**

$$\Delta E = W + Q$$

This gives us a more general statement of conservation of energy:

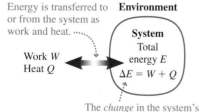

Energy is transferred to or from the system as work and heat.

Environment

System
Total energy E
$\Delta E = W + Q$

Work W
Heat Q

The *change* in the system's energy equals the amount of work done or heat transferred.

FIGURE 10.6 The law of conservation of energy.

Equation 10.4 is the fullest expression of the law of conservation of energy. It's usually called the **first law of thermodynamics,** but it's really just a restatement of the law of conservation of energy to include the possibility of energy transfers. In this chapter we'll refer to it simply as the law of conservation of energy.

NOTE ▶ It's important to realize that even when the system is not isolated, energy is conserved overall. The energy transferred to the system as, say, work increases the energy of the system. But this energy is *removed* from the environment, so that the total energy of system *plus* environment is still conserved. ◀

Systems and Conservation of Energy

To apply the law of conservation of energy, you need to carefully define which objects make up the system and which belong to the environment. This choice will affect how we analyze the various energy transfers and tranformations that occur. In doing so, we need to make a distinction between two classes of forces. **Internal forces** are forces between objects *within* the system. If a weightlifter and barbell are both part of the system, the forces $\vec{F}_{\text{weightlifter on barbell}}$ and $\vec{F}_{\text{barbell on weightlifter}}$ are both internal forces. Internal forces are responsible for energy transformations within the system. Because they are internal to the system, however, **internal forces cannot do work on the system** and thereby change its energy. **External forces** act on the system, but their agent is part of the *environment*. **External forces *can* do work on the system,** transferring energy in or out of it. Whether a given force is an internal or external force depends on the choice of what's included in the system. The following table shows some choices for a crane accelerating a heavy ball upward.

Airplanes are assisted in takeoff from aircraft carriers by a steam-powered catapult under the flight deck. The force of this catapult does work W on the plane, leading to a large increase ΔK in the plane's kinetic energy.

Different choices of the system

The ball only	Ball + earth	Ball + earth + crane
System: The ball only	Ball + earth	Ball + earth + crane
Internal forces: None	$\vec{w}$	$\vec{T}, \vec{w}$, many internal forces of crane
External forces: $\vec{T}, \vec{w}$	$\vec{T}$	None
System energies: K	K, U_g	K, U_g, E_{chem}
Energy analysis: Tension does positive work and the weight does negative work, but since $T > w$ the net work is positive. This work serves to increase the only energy of the system, its kinetic energy. Notice that since the earth is *not* part of the system, the system has no gravitational potential energy.	The weight force is now an *internal* force. That is, it is an interaction force between two objects—the ball and the earth—that are part of the system. The tension force is still an *external* force that does work on the system. This work increases the gravitational potential energy and the kinetic energy of the system.	Now all the forces are internal, and no work is done on the system: The system is *isolated*. With this choice of system, the increased potential and kinetic energy of the ball come from an energy *transformation* from the chemical energy of the crane's fuel.
Energy equation: $\Delta K = W$	$\Delta K + \Delta U_g = W$	$\Delta K + \Delta U_g + \Delta E_{\text{chem}} = 0$

There are evidently many possible choices of the system for a given situation. However, certain choices can make problem solving using the law of energy conservation easier. For the crane above, we'd probably choose the second system consisting of the ball and the earth, since it is a good balance between reducing the number of external forces and having only simple system energies such as K

and U_g. The third choice would be hard to work with, since the many complicated internal forces are difficult to calculate. Tactics Box 10.1 gives some suggestions on how to make a good choice for the system.

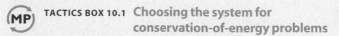

TACTICS BOX 10.1 Choosing the system for conservation-of-energy problems

Exercise 6

The system should include all of the objects identified as follows:

❶ If the speed of an object or objects is changing, the system should include these moving objects because their kinetic energy is changing.

❷ If the height of an object or objects is changing, the system should include the raised object(s) *plus* the earth. This is because potential energy is stored via the gravitational interaction of the earth and object(s).

❸ If the compression or extension of a spring is changing, the system should include the spring because elastic potential energy is stored in the spring itself.

❹ If kinetic or rolling friction is present, the system should include the moving object and the surface on which it slides or rolls. This is because thermal energy is created in both the moving object and the surface, and we want this thermal energy to all be within the system.

Working with Energy Transformations

The law of conservation of energy applies to every form of energy, from kinetic to chemical to nuclear. For the rest of this chapter, however, we'll narrow our focus a bit and only concern ourselves with the forms of energy typically transformed during the motion of ordinary objects. These energies are the kinetic energy K, the potential energy U (which includes both U_g and U_s), and thermal energy E_{th}. The sum of the kinetic and potential energy, $K + U = K + U_g + U_s$, is called the **mechanical energy** of the system. We'll also limit our analysis to energy transfers in the form of work W. In Chapter 11 we'll expand our scope to include other forms of energy listed in the earlier table, as well as energy transfers as heat Q.

The fact that energy is conserved can be a powerful tool for analyzing the dynamics of moving objects. To see how we can apply the law of conservation of energy to dynamics problems, let's use the fact that the change in any quantity is its final value minus its initial value so that, for example, $\Delta K = K_f - K_i$. Then we can write the law of conservation of energy, Equation 10.4 (with $Q = 0$), as

$$(K_f - K_i) + (U_f - U_i) + \Delta E_{th} = W \tag{10.5}$$

NOTE ▶ We don't rewrite ΔE_{th} as $(E_{th})_f - (E_{th})_i$ because the initial thermal energy of an object is typically unknown. Only the *change* in E_{th} can be measured. ◀

Rearranging, we have

$$\underbrace{K_i + U_i}_{} + \underbrace{W}_{} = \underbrace{K_f + U_f + \Delta E_{th}}_{} \tag{10.6}$$

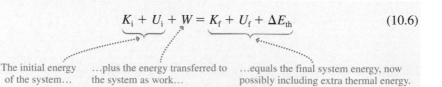

The initial energy of the system...   ...plus the energy transferred to the system as work...   ...equals the final system energy, now possibly including extra thermal energy.

If no external forces do work on the system, $W = 0$ in Equation 10.6 and the system is *isolated*. **If no kinetic friction is present,** ΔE_{th} **will be zero and mechanical energy will be conserved.** Equation 10.6 then becomes the **law of conservation of mechanical energy:**

$$K_i + U_i = K_f + U_f \tag{10.7}$$

Equations 10.6 and 10.7 summarize what we have learned about the conservation of energy, and they will be the basis of our strategy for solving problems using the law of conservation of energy. Much of the rest of this chapter will be concerned with finding quantitative expressions for the different forms of energy in the system and discussing the important question of what to include in the system. We'll use the following Problem-Solving Strategy as we further develop these ideas.

Spring into action BIO A locust can jump as far as one meter, an impressive distance for such a small animal. To make such a jump, its legs must extend much more rapidly than muscles can ordinarily contract. Thus, instead of using its muscles to make the jump directly, the locust uses them to more slowly stretch an internal "spring" near its knee joint. This stores elastic potential energy in the spring. When the muscles relax, the spring is suddenly released, and its energy is rapidly converted into kinetic energy of the insect.

(MP) **PROBLEM-SOLVING STRATEGY 10.1   Conservation of energy problems**

PREPARE   Choose what to include in your system (see Tactics Box 10.1). Draw a before-and-after visual overview, as outlined in Tactics Box 9.1. Note known quantities, and determine what quantity you're trying to find. If the system is isolated and if there is no friction, your solution will be based on Equation 10.7, otherwise you should use Equation 10.6.

Identify which mechanical energies in the system are changing:

- If the *speed* of the object is changing, include K_i and K_f in your solution.
- If the *height* of the object is changing, include $(U_g)_i$ and $(U_g)_f$.
- If the *length* of a spring is changing, include $(U_s)_i$ and $(U_s)_f$.
- If kinetic friction is present, ΔE_{th} will be positive. Some kinetic or potential energy will be transformed into thermal energy.

If an external force acts on the system, you'll need to include the work W done by this force in Equation 10.6.

SOLVE   Depending on the problem, you'll need to calculate initial and/or final values of these energies and insert them into Equation 10.6 or 10.7. Then you can solve for the unknown energies, and from these any unknown speeds (from K), positions (from U), or displacements or forces (from W).

ASSESS   Check the signs of your energies. Kinetic energy, as we'll see, is always positive. In the systems we'll study in this chapter, thermal energy can only increase, so that its change is positive. In Chapters 11 and 12 we'll study systems for which the thermal energy can decrease.

10.4 Work

We've already discussed work as the transfer of energy between a system and its environment by the application of forces on the system. We also noted that in order for energy to be transferred in this way, the system must undergo a displacement—it must *move*—during the time that the force is applied. Let's further investigate the relationship between work, force, and displacement. We'll find that there is a simple expression for work, which we can then use to quantify other kinds of energy as well.

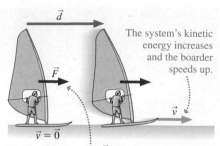

The system's kinetic energy increases and the boarder speeds up.

The force of the wind $\vec{F}$ does work on the system.

FIGURE 10.7 The force of the wind does work on the system, increasing its kinetic energy K.

Consider a system consisting of a windsurfer at rest, as shown on the left in Figure 10.7. Let's assume that there is no friction between his board and the water. Initially the system has no kinetic energy. But if a force from outside the system, such as the force due to the wind, begins to act on the system, the surfer will begin to speed up, and his kinetic energy will increase. In terms of energy transfers, we would say that the energy of the system has increased because of the work done on the system by the force of the wind.

What determines how much work is done by the force of the wind? First, we note that the greater the distance over which the wind pushes the surfer, the faster the surfer goes, and the more his kinetic energy increases. This implies a greater transfer of energy. So **the larger the displacement, the greater the work done.** Second, if the wind pushes with a stronger force, the surfer speeds up more rapidly, and the change in his kinetic energy is greater than with a weaker force. **The stronger the force, the greater the work done.**

This experiment suggests that the amount of energy transferred into a system by a force $\vec{F}$—that is, the amount of work done by $\vec{F}$—depends on both the magnitude F of the force *and* the displacement d of the system. Many experiments of this kind have established that the amount of work done by $\vec{F}$ is *proportional* to both F and d. For the simplest case described above, where the force $\vec{F}$ is constant and points in the direction of the object's displacement, the expression for the work done is found to be

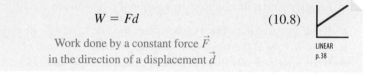

$$W = Fd \qquad\qquad (10.8)$$

Work done by a constant force $\vec{F}$
in the direction of a displacement $\vec{d}$

LINEAR
p. 38

The unit of work, that of force multiplied by distance, is N · m. This unit is so important that it has been given its own name, the **joule** (rhymes with *tool*). We define:

$$1 \text{ joule} = 1 \text{ J} = 1 \text{ N} \cdot \text{m}$$

Since work is simply energy being transferred, **the joule is the unit of *all* forms of energy.** Note that work is a *scalar* quantity.

EXAMPLE 10.1 Work done in pushing a crate

Sarah pushes a heavy crate 3.0 m along the floor at a constant speed. She pushes with a constant horizontal force of magnitude 70 N. How much work does Sarah do on the crate?

PREPARE We begin with the visual overview in Figure 10.8. Sarah pushes with a constant force in the direction of the crate's motion, so we can use Equation 10.8 to find the work done.

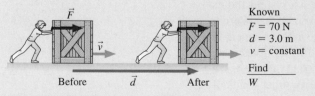

Known
$F = 70 \text{ N}$
$d = 3.0 \text{ m}$
$v = \text{constant}$

Find
W

FIGURE 10.8 Sarah pushing a crate.

SOLVE The work done by Sarah is given by

$$W = Fd = (70 \text{ N})(3.0 \text{ m}) = 210 \text{ J}$$

ASSESS Since the crate moves at a constant speed, it must be in dynamic equilibrium with $\vec{F}_{net} = \vec{0}$. This means that a friction force (not shown) must act opposite to Sarah's push. If friction is present, Tactics Box 10.1 suggests taking the crate *and* the floor as the system. The work Sarah does represents energy transferred *into* the system. In this case, the work increases the thermal energy in the crate and the part of the floor along which it slid. Contrast this with the windsurfer, where work increased the windsurfer's kinetic energy. Both situations are consistent with the energy model shown in Figure 10.4, which you should review at this point.

Force at an Angle to the Displacement

Pushing a crate in the same direction as the crate's displacement is the most efficient way to transfer energy into the system, and so the largest possible amount of work is done. Less work is done if the force acts at an angle to the displacement. To see this, consider the kite buggy of Figure 10.9a, pulled along a horizontal path by the angled force of the kite string $\vec{F}$. As shown in Figure 10.9b, we can break $\vec{F}$ into a component $F_\perp$ perpendicular to the motion, and a component $F_\parallel$ parallel to the motion. Only the parallel component acts to accelerate the rider and increase his kinetic energy, so only the parallel component does work on the rider. From Figure 10.9b, we see that if the angle between $\vec{F}$ and the displacement is θ, then the parallel component is $F_\parallel = F\cos\theta$. So when the force acts at an angle θ to the direction of the displacement, we have

$$W = F_\parallel d = Fd\cos\theta \qquad (10.9)$$

Work done by a constant force $\vec{F}$ at an angle θ to the displacement $\vec{d}$

Notice that this more general definition of work agrees with Equation 10.8 if $\theta = 0°$.

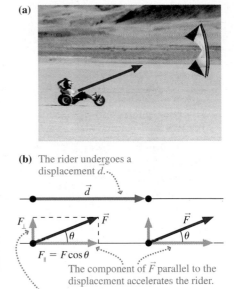

(a)

(b) The rider undergoes a displacement $\vec{d}$.

$F_\parallel = F\cos\theta$

The component of $\vec{F}$ parallel to the displacement accelerates the rider.

The component of $\vec{F}$ perpendicular to the displacement only pulls up on the rider. It doesn't accelerate him.

FIGURE 10.9 Finding the work done when the force is at an angle to the displacement.

CONCEPTUAL EXAMPLE 10.2 **Work done by a parachute**

A drag racer is slowed by a parachute. What is the sign of the work done?

REASON The drag force on the drag racer is shown in Figure 10.10, along with the dragster's displacement as it slows. The force points in the opposite direction to the displacement, so that the angle θ in Equation 10.9 is 180°. Then $\cos\theta = \cos(180°) = -1$. Since F and d in Equation 10.9 are magnitudes, and hence positive, this means that the work $W = Fd\cos\theta = -Fd$ done by the drag force is *negative*.

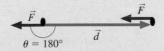

$\theta = 180°$ $\qquad \vec{d}$

FIGURE 10.10 The force acting on a drag racer.

ASSESS Applying Equation 10.4, the law of conservation of energy, to this situation, we have

$$\Delta K = W$$

because the only system energy that changes is the racer's kinetic energy K. Since the kinetic energy is decreasing, its change ΔK is negative. This agrees with the sign of W. This example illustrates the general principle that **negative work represents a transfer of energy out of the system.**

Tactics Box 10.2 shows how to calculate the work done by a force at any angle to the direction of motion. The system illustrated is a block sliding on a frictionless horizontal surface, so that only the kinetic energy is changing. However, the same relationships hold for any object undergoing a displacement.

The quantities F and d are always positive, so **the sign of W is determined entirely by the angle θ between the force and the displacement.** Note that

Activ
Physics
ONLINE
5.1

Equation 10.9, $W = Fd\cos\theta$, is valid for any angle θ. In three special cases, $\theta = 0°$, $\theta = 90°$, and $\theta = 180°$, however, there are simple versions of Equation 10.9 that you can use. These are noted in Tactics Box 10.2.

(MP) **TACTICS BOX 10.2** Calculating the work done by a constant force   ✐ Exercises 9,11,12

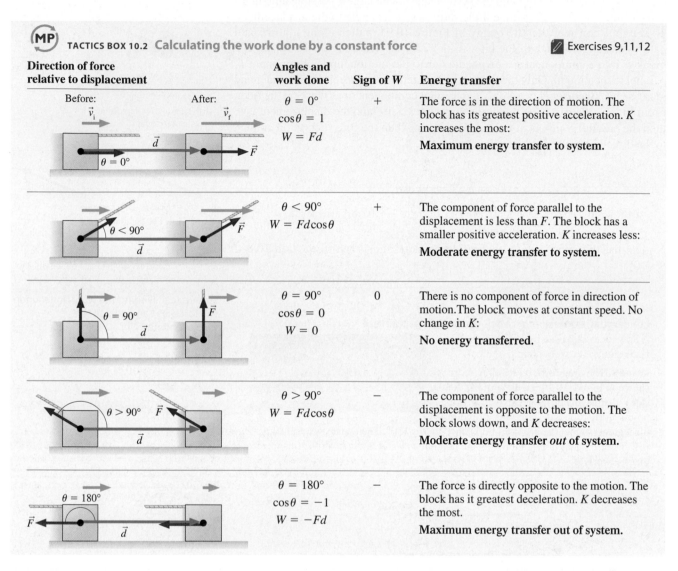

Direction of force relative to displacement		Angles and work done	Sign of W	Energy transfer
Before: $\vec{v}_i$ After: $\vec{v}_f$ $\vec{d}$ $\theta = 0°$ $\vec{F}$		$\theta = 0°$ $\cos\theta = 1$ $W = Fd$	+	The force is in the direction of motion. The block has its greatest positive acceleration. K increases the most: **Maximum energy transfer to system.**
$\theta < 90°$ $\vec{d}$ $\vec{F}$		$\theta < 90°$ $W = Fd\cos\theta$	+	The component of force parallel to the displacement is less than F. The block has a smaller positive acceleration. K increases less: **Moderate energy transfer to system.**
$\theta = 90°$ $\vec{d}$ $\vec{F}$		$\theta = 90°$ $\cos\theta = 0$ $W = 0$	0	There is no component of force in direction of motion. The block moves at constant speed. No change in K: **No energy transferred.**
$\theta > 90°$ $\vec{F}$ $\vec{d}$ $\vec{F}$		$\theta > 90°$ $W = Fd\cos\theta$	−	The component of force parallel to the displacement is opposite to the motion. The block slows down, and K decreases: **Moderate energy transfer *out* of system.**
$\theta = 180°$ $\vec{F}$ $\vec{d}$ $\vec{F}$		$\theta = 180°$ $\cos\theta = -1$ $W = -Fd$	−	The force is directly opposite to the motion. The block has it greatest deceleration. K decreases the most. **Maximum energy transfer out of system.**

EXAMPLE 10.2 Work done in pulling a suitcase

A strap inclined upward at a 45° angle pulls a suitcase through the airport. The tension in the strap is 20 N. How much work does the tension do if the suitcase is pulled 100 m at a constant speed?

PREPARE Figure 10.11 shows a visual overview. Since the case moves at a constant speed, there must be a rolling friction force acting to the left. Tactics Box 10.1 suggests in this case that we take as our system the suitcase and the floor upon which it rolls.

SOLVE We can use Equation 10.9 to find that the tension does work

$$W = Td\cos\theta = (20 \text{ N})(100 \text{ m})\cos 45° = 1400 \text{ J}$$

ASSESS Because a person is pulling on the other end of the strap, causing the tension, we would say informally that the person does 1400 J of work on the suitcase. This work represents

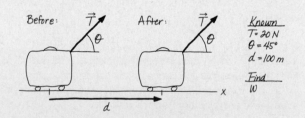

FIGURE 10.11 A suitcase pulled by a strap.

energy transferred into the suitcase/floor system. Since the suitcase moves at a constant speed, the system's kinetic energy doesn't change. Thus the work goes entirely into increasing the *thermal* energy E_{th} of the suitcase and the floor.

If several forces act on an object that undergoes a displacement, each does work on the object. The **total (or net) work** W_{total} is the sum of the work done by each force. The total work represents the total energy transfer *to* the system from the environment (if $W_{total} > 0$) or *from* the system to the environment (if $W_{total} < 0$).

Forces That Do No Work

The fact that a force acts on an object doesn't mean that the force will do work on the object. The table below shows three common cases where a force does no work.

Forces that do no work

If the object undergoes no displacement while the force acts, no work is done.

This can sometimes seem counterintuitive. The weightlifter struggles mightily to hold the barbell over his head. But during the time the barbell remains stationary, he does no work on it because its displacement is zero. But why then is it so hard for him to hold it there? We'll see in Chapter 11 that it takes a rapid conversion of his internal chemical energy to keep his arms extended under this great load.

A force perpendicular to the displacement does no work.

The woman exerts only a vertical force on the briefcase she's carrying. This force has no component in the direction of the displacement, so the briefcase moves at a constant velocity and its kinetic energy remains constant. Since the energy of the briefcase doesn't change, it must be that no energy is being transferred to it as work.

(This is the case where $\theta = 90°$ in Tactics Box 10.2.)

If the part of the object on which the force acts undergoes no displacement, no work is done.

Even though the wall pushes on the skater with a normal force $\vec{n}$ and she undergoes a displacement $\vec{d}$, the wall does no work on her, because the point of her body on which $\vec{n}$ acts—her hands—undergoes no displacement. This makes sense: How could energy be transferred as work from an inert, stationary object? So where does her kinetic energy come from? This will be the subject of much of Chapter 11. Can you guess?

STOP TO THINK 10.2 Which force does the most work?

A. The 10 N force.
B. The 8 N force.
C. The 6 N force.
D. They all do the same amount of work.

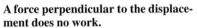

10.5 Kinetic Energy

We've already qualitatively discussed kinetic energy, an object's energy of motion. Let's now use what we've learned about work, and some simple kinematics, to find a quantitative expression for kinetic energy. Consider the system consisting of a car being pulled by a tow rope as in Figure 10.12. The rope pulls with a constant force $\vec{F}$ while the car undergoes a displacement $\vec{d}$, so that the force does work $W = Fd$ on the car. If we ignore friction and drag, the work done by $\vec{F}$ will

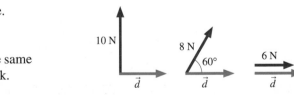

FIGURE 10.12 The work done by the tow rope increases the car's kinetic energy.

be transferred entirely into the car's energy of motion—its kinetic energy. In this case, the law of conservation of energy, Equation 10.6, reads

$$K_i + W = K_f$$

or

$$W = K_f - K_i \qquad (10.10)$$

Using kinematics, we can find another expression for the work done, in terms of the car's initial and final speeds. Recall from Chapter 2 the kinematic equation relating an object's displacement and its change in velocity:

$$v_f^2 = v_i^2 + 2a\Delta x$$

Applied to the motion of our car, $\Delta x = d$ is the car's displacement and, from Newton's second law, the acceleration is $a = F/m$. Thus we can write

$$v_f^2 = v_i^2 + \frac{2Fd}{m} = v_i^2 + \frac{2W}{m}$$

where we have replaced Fd with the work W. If we now solve for the work, we find

$$W = \frac{1}{2}m(v_f^2 - v_i^2) = \frac{1}{2}mv_f^2 - \frac{1}{2}mv_i^2$$

If we compare this result with Equation 10.10, we see that

$$K_f = \frac{1}{2}mv_f^2 \qquad \text{and} \qquad K_i = \frac{1}{2}mv_i^2$$

6.1 Act|v
Phys|cs
ONLINE

In general, then, an object moving at a speed v has kinetic energy

$$K = \frac{1}{2}mv^2 \qquad (10.11)$$

QUADRATIC
p. 50

Kinetic energy of an object of mass m moving with speed v

From Equation 10.11, the units of kinetic energy are mass times speed squared, or $\text{kg} \cdot (\text{m/s})^2$. But

$$1\,\text{kg} \cdot (\text{m/s})^2 = \underbrace{1\,\text{kg} \cdot (\text{m/s}^2)}_{1\,\text{N}} \cdot \text{m} = 1\,\text{N} \cdot \text{m} = 1\,\text{J}$$

We see that the units of kinetic energy are the same as those of work, as they must be. Table 10.1 gives some approximate kinetic energies. Everyday kinetic energies range from a tiny fraction of a fraction of a joule to nearly a million joules for a speeding car.

TABLE 10.1 Some approximate kinetic energies

Object	Kinetic energy
Walking ant	1×10^{-8} J
Penny dropped 1 m	2.5×10^{-3} J
Person walking	70 J
100 mph fastball	150 J
Bullet	5000 J
Car, 60 mph	5×10^5 J
Supertanker	2×10^{10} J

CONCEPTUAL EXAMPLE 10.3 **Kinetic energy changes for a car**

Compare the increase in a 1000 kg car's kinetic energy as it speeds up by 5.0 m/s starting from 5.0 m/s, to its increase in kinetic energy as it speeds up by 5.0 m/s starting from 10 m/s.

REASON The change in the car's kinetic energy in going from 5 m/s to 10 m/s is

$$\Delta K_{5\to 10} = \frac{1}{2}mv_f^2 - \frac{1}{2}mv_i^2$$

This gives

$$\Delta K_{5\to 10} = \frac{1}{2}(1000\,\text{kg})(10\,\text{m/s})^2 - \frac{1}{2}(1000\,\text{kg})(5.0\,\text{m/s})^2$$

$$= 3.8 \times 10^4\,\text{J}$$

while

$$\Delta K_{10\to 15} = \frac{1}{2}(1000\,\text{kg})(15\,\text{m/s})^2 - \frac{1}{2}(1000\,\text{kg})(10\,\text{m/s})^2$$

$$= 6.3 \times 10^4\,\text{J}$$

Even though the increase in the car's *speed* was the same in both cases, the increase in kinetic energy is substantially larger in the second case.

ASSESS Kinetic energy depends on the *square* of the speed v. If we plot the kinetic energy of the car as in Figure 10.13, we see that the energy of the car increases rapidly with speed. We can also see graphically why the change in K for a fixed 5 m/s change in v is greater at high speeds than at low speeds. In part this is why it's harder to accelerate your car at high speeds than at low speeds.

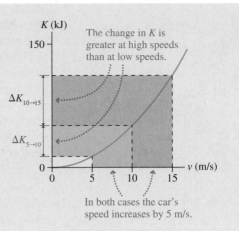

FIGURE 10.13 The kinetic energy increases as the *square* of the speed.

EXAMPLE 10.3 Speed of a bobsled after pushing

A two-man bobsled has a mass of 390 kg. Starting from rest, the two racers push the sled for the first 50 m with a net force of 270 N. Neglecting friction, what is the sled's speed at the end of the 50 m?

PREPARE This is the first example where we fully use Problem-Solving Strategy 10.1. We start by identifying the bobsled as the system; the two racers pushing the sled are part of the environment. The racers do work on the system by pushing it with force $\vec{F}$. Because the speed of the sled changes, we'll need to include kinetic energy. Neither U_g nor U_s changes, so we won't need to consider these energies. Figure 10.14 lists the known quantities and the quantity (v_f) that we want to find.

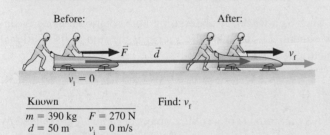

Known	Find: v_f
$m = 390$ kg $F = 270$ N	
$d = 50$ m $v_i = 0$ m/s	

FIGURE 10.14 The work done by the pushers increases the sled's kinetic energy.

SOLVE With only kinetic energy changing, the conservation of energy equation, Equation 10.6, is

$$K_i + W = K_f$$

Using our expressions for kinetic energy and work, this becomes

$$\frac{1}{2}mv_i^2 + Fd = \frac{1}{2}mv_f^2$$

Because $v_i = 0$, the energy equation reduces to

$$Fd = \frac{1}{2}mv_f^2$$

We can solve for the final speed to get

$$v_f = \sqrt{\frac{2Fd}{m}} = \sqrt{\frac{2(270 \text{ N})(50 \text{ m})}{390 \text{ kg}}} = 8.3 \text{ m/s}$$

ASSESS We solved this problem using the concept of energy conservation. In this case, we could also have solved it using Newton's second law and kinematics. However, we'll soon see that energy conservation can solve problems that would be very difficult for us to solve using Newton's laws alone.

STOP TO THINK 10.3 Rank in order, from greatest to least, the kinetic energies of the sliding pucks.

FIGURE 10.15 The large rotating blades of a windmill have a great deal of kinetic energy.

Rotational Kinetic Energy

We've just found an expression for the kinetic energy of an object moving along a line or some other path. This energy is called **translational kinetic energy.** Consider now an object rotating about a fixed axis, such as the windmill blades in Figure 10.15. Although the blades have no overall translational motion, each

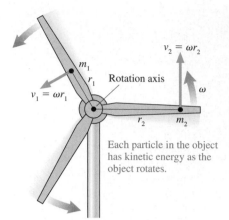

FIGURE 10.16 Rotational kinetic energy is due to the circular motion of the particles.

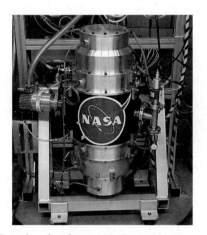

Rotational recharge The International Space Station (ISS) gets its electrical power from solar panels. But during each 92-min orbit, the ISS is in the earth's shadow for 30 min. The batteries that currently provide power during these blackouts need periodic replacement, which is very expensive in space. A promising new technology would replace the batteries with a *flywheel*—a cylinder rotating at very high angular speed. Energy from the solar cells is used to speed up the flywheel, storing energy as rotational kinetic energy, which can then be converted back into electrical energy when the ISS is in shadow.

particle in the blade is moving and hence has kinetic energy. Adding up the kinetic energy for each particle that makes up the blades, we find that the blades have **rotational kinetic energy,** the kinetic energy due to rotation.

Figure 10.16 shows two of the particles making up a windmill blade that rotates with angular velocity ω. Recall from Section 7.2 that a particle moving with angular velocity ω in a circle of radius r has a speed $v = \omega r$. Thus particle 1, which rotates in a circle of radius r_1, moves with speed $v_1 = r_1\omega$. Particle 2, which rotates in a circle with a larger radius r_2, moves with a larger speed $v_2 = r_2\omega$. The object's rotational kinetic energy is the sum of the kinetic energies of *all* of the particles:

$$K_{\text{rot}} = \frac{1}{2}m_1v_1^2 + \frac{1}{2}m_2v_2^2 + \cdots$$

$$= \frac{1}{2}m_1r_1^2\omega^2 + \frac{1}{2}m_2r_2^2\omega^2 + \cdots = \frac{1}{2}\left(\sum mr^2\right)\omega^2$$

You will recognize the term in parentheses as our old friend, the moment of inertia I. Thus the rotational kinetic energy is

$$K_{\text{rot}} = \frac{1}{2}I\omega^2 \qquad (10.12)$$

QUADRATIC
p.50

Rotational kinetic energy of object with moment of inertia I and angular velocity ω

NOTE ▶ Rotational kinetic energy is *not* a new form of energy. This is the familiar kinetic energy of motion, only now expressed in a form that is especially convenient for rotational motion. Comparison with the familiar $\frac{1}{2}mv^2$ shows again that the moment of inertia I is the rotational equivalent of mass. ◀

A rolling object, such as a wheel, is undergoing both rotational *and* translational motions. Consequently, its total kinetic energy is the sum of its rotational and translational kinetic energies:

$$K = K_{\text{trans}} + K_{\text{rot}} = \frac{1}{2}mv^2 + \frac{1}{2}I\omega^2$$

Recall from Section 6.3 that v and ω of a rolling object of radius R are related by $\omega = v/R$. Thus we can write the kinetic energy of a rolling object as

$$K_{\text{rolling}} = \frac{1}{2}mv^2 + \frac{1}{2}I\left(\frac{v}{R}\right)^2 = \frac{1}{2}\left(m + \frac{I}{R^2}\right)v^2 \qquad (10.13)$$

This illustrates the important fact that **the kinetic energy of a rolling object is always greater than that of a nonrotating object moving at the same speed.**

EXAMPLE 10.4 Kinetic energy of a bicycle

Bike 1 has a 10.0 kg frame and 1.00 kg wheels, while bike 2 has a 9.00 kg frame and 1.50 kg wheels. Both bikes thus have the same 12.0 kg total mass. What is the kinetic energy of each bike when they are ridden at 12.0 m/s? Model each wheel as a hoop of radius 35.0 cm.

PREPARE Each bike's frame has only translational kinetic energy $K_{\text{frame}} = \frac{1}{2}Mv^2$, where M is the mass of the frame. The kinetic energy of each rolling wheel is given by Equation 10.13. From

Table 7.2, we find that I for a hoop is mR^2, where m is the mass of one wheel.

SOLVE From Equation 10.13 the kinetic energy of each rolling wheel is

$$K_{\text{wheel}} = \frac{1}{2}\left(m + \frac{mR^2}{R^2}\right)v^2 = \frac{1}{2}(2m)v^2 = mv^2$$

Then the total kinetic energy of a bike is

$$K = K_{\text{frame}} + 2K_{\text{wheel}} = \frac{1}{2}Mv^2 + 2mv^2$$

The factor of 2 in the second term occurs because each bike has two wheels. Thus the kinetic energies of the two bikes are

$$K_1 = \frac{1}{2}(10.0 \text{ kg})(12.0 \text{ m/s})^2 + 2(1.00 \text{ kg})(12.0 \text{ m/s})^2$$
$$= 1010 \text{ J}$$

$$K_2 = \frac{1}{2}(9.00 \text{ kg})(12.0 \text{ m/s})^2 + 2(1.50 \text{ kg})(12.0 \text{ m/s})^2$$
$$= 1080 \text{ J}$$

The kinetic energy of bike 2 is about 7% higher than that of bike 1. Note that the radius of the wheels was not needed in this calculation.

ASSESS As the cyclists on these bikes accelerate from rest to 12 m/s, they must convert some of their internal chemical energy into the kinetic energy of the bikes. Racing cyclists want to use as little of their own energy as possible. Although both bikes have the same total mass, the one with the lighter wheels will take less energy to get it moving. Shaving a little extra weight off your wheels is more useful than taking that same weight off your frame.

It's important that racing bike wheels are as light as possible.

10.6 Potential Energy

When two or more objects in a system interact, it is sometimes possible to *store* energy in that system in a way that the energy can be easily recovered. For instance, the earth and a ball interact by the gravitational force between them. If the ball is lifted up into the air, energy is stored in the ball-earth system, energy that can later be recovered as kinetic energy when the ball is released and falls. Similarly, a spring is a system made up of countless atoms that interact via their atomic "springs." If we push a box against a spring, energy is stored that can be recovered when the spring later pushes the box across the table. This sort of stored energy is called **potential energy,** since it has the *potential* to be converted into other forms of energy such as kinetic or thermal energy.

The forces due to gravity and springs are special in that they allow for the storage of energy. Other interaction forces do not. When a crate is pushed across the floor, the crate and the floor interact via the force of friction, and the work done on the system is converted into thermal energy. But this energy is *not* stored up for later recovery—it slowly diffuses into the environment and cannot be recovered.

Interaction forces that can store useful energy are called **conservative forces.** The name comes from the important fact, which we'll soon look at in detail, that when only conservative forces act, the mechanical energy of a system is *conserved.* Gravity and elastic forces are conservative forces, and later we'll see that the electric force is a conservative force as well. Friction, on the other hand, is a **nonconservative force.** When two objects interact via a friction force, energy is not stored. It is usually transformed into thermal energy.

Let's look more closely at the potential energies associated with the two conservative forces—gravity and springs—that we'll study in this chapter.

Gravitational Potential Energy

To find an expression for gravitational potential energy, let's consider the system of the book and the earth shown in Figure 10.17a on the next page. The book is lifted at a constant speed from its initial position at y_i to a final height y_f.

We can analyze this situation using the approach of Problem-Solving Strategy 10.1. The lifting force of the hand is external to the system and so does work W on the system, increasing its energy. The book is lifted at a constant speed, so its kinetic energy doesn't change. Because there's no friction, the book's thermal energy doesn't change either. Thus the work done goes entirely into increasing the gravitational potential energy of the system. The law of conservation of energy, Equation 10.6, then reads

(a)

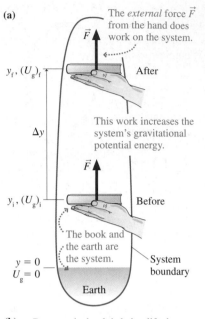

The *external* force $\vec{F}$ from the hand does work on the system.

$y_f, (U_g)_f$ After

This work increases the system's gravitational potential energy.

Δy

$y_i, (U_g)_i$ Before

The book and the earth are the system.

$y = 0$
$U_g = 0$

System boundary

Earth

(b)   Because the book is being lifted at a constant speed, it is in dynamic equilibrium with $\vec{F}_{net} = \vec{0}$. Thus $F = w = mg$.

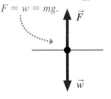

$\vec{F}$

$\vec{w}$

FIGURE 10.17 Lifting a book increases its gravitational potential energy.

The initial gravitational potential energy. . .   . . .plus the energy put into the system as work. . .

$$(U_g)_i + W = (U_g)_f \tag{10.14}$$

. . .equals the final gravitational potential energy.

The work done is $W = F\Delta y$, where $\Delta y = y_f - y_i$ is the vertical distance that the book is lifted. From the free-body diagram of Figure 10.17b, we see that $F = mg$. This gives $W = mg\Delta y$, so that

$$(U_g)_i + mg\Delta y = (U_g)_f$$

or

$$(U_g)_f = (U_g)_i + mg\Delta y \tag{10.15}$$

Since our final height was greater than our initial height, Δy is positive and $(U_g)_f > (U_g)_i$. **The higher the object is lifted, the greater the gravitational potential energy in the object/earth system.**

Equation 10.15 gives the final gravitational potential energy $(U_g)_f$ in terms of its initial value $(U_g)_i$. But what is the value of $(U_g)_i$? We can gain some insight by writing Equation 10.15 in terms of energy *changes*. We have

$$(U_g)_f - (U_g)_i = mg\Delta y$$

or

$$\Delta U_g = mg\Delta y$$

For example, if we lift a 1.5 kg book up by 2.0 m, we increase its gravitational potential energy by $\Delta U_g = (1.5 \text{ kg})(9.8 \text{ m/s}^2)(2.0 \text{ m}) = 29.4$ J. This increase is *independent* of the book's starting height: We would get the *same* increase whether we lifted the book 2.0 m starting at sea level or starting at the top of Mount Everest. If we then dropped the book 2.0 m, we would recover the same 29.4 J as kinetic energy, whether in Miami or on Everest. This illustrates an important general fact about *every* form of potential energy: **Only *changes* in potential energy are significant.**

Because of this fact, we are free to choose a *reference level* where we define U_g to be zero. Our expression for U_g is particularly simple if we choose this reference level to be at $y = 0$. We then have

$$U_g = mgy \tag{10.16}$$

Gravitational potential energy of an object of mass m at a height y (assuming $U_g = 0$ when the object is at $y = 0$)

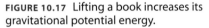

LINEAR
p. 38

NOTE ▶ We've emphasized that gravitational potential energy is an energy of the earth-object *system*. In solving problems using the law of conservation of energy, you'll need to include the earth as part of your system. For simplicity, we'll usually speak of "the gravitational potential energy of the ball," but what we really mean is the potential energy of the earth-ball system. ◀

EXAMPLE 10.5 Hitting the bell

At the county fair, Katie tries her hand at the ring-the-bell attraction, as shown in Figure 10.18. She swings the mallet hard enough to give the ball an initial upward speed of 8.0 m/s. Will the ball ring the bell, 3.0 m from the bottom?

PREPARE As discussed above and in Tactics Box 10.1, we'll choose the ball *and* the earth as the system. Figure 10.18 shows the visual overview. If we assume that the track along which the ball moves is frictionless, then only the *mechanical* energy of the system changes. The only force on the ball after it leaves the

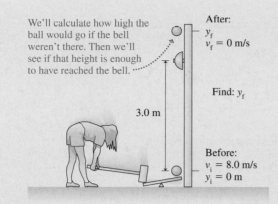

We'll calculate how high the ball would go if the bell weren't there. Then we'll see if that height is enough to have reached the bell.

After:
y_f
$v_f = 0$ m/s

Find: y_f

3.0 m

Before:
$v_i = 8.0$ m/s
$y_i = 0$ m

FIGURE 10.18 Before-and-after visual overview of the ring-the-bell attraction.

bottom lever is gravity, but gravity is an *internal* force due to our choice of the ball *plus* the earth as the system. This means that the gravitational interaction is included as gravitational potential energy rather than as external work. Since no *external* forces do work on the earth-ball system, the system is isolated. We can then use the law of conservation of mechanical energy, Equation 10.7.

SOLVE Equation 10.7 tells us that $K_i + (U_g)_i = K_f + (U_g)_f$. We can use our expressions for kinetic and potential energy to write this as

$$\frac{1}{2}mv_i^2 + mgy_i = \frac{1}{2}mv_f^2 + mgy_f$$

Let's ignore the bell for the moment and figure out how far the ball would rise if there were nothing in its way. We know that the ball starts at $y_i = 0$ m and that its speed v_f at the highest point is zero. Thus the energy equation simplifies to

$$mgy_f = \frac{1}{2}mv_i^2$$

This is easily solved for the height y_f:

$$y_f = \frac{v_i^2}{2g} = \frac{(8.0 \text{ m/s})^2}{2(9.8 \text{ m/s}^2)} = 3.3 \text{ m}$$

This is higher than the point where the bell sits, so the ball would actually hit it on the way up.

ASSESS Notice that the mass canceled and wasn't needed, a fact about free fall that you should remember from Chapter 2.

An important conclusion from Equation 10.16 is that **gravitational potential energy depends only on the height of the object above the reference level $y = 0$, not on the object's horizontal position.** Consider carrying a briefcase while walking on level ground at a constant speed. As shown in the table on page 309, the force of your hand on the briefcase is *vertical* and hence *perpendicular* to the displacement. No work is done on the briefcase and consequently its gravitational potential energy remains constant as long as its height above the ground doesn't change as you walk.

This idea can be applied to more complicated cases, such as the 51 kg hiker in Figure 10.19. His gravitational potential energy depends *only* on his height y above the reference level, so it's the same value $U_g = mgy = 50$ kJ at any point on path A where he is at a height $y = 100$ m above the reference level. If he had instead taken path B, his gravitational potential energy at 100 m would be the same 50 kJ. It doesn't matter *how* he gets to 100 m, his potential energy at that height will be the same. This demonstrates an important aspect of all potential energies: **The potential energy depends only on the *position* of the object and not on the path the object took to get to that position.** This fact will allow us to use the law of conservation of energy to easily solve a variety of problems that would be very difficult to solve using Newton's laws alone, because we won't need to know the details of the path of the object—just its starting and ending points.

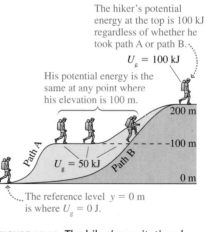

The hiker's potential energy at the top is 100 kJ regardless of whether he took path A or path B.

$U_g = 100$ kJ

His potential energy is the same at any point where his elevation is 100 m.

200 m

Path A   Path B
$U_g = 50$ kJ

100 m

0 m

The reference level $y = 0$ m is where $U_g = 0$ J.

FIGURE 10.19 The hiker's gravitational potential energy depends only on his height above the $y = 0$ reference level.

EXAMPLE 10.6 Speed at the bottom of a water slide

Still at the county fair, Katie tries the water slide, whose shape is shown in Figure 10.20. The starting point is 9.0 m above the ground. She pushes off with an initial speed of 2.0 m/s. If the slide is frictionless, how fast will Katie be traveling at the bottom?

PREPARE Figure 10.20 on the next page shows a visual overview of the slide. Because there is no friction, Tactics Box 10.1 suggests that we take as our system Katie (the moving object) and the earth. With this choice of system, the only energies in the system are kinetic and gravitational potential energy. Note that the slope of the slide is not constant, so Katie's

Continued

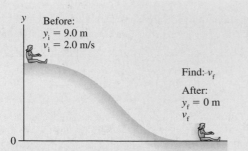

FIGURE 10.20 Before-and-after visual overview of Katie on the water slide.

acceleration will not be constant either. Thus we can't use constant-acceleration kinematics to find her speed. But we *can* use the law of conservation of energy to easily solve for her speed. Because there is no friction, the mechanical energy is conserved.

SOLVE Conservation of mechanical energy gives

$$K_i + (U_g)_i = K_f + (U_g)_f$$

or

$$\frac{1}{2}mv_i^2 + mgy_i = \frac{1}{2}mv_f^2 + mgy_f$$

Taking $y_f = 0$ m we have

$$\frac{1}{2}mv_i^2 + mgy_i = \frac{1}{2}mv_f^2$$

which we can solve to get

$$v_f = \sqrt{v_i^2 + 2gy_i}$$

$$= \sqrt{(2.0 \text{ m/s})^2 + 2(9.8 \text{ m/s}^2)(9.0 \text{ m})} = 13 \text{ m/s}$$

ASSESS It is important to realize that the *shape* of the slide does not matter because gravitational potential energy depends only on the *height* above a reference level. **In sliding down any (frictionless) slide of the same height, your speed at the bottom would be the same.**

STOP TO THINK 10.4 Rank in order, from largest to smallest, the gravitational potential energies of identical balls 1 to 4.

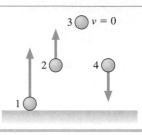

Elastic Potential Energy

Energy can also be stored in a compressed or extended spring as **elastic** (or **spring**) **potential energy** U_s. We can find out how much energy is stored in a spring by using an external force to slowly compress the spring. This external force does work on the spring, transferring energy to the spring. Since only the elastic potential energy of the spring is changing, the law of conservation of energy reads

$$W = \Delta U_s \quad (10.17)$$

That is, we can find out how much elastic potential energy is stored in the spring by calculating the amount of work needed to compress the spring.

Figure 10.21 shows a spring being compressed by a hand. In Section 8.3 we found that the force that the spring will exert on the hand is equal to $-k\Delta x$, where Δx is the displacement of the end of the spring from its equilibrium position and k is the spring constant. In Figure 10.21 we have set the origin of our coordinate system at the equilibrium position. The displacement from equilibrium Δx is therefore equal to x, and the spring force is then $-kx$. By Newton's third law, this means that the force that the hand exerts on the spring is equal to $+kx$.

As we compress the end of the spring from its equilibrium position to a final displacement x, the force we apply increases from zero to kx. This is not a constant force, so we can't use Equation 10.8, $W = Fd$, to find the work done, because this equation is valid only for a constant force. However, it seems reasonable that we could calculate the work by using the *average* force in Equation 10.8. Because the force varies from $F_i = 0$ to $F_f = kx$, the average force used to compress the spring is

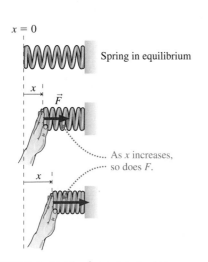

FIGURE 10.21 The force required to compress a spring is not constant.

$$F_{avg} = \frac{1}{2}(F_f + F_i) = \frac{1}{2}(kx + 0) = \frac{1}{2}kx$$

Thus the work done by the hand is

$$W = F_{avg}d = F_{avg}x = \left(\frac{1}{2}kx\right)x = \frac{1}{2}kx^2$$

This work is stored as potential energy in the spring, so we can use Equation 10.17 to find that the elastic potential energy increases by

$$\Delta U_s = \frac{1}{2}kx^2$$

Just as in the case of gravitational potential energy, we have found an expression for the *change* in U_s, not U_s itself. Again, we are free to set $U_s = 0$ at any convenient spring extension. An obvious choice is to set $U_s = 0$ at the point where the spring is in equilibrium, neither compressed nor stretched; that is, at $x = 0$. With this choice we have

$$U_s = \frac{1}{2}kx^2 \qquad (10.18)$$

Elastic potential energy of a spring displaced a distance x from equilibrium (assuming $U_s = 0$ when the end of the spring is at $x = 0$)

QUADRATIC
p. 50

NOTE ▶ Since U_s depends on the *square* of the displacement x, U_s is the same whether x is positive (the spring is compressed as in Figure 10.21) or negative (the spring is stretched). ◀

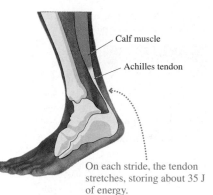

Calf muscle

Achilles tendon

On each stride, the tendon stretches, storing about 35 J of energy.

Spring in your step BIO As you run, you lose some of your mechanical energy each time your foot strikes the ground; this energy is transformed into unrecoverable thermal energy. Luckily, about 35% of the decrease of your mechanical energy when your foot lands is stored as elastic potential energy in the stretchable Achilles tendon of the lower leg. On each plant of the foot the tendon is stretched, storing some energy. The tendon springs back as you push off the ground again, helping to propel you forward. This recovered energy reduces the amount of internal chemical energy you use, increasing your efficiency.

EXAMPLE 10.7 Speed of a spring-launched ball

A spring-loaded toy gun is used to launch a 10 g plastic ball. The spring, which has a spring constant of 10 N/m, is compressed by 10 cm as the ball is pushed into the barrel. When the trigger is pulled, the spring is released and shoots the ball back out. What is the ball's speed as it leaves the barrel? Assume that friction is negligible.

PREPARE Assume the spring obeys Hooke's law $F = -kx$, and is massless so that it has no kinetic energy of its own. Using Tactics Box 10.1 we choose the system to be the spring and the ball. There's no friction, hence the system's mechanical energy $K + U_s$ is conserved.

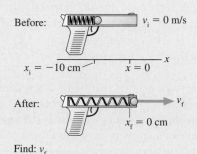

Before: $v_i = 0$ m/s

$x_i = -10$ cm $x = 0$

After: v_f

$x_f = 0$ cm

Find: v_f

FIGURE 10.22 The before-and-after visual overview of a ball being shot out of a spring-loaded toy gun.

Figure 10.22 shows a before-and-after visual overview. The compressed spring will push on the ball until the spring has returned to its equilibrium length. We have chosen the origin of the coordinate system at the equilibrium position of the free end of the spring, making $x_i = -10$ cm and $x_f = 0$ cm.

SOLVE The energy conservation equation is $K_i + (U_s)_i = K_f + (U_s)_f$. We can use the elastic potential energy of the spring, Equation 10.18, to write this as

$$\frac{1}{2}mv_i^2 + \frac{1}{2}kx_i^2 = \frac{1}{2}mv_f^2 + \frac{1}{2}kx_f^2$$

We know that $x_f = 0$ m and $v_i = 0$ m/s, so this simplifies to

$$\frac{1}{2}mv_f^2 = \frac{1}{2}kx_i^2$$

It is now straightforward to solve for the ball's speed:

$$v_f = \sqrt{\frac{kx_i^2}{m}} = \sqrt{\frac{(10 \text{ N/m})(-0.10 \text{ m})^2}{0.010 \text{ kg}}} = 3.2 \text{ m/s}$$

ASSESS This is *not* a problem that we could have easily solved with Newton's laws. The acceleration is not constant, and we have not learned how to handle the kinematics of nonconstant acceleration. But with conservation of energy—it's easy!

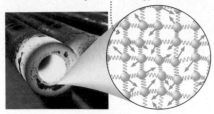

Hot object: Fast-moving molecules have lots of kinetic and elastic potential energy.

Cold object: Slow-moving molecules have little kinetic and elastic potential energy.

FIGURE 10.23 A molecular view of thermal energy.

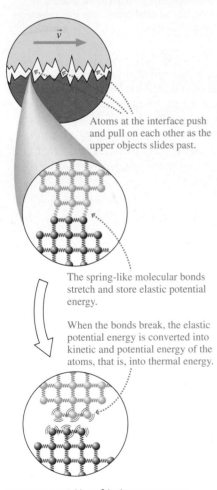

Atoms at the interface push and pull on each other as the upper objects slides past.

The spring-like molecular bonds stretch and store elastic potential energy.

When the bonds break, the elastic potential energy is converted into kinetic and potential energy of the atoms, that is, into thermal energy.

FIGURE 10.24 How friction causes an increase in thermal energy.

> **STOP TO THINK 10.5**   A spring-loaded gun shoots a plastic ball with a speed of 4 m/s. If the spring is compressed twice as far, the ball's speed will be
>
> A. 2 m/s.   B. 4 m/s.   C. 8 m/s.   D. 16 m/s.

10.7 Thermal Energy

We noted earlier that thermal energy is related to the microscopic motion of the molecules of an object. As Figure 10.23 shows, the molecules in a hot object jiggle around their average positions more than the molecules in a cold object. This has two consequences. First, each atom is on average moving faster in the hot object. This means that each atom has a higher *kinetic energy*. Second, each atom in the hot object tends to stray further from its equilibrium position, leading to a greater stretching or compressing of the spring-like molecular bonds. This means that each atom has on average a higher *potential energy*. The potential energy stored in any one bond and the kinetic energy of any one atom are both exceedingly small, but there are incredibly many bonds and atoms. The sum of all these microscopic potential and kinetic energies is what we call **thermal energy.**

Is this microscopic energy worth worrying about? To see, consider a 500 g (≈ 1 lb) iron ball moving at the respectable speed of $v_{ball} = 20$ m/s (≈ 45 mph). Its kinetic energy is $K = \frac{1}{2}mv_{ball}^2 = 100$ J.

How fast do the atoms jiggle about their equilibrium positions? This speed depends on the temperature, but at room temperature it's very high—roughly 500 m/s. So each atom is on average traveling in a straight line at 20 m/s, but jiggling about this average motion at a speed of 500 m/s! This a factor of 25 times faster. And since kinetic energy is proportional to the *square* of the speed, the kinetic energy due to the microscopic motion is about 625 times greater than that due to the overall motion. And it turns out that the microscopic potential energy is just as large. Thus the ball that has an ordinary kinetic energy of 100 J has an internal thermal energy of $2 \times 625 \times 100$ J = 125,000 J!

Transforming Mechanical Energy into Thermal Energy

Consider a snowboarder sliding on level snow. After a while, he will glide to a stop because of the friction force of the snow on his board. We can analyze this using the law of conservation of energy. Following Tactics Box 10.1 we'll take the system to be the boarder *plus* the snow. Then there are no forces external to the system that do work on it and, since he's moving horizontally, his potential energy doesn't change. Then the law of conservation of energy is $K_i = K_f + \Delta E_{th}$, or $\Delta E_{th} = K_i - K_f$. He's slowing to a stop, so $K_i > K_f$ and ΔE_{th} is positive. The system's thermal energy *increases* as kinetic energy is transformed into thermal energy.

This increase in thermal energy is a general feature of any system where friction between sliding objects is present: **When two objects slide against each other with friction present, mechanical energy is always transformed into thermal energy.** An atomic-level explanation is shown in Figure 10.24.

The presence of friction has two important consequences for our conservation of energy Problem-Solving Strategy 10.1:

1. As stated in Tactics Box 10.1, we must include in the system not only the moving object but also the surface against which it slides. This is because the thermal energy generated by friction resides in *both* object and surface (as in Figure 10.24), and it is usually impossible to tell what fraction resides

in each. By choosing both object and surface to be in the system, we know that *all* the thermal energy ends up in the system.

2. In addition to the mechanical energy $K + U$ we now must include ΔE_{th} in the conservation of energy equation.

TRY IT YOURSELF

Agitating atoms Vigorously rub a some-what soft object such as a blackboard eraser on your desktop for about 10 seconds. If you then pass your fingers over the spot where you rubbed, you'll feel a distinct warm area. Congratulations: you've just sct some 100,000,000,000,000,000,000,000 atoms into motion!

EXAMPLE 10.8 Thermal energy created sledding down a hill

George jumps onto his sled and starts from rest at the top of a 5.0-m-high hill. His speed at the bottom is 8.0 m/s. The mass of George and the sled is 55 kg. How much thermal energy was produced in this process?

PREPARE Figure 10.25 shows the before-and-after visual overview. The statement of the problem implies that thermal energy will be generated, so following Tactics Box 10.1, we'll take the system to include both George and the sled *and* the slope. Because his height is changing, his gravitational potential energy is changing and we'll need to include the earth in the system as well. No forces act from outside this system, so the work W is zero.

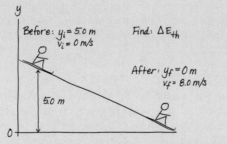

FIGURE 10.25 Visual overview of George sliding down the hill.

SOLVE Here the law of conservation of energy reads

$$K_i + (U_g)_i = K_f + (U_g)_f + \Delta E_{th}$$

In terms of positions and speeds,

$$\frac{1}{2}mv_i^2 + mgy_i = \frac{1}{2}mv_f^2 + mgy_f + \Delta E_{th}$$

Because $v_i = 0$ m/s and $y_f = 0$ m, this simplifies to

$$mgy_i = \frac{1}{2}mv_f^2 + \Delta E_{th}$$

from which we have

$$\Delta E_{th} = mgy_i - \frac{1}{2}mv_f^2$$

$$= (55 \text{ kg})(9.8 \text{ m/s}^2)(5.0 \text{ m}) - \frac{1}{2}(55 \text{ kg})(8.0 \text{ m/s})^2$$

$$= 940 \text{ J}$$

ASSESS The change in E_{th} is positive, as it must be. This extra thermal energy resides in the sled and all along the slope where George slid. You should be able to show that about 35% of George's original gravitational potential energy was transformed into thermal energy as he slid down the hill.

10.8 Further Examples of Conservation of Energy

In this section, we'll tie together what we've learned about using the law of conservation of energy to solve dynamics problems. In each, we use the key idea of setting the "before" energy equal to the "after" energy.

Activ Physics ONLINE   5.2–5.7, 7.11–7.13

EXAMPLE 10.9 Where will the sled stop?

A sledder, starting from rest, slides down a 10-m-high hill. At the bottom of the hill is a long horizontal patch of rough snow. The hill is nearly frictionless, but the coefficient of friction between the sled and the rough snow at the bottom is $\mu_k = 0.30$. How far will the sled slide along the rough patch?

PREPARE A picture of the sledder is shown in Figure 10.26. We'll break this problem into Part A, his motion down the hill; and Part B, his motion along the ice. We know how to use conservation of energy to find his speed at the bottom of the hill. Along the rough patch, however, we'll use kinematics and Newton's laws, as studied in Chapters 2 and 5, to find how far he slides.

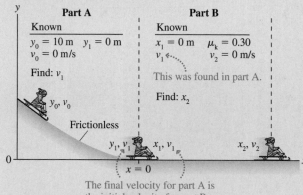

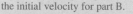

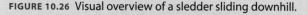

FIGURE 10.26 Visual overview of a sledder sliding downhill.

SOLVE We'll solve Part A first and find the sled's speed v_f at the bottom. The hill is frictionless, so mechanical energy is conserved and we have

$$mgy_i + \frac{1}{2}mv_i^2 = mgy_f + \frac{1}{2}mv_f^2$$

Since $v_i = 0$ m/s and $y_f = 0$ m, this reduces to

$$mgy_i = \frac{1}{2}mv_f^2$$

so that

$$v_f = \sqrt{2gy_i} = \sqrt{2(9.8 \text{ m/s}^2)(10 \text{ m})} = 14.0 \text{ m/s}$$

On the rough patch in Part B, where the only horizontal force is the kinetic friction force f_k pointing to the left, the sled's acceleration is

$$a = -\frac{f_k}{m} = -\frac{\mu_k n}{m} = -\frac{\mu_k mg}{m}$$

$$= -\mu_k g = -(0.30)(9.8 \text{ m/s}^2) = -2.94 \text{ m/s}^2$$

The negative acceleration indicates that the sled is slowing down, as expected.

We now use kinematics to find how far the sled slides. We know the acceleration a, as well as the initial and final velocities along the horizontal patch, and we want to know the final position x_2. This suggests using the kinematic equation

$$v_f^2 = v_i^2 + 2a(x_f - x_i)$$

to find the final position. For the motion of Part B, the final velocity $v_f = v_2 = 0$ m/s, the initial velocity is $v_1 = 14.0$ m/s, and the initial position is $x_i = x_1 = 0$ m. We can then solve for the final position in the kinematic equation to get

$$x_2 = -\frac{v_1^2}{2a} = -\frac{(14.0 \text{ m/s})^2}{-5.88 \text{ m/s}^2} = 33 \text{ m}$$

ASSESS When friction is present, mechanical energy is *not* conserved: Some of the mechanical energy of the system is inevitably transformed into thermal energy. Thus we cannot use the law of conservation of mechanical energy for such problems. Instead, we'll need to use Newton's laws and kinematics to find how far objects slide. As in this example, however, there will often be a part of the problem with no friction that we *can* solve using the law of conservation of mechanical energy.

EXAMPLE 10.10 Who wins the great downhill race?

Figure 10.27 shows a contest in which a sphere, a cylinder, and a circular hoop, each with mass M and radius R, are placed at height h on a slope of angle θ. All three are simultaneously released from rest and roll down the ramp without slipping. Which one will win the race to the bottom of the hill?

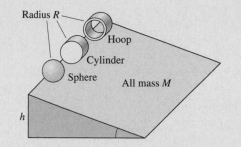

FIGURE 10.27 Which will win the downhill race?

PREPARE With no sliding friction, the total mechanical energy is conserved. However, the kinetic energy of each object must include a contribution from its rotational kinetic energy.

SOLVE Conservation of energy tells us that the gravitational potential energy $(U_g)_i = Mgh$ at the top will be transformed into an equal amount of kinetic energy K_f at the bottom. Thus

$$(U_g)_i = Mgh = K_f = \frac{1}{2}\left(M + \frac{I}{R^2}\right)v^2$$

where we used Equation 10.13 for the total (translational plus rotational) kinetic energy. The speed at the bottom is then

$$v = \sqrt{\frac{2Mgh}{M + \dfrac{I}{R^2}}}$$

Table 7.2 gives the moment of inertia for each of the three shapes. We have

Shape	Moment of Inertia I	$v = \sqrt{2Mgh/(M + I/R^2)}$
Sphere	$\frac{2}{5}MR^2$	$v = \sqrt{\frac{10}{7}gh} = 1.19\sqrt{gh}$
Cylinder	$\frac{1}{2}MR^2$	$v = \sqrt{\frac{4}{3}gh} = 1.15\sqrt{gh}$
Hoop	MR^2	$v = \sqrt{gh}$

The sphere has the largest speed at the bottom, a full 19% faster than the hoop. Because the sphere always travels faster than the hoop or the cylinder, it will win the race, followed by the cylinder and the hoop.

ASSESS All the objects have the same kinetic energy at the bottom, because they all started with the same energy, Mgh, at the top. But the object with the smallest I will have the smallest *rotational* kinetic energy at the bottom, and hence the largest *translational* kinetic energy and the largest v. An ordinary sliding object (no rotation) reaches the bottom with speed $v = \sqrt{2gh} = 1.41\sqrt{gh}$. This is significantly faster than any of the rolling objects. The sliding object is faster because *all* its kinetic energy is translational—and it's the translational motion that gets you down the hill.

10.9 Energy in Collisions

In Chapter 9 we studied collisions between two objects. We found that if no external forces are acting on the objects, the total *momentum* of the objects will be conserved. Now we wish to study what happens to *energy* in collisions. The energetics of collisions are important in many applications in biokinetics, such as designing safer automobiles and bicycle helmets.

Let's first re-examine a perfectly inelastic collision. We studied just such a collision in Example 9.8. Recall that in such a collision the two objects stick together and then move with a common final velocity. What happens to the energy?

Activ Physics   6.2, 6.5, 6.8, 6.9

EXAMPLE 10.11 Energy transformations in a perfectly inelastic collision

Figure 10.28 shows two air track gliders that are pushed toward each other, collide, and stick together. In Example 9.8, we used conservation of momentum to find the final velocity shown in Figure 10.28 from given initial velocities. Compare the initial and final *mechanical energies* of the system.

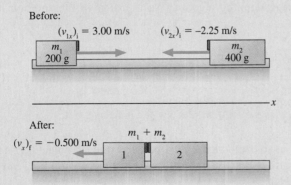

FIGURE 10.28 Initial and final velocities in a completely inelastic collision.

PREPARE We'll choose our system to be the two gliders. Because the tracks are horizontal, there is no change in potential energy. Thus the law of conservation of energy, Equation 10.6, reads $K_i = K_f + \Delta E_{th}$. The total energy before the collision must equal the total energy after, but the total *mechanical* energies need not be equal.

SOLVE The initial kinetic energy is

$$K_i = \frac{1}{2}m_1(v_{1x})_i^2 + \frac{1}{2}m_2(v_{2x})_i^2$$

$$= \frac{1}{2}(0.200 \text{ kg})(3.00 \text{ m/s})^2 + \frac{1}{2}(0.400 \text{ kg})(-2.25 \text{ m/s})^2$$

$$= 1.91 \text{ J}$$

Because the gliders stick together and move as a single object with mass $m_1 + m_2$, the final kinetic energy is

$$K_f = \frac{1}{2}(m_1 + m_2)(v_x)_f^2$$

$$= \frac{1}{2}(0.600 \text{ kg})(-0.500 \text{ m/s})^2 = 0.0750 \text{ J}$$

From the conservation of energy equation above, we find that the thermal energy increases by

$$\Delta E_{th} = K_i - K_f = 1.91 \text{ J} - 0.075 \text{ J} = 1.84 \text{ J}$$

This amount of the initial kinetic energy is transformed into thermal energy during the impact of the collision.

ASSESS About 96% of the initial kinetic energy is transformed into thermal energy. This is typical of many real-world collisions.

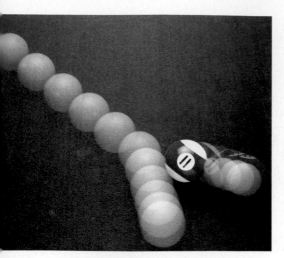

In a collision between a cue ball and a stationary ball, the mechanical energy of the balls is almost perfectly conserved.

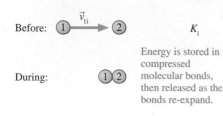

FIGURE 10.29 A perfectly elastic collision.

Elastic Collisions

Figure 9.1 showed a collision of a tennis ball with a racket. The ball is compressed and the racket strings stretch as the two collide, then the ball expands and the strings relax as the two are pushed apart. In the language of energy, the kinetic energy of the objects is transformed into the elastic potential energy of the ball and strings, then back into kinetic energy as the two objects spring apart. If *all* of the kinetic energy is stored as elastic potential energy, and then *all* of the elastic potential energy is transformed back into the post-collision kinetic energy of the objects, then mechanical energy is conserved. A collision for which mechanical energy is conserved is called a **perfectly elastic collision.**

Needless to say, most real collisions fall somewhere between perfectly elastic and perfectly inelastic. A rubber ball bouncing on the floor might "lose" 20% of its kinetic energy on each bounce and return to only 80% of the height of the previous bounce. But collisions between two very hard objects, such as two pool balls or two steel balls, come close to being perfectly elastic. And collisions between microscopic particles, such as atoms or electrons, can be perfectly elastic.

Figure 10.29 shows a head-on, perfectly elastic collision of a ball of mass m_1, having initial velocity $(v_{1x})_i$, with a ball of mass m_2 that is initially at rest. The balls' velocities after the collision are $(v_{1x})_f$ and $(v_{2x})_f$. These are velocities, not speeds, and have signs. Ball 1, in particular, might bounce backward and have a negative value for $(v_{1x})_f$.

The collision must obey two conservation laws: conservation of momentum (obeyed in any collision) and conservation of mechanical energy (because the collision is perfectly elastic). Although the energy is transformed into potential energy during the collision, the mechanical energy before and after the collision is purely kinetic energy. Thus

momentum conservation: $m_1(v_{1x})_i = m_1(v_{1x})_f + m_2(v_{2x})_f$

energy conservation: $\frac{1}{2}m_1(v_{1x})_i^2 = \frac{1}{2}m_1(v_{1x})_f^2 + \frac{1}{2}m_2(v_{2x})_f^2$

Momentum conservation alone is not sufficient to analyze the collision because there are two unknowns: the two final velocities. That is why we did not consider perfectly elastic collisions in Chapter 9. Energy conservation gives us another condition. The complete solution of these two equations involves straightforward but rather lengthy algebra. We'll just give the solution here, which is:

$$(v_{1x})_f = \frac{m_1 - m_2}{m_1 + m_2}(v_{1x})_i \qquad (v_{2x})_f = \frac{2m_1}{m_1 + m_2}(v_{1x})_i \qquad (10.19)$$

Perfectly elastic collision with object 2 initially at rest

Equations 10.19 allow us to compute the final velocity of each object. Let's look at a common and important example: a perfectly elastic collision between two objects of equal mass.

EXAMPLE 10.12 Velocities in an air hockey collision

On an air hockey table, a moving puck, traveling at 2.3 m/s, makes a head-on collision with a stationary puck. What are the final velocities of each of the pucks?

PREPARE The before-and-after visual overview is shown Figure 10.30. We've sketched in final velocities in the picture, but we don't really know yet which way the pucks will move. Because one puck was initially at rest, we can use Equation 10.19 to find the final velocities of the pucks. The pucks are identical, so we have $m_1 = m_2 = m$.

Before: $(v_{1x})_i = 2.3$ m/s $(v_{2x})_i = 0$ m/s

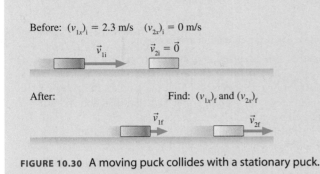

After: Find: $(v_{1x})_f$ and $(v_{2x})_f$

FIGURE 10.30 A moving puck collides with a stationary puck.

SOLVE We use Equation 10.19 with $m_1 = m_2 = m$ to get

$$(v_{1x})_f = \frac{m - m}{m + m}(v_{1x})_i = 0 \text{ m/s}$$

$$(v_{2x})_f = \frac{2m}{m + m}(v_{1x})_i = (v_{1x})_i = 2.3 \text{ m/s}$$

The incoming puck stops dead, and the initially stationary puck goes off with the same velocity that the incoming one had.

ASSESS You can see that momentum and energy are conserved: the incoming puck's momentum and energy are completely transferred to the outgoing puck. If you've ever played pool, you've probably seen this sort of collision when you hit a ball head-on with the cue ball: the cue ball stops and the other ball picks up the cue ball's velocity.

Other cases where the colliding objects are of unequal mass will be treated in the end-of-chapter problems.

Forces in Collisions

The collision between two pool balls occurs very quickly, and the forces are typically very large and difficult to calculate. Fortunately, by using the concepts of momentum and energy conservation we can often calculate the final velocities of the balls without having to know the forces between them. There are collisions, however, where knowing the forces involved is of critical importance. The following example shows how a helmet helps protect the head from the large forces involved in a bicycle accident.

EXAMPLE 10.13 Protecting your head

A bike helmet is basically a shell of hard, crushable foam 3.0 cm thick. In testing, the helmet is strapped onto a 5.0 kg headform that is dropped from a height of 2.0 m onto a hard anvil. What force is encountered by the head in such a fall?

The foam inside a bike helmet is designed to crush upon impact.

PREPARE A visual overview of the test is shown in Figure 10.31. We can use the law of conservation of energy, Equation 10.6, to estimate the force on the headform. We'll choose the headform and the earth to be the system; the foam in the helmet will be part of the environment. We make this choice so that the force on the headform due to the foam is an *external* force that does work W on the headform.

SOLVE The headform starts at initial height $y_i = 2.0$ m above the anvil and ends at rest with the foam fully crushed. Then the law of conservation of energy is

$$K_i + (U_g)_i + W = K_f + (U_g)_f$$

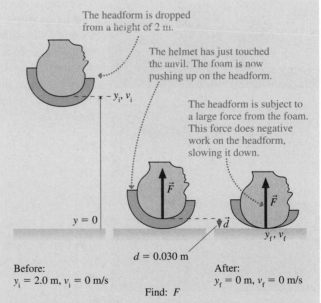

The headform is dropped from a height of 2 m.

The helmet has just touched the anvil. The foam is now pushing up on the headform.

The headform is subject to a large force from the foam. This force does negative work on the headform, slowing it down.

Before:
$y_i = 2.0$ m, $v_i = 0$ m/s

After:
$y_f = 0$ m, $v_f = 0$ m/s

Find: F

$d = 0.030$ m

FIGURE 10.31 The foam in the helmet does negative work on the headform.

Continued

In words, this states that the initial energy of the system, plus the energy transferred to the system as work, equals the final energy of the system.

Since the headform starts and ends at rest, both K_i and K_f are zero. Taking our reference height $y = 0$ at the anvil, $(U_g)_f$ is zero as well. Since $(U_g)_i = mgy_i$, conservation of energy gives simply $mgy_i + W = 0$, or

$$mgy_i = -W$$

As the foam is crushed, it pushes up on the headform with force $\vec{F}$, doing work on it. This force is directed *opposite* to the displacement $\vec{d}$ of the headform, so the work done is *negative*—kinetic energy is being *removed* from the headform, slowing it down. The work done is $-Fd$, if we assume the force is relatively constant, so we have

$$mgy_i = -(-Fd)$$

or

$$F = \frac{mgy_i}{d} = \frac{(5.0 \text{ kg})(9.8 \text{ m/s})(2.0 \text{ m})}{0.030 \text{ m}} = 3300 \text{ N}$$

This is the force that acts on the head to bring it to a halt in only 3 cm. More important from the perspective of possible brain injury is the head's *acceleration*

$$a = \frac{F}{m} = \frac{3300 \text{ N}}{5.0 \text{ kg}} = 660 \text{ m/s}^2 = 67g$$

where g is the acceleration due to gravity.

ASSESS The accepted threshold for serious brain injury is around $300g$, so this helmet would protect the rider in all but the most serious accidents. Without the helmet, the rider's head would come to a stop in a much smaller distance and thus be subjected to a much larger acceleration.

It's also interesting to ask where the original energy of the headform went. The work on it was negative, indicating a transfer of energy from the headform to the environment—the foam. As the foam crushes, there is a great deal of internal friction and rubbing between parts of the foam. This causes the foam to get warmer, increasing its thermal energy. This increase must be exactly equal to the energy lost by the headform.

10.10 Power

Both these cars take about the same energy to reach 60 mph, but the race car gets there in a much shorter time, so its *power* is much greater.

We've now studied how energy can be transformed from one kind to another and how it can be transferred between the environment and the system as work. In many situations we would like to know *how quickly* the energy is transformed or transferred. Is a transfer of energy very rapid, or does it take place over a long time? In passing a truck, your car needs to transform a certain amount of the chemical energy in its fuel into kinetic energy. It makes a *big* difference whether your engine can do this in 20 s or 60 s!

The question "How quickly?" implies that we are talking about a *rate*. For example, the velocity of an object—how fast it is going—is the *rate of change* of position. So when we raise the issue of how fast the energy is transformed, we are talking about the *rate of transformation* of energy. Suppose in a time interval Δt an amount of energy ΔE is transformed from one form to another. The rate at which this energy is transformed is called the **power** P, and it is defined as

$$P = \frac{\Delta E}{\Delta t} \tag{10.20}$$

Power when amount of energy ΔE is transformed in time interval Δt

The unit of power is the **watt,** which is defined as 1 watt = 1 W = 1 J/s.

Power also measures the rate at which energy is transferred into or out of a system as work W. If work W is done in time interval Δt, the rate of energy *transfer* is

$$P = \frac{W}{\Delta t} \tag{10.21}$$

Power when amount of work W is done in time interval Δt

A force that is doing work (i.e., transferring energy) at a rate of 3 J/s has an "output power" of 3 W. A system gaining energy at the rate of 3 J/s is said to "consume" 3 W of power. Common prefixes used with power are mW (milliwatts), kW (kilowatts), and MW (megawatts).

The English unit of power is the *horsepower*. The conversion factor to watts is

$$1 \text{ horsepower} = 1 \text{ hp} = 746 \text{ W}$$

Many common appliances, such as motors, are rated in hp.

We can express Equation 10.21 in a different form. If in the time interval Δt an object undergoes a displacement Δx, the work done by a force acting on the object is $W = F\Delta x$. Then Equation 10.21 can be written

$$P = \frac{W}{\Delta t} = \frac{F\Delta x}{\Delta t} = F\frac{\Delta x}{\Delta t} = Fv$$

The rate at which energy is transferred to an object as work—the power—is the product of the force that does the work, and the velocity of the object:

$$P = Fv \qquad (10.22)$$

Rate of energy transfer due to a force F acting on an object moving at velocity v

LINEAR
p. 38

EXAMPLE 10.14 Power to pass a truck

You are behind a 1500 kg truck traveling at 60 mph (27 m/s). To pass it, you speed up to 75 mph (34 m/s) in 6.0 s. What power is required to do this?

PREPARE Your car is undergoing an energy transformation from the chemical energy of your fuel to the kinetic energy of the car. We can calculate the amount of energy transformed by finding the change ΔK in the kinetic energy.

SOLVE We have

$$K_i = \frac{1}{2}mv_i^2 = \frac{1}{2}(1500 \text{ kg})(27 \text{ m/s})^2 = 5.47 \times 10^5 \text{ J}$$

$$K_f = \frac{1}{2}mv_f^2 = \frac{1}{2}(1500 \text{ kg})(34 \text{ m/s})^2 = 8.67 \times 10^5 \text{ J}$$

so that

$$\Delta K = K_f - K_i$$
$$= 8.67 \times 10^5 \text{ J} - 5.47 \times 10^5 \text{ J} = 3.20 \times 10^5 \text{ J}$$

To transform this amount of energy in 6 s, the power required is

$$P = \frac{\Delta K}{\Delta t} = \frac{3.20 \times 10^5 \text{ J}}{6.0 \text{ s}} = 53,000 \text{ W} = 53 \text{ kW}$$

This is about 71 hp. This power is in addition to the power needed to overcome drag and friction and cruise at 60 mph, so the total power required from the engine will be even greater than this.

ASSESS You use a large amount of energy to perform a simple driving maneuver such as this. 3.20×10^5 J is enough energy to lift an 80 kg person 410 m in the air—the height of a tall skyscraper. And 53 kW would lift him there in only 6 s!

STOP TO THINK 10.6 Four students run up the stairs in the time shown. Rank in order, from largest to smallest, their power outputs P_A to P_D.

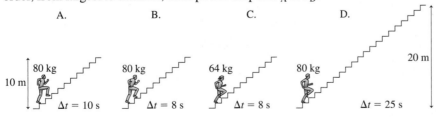

SUMMARY

The goal of Chapter 10 has been to learn about energy and how to solve problems using the law of conservation of energy.

GENERAL PRINCIPLES

General Energy Model

Within a system, energy can be **transformed** between various forms.

Energy can be **transferred** into or out of a system in two basic ways:

- **Work:** The transfer of energy by mechanical forces.

- **Heat:** The nonmechanical transfer of energy from a hotter to a colder object.

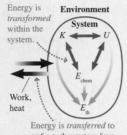

Energy is *transformed* within the system.

Environment
System
$K \leftrightarrow U$
E_{chem}
E_{th}

Work, heat

Energy is *transferred* to or from the system from or to the environment.

Solving Energy Conservation Problems

PREPARE Choose your system (Tactics Box 10.1). Decide what forms of energy are changing. If there is friction, then thermal energy will be created. Check for external forces that will do work on your system.

SOLVE Use Equation 10.6 to relate the initial energy of your system, plus the work done, to the final energy of the system:

$$K_i + U_i + W = K_f + U_f + \Delta E_{th}$$

ASSESS Kinetic energy is always positive. The *change* in thermal energy should be positive.

Law of Conservation of Energy

Isolated system: No energy is transferred into or out of the system. Each form of energy within the system can change, but the total change in energy is zero. Energy of the system is *conserved:*

The change in the system's energy is zero.

$$\Delta K + \Delta U_g + \Delta U_s + \Delta E_{th} + \Delta E_{chem} + \ldots = 0$$

Nonisolated system: Energy can be exchanged with the environment as work or heat. The energy of the system changes by the amount of work done or heat transferred:

$$\Delta K + \Delta U_g + \Delta U_s + \Delta E_{th} + \Delta E_{chem} + \ldots = W + Q$$

The system's energy changes by the amount of work done and heat transferred.

Systems with mechanical and thermal energy only: The initial mechanical energy, plus the work done, equals the final mechanical energy plus additional thermal energy:

$$K_i + U_i + W = K_f + U_f + \Delta E_{th}$$

In terms of energy changes, this can be written

$$\Delta K + \Delta U_g + \Delta U_s + \Delta E_{th} = W$$

IMPORTANT CONCEPTS

Mechanical energy is the sum of a system's kinetic and potential energies:

Mechanical energy $= K + U = K + U_g + U_s$

Kinetic energy is an energy of motion

$$K = \frac{1}{2}mv^2 + \frac{1}{2}I\omega^2$$

Translational ····· ····· Rotational

Potential energy is energy stored in a system of interacting objects

- **Gravitational potential energy:** $U_g = mgy$

- **Elastic potential energy:** $U_s = \frac{1}{2}kx^2$

Thermal energy is the sum of the microscopic kinetic and potential energy of all the molecules in an object. The hotter an object, the more thermal energy it has. When kinetic (sliding) friction is present, mechanical energy will be transformed into thermal energy.

Work is the process by which energy is transferred to or from a system by the application of mechanical forces.

If a particle moves through a displacement $\vec{d}$ while acted upon by a constant force $\vec{F}$, the force does work

$$W = F_{\parallel}d = Fd\cos\theta$$

Only the component of the force parallel to the

APPLICATIONS

Perfectly elastic collisions
Both mechanical energy and momentum are conserved.

Object 2 initially at rest

Before: ①→ ② K_i

$(v_{1x})_i$

After: ①→ ②→ $K_f = K_i$
$(v_{1x})_f$ $(v_{2x})_f$

$$(v_{1x})_f = \frac{m_1 - m_2}{m_1 + m_2}(v_{1x})_i$$

$$(v_{2x})_f = \frac{2m_1}{m_1 + m_2}(v_{1x})_i$$

Power is the rate at which energy is transformed . . .

$$P = \frac{\Delta E}{\Delta t}$$ ····· Amount of energy transformed
····· Time required to transform it

. . . or at which work is done.

$$P = \frac{W}{\Delta t}$$ ····· Amount of work done
····· Time required to do work

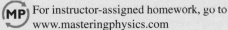 For instructor-assigned homework, go to
www.masteringphysics.com

Problem difficulty is labeled as I (straightforward) to IIIII (challenging).

Problems labeled ✎ can be done on a Workbook Dynamics Worksheet; INT integrate significant material from earlier chapters; BIO are of biological or medical interest.

QUESTIONS

Conceptual Questions

1. The brake shoes of your car are made of a material that can tolerate very high temperatures without being damaged. Why is this so?
2. When you pound a nail with a hammer, the nail gets quite warm. Describe the energy transformations that lead to the addition of thermal energy in the nail.

For Questions 3 through 8, give a specific example of a system with the energy transformation shown. In these questions, W is the work done on the system by the environment, and K and U are the kinetic and potential energies of the system, respectively.

3. $W \to K$ with $\Delta U = 0$.
4. $W \to U$ with $\Delta K = 0$.
5. $K \to U$ with $W = 0$.
6. $K \to W$ with $\Delta U = 0$.
7. $U \to K$ with $W = 0$.
8. $U \to W$ with $\Delta K = 0$.
9. A ball of putty is dropped from a height of 2 m onto a hard floor, where it sticks. What object or objects need to be included within the system if the system is to be isolated during this process?
10. A 0.5 kg mass on a 1-m-long string swings in a circle on a horizontal, frictionless table at a steady speed of 2 m/s. How much work does the tension in the string do on the mass during one revolution? Explain.
11. Particle A has less mass than particle B. Both are pushed forward across a frictionless surface by equal forces for 1 s. Both start from rest.
    a. Compare the amount of work done on each particle. That is, is the work done on A greater than, less than, or equal to the work done on B? Explain.
    b. Compare the impulses delivered to particles A and B. Explain.
    c. Compare the final speeds of particles A and B. Explain.
12. The meaning of the word "work" is quite different in physics from its everyday usage. Give an example of an action a person could do that "feels like work" but that does not involve any work as we've defined it in this chapter.
13. To change a tire, you need to use a jack to raise one corner of your car. While doing so, you happen to notice that pushing the jack handle down 20 cm raises the car only 0.2 cm. Use energy concepts to explain why the handle must be moved so far to raise the car by such a small amount.

Questions 14 through 17 refer to a weightlifter raising a barbell from the floor to above his head. Describe the energy transformations that occur if the system is chosen as specified in the question. Use the notation of Section 10.2 for the various forms of energy and energy transfer.

14. The system is the barbell alone.
15. The system is the weightlifter alone.
16. The system is the barbell plus the earth.
17. The system is the barbell plus the earth plus the weightlifter.

In Questions 18 through 20, imagine yourself doing a chin-up. You start from rest with your arms extended above your head, and end at rest with your elbows bent and your hands still gripping the bar. Describe the energy transformations that occur if the system is chosen as specified in the question. Use the notation of Section 10.2 for the various forms of energy and energy transfer.

18. The system is you alone.
19. The system is you plus the chin-up bar.
20. The system is you plus the chin-up bar plus the earth.
21. One kilogram of matter contains approximately 10^{17} J of nuclear energy. Why don't we need to include this energy when we study ordinary energy transformations?
22. A roller coaster car rolls down a frictionless track, reaching speed v_f at the bottom.
    a. If you want the car to go twice as fast at the bottom, by what factor must you increase the height of the track?
    b. Does your answer to part a depend on whether the track is straight or not? Explain.
23. A spring gun shoots out a plastic ball at speed v_i. The spring is then compressed twice the distance it was on the first shot.
    a. By what factor is the spring's potential energy increased?
    b. By what factor is the ball's speed increased? Explain.
24. Sandy and Chris stand on the edge of a cliff and throw identical mass rocks at the same speed. Sandy throws her rock horizontally while Chris throws his upward at an angle of 45° to the horizontal. Are the rocks moving at the same speed when they hit the ground, or is one moving faster than the other? If one is moving faster, which one? Explain.
25. If you allow a can of chicken broth to join the rolling-object race discussed in Example 10.10, it wins handily. A can of tomato paste, on the other hand, ties with the cylinder. Why? **Hint:** Try to picture how the stuff inside each can moves as the can rolls.
26. A solid cylinder and a cylindrical shell have the same mass, same radius, and turn on frictionless, horizontal axles. (The cylindrical shell has light-weight spokes connecting the shell to the axle.) A rope is wrapped around

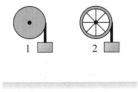

FIGURE Q10.26

each cylinder and tied to a block. The blocks have the same mass and are held the same height above the ground as shown in Figure Q10.26. Both blocks are released simultaneously. The ropes do not slip. Which block hits the ground first? Or is it a tie? Explain.

27. You are much more likely to be injured if you fall and your
BIO head strikes the ground than if your head strikes a gymnastics
pad. Use energy and work concepts to explain why this is so.

Multiple-Choice Questions

28. ‖ If you walk up a flight of stairs at constant speed, gaining ver-
tical height h, the work done on you (the system, of mass m) is
A. $+mgh$, by the normal force of the stairs.
B. $-mgh$, by the normal force of the stairs.
C. $+mgh$, by the gravitational force of the earth.
D. $-mgh$, by the gravitational force of the earth.

29. | You and a friend each carry a 15 kg suitcase up two flights of
stairs, walking at a constant speed. Take each suitcase to be the
system. Suppose you carry your suitcase up the stairs in 30 s
while your friend takes 60 s. Which of the following is true?
A. You did more work, but both of you expended the same
power.
B. You did more work and expended more power.
C. Both of you did the same work, but you expended more
power.
D. Both of you did the same work, but you expended less power.

30. | A woman uses a pulley and a rope to raise a 20 kg weight to
a height of 2 m. If it takes 4 s to do this, about how much
power is she supplying?
A. 100 W   B. 200 W   C. 300 W   D. 400 W

31. | A hockey puck sliding along frictionless ice with speed v to
the right collides with a horizontal spring and compresses it
by 2.0 cm before coming to a momentary stop. What will be
the spring's maximum compression if the same puck hits it at
a speed of $2v$?
A. 2.0 cm   B. 2.8 cm   C. 4.0 cm
D. 5.6 cm   E. 8.0 cm

32. ‖ A block slides down a smooth ramp, starting from rest at a
height h. When it reaches the bottom it's moving at speed v_i.
It then continues to slide up a second smooth ramp. At what
height is its speed equal to $v_i/2$?
A. $h/4$ B. $h/2$ C. $3h/4$ D. $2h$

33. ‖ A wrecking ball is suspended from a 5.0-m-long cable that
makes a 30° angle with the vertical. The ball is released and
swings down. What is the ball's speed at the lowest point?
A. 7.7 m/s   B. 4.4 m/s   C. 3.6 m/s   D. 3.1 m/s

PROBLEMS

Section 10.4 Work

1. | During an etiquette class, you walk slowly and steadily at
0.20 m/s for 2.5 m with a 0.75 kg book balanced on top of
your head. How much work does your head do on the book?

2. | A 2.0 kg book is lying on a 0.75-m-high table. You pick it
up and place it on a bookshelf 2.3 m above the floor.
a. How much work does gravity do on the book?
b. How much work does your hand do on the book?

3. ‖ The two ropes seen in Figure P10.3 are used to lower a
255 kg piano exactly 5 m from a second-story window to the
ground. How much work is done by each of the three forces?

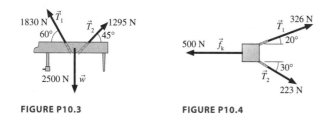

FIGURE P10.3   **FIGURE P10.4**

4. ‖‖ The two ropes shown in the bird's-eye view of Figure P10.4
are used to drag a crate exactly 3 m across the floor. How
much work is done by each of the ropes on the crate?

5. ‖‖ a. At the airport, you ride a "moving sidewalk" that carries
you horizontally for 25 m at 0.70 m/s. Assuming that you
were moving at 0.70 m/s before stepping onto the moving sidewalk and continue at 0.70 m/s afterward, how
much work does the moving sidewalk do on you? Your
mass is 60 kg.
b. An escalator carries you from one level to the next in the
airport terminal. The upper level is 4.5 m above the
lower level, and the length of the escalator is 7.0 m.

How much work does the up escalator do on you when
you ride it from the lower level to the upper level?
c. How much work does the down escalator do on you
when you ride it from the upper level to the lower level?

6. | A boy flies a kite with the string at a 30° angle to the hori-
zontal. The tension in the string is 4.5 N. How much work
does the string do on the boy if the boy
a. stands still?
b. walks a horizontal distance of 11 m away from the kite?
c. walks a horizontal distance of 11 m toward the kite?

Section 10.5 Kinetic Energy

7. | Which has the larger kinetic energy, a 10 g bullet fired at
500 m/s or a 10 kg bowling ball sliding at 10 m/s?

8. | At what speed does a 1000 kg compact car have the same
kinetic energy as a 20,000 kg truck going 25 km/hr?

9. ‖ An oxygen atom is four times as massive as a helium atom.
In an experiment, a helium atom and an oxygen atom have the
same kinetic energy. What is the ratio v_{He}/v_O of their speeds?

10. ‖‖ Sam's job at the amusement park is to slow down and bring
to a stop the boats in the log ride. If a boat and its riders have a
mass of 1200 kg and the boat drifts in at 1.2 m/s, how much
work does Sam do to stop it?

11. ‖ A 20 g plastic ball is moving to the left at 30 m/s. How
much work must be done on the ball to cause it to move to the
right at 30 m/s?

12. | The turntable in a microwave oven has a moment of inertia
of 0.040 kg · m² and is rotating once every 4.0 s. What is its
kinetic energy?

13. | An energy storage system based on a flywheel (a rotating
disk) can store a maximum of 4.0 MJ when the flywheel is
rotating at 20,000 revolutions per minute. What is the moment
of inertia of the flywheel?

Section 10.6 Potential Energy

14. | The lowest point in Death Valley is 85.0 m below sea level. The summit of nearby Mt. Whitney has an elevation of 4420 m. What is the change in gravitational potential energy of an energetic 65.0 kg hiker who makes it from the floor of Death Valley to the top of Mt. Whitney?

15. | a. What is the kinetic energy of a 1500 kg car traveling at a speed of 30 m/s ($\approx$65 mph)?
    b. From what height should the car be dropped to have this same amount of kinetic energy just before impact?
    c. Does your answer to part b depend on the car's mass?

16. || A boy reaches out of a window and tosses a ball straight up with a speed of 10 m/s. The ball is 20 m above the ground as he releases it. Use conservation of energy to find
    a. The ball's maximum height above the ground.
    b. The ball's speed as it passes the window on its way down.
    c. The speed of impact on the ground.

17. || a. With what minimum speed must you toss a 100 g ball straight up to just barely hit the 10-m-high ceiling of the gymnasium if you release the ball 1.5 m above the floor? Solve this problem using energy.
    b. With what speed does the ball hit the floor?

18. || What minimum speed does a 100 g puck need to make it to the top of a frictionless ramp that is 3.0 m long and inclined at 20°?

19. || A car is parked at the top of a 50-m-high hill. It slips out of gear and rolls down the hill. How fast will it be going at the bottom? (Ignore friction.)

20. || A pendulum is made by tying a 500 g ball to a 75-cm-long string. The pendulum is pulled 30° to one side, then released.
    a. What is the ball's speed at the lowest point of its trajectory?
    b. To what angle does the pendulum swing on the other side?

21. ||| A 1500 kg car is approaching the hill shown in Figure P10.21 at 10 m/s when it suddenly runs out of gas.
    a. Can the car make it to the top of the hill by coasting?
    b. If your answer to (a) is yes, what is the car's speed after coasting down the other side?

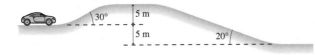

FIGURE P10.21

22. | How much energy can be stored in a spring with a spring constant of 500 N/m if its maximum possible stretch is 20 cm?

23. | How far must you stretch a spring with $k = 1000$ N/m to store 200 J of energy?

24. | A student places her 500 g physics book on a frictionless table. She pushes the book against a spring, compressing the spring by 4.00 cm, then releases the book. What is the book's speed as it slides away? The spring constant is 1250 N/m.

25. | A 10 kg runaway grocery cart runs into a spring with spring constant 250 N/m and compresses it by 60 cm. What was the speed of the cart just before it hit the spring?

26. | As a 15,000 kg jet lands on an aircraft carrier, its tail hook snags a cable to slow it down. The cable is attached to a spring with spring constant 60,000 N/m. If the spring stretches 30 m to stop the plane, what was the plane's landing speed?

27. |||| The elastic energy stored in your tendons can contribute up

BIO   to 35% of your energy needs when running. Sports scientists

have studied the change in length of the knee extensor tendon in sprinters and nonathletes. They find (on average) that the sprinters' tendons stretch 41 mm, while nonathletes' stretch only 33 mm. The spring constant for the tendon is the same for both groups, 33 N/mm. What is the difference in maximum stored energy between the sprinters and the nonathletes?

28. || You're driving at 35 km/hr when the road suddenly descends 15 m into a valley. You take your foot off the accelerator and coast down the hill. Just as you reach the bottom you see the police officer hiding behind the speed limit sign that reads "70 km/hr." Are you going to get a speeding ticket?

29. || Your friend's Frisbee has become stuck 16 m above the ground in a tree. You want to dislodge the Frisbee by throwing a rock at it. The Frisbee is stuck pretty tight, so you figure the rock needs to be traveling at least 5.0 m/s when it hits the Frisbee. If you release the rock 2.0 m above the ground, with what minimum speed must you throw it?

Section 10.7 Thermal Energy

30. | A 1500 kg car traveling at 20 m/s skids to a halt.
    a. What energy transfers and transformations occur during the skid?
    b. What is the change in the thermal energy of the car and the road surface?

31. || A 20 kg child slides down a 3.0-m-high playground slide. She starts from rest, and her speed at the bottom is 2.0 m/s.
    a. What energy transfers and transformations occur during the slide?
    b. What is the change in the thermal energy of the slide and the seat of her pants?

32. ||| A fireman of mass 80 kg slides down a pole. When he reaches the bottom, 4.2 m below his starting point, his speed is 2.2 m/s. By how much has thermal energy increased during his slide?

Section 10.9 Energy in Collisions

33. | A 50 g marble moving at 2.0 m/s strikes a 20 g marble at rest. What is the speed of each marble immediately after the collision? Assume the collision is perfectly elastic.

34. | Ball 1, with a mass of 100 g and traveling at 10 m/s, collides head-on with ball 2, which has a mass of 300 g and is initially at rest. What are the final velocities of each ball if the collision is (a) perfectly elastic? (b) perfectly inelastic?

35. || A proton is traveling to the right at 2.0×10^7 m/s. It has a head-on, perfectly elastic collision with a stationary carbon atom. The mass of the carbon atom is 12 times the mass of the proton. What are the speed and direction of each after the collision?

36. | Two balls undergo a perfectly elastic head-on collision, with one ball initially at rest. If the incoming ball has a speed of 200 m/s, what are the final speed and direction of each ball if
    a. the incoming ball is *much* more massive than the stationary ball?
    b. the stationary ball is *much* more massive than the incoming ball?

37. |||| Derive Equations 10.19 for the final speeds of two objects undergoing a perfectly elastic collision, with one object initially stationary.

Section 10.10 Power

38. | a. How much work does an elevator motor do to lift a 1000 kg elevator a height of 100 m?
    b. How much power must the motor supply to do this in 50 s at constant speed?

39. ‖ a. How much work must you do to push a 10 kg block of steel across a steel table at a steady speed of 1.0 m/s for 3.0 s? The coefficient of kinetic friction for steel on steel is 0.60.
    b. What is your power output while doing so?

40. | Which consumes more energy, a 1.2 kW hair dryer used for 10 min or a 10 W night light left on for 24 hr?

41. ‖‖ A 1000 kg sports car accelerates from 0 to 30 m/s in 10 s. What is the average power of the engine?

42. ‖‖ In just 0.30 s, you compress a spring (spring constant 5000 N/m), which is initially at its equilibrium length, by 4.0 cm. What is your average power output?

43. ‖‖‖ In the winter sport of curling, players give a 20 kg stone a push across a sheet of ice. A curler accelerates a stone to a speed of 3.0 m/s over a time of 2.0 s.
    a. How much force does the curler exert on the stone?
    b. What average power does the curler use to bring the stone up to speed?

44. ‖‖‖ A 710 kg car drives at a constant speed of 23 m/s. It is subject to a drag force of 500 N. What power is required from the car's engine to drive the car
    a. on level ground?
    b. up a hill with a slope of 2.0°?

45. ‖ An elevator weighing 2500 N ascends at a constant speed of 8.0 m/s. How much power must the motor supply to do this?

General Problems

46. ‖‖ A 2.3 kg box, starting from rest, is pushed up a ramp by a 10 N force parallel to the ramp. The ramp is 2.0 m long and tilted at 17°. The speed of the box at the top of the ramp is 0.80 m/s. Consider the system to be the box + ramp + earth.
    a. How much work W does the force do on the system?
    b. What is the change ΔK in the kinetic energy of the system?
    c. What is the change ΔU_g in the gravitational potential energy of the system?
    d. What is the change ΔE_{th} in the thermal energy of the system?

47. ‖ A 55 kg skateboarder wants to just make it to the upper edge of a "half-pipe" with a radius of 3.0 m, as shown in Figure P10.47. What speed v_i does he need at the bottom if he is to coast all the way up?

FIGURE P10.47

    a. First do the calculation treating the skateboarder and board as a point particle, with the entire mass nearly in contact with the half-pipe.
    b. More realistically, the mass of the skateboarder in a deep crouch might be thought of as concentrated 0.75 m from the half-pipe. Assuming he remains in that position all the way up, what v_i is needed to reach the upper edge ?

48. ‖ Fleas have remarkable jumping ability. If a 0.50 mg flea jumps straight up, it will reach a height of 40 cm if there is no air resistance. In reality, air resistance limits the height to 20 cm.
    a. What is the flea's kinetic energy as it leaves the ground?

b. At its highest point, what fraction of the initial kinetic energy has been converted to potential energy?

49. ‖‖‖ A marble slides without friction in a *vertical* plane around the inside of a smooth, 20-cm-diameter horizontal pipe. The marble's speed at the bottom is 3.0 m/s; this is fast enough so that the marble makes a complete loop, never losing contact with the pipe. What is its speed at the top?

50. ‖‖‖ A 20 kg child is on a swing that hangs from 3.0-m-long chains, as shown in Figure P10.50. What is her speed v_i at the bottom of the arc if she swings out to a 45° angle before reversing direction?

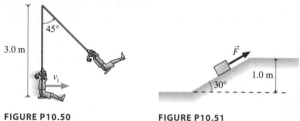

FIGURE P10.50   **FIGURE P10.51**

51. | Suppose you lift a 20 kg box by a height of 1.0 m.
    a. How much work do you do in lifting the box?
    Instead of lifting the box straight up, suppose you push it up a 1.0-m-high ramp that makes a 30° degree angle with the horizontal, as shown in Figure P10.51. Being clever, you choose a ramp with no friction.
    b. How much force F is required to push the box straight up the slope at a constant speed?
    c. How long is the ramp?
    d. Use your force and distance results to calculate the work you do in pushing the box up the ramp. How does this compare to your answer to part a?

52. ‖ A cannon tilted up at a 30° angle fires a cannon ball at 80 m/s from atop a 10-m-high fortress wall. What is the ball's impact speed on the ground below? Ignore air resistance.

53. ‖‖‖ The sledder shown in Figure P10.53 starts from the top of a frictionless hill and slides down into the valley. What initial speed v_i does the sledder need to just make it over the next hill?

FIGURE P10.53

54. ‖‖‖ A 100 g granite cube slides down a frictionless 40° incline. At the bottom, just after it exits onto a horizontal table, the granite collides with a 200 g steel cube at rest. How high above the table should the granite cube be released to give the steel cube a speed of 150 cm/s?

55. ‖ A 50 g ice cube can slide without friction up and down a 30° slope. The ice cube is pressed against a spring at the bottom of the slope, compressing the spring 10 cm. The spring constant is 25 N/m. When the ice cube is released, what distance will it travel up the slope before reversing direction?

56. ‖‖‖‖ In a physics lab experiment, a spring clamped to the table shoots a 20 g ball horizontally. When the spring is compressed 20 cm, the ball travels horizontally 5.0 m and lands on the floor 1.5 m below the point at which it left the spring. What is the spring constant?

57. | The desperate contestants on a TV survival show are very hungry. The only food they can see is some fruit hanging on a branch high in a tree. Fortunately, they have a spring they can use to launch a rock. The spring constant is 1000 N/m, and they can compress the spring a maximum of 30 cm. All the rocks on the island seem to have a mass of 400 g.
   a. With what speed does the rock leave the spring?
   b. If the fruit hangs 15 m above the ground, will they feast or go hungry?

58. | The maximum energy a bone can absorb without breaking BIO is surprisingly small. For a healthy human of mass 60 kg, experimental data show that the leg bones can absorb about 200 J.
   a. From what maximum height could a person jump and land rigidly upright on both feet without breaking his legs? Assume that all the energy is absorbed in the leg bones in a rigid landing.
   b. People jump from much greater heights than this; explain how this is possible.
   Hint: Think about how people land when they jump from greater heights.

59. ‖ In an amusement park water slide, people slide down an essentially frictionless tube. They drop 3.0 m and exit the slide, INT moving horizontally, 1.2 m above a swimming pool. What horizontal distance do they travel from the exit point before hitting the water? Does the mass of the person make any difference?

60. ‖ The 5.0-m-long rope in Figure P10.60 hangs vertically from a tree right at the edge of a ravine. A woman wants to use the rope to swing to the other side of the ravine. She runs as fast as she can, grabs the rope, and swings out over the ravine.

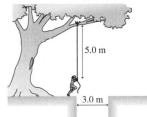

FIGURE P10.60

   a. As she swings, what energy conversion is taking place?
   b. When she's directly over the far edge of the ravine, how much higher is she than when she started?
   c. Given your answers to parts a and b, how fast must she be running when she grabs the rope in order to swing all the way across the ravine?

61. ‖‖ You have been asked to design a "ballistic spring system" to measure the speed of bullets. A bullet of mass m is fired INT into a block of mass M. The block, with the embedded bullet, then slides across a frictionless table and collides with a horizontal spring whose spring constant is k. The opposite end of the spring is anchored to a wall. The spring's maximum compression d is measured.
   a. Find an expression for the bullet's initial speed v_B in terms of m, M, k, and d.
   Hint: This is a two-part problem. The bullet's collision with the block is an inelastic collision. What quantity is conserved in an inelastic collision? Subsequently the block hits a spring on a frictionless surface. What quantity is conserved in this collision?
   b. What was the speed of a 5.0 g bullet if the block's mass is 2.0 kg and if the spring, with $k = 50$ N/m, was compressed by 10 cm?
   c. What fraction of the bullet's initial kinetic energy is "lost"? Where did it go?

62. ‖ A new event, shown in Figure P10.62, has been INT proposed for the Winter Olympics. An athlete will sprint 100 m, starting from rest, then leap onto a 20 kg bobsled. The person

FIGURE P10.62

and bobsled will then slide down a 50-m-long ice-covered ramp, sloped at 20°, and into a spring with a carefully calibrated spring constant of 2000 N/m. The athlete who compresses the spring the farthest wins the gold medal. Lisa, whose mass is 40 kg, has been training for this event. She can reach a maximum speed of 12 m/s in the 100 m dash.
   a. How far will Lisa compress the spring?
   b. The Olympic committee has very exact specifications about the shape and angle of the ramp. Is this necessary? If the committee asks your opinion, what factors about the ramp will you tell them are important?

63. ‖‖ A 20 g ball is fired horizontally with initial speed v_i toward a 100 g ball that is hanging motionless from a 1.0-m-long string. The balls undergo a head-on, perfectly elastic collision, after which the 100 g ball swings out to a maximum angle $\theta_{max} = 50°$. What was v_i?

64. | A 70 kg human sprinter can accelerate from rest to 10 m/s BIO in 3.0 s. During the same interval, a 30 kg greyhound can accelerate from rest to 20 m/s. Compute (a) the change in kinetic energy and (b) the average power output for each.

65. ‖ A 50 g ball of clay traveling at speed v_i hits and sticks to a INT 1.0 kg block sitting at rest on a frictionless surface.
   a. What is the speed of the block after the collision?
   b. Show that the mechanical energy is *not* conserved in this collision. What percentage of the ball's initial kinetic energy is "lost"? Where did this kinetic energy go?

66. ‖ A package of mass m is released from rest at a warehouse INT loading dock and slides down a 3.0-m-high frictionless chute to a waiting truck. Unfortunately, the truck driver went on a break without having removed the previous package, of mass $2m$, from the bottom of the chute as shown in Figure P10.66.
   a. Suppose the packages stick together. What is their common speed after the collision?
   b. Suppose the collision between the packages is perfectly elastic. To what height does the package of mass m rebound?

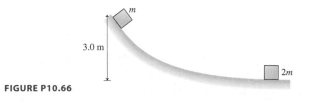

FIGURE P10.66

67. ‖‖‖ A 50 kg sprinter, starting from rest, runs 50 m in 7.0 s at INT constant acceleration.
   a. What is the magnitude of the horizontal force acting on the sprinter?
   b. What is the sprinter's average power output during the first 2.0 s of his run?
   c. What is the sprinter's average power output during the final 2.0 s?

68. ||| Bob can throw a 500 g rock with a speed of 30 m/s. He
INT moves his hand forward 1.0 m while doing so.
   a. How much force, assumed to be constant, does Bob apply to the rock?
   b. How much work does Bob do on the rock?
69. |||| A 2.0 hp electric motor on a water well pumps water from 10 m below the surface. The density of water is 1.0 kg per L. How many liters of water can the motor pump in 1 hr?
70. | The human heart has to pump the average adult's 6.0 L of
BIO blood through the body every minute. The heart must do work to overcome frictional forces that resist the blood flow. The average blood pressure is 1.3×10^4 N/m².
   a. Compute the work done moving the 6.0 L of blood completely through the body, assuming the blood pressure always takes its average value.
   b. What power output must the heart have to do this task once a minute?
   Hint: When the heart contracts, it applies force to the blood. Pressure is just force/area, so we can write work = (pressure)(area)(distance). But (area)(distance) is just the blood volume passing through the heart.

Passage Problems

Tennis Ball Testing

A tennis ball bouncing on a hard surface compresses and then rebounds. The details of the rebound are specified in tennis regulations. Tennis balls, to be acceptable for tournament play, must have a mass of 57.5 g. When dropped from a height of 2.5 m onto a concrete surface, a ball must rebound to a height of 1.4 m. During impact, the ball compresses by approximately 6 mm.

71. | How fast is the ball moving when it hits the concrete surface? (Ignore air resistance.)
   A. 5 m/s   B. 7 m/s   C. 25 m/s   D. 50 m/s
72. | If the ball accelerates uniformly when it hits the floor, what is its approximate acceleration as it comes to rest before rebounding?
   A. 1000 m/s² B. 2000 m/s² C. 3000 m/s² D. 4000 m/s²
73. | The ball's kinetic energy just after the bounce is less than just before the bounce. In what form does this lost energy end up?
   A. Elastic potential energy
   B. Gravitational potential energy
   C. Thermal energy
   D. Rotational kinetic energy
74. | By what percent does the kinetic energy decrease?
   A. 35%   B. 45%   C. 55%   D. 65%
75. | When a tennis ball bounces from a racket, the ball loses approximately 30% of its kinetic energy to thermal energy. A ball that hits a racket at a speed of 10 m/s will rebound with approximately what speed?
   A. 8.5 m/s B. 7.0 m/s C. 4.5 m/s D. 3.0 m/s

Work and Power in Cycling

When you ride a bicycle at constant speed, almost all of the energy you expend goes into the work you do against the drag force of the air. In this problem, assume that *all* of the energy expended goes into working against drag. As we saw in Section 5.7, the drag force on an object is approximately proportional to the square of its speed with respect to the air. For this problem, assume that $F \propto v^2$ exactly and that the air is motionless with respect to the ground unless noted otherwise. Suppose a cyclist and her bicycle have a combined mass of 60 kg and she is cycling along at a speed of 5 m/s.

76. | If the drag force on the cyclist is 10 N, how much energy does she use in cycling 1 km?
   A. 6 kJ   B. 10 kJ   C. 50 kJ   D. 100 kJ
77. | Under these conditions, how much power does she expend as she cycles?
   A. 10 W   B. 50 W   C. 100 W   D. 200 W
78. | If she doubles her speed to 10 m/s, how much energy does she use in cycling 1 km?
   A. 20 kJ B. 40 kJ C. 400 kJ D. 400 kJ
79. | How much power does she expend when cycling at that speed?
   A. 100 W B. 200 W C. 400 W D. 1000 W
80. | Upon reducing her speed back down to 5 m/s, she hits a headwind of 5 m/s. How much power is she expending now?
   A. 100 W B. 200 W C. 500 W D. 1000 W

STOP TO THINK ANSWERS

Stop to Think 10.1: D. Since the child slides at a constant speed, his kinetic energy doesn't change. But his gravitational potential energy at the top of the slide decreases as he descends, and is transformed into thermal energy in the slide and his bottom.

Stop to Think 10.2: C. $W = Fd\cos\theta$. The 10 N force at 90° does no work at all. $\cos 60° = \frac{1}{2}$, so the 8 N force does less work than the 6 N force.

Stop to Think 10.3: B > D > A = C. $K = (1/2)mv^2$. Using the given masses and velocities, we find $K_A = 2.0$ J, $K_B = 4.5$ J, $K_C = 2.0$ J, $K_D = 4.0$ J.

Stop to Think 10.4: $(U_g)_3 > (U_g)_2 = (U_g)_4 > (U_g)_1$. Gravitational potential energy depends only on height, not speed.

Stop to Think 10.5: C. U_s depends on x^2, so doubling the compression increases U_s by a factor of 4. All the potential energy is converted to kinetic energy, so K increases by a factor of 4. But K depends on v^2, so v increases by only a factor of $\sqrt{4} = 2$.

Stop to Think 10.6: $P_B > P_A = P_C > P_D$. The power here is the rate at which each runner's internal chemical energy is converted into gravitational potential energy. The change in gravitational potential energy is $mg\Delta y$, so the power is $mg\Delta y/\Delta t$. For runner A, the ratio $m\Delta y/\Delta t$ equal (80 kg)(10 m)/(10 s) = 80 kg · m/s. For C, it's the same. For B, it's 100 kg · m/s, while for D the ratio is 64 kg · m/s.

11

USING ENERGY

The strange masks worn by the fighters have a very simple purpose: to measure how much oxygen their bodies are using. How can this tell us how efficiently their bodies are using energy?

Looking Ahead ▶▶

The goal of Chapter 11 is to learn more about energy transformations and transfers, the laws of thermodynamics, and theoretical and practical limitations on energy use. In this chapter, you will learn to:

- ▶ Use the concept of efficiency.
- ▶ Discuss how energy is used in the body.
- ▶ Understand the related concepts of heat, temperature, and thermal energy.
- ▶ Apply the first and second laws of thermodynamics.
- ▶ Understand the concept of entropy and the limits it places on energy transformations.

Looking Back ◀◀

This chapter is a companion chapter to Chapter 10. All of the concepts that you learned in that chapter will be used here. Please review especially:

- ◀ Section 10.2 Basic energy concepts; forms of energy.
- ◀ Section 10.3 The law of conservation of energy.
- ◀ Section 10.7 Thermal energy.

Chapter 10 introduced the concept of energy, and in it you learned to solve problems in which energy is transformed from one form to another. For example, you now know how to calculate the energy required to lift a box a certain distance off the floor.

But how much energy would your body actually *use* to complete this task? We will find that the energy used by your body is more than the change in potential energy of the box. Why is this? As you lift the box, where does the energy come from? If you put the box back on the floor, where has the energy gone? Is it "lost"? In this chapter, we will consider just such questions of energy transformations and efficiency. Answering these questions will lead us to learn more about thermal energy, temperature, and heat.

11.1 Transforming Energy

As we saw in Chapter 10, energy can't be created or destroyed; it can only be converted from one form to another. When we say we are *using energy,* we mean that we are transforming it, such as transforming the chemical energy of food into the kinetic energy of your body. Let's revisit the idea of energy transformations, considering some realistic situations that have interesting theoretical and practical limitations.

Energy transformations

Light energy hitting a solar cell on top of these walkway lights is converted to electric energy and then stored as chemical energy in a battery. At night, the battery's chemical energy is converted to electric energy that is then converted to light energy in a light-emitting diode.

Light energy is absorbed by photosynthetic pigments in soybean plants, which use this energy to create concentrated chemical energy. The soybeans are harvested and their oil is used to make candles. When the candle burns, the stored chemical energy is transformed into light energy and thermal energy.

A wind turbine converts the translational kinetic energy of moving air into electric energy. Miles away, this energy can be used by a fan, which transforms the electric energy back into the kinetic energy of moving air.

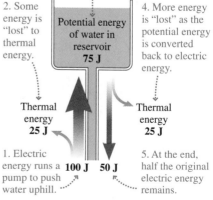

3. Less than 100 J is stored.

2. Some energy is "lost" to thermal energy.

Potential energy of water in reservoir
75 J

4. More energy is "lost" as the potential energy is converted back to electric energy.

Thermal energy
25 J

Thermal energy
25 J

1. Electric energy runs a pump to push water uphill.

100 J   **50 J**

5. At the end, half the original electric energy remains.

FIGURE 11.1 Energy transformations for a pumped storage system.

In each of the processes in the table, energy is transformed from one form to another, ending up in the same form as it began. But some energy appears to have been "lost" along the way. The garden lights shine with a glow that is much less bright than the sun that shined on them! No energy is really lost, because energy is conserved; it is merely converted to other forms that are less useful to us.

Let's look at a practical example. Electric power companies sometimes use excess electric energy to pump water uphill. They can then reclaim this energy when demand is greater by letting the water generate electricity as it flows back down. Figure 11.1 shows the energy transformations in this process, assuming 100 J of electric energy at the start.

At the end of the process, only 50 J of electric energy is recovered. No energy is actually lost, but when other forms of energy are transformed to thermal energy, this change is *irreversible:* We cannot easily recover this thermal energy and convert it back to electric energy. **The energy isn't lost, but it is lost to our use.** In order to look at losses in energy transformations in more detail, we will define the notion of *efficiency.*

Efficiency

To get 50 J of energy out of your pumped storage plant, you would actually need to *pay* to put in 100 J of energy, as we saw in Figure 11.1. Because only 50% of the energy is returned as useful energy, we can say that this plant has an efficiency of 50%. Generally, we can define efficiency as

$$e = \frac{\text{what you get}}{\text{what you had to pay}} \qquad (11.1)$$

General definition of efficiency

The larger the energy losses in a system, the lower its efficiency.

The reductions in efficiency can arise from two different sources:

- **Process limitations.** In many cases that we will consider, the energy losses are due to practical details of an energy transformation process. You could, in principle, design a process that entailed smaller losses.
- **Fundamental limitations.** In other cases that we will consider, energy losses are due to physical laws that cannot be circumvented. The best process that could theoretically be designed will have less than 100% efficiency. As we will see, these limitations result from the difficulty of transforming thermal energy into other forms of energy.

As you walk up a flight of stairs, you increase your body's potential energy by about 1800 J. But a measurement of the energy used by your body to climb the stairs would show that your body uses 7200 J to complete this task. The 1800 J increase in potential energy is "what you get"; the 7200 J your body uses is "what you had to pay." We can use Equation 11.1 to compute the efficiency for this action to be

$$e = \frac{1800 \text{ J}}{7200 \text{ J}} = 0.25 = 25\%$$

This low efficiency is due to *process limitations*. The efficiency is limited by the biochemistry of how your food is digested and the biomechanics of how you move. This is not a fundamental limit; the process could be made more efficient. For instance, changing the angle of the stairs can allow a person to climb the same height with a smaller total energy expenditure.

The electric energy you use daily must be generated from other sources, in most cases the chemical energy in coal or other fossil fuels. The chemical energy is converted to thermal energy by burning, and the resulting thermal energy is converted to electric energy. A typical power plant cycle is shown in Figure 11.2. "What you get" is the energy output, the 35 J of electric energy. "What you had to pay" is the energy input, the 100 J of chemical energy, giving an efficiency of

$$e = \frac{35 \text{ J}}{100 \text{ J}} = 0.35 = 35\%$$

This low efficiency turns out to be largely due to a *fundamental limitation*. Thermal energy cannot be transformed to other forms of energy with 100% efficiency. A 35% efficiency is close to the theoretical maximum, and no power plant could be designed that would do better than this maximum. In later sections, we will explore the fundamental properties of thermal energy that make this so.

This idea of efficiency will form a thread that will run through this chapter. We will take some time to explore how to solve problems using this concept.

These cooling towers release thermal energy from a coal-fired power plant.

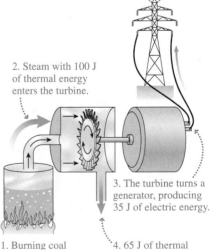

2. Steam with 100 J of thermal energy enters the turbine.

3. The turbine turns a generator, producing 35 J of electric energy.

1. Burning coal produces 100 J of thermal energy.

4. 65 J of thermal energy is exhausted into the environment.

FIGURE 11.2 Energy transformations in a coal-fired power plant.

PREPARE There are two key components to define as we prepare to compute efficiency:

❶ Choose what energy to count as "what you get." This could be the useful energy output of an engine or process or the work that is done in completing a process. For example, when you climb a flight of stairs, "what you get" is your change in potential energy.

❷ "What you had to pay" will generally be the total energy input needed for an engine, task, or process. For example, when you run your air conditioner, "what you had to pay" is the electric energy input.

SOLVE You may need to do additional calculations:

■ Compute values for "what you get" and "what you had to pay."
■ Be certain that all energy values are in the same units.

After this, compute the efficiency using $e = \frac{\text{what you get}}{\text{what you had to pay}}$.

ASSESS Check your answer to see if it is reasonable, given what you know about typical efficiencies for the process under consideration.

A Case Study in Energy Efficiency

We will now look at a practical example that involves energy transformations and energy transfers by work and heat: the energy transformations and transfers in moving a car along the road. How efficient is your car as you drive down the highway?

Why does your car use energy to move forward at all? In the absence of external forces, your car would continue at a constant speed once it was moving. But if you are driving at highway speed and take your foot off the gas, your car will slow down; this is almost entirely due to one external force, the drag force on the car. The car must apply a forward force to move through the air; it has to do work to move forward.

CONCEPTUAL EXAMPLE 11.1 Energy transformations and transfers for a car

A car is traveling along a level road at a steady 65 mph (approximately 30 m/s). What energy transformations and transfers are taking place?

REASON We will take the car as our system. Let's write the conservation of energy equation for a system that isn't isolated, Equation 10.4, including all types of energy which might apply to the car:

$$\Delta K + \Delta U_g + \Delta E_{th} + \Delta E_{chem} = W + Q$$

The car is moving at a constant speed, so the kinetic energy is not changing. The road is level, so there is no change in potential energy. If the car is running well, its temperature will be constant; its thermal energy isn't changing. Chemical energy does change, decreasing as fuel is burned. The equation thus reduces to

$$\Delta E_{chem} = W + Q$$

Chemical energy in the car's fuel is being used to do work because of the drag force and to transfer energy as heat to the environment. These are the only significant energy transfers taking place at constant speed.

ASSESS The car has kinetic energy, potential energy and thermal energy, but these forms of energy are all constant. Only the chemical energy changes.

We can now use what we know about the drag force to compute the minimum amount of energy necessary to move a car forward. We can take this as "what you get"; it is the minimum amount of energy that could be used to move the car forward. We can then use some real numbers to compute "what you had to pay"—the energy actually used to move the car forward. Once we have these two pieces of information, we can calculate an efficiency.

EXAMPLE 11.1 Minimum power for a car on a freeway
A Chevrolet Corvette has a drag coefficient of 0.30; the area of the front of the car is 1.8 m². What minimum power does this car need to move at a steady speed of 30 m/s on level ground?

PREPARE In Conceptual Example 11.1, we saw that the car was doing work because of drag. The car is pushing on the atmosphere with a force F equal in magnitude to the drag force D; this is the force that is doing work on the environment.

SOLVE In Chapter 5, we saw that the drag force on an object moving at a speed v is given by $D = \frac{1}{2}C\rho Av^2$, where C is the drag coefficient, ρ the density of air (approximately 1.2 kg/m²), A the area of the front of the car, and v the speed. Thus the drag force on the car moving at 30 m/s is

$$D = (\tfrac{1}{2})(0.30)(1.2 \text{ kg/m}^3)(1.8 \text{ m}^2)(30 \text{ m/s})^2 = 290 \text{ N}$$

Now that we know the magnitude of the force we can compute the necessary power by using Equation 10.22:

$$P = Fv = (290 \text{ N})(30 \text{ m/s}) = 8700 \text{ W}$$

ASSESS We can get a better idea of the size of this power by converting to horsepower:

$$P = 8700 \text{ W}\left(\frac{1 \text{ hp}}{746 \text{ W}}\right) = 12 \text{ hp}.$$

A Corvette will likely have a 350 hp engine. The power necessary to drive at highway speeds is just over 3% of the power available from the engine! The extra power is used to provide power for rapid acceleration (as we saw in Example 10.14) and for climbing hills. And, as we'll explore more later, quite a bit of the engine's power is "wasted" as heat energy transferred to the environment.

EXAMPLE 11.2 The efficiency of an automobile
The Corvette of the previous example is rated at 25 miles per gallon of gas at highway speeds. What is the efficiency of a Corvette driven 25 miles (40 km) at 30 m/s? (A gallon of gas contains 1.4×10^8 J of chemical energy.)

PREPARE We can use the steps outlined in Problem-Solving Strategy 11.1 for computing efficiency. "What you get" is the minimum energy to move the car forward as in the previous example; "what you had to pay" is the energy in one gallon of gas—the energy the car actually uses to drive this distance.

SOLVE As we saw in the previous example, the power necessary to overcome drag is 8700 W. The minimum energy needed to travel 40 km, which we will label ΔE_{min}, is

$$\Delta E_{\text{min}} = P\Delta t = P\left(\frac{\Delta x}{v}\right) = (8700 \text{ W})\left(\frac{40,000 \text{ m}}{30 \text{ m/s}}\right)$$
$$= 11.6 \times 10^6 \text{ J}$$

To drive 25 miles the car will use one gallon of gas, with energy 1.4×10^8 J. We can now use Equation 11.1 to find the efficiency:

$$e = \frac{\text{what you get}}{\text{what you had to pay}} = \frac{11.6 \times 10^6 \text{ J}}{1.4 \times 10^8 \text{ J}} = 0.083 = 8.3\%$$

ASSESS Only 8.3% of the chemical energy from the gasoline is used to overcome drag. The rest of the energy—about 92%—is "lost" as heat to the environment.

STOP TO THINK 11.1   Walking or running up a flight of stairs takes the same amount of energy. Which is more efficient?

A. Walking up the stairs.
B. Running up the stairs.
C. The efficiency is the same.

11.2 Energy in the Body: Energy Inputs

In this section, we will look at energy in the body, which will give us the opportunity to explore a number of different energy transformations and transfers in another practical context. Figure 11.3 shows the body considered as the system for energy analysis. The chemical energy in food provides the necessary energy input for your body to function. It is this energy that is used for energy transfers with the environment.

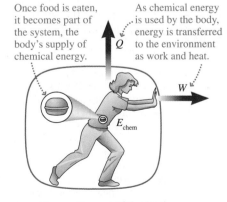

Once food is eaten, it becomes part of the system, the body's supply of chemical energy.

As chemical energy is used by the body, energy is transferred to the environment as work and heat.

Q

W

E_{chem}

FIGURE 11.3 Energy of the body, considered as the system.

Getting Energy from Food

Suppose you do a very simple exercise: lifting a mass and increasing its potential energy. Where did the energy come from to do this? At some point, the energy came from the food you ate, but what were the intermediate steps? The chemical energy in food is made available to the cells in the body by a two-step process. First, the digestive system breaks down food into simpler molecules such as glucose, a simple sugar, or long chains of glucose molecules called glycogen. These molecules are delivered via the bloodstream to cells in the body, where they are metabolized by combining with oxygen to produce simpler molecules, as in Equation 11.2.

Glucose from the digestion of food combines with oxygen that is breathed in to produce... ...carbon dioxide, which is exhaled; water, which can be used by the body; and energy.

$$C_6H_{12}O_6 + 6O_2 \longrightarrow 6CO_2 + 6H_2O + energy \qquad (11.2)$$

Glucose   Oxygen   Carbon dioxide   Water

This metabolism releases energy, some of which is stored in a molecule called ATP, or adenosine triphosphate. Cells in the body use this ATP to do all the work of life: Muscle cells use it to contract, nerve cells use it to produce electrical signals, and so on.

Oxidation reactions like those in Equation 11.2 "burn" the fuel that you obtain by eating. The oxidation of 1 g of glucose (or any other carbohydrate) will release approximately 17 kJ of energy. Table 11.1 compares the energy content of carbohydrates and other foods to other familiar sources of chemical energy.

It is possible to measure the chemical energy content of food by burning it. Burning food may seem quite different from metabolizing it, but if glucose is burned, the chemical formula for the reaction is exactly that of Equation 11.2; the two reactions are the same. Burning food transforms all of its chemical energy to thermal energy, which can be easily measured. Thermal energy is often measured in units of **calories (cal)** rather than in joules; 1.00 calorie is equivalent to 4.19 joules.

NOTE ▶ The energy content of food is usually quoted in Calories (Cal) (with a capital "c"); one Calorie (also called a "food calorie") is equal to 1000 calories or 1 kcal. If a candy bar contains 230 Cal, this means that, if burned, it would produce 230,000 cal (or 962 kJ) of thermal energy. ◀

TABLE 11.1 Energy in fuels

Fuel	Energy in 1 g of fuel (in kJ)
Hydrogen	121
Gasoline	44
Fat (in food)	38
Coal	27
Carbohydrates (in food)	17
Wood chips	15

Counting calories BIO Most foods burn quite well, as this photo of corn chips illustrates. You could set food on fire to measure its energy content, but this isn't really necessary. The chemical energies of the basic components of food (carbohydrates, proteins, fats) have been carefully measured—by burning—in a device called a *calorimeter.* Foods are analyzed to determine their composition, and their chemical energy can then be calculated.

EXAMPLE 11.3 Energy in food

A 12 oz can of soda contains approximately 40 g (or a bit less than 1/4 cup) of sugar, a simple carbohydrate. How much chemical energy in joules does this sugar contain? How many Calories is this?

SOLVE 1 g of sugar contains 17 kJ of energy; 40 g contains:

$$40 \text{ g} \times \frac{17 \times 10^3 \text{ J}}{1 \text{ g}} = 68 \times 10^4 \text{ J} = 680 \text{ kJ}$$

Converting to Calories, we get:

$$680 \text{ kJ} = 6.8 \times 10^5 \text{ J} = (6.8 \times 10^5 \text{ J})\frac{1.00 \text{ cal}}{4.19 \text{ J}}$$
$$= 1.6 \times 10^5 \text{ cal} = 160 \text{ Cal}$$

ASSESS 160 Calories is a typical value for the energy content of a 12 oz can of soda (check the nutrition label on one to see!), so this calculation seems reasonable.

TABLE 11.2 Energy content of foods

Food	Energy content in Cal	Energy content in kJ
Fried egg	100	420
Large apple	125	525
Slice of pizza	300	1260
Slice of apple pie	400	1680
Fast-food meal: Burger, fries, drink, large size	1350	5670

The first item on the nutrition label on packaged foods is Calories—a measure of the chemical energy in the food. (In Europe, where SI units are standard, you will also find the energy content listed in kJ.) The energy content of some common foods is given in Table 11.2.

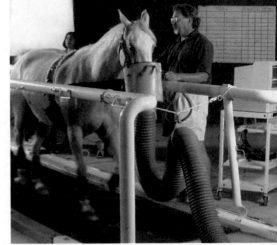

A respiratory apparatus can be used to measure metabolic energy use of animals as well as humans, in this case the energy used by a horse in walking.

11.3 Energy in the Body: Energy Outputs

Your body uses energy in many ways. Even at rest, your body uses energy for tasks such as building and repairing tissue, digesting food, and keeping warm. The number of joules used per second (that is, the power in watts) by different tissues in the resting body is given in Table 11.3.

Even at rest, your body uses energy at the rate of approximately 100 W. This energy comes from chemical energy in your body's stores, and is ultimately converted entirely to thermal energy, which is transferred as heat to the environment. If 100 people are in a lecture hall, they are adding thermal energy to the room at a rate of 10,000 W, and the air conditioning must be designed to take account of this!

Energy Use in Activities

Your body stores very little energy as ATP; as your body uses energy your cells must continually metabolize carbohydrates to produce it. This requires oxygen, as we saw in Equation 11.2. Physiologists can precisely measure the body's energy use by measuring how much oxygen the body is taking up with a respiratory apparatus, as seen in the photograph at the start of the chapter. The spirometer measures the body's *total metabolic energy use*—all of the energy used by the body while performing an activity. This total will include all of the body's basic processes such as breathing plus whatever additional energy is needed to perform the activity.

The metabolic energy used in an activity depends on an individual's size, level of fitness, and other variables. But we can make reasonable estimates for the power used in various activities for a typical individual. Some values are given in Table 11.4.

Your body uses energy above the resting rate to perform different activities. For instance, typing requires 125 W, as seen in Table 11.4. This corresponds to 100 W for basic functions plus 25 W for the work of typing.

Suppose you climb a set of stairs at a constant speed, as in Figure 11.4 on the next page. What is your body's efficiency for this process? To answer this question we will take the system to be your body together with the earth, so that we can consider gravitational potential energy.

TABLE 11.3 Energy usage at rest

Organ	Resting power (W) of 68 kg individual
Liver	26
Brain	19
Heart	7
Kidneys	11
Skeletal muscle	18
Remainder of body	19
Total	**100**

TABLE 11.4 Metabolic power use during activities by a 68 kg (150 lb) individual

Activity	Metabolic power (W) of 68 kg individual
Typing	125
Ballroom dancing	250
Walking at 5 km/hr	380
Cycling at 15 km/hr	480
Swimming at a fast crawl	800
Running at 15 km/hr	1150

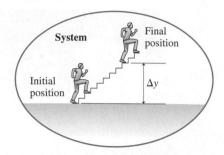

FIGURE 11.4 Pictorial representation of climbing a set of stairs.

Given how we have defined the system, there are no external forces that do work and there is no heat input—the system is isolated. Any energy transformations are internal to the system. We can use Equation 10.3 for the energy changes in the system, using the only forms of energy that change in this situation: gravitational potential, chemical (in your body), and thermal energy; kinetic energy does not change because your speed is constant:

$$\Delta U_g + \Delta E_{th} + \Delta E_{chem} = 0 \qquad (11.3)$$

Thermal energy and gravitational potential energy are increasing, so ΔE_{th} and ΔU_g are positive; chemical energy is being used, so ΔE_{chem} is a negative number. We can get a better feeling about what is happening by rewriting Equation 11.3 as

$$|\Delta E_{chem}| = \Delta U_g + \Delta E_{th}$$

The *magnitude* of the change in the chemical energy is equal to the sum of the changes in the gravitational potential and thermal energies. Chemical energy from your body is converted into potential energy and thermal energy; in the final position, you are at a greater height, and your body is slightly warmer!

Earlier in the chapter, we noted that the efficiency for stair climbing is about 25%. Let's see where that number comes from.

1. *What you get.* "What you get" is the change in potential energy: You have raised your body to the top of the stairs. If you climb a flight of stairs of vertical height Δy, we can easily compute the increase in potential energy $\Delta U_g = mg\Delta y$. Assuming a mass of 68 kg and a change in height of 2.7 m (about 9 ft, a reasonable value for a flight of stairs) we compute (to two significant figures)

$$\Delta U_g = (68 \text{ kg})(9.8 \text{ m/s}^2)(2.7 \text{ m}) = 1800 \text{ J}$$

2. *What you had to pay.* The cost is the metabolic energy the body used in completing the task. Physiologists can measure directly how much energy $|\Delta E_{chem}|$ your body uses to perform a task. A typical value for climbing a flight of stairs is

$$|E_{chem}| = 7200 \text{ J}$$

Given the definition of efficiency of Equation 11.2, we can compute an efficiency for climbing the stairs as

$$e = \frac{\Delta U_g}{|\Delta E_{chem}|} = \frac{1800 \text{ J}}{7200 \text{ J}} = 0.25 = 25\%$$

For the types of activities we will consider in this chapter, such as running, walking, and cycling, the body's efficiency is typically in the range of 20–30%.

In the above example, 25% of the energy used by your body ends up as increased potential energy; the other 75% ends up as thermal energy. Thermal energy losses arise from several sources, including the metabolism of glucose or glycogen to produce ATP, the use of ATP to make muscle cells contract, and from basic body processes that continue to be performed. Whenever you exercise, most

High heating costs? BIO The daily energy use of mammals is much higher than that of reptiles, largely because mammals use energy to maintain a constant body temperature. A 40 kg timber wolf uses approximately 19,000 kJ during the course of a day. A Komodo dragon, a reptilian predator of the same size, uses only 2100 kJ.

of the energy you use is converted to thermal energy; that's why you warm up when you work out!

EXAMPLE 11.4 How many flights?

If you have consumed a 12 oz can of soda, how many flights of stairs could you climb on the chemical energy it contains? Assume a mass of 68 kg and that a flight of stairs has a vertical height of 2.7 m.

PREPARE In Example 11.3, we calculated that the soda contains a chemical energy of 680 kJ. If the body uses this energy to climb stairs, we can assume that 25% of the energy is transformed to increased potential energy:

$$\Delta U_g = (0.25)(680 \times 10^3 \text{ J}) = 1.7 \times 10^5 \text{ J}$$

SOLVE Since $\Delta U_g = mg\Delta y$, the height gained is

$$\Delta y = \frac{\Delta U_g}{mg} = \frac{1.7 \times 10^5 \text{ J}}{(68 \text{ kg})(9.8 \text{ m/s}^2)} = 255 \text{ m}$$

If we assume that each flight has a height of 2.7 m, this gives

$$\text{Number of flights} = \frac{255 \text{ m}}{2.7 \text{ m}} \cong 94 \text{ flights}$$

ASSESS This is almost enough to get to the top of the Empire State Building—all fueled by one can of soda! This is a remarkable result.

The metabolic power values given in Table 11.4 represent the energy *used by the body* while these activities are being performed. Given that the body's efficiency is only 20–30%, the body's actual *useful power output* is quite a bit less than this. The table's value for cycling at 15 km/hr (a bit less than 10 mph) is 480 W. If we assume that the efficiency for cycling is approximately 25%, the actual power going to forward propulsion will only be 120 W. An elite racing cyclist going more than twice as fast as 15 km/hr might be using about 300 W for forward propulsion. This is a surprisingly low figure, as noted in Figure 11.5.

The energy you use per second running is proportional to your speed; running twice as fast takes approximately twice as much power. But running twice as fast takes you twice as far in the same time, so the energy you use to run a certain distance doesn't depend on how fast you run! Running a marathon takes approximately the same amount of energy whether you complete it in 2 hours, 3 hours or 4 hours; it is only the power that varies.

NOTE ▶ It is important to remember the distinction between the metabolic energy used to perform a task and the work done in a physics sense; these values can be quite different. Your muscles use power when applying a force, even when there is no motion. Holding a weight above your head involves no external work, but it clearly takes metabolic power to keep the weight in place! ◀

FIGURE 11.5 Amazingly, a racing cyclist moving at 35 km/hr uses about the same power as an electric scooter moving at 5 km/hr.

CONCEPTUAL EXAMPLE 11.2 Energy in weightlifting

A weightlifter lifts a 50 kg bar from the floor to a position over his head and back to the floor again 10 times in succession. At the end of this exercise, what energy transformations have taken place?

REASON We will take the system to be the weightlifter plus the bar. The environment does no work on the system, and we will assume that the time is short enough that no heat is transferred from the system to the environment. The bar has returned to its starting position and is not moving, so there has been no change in potential or kinetic energy. The equation for energy conservation is thus

$$\Delta E_{\text{chem}} + \Delta E_{\text{th}} = 0$$

Each time the bar is raised or lowered, the muscles use chemical energy. Ultimately, all of this energy is transformed to thermal energy.

ASSESS Most exercises in the gym—lifting weights, running on a treadmill—involve only the transformation of chemical energy to thermal energy.

EXAMPLE 11.5 Energy usage for a cyclist

A cyclist pedals for 20 min at a speed of 15 km/hr. How much metabolic energy is required?

PREPARE Table 11.4 gives a value of 480 W for the power used in cycling at a speed of 15 km/hr. This is the energy that is used per second.

SOLVE The cyclist uses energy at this rate for 20 min, or 1200 s. Power is the rate of using energy, $P = \Delta E/\Delta t$, so the energy required is

$$\Delta E = P\Delta t = (480 \text{ J/s})(1200 \text{ s}) = 580 \text{ kJ}$$

ASSESS How much energy is 580 kJ? A look at Table 11.2 shows that this is slightly more than the amount of energy available in a large apple, and only 10% of the energy available in a large fast-food meal. If you eat such a meal and plan to "work it off" by cycling, you should plan on cycling at a pretty good clip for a bit over 3 hr.

Energy Storage

The body gets energy from food; if this energy is not used, it will be stored. A small amount of energy needed for immediate use is stored as ATP. A larger amount of energy is stored as chemical energy of glycogen and glucose in muscle tissue and the liver. A healthy adult might store 400 g of these carbohydrates, which is a little more carbohydrate than is typically consumed in one day.

If the energy input from food continuously exceeds the energy outputs of the body, this energy will be stored in the form of fat under the skin and around the organs. From an energy point of view, gaining weight is simply explained!

EXAMPLE 11.6 Running out of fuel

The body stores about 400 g of carbohydrates. Approximately how far could a 68 kg runner travel on this stored energy?

PREPARE Table 11.1 gives a value of 17 kJ per g of carbohydrate. The 400 g of carbohydrates in the body contain an energy of

$$E_{\text{chem}} = (400 \text{ g})(17 \times 10^3 \text{ J/g}) = 6.8 \times 10^6 \text{ J}$$

SOLVE Table 11.4 gives the power used in running at 15 km/hr as 1150 W. The time that the stored chemical energy will last at this rate is

$$\Delta t = \frac{\Delta E_{\text{chem}}}{P} = \frac{6.8 \times 10^6 \text{ J}}{1150 \text{ W}} = 5.91 \times 10^3 \text{ s} = 1.64 \text{ hr}$$

And the distance that can be covered during this time at 15 km/hr is

$$\Delta x = v\Delta t = (15 \text{ km/hr})(1.64 \text{ hr}) = 25 \text{ km}$$

to two significant figures.

ASSESS A marathon is longer than this—just over 42 km. Even with "carbo loading" before the event (eating high-carbohydrate meals), many marathon runners "hit the wall" before the end of the race as they reach the point where they have exhausted their store of carbohydrates. The body has other energy stores (in fats, for instance), but the rate that they can be drawn on is much lower.

Energy and Locomotion

When you walk at a constant speed on level ground, your kinetic energy is constant. Your potential energy is also constant. So why does your body need energy to walk? Where does this energy go?

We use energy to walk because of mechanical inefficiencies in our gait. Figure 11.6 shows how the speed of your foot typically changes during each stride. The kinetic energy of your leg and foot increases, only to go to zero at the end of the stride. The kinetic energy is mostly transformed into thermal energy in your muscles and in your shoe. This thermal energy is lost; it can't be used for making further strides.

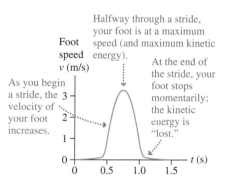

FIGURE 11.6 Human locomotion analysis.

Footwear can be designed to minimize the loss of kinetic energy to thermal energy. A spring in the sole of the shoe can store potential energy, which can be returned to kinetic energy during the next stride. Such a spring will make the collision with the ground more elastic. We saw in Chapter 10 that the tendons in the ankle do store a certain amount of energy during a stride; very stout tendons in the legs of kangaroos store energy even more efficiently. Their peculiar hopping gait is quite efficient at high speeds.

STOP TO THINK 11.2 A person is running at a constant speed on level ground. Chemical energy in the runner's body is being transformed to other forms of energy; most of the chemical energy is transformed to

A. Kinetic energy.   B. Potential energy.   C. Thermal energy.

11.4 Thermal Energy and Temperature

We have frequently spoken of the energy that is transformed to thermal energy as being "lost." Regardless of whether "the system" is a car, a power plant, or your body, this thermal energy is exhausted into the environment as heat. But why is thermal energy in your body and other systems not converted to other forms of energy and used for practical purposes? To continue our study of how energy is used, we need to understand a bit more about one particular kind of energy—thermal energy.

What do you mean when you say something is "hot"? Do you mean that it has a high temperature? Or do you mean that it has a lot of thermal energy? Perhaps both of these definitions are the same? Let's give some thought to the definitions of temperature and thermal energy, and the relationship between them.

Where do you wear the weights? BIO
If you wear a backpack with a mass equal to 1% of your body mass, your energy expenditure for walking will increase by 1%. But if you wear ankle weights with a combined mass of 1% of your body mass, the increase in energy expenditure is 6%, because you must repeatedly accelerate this extra mass. If you want to "burn more fat," wear the weights on your ankles, not on your back!

Measuring Temperature

We are all familiar with the idea of temperature. You hear the word used nearly every day. But just what is temperature a measure of? Velocity is a measure of how fast a system moves. What physical property of the system have you determined if you measure its temperature?

Let's start by looking at how you measure temperature. This is what a *thermometer* does. In a common glass-tube thermometer, for example, a small volume of mercury or alcohol expands or contracts when placed in contact with a "hot" or "cold" object. The object's temperature is determined by the height of the column of liquid.

A thermometer needs a *temperature scale* to be a useful measuring device. In 1742, the Swedish astronomer Anders Celsius sealed mercury into a small capillary tube and observed how it moved up and down the tube as the temperature changed. He selected two temperatures that anyone could reproduce, the freezing and boiling points of pure water, and labeled them 0 and 100. He then marked off the glass tube into one hundred equal intervals between these two reference points. By doing so, he invented the temperature scale that we today call the *Celsius scale*. The units of the Celsius temperature scale are "degrees Celsius," which we abbreviate as °C. The *Fahrenheit scale,* still widely used in the United States, is related to the Celsius scale by

$$T_C = \frac{5}{9}(T_F - 32°) \tag{11.4}$$

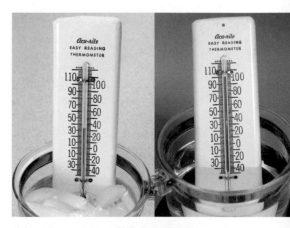

Thermal expansion of the liquid in the thermometer pushes it higher when immersed in hot water than in ice water.

1. The gas is made of a large number N of atoms, each of mass m, all moving randomly.

2. The atoms in the gas are quite far from each other and interact only rarely when they collide.

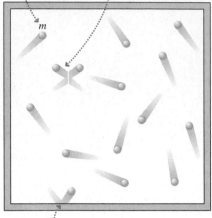

m

3. The collisions of the atoms with each other (and with walls of the container) are elastic; no energy is lost in these collisions.

FIGURE 11.7 Motion of atoms in an ideal gas.

3. The temperature is also increased.

Before   After

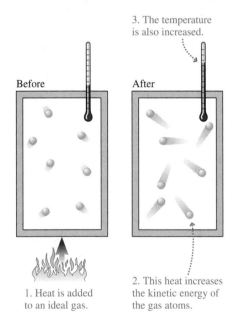

1. Heat is added to an ideal gas.

2. This heat increases the kinetic energy of the gas atoms.

FIGURE 11.8 Heating an ideal gas.

8.1–8.3   Activ Physics

An Atomic View of Thermal Energy and Temperature

When you add heat to an object, you increase its thermal energy; you also increase its temperature. In Chapter 10, we noted that the thermal energy of an object is the sum of the kinetic and potential energy of its atoms or molecules. Now we extend our discussion to include the concept of temperature. Let's start with the simplest possible system, a gas of atoms. In fact, we will make the system even simpler by defining the properties of what we will term an **ideal gas** in Figure 11.7. This is a rather simple model, but it gives a remarkably accurate description of the behavior of many real gases.

Because the ideal-gas atoms do not interact with each other (except for collisions), the system has no potential energy. The only internal energy in the ideal gas is the translational kinetic energy of the individual atoms. Thus, **the thermal energy of an ideal gas is equal to the total kinetic energy of the atoms that make up the gas.**

Adding heat to an ideal gas, as in Figure 11.8, will increase the thermal energy of the gas; the atoms will move faster. Adding heat to the gas will also increase its temperature. So temperature must be related to the kinetic energy of the individual atoms, but how? We can get one hint from the following fact: The temperature of a system doesn't depend on the size of the system. If you have two glasses of water, each at a temperature of 20°C, and you pour them together, you will have a larger volume of water at the same temperature of 20°C. There are more atoms, and therefore there is more *total* thermal energy, but the temperature has not changed. Even though there are more atoms, the *average* energy per atom is the same. It is this average kinetic energy of the atoms that is related to the temperature. **The temperature of an ideal gas is a measure of the average kinetic energy of the atoms that make up the gas.**

Let's consider the average kinetic energy of atoms in the gas. An atom of mass m and velocity v has kinetic energy $K = \frac{1}{2}mv^2$. The average kinetic energy of an atom is computed as the sum of the kinetic energies of all of the atoms divided by the number of atoms:

$$K_{\text{avg}} = \frac{\sum \frac{1}{2}mv^2}{N} = \frac{1}{2}m\frac{\sum v^2}{N} = \frac{1}{2}m(v^2)_{\text{avg}} \tag{11.5}$$

The quantity $\sum v^2/N$ that appears in Equation 11.5 is the average of the speed squared, not the square of the average speed. If we want to know how fast a typical atom is moving, we should look at the square root of this average:

$$\text{Speed of a typical atom} = \sqrt{(v^2)_{\text{avg}}} = v_{\text{rms}} \tag{11.6}$$

The quantity that we've defined as v_{rms} is called the **root mean square speed,** often referred to as the *rms speed.* Rewriting Equation 11.5 in terms of v_{rms} gives the average kinetic energy per atom:

$$K_{\text{avg}} = \frac{1}{2}mv_{\text{rms}}^2 \tag{11.7}$$

If the gas consists of N atoms, the total kinetic energy of the atoms is just NK_{avg}. The total kinetic energy of the atoms in the gas is the thermal energy of the gas, so we find that

$$E_{\text{th}} = \frac{1}{2}Nmv_{\text{rms}}^2$$

If we can establish a connection between K_{avg} and the temperature T, which we will do in the next section, then we'll know how both the thermal energy E_{th} and the typical atomic speed are related to temperature.

The Meaning of Temperature on the Kelvin Scale

When the Celsius temperature scale was defined, zero degrees was fixed to be the freezing point of water. The Fahrenheit scale has a zero point that was set as the temperature of an ice-salt mix. On both of these scales the choice of zero is arbitrary, and both scales allow temperatures that are negative or "below zero."

If we use the average kinetic energy of atoms as the basis for our definition of temperature, our temperature scale will have a very natural zero—the point at which kinetic energy is zero. Kinetic energy is always positive, so the zero on our temperature scale will be an **absolute zero;** no temperature below this is possible.

This is how zero is defined on the temperature scale called the *Kelvin scale:* **Zero degrees is the point at which the kinetic energy of atoms is zero.** All temperatures on the Kelvin scale are positive, so it is often called an *absolute temperature scale*. The units of the Kelvin temperature scale are "kelvin" (not degrees kelvin!), which is abbreviated K.

The spacing between divisions for the Kelvin scale is the same as that of the Celsius scale; the only difference is the position of the zero point. It is found experimentally that absolute zero—the temperature at which atoms would cease moving—is $-273°C$. Thus the Kelvin scale is obtained from the Celsius scale by shifting the zero point by 273 divisions. For this reason, the conversion between Celsius and Kelvin temperatures is quite straightforward:

$$T = T_C + 273 \qquad (11.8)$$

Thus the freezing point of water at $0°C$ is $T = 0 + 273 = 273$ K. A $30°C$ warm summer day is $T = 303$ K on the Kelvin scale. A side-by-side comparison of these scales is given in Figure 11.9.

> **NOTE** ▶ We will use the symbol T for temperature in kelvin. We will use subscripts to denote other scales. In the equations in this chapter and the rest of the text, *T must be interpreted as a temperature in kelvin.* ◀

Optical molasses It isn't possible to reach absolute zero, where the atoms would be still, but it is possible to get quite close by slowing the atoms down directly. These crossed laser beams produce what is known as "optical molasses." As we will see in Chapter 28, light is made of photons, which carry energy and momentum. Interactions of the atoms of a diffuse gas with the photons cause the atoms to slow down. Atoms can be slowed to very slow speeds that correspond to a temperature as cold as 5×10^{-10} K!

EXAMPLE 11.7 Temperature scales

The coldest temperature ever measured on earth was $-129°F$, in Antarctica. What is this in $°C$ and K?

SOLVE We use Equation 11.4 to convert the temperature to the Celsius scale:

$$T_C = \frac{5}{9}(-129° - 32°) = -89°C$$

We can then use Equation 11.8 to convert this to kelvin:

$$T = -89 + 273 = 184 \text{ K}$$

ASSESS This is quite cold, but quite a bit warmer than the coldest temperatures achieved in the laboratory.

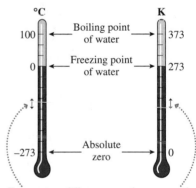

Temperature *differences* are the same on the Celsius and Kelvin scales. The temperature difference between the freezing point and boiling point of water is $100°C$ or 100 K.

FIGURE 11.9 Celsius and Kelvin temperature scales.

We noted above that the temperature of an ideal gas is a measure of the average kinetic energy of the atoms. It can be shown that temperature on the Kelvin scale is related to the average kinetic energy per atom by

$$T = \frac{2}{3}\frac{K_{avg}}{k_B} \qquad (11.9)$$

Temperature of an ideal gas in terms of the average kinetic energy per atom

LINEAR
p. 38

where k_B is a constant known as **Boltzmann's constant.** Its value is

$$k_B = 1.38 \times 10^{-23} \text{ J/K}$$

Martian airsicles The atmosphere of Mars is mostly carbon dioxide. At night, the temperature may drop so low that the molecules in the atmosphere will slow down enough to stick together—the atmosphere actually freezes. The frost on the surface in this image from the Viking 2 lander is composed partially of frozen carbon dioxide.

We can rewrite Equation 11.9 to express the average kinetic energy of the atoms in an ideal gas at a temperature T as

$$K_{avg} = \frac{3}{2}k_B T \tag{11.10}$$

Interestingly, the average kinetic energy per atom depends *only* on the temperature, not on the atom's mass. For example, the average kinetic energy of an atom (of any gas) at room temperature (20°C, or 293 K) is

$$K_{avg} = \frac{3}{2}(1.38 \times 10^{-23} \text{ J/K})(293 \text{ K}) = 6.07 \times 10^{-21} \text{ J}$$

Combining Equation 11.10 with our definition of the root-mean-square speed of the atoms in a gas in Equation 11.7, we find that

$$v_{rms} = \sqrt{\frac{3k_B T}{m}} \tag{11.11}$$

The rms speed is proportional to the *square root* of the temperature. This is a new mathematical form that we will see again, so we will take a look at its properties.

⟋ **Square-root relationships**   ✐ Exercise 12   (MP)

Two quantities are said to have a **square-root relationship** if y is proportional to the square root of x. We write the mathematical relationship as

$$y = A\sqrt{x}$$

y is proportional to the square root of x

The graph of a square-root relationship is a parabola that has been rotated by 90°.

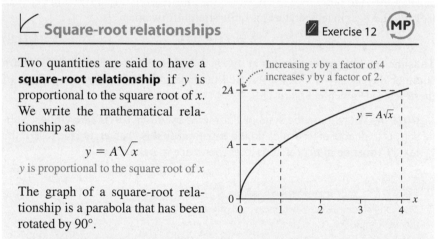

Increasing x by a factor of 4 increases y by a factor of 2.

$y = A\sqrt{x}$

SCALING Suppose we quadruple x, so that $x_f = 4x_i$. Then y will change to

$$y_f = A\sqrt{x_f} = A\sqrt{4x_i} = 2A\sqrt{x_i} = 2y_i$$

Increasing x by a factor of 4 causes y to increase by a factor of $\sqrt{4}$, or 2. Similarly, if we *decrease* x by a factor of 4, y will decrease by a factor of 2. Generally, we can say that:

Changing x by a factor C changes y by a factor $\sqrt{C}$.

We have been treating ideal gases made of atoms, but our results are equally valid for real gases made of atoms or molecules. From Equation 11.11 we see that **a higher temperature corresponds to faster atomic or molecular speeds.** At a given temperature, the speed of atoms or molecules in a gas varies with the atomic or molecular mass. A gas with lighter atoms will have faster atoms, on average, than a gas with heavier atoms.

The masses of atoms and molecules in chemistry are measured in **atomic mass units,** denoted u; this is the atomic mass that you will find in a periodic table. The atomic mass of a hydrogen atom is 1 u, so that of an H_2 molecule is 2 u. However, the equations in this chapter require that the masses of atoms and molecules be in kg. The conversion factor is

$$1 \text{ u} = 1.66 \times 10^{-27} \text{ kg}$$

EXAMPLE 11.8 Speed of nitrogen molecules

Nitrogen gas consists of molecules, N_2. At 20°C, what is the root-mean-square speed of the molecules of nitrogen?

PREPARE In the periodic table, you can find that the mass of a nitrogen atom is 14 u. A N_2 molecule consists of two atoms, so its mass is 28 u. Thus the molecular mass in SI units (i.e., kg) is

$$m = 28 \text{ u} \times \frac{1.66 \times 10^{-27} \text{ kg}}{1 \text{ u}} = 4.65 \times 10^{-26} \text{ kg}$$

Convert the temperature to kelvin:

$$T = 20 + 273 = 293 \text{ K}$$

SOLVE We compute v_{rms} for the nitrogen atoms using Equation 11.11:

$$v_{rms} = \sqrt{\frac{3k_B T}{m}} = \sqrt{\frac{3(1.38 \times 10^{-23} \text{ J/K})(293 \text{ K})}{4.65 \times 10^{-26} \text{ kg}}} = 510 \text{ m/s}$$

ASSESS This is *very* fast, just a bit faster than the speed of sound in air. This result makes sense, however: Sound waves must be carried by moving air molecules, so we would expect the speed of sound to be close to the speed of the molecules.

Computing Thermal Energy

The speeds of gas atoms are quite high. The associated kinetic energy of the atoms in a gas will be quite high as well. The thermal energy of an ideal gas consisting of N atoms can be computed using the above result for the average kinetic energy:

$$E_{th} = NK_{avg} = \frac{3}{2}Nk_B T \qquad (11.12)$$

LINEAR
p. 38

Thermal energy of an ideal gas of N atoms

Thus, **thermal energy is directly proportional to temperature.** Consequently, a change in the thermal energy of an ideal gas is proportional to a change in temperature:

$$\Delta E_{th} = \frac{3}{2}Nk_B \Delta T \qquad (11.13)$$

This relationship between a change in temperature and a change in thermal energy is for an ideal gas, but solids, liquids, and other gases all follow similar rules, as we will see in the next chapter. For now, we will simply note two important conclusions that apply to any substance:

1. The thermal energy of a substance is proportional to the number of atoms. A gas with more atoms has more thermal energy than a gas at the same temperature with fewer atoms.
2. A change in temperature causes a proportional change in the substance's thermal energy. A larger temperature change causes a larger change in thermal energy.

EXAMPLE 11.9 Energy change in heating a gas

100 cm³ of gas at 1 atmosphere pressure and 20°C contains 2.5×10^{21} molecules. (You'll learn how to show this in the next chapter.) By how much does its thermal energy change if it is heated to 30°C?

SOLVE The temerature *change* is $\Delta T = 10$ K. Using Equation 11.13, we find

$$\Delta E_{th} = \frac{3}{2}Nk_B \Delta T = \frac{3}{2}(2.5 \times 10^{21})(1.38 \times 10^{-23} \text{ J/K})(10 \text{ K}) = 0.52 \text{ J}$$

ASSESS Although the number of molecules in a typical gas is vast, a typical change in thermal energy is quite comparable to the values we've been calculating for the kinetic and potential energies of macroscopic objects.

Is it cold in space? The space shuttle orbits in the upper thermosphere, about 300 km above the surface of the earth. There is still a trace of atmosphere left at this altitude, and it has quite a high temperature—over 1000°C. Although the average speed of the air molecules here is quite high, there are so few air molecules present that the thermal energy is extremely low.

Consider gases of the following atoms or molecules, all at the same temperature. Which atoms or molecules have the fastest average speed?

A. H_2 B. He   C. N_2 D. O_2 E. All the same.

11.5 Heat and the First Law of Thermodynamics

In Chapter 10, we saw that a system could exchange energy with the environment through two different means: Work and heat. Work was treated in some detail in Chapter 10; now it is time to look at the transfer of energy by heat. This will begin our study of a topic called *thermodynamics,* the study of thermal energy and heat and their relationships to other forms of energy and energy transfer.

What Is Heat?

Heat is a more elusive concept than work. We use the word "heat" very loosely in the English language, often as synonymous with *hot.* We might say, on a very hot day, that "This heat is oppressive." If your apartment is cold you may say, "Turn up the heat." These expressions date to a time long ago when it was thought that heat was a *substance* with fluid-like properties.

If you place a hot object and a cold object together, they evolve toward a common final temperature. Common sense suggests that "something" flows from the hot object to the cold until equilibrium is achieved. This "heat fluid" was called *caloric.* The notion that objects somehow "contain" heat, and that heat can move around, lingers on in expressions like *heat flow, heat loss,* and *heat capacity*

One of the first to disagree with this notion, in the late 1700s, was Benjamin Thompson, known by his title Count Rumford. While watching the hot metal chips thrown off during the boring of cannons, he began to think about heat. If caloric is a substance, the cannon and borer should eventually run out of caloric and the heat generation ought to decrease with time. But it does not. Rumford noted that the heat generation appears to be "inexhaustible," which is not consistent with the idea of heat as a substance. He concluded that heat is not a substance—it is *motion!*

The British physicist James Joule carried out careful experiments in the 1840s to learn how it is that systems change their temperature. Using experiments like those shown in Figure 11.10, Joule found that you can raise the temperature of a beaker of water by two entirely different means:

1. Heating it with a flame, or
2. Doing work on it with a rapidly spinning paddle wheel.

Surprisingly, the final state of the water is *exactly* the same in both cases. This implies that heat and work are essentially equivalent to each other—**heat and work are simply two different ways of transferring energy to or from a system.**

Thermal energy is transferred between a system and the environment as a consequence of a temperature difference between them. When you place a pan of water on the stove, thermal energy is transferred *from* the hotter flame *to* the cooler water. If you place the water in a freezer, thermal energy is transferred from the warmer water to the colder air in the freezer.

Energy is transferred when the faster molecules in the hotter object collide with the slower molecules in the cooler object. On average, these collisions cause

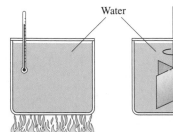

Water

The flame heats the water.

The spinning paddle does work on the water.

FIGURE 11.10 Joule's experiments to show the equivalence of heat and work.

TRY IT YOURSELF

Energetic cooking We can do a modern version of Joule's experiment in the kitchen. Next time you mix food in a blender, notice how it actually warms the food! As the blades rotate, friction increases the thermal energy of the food in the blender. A blender uses about as much power as a microwave, and most of this energy ends up as thermal energy in the food, so the temperature rises as the food is chopped and blended.

the faster molecules to lose energy and the slower molecules to gain energy. The net result is that energy is transferred from the hotter object to the colder object. The process itself, whereby energy is transferred between the system and the environment via atomic-level collisions, is called a *thermal interaction*. **Heat is the energy transfer during a thermal interaction.**

An Atomic Model of Heat

Let's look at a thermal interaction in some detail, considering what happens at an atomic level. Figure 11.11 shows a rigid, insulated container that is divided into two sections by a very thin membrane. Each side is filled with a gas of the same kind of atoms. The left side, which we'll call system 1, has N_1 atoms at an initial temperature T_{1i}. System 2 on the right has N_2 atoms at an initial temperature T_{2i}. We imagine the membrane to be so thin that atoms can collide at the boundary as if the membrane were not there, yet it is a barrier that prevents atoms from moving from one side to the other.

Suppose that system 1 is initially at a higher temperature: $T_{1i} > T_{2i}$. This means that the atoms in system 1 will have a higher average kinetic energy. Figure 11.12 shows a fast atom and a slow atom approaching the barrier from opposite sides. They undergo a perfectly elastic collision at the barrier. Although no net energy is lost in a perfectly elastic collision, the faster atom loses energy while the slower one gains energy. In other words, there is an energy *transfer* from the faster atom's side to the slower atom's side.

Because the atoms in system 1 are, on average, more energetic than the atoms in system 2, *on average* the collisions transfer energy from system 1 to system 2. This is not true for every collision. Sometimes a fast atom in system 2 collides with a slow atom in system 1, transferring energy from 2 to 1. But the net energy transfer, from all collisions, is from the warmer system 1 to the cooler system 2. This transfer of energy is heat; **thermal energy is transferred from the faster moving atoms on the warmer side to the slower moving atoms on the cooler side.**

This transfer will continue until a stable situation is reached. This is a situation we call **thermal equilibrium.** How do the systems "know" when they've reached thermal equilibrium? Energy transfer continues until the atoms on both sides of the barrier have the *same average kinetic energy*. Once the average kinetic energies are the same, individual collisions will still transfer energy from one side to the other. But since both sides have atoms with the same average kinetic energies, the amount of energy transferred from 1 to 2 will equal that transferred from 2 to 1. Once the average kinetic energies are the same, there will be no more net energy transfer.

An important result from the previous section is that the average kinetic energy of the atoms in a system is directly proportional to the system's temperature. If two systems exchange energy until their atoms have the same average kinetic energy, we can say that

$$T_{1f} = T_{2f} = T_f$$

That is, heat is transferred until the two systems reach a common final temperature; we call this final state **thermal equilibrium.** We considered a rather artificial system in this case, but the result is quite general: **Two systems placed in thermal contact will transfer thermal energy until their final temperatures are the same.** This process is illustrated in Figure 11.13.

Heat is a transfer of energy. The sign that we use for transfers is defined in Figure 11.14 on the next page. In the process of Figure 11.13, Q_1 is negative as system 1 loses energy; Q_2 is positive as system 2 gains energy. Because no energy

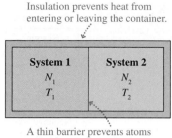

Insulation prevents heat from entering or leaving the container.

A thin barrier prevents atoms from moving from system 1 to 2 but still allows them to collide.

FIGURE 11.11 Two gases are separated by a thin barrier through which they can thermally interact.

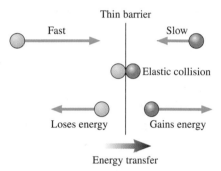

FIGURE 11.12 Collisions at a barrier transfer energy from faster molecules to slower molecules.

Collisions transfer energy from the warmer system to the cooler as more energetic atoms lose energy to less energetic atoms. This energy transfer is heat.

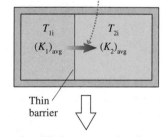

Thermal equilibrium occurs when the systems have the same average translational kinetic energy and thus the same temperature.

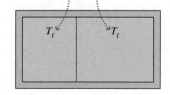

FIGURE 11.13 Two systems in thermal contact exchange thermal energy.

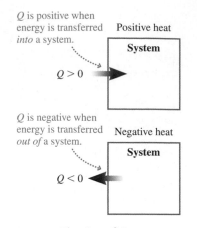

FIGURE 11.14 The sign of Q.

escapes from the container, all of the energy that was lost by system 1 was gained by system 2

$$Q_2 = -Q_1$$

as required by energy conservation. That is, the heat energy lost by one system is gained by the other.

The above argument was made for gases, but it is completely general. We could make a similar analysis for different gases, solids, or liquids. In each case, the final result is the same: **Heat is transferred from hot to cold.**

The First Law of Thermodynamics

In Chapter 10, we looked at several versions of the law of conservation of energy. The most general statement was the law of conservation of energy for systems that were not isolated, Equation 10.4:

$$\Delta K + \Delta U_g + \Delta U_s + \Delta E_{th} + \Delta E_{chem} + \cdots = W + Q$$

In Chapter 10, we focused on systems where the potential and kinetic energy could change, such as a sled moving down a hill. Earlier in this chapter, we looked at the body, where the chemical energy changes. Now let's consider systems in which only the thermal energy changes. That is, we will consider systems that aren't moving, that aren't changing chemically, but whose temperatures can change. Such systems are the province of what is called **thermodynamics.** The question of how to keep your house cool in the summer is a question of thermodynamics. If thermal energy is transferred into your house, the temperature will rise; if thermal energy is transferred out, the temperature will drop.

When the law of conservation of energy was introduced, we noted that it was sometimes called the first law of thermodynamics. This is true, but a more common statement of the first law of thermodynamics applies to systems in which only thermal energy changes take place.

First law of thermodynamics For systems in which only the thermal energy changes, the change in thermal energy is equal to the energy transferred into or out of the system as work W and/or heat Q:

$$\Delta E_{th} = W + Q \qquad (11.14)$$

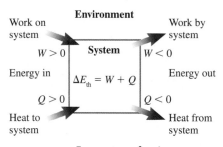

FIGURE 11.15 Energy transfers in a thermodynamic system.

Only work and heat, two ways of transferring energy between a system and the environment, cause the system's energy to change. In thermodynamic systems, the only energy change will be a change in thermal energy. Whether this energy increases or decreases depends on the signs of W and Q. In Chapter 10, we noted that when $W > 0$, work is done on the system, increasing its energy, and when $W < 0$ work is done by the system, decreasing its energy. A similar sign convention holds for heat, as we saw in Figure 11.14. The possible energy transfers between a system and the environment are illustrated in Figure 11.15.

CONCEPTUAL EXAMPLE 11.3 Compressing a gas

Suppose a gas is in an insulated container, so that no heat energy can escape. If a piston is used to compress the gas, what happens to the temperature of the gas?

REASON The piston applies a force to the gas, and there is a displacement. This means that work is done on the gas by the piston ($W > 0$). No thermal energy can be exchanged with the environment, meaning $Q = 0$. Since energy is transferred into the system, the thermal energy of the gas must increase. This means that the temperature must increase as well.

ASSESS When you use a bike pump to inflate a tire, you may have noticed that the pump and the tire get warm. This temperature increase is due to the warming of the air by the compression.

EXAMPLE 11.10 Work and heat in an ideal gas

Suppose a container holds 5.0×10^{22} molecules of an ideal gas. 50 J of work are done on the gas by a piston that compresses it. The temperature of the gas changes by 30°C during this process. How much heat is transferred to or from the environment?

SOLVE For an ideal gas, the change in thermal energy is proportional to the change in temperature, as we see in Equation 11.13. We compute the change in thermal energy to be

$$\Delta E_{th} = \frac{3}{2} N k_B \Delta T$$

$$= \left(\frac{3}{2}\right)(5.0 \times 10^{22})(1.38 \times 10^{-23} \text{ J/K})(30 \text{ K}) = 31 \text{ J}$$

The first law of thermodynamics, Equation 11.14, tells us that the change in thermal energy ΔE_{th} of the gas is equal to the sum of W and Q. W is known, and we have just calculated ΔE_{th}. Combining, we get

$$Q = \Delta E_{th} - W = 31 \text{ J} - 50 \text{ J} = -19 \text{ J}$$

ASSESS Q is negative, meaning that energy is transferred from the hot gas to the cooler environment in this process.

Energy-Transfer Diagrams

Suppose you drop a hot rock into the ocean. Heat is transferred from the rock to the ocean until the rock and ocean are at the same temperature. Although the ocean warms up ever so slightly, ΔT_{ocean} is so small as to be completely insignificant. For all practical purposes, the ocean is infinite and unchangeable.

An **energy reservoir** is an object or a part of the environment so large that its temperature does not noticeably change when heat is transferred between the system and the reservoir. A reservoir at a higher temperature than the system is called a *hot reservoir*. A vigorously burning flame is a hot reservoir for small objects placed in the flame. A reservoir at a lower temperature than the system is called a *cold reservoir*. The ocean is a cold reservoir for the hot rock. We will use T_H and T_C to designate the temperatures of the hot and cold reservoirs.

Heat energy is transferred between a system and a reservoir if they have different temperatures. We will define

Q_H = amount of heat transferred to or from a hot reservoir

Q_C = amount of heat transferred to or from a cold reservoir

By definition, Q_H and Q_C are *positive* quantities.

Figure 11.16a shows a heavy copper bar placed between a hot reservoir (at temperature T_H) and a cold reservoir (at temperature T_C). Heat Q_H is transferred from the hot reservoir into the copper and heat Q_C is transferred from the copper to the cold reservoir. Figure 11.16b is an **energy-transfer diagram** for this process. The hot reservoir is always drawn at the top, the cold reservoir at the bottom, and the system—the copper bar in this case—between them. The reservoirs and the system are connected by "pipes" that show the energy transfers. Figure 11.16b shows heat Q_H being transferred into the system and Q_C being transferred out.

Figure 11.17 illustrates an important fact about heat transfers that we have seen: Spontaneous transfers go in one direction only. The fact that heat is not transferred from a colder object to a hotter object is an important result that will have significant practical implications, as we will soon discover.

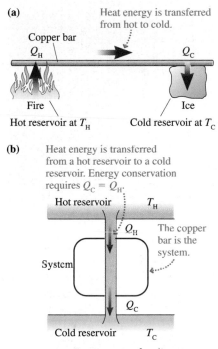

(a) Heat energy is transferred from hot to cold.

Copper bar

Q_H Q_C

Fire   Ice

Hot reservoir at T_H Cold reservoir at T_C

(b) Heat energy is transferred from a hot reservoir to a cold reservoir. Energy conservation requires $Q_C = Q_H$.

Hot reservoir T_H

Q_H The copper bar is the system.

System

Q_C

Cold reservoir T_C

FIGURE 11.16 Energy-transfer diagrams.

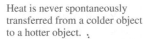

Heat is never spontaneously transferred from a colder object to a hotter object.

Hot reservoir T_H

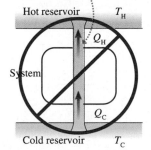

Q_H

System

Q_C

Cold reservoir T_C

FIGURE 11.17 An impossible energy transfer.

CONCEPTUAL EXAMPLE 11.4 Energy transfers and the body

Why—in physics terms—is it more taxing on the body to exercise in very hot weather?

REASON Your body continuously converts chemical energy to thermal energy, as we have seen. In order to maintain a constant body temperature, your body must continuously transfer heat to the environment. This is a simple matter in cool weather when heat is spontaneously transferred to the environment, but when the air temperature is higher than your body temperature, your body cannot cool itself this way and must use other mechanisms to transfer this energy, such as perspiring. These mechanisms require additional energy expenditure.

ASSESS Strenuous exercise in hot weather can easily lead to a rise in body temperature if the body cannot exhaust heat quickly enough.

STOP TO THINK 11.4 The radiator in an automobile is at a higher temperature than the air around it. Considering the radiator as the system, we can say that

A. $Q > 0$ B. $Q = 0$ C. $Q < 0$

Falling water turns a waterwheel.

11.6 Heat Engines

In the early stages of the industrial revolution, most of the energy needed to run mills and factories came from water power. Water in a high reservoir will naturally flow downhill. A waterwheel can be used to harness this natural flow of water to produce some useful energy. That is, some of the potential energy lost by the water as it flows downhill can be converted into other forms.

It is possible to do something similar with heat. Thermal energy is naturally transferred from a hot reservoir to a cold; it is possible to take some of this energy as it is transferred and convert it to other forms. This is the job of a device known as a **heat engine.**

A simple example of a heat engine is shown in Figure 11.18a. Here a *Peltier device*—a device that produces a voltage if there is a temperature difference between its two sides—is sandwiched between two aluminum vanes that sit in cups of water. The cup of water on the left is hot; the cup on the right is cold. The temperature difference produces a voltage that drives the fan.

Figure 11.18b is a diagram that describes the operation of this device in thermodynamic terms. The Peltier device is the system. Heat energy is transferred through the device from the hot water to the cold water. As the transfer occurs, some of this energy is converted into electric energy to run the fan. The cold-water reservoir is crucial for the operation of the device. The heat energy that is converted to electric energy to run the fan comes from the hot water, but the fan won't run unless some heat energy is transferred into the cold water. Figure 11.19 is a thermodynamic picture of this heat engine—or any heat engine.

This schematic diagram of Figure 11.19 illustrates everything important about the basic physics of a heat engine. It takes in energy as heat from the hot reservoir, turns some of it into useful work, and exhausts the balance as waste heat in the cold reservoir. Any heat engine has exactly the same schematic.

Efficiency of a Heat Engine

We assume, for a heat engine, that the engine's thermal energy doesn't change. This means that there is no net energy transfer into or out of the heat engine. Because energy is conserved, we can say that the useful work extracted is equal to

(a)

Peltier device

(b)

W_{out} ← Energy is extracted to run the fan.

Heat energy is transferred from the hot water.

System

Q_H

Q_C

Excess heat energy is transferred into the cold water.

FIGURE 11.18 A simple heat engine.

the difference between the heat energy transferred from the hot reservoir and the heat exhausted into the cold reservoir:

$$W_{out} = Q_H - Q_C$$

The energy input to the engine is Q_H and the energy output is W_{out}.

NOTE ► We earlier defined Q_H and Q_C to be positive quantities. We also define the energy output of a heat engine W_{out} to be a positive quantity. For heat engines, the directions of the transfers will always be clear, and we will take all of these basic quantities to be positive. ◄

We can use the definition of efficiency, from earlier in the chapter, to compute the heat engine's efficiency:

$$e = \frac{\text{what you get}}{\text{what you had to pay}} = \frac{W_{out}}{Q_H} = \frac{Q_H - Q_C}{Q_H} \qquad (11.15)$$

The heat energy that is not converted to work ends up in the cold reservoir as waste heat.

Why should we waste energy this way? Why don't we make a heat engine like the one shown in Figure 11.20 that converts 100% of the heat into useful work? The surprising answer is that we can't. No heat engine can operate without exhausting some fraction of the heat into a cold reservoir. This isn't a limitation on our engineering abilities. As we'll see, it's a fundamental law of nature.

The maximum possible efficiency of a heat engine is fixed by the *second law of thermodynamics,* which we will explore in detail in Section 11.8. We will not do a detailed derivation, but simply note that the second law gives the maximum efficiency of a heat engine as

$$e_{max} = 1 - \frac{T_C}{T_H} \qquad (11.16)$$

Theoretical maximum efficiency of a heat engine

The maximum efficiency of any heat engine is therefore fixed by the ratio of the temperatures of the hot and cold reservoirs. It is possible to increase the efficiency of a heat engine by increasing the temperature of the hot reservoir or decreasing the temperature of the cold reservoir. The efficiency of Equation 11.16 is also called the *Carnot efficiency,* after a particular heat engine that achieves this maximum possible efficiency.

NOTE ► The temperatures in Equation 11.16 *must* be in K, not °C. ◄

Because $e_{max} < 1$, the work done is always less than the heat input ($W_{out} < Q_H$). Consequently, there *must* be heat Q_C exhausted to the cold reservoir. That is why the heat engine of Figure 11.20 is impossible. Think of a heat engine as analogous to a water wheel: The engine "siphons off" some of the energy that is spontaneously flowing from hot to cold, but it can't completely shut off the flow.

The fact that heat engine efficiency is limited—and that heat must be exhausted into a cold reservoir—is of tremendous practical importance. Most of the energy that you use daily comes from the conversion of chemical energy into thermal energy and the subsequent conversion of that energy into other forms. Let's look at some common examples of heat engines.

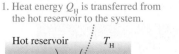
1. Heat energy Q_H is transferred from the hot reservoir to the system.

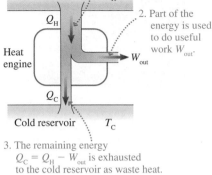
2. Part of the energy is used to do useful work W_{out}.

3. The remaining energy $Q_C = Q_H - W_{out}$ is exhausted to the cold reservoir as waste heat.

FIGURE 11.19 The operation of a heat engine.

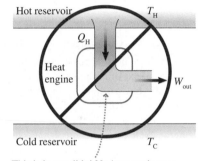

This is impossible! No heat engine can convert 100% of heat into useful work.

FIGURE 11.20 A perfect (and impossible!) heat engine.

Heat engines

Most of the electricity that you use was generated by heat engines. Coal or other fossil fuels are burned to produce the thermal energy of steam. The steam does work by spinning a turbine attached to a generator, which produces electricity. At the end of the proccess, approximately 65% of the thermal energy is deposited in a cold reservoir, usually a lake or a river.

Your car gets the energy it needs to run from the chemical energy in gasoline. The gasoline is burned; the resulting hot gases are the hot reservoir. Some of this thermal energy is converted to kinetic energy of the moving vehicle and electric energy to run systems in the car, but more than 90% may be lost as heat to the environment—the cold reservoir, as we saw earlier.

There are many small, simple heat engines that are part of things you use daily. This fan, which can be put on top of a wood stove, uses the thermal energy of the stove to provide power to drive air around the room. Where are the hot and cold reservoirs in this device?

EXAMPLE 11.11 The efficiency of a nuclear power plant
Energy from nuclear reactions in the core or a nuclear reactor produces high-pressure steam at a temperature of 290°C. After the steam is used to spin a turbine, it is condensed (by using cooling water from a nearby river) back to water at 20°C. The excess heat is deposited in the river. The water is then reheated, and the cycle beings again. What is the maximum possible efficiency that this plant could achieve?

PREPARE A nuclear power plant is a heat engine, with energy transfers as illustrated in Figure 11.19. Q_H is the energy in the steam. T_H is the temperature of the steam, 290°C. The steam is cooled and condensed and the heat Q_C exhausted to the river. The river is the cold reservoir, so T_C is 20°C.

In kelvin, these temperatures are:

$$T_H = 290°C = 563 \text{ K}, \; T_C = 20°C = 293 \text{ K}$$

SOLVE We use Equation 11.16 to compute the maximum possible efficiency:

$$e_{max} = 1 - \frac{T_C}{T_H} = 1 - \frac{293 \text{ K}}{563 \text{ K}} = 0.479 \approx 48\%$$

ASSESS This is the maximum possible efficiency. There are practical limitations as well that limit real power plants, whether nuclear or coal- or gas-fired, to an efficiency $e \approx 0.35$. This means that 65% of the energy from the fuel is exhausted as waste heat into a river or lake, where it may cause problematic warming in the local environment.

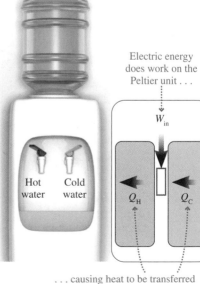

Electric energy does work on the Peltier unit . . .

W_{in}

Hot water   Cold water

Q_H Q_C

. . . causing heat to be transferred from the cold water to the hot water.

FIGURE 11.21 A heat pump provides hot and cold water.

STOP TO THINK 11.5   Which of the following changes (there may be more than one) would increase the maximum theoretical efficiency of a heat engine?

A. Increase T_H.   B. Increase T_C.   C. Decrease T_H.   D. Decrease T_C.

11.7 Heat Pumps

As the inside of your refrigerator is colder than the air in your kitchen, heat energy will be transferred from the room to the inside of the refrigerator, warming it. This happens every time you open the door and as heat "leaks" through the walls. But you want the inside of the refrigerator to stay cool. To keep it cool, you need some way to "pump" this heat back out to the warmer room. Transferring heat energy from a cold reservoir to a hot reservoir—the opposite of the natural direction—is the job of a **heat pump.**

The Peltier device that we saw in the last section can be used as a heat pump. If an electric current is put through the device, a temperature difference develops between the two sides. As time goes by, the cold side gets colder and the hot side gets hotter. A practical application of such a heat pump is the water cooler/heater shown in Figure 11.21. The energy that is removed from the cold water ends up as increased thermal energy in the hot water. A single process cools the cold water and heats the hot water.

Heat pumps, like heat engines, are very practical devices. You probably have at least one heat pump in your house and one in your car—a refrigerator and an air conditioner. Each transfers heat energy from someplace cold to someplace warm.

The heat pump in a refrigerator transfers heat from the cold air *inside* the refrigerator to the warmer air in the room. Coils inside the refrigerator that are colder than the inside air take in thermal energy. This energy is transferred to warm coils *outside* the refrigerator that transfer heat to the room. In the antique refrigerator shown here, the cold coils are in the metal trays at the top of the cabinet, the warm coils outside the cabinet in the cylindrical unit on top. When the refrigerator is running, these top coils are quite warm. If you look closely at your refrigerator, you will find warm coils or a warm air exhaust that transfers heat to the room, heat that was removed from the inside.

An air conditioner works similarly, transferring heat from the cool air of a house (or a car) to the warmer air outside. You have probably seen room air conditioners that are single units meant to fit in a window. The side of the air conditioner that faces the room is the cool side; the other side of the air conditioner is warm. Heat is pumped from the cool side to the warm side, cooling the room.

Conversely, you can also use a heat pump to *warm* your house in the winter by moving thermal energy from outside your house to inside. A unit outside the house takes in heat energy that is pumped to a unit inside the house, warming the inside air.

Water doesn't spontaneously flow from a lower to a higher elevation. Nonetheless, you can force water to flow uphill by doing work on it with a pump. Similarly, heat doesn't spontaneously flow from cold to hot, but we can transfer heat energy from a cold reservoir to a hot reservoir by doing work. That's what a heat pump does, as shown in Figure 11.22.

Energy must be conserved, so the heat deposited in the hot side must equal the sum of the heat removed from the cold side and the work input:

$$Q_H = Q_C + W_{in}$$

For heat pumps, rather than compute efficiency, we compute an analogous quantity called the **coefficient of performance COP.**

There are two different ways that one can use a heat pump, as noted above. A refrigerator uses a heat pump for cooling, removing heat from a cold reservoir to keep it cold. You can also use a heat pump for heating, moving heat from a cold reservoir to a hot reservoir, to keep it warm. In both cases, we must do work to make this happen.

If we use a heat pump for cooling, we define the coefficient of performance as

$$\text{COP} = \frac{\text{what you get}}{\text{what you had to pay}} = \frac{\text{energy removed from the cold reservoir}}{\text{work required to perform the transfer}} = \frac{Q_C}{W_{in}}$$

A calculation similar to that for a heat engine can be done to show that the greater the temperature difference between the hot and cold, the more work is required to transfer the energy. The maximum possible coefficient of performance is related to the temperatures of the hot and cold reservoirs as

$$\text{COP}_{max} = \frac{T_C}{T_H - T_C} \qquad (11.17)$$

Theoretical maximum coefficient of performance
of a heat pump used for cooling

A refrigerator from 1934.

The amount of heat exhausted to the hot reservoir is larger than the amount of heat extracted from the cold reservoir.

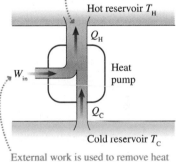

External work is used to remove heat from a cold reservoir and exhaust heat to a hot reservoir.

FIGURE 11.22 The operation of a heat pump.

Hot or cold lunch? Small coolers like this use Peltier devices that run off the 12 V electrical system of your car. They can transfer heat from the interior of the cooler to the outside, keeping food or drinks cool. But Peltier devices, like some other heat pumps, are reversible; switching the direction of current reverses the direction of heat transfer. A flick of a switch will cause heat to be transferred into the interior, keeping your lunch warm!

If we use a heat pump for heating, we define the coefficient of performance as

$$\text{COP} = \frac{\text{what you get}}{\text{what you had to pay}} = \frac{\text{energy added to the hot reservoir}}{\text{work required to perform the transfer}} = \frac{Q_H}{W_{in}}$$

In this case, the maximum possible coefficient of performance is

$$\text{COP}_{max} = \frac{T_H}{T_H - T_C} \qquad (11.18)$$

Theoretical maximum coefficient of performance
of a heat pump used for heating

In both cases, **a larger coefficient of performance means a more efficient heat pump.** Unlike the efficiency of a heat engine, which must be less than 1, the COP of a heat pump can be—and usually is—greater than 1. The following example shows that the COP can be quite high for typical temperatures.

EXAMPLE 11.12 Coefficient of performance of a refrigerator
The inside of your refrigerator is approximately 0°C. Heat from the inside of your refrigerator is deposited into the air in your kitchen, which has a temperature of approximately 20°C. At these operating temperatures, what is the maximum possible coefficient of performance of your refrigerator?

PREPARE The temperatures of the hot side and the cold side must be expressed in kelvin:

$$T_H = 20°C = 293 \text{ K}, T_C = 0°C = 273 \text{ K}$$

SOLVE We use Equation 11.17 to compute the maximum coefficient of performance:

$$\text{COP}_{max} = \frac{T_C}{T_H - T_C} = \frac{273 \text{ K}}{293 \text{ K} - 273 \text{ K}} = 13.6$$

ASSESS A coefficient of performance of 13.6 means that we pump 13.6 J of heat for an energy cost of 1 J. Due to practical limitations, the coefficient of performance of an actual refrigerator is typically ≈5. Other factors affect the overall efficiency of the appliance, including how well insulated it is.

CONCEPTUAL EXAMPLE 11.5 Keeping your cool?
It's a hot day, and your apartment is rather warm. Your roommate suggests cooling off the apartment by keeping the door of the refrigerator open. Will this help the situation?

REASON It's time for a physics lesson for your roommate! What you want to do is remove heat from the room. Your refrigerator, a heat pump, is designed to transfer heat from its inside to the outside. If you leave the door open, all parts of the refrigerator are exposed to the room air. It simply transfers heat from one part of the room to another—it won't make the room cooler.

ASSESS An air conditioner is a heat pump too. It must have a hot side that is outside the house, so that there is a net transfer of heat from inside to outside.

STOP TO THINK 11.6 Which of the following changes would allow your refrigerator to use less energy to run? (There may be more than one correct answer.)

A. Increasing the temperature inside the refrigerator.
B. Increasing the temperature of the kitchen.
C. Decreasing the temperature inside the refrigerator.
D. Decreasing the temperature of the kitchen.

11.8 Entropy and the Second Law of Thermodynamics

Throughout the chapter, we have noticed certain trends and certain limitations in energy transformations and transfers. Heat is transferred spontaneously from hot to cold, not from cold to hot. Once energy is transformed to thermal energy it is (in some sense) "lost." The transfer of heat from hot to cold is an example of an **irreversible** process, a process that can only happen in one direction. Why are some processes irreversible? The spontaneous transfer of heat from cold to hot would not violate any law of physics that we have seen to this point, but it is never observed. There must be another law of physics that prevents it. In this section we will explore the basis for this law, which we will call the **second law of thermodynamics.**

Reversible and Irreversible Processes

Stirring the cream in your coffee mixes the cream and coffee together. No amount of stirring ever unmixes them. If you shake a jar that has red marbles on the top and blue marbles on the bottom, the two colors are quickly mixed together. No amount of shaking ever separates them again. If you watched a movie of someone shaking a jar and saw the red and blue marbles separating, you would be certain that the movie was running backward. In fact, a reasonable definition of an irreversible process is one for which a backward-running movie shows a physically impossible process. In Figure 11.23, we see two possible movies of a particular physical process: the collision of a car with a barrier, one forward, the other backward.

The backward movie of Figure 11.23b is obviously wrong. But what has been violated in the backward movie? To have the crumpled car spring away from the wall would not violate any laws of physics we have so far discussed.

At a microscopic level, collisions are completely reversible. In Figure 11.24 we see two possible movies of a collision between two gas molecules, one forward, the other backward. You cannot tell, just by looking at the two movies, which is really going forward and which is being played backward. Nothing in either collision looks wrong, and no measurements you might make on either would reveal any violations of Newton's laws. **Interactions at the molecular level are reversible processes.**

If microscopic motions are all reversible, how can macroscopic phenomena such as the car crash end up being irreversible? If reversible collisions can cause heat to be transferred from hot to cold, why do they never cause heat to be transferred from cold to hot? There must be something at work that can distinguish the past from the future.

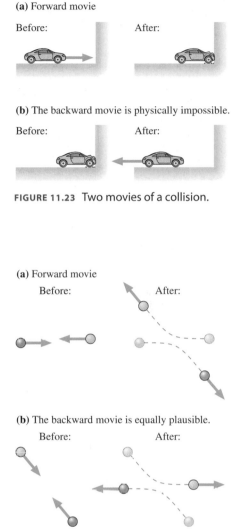

(a) Forward movie

Before: After:

(b) The backward movie is physically impossible.

Before: After:

FIGURE 11.23 Two movies of a collision.

(a) Forward movie

Before: After:

(b) The backward movie is equally plausible.

Before: After:

FIGURE 11.24 Molecular collisions are reversible.

Which Way to Equilibrium?

Stated another way, how do two systems initially at different temperatures "know" which way to go to reach equilibrium? Perhaps an analogy will help.

Figure 11.25 shows two boxes, numbered 1 and 2, containing identical balls. Box 1 starts with more balls than box 2, so $N_{1i} > N_{2i}$. Once every second, one ball is chosen at random and moved to the other box. This is a reversible process because a ball can move from box 2 to box 1 just as easily as from box 1 to box 2. What do you expect to see if you return several hours later?

Because balls are chosen at random, and because $N_{1i} > N_{2i}$, it's initially more likely that a ball will move from box 1 to box 2 than from box 2 to box 1. Sometimes a ball will move "backward" from box 2 to box 1, but overall there's a net movement of balls from box 1 to box 2. The system will evolve until $N_1 \approx N_2$. This is a stable situation—equilibrium!—with an equal number of balls moving in both directions.

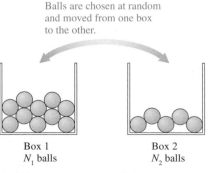

Balls are chosen at random and moved from one box to the other.

Box 1
N_1 balls

Box 2
N_2 balls

FIGURE 11.25 Moving balls between boxes.

But couldn't it go the other way, with N_1 getting even larger while N_2 decreases? In principle, any arrangement of the balls is possible. But certain arrangements are more likely. Each ball is equally likely to be in either box. With four balls, the odds are 1 in 2^4 or 1 in 16 that, at a randomly chosen instant of time, you would find all the balls in box 1. Were you to do so, you wouldn't find that to be terribly surprising.

With 10 balls, the probability that all of the balls are in the left box is about 1 in 1000. With 100 balls, the probability has dropped to about 1 in 1,000,000,000,000,000,000,000,000,000,000, or 1 in 10^{30}. The larger the number of balls, the more unlikely these extreme arrangements become. For the case of Figure 11.25, though each transfer is reversible, **the statistics of large numbers make it overwhelmingly likely that the system will evolve toward a state in which $N_1 \approx N_2$.** For the 10^{23} or so particles in a realistic system of atoms or molecules, the numbers are so astronomical that we will never see a state that deviates appreciably from the equilibrium case.

In Section 11.5 we considered the exchange of energy between two systems via collisions. Each collision is reversible, just as likely to transfer energy from 1 to 2 as from 2 to 1. But it is overwhelmingly likely that the net result of many, many collisions will be to transfer energy from system 1 to system 2 until the two temperatures are the same—in other words, for heat energy to be transferred from hot to cold.

A system reaches thermal equilibrium not by any plan or by outside intervention, but simply because **equilibrium is the most probable state in which to be.** It is possible that the system will move away from equilibrium, with heat moving from cold to hot, but this is remotely improbable in any realistic system. The consequence of having a vast number of random events is that the system evolves in one direction, toward equilibrium, and not the other. Reversible microscopic events lead to irreversible macroscopic behavior because some macroscopic states are vastly more probable than others.

Order, Disorder, and Entropy

Figure 11.26 shows microscopic views of three containers of gas. The top diagram shows a group of atoms arranged in a very ordered pattern. This is a highly ordered and nonrandom system, with each atom's position precisely specified. Contrast this with the system on the bottom, in which there is no order at all. It is extremely improbable that the atoms in a container would *spontaneously* arrange themselves into the ordered pattern of the top picture; even a small change in the pattern is quite noticeable. By contrast, there are a vast number of arrangements like the one on the bottom that randomly fill the container. A small change in the pattern is hard to detect.

Scientists and engineers use the term **entropy** to quantify the probability that a certain state of a system will occur. The ordered arrangement of the top system, which has a very small probability of spontaneous occurrence, has a very low entropy. The entropy of the randomly filled container is high; it has a large probability of occurrence. The entropy of the middle picture is somewhere in between. It is often said that entropy measures the amount of *disorder* in a system. The entropy in Figure 11.26 increases as you move from the ordered system on the top to the disordered system on the bottom.

Suppose you ordered the atoms in Figure 11.26 as they are in the top diagram, and then you let them go. After you release them, they will move, as gas atoms do, in a random manner. After some time, you would expect their arrangement to appear as in the bottom diagram. In fact, the series of diagrams from top to bottom could be thought of as a series of movie frames showing the evolution of the

The arrangement is unlikely; all of the particles must be precisely arranged. The movement of one particle is easy to notice. ⋯

A progression of states of a system of particles. The top state is ordered; the bottom is not.

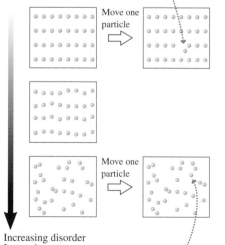

Increasing disorder
Increasing entropy
Increasing probability

The movement of one particle is hard to spot. Many arrangements have a similar appearance, so an arrangement like this is quite likely.

FIGURE 11.26 Ordered and disordered arrangements of atoms in a gas.

positions of the atoms after they are released. The system of atoms would be spontaneously moving toward a state of higher entropy. You would correctly expect this process to be irreversible. That is, the particles will never spontaneously recreate their initial ordered state.

Two thermally interacting systems with different temperatures have a low entropy. These systems are ordered in the sense that the faster atoms are on one side of the barrier, the slower atoms on the other. The most random possible distribution of energy, and hence the least ordered system, corresponds to the situation where the two systems are in thermal equilibrium with equal temperatures. **Entropy increases as two systems with initially different temperatures move toward thermal equilibrium.** Entropy would decrease if heat energy moved from cold to hot, making the hot system hotter and the cold system colder.

The Second Law of Thermodynamics

The fact that macroscopic systems evolve irreversibly toward equilibrium is a statement about nature that is not contained in any of the laws of physics we have encountered. It is a new law of physics, one known as the **second law of thermodynamics.**

The formal statement of the second law of thermodynamics is given in terms of entropy:

> **Second law of thermodynamics** The entropy of an isolated system never decreases. The entropy either increases, until the system reaches equilibrium, or, if the system began in equilibrium, stays the same.

NOTE ▶ The qualifier "isolated" is crucial. We can order the system by reaching in from the outside, perhaps using little tweezers to place atoms in a lattice. Similarly, we can transfer heat from cold to hot by using a refrigerator. The second law is about what a system can or cannot do *spontaneously,* on its own, without outside intervention. ◀

The second law of thermodynamics tells us that an isolated system evolves such that

- Order turns into disorder and randomness.
- Information is lost rather than gained.
- The system "runs down" as other forms of energy are transformed to thermal energy.

An isolated system never spontaneously generates order out of randomness. It is not that the system "knows" about order or randomness, but rather that there are vastly more states corresponding to randomness than there are corresponding to order. As collisions occur at the microscopic level, the laws of probability dictate that the system will, on average, move inexorably toward the most probable and thus most random macroscopic state.

Entropy and Thermal Energy

Suppose we have a very cold, moving baseball, with essentially no thermal energy, and a stationary helium balloon at room temperature, as shown in Figure 11.27. The atoms in the baseball and the atoms in the balloon are all moving, but there is a big difference in their motions. The atoms in the baseball are all moving in the same direction at the same speed, but the atoms in the balloon are moving in random directions. The ordered, organized motion of the baseball has

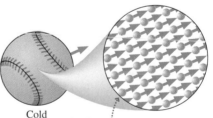

Cold baseball   The molecules in the baseball all move in the same direction at the same speed. This ordered motion is the ball's kinetic energy.

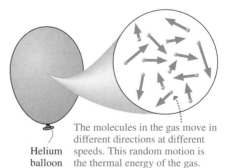

Helium balloon   The molecules in the gas move in different directions at different speeds. This random motion is the thermal energy of the gas.

FIGURE 11.27 Kinetic energy and thermal energy compared.

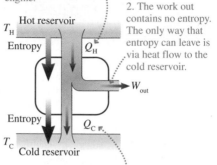

1. Heat energy Q_H comes into the system from the hot reservoir. Entropy is added to the heat engine.

2. The work out contains no entropy. The only way that entropy can leave is via heat flow to the cold reservoir.

3. Heat energy Q_C goes out of the heat engine into the cold reservoir. Entropy is removed from the heat engine.

FIGURE 11.28 A diagram of a heat engine, with entropy changes noted.

low entropy while the disorganized, random motions of the gas atoms—what we have called thermal energy—has high entropy. You can see that a conversion of macroscopic kinetic energy to thermal energy means an increase in entropy.

Thermal energy has entropy; other forms of energy do not. We saw, in section 11.5, that the conversion of other forms of energy to thermal energy was irreversible. Now, we can explain why: **When another form of energy is converted to thermal energy, there is an increase in entropy. The process will not spontaneously reverse.**

This also explains why it is hard to make a heat engine, a device that turns thermal energy into other forms of energy. We have redrawn our schematic diagram of a heat engine in Figure 11.28, adding arrows representing entropy. Now we can see why heat must be exhausted into the cold reservoir: We need to get rid of entropy! Each unit of entropy to be exhausted requires heat to carry it out, and this limits the efficiency of the engine. Some heat must be "wasted." The best we can do is to exhaust as little heat as we can while getting rid of entropy. Although the full details are beyond the scope of this book, the maximum efficiencies given in equations 11.16, 11.17, and 11.18 can be derived by assuming that we exhaust as little heat as possible to get rid of entropy. Efficiencies or coefficients of performance higher than these would violate the second law of thermodynamics.

Alternative Statements of the Second Law of Thermodynamics

Our formal statement of the second law dealt with entropy. The fact that the entropy of an isolated system always increases has other consequences. Some of these consequences can, in fact, be used as alternative, informal statements of the second law of thermodynamics. We will make a table of these alternative statements as a way of summarizing what we have learned about thermodynamics.

TABLE 11.5 Alternative statements of the second law of thermodynamics

Phenomenon	Related Statement of the Second Law
Heat energy spontaneously flows from hot to cold.	When two systems at different temperatures interact, heat energy is transferred spontaneously from the hotter to the colder system, never from the colder to the hotter.
Entropy considerations limit the possible efficiency of a heat engine.	It is not possible to make a heat engine that converts thermal energy into an equivalent amount of work.
It takes energy to move heat from a cold object to a warm object.	It is not possible to make a heat pump that moves heat from a cold object to a hot object without an external energy input.
The entropy of an isolated system will never spontaneously decrease.	The time direction in which the entropy of an isolated system increases is "the future."

The last of the statements of the second law is perhaps the most compelling. The irreversible evolution of systems from less-likely states to more-likely states is what gives us a direction of time. As we noted, stirring blends your coffee and cream, it never unmixes them. Friction causes an object to stop while increasing its thermal energy; the random atomic motions of thermal energy never spontaneously organize themselves into a macroscopic motion of an entire object. A plant in a sealed, darkened jar dies and decomposes to carbon and various gases; the gases and carbon never spontaneously assemble themselves into a flower. These are all examples of irreversible processes. They each show a clear direction of time, a distinct difference between past and future.

Which of the following processes does not involve a change in entropy?

A. An electric heater raises the temperature of a cup of water by 20°C.
B. A ball rolls up a ramp, decreasing in speed as it rolls higher.
C. A basketball is dropped from 2 m and bounces until it comes to rest.
D. The sun shines on a black surface and warms it.

11.9 Systems, Energy, and Entropy

Over the past two chapters, we have learned a good deal about energy and how it is used. By introducing the concept of entropy, we were able to see how limits on our ability to use energy come about. We will close this chapter, and this part of the book, by considering a few final questions that bring these different pieces together.

The Conservation of Energy and Energy Conservation

We have all heard, for many years, that it is important to "conserve energy." We are asked to turn off lights when we leave rooms, to drive our cars less, to turn down our thermostats. But this brings up an interesting question: If we have a law of conservation of energy, which states that energy can't be created or destroyed, what do we really mean by "conserving energy?" If energy can't be created or destroyed, how can there be an "energy crisis"?

In fact, we aren't, as a society or as a planet, running out of energy. We can't! What we *can* run out of is high-quality sources of energy, forms of energy that can easily be converted into other forms of energy. Oil is a good example. Oil contains chemical energy, in large quantity. It is a liquid, so is easily transported, and it is easily burned in a host of devices to generate heat, electricity, or motion. When we burn oil, we don't use up its energy—we simply convert its chemical energy into thermal energy. As we do this, we decrease the amount of high-quality chemical energy in the world and increase the supply of thermal energy. And, as we have seen, thermal energy is less useful, as it is difficult to convert it to other forms.

Entropy and Life

The second law predicts that systems will run down, that order will evolve toward disorder and randomness, and that complexity will give way to simplicity. But just look around you!

- Plants grow from simple seeds to complex entities.
- Single-celled fertilized eggs grow into complex adult organisms.
- Over the last billion years or so, life has evolved from simple unicellular organisms to very complex forms.

Everywhere we look, it seems, the second law is being violated. How can this be?

There is an important qualification in the second law of thermodynamics: It applies only to *isolated systems,* systems that do not exchange energy with their environment. The situation is entirely different if energy is transferred into or out of the system.

Your body is not an isolated system. Every day, you take in chemical energy in the food you eat. As you use this energy, most of it ends up as thermal energy that you exhaust as heat into the environment—taking away the entropy that natural processes in your body creates. An energy and entropy diagram of this situation is given in Figure 11.29. It is this continual exchange of energy (and entropy) with the environment that makes your life—and all life—possible.

Sealed, but not isolated BIO This glass container is a completely sealed system containing living organisms, shrimp and algae. But the organisms will live and grow for many years. The reason this is possible is that the glass sphere, though sealed, is not an *isolated* system. Energy can be transferred in and out as light and heat. If the container were to be placed in a darkened room, the organisms would quickly perish.

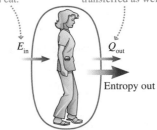

Chemical energy comes into your body in the food you eat.

Energy leaves your body mostly as heat, meaning entropy is transferred as well.

E_{in} Q_{out}

Entropy out

FIGURE 11.29 Thermodynamic view of the body.

SUMMARY

The goal of Chapter 11 has been to learn more about energy transformations and transfers, the laws of thermodynamics, and theoretical and practical limitations on energy use.

GENERAL PRINCIPLES

Energy and Efficiency

When energy is transformed from one form to another, some may be "lost," usually to thermal energy, due to practical or theoretical constraints. This limits the efficiency of processes. We define **efficiency** as:

$$e = \frac{\text{what you get}}{\text{what you had to pay}}$$

Entropy and Irreversibility

Systems move toward more probable states. These states have higher entropy—more disorder. This change is irreversible. Changing other forms of energy to thermal energy is irreversible.

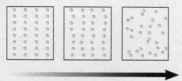

Increasing probability
Increasing entropy

The Laws of Thermodynamics

The **first law of thermodynamics** is the law of conservation of energy for systems in which only thermal energy changes:

$$\Delta E_{th} = W + Q$$

The **second law of thermodynamics** can be stated in a few different ways:

- The entropy of an isolated system always increases.
- Heat energy spontaneously flows only from hot to cold.
- No heat engine can be 100% efficient.
- "The future" is the time direction of entropy increase.

Heat is the transfer of energy via a thermal interaction. Energy will be transferred until thermal equilibrium is reached.

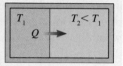

IMPORTANT CONCEPTS

Thermal energy
- For a gas, the thermal energy is the **total kinetic energy** of motion of the atoms.

$$E_{th} = NK_{avg} = \frac{1}{2}Nmv_{rms}^2$$

- Thermal energy is random kinetic energy and thus has entropy.

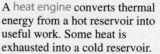

Temperature
- For a gas, temperature is proportional to the **average kinetic energy** of the motion of the atoms.

$$T = \frac{2}{3}\frac{K_{avg}}{k_B}$$

- Two systems are in **thermal equilibrium** if they are at the same temperature. No heat energy is transferred at thermal equilibrium.

A heat engine converts thermal energy from a hot reservoir into useful work. Some heat is exhausted into a cold reservoir.

$$e_{max} = 1 - \frac{T_C}{T_H}$$

A heat pump uses an energy input to transfer heat from a cold side to a hot side. When used for cooling:

$$COP_{max} = \frac{T_C}{T_H - T_C}$$

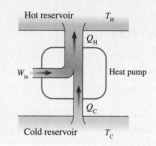

APPLICATIONS

Efficiencies

Energy in the body
Cells in the body metabolize chemical energy in food. Efficiency for most actions is about 25%.

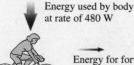

Energy used by body at rate of 480 W

Energy for forward propulsion at rate of 120 W

Power plants
A typical power plant converts about 1/3 of the energy input into useful work. The rest is exhausted as waste heat.

Chemical energy in

Waste heat

Useful work out

Temperature scales

Zero on the **Kelvin temperature scale** is the temperature at which the kinetic energy of atoms is zero. This is **absolute zero**. The conversion from °C to K is

$$T = T_C + 273$$

▶ All temperatures in equations must be in kelvin. ◀

MP For instructor-assigned homework, go to www.masteringphysics.com

Problem difficulty is labeled as | (straightforward) to ||||| (challenging).

Problems labeled can be done on a Workbook Dynamics Worksheet; INT integrate significant material from earlier chapters; BIO are of biological or medical interest.

QUESTIONS

Conceptual Questions

1. Rub your hands together vigorously. What happens? Discuss the energy transfers and transformations that take place.
2. Write a few sentences describing the energy transformations that occur from the time moving water enters a hydroelectric plant until you see some water being pumped out of a nozzle in a public fountain. Use the "Energy transformations" table on page 40 as an example.
3. BIO Describe the energy transfers and transformations that occur from the time you sit down to breakfast until you've completed a fast bicycle ride.
4. BIO According to Table 11.4, cycling at 15 km/hr requires less metabolic energy than running at 15 km/hr. Suggest reasons why this is the case.
5. For most automobiles, the number of miles per gallon decreases as highway speed increases. Fuel economy drops as speeds increase from 55 to 65 mph, then decreases further as speeds increase to 75 mph. Explain why this is the case.
6. When the space shuttle returns to earth, its surfaces get very hot as it passes through the atmosphere at high speed.
   a. Has the space shuttle been heated? If so, what was the source of the heat? If not, why is it hot?
   b. Energy must be conserved. What happens to the space shuttle's initial kinetic energy?
7. One end of a short aluminum rod is in a campfire and the other end is in a block of ice, as shown in Figure Q11.7. If 100 J of energy are transferred from the fire to the rod, and if the temperature at every point in the rod has reached a steady value, how much energy goes from the rod into the ice?

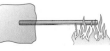

FIGURE Q11.7

8. Two blocks of copper, 1 kg and 3 kg, are at the same temperature. Which block has more thermal energy? If the blocks are placed in thermal contact, will the thermal energy of the blocks change? If so, how?
9. If the temperature T of an ideal gas doubles, by what factor does the average kinetic energy of the atoms change?
10. A bottle of helium gas and a bottle of argon gas contain equal numbers of atoms at the same temperature. Which bottle, if either, has the greater total thermal energy?

For Questions 11 through 16, give a specific example of a process that has the energy changes and transfers described. (For example, if the question states "$\Delta E_{th} > 0, W = 0$," you are to describe a process that has an increase in thermal energy and no transfer of energy by work. You could write "Heating a pan of water on the stove.")

11. $\Delta E_{th} < 0, W = 0$
12. $\Delta E_{th} > 0, Q = 0$
13. $\Delta E_{th} < 0, Q = 0$
14. $\Delta E_{th} > 0, W \neq 0, Q \neq 0$
15. $\Delta E_{th} < 0, W \neq 0, Q \neq 0$
16. $\Delta E_{th} = 0, W \neq 0, Q \neq 0$
17. A drop of green ink falls into a beaker of clear water. First *describe* what happens. Then *explain* the outcome in terms of entropy.
18. If you hold a rubber band loosely between two fingers and let it rest against your lips, and then stretch the rubber band, the sensitive skin of your lips will detect a small temperature increase. The rubber band soon returns to its original temperature. What are the signs of W and Q for this process?
19. The exterior unit of a heat pump designed for heating is sometimes buried underground in order to use the earth as a thermal reservoir. Why is it worthwhile to bury the heat exchanger, even if the underground unit costs more to purchase and install than one above ground?
20. Assuming improved materials and better processes, can engineers ever design a heat engine that exceeds the maximum efficiency indicated by Equation 11.16? If not, why not?

Multiple-Choice Questions

21. | A person is walking on level ground at constant speed. What energy transformation is taking place?
   A. Chemical energy is being transformed to thermal energy.
   B. Chemical energy is being transformed to kinetic energy.
   C. Chemical energy is being transformed to kinetic energy and thermal energy.
   D. Chemical energy and thermal energy are being transformed to kinetic energy.
22. || A person walks 1 km, turns around, and runs back to where he started. Compare the energy used and the power during the two segments:
   A. The energy used and the power are the same for both segments.
   B. The energy used while walking is greater, the power while running is greater.
   C. The energy used while running is greater, the power while running is greater.
   D. The energy used is the same for both segments, the power while running is greater.
23. | The temperature of the air in a basketball increases as it is pumped up. This means that:
   A. The total kinetic energy of the air is increasing and the average kinetic energy of each atom is decreasing.
   B. The total kinetic energy of the air is increasing and the average kinetic energy of each atom is increasing.
   C. The total kinetic energy of the air is decreasing and the average kinetic energy of each atom is decreasing.
   D. The total kinetic energy of the air is decreasing and the average kinetic energy of each atom is increasing.

24. | The thermal energy of a container of helium gas is halved. What happens to the temperature, in kelvin?
    A. It decreases to one-fourth its initial value.
    B. It decreases to one-half its initial value.
    C. It stays the same.
    D. It increases to twice its initial value.

25. | An inventor approaches you with a device that he claims will take 100 J of thermal energy input and produce 200 J of electricity. You decide not to invest your money because this device would violate
    A. the first law of thermodynamics.
    B. the second law of thermodynamics.
    C. both the first and second laws of thermodynamics.

26. | While keeping your food cold, your refrigerator transfers energy from the inside to the surroundings. Thus thermal energy goes from a colder object to a warmer one. What can you say about this?
    A. It is a violation of the second law of thermodynamics.
    B. It is not a violation of the second law of thermodynamics because refrigerators can have efficiency of 100%.
    C. It is not a violation of the second law of thermodynamics because the second law doesn't apply to refrigerators.
    D. The second law of thermodynamics applies in this situation, but it is not violated because the energy did not spontaneously go from cold to hot.

PROBLEMS

Section 11.1 Transforming Energy

1. | A 20%-efficient engine accelerates a 1500 kg car from rest to 15 m/s. How much energy is transferred to the engine by burning gasoline?

2. | A 60% efficient device uses chemical energy to generate 600 J of electric energy.
   a. How much chemical energy is used?
   b. A second device uses twice as much chemical energy to generate half as much electric energy? What is its efficiency?

3. | A typical photovoltaic cell delivers 4.0×10^{-3} W of electric energy when illuminated with 1.2×10^{-1} W of light energy. What is the efficiency of the cell?

4. | A 15 W compact fluorescent bulb and a 75 W incandescent bulb each produce 3.0 W of visible light energy. What are the efficiencies of these two types of bulbs for converting electric energy into light?

FIGURE P11.4 Compact fluorescent and incandescent bulbs.

Section 11.2 Energy in the Body: Energy Inputs

5. ‖ A fast-food hamburger (with cheese and bacon) contains 1000 Calories. (a) What is the burger's energy in joules? (b) If all this energy is used to lift a 10 kg mass, how high can it be lifted?

6. ‖ In an average human, basic life processes require energy to BIO be supplied at a steady rate of 100 W. What daily energy intake, in Calories, is required to maintain these basic processes? This is the minimum daily caloric intake needed to avoid starvation.

7. | An "energy bar" contains 6.0 g of fat. How much energy is BIO this in joules? In calories? In Calories?

8. | An "energy bar" contains 22 g of carbohydrates. How BIO much energy is this in joules? In calories? In Calories?

Section 11.3 Energy in the Body: Energy Outputs

9. ‖ An "energy bar" contains 22 g of carbohydrates. If the BIO energy bar was his only fuel, how far could a 68 kg person walk at 5.0 km/hr?

10. ‖ Suppose your body was able to use the chemical energy in BIO gasoline. How far could you pedal a bicycle at 15 km/hr on the energy in 1 gal of gas? (1 gal of gas has a mass of 3.2 kg.)

11. ‖‖ The label on a candy bar says 400 Calories. Assuming a BIO typical efficiency for energy use by the body, if a 60 kg person were to use the energy in this candy bar to climb stairs, how high could she go?

12. ‖ A weightlifter curls a 30 kg bar, raising it each time a dis- BIO tance of 0.60 m. How many times must he repeat this exercise to burn off the energy in one slice of pizza?

13. ‖‖ A weightlifter works out at the gym each day. Part of her BIO routine is to lie on her back and lift a 40 kg barbell straight up from chest height to full arm extension, a distance of 0.50 m.
    a. How much work does the weightlifter do to lift the barbell one time?
    b. If the weightlifter does 20 repetitions a day, what total energy does she expend on lifting? Assume 25% efficiency.
    c. How many 400 Calorie donuts can she eat a day to supply that energy?

Section 11.4 Thermal Energy and Temperature

14. | What are the rms speeds of helium and argon atoms in a gas at 1000°C?

15. | 30 m/s is a typical highway speed for a car. At what temperature do the molecules of nitrogen gas have an rms speed of 30 m/s?

16. ‖ A gas consists of a mixture of neon and argon. The rms speed of the neon atoms is 400 m/s. What is the rms speed of the argon atoms?

17. ‖ At what temperature do hydrogen molecules have the same rms speed as nitrogen molecules at 100°C?

18. ‖ At what temperature, in °C, is the rms speed of oxygen molecules (a) half and (b) twice its value at 0°C?

Section 11.5 Heat and the First Law of Thermodynamics

19. | 500 J of work are done on a system in a process that decreases the system's thermal energy by 200 J. How much energy is transferred to or from the system as heat?

20. | 600 J of heat energy are transferred to a system that does 400 J of work. By how much does the system's thermal energy change?

21. | 300 J of energy are transferred to a system in the form of heat while the thermal energy increases by 150 J. How much work is done on or by the system?
22. | 10 J of heat are removed from a gas sample while it is being compressed by a piston that does 20 J of work. What is the change in the thermal energy of the gas? Does the temperature of the gas increase or decrease?

Section 11.6 Heat Engines

23. ⦀ A heat engine extracts 55 kJ from the hot reservoir and exhausts 40 kJ into the cold reservoir. What are (a) the work done and (b) the efficiency?
24. | A heat engine does 20 J of work while exhausting 30 J of waste heat. What is the engine's efficiency?
25. | A heat engine does 200 J of work while exhausting 600 J of heat to the cold reservoir. What is the engine's efficiency?
26. | A heat engine with an efficiency of 40% does 100 J of work. How much heat is (a) extracted from the hot reservoir and (b) exhausted into the cold reservoir?
27. ‖ a. At what cold-reservoir temperature (in °C) would an engine operating at maximum theoretical efficiency with a hot-reservoir temperature of 427°C have an efficiency of 60%?
   b. If another engine, operating at maximum theoretical efficiency with a hot-reservoir temperature of 400°C, has the same efficiency, what is its cold-reservoir temperature?
28. ‖ A heat engine operating between energy reservoirs at 20°C and 600°C has 30% of the maximum possible efficiency. How much energy does this engine extract from the hot reservoir to do 1000 J of work?

Section 11.7 Heat Pumps

29. | A refrigerator takes in 20 J of work and exhausts 50 J of heat. What is the refrigerator's coefficient of performance?
30. | Air conditioners are rated by their coefficient of performance at 80°F inside temperature and 95°F outside temperature. An efficient but realistic air conditioner has a coefficient of performance of 3.2. What is the maximum possible coefficient of performance?
31. | 50 J of work are done on a refrigerator with a coefficient of performance of 4.0. How much heat is (a) extracted from the cold reservoir and (b) exhausted to the hot reservoir?
32. ‖ Find the maximum possible coefficient of performance for a heat pump used to heat a house in a northerly climate in winter. The inside is kept at 20°C while the outside is −20°C.

Section 11.8 Entropy and the Second Law of Thermodynamics

33. ‖ Which, if any, of the heat engines in Figure P11.33 at right violate (a) the first law of thermodynamics or (b) the second law of thermodynamics? Explain.
34. ‖ Which, if any, of the refrigerators in Figure P11.34 at right violate (a) the first law of thermodynamics or (b) the second law of thermodynamics? Explain.
35. ‖ Draw all possible distinct arrangements in which three balls (labeled A, B, C) are placed into two different boxes (1 and 2), as in Figure 11.25. If all arrangements are equally likely, what is the probability that all three will be in box 1?

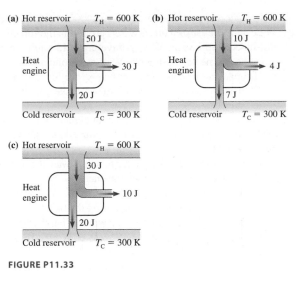

FIGURE P11.33

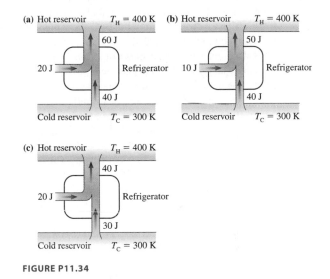

FIGURE P11.34

General Problems

36. ‖ How many slices of pizza must you eat to walk for 1.0 hr at
BIO a speed of 5.0 km/hr? (Assume your mass is 68 kg.)
37. | A 60 kg person climbs to the top of a 500-m-high hill. If the energy necessary to do this were provided by burning coal, how much coal would be needed?
38. ‖ For how long would a 68 kg person have to swim at a fast
BIO crawl to use all the energy available in a typical fast food meal of burger, fries, and a drink?
39. | a. How much metabolic energy is required for a 68 kg per-
BIO son to run at a speed of 15 km/hr for 20 min?
   b. How much metabolic energy is required for this person to walk at a speed of 5.0 km/hr for 60 min? Compare your result to your answer to part a.
   c. Compare your results of parts a and b to the result of Example 11.5. Of these three modes of human motion, which is the most efficient?

40. ||| To a good approximation, the only external force that does
BIO work on a cyclist moving on level ground is the force of air
resistance. Suppose a cyclist is traveling at 15 km/hr on level
ground. Assume he is using 480 W of metabolic power.
   a   Estimate the amount of power he uses for forward motion.
   b. How much force must he exert to overcome the force of air
       resistance?

41. || The winning time for the 2005 annual race up 86 floors of
BIO the Empire State Building was 10 min and 49 s. The winner's
INT mass was 60 kg.
   a. If each floor was 3.7 m high, what was the winner's
      change in gravitational potential energy?
   b. If the efficiency in climbing stairs is 25%, what total
      energy did the winner expend during the race?
   c. How many food Calories did the winner "burn" in the race?
   d. Of those Calories, how many were converted to thermal
      energy?
   e. What was the winner's metabolic power in watts during
      the race up the stairs?

42. ||| Championship swimmers take about 22 s and about 30 arm
BIO strokes to move through the water in a 50 m freestyle race.
INT a. From Table 12.4, a swimmer's metabolic power is 800 W.
      If the efficiency for swimming is 25%, how much energy
      is expended moving through the water in a 50 m race?
   b. If half the energy is used in arm motion and half in leg
      motion, what is the energy expenditure per arm stroke?
   c. Model the swimmer's hand as a paddle. During one arm
      stroke, the paddle moves halfway around a 90-cm-radius
      circle. If all the swimmer's forward propulsion during an
      arm stroke comes from the hand pushing on the water and
      none from the arm (somewhat of an oversimplification),
      what is the average force of the hand on the water?

43. | Helium remains very nearly an ideal gas to temperatures as
low as 5 K. What are the root-mean-square speed and the
average kinetic energy of helium atoms at this temperature?

44. | The temperature at the center of the sun is $\approx 2 \times 10^7$ K.
What are (a) the average kinetic energy and (b) the rms speed
of a proton in the center of the sun?

45. || From what height must an oxygen molecule fall from rest
in a vacuum so that its kinetic energy at the bottom equals the
INT average kinetic energy of an oxygen molecule at 300 K?

46. | The surface of the sun consists
mostly of hydrogen *atoms* (not molecules) at a temperature of 6000 K. What
are (a) the average translational kinetic
energy per atom and (b) the rms speed
of the atoms?

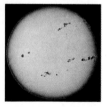

47. | What is the total translational kinetic
energy at 0°C of 1.0×10^{21} molecules
of (a) hydrogen and (b) oxygen?

48. || A container holding argon atoms changes temperature by
20°C when 30 J of heat are removed. How many atoms are in
the container?

49. | A heat engine with a high-temperature reservoir at 400 K
has an efficiency of 0.20. What is the maximum possible temperature of the cold reservoir?

50. || A heat engine does 10 J of work and exhausts 15 J of waste
heat.
   a. What is the engine's efficiency?
   b. If the cold-reservoir temperature is 20°C, what is the mini-
      mum possible temperature in °C of the hot reservoir?

51. || The heat exhausted to the cold reservoir of an engine oper-
ating at maximum theoretical efficiency is two-thirds the heat
extracted from the hot reservoir. What is the temperature ratio
T_C/T_H?

52. || An engine operating at maximum theoretical efficiency
whose cold-reservoir temperature is 7°C is 40% efficient. By
how much should the temperature of the hot reservoir be
increased to raise the efficiency to 60%?

53. | An ideal heat pump used for heating has a coefficient of
performance of 10 and a cold-reservoir temperature of 0°C.
What is the hot-reservoir temperature in °C?

54. || The coefficient of performance of a refrigerator is 5.0.
   a. If the compressor uses 10 J of energy, how much heat is
      exhausted to the hot reservoir?
   b. If the hot-reservoir temperature is 27°C, what is the lowest
      possible temperature in °C of the cold reservoir?

55. || An engineer claims to have measured the characteristics of
a heat engine that takes in 100 J of thermal energy and produces 50 J of useful work. Is this engine possible? If so, what
is the smallest possible ratio of the temperatures (in kelvin) of
the hot and cold reservoirs?

56. || A 32% efficient electric power plant produces 900 MJ of
electric energy per second and discharges waste heat into 20°C
ocean water. Suppose the waste heat could be used to heat
homes during the winter instead of being discharged into the
ocean. A typical American house requires an average 20 kW
for heating. How many homes could be heated with the waste
heat of this one power plant?

57. ||| A typical coal-fired power plant burns 300 metric tons of
coal *every hour* to generate 2.7×10^6 MJ of electric energy.
1 metric ton = 1000 kg; 1 metric ton of coal has a volume of
1.5 m³. The heat of combustion of coal is 28 MJ/kg. Assume
that *all* heat is transferred from the fuel to the boiler and that
all the work done in spinning the turbine is transformed into
electric energy.
   a. Suppose the coal is piled up in a 10 m × 10 m room. How
      tall must the pile be to operate the plant for one day?
   b. What is the power plant's efficiency?

58. || Each second, a nuclear power plant generates 2000 MJ of
thermal energy from nuclear reactions in the reactor's core.
This energy is used to boil water and produce high-pressure
steam at 300°C. The steam spins a turbine, which produces
700 MJ of electric power, then the steam is condensed and
the water is cooled to 30°C before starting the cycle again.
   a. What is the maximum possible efficiency of the plant?
   b. What is the plant's actual efficiency?

59. | The surface waters of tropical oceans are at a temperature
of 27°C while water at a depth of 1200 m is at 3°C. It has been
suggested these warm and cold waters could be the heat reservoirs for a heat engine, allowing us to do work or generate
electricity from the thermal energy of the ocean. What is the
maximum efficiency possible of such a heat engine?

60. || The light energy that falls on a square meter of ground over
the course of a typical sunny day is about 20 MJ. The average
rate of electric energy consumption in one house is 1.0 kW.
   a. On average, how much energy does one house use during
      each 24 hr day?
   b. If light energy to electric energy conversion using solar
      cells is 5% efficient, how many square miles of land must
      be covered with solar cells to supply the electrical energy
      for 250,000 houses? Assume there is no cloud cover.

Passage Problems

Kangaroo Locomotion BIO

Kangaroos have very stout tendons in their legs that can be used to store energy. When a kangaroo lands on its feet, the tendons stretch, transforming kinetic energy of motion to elastic potential energy. Much of this energy can be transformed  back into kinetic energy as the kangaroo takes another hop. The kangaroo's peculiar hopping gait is not very efficient at low speeds but is quite efficient at high speeds.

Figure P11.61 shows the energy cost of human and kangaroo locomotion. The graph shows oxygen uptake (in mL/s) per kg of body mass, allowing a direct comparison between the two species.

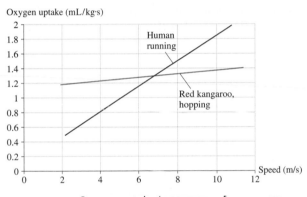

FIGURE P11.61 Oxygen uptake (a measure of energy use per second) for a running human and a hopping kangaroo.

For humans, the energy used per second (i.e., power) is proportional to the speed. That is, the human curve nearly passes through the origin, so running twice as fast takes approximately twice as much power. For a hopping kangaroo, the graph of energy use has only a very small slope. In other words, the energy used per second changes very little with speed. Going faster requires very little additional power. Treadmill tests on kangaroos and observations in the wild have shown that they do not become winded at any speed at which they are able to hop. No matter how fast they hop, the necessary power is approximately the same.

61. ‖ A person runs 1 km. How does his speed affect the total energy needed to cover this distance?
    A. A faster speed requires less total energy.
    B. A faster speed requires more total energy.
    C. The total energy is about the same for a fast speed and a slow speed.

62. ‖ A kangaroo hops 1 km. How does its speed affect the total energy needed to cover this distance?
    A. A faster speed requires less total energy.
    B. A faster speed requires more total energy.
    C. The total energy is about the same for a fast speed and a slow speed.

63. ‖ At a speed of 4 m/s,
    A. A running human is more efficient than an equal-mass hopping kangaroo.
    B. A running human is less efficient than an equal-mass hopping kangaroo.
    C. A running human and an equal-mass hopping kangaroo have about the same efficiency.

64. ‖ At approximately what speed would a human use half the power of an equal-mass kangaroo moving at the same speed?
    A. 3 m/s   B. 4 m/s   C. 5 m/s   D. 6 m/s

65. ‖ At approximately what speed would a human use twice the power of a kangaroo of half the mass moving at the same speed?
    A. 3 m/s   B. 5 m/s   C. 7 m/s   D. 9 m/s

STOP TO THINK ANSWERS

Stop to Think 11.1: C. In each case, "what you get" is the same because the potential energy change is the same; "What you had to pay" is the same as well.

Stop to Think 11.2: C. As the body uses chemical energy from food, approximately 75% is transformed to thermal energy. Also, kinetic energy of motion of the legs and feet is transformed to thermal energy with each stride. Most of the chemical energy is transformed to thermal energy.

Stop to Think 11.3: A. The temperatures are the same, so all atoms have the same average kinetic energy. Hydrogen molecules, with the smallest mass, with thus have the highest typical speed.

Stop to Think 11.4: C. The radiator is at a higher temperature than the surrounding air. Thermal energy is transferred out of the system to the environment, so $Q < 0$.

Stop to Think 11.5: A, D. The efficiency is fixed by the ratio of T_H to T_C. Increasing this ratio increases efficiency; the heat engine will be more efficient with a hotter hot reservoir or a colder cold reservoir.

Stop to Think 11.6: A, D. The closer the temperatures of the hot and cold reservoirs, the more efficient the heat pump can be. (It is also true that having the two temperatures be closer will cause less thermal energy to "leak" out.) Any change that makes the two temperatures closer will allow the refrigerator to use less energy to run.

Stop to Think 11.7: B. In this case, kinetic energy is transformed to potential energy; there is no entropy change. In the other cases, energy is transformed to thermal energy, meaning entropy increases.

Conservation Laws

In Part II we have discovered that we don't need to know all the details of an interaction to relate the properties of a system "before" the interaction to those "after" the interaction. We also found two important quantities, momentum and energy, that are often conserved. Momentum and energy are characteristics of a system.

Momentum and energy have conditions under which they are conserved. The total momentum $\vec{P}$ and the total energy E are conserved for an *isolated system*. Of course, not all systems are isolated. For both momentum and energy, it was useful to develop a *model* of a system interacting with its environment. Interactions within the system do not change $\vec{P}$ or E. The kinetic, potential, and thermal energy *within* the system can be transformed without changing E. Interactions between the system and the environment *do* change the system's momentum and energy. In particular,

- Impulse is the transfer of momentum to or from the system: $\Delta \vec{p} = \vec{J}$.

- Work is the transfer of energy to or from the system in a mechanical interaction: $\Delta E = W$.
- Heat is the energy transferred to or from the system in a thermal interaction: $\Delta E = Q$.

The laws of conservation of momentum and energy, when coupled with the laws of Newtonian mechanics of Part I, form a powerful set of tools for analyzing motion. But energy is a concept that can be used for much more than the study of motion; it can be used to analyze how your body uses food, how the sun shines, and a host of other problems.

The study of the transformation of energy from one form to another reveals certain limits. Thermal energy is different from other forms of energy. A transformation of energy from another form to thermal energy is *irreversible*. The study of heat and thermal energy thus led us to the discipline of thermodynamics, a subject we will take up further in Part III as we look at properties of matter.

KNOWLEDGE STRUCTURE II Conservation Laws

BASIC GOALS	How is the system "after" an interaction related to the system "before"? What quantities are conserved, and under what conditions? Why are some energy changes more efficient than others?
GENERAL PRINCIPLES	**Law of Conservation of Momentum** For an isolated system, $\vec{P}_f = \vec{P}_i$ **Law of Conservation of Energy** For an isolated system, there is no change in the system's energy: $$\Delta K + \Delta U_g + \Delta U_s + \Delta E_{th} + \Delta E_{chem} + \cdots = 0$$ Energy can be exchanged with the environment as work or heat: $$\Delta K + \Delta U_g + \Delta U_s + \Delta E_{th} + \Delta E_{chem} + \cdots = W + Q$$ **Laws of Thermodynamics** First law: If only thermal energy changes, $\Delta E_{th} = W + Q$ Second law: The entropy of an isolated system always increases.
BASIC PROBLEM-SOLVING STRATEGY	Draw a visual overview for the system "before" and "after", then use the conservation of momentum or energy equations to relate the two. If necessary, calculate impulse and/or work.

Momentum and impulse

In a collision, the total momentum

$$\vec{P} = \vec{p}_1 + \vec{p}_2 = m_1\vec{v}_1 + m_2\vec{v}_2$$

before and after is equal.

Before: m_1 ①$\xrightarrow{(v_{1x})_i}$ $\xleftarrow{(v_{2x})_i}$ ② m_2

After: $\xleftarrow{(v_{1x})_f}$ ① ② $\xrightarrow{(v_{2x})_f}$

A force can change the momentum of an object. The change is the **impulse:** $\Delta p_x = J_x$

$J_x =$ area under force curve

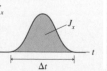

Basic model of energy

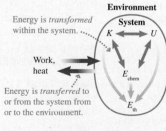

Energy is *transformed* within the system.

Work, heat

Energy is *transferred* to or from the system from or to the environment.

Work $W = F_{\parallel}d$ is done by the component of a force parallel to a displacement.

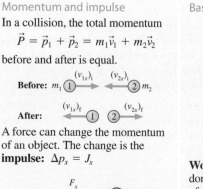

$F_{\parallel} = F\cos\theta$

Limitations on energy transfers and transformations

Thermal energy is random kinetic energy. Changing other forms of energy to thermal energy is **irreversible.**

When transforming energy from one form to another, some may be "lost" as thermal energy. This limits efficiency:

Efficiency: $e = \dfrac{\text{what you get}}{\text{what you had to pay}}$

A heat engine can convert thermal energy to other forms. The efficiency must be less than 100%.

Order out of Chaos

The second law of thermodynamics specifies that "the future" is the direction of entropy increase. But, as we have seen, this doesn't mean that systems must invariably become more random. You don't need to look far to find examples of systems that spontaneously evolve to a state of greater order.

A snowflake is a perfect example. As water freezes, the random motion of water molecules is transformed into the orderly arrangement of a crystal. The entropy of the snowflake is less than that of the water vapor from which it formed. Has the second law of thermodynamics been turned on its head?

The entropy of the water molecules in the snowflake certainly decreases, but the water doesn't freeze as an isolated system. For it to freeze, heat energy must be transferred from the water to the surrounding air. The entropy of the air increases by *more* than the entropy of the water decreases. Thus the *total* entropy of the water + air system increases when a snowflake is formed, just as the second law predicts. If the system isn't isolated, its entropy can decrease without violating the second law as long as the entropy increases somewhere else.

Systems that become *more* ordered as time passes, and in which the entropy decreases, are called *self-organizing systems*. These systems can't be isolated. It is common in self-organizing systems to find a substantial flow of energy *through* the system. Your body takes in chemical energy from food, makes use of that energy, and then gives waste heat back to the environment. It is this energy flow that allows systems to develop a high degree of order and a very low entropy. The entropy of the environment undergoes a significant *increase* so as to let selected subsystems decrease their entropy and become more ordered.

Self-organizing systems don't violate the second law of thermodynamics, but this fact doesn't really explain their existence. If you toss a coin, no law of physics says that you can't get heads 100 times in a row—but you don't expect this to happen. Can we show that self-organization isn't just possible, but likely?

Let's look at a simple example. Suppose you heat a shallow dish of oil at the bottom, while holding the temperature of the top constant. When the temperature difference between the top and the bottom of the dish is small, heat is transferred from the bottom to the top by conduction. But convection begins when the temperature difference becomes large enough. The pattern of convection needn't be random, though; it can develop in a stable, highly ordered pattern, as we see in the figure. Convection is a much more efficient means of transferring energy than conduction, so the rate of transfer is *increased* as a result of the development of these ordered *convection cells*.

The development of the convection cells is an example of self-organization. The roughly 10^{23} molecules in the fluid had been moving randomly but now have begun behaving in a very orderly fashion. But there is more to the story. The convection cells transfer energy from the hot lower side of the dish to the cold upper side. This hot-to-cold energy transfer increases the entropy of the surrounding environment, as we have seen. In becoming more organized, the system has become more effective at transferring heat, resulting in a greater rate of entropy increase! Order has arisen out of disorder in the system, but the net result is a more rapid increase of the disorder of the universe.

Convection cells are thus a thermodynamically favorable form of order. We should expect this, because convection cells aren't confined to the laboratory. We see them in the sun, where they transfer energy from lower levels to the surface, and in the atmosphere of the earth, where they give rise to some of our most dramatic weather.

Self-organizing systems are a very active field of research in physical and biological sciences. The 1977 Nobel Prize in chemistry was awarded to the Belgian scientist Ilya Prigogine for his studies of *nonequilibrium thermodynamics,* the basic science underlying self-organizing systems. Prigogine and others have shown how energy flow through a system can, when the conditions are right, "bring order out of chaos." And this spontaneous ordering is not just possible—it can be probable. The existence and evolution of self-organizing systems, from thunderstorms to life on earth, might just be nature's preferred way of increasing entropy in the universe.

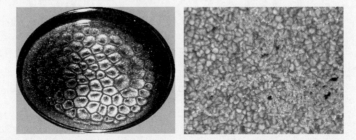

Convection cells in a shallow dish of oil heated from below (left) and in the sun (right). In both, warmer fluid is rising (lighter color) and cooler fluid is sinking (darker color).

PROPERTIES OF MATTER

Beyond the Particle Model

The first 11 chapters of this book have made extensive use of the *particle model* in which we represent objects as point masses. The particle model is especially useful for describing how discrete objects move through space and how they interact with each other. Whether a ball is made of metal or wood is irrelevant to calculating its trajectory.

But there are many situations where the distinction between metal and wood is crucial. If you toss a metal ball and a wood ball into a pond, one sinks and the other floats. If you stir a pan on the stove with a metal spoon, it will quickly get too hot to hold unless it has a wooden handle.

Wood and metal have different physical properties. So do air and water. Our goal in Part III is to describe and understand the similarities and differences of different materials. To do so, we must go beyond the particle model and dig deeper into the nature of matter.

Macroscopic Physics

In Part III, we will be concerned with systems that are solids, liquids, or gases. Although it is sometimes useful to think of these systems as collections of particle-like atoms, properties such as pressure, temperature, specific heat, and viscosity are characteristics of the system as a whole, not of individual particles. Solids, liquids, and gases are often called *macroscopic* systems, the prefix *macro* (the opposite of *micro*) meaning "large."

The questions we will be asking are different from those we asked about particles. For example,

- How do the temperature and pressure of a system change if you heat it? Why do some materials respond quickly, others slowly?
- What are the mechanisms by which a system exchanges heat energy with its environment? Why does blowing on a cup of hot coffee cause it to cool off?
- Why are there three phases of matter—solids, liquids, gases? What happens during a phase change?
- Why do some objects float while others, with the same mass, sink? What keeps a massive steel ship afloat?
- What are the laws of motion of a flowing liquid? How do they differ from the laws governing the motion of a particle?

Both Newton's laws and the law of conservation of energy will remain important tools—they are, after all, the basic laws of physics—but we'll have to learn how they apply to macroscopic systems.

It should come as no surprise that an understanding of macroscopic systems and their properties is essential for understanding the world around us. Biological systems, from cells to ecosystems, are macroscopic systems exchanging energy with their environment. On a larger scale, energy transport on earth gives us weather, and the special physical properties of water help explain why the earth is hospitable for life.

Individual bees do not have the ability to regulate their body temperature. But a colony of bees, working together, can very precisely regulate the temperature of their hive. How can the bees use the heat generated by their muscles, the structure of the hive, and the evaporation of water to achieve this control? In Part III, we will learn about the flows of matter and energy that drive such processes.

12

THERMAL PROPERTIES OF MATTER

This image of an elephant shows its surface temperatures. Red represents warm patches (such as the trunk) and blue cool patches (such as the ears). The image shows energy that the elephant *radiates*. Why does an elephant (or any other warm object) radiate energy?

Looking Ahead ▶▶

The goal of Chapter 12 is to use the atomic model of matter to explain and explore many macroscopic phenomena associated with heat, temperature, and the properties of matter. In this chapter, you will learn to:

▶ Use the atomic model of matter to explain the thermal expansion of solids and liquids, changes of phase, pressure in gases, and the transfer of heat.

▶ Apply the ideal-gas law to gases undergoing changes in pressure, volume, and temperature.

▶ Solve problems involving heat transfer between systems.

▶ Understand the heat-transfer processes of conduction, convection, and radiation.

Looking Back ◀◀

This chapter will make extensive use of the concepts of energy and heat from Chapters 10 and 11. Please review:

◀ Section 10.7 The nature of thermal energy.

◀ Section 11.4 The connection between thermal energy and temperature, the nature of thermal energy and temperature for ideal gases.

◀ Section 11.5 Heat and the first law of thermodynamics.

Elephants, humans, and all other animals convert the chemical energy of food to thermal energy. This is the process of metabolism. But if an animal had no means of getting rid of excess thermal energy, its temperature would soon rise to dangerous levels.

Chapter 11 introduced the concepts of thermal energy, temperature, and heat, but many unanswered questions remain. For example:

■ How is heat transferred to or from a system? How can an elephant keep cool even in the hot African savanna?

■ What are the consequences of heating or cooling a system? Changing the temperature is one obvious possibility, but are there others? And what is the connection between the amount of heat transferred to a system and the amount by which its temperature changes?

■ How do the properties of matter depend on temperature? Whether a system is a solid or liquid depends on its temperature. So does the pressure of a gas.

This chapter will cover a very wide range of topics as it addresses such questions, but there is a central organizing theme that will help us see the connections among the diverse phenomena. Namely, we will use the *atomic model of matter* throughout the chapter to predict, explain, and understand the fascinating interplay of matter and heat in our world.

12.1 The Atomic Model of Matter

We have, throughout this book, used a model in which we represent matter as being made of particle-like atoms. This atomic model has been invaluable in explaining friction, elastic forces, and the nature of thermal energy. We now want to use this atomic model to understand the thermal properties of matter.

Each element and most compounds can exist as a solid, liquid, or gas. These three **phases** of matter are familiar from everyday experience. An atomic view of the three phases is shown in Figure 12.1.

- A **gas** is a system in which each particle moves freely through space until, on occasion, it collides with another particle or the wall of its container.
- In a **liquid,** weak bonds permit motion while keeping the particles close together.
- A rigid **solid** has a definite shape, and can be compressed or deformed only slightly, as we saw in Chapter 8. It consists of atoms connected by spring-like molecular bonds.

Our atomic model makes some simplifications that are worth noting. The basic particles of the gas in Figure 12.1 are drawn as simple spheres; no mention is made of the nature of the particles. The balloon might contain either helium (in which the basic particles are helium atoms) or air (in which the basic particles are nitrogen and oxygen molecules). A helium atom and a nitrogen molecule are quite different from each other, but many of the properties of the gas as a whole do not depend on the nature of the particles—a gas of helium atoms or oxygen molecules may behave identically. In such cases, we will simply refer to gas *particles,* which may be either atoms or molecules.

The basic particles in the liquid water in Figure 12.1 are water molecules. We ignore the structure of the molecules, so when we speak of the bonds that hold the particles in the liquid state, we are referring to the relatively weak intermolecular bonds among the water molecules, not the strong bonds between the hydrogen and oxygen atoms that form the molecules. The basic particles of the gold bars in Figure 12.1 are gold atoms. The bonds that hold these atoms together are the bonds that give the solid its structure. But there are solids that are composed of molecules that are held together by bonds between the molecules; water ice is one such example. In this case, the particles are the molecules, and the bonds that form the solid are the bonds between the molecules.

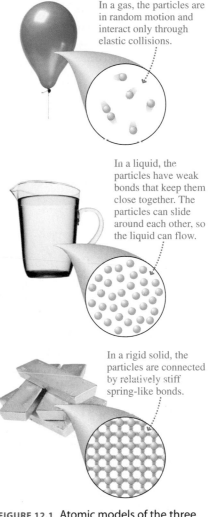

In a gas, the particles are in random motion and interact only through elastic collisions.

In a liquid, the particles have weak bonds that keep them close together. The particles can slide around each other, so the liquid can flow.

In a rigid solid, the particles are connected by relatively stiff spring-like bonds.

FIGURE 12.1 Atomic models of the three phases of matter: solid, liquid, and gas.

Atomic Mass and Atomic Mass Number

Before we see how the atomic model explains the thermal properties of matter, we need to remind you of some "atomic accounting." Recall that atoms of different elements have different masses. The mass of an atom is determined primarily by its most massive constituents, the protons and neutrons in its nucleus. The *sum* of the number of protons and neutrons is called the **atomic mass number** A:

$$A = \text{number of protons} + \text{number of neutrons}$$

TABLE 12.1 Some atomic mass numbers

Element	Symbol	A
Hydrogen	^{1}H	1
Helium	^{4}He	4
Carbon	^{12}C	12
Nitrogen	^{14}N	14
Oxygen	^{16}O	16
Neon	^{20}Ne	20
Aluminum	^{27}Al	27
Argon	^{40}Ar	40
Lead	^{207}Pb	207

A, which by definition is an integer, is written as a leading superscript on the atomic symbol. For example, the primary isotope of carbon, with six protons (which makes it carbon) and six neutrons, has $A = 12$ and is written ^{12}C. The radioactive isotope ^{14}C, used for carbon dating of archeological finds, contains six protons and eight neutrons.

The **atomic mass** scale is established by defining the mass of ^{12}C to be exactly 12 u, where u is the symbol for the *atomic mass unit*. That is, $m(^{12}$C$) = 12$ u. In kg, the atomic mass unit is

$$1 \text{ u} = 1.66 \times 10^{-27} \text{ kg}$$

Atomic masses are all very nearly equal to the integer atomic mass number A. For example, the mass of ^{1}H, with $A = 1$, is $m = 1.0078$ u. For our present purposes, it will be sufficient to use the integer atomic mass numbers as the values of the atomic mass. That is, we'll use $m(^1$H$) = 1$ u, $m(^4$He$) = 4$ u, and $m(^{16}$O$) = 16$ u. For molecules, the **molecular mass** is the sum of the atomic masses of the atoms forming the molecule. Thus the molecular mass of the diatomic molecule O_2, the constituent of oxygen gas, is $m(O_2) = 2m(^{16}$O$) = 32$ u.

> **NOTE** ▶ An element's atomic mass number is *not* the same as its atomic number. The *atomic number,* which gives the element's position in the periodic table, is the number of protons. ◀

Table 12.1 shows the atomic mass numbers of some of the elements that we'll use for examples and homework problems. A complete periodic table, including atomic masses, is found in Appendix B.

The Definition of the Mole

TABLE 12.2 Monatomic and diatomic gases

Monatomic	Diatomic
Helium (He)	Hydrogen (H_2)
Neon (Ne)	Nitrogen (N_2)
Argon (Ar)	Oxygen (O_2)

One way to specify the amount of substance in a system is to give its mass. Another way, one connected to the number of atoms, is to measure the amount of substance in *moles*. By definition, one **mole** of matter is the amount of substance containing as many basic particles as there are atoms in 12 g of ^{12}C. Many decades of ingenious experiments have determined that there are 6.02×10^{23} atoms in 12 g of ^{12}C, so we can say that **1 mole of substance, abbreviated 1 mol, is 6.02×10^{23} basic particles.**

The basic particle depends on the substance, as noted in the previous section. Helium is a **monatomic gas,** meaning that the basic particle is the helium atom. Thus 6.02×10^{23} helium atoms are 1 mol of helium. But oxygen gas is a **diatomic gas** because the basic particle is the two-atom diatomic molecule O_2. 1 mol of oxygen gas contains 6.02×10^{23} *molecules* of O_2 and thus $2 \times 6.02 \times 10^{23}$ oxygen atoms. Table 12.2 lists the monatomic and diatomic gases that we will use for examples and problems.

The number of basic particles per mole of substance is called **Avogadro's number,** N_A. The value of Avogadro's number is thus

$$N_A = 6.02 \times 10^{23} \text{ mol}^{-1}$$

The number n of moles in a substance containing N basic particles is

$$n = \frac{N}{N_A} \tag{12.1}$$

Moles of a substance in terms of the number of basic particles

One mole of helium, sulfur, copper, and mercury.

The **molar mass** of a substance M_{mol} is the mass *in grams* of 1 mol of substance. By definition, the molar mass of ^{12}C is $M_{\text{mol}}(^{12}$C$) = 12$ g/mol. For other substances, the numerical value of the molar mass equals the numerical value of the atomic or molecular mass. That is, the molar mass of He, with $m = 4$ u, is $M_{mol}(\text{He}) = 4$ g/mol and the molar mass of diatomic O_2 is $M_{mol}(O_2) = 32$ g/mol.

You can use the molar mass to determine the number of moles. In one of the few instances where the proper units are *grams* rather than kilograms, the number of moles contained in a system of mass M consisting of atoms or molecules with molar mass M_{mol} is

$$n = \frac{M \text{ (in grams)}}{M_{mol}} \qquad (12.2)$$

Moles of a substance in terms of its mass

EXAMPLE 12.1 Determining quantities of oxygen

A system contains 100 g of oxygen. How many moles does it contain? How many molecules?

SOLVE The diatomic oxygen molecule O_2 has a mass $M_{mol} = 32$ g/mol. From Equation 12.2,

$$n = \frac{100 \text{ g}}{32 \text{ g/mol}} = 3.1 \text{ mol}$$

Each mole contains N_A molecules, so the total number is $N = nN_A = 1.9 \times 10^{24}$ molecules.

Volume

An important property that characterizes a macroscopic system is its volume V, the amount of space the system occupies. The SI unit of volume is m^3. Nonetheless, both cm^3 and, to some extent, liters (L) are widely used metric units of volume. In most cases, you *must* convert these to m^3 before doing calculations.

While it is true that $1 \text{ m} = 100 \text{ cm}$, it is *not* true that $1 \text{ m}^3 = 100 \text{ cm}^3$. Figure 12.2 shows that the volume conversion factor is $1 \text{ m}^3 = 10^6 \text{ cm}^3$. A liter is 1000 cm^3, so $1 \text{ m}^3 = 10^3 \text{ L}$. A milliliter (1 mL) is the same as 1 cm^3.

Subdivide the 1 m × 1 m × 1 m cube into little cubes 1 cm on a side. You will get 100 subdivisions along each edge.

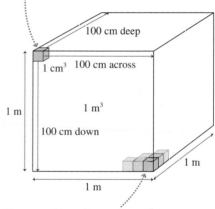

There are $100 \times 100 \times 100 = 10^6$ little 1 cm^3 cubes in the big 1 m^3 cube.

FIGURE 12.2 There are 10^6 cm^3 in 1 m^3.

STOP TO THINK 12.1 Which system contains more atoms: 5 mol of helium ($A = 4$) or 1 mol of neon ($A = 20$)?

A. Helium.   B. Neon.   C. They have the same number of atoms.

12.2 Thermal Expansion

In our introduction to temperature in Chapter 11, we noted that temperature can be measured with a thermometer. The accompanying figure showed an alcohol thermometer, a glass tube with a bulb of alcohol at the bottom. As the temperature of the alcohol increases, so does its volume; this increase in volume—an example of **thermal expansion**—is responsible for the liquid's motion up the column.

Our atomic model gives a very good way to visualize thermal expansion. Suppose a cube of material, as shown in Figure 12.3a, is initially at temperature T_i. The edge of the cube has length L_i, and the cube's volume is V_i. We then heat the cube, increasing its thermal energy. This increases its temperature to T_f, as shown in Figure 12.3b. The atoms now jiggle around faster, so the bonds are "stretched." This increases the average distance between atoms, and so the cube's volume increases to V_f.

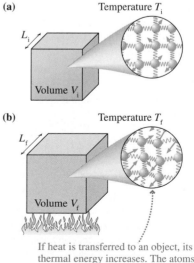

If heat is transferred to an object, its thermal energy increases. The atoms and molecules jiggle around at higher speeds and move farther apart.

FIGURE 12.3 Increasing the temperature of a solid causes it to expand.

Expanding spans A long steel bridge will slightly increase in length on a hot day and decrease on a cold day. Thermal expansion joints let the bridge's length change without causing the roadway to buckle.

For most substances, the change in volume $\Delta V = V_f - V_i$ is linearly related to the change in temperature $\Delta T = T_f - T_i$ by

$$\Delta V = \beta V_i \Delta T \qquad (12.3)$$

Volume thermal expansion

LINEAR
p. 38

The constant β is known as the **coefficient of volume expansion.** Its value depends on the nature of the bonds in a material; it can vary quite a bit from one substance to another. Because ΔT is measured in K, the units of β are K^{-1}.

As a solid's volume increases, each of its linear dimensions increases as well. We can write a similar expression for this linear thermal expansion. If an object of initial length L_i undergoes a temperature change ΔT, its length changes to L_f. The change in length, $\Delta L = L_f - L_i$, is given by

$$\Delta L = \alpha L_i \Delta T \qquad (12.4)$$

Linear thermal expansion

The constant α is the **coefficient of linear expansion.** Note that Equations 12.3 and 12.4 apply equally well to thermal *contractions,* in which case both ΔT and ΔV (or ΔL) are negative.

NOTE ▶ The expressions for thermal expansion are only approximate expressions that apply over a limited range of temperatures. They are what we call **empirical formulas;** they are a good fit to measured data, but they do not represent any underlying fundamental law. There are materials that do *not* closely follow Equations 12.3 or 12.4; water at low temperatures is one example, as we will see. ◄

Values of α and β for some common materials are listed in Table 12.3 The volume expansion of a liquid can be measured but, because a liquid can change shape, we don't assign a coefficient of linear expansion to a liquid. While α and β do vary slightly with temperature, the room-temperature values in Table 12.3 are quite adequate for all problems in this book.

TABLE 12.3 Coefficients of linear and volume thermal expansion at 20°C

Substance	Linear α (K^{-1})	Volume β (K^{-1})
Aluminum	23×10^{-6}	69×10^{-6}
Glass	9×10^{-6}	27×10^{-6}
Iron or steel	12×10^{-6}	36×10^{-6}
Concrete	12×10^{-6}	36×10^{-6}
Ethyl alcohol		1100×10^{-6}
Water		210×10^{-6}
Air (and other gases)		3400×10^{-6}

EXAMPLE 12.2 How much closer to space?

The height of the Space Needle, a steel observation tower in Seattle, is 180 meters on a 0°C winter day. How much taller is it on a hot summer day when the temperature is 30°C?

PREPARE The increase in temperature is

$$\Delta T = T_f - T_i = 30°C - 0°C = 30°C = 30 \text{ K}$$

NOTE ▶ In practice, ΔT is usually measured in °C. But the Kelvin and the Celsius temperature scales have the same step size, so ΔT in K has exactly the same numerical value as ΔT in °C. Thus

■ You do not need to convert temperatures from °C to K if you only need a temperature *change* ΔT.
■ You do need to convert anytime you need the actual temperature T. ◄

SOLVE The coefficient of linear expansion is given in Table 12.3; we can use this value in Equation 12.4 to compute the increase in height:

$$\Delta L = \alpha L_i \Delta T = (12 \times 10^{-6} \text{ K}^{-1})(180 \text{ m})(30 \text{ K}) = 0.065 \text{ m}$$

ASSESS Compared to 180 m, an expansion of 6.5 cm is small—not something you would easily notice—but not negligible. The thermal expansion of structural elements in towers and bridges must be accounted for in the design to avoid damaging stresses. When designers failed to properly account for thermal stresses in marble panels cladding the Amoco Building in Chicago, all 43,000 panels had to be replaced, at great cost.

CONCEPTUAL EXAMPLE 12.1 **What happens to the hole?**

A metal plate has a circular hole in it. As the plate is heated, does the hole get larger or smaller?

REASON As the plate expands, you might think the hole would shrink (with the metal expanding into the hole). But suppose we took a metal plate and simply drew a circle where a hole could be cut. On heating, the plate and the marked area both expand, as we see in Figure 12.4. We could cut a hole on the marked line before or after heating; the size of the hole would be larger in the latter case. Therefore, the size of the hole must expand as the plate expands.

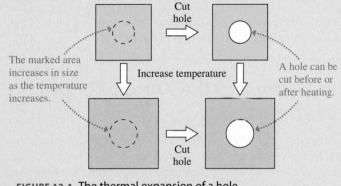

FIGURE 12.4 The thermal expansion of a hole.

ASSESS This somewhat counterintuitive result has many practical applications, as you can see from the "Try It Yourself" exercise on loosening a jar lid.

Special Properties of Water and Ice

Water, the most important molecule for life, differs from other liquids in many important ways. If you cool a sample of water toward its freezing point of 0°C, you would expect its volume to decrease. Figure 12.5a shows a graph of the volume of a mole of water versus temperature. As the water is cooled, you can see that the volume indeed decreases—to a point. Once the water is cooled below 4°C, further cooling results in an *increase* in volume.

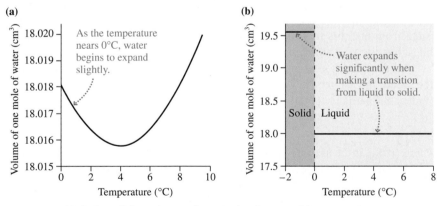

FIGURE 12.5 Variation of the volume of one mole of water with temperature.

This anomalous behavior of water is a result of the charge distribution on the water molecules, which determines how nearby molecules interact with each other. As water approaches the freezing point, molecules begin to form clusters that are more strongly bound. Water molecules must actually get a bit farther apart to form such clusters, meaning the volume increases. Freezing the water results in an even greater increase in volume, as illustrated in Figure 12.5b.

Thermal expansion to the rescue If you have a stubborn lid on a glass jar, try this: Put the lid under very hot water for a short time. Heating the lid and the jar makes them both expand, but, as you can see from the data in Table 12.3, the steel lid—and the opening in the lid that fits over the glass jar—expands by more than the glass jar. The jar lid is now looser, and can be more easily removed.

This expansion on freezing has important consequences. In most materials, the solid phase is denser than the liquid phase and so the solid material sinks. Because water *expands* as it freezes, ice is less dense than liquid water and floats. Not only do ice cubes float in your cold drink, a lake freezes by forming ice on the top rather than at the bottom. This layer of ice then insulates the water below from much colder air above. Thus most lakes do not freeze solid, allowing aquatic life to survive even the harshest winters.

> **STOP TO THINK 12.2**   An aluminum ring is tight around a solid iron rod. If we wish to loosen the ring to remove it from the rod, we should
>
> A. Increase the temperature of the ring and rod.
> B. Decrease the temperature of the ring and rod.

12.3 Pressure and the Kinetic Theory of an Ideal Gas

Solids and liquids are nearly incompressible because the atomic particles are in close contact with each other. Gases differ from solids or liquids in that gases are highly compressible. That is, a gas can expand or contract not only in response to temperature but also to squeezing the particles closer together or pulling them farther apart. In Chapter 11, we introduced an atomic-level model of an *ideal gas,* reviewed in Figure 12.6. Our goal in this section is to further develop this model of an ideal gas. The explanation of properties of the gas as a whole in terms of randomly moving particles is called **kinetic theory.**

Temperature

We had our first taste of kinetic theory in Chapter 11, when we expressed the temperature of an ideal gas in terms of the motion of individual particles of the gas. To see that the particles are moving with different speeds, Figure 12.7 shows data from an experiment to measure the molecular speeds in nitrogen gas at 20°C. The results are presented as a *histogram,* a bar chart in which the height of the bar indicates what percentage of the molecules have a speed in the range of speeds shown below the bar. For example, 16% of the molecules have speeds in the range from 600 m/s to 700 m/s. The most probable speed, as judged from the tallest bar, is ≈550 m/s. This is really fast, ≈1200 mph!

Even though different molecules have different speeds, we can define an average speed. In Chapter 11, we defined the root-mean-square speed, or rms speed, of ideal gas particles to be

$$v_{rms} = \sqrt{(v^2)_{avg}}$$

where $(v^2)_{avg}$ is the average value of the squares of all the molecular velocities.

In terms of the rms speed, the average kinetic energy of the gas particles is

$$K_{avg} = \tfrac{1}{2}m(v^2)_{avg} = \tfrac{1}{2}mv_{rms}^2 \tag{12.5}$$

An important result from Chapter 11 was that the average kinetic energy of the gas particles is directly proportional to the absolute temperature T:

$$K_{avg} = \tfrac{3}{2}k_B T \tag{12.6}$$

where $k_B = 1.38 \times 10^{-23}$ J/K is Boltzmann's constant. Indeed, this equation is really the definition of temperature—temperature measures the average kinetic energy of the particles in a system. So the important concept of temperature is intimately connected to the random motions of the particles in the gas.

1. The gas is made of a large number N of particles of mass m, each moving randomly.

2. The particles are quite far from each other and interact only rarely when they collide.

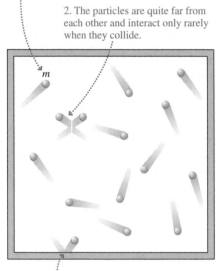

m

3. The collisions of the particles with each other (and with walls of the container) are elastic; no energy is lost in these collisions.

FIGURE 12.6 The ideal-gas model.

The height of a bar represents the percent of molecules with speed in the range on the horizontal axis.

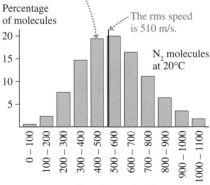

Percentage of molecules

The rms speed is 510 m/s.

N₂ molecules at 20°C

Speed range (m/s)

FIGURE 12.7 The distribution of molecular speeds in nitrogen gas at 20°C.

Pressure

Everyone has some sense of the concept of *pressure*. If you get a hole in your bicycle tire, the higher-pressure air inside comes squirting out. It's hard to get the lid off a vacuum-sealed jar because of the low pressure inside. But just what is pressure?

Let's take a kinetic theory view of pressure, defining it in terms of the motion of particles of a gas. Suppose we have a sample of gas in a container with rigid walls. As the particles in a gas move around, they sometimes collide with and bounce off the walls. As Figure 12.8 shows, each collision exerts a force on the wall. The force due to any one collision is small, but there are an exceedingly large number of collisions every second. The net force on a surface can be quite substantial.

The force on a surface is proportional to the area. Doubling the surface area will double the number of collisions and thus double the force. Rather than the force itself, a more useful quantity is the force-to-area ratio F/A. This ratio is a property of the gas itself, independent of the surface, and it's what we call the gas **pressure:**

$$p = \frac{F}{A} \qquad (12.7)$$

Definition of pressure in a gas as the force-to-area ratio

There are an enormous number of collisions of particles against the wall every second.

Each collision exerts a tiny force on the wall. The net force due to all the collisions causes the gas to have a pressure.

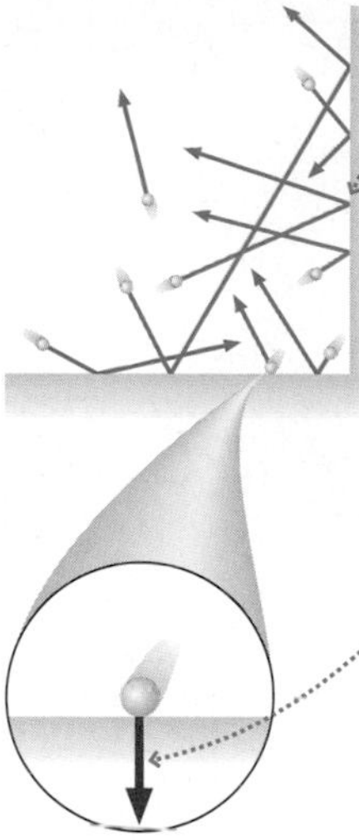

FIGURE 12.8 The pressure in a gas is due to the net force of the particles colliding with the walls.

Notice that pressure is a scalar, not a vector. You can see from Equation 12.7 that a gas exerts a force of magnitude

$$F = pA \qquad (12.8)$$

on a surface of area A. The force is *perpendicular* to the surface.

NOTE ▶ Pressure itself is *not* a force, even though we sometimes talk informally about "the force exerted by the pressure." The correct statement is that the *gas* exerts a force on a surface. ◀

From its definition, you can see that pressure has units of N/m^2. The SI unit of pressure is the **pascal,** defined as

$$1 \text{ pascal} = 1 \text{ Pa} = 1 \frac{N}{m^2}$$

This unit is named for the 17th-century French scientist Blaise Pascal, who was one of the first to study gases. A pascal is a very small pressure, so we usually see pressures in kilopascals, where $1 \text{ kPa} = 1000 \text{ Pa}$.

The total force on the surface of your body due to the pressure of the atmosphere is over 40,000 pounds. Why doesn't this enormous force crush you? To answer this question, let's look at a surface with pressure on both sides. Figure 12.9a shows a situation where the pressure is the same on both sides of a surface. The forces pushing on both sides may be quite large, but they exactly balance and there's no net force. The pressure of the atmosphere is pushing in on your skin, but it is balanced by an equal pressure inside your body pushing out.

Net forces are exerted only where there is a pressure *difference* Δp, as in Figure 12.9b. The net force on a surface of area A is

$$F_{net} = F_2 - F_1 = p_2 A - p_1 A = A(p_2 - p_1) = A \Delta p$$

(a) Equal pressures   **(b)** Unequal pressures

p_1 $p_2 = p_1$ p_1 $p_2 > p_1$

F_1 F_2 F_1 F_2

The forces on both sides are equal. There is no net force.

Now there is a net force because of the pressure difference.

FIGURE 12.9 The net force depends on the pressure difference.

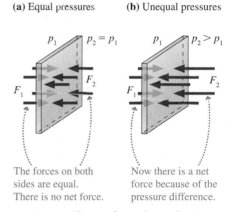

▶ **Too little pressure too fast** BIO This photo shows a rockfish, a popular game fish that can be caught at depths up to 1000 ft. The great pressure at these depths is balanced by an equally large pressure inside the fish's swim bladder, a gas-filled organ that we will learn more about in Chapter 13. If the fish is hooked and rapidly raised to the lower pressure of the surface, the pressure inside the swim bladder is suddenly much larger than the pressure outside, causing the swim bladder to expand dramatically, damaging nearby organs. When reeled in from a great depth, these fish seldom survive.

A tire-pressure gauge reads the gauge pressure p_g, not the absolute pressure p. Zero gauge pressure means the pressure inside the tire is the same as the pressure outside the tire—and the tire is flat.

This is the force that holds the lid on a vacuum-sealed jar, where the pressure inside is less than the pressure outside. To remove the lid, you have to exert a force larger than the net force due to the pressure difference.

Decreasing the number of molecules in a container decreases the pressure because there are fewer collisions with the walls. The pressure in a completely empty container would be $p = 0$ Pa. This is a called a *perfect vacuum*. A perfect vacuum cannot be achieved because it's impossible to remove every molecule from a region of space. In practice, a **vacuum** is an enclosed space in which $p \ll 1$ atm. Using $p = 0$ Pa is then a good approximation.

Measuring Pressure

The most important pressure for our daily lives is the pressure of the atmosphere. The pressure of the atmosphere varies with altitude and the weather, but the global average pressure at sea level, called the *standard atmosphere,* is

$$1 \text{ standard atmosphere} = 1 \text{ atm} = 101{,}300 \text{ Pa} = 101.3 \text{ kPa}$$

Just as we measured acceleration in units of g, we will often measure pressure in units of atm.

In the United States, where force is measured in pounds and area in square inches, pressure is often pounds per square inch, or psi. When you measure pressure in the tires on your car or bike, you probably use a gauge that reads in psi. The conversion factor is

$$1 \text{ atm} = 14.7 \text{ psi}$$

Because the effects of pressure depend on pressure differences, many common gauges measure not the actual or *absolute pressure p* but what is called the **gauge pressure.** The gauge pressure, denoted p_g, is the pressure *in excess* of 1 atm. That is,

$$p_g = p - 1 \text{ atm}$$

You need to add 1 atm = 101.3 kPa to the reading of a pressure gauge to find the absolute pressure p you need for doing most calculations in this chapter.

EXAMPLE 12.3   Finding the force due to a pressure difference

Patients suffering from decompression sickness and other conditions may be treated in a hyperbaric oxygen chamber filled with oxygen at greater than atmospheric pressure. A cylindrical chamber with flat end plates of diameter 0.75 m is filled with oxygen to a gauge pressure of 27 kPa. What is the net force on the end plate of the cylinder?

PREPARE There is a net force on the end plate because of the pressure *difference* between the inside and outside. 27 kPa is the pressure in excess of 1 atm. If we assume the pressure out-side is 1 atm, then 27 kPa is Δp, the pressure difference across the surface.

SOLVE The end plate has area $A = \pi(0.75 \text{ m}/2)^2 = 0.442 \text{ m}^2$. The pressure difference causes a net force

$$F_{\text{net}} = A\,\Delta p = (0.442 \text{ m}^2)(27{,}000 \text{ Pa}) = 12 \text{ kN}$$

ASSESS This is a large force—more than the weight of one ton. Even modest pressure differences can cause substantial forces on a large enough area. It's remarkable to think that this large force is due to the collisions of individual molecules with the ends of the cylinder.

From Collisions to Pressure and the Ideal-Gas Law

We can use the fact that the pressure in a gas is due to the collision of particles with the walls to make some qualitative predictions. Figure 12.10 presents a few such predictions.

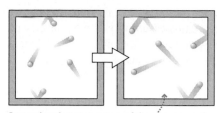

Increasing the temperature of the gas means the particles move at higher speeds. They hit the walls more often and with more force, so there is more pressure.

■ Increasing the temperature leads to an increase in pressure.

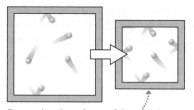

Decreasing the volume of the container means more frequent collisions with the walls of the container, and thus more pressure.

■ Decreasing the volume leads to an increase in pressure.

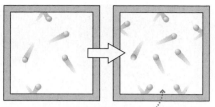

Increasing the number of particles in the container means more frequent collisions with the walls of the container, and thus more pressure.

■ Increasing the number of particles leads to an increase in pressure.

FIGURE 12.10 A kinetic theory model of pressure changes.

We see that temperature, volume, and the number of particles all affect the gas pressure in a container. To analyze pressure, let's start by considering a single collision of a particle with a wall. Figure 12.11 shows a particle moving parallel to one side of the box as it collides with a wall. Collisions in an ideal gas are perfectly elastic, so no energy is lost in the collision. Thus the particle rebounds from the wall with its velocity changed from v to $-v$.

We looked at collisions in Chapter 9. Recall that the average force exerted during a collision of duration Δt is related to the particle's change in momentum by

$$(F_{\text{avg}})_x \, \Delta t = \Delta p_x = m(v_x)_f - m(v_x)_i$$

With $(v_x)_i = v$ and $(v_x)_f = -v$ the average force on the particle is

$$(F_{\text{molecule}})_{\text{avg}\,x} = \frac{m(v_x)_f - m(v_x)_i}{\Delta t} = -m\frac{2v}{\Delta t}$$

The force on the particle is negative because it points to the left. According to Newton's third law, the wall experiences the equal but opposite force

$$(F_{\text{wall}})_{\text{avg}\,x} = +m\frac{2v}{\Delta t} \tag{12.9}$$

as a result of this single collision. Not surprisingly, faster particles exert larger forces when they rebound from the wall.

It is the sum of many such collisions that leads to the macroscopic phenomenon of pressure. To find the pressure, we need to multiply the force due to one collision by the total number N_{coll} of collisions that occur during the time interval Δt. If there are N particles moving with speed v in a box of volume V, it can be shown that the number of collisions is

$$N_{\text{coll}} = \frac{1}{2}\frac{N}{V}Av \, \Delta t$$

Although we've not proved this result, it makes sense that increasing the number of particles N or increasing the size of the wall A will increase the number of collisions, whereas increasing the volume of the box (which spreads the particles out) will decrease the number of collisions.

The net force on the wall is found by combining this expression for N_{coll} with Equation 12.9. Δt cancels, giving

$$F_{\text{net}} = N_{\text{coll}}(F_{\text{wall}})_{\text{avg}\,x} = \left(\frac{1}{2}\frac{N}{V}Av\Delta t\right)\left(\frac{2mv}{\Delta t}\right) = \frac{N}{V}mv^2A$$

FIGURE 12.11 A particle colliding with the wall exerts a force on it.

Before: $(v_x)_i = v$
After: $(v_x)_f = -v$
Wall of area A

Pumping up a bicycle tire adds more particles to the fixed volume of the tire, increasing the pressure.

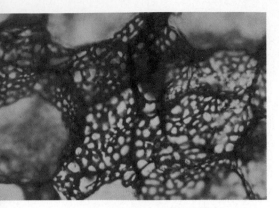

Gas exchange in the lungs BIO When you draw a breath, how does the oxygen get into your bloodstream? Kinetic theory provides the answer. The above picture is a highly magnified view of the alveoli, air sacs in the lungs, that are surrounded by capillaries, fine blood vessels. The thin membranes of the alveoli and the capillaries are permeable—small molecules can move across them. If a permeable membrane separates two regions of space having different concentrations of a molecule, the rapid motion of the molecules causes a net transport of molecules in the direction of lower concentration. This transport—entirely due to the motion of the molecules—is known as **diffusion.** In the lungs, the higher concentration of oxygen in the alveoli drives the diffusion of oxygen into the blood in the capillaries. At the same time, carbon dioxide diffuses from the blood into the alveoli. Diffusion is a rapid and effective means of transport in this case because the membranes are thin. Diffusion won't work to get oxygen from the lungs to other parts of the body, because diffusion is slow over large distances, so the oxygenated blood must be pumped throughout the body.

This result is based on the assumption that all particles have the same speed and are moving parallel to the x-axis. We can relax this assumption if we replace v by v_{rms} and, because there are three dimensions, divide by 3. Doing so, we find that the force exerted on the wall by a vast number of collisions is

$$F_{net} = \frac{1}{3}\frac{N}{V}mv_{rms}^2 A$$

Dividing F_{net} by A, we finally have an expression for the pressure in the gas:

$$p = \frac{F_{net}}{A} = \frac{1}{3}\frac{N}{V}mv_{rms}^2 = \frac{2}{3}\frac{N}{V}\left(\frac{1}{2}mv_{rms}^2\right)$$

But $\frac{1}{2}mv_{avg}^2$ is the average kinetic energy per particle in the gas, so we're left with

$$p = \frac{2}{3}\frac{N}{V}K_{avg} \tag{12.10}$$

The pressure of an ideal gas depends on just two things: the number of particles per cubic meter (N/V) and the average kinetic energy of the particles. Surprisingly, it does not depend on the nature of the particles.

As remarkable as this result is, we can take it one more step and "discover" a law you have likely seen in a previous course. If we replace K_{avg} with the temperature expression of Equation 12.6 and take the V to the other side of the equation, we end up with

$$pV = Nk_B T \tag{12.11}$$

Ideal-gas law, version 1

Remarkably, we've derived the **ideal-gas law:** The pressure of a gas multiplied by its volume is equal to the number of gas particles multiplied by Boltzmann's constant multiplied by the temperature. The derivation was longer than we are accustomed to in this text, but it was worthwhile in this particular case. It is not only the final result that is important, but also the process behind it: **By considering the forces of individual particles colliding with the wall of a container (a microscopic picture), we have derived an equation that relates the pressure, volume, and temperature of a gas (a macroscopic picture).** The ideal-gas law is not simply an empirical finding but is a direct consequence of the atomic model of matter.

Equation 12.11 is written in terms of the number N of particles in the gas, whereas the ideal-gas law is stated in chemistry in terms of the number n of moles. But the change is easy to make. The number of particles is $N = nN_A$, so we can rewrite Equation 12.11 as

$$pV = nN_A k_B T = nRT \tag{12.12}$$

Ideal-gas law, version 2

where $R = N_A k_B = 8.31$ J/mol · K is the *gas constant*. The units may seem a little unusual, but the product of Pa and m^3, the units of pV, is equivalent to J.

You may have learned to work gas problems using units of atmospheres and liters. To do so, you had a different numerical value of R that was expressed in those units. In physics we always work gas problem in SI units. Let's review the meaning and the units of the various quantities in the ideal-gas law:

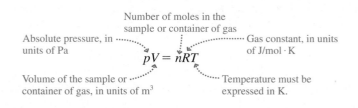

Number of moles in the sample or container of gas

Absolute pressure, in units of Pa

Gas constant, in units of J/mol · K

$$pV = nRT$$

Volume of the sample or container of gas, in units of m^3

Temperature must be expressed in K.

EXAMPLE 12.4 Finding the volume of a mole
What volume is occupied by one mole of an ideal gas at a pressure of 1.00 atm and a temperature of 0°C?

PREPARE The first step in ideal-gas law calculations is to convert all quantities to SI units:

$$p = 1.00 \text{ atm} = 101.3 \times 10^3 \text{ Pa}$$

$$T = 0 + 273 = 273 \text{ K}$$

SOLVE We use the ideal-gas law equation to compute

$$V = \frac{nRT}{p} = \frac{(1.00 \text{ mol})(8.31 \text{ J/mol} \cdot \text{K})(273 \text{ K})}{101.3 \times 10^3 \text{ Pa}}$$

$$= 0.0224 \text{ m}^3 = 22.4 \text{ L}$$

ASSESS When we do calculations using gases, it will be useful to keep this volume in mind to see if our answers make physical sense.

NOTE ▶ The conditions of this example, 1 atm pressure and 0°C, are called *standard temperature and pressure,* abbreviated **STP.** They are a common reference point for many gas processes. ◀

Because 1 m^3 = 1000 L, this example tells us that the volume of one mole of gas at STP is 22.4 liters, a value you might recall from chemistry.

12.4 Ideal-Gas Processes

The ideal-gas law provides a connection between the pressure, volume, and temperature of a gas. Changing one of these variables causes one or both of the others to change. For example, suppose you measure the pressure in the tires on your car on a cold morning. How much will the tire pressure increase after you've driven at highway speeds for 30 minutes and the tires are much warmer? We will solve a problem like this later in the chapter, but, for now, note these properties of this process:

Activ Physics ONLINE 8.4–8.6, 8.8–8.11

- The quantity of gas is fixed. No air is added to or removed from the tire. All of the processes we will consider will involve a fixed quantity of gas.
- There is a well-defined initial state. The initial values of pressure, volume, and temperature will be designated p_i, V_i, T_i.
- There is a well-defined final state in which pressure, volume, and temperature have values p_f, V_f, T_f.

For gases in sealed containers, the number of moles (and number of molecules) does not change. In that case, the ideal-gas law can be written

$$\frac{pV}{T} = nR = \text{constant}$$

The values of the variables in the initial and final states are then related by

$$\frac{p_f V_f}{T_f} = \frac{p_i V_i}{T_i} \tag{12.13}$$

Initial and final states for an ideal gas in a sealed container

(a)   Each state of an ideal gas is represented as a point on a pV diagram.

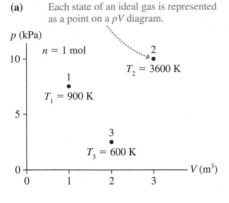

(b)   A process that changes the gas from one state to another is represented by a trajectory on a pV diagram.

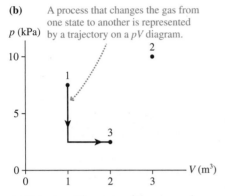

FIGURE 12.12 The state of the gas and ideal-gas processes can be shown on a pV diagram.

This before-and-after relationship between the two states, reminiscent of a conservation law, will be valuable for many problems.

NOTE ▶ Because pressure and volume appear on both sides of the equation, this is a rare case of an equation for which we may use any units we wish, not just SI units. However, temperature *must* be in K. The difference is because unit-conversion factors for pressure and volume are multiplicative factors, and the same factor on both sides of the equation cancels. But the conversion from K to °C is an *additive* factor, and additive factors in the denominator don't cancel. ◀

pV Diagrams

It will be useful to represent ideal-gas processes on a graph called a **pV diagram.** The important idea behind a pV diagram is that each point on the graph represents a single, unique state of the gas. This may seem surprising, because a point on a graph specifies only the value of pressure and volume. But knowing p and V, and assuming that n is known for a sealed container, we can find the temperature from the ideal-gas law. Thus each point on a pV diagram actually represents a triplet of values (p, V, T) specifying the state of the gas.

For example, Figure 12.12a is a pV diagram showing three states of a system consisting of 1 mol of gas. The value of p and V can be read from the axes, then the temperature at that point calculated from the ideal-gas law. An ideal-gas process—a process that changes the state of the gas by, for example, heating it or compressing it—can be represented as a "trajectory" in the pV diagram. Figure 12.12b shows one possible process by which the gas of Figure 12.12a is changed from state 1 to state 3.

Constant-Volume Processes

Suppose you have a gas in the closed, rigid container shown in Figure 12.13a. Warming the gas will raise its pressure without changing its volume. This is an example of a **constant-volume process.** $V_f = V_i$ for a constant-volume process.

Because the value of V doesn't change, this process is shown as the vertical line i → f on the pV diagram of Figure 12.13b. **A constant-volume process appears on a pV diagram as a vertical line.**

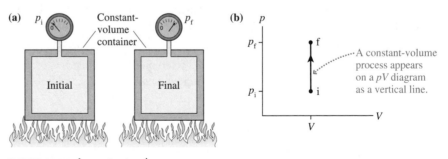

FIGURE 12.13 A constant-volume process.

EXAMPLE 12.5 Computing tire pressure on a hot day

The pressure in a car tire is 30.0 psi on a cool morning when the air temperature is 0°C. After the day warms up and bright sun shines on the black tire, the temperature of the air inside the tire reaches 30°C. What is the tire pressure at this temperature?

PREPARE A tire is (to a good approximation) a sealed container with constant volume; this is a constant-volume process. The measured tire pressure is a gauge pressure, but the ideal-gas law

requires an absolute pressure. We must correct for this. The initial pressure is

$$p_i = (p_g)_i + 1.00 \text{ atm} = 30.0 \text{ psi} + 14.7 \text{ psi} = 44.7 \text{ psi}$$

Temperatures must be in kelvin, so we convert:

$$T_i = 0°C + 273 = 273 \text{ K}$$
$$T_f = 30°C + 273 = 303 \text{ K}$$

SOLVE The gas is in a sealed container, so we can use the ideal-gas law as given in Equation 12.13 to solve for the final pressure. In this equation, we divide both sides by V_f, and then cancel the ratio of the two volumes, which is equal to 1 for this constant-volume process:

$$p_f = p_i \frac{V_i}{V_f} \frac{T_f}{T_i} = p_i \frac{T_f}{T_i}$$

The units for p_f will be the same as those for p_i, so we can keep the initial pressure in psi. The pressure at the higher temperature is

$$p_f = 44.7 \text{ psi} \times \frac{303 \text{ K}}{273 \text{ K}} = 49.6 \text{ psi}$$

This is an absolute pressure, but the problem asks for the measured pressure in the tire—a gauge pressure. Converting to gauge pressure gives

$$(p_g)_f = p_f - 1.00 \text{ atm} = 49.6 \text{ psi} - 14.7 \text{ psi} = 34.9 \text{ psi}$$

ASSESS This is a significant change in pressure. You may find that you need to add air to your tires in the winter, when the temperature drops.

Constant-Pressure Processes

Many gas processes take place at a constant, unchanging pressure. A constant-pressure process is also called an **isobaric process.** $p_f = p_i$ for an isobaric process.

Figure 12.14a shows one method of changing the state of a gas while keeping the pressure constant. A cylinder of gas has a tight-fitting, massless piston that can slide up and down but seals the container so that no atoms enter or escape. The mass on top applies a constant downward force Mg on the piston. The air also presses down on the piston. The gas pressure is determined by the requirement that the gas must support both the mass on top of the piston and the air pressing inward. This pressure is independent of the temperature of the gas or the height of the piston, so it stays constant as long as M is unchanged. Figure 12.14b shows this trajectory for the isobaric process. Because the pressure does not change **a constant-pressure process appears on a pV diagram as a horizontal line.**

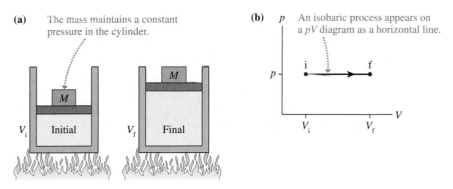

FIGURE 12.14 A constant-pressure (isobaric) process.

EXAMPLE 12.6 A constant-pressure compression
A gas occupying 50.0 cm³ at 50°C is cooled at constant pressure until the temperature is 10°C. What is its final volume?

PREPARE This is a sealed system, so we can use Equation 12.13. The pressure of the gas doesn't change, so this is an isobaric process with $p_i/p_f = 1$.

The temperatures must be in kelvin, so we convert:

$$T_i = 50°C + 273 = 323 \text{ K}$$

and

$$T_f = 10°C + 273 = 283 \text{ K}$$

SOLVE We can use the ideal-gas law for a sealed container to solve for V_f:

$$V_f = V_i \frac{p_i}{p_f} \frac{T_f}{T_i} = 50.0 \text{ cm}^3 \times 1 \times \frac{283 \text{ K}}{323 \text{ K}} = 43.8 \text{ cm}^3$$

ASSESS In this solution and the previous one, we have not converted pressure and volume units as these multiplicative factors will cancel. But we did convert temperature to kelvin as this is an *additive* factor that does *not* cancel.

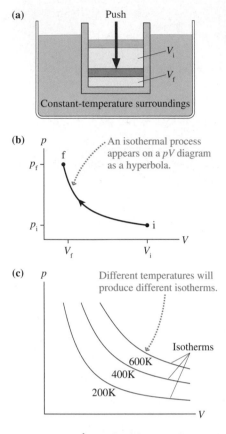

(a) Push

Constant-temperature surroundings

V_i

V_f

(b) p

An isothermal process appears on a pV diagram as a hyperbola.

p_f — f

p_i — i

V_f V_i — V

(c) p

Different temperatures will produce different isotherms.

Isotherms

600K

400K

200K

V

FIGURE 12.15 A constant-temperature (isothermal) process.

Constant-Temperature Process

A constant-temperature process is also called an **isothermal process.** For an isothermal process $T_f = T_i$. One possible isothermal process is illustrated in Figure 12.15. A piston is being pushed down to compress a gas, but the gas cylinder is submerged in a large container of liquid that is held at a constant temperature. If the piston is pushed *slowly,* then heat energy transfer through the walls of the cylinder will keep the gas at the same temperature as the surrounding liquid. This would be an *isothermal compression.* The reverse process, with the piston slowly pulled out, would be an *isothermal expansion.*

Representing an isothermal process on the pV diagram is a little more complicated than the two previous processes because both p and V change. As long as T remains fixed, we have the relationship

$$p = \frac{nRT}{V} = \frac{\text{constant}}{V} \qquad (12.14)$$

INVERSE
p. 118

Because there is an inverse relationship between p and V, the graph of an isothermal process is a *hyperbola*.

The process shown as i → f in Figure 12.15b represents the *isothermal compression* shown in Figure 12.15a. An *isothermal expansion* would move in the opposite direction along the hyperbola. The graph of an isothermal process is known as an **isotherm.**

The location of the hyperbola depends on the value of T. If we raise the constant temperature of the process in Figure 12.15a, the isotherm will move farther from the origin of the pV diagram. Figure 12.15c shows three isotherms for this process at three different temperatures. A gas undergoing an isothermal process will move along the isotherm having the appropriate temperature.

EXAMPLE 12.7 Compressing air in the lungs

A snorkeler takes a deep breath at the surface, filling his lungs with 4.0 L of air. He then descends to a depth of 5.0 m, where the pressure is 1.5 atm. At this depth, what is the volume of air in the snorkeler's lungs?

PREPARE The pressure inside the snorkeler's lungs equals the pressure of the surrounding water, since otherwise his lungs would collapse. Further, the air stays at body temperature, making this an isothermal process with $T_f = T_i$.

SOLVE The ideal-gas law for a sealed container (the lungs) gives

$$V_f = V_i \frac{p_i}{p_f} \frac{T_f}{T_i} = 4.0 \text{ L} \times \frac{1.0 \text{ atm}}{1.5 \text{ atm}} = 2.7 \text{ L}$$

ASSESS The air has a smaller volume at the greater pressure, as we would expect.

Work

When the spark plug fires in a cylinder of your car, it ignites the gaseous fuel-air mixture inside. The hot gas expands, pushing the piston out and, through various mechanical linkages, turning the wheels of your car. As the gas expands, it does *work* on the piston. Similarly, the gas in Figure 12.14a does work as it lifts the mass on the piston. pV diagrams are particularly useful for calculating the work done by a gas.

You learned in Chapter 10 that the work done by a constant force F in pushing an object distance d is $W = Fd$. Let's apply this idea to a gas. Figure 12.16a shows a gas cylinder sealed at one end by a movable piston. Force $\vec{F}_{gas}$ is due to the gas pressure and has magnitude $F_{gas} = pA$. Force $\vec{F}_{ext}$, perhaps a force applied by a piston rod, is equal in magnitude and opposite in direction to $\vec{F}_{gas}$. The gas pressure would blow the piston out if the external force weren't there!

Suppose the gas expands at constant pressure, pushing the piston outward from x_i to x_f, a distance $d = x_f - x_i$, as shown in Figure 12.16b. As it does, the force due to the gas pressure does work

$$W_{gas} = F_{gas}d = (pA)(x_f - x_i) = p(x_fA - x_iA)$$

But x_iA is the cylinder's initial volume V_i (recall that the volume of a cylinder is length times the area of the base), and x_fA is the final volume V_f. Thus the work done is

$$W_{gas} = p(V_f - V_i) = p\,\Delta V \qquad (12.15)$$

Work done by a gas in an isobaric process

where ΔV is the *change* in volume.

Equation 12.15 has a particularly simple interpretation on a pV diagram. As Figure 12.17a shows, $p\,\Delta V$ is the "area under the pV graph" between V_i and V_f. Although we've shown this result only for an isobaric process, it turns out to be true for all ideal-gas processes. That is, as Figure 12.17b shows,

$$W_{gas} = \text{area under the } pV \text{ graph between } V_i \text{ and } V_f$$

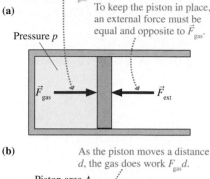

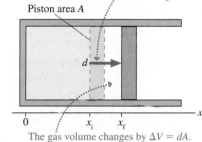

FIGURE 12.16 The expanding gas does work on the piston.

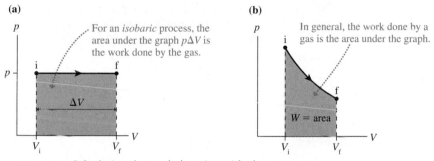

FIGURE 12.17 Calculating the work done in an ideal-gas process.

Thus the work done by a gas can be determined from the geometry of the pV diagram. There are a few things we need to clarify about this determination:

- In order for a gas to do work, the volume must change. Thus no work is done in a constant-volume process.
- The work done by a gas $W_{gas} = p\Delta V$ is positive when the gas expands ($\Delta V > 0$) but negative (meaning work is done on the gas) when the gas is compressed ($\Delta V < 0$).
- The simple relationship of Equation 12.15 applies only to constant-pressure processes. For any other ideal-gas process, you must use the geometry of the pV diagram to calculate the area under the graph.
- To calculate work, pressure must be in Pa and volume in m^3. The product of Pa (i.e., N/m^2) and m^3 is joules, the units of work and energy.
- W_{gas} is *not* the work W that appears in the first law of thermodynamics. The first-law work is work done *on* the system, whereas W_{gas} is work done *by* the system. But $W = -W_{gas}$, so the first law of thermodynamics can be written

$$\Delta E_{th} = Q - W_{gas} \qquad (12.16)$$

EXAMPLE 12.8 Finding the work done by a gas

0.0160 mol of nitrogen follow the ideal-gas process shown in Figure 12.18.

a. What are the initial and final temperatures of the gas?
b. How much work is done by the gas?

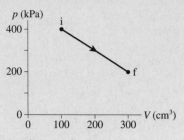

FIGURE 12.18 pV diagram for Example 12.8.

PREPARE We can use the ideal-gas law to determine the temperatures, but first we must convert the volumes to SI units:

$$V_i = 100 \text{ cm}^3 \times \frac{1 \text{ m}^3}{10^6 \text{ cm}^3} = 1.00 \times 10^{-4} \text{ m}^3$$

$$V_f = 300 \text{ cm}^3 \times \frac{1 \text{ m}^3}{10^6 \text{ cm}^3} = 3.00 \times 10^{-4} \text{ m}^3$$

This is not a constant-pressure process, so we will need to calculate the area under the graph to determine the amount of work done by the gas as it expands.

SOLVE a. The initial and final temperatures are

$$T_i = \frac{p_i V_i}{nR} = \frac{(4.00 \times 10^5 \text{ Pa})(1.00 \times 10^{-4} \text{ m}^3)}{(0.0160 \text{ mol})(8.31 \text{ J/mol} \cdot \text{K})} = 301 \text{ K}$$

$$T_f = \frac{p_f V_f}{nR} = \frac{(2.00 \times 10^5 \text{ Pa})(3.00 \times 10^{-4} \text{ m}^3)}{(0.0160 \text{ mol})(8.31 \text{ J/mol} \cdot \text{K})} = 451 \text{ K}$$

b. We analyze the area under the graph as a triangle on top of a rectangle, with areas computed as in Figure 12.19. Notice that the areas are in J because they are the product of Pa and m³.

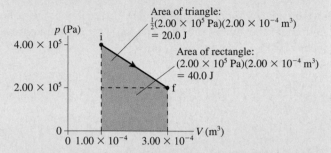

FIGURE 12.19 The work done by the gas is the total area under the graph.

The total work is

$$W_{gas} = \text{area of triangle} + \text{area of rectangle}$$
$$= 20.0 \text{ J} + 40.0 \text{ J} = 60.0 \text{ J}$$

ASSESS The gas did work on the piston as it expanded. In a practical situation, this work would be transferred via gears and levers to a machine that performs some useful function.

Adiabatic Processes

The last ideal-gas process we'll look at is one in which no heat energy is transferred between the gas and the environment and thus $Q = 0$. This is called an **adiabatic process.** There are two ways an adiabatic process can come about: Either the system is thermally insulated, preventing heat transfer, or, more commonly, the process happens too quickly for a transfer to occur. A rapid compression or expansion of a gas doesn't allow time for heat transfer, so a rapid compression is called an *adiabatic compression* and a rapid expansion is an *adiabatic expansion.*

If you've ever pumped up a bicycle tire with a hand pump, you may have noticed that the pump gets quite warm. Why? Pressing the pump handle causes a *rapid* compression of the air inside—an adiabatic process. For an adiabatic process, with $Q = 0$, the first law of thermodynamics, Equation 12.16, is $\Delta E_{th} = -W_{gas}$. You do work *on* the gas to compress it, so the work done *by* the gas is negative ($W_{gas} < 0$). Thus the thermal energy of the gas increases ($\Delta E_{th} > 0$) and the temperature goes up.

Similarly, a gas that does work on a piston by expanding adiabatically ($W_{gas} > 0$) gets colder as its thermal energy decreases. Thus **an adiabatic expansion lowers the temperature of a gas** and **an adiabatic compression raises the temperature of a gas.** You can use an adiabatic process to change the gas temperature without using heat! An adiabatic processes uses work, rather than heat, to change the speeds of the molecules and thus change the temperature.

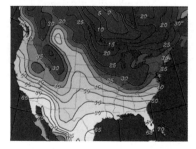

Warm mountain winds This image shows surface temperatures (in °F) in North America on a winter day in 2003. Notice the bright green area of unseasonably warm temperatures extending north and west from the center of the continent. This warm area was produced by a Chinook wind. (Similar winds in other areas of the world have other names, such as the Santa Ana winds in southern California.) On this day, strong westerly winds were blowing down off the Rocky Mountains. The air was rapidly compressed as it descended to lower regions of higher pressure in a very short time. This rapid compression was an adiabatic process that significantly increased the air temperature.

Adiabatic processes have many important applications. In the atmosphere, large masses of air that move across the planet transfer heat quite slowly, so adiabatic expansions and compressions often lead to changes in temperature, as we see in the discussion of the Chinook wind.

CONCEPTUAL EXAMPLE 12.2 Adiabatic curves on a *pV* diagram

Figure 12.20 shows the *pV* diagram of a gas undergoing an isothermal compression from point 1 to point 2. Roughly how would the *pV* diagram look if the gas were compressed from point 1 to the same final pressure by a rapid adiabatic compression?

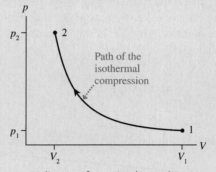

FIGURE 12.20 *pV* diagram for an isothermal compression.

REASON An adiabatic compression increases the temperature of the gas as the work done on the gas is transformed into thermal

energy. Consequently, as seen in Figure 12.21, the curve of the adiabatic compression cuts across the isotherms to end on a higher-temperature isotherm when the gas pressure reaches p_2.

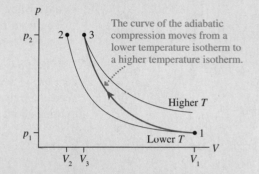

FIGURE 12.21 *pV* diagram for an adiabatic compression.

ASSESS The temperature at the final point of an adiabatic compression is higher than at the starting point. The final volume for the adiabatic compression is therefore larger than that for the isothermal compression, in agreement with the diagram as drawn. Similarly, an adiabatic expansion ends on a lower-temperature isotherm.

STOP TO THINK 12.3 What is the ratio T_f/T_i for this process?

A. 1/4
B. 1/2
C. 1 (no change)
D. 2
E. 4
F. There is not enough information to decide.

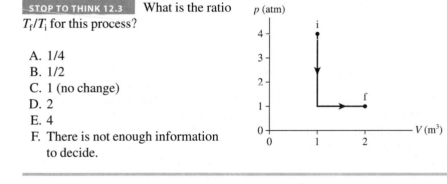

12.5 Specific Heat and Heat of Transformation

If you hold a glass of cold water in your hand, the heat from your hand will increase the temperature of the water. Similarly, the heat from your hand will melt an ice cube; this is what we will call a *phase change*. Increasing the temperature and changing the phase are two possible *consequences* of heating a system, and they form an important part of our understanding of the thermal properties of matter.

Specific Heat

Adding 4190 J of heat energy to water raises its temperature by 1 K. If you are fortunate enough to have 1 kg of gold, you need only 129 J of heat to raise its temperature by 1 K. The amount of heat that raises the temperature of 1 kg of a

TABLE 12.4 Specific heats of solids and liquids

Substance	c (J/kg · K)
Solids	
Lead	128
Gold	129
Copper	385
Iron	449
Aluminum	900
Water ice	2090
Mammalian body	3400
Liquids	
Mercury	140
Ethyl alcohol	2400
Water	4190

substance by 1 K is called the **specific heat** of that substance. The symbol for specific heat is c. Water has specific heat $c_{water} = 4190$ J/kg · K, and the specific heat of gold is $c_{gold} = 129$ J/kg · K. Specific heat depends only on the material from which an object is made. Table 12.4 provides some specific heats for common liquids and solids.

NOTE ▶ *Heat* is energy—specifically, the energy transferred between a hotter object and a colder object. *Specific heat,* the property of an object, does not use the word *heat* in the same way. Specific heat is an old idea, dating back to the days when heat was thought to be a substance that was contained in objects. The term has continued in use even though our understanding of heat has changed. ◀

If heat c is required to raise the temperature of 1 kg of substance by 1 K, then heat Mc is needed to raise the temperature of mass M by 1 K and $(Mc) \Delta T$ is needed to raise the temperature of mass M by ΔT. In other words, the heat needed to bring about a temperature change ΔT is

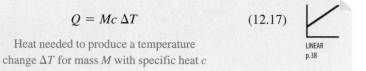

$$Q = Mc \, \Delta T \qquad (12.17)$$

Heat needed to produce a temperature change ΔT for mass M with specific heat c

LINEAR
p.38

Q can be either positive (temperature goes up) or negative (temperature goes down).

Because $\Delta T = Q/Mc$, it takes more heat energy to change the temperature of a substance with a large specific heat than to change the temperature of a substance with a small specific heat. Water, with a very large specific heat, is slow to warm up and slow to cool down. This is fortunate for us. The large "thermal inertia" of water is essential for the biological processes of life.

Hurricane season During a cold night, the large specific heat of water prevents the temperature of a lake from dropping nearly as much as that of the surrounding air. Early in the morning, water vapor evaporating from the warm lake quickly condenses in the colder air, forming mist. The thermal inertia of the oceans is even greater. Ocean temperatures in the northern hemisphere don't reach their maximum values until the late summer or early fall. This is why hurricanes, which require warm ocean water to form, happen in the fall but not the spring.

EXAMPLE 12.9 Energy to run a fever
A 70 kg student catches the flu, and his body temperature increases from 37.0°C (98.6°F) to 39.0°C (102.2°F). How much energy is required to raise his body's temperature?

SOLVE Raising the temperature of the body uses energy supplied internally from the chemical reactions of the body's metabolism, which transfer heat to the body. The specific heat of the body is given in Table 12.4 as 3400 J/kg · K. We can use Equation 12.17 to find the necessary heat energy:

$$Q = Mc \, \Delta T = (70 \text{ kg})(3400 \text{ J/kg} \cdot \text{K})(2.0 \text{ K}) = 4.8 \times 10^5 \text{ J}$$

ASSESS Raising the temperature of an object as large as the body will take a good deal of energy, so the size of this result seems reasonable. Looking back to Chapter 11, we see that this is approximately the energy in a large apple, or the amount of energy required to walk one mile.

Phase Changes

The temperature inside the freezer compartment of a refrigerator is typically about −20°C. Suppose you were to remove a few ice cubes from the freezer and place them in a sealed container with a thermometer. Then, as Figure 12.22a shows, you put a steady flame under the container. We'll assume that the heating is done slowly so that the inside of the container always has a single, uniform temperature.

Figure 12.22b shows the temperature as a function of time. After steadily rising from the initial −20°C, the temperature remains fixed at 0°C for an extended period of time. This is the interval of time during which the ice melts. As it's melting, the ice temperature is 0°C and the liquid water temperature is 0°C. Even though the system is being heated, the temperature doesn't begin to rise until all the ice has melted. The energy that is being added to the system is being used to break bonds between the water molecules, changing the form of the water from solid to liquid.

The thermal energy of a solid is the kinetic energy of the vibrating atoms plus the potential energy of stretched and compressed molecular bonds. If you heat a solid, at some point the thermal energy gets so large that the molecular bonds begin to break, allowing the atoms to move around—the solid begins to melt. Continued heating will produce further melting, as the energy is used to break more bonds. The temperature will not rise until all of the bonds are broken. If you wish to return the liquid to a solid, this energy must be removed.

NOTE ▶ In everyday language, the three phases of water are called *ice, water,* and *steam.* That is, the term *water* implies the liquid phase. Scientifically, these are the solid, liquid, and gas phases of the compound called *water.* When we are working with different phases of water, we'll use the term *water* in the scientific sense of a collection of H_2O molecules. We'll say either *liquid* or *liquid water* to denote the liquid phase. ◀

(a)

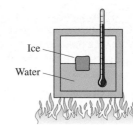

(b)

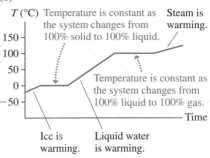

FIGURE 12.22 The temperature as a function of time as water is transformed from solid to liquid to gas.

CONCEPTUAL EXAMPLE 12.3 Strategy for cooling a drink

If you have a warm soda that you wish to cool, would it be more effective to add 25 g of liquid water at 0°C or 25 g of water ice at 0°C?

REASON If you add liquid water at 0°C, heat will be transferred from the soda to the water, raising the temperature of the water and lowering that of the soda. If you add water ice at 0°C, heat first will be transferred from the soda to the ice to melt it, transforming the 0°C ice to 0°C liquid water, then will be transferred to the liquid water to raise its temperature. Thus more thermal energy will be removed from the soda, giving it a lower final temperature, if ice is used rather than liquid water.

The temperature at which a solid becomes a liquid or, if the thermal energy is reduced, a liquid becomes a solid is called the **melting point** or the **freezing point.** Melting and freezing are *phase changes.* A system at the melting point is in **phase equilibrium,** meaning that any amount of solid can coexist with any amount of liquid. Raise the temperature ever so slightly and the entire system soon becomes liquid. Lower it slightly and it all becomes solid.

You can see another region of phase equilibrium of water in Figure 12.22b at 100°C. This is a phase equilibrium between the liquid phase and the gas phase, and any amount of liquid can coexist with any amount of gas at this temperature. As heat is added to the system, the temperature stays the same. The added energy is used to break bonds between the liquid molecules, allowing them to move into the gas phase. The temperature at which a gas becomes a liquid or, if the thermal energy is increased, a liquid becomes a gas is called the **condensation point** or the **boiling point.**

NOTE ▶ Liquid water becomes solid ice at 0°C, but that doesn't mean the temperature of ice is always 0°C. Ice reaches the temperature of its surroundings. If the air temperature in a freezer is −20°C, then the ice temperature is −20°C, Likewise, steam can be heated to temperatures above 100°C. That doesn't happen when you boil water on the stove because the steam escapes, but steam can be heated far above 100°C in a sealed container. ◀

Tin pest Some substances have multiple solid phases in which the atoms are arranged in lattices with different structures. Such substances can undergo a phase change between these solid phases; the atoms rearrange themselves, and the properties of the substance can change. For example, the element tin slowly changes from the usual "white" phase to a brittle "gray" phase at temperatures below 13°C. The tin sample in this photo was kept below this transition temperature for a year; the resulting "tin pest" is quite obvious. Some historians speculate that this phase change destroyed the tin buttons on the coats of Napoleon's troops during their winter invasion of Russia in 1812, contributing to their defeat as they froze in the bitter cold.

CONCEPTUAL EXAMPLE 12.4 Fast or slow boil?

You are cooking pasta on the stove; the water is at a slow boil. Will the pasta cook more quickly if you turn up the burner on the stove so that the water is at a fast boil?

REASON Water boils at 100°C; no matter how vigorously the water is boiling, the temperature is the same. It is the tempera-ture of the water that determines how fast the cooking takes place. Adding heat at a faster rate will make the water boil away more rapidly, but will not change the temperature—and will not alter the cooking time.

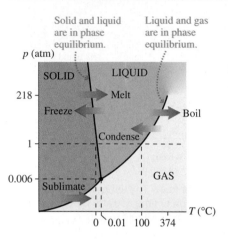

FIGURE 12.23 Phase diagram (not to scale) for water.

A **phase diagram** is used to show how the phases and phase changes of a substance vary with both temperature and pressure. Figure 12.23 shows the phase diagram for water. You can see that the diagram is divided into three regions corresponding to the solid, liquid, and gas phases. The boundary lines separating the regions indicate the phase transitions. The system is in phase equilibrium at a pressure-temperature point that falls on one of these lines.

Phase diagrams contain a great deal of information. Notice on the diagram that the dotted line at $p = 1$ atm crosses the solid-liquid boundary at 0°C and the liquid-gas boundary at 100°C. These well-known melting and boiling point temperatures of water apply only at standard atmospheric pressure. These temperatures only apply for pure water; if the water contains dissolved substances, the melting point is lowered and the boiling point is raised.

The variation of melting and boiling points with pressure has practical consequences. If you boil water in Denver, the "Mile-High City," where the altitude results in atmospheric pressure significantly less than that at sea level, you will find that the boiling point is several degrees below 100°C. It takes a bit longer to cook pasta in Denver than in coastal Miami. A *pressure cooker* works by allowing the pressure inside to exceed 1 atm. This raises the boiling point, so foods in boiling water are at a temperature $> 100°C$ and cook faster.

Heat of Transformation

In Figure 12.22b, the phase changes appeared as horizontal line segments on the graph. During these segments, heat is being transferred to the system but the temperature isn't changing. The thermal energy continues to increase during a phase change, but, as noted, the additional energy goes into breaking molecular bonds rather than speeding up the molecules. **A phase change is characterized by a change in thermal energy without a change in temperature.**

The amount of heat energy that causes 1 kg of a substance to undergo a phase change is called the **heat of transformation** of that substance. For example, laboratory experiments show that 333,000 J of heat are needed to melt 1 kg of ice at 0°C. The symbol for heat of transformation is L. The heat required for the entire system of mass M to undergo a phase change is

$$Q = ML \tag{12.18}$$

Heat of transformation is a generic term that refers to any phase change. Two specific heats of transformation are the **heat of fusion** L_f, the heat of transformation between a solid and a liquid, and the **heat of vaporization** L_v, the heat of transformation between a liquid and a gas. The heat needed for these phase changes is

$$Q = \begin{cases} \pm ML_f & \text{Heat needed to melt/freeze mass } M \\ \pm ML_v & \text{Heat needed to boil/condense mass } M \end{cases} \tag{12.19}$$

where the $\pm$ indicates that heat must be *added* to the system during melting or boiling but *removed* from the system during freezing or condensing. **You must explicitly include the minus sign when it is needed.**

Frozen frogs BIO It seems impossible, but common wood frogs survive the winter with much of their bodies frozen. They use the principle that the freezing point of water decreases when there are substances dissolved in it; this is also the principle behind the antifreeze in your car's radiator. Although the liquid water *between* cells in the frogs' bodies freezes, the water *inside* their cells remains liquid because of high concentrations of dissolved glucose. This prevents the cell damage that accompanies the freezing and subsequent thawing of tissues. When spring arrives, the frogs thaw and appear no worse for their winter freeze.

Table 12.5 lists some heats of transformation. Notice that the heat of vaporization is always much larger than the heat of fusion. Our atomic model explains this. Melting breaks just enough molecular bonds to allow the system to lose rigidity and flow. Even so, the molecules in a liquid remain close together and loosely bonded. Vaporization breaks all bonds completely and sends the molecules flying apart. This process requires a larger increase in the thermal energy and thus a larger quantity of heat.

TABLE 12.5 Melting/boiling temperatures and heats of transformation at standard atmospheric pressure

Substance	T_m (°C)	L_f (J/kg)	T_b (°C)	L_v (J/kg)
Nitrogen (N_2)	−210	0.26×10^5	−196	1.99×10^5
Ethyl alcohol	−114	1.09×10^5	78	8.79×10^5
Mercury	−39	0.11×10^5	357	2.96×10^5
Water	0	3.33×10^5	100	22.6×10^5
Lead	328	0.25×10^5	1750	8.58×10^5

Lava—molten rock—undergoes a phase change from liquid to solid when it contacts liquid water; the transfer of heat to the water causes the water to undergo a phase change from liquid to gas.

EXAMPLE 12.10 Melting a popsicle

A girl eats a 45 g frozen popsicle that was taken out of a −10°C freezer. How much energy does her body use to bring the popsicle up to body temperature?

PREPARE We can assume that the popsicle is pure water. Normal body temperature is 37°C. The specific heats of ice and liquid water are given in Table 12.4; the heat of fusion of water is given in Table 12.5.

SOLVE There are three parts to the problem: The popsicle must be warmed to 0°C, then melted, and then the resulting water must be warmed to body temperature. The heat needed to warm the frozen water by $\Delta T = 10°C = 10$ K to the melting point is

$$Q_1 = Mc_{ice}\Delta T = (0.045 \text{ kg})(2090 \text{ J/kg} \cdot \text{K})(10 \text{ K}) = 940 \text{ J}$$

Note that we use the specific heat of water ice, not liquid water, in this equation. Melting 45 g of ice requires heat

$$Q_2 = ML = (0.045 \text{ kg})(3.33 \times 10^5 \text{ J/kg}) = 15,000 \text{ J}$$

The liquid water must now be warmed to body temperature; this requires heat

$$Q_3 = Mc_{water}\Delta T = (0.045 \text{ kg})(4190 \text{ J/kg} \cdot \text{K})(37 \text{ K})$$
$$= 7000 \text{ J}$$

The total energy is the sum of these three values, $Q_{total} = 23,000$ J.

ASSESS More energy is needed to melt the ice than to warm the water, as we would expect. A commercial popsicle has 40 Calories, which is about 170 kJ. Roughly 15% of the chemical energy in this frozen treat is used to bring it up to body temperature!

Evaporation

Water boils at 100°C. But individual molecules of water can move from the liquid phase to the gas phase at lower temperatures. This process is known as **evaporation.** Water evaporates as sweat from your skin at a temperature well below 100°C. We can use our atomic model to explain this.

In gases, we have seen that there is a variation in particle speeds. The same principle applies to water and other liquids. At any temperature, some molecules will be moving fast enough to go into the gas phase. And they will do so, carrying away thermal energy as they go. The molecules that leave the liquid are the ones that have the highest kinetic energy, so evaporation reduces the average kinetic energy (and thus the temperature) of the liquid left behind.

The evaporation of water, both from sweat and in moisture you exhale, is one of the body's methods of exhausting the excess heat of metabolism to the environment, allowing you to maintain a steady body temperature. The heat to evaporate a mass M of water is $Q = ML_v$, and this amount of heat is removed from your body. However, the heat of vaporization L_v is a little larger than the value for boiling. At a skin temperature of 30°C, the heat of vaporization of water is $L_v = 24 \times 10^5$ J/kg, or 6% larger than the value at 100°C in Table 12.5.

Keeping your cool BIO Humans (and cattle and horses) have sweat glands, so we can perspire to moisten our skin, allowing evaporation to cool our bodies. Animals that do not perspire can also use evaporation to keep cool. Dogs, goats, rabbits, and even birds pant, evaporating water from their respiratory passages. Elephants spray water on their skin; other animals may lick their fur.

EXAMPLE 12.11 Computing heat loss by perspiration
The human body can produce approximately 30 g of perspiration per minute. At what rate is it possible to exhaust heat by the evaporation of perspiration?

SOLVE The evaporation of 30 g of perspiration at normal body temperature requires a heat energy

$$Q = (0.030 \text{ kg})(24 \times 10^5 \text{ J/kg}) = 7.2 \times 10^4 \text{ J}$$

This is the heat lost per minute; the rate of heat loss is

$$\frac{Q}{\Delta t} = \frac{7.2 \times 10^4 \text{ J}}{60 \text{ s}} = 1200 \text{ W}$$

ASSESS Given the metabolic power required for different activities, as listed in Chapter 11, this rate of heat removal is quite sufficient to keep the body cool even when exercising in hot weather—as long as the person drinks enough water to keep up this rate of perspiration!

STOP TO THINK 12.4   1 kg of barely molten lead, at 328°C, is poured into a large beaker holding liquid water right at the boiling point, 100°C. The mass of the water that will be boiled away as the lead solidifies is

A. 0 kg   B. <1 kg   C. 1 kg   D. >1 kg

12.6 Calorimetry

At one time or another you've probably put an ice cube into a hot drink to cool it quickly. You were engaged, in a somewhat trial-and-error way, in a practical aspect of heat transfer known as **calorimetry,** the quantitative measurement of the heat transferred between systems or evolved in reactions. You know that heat will be transferred from the hot drink into the cold ice cube, reducing the temperature of the drink, as shown in Figure 12.24.

Let's make this qualitative picture more precise. Figure 12.25 shows two systems thermally interacting with each other but isolated from everything else. Suppose they start at different temperatures T_1 and T_2. As you know, heat energy will be transferred from the hotter to the colder system until they reach a common final temperature T_f. The systems will then be in thermal equilibrium and the temperature will not change further. In the cooling coffee example of Figure 12.24, the coffee is System 1, the ice is System 2, and the insulating barrier is the mug.

The insulation prevents any heat energy from being transferred to or from the environment, so energy conservation tells us that any energy leaving the hotter system must enter the colder system. That is, the systems *exchange* energy with no net loss or gain. The concept is straightforward, but to state the idea mathematically we need to be careful with signs.

Let Q_1 be the energy transferred to system 1 as heat. Q_1 is positive if energy *enters* system 1, negative if energy *leaves* system 1. Similarly, Q_2 is the energy transferred to system 2. The fact that the systems are merely exchanging energy can be written $|Q_1| = |Q_2|$. That is, the energy *lost* by the hotter system is the energy *gained* by the colder system. But Q_1 and Q_2 have opposite signs, so $Q_1 = -Q_2$. No energy is exchanged with the environment, hence it makes more sense to write this relationship as

$$Q_{net} = Q_1 + Q_2 = 0 \tag{12.20}$$

This idea is easily extended. No matter how many systems interact thermally, Q_{net} must equal zero.

NOTE ▶ The signs are very important in calorimetry problems. ΔT is always $T_f - T_i$, so ΔT and Q are negative for any system whose temperature decreases. The proper sign of Q for any phase change must be supplied *by you*, depending on the direction of the phase change. ◀

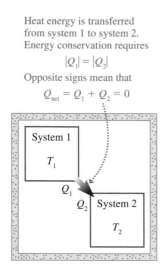

FIGURE 12.24 Cooling a hot drink with ice.

Heat energy is transferred from system 1 to system 2. Energy conservation requires

$$|Q_1| = |Q_2|$$

Opposite signs mean that

$$Q_{net} = Q_1 + Q_2 = 0$$

System 1
T_1
Q_1
Q_2 System 2
T_2

FIGURE 12.25 Two systems interact thermally. In this particular example, where heat is being transferred from system 1 to system 2, $Q_2 > 0$ and $Q_1 < 0$.

(MP) **PROBLEM-SOLVING STRATEGY 12.1** Calorimetry problems

PREPARE Identify the interacting systems. Assume that they are isolated from the larger environment. List known information and identify what you need to find. Convert all quantities to SI units.

SOLVE The statement of energy conservation is

$$Q_{net} = Q_1 + Q_2 + \cdots = 0$$

■ For systems that undergo a temperature change, $Q_{\Delta T} = Mc(T_f - T_i)$. Be sure to have the temperatures T_i and T_f in the correct order.

■ For systems that undergo a phase change, $Q_{phase} = \pm ML$. Supply the correct sign by observing whether energy enters or leaves the system during the transition.

■ Some systems may undergo a temperature change *and* a phase change. Treat the changes separately. The heat energy is $Q = Q_{\Delta T} + Q_{phase}$.

ASSESS The final temperature should be in the middle of the initial temperatures. A T_f that is higher or lower than all initial temperatures is an indication that something is wrong, usually a sign error.

EXAMPLE 12.12 Using calorimetry to identify a metal

200 g of an unknown metal is heated to 200.0°C then dropped into 50.0 g of water at 20.0°C in an insulated container. The water temperature rises within a few seconds to 39.7°C, then changes no further. Identify the metal.

PREPARE The metal and the water interact thermally; there are no phase changes. We know all the initial and final temperatures. We will label the temperatures as follows: The initial temperature of the metal is T_m; the initial temperature of the water is T_w. The common final temperature is T_f. For water, $c_w = 4190$ J/kg · K is known from Table 12.4. Only the specific heat c_m of the metal is unknown.

SOLVE Energy conservation requires $Q_w + Q_m = 0$. Using $Q = Mc(T_f - T_i)$ for each, we have

$$Q_w + Q_m = M_w c_w(T_f - T_w) + M_m c_m(T_f - T_m) = 0$$

This is solved for the unknown specific heat:

$$c_m = \frac{-M_w c_w(T_f - T_w)}{M_m(T_f - T_m)}$$

$$= \frac{-(0.0500 \text{ kg})(4190 \text{ J/kg} \cdot \text{K})(39.7°C - 20.0°C)}{(0.200 \text{ kg})(39.7°C - 200.0°C)}$$

$$= 129 \text{ J/kg} \cdot \text{K}$$

Referring to Table 12.4, we find we have either 200 g of gold or, if we made an ever-so-slight experimental error, 200 g of lead!

ASSESS The temperature of the unknown metal changed much more than the temperature of the water. That is, the unknown metal had much less thermal inertia, and thus, as we found, its specific heat should be much less than that of water.

EXAMPLE 12.13 Calorimetry with a phase change

Your 500 mL soda is at 20°C, room temperature, so you add 100 g of ice from the −20°C freezer. Does all the ice melt? If so, what is the final temperature? If not, what fraction of the ice melts? Assume that you have a well-insulated cup.

PREPARE We need to distinguish between two possible outcomes. If all the ice melts, then $T_f > 0°C$. It's also possible that the soda will cool to 0°C before all the ice has melted, leaving the ice and liquid in equilibrium at 0°C. We need to distinguish between these before knowing how to proceed. All

the initial temperatures, masses, and specific heats are known. The final temperature of the combined soda + ice system is unknown.

SOLVE Let's first calculate the heat needed to melt all the ice and leave it as liquid water at 0°C. To do so, we must warm the ice to 0°C, then change it to water. The heat input for this two-stage process is

$$Q_{melt} = M_{ice} c_{ice}(20 \text{ K}) + M_{ice} L_f = 37,000 \text{ J}$$

where L_f is the heat of fusion of water.

Continued

Q_{melt} is a *positive* quantity because we must *add* heat to melt the ice. Next, let's calculate how much heat energy will leave the soda if it cools all the way to 0°C. The volume is $V = 500$ mL $= 5.00 \times 10^{-4}$ m³ and its density is that of water, $\rho = 1000$ kg/m³, thus the mass is $M_{soda} = \rho V = 0.500$ kg. The heat is

$$Q_{cool} = M_{soda}c_{water}(-20 \text{ K}) = -42{,}000 \text{ J}$$

where $\Delta T = -20$ K because the temperature decreases. Because $|Q_{cool}| > Q_{melt}$, the soda has sufficient energy to melt all the ice. Hence the final state will be all liquid at $T_f > 0$. (Had we found $|Q_{cool}| < Q_{melt}$, then the final state would have been an ice-liquid mixture at 0°C.)

Energy conservation requires $Q_{ice} + Q_{soda} = 0$. The heat Q_{ice} consists of three terms: warming the ice to 0°C, melting the ice to water at 0°C, then warming the 0°C water to T_f. The mass will still be M_{ice} in the last of these steps, because it is the "ice system," but we need to use the specific heat of *liquid water*. Thus

$$Q_{ice} + Q_{soda} = [M_{ice}c_{ice}(20 \text{ K}) + M_{ice}L_f$$
$$+ M_{ice}c_{water}(T_f - 0°C)]$$
$$+ M_{soda}c_{water}(T_f - 20°C) = 0$$

We've already done part of the calculation, allowing us to write

$$37{,}000 \text{ J} + M_{ice}c_{water}(T_f - 0°C) + M_{soda}c_{water}(T_f - 20°C) = 0$$

Solving for T_f gives

$$T_f = \frac{20M_{soda}c_{water} - 37{,}000}{M_{ice}c_{water} + M_{soda}c_{water}} = 1.9°C$$

ASSESS A good deal of ice has been put in the soda, so it ends up being cooled nearly to the freezing point, as we might expect.

STOP TO THINK 12.5   1 kg of lead at 100°C is dropped into a container holding 1 kg of water at 0°C. Once the lead and water reach thermal equilibrium, the final temperature is

A. <50°C.   B. 50°C.   C. >50°C.

12.7 Thermal Properties of Gases

We've examined the thermal properties of solids and liquids—their thermal expansion and their specific heat. We now turn to the thermal properties of gases—what happens when they are heated? Gases are harder to characterize than solids or liquids because the heat required to cause a specified temperature change depends on the *process* by which the gas changes state.

Just as for solids and liquids, heating a gas changes its temperature. But by how much? Figure 12.26 shows two isotherms on the pV diagram for a gas. Processes A and B, which start on the T_i isotherm and end on the T_f isotherm, have the *same* temperature change $\Delta T = T_f - T_i$, so we might expect them both to require the same amount of heat. But it turns out that process A, which takes place at constant volume, requires *less* heat than does process B, which occurs at constant pressure. The reason is that work is done in process B but not in process A.

It is useful to define two different versions of the specific heat of gases, one for constant volume processes and one for constant pressure processes. We will define these as *molar* specific heats because we usually do gas calculations using moles instead of mass. The quantity of heat needed to change the temperature of n moles of gas by ΔT is, for a constant volume process

$$Q = nC_V \, \Delta T \tag{12.21}$$

and for a constant pressure process

$$Q = nC_P \, \Delta T \tag{12.22}$$

C_V is the **molar specific heat at constant volume** and C_P is the **molar specific heat at constant pressure.** Table 12.6 gives the values of C_V and C_P for a few common monatomic and diatomic gases. The units are J/mol · K. The values for air are essentially equal to those for N_2.

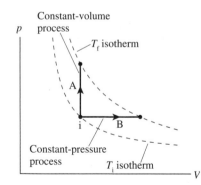

FIGURE 12.26 Processes A and B have the same ΔT and the same ΔE_{th}, but they require different amounts of heat.

It's interesting that all the monatomic gases in Table 12.6 have the same value for C_P and for C_V. The same is very nearly true for the diatomic gases. Why should this be? We found in Chapter 11 that the average kinetic energy of the atoms in a gas is $K_{avg} = \frac{3}{2}k_B T$, a fact we used earlier in the kinetic-theory derivation of the pressure of an ideal gas. For an ideal monatomic gas, the thermal energy E_{th} is simply the sum total of the kinetic energies of all the atoms; there is no other way a monatomic gas can store energy. Thus the thermal energy of a gas of N atoms is

$$E_{th} = N\left(\frac{3}{2}k_B T\right) = \frac{3}{2}n(N_A k_B)T = \frac{3}{2}nRT$$

In arriving at this last expression, we first related the number of atoms to the number of moles by $N = nN_A$, then used the definition of the gas constant, $R = N_A k_B$.

If the temperature of a monatomic gas changes by ΔT, its thermal energy changes by

$$\Delta E_{th} = \frac{3}{2}nR\Delta T \qquad (12.23)$$

If we keep the volume of the gas constant so that no work is done, this energy change can come only from heat:

$$Q = \frac{3}{2}nR\Delta T \qquad (12.24)$$

Comparing Equation 12.24 with the definition of molar specific heat in Equation 12.21, we see that the molar specific heat at constant volume must be

$$C_V(\text{monatomic gas}) = \frac{3}{2}R = 12.5 \text{ J/mol} \cdot \text{K} \qquad (12.25)$$

This theoretical prediction of kinetic theory is exactly the measured value of C_V in Table 12.6. **The perfect agreement of theory and experiment is strong evidence that the atomic model of gases that we have been using is correct.**

The molar specific heat at constant pressure differs because in an isobaric (constant-pressure) process, such as process B in Figure 12.26, the gas does work. The expression for ΔE_{th} in Equation 12.23 is still valid, but now, according to the first law of thermodynamics, $\Delta E_{th} = Q - W_{gas}$. Thus

$$Q = \Delta E_{th} + W_{gas} = \frac{3}{2}nR\Delta T + p\Delta V \qquad (12.26)$$

where we used the fact, discovered earlier, that the work done in an isobaric process is $W_{gas} = p\Delta V$.

The ideal-gas law, $pV = nRT$, implies that if p is constant and only V and T change, then $p\Delta V = nR\Delta T$. Using this result in Equation 12.26 gives the heat needed to change the temperature by ΔT in a constant-pressure process:

$$Q = \frac{3}{2}nR\Delta T + nR\Delta T = \frac{5}{2}nR\Delta T$$

A comparison with the definition of molar specific heat shows that

$$C_P(\text{monatomic gas}) = \frac{5}{2}R = 20.8 \text{ J/mol} \cdot \text{K} \qquad (12.27)$$

Checking the data in Table 12.6 shows that the theory-experiment agreement remains perfect.

TABLE 12.6 Molar specific heats of gases (J/mol · K) at 20°C

Gas	C_P	C_V
Monatomic Gases		
He	20.8	12.5
Ne	20.8	12.5
Ar	20.8	12.5
Diatomic Gases		
H_2	28.7	20.4
N_2	29.1	20.8
O_2	29.2	20.9
Triatomic Gases		
Water vapor	33.3	25.0

The thermal energy of a diatomic gas is the sum of the translational kinetic energy of the molecules . . .

. . . plus the rotational kinetic energy of the molecules.

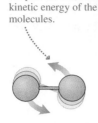

FIGURE 12.27 Thermal energy of a diatomic gas.

The thermal energy of a monatomic gas consists exclusively of the translational kinetic energy of the atoms; heating a monatomic gas means that the atoms move faster. Air is primarily made of the diatomic gases N_2 and O_2. The thermal energy of a diatomic gas has more possible forms than for a monatomic gas, as shown in Figure 12.27. Heating a diatomic gas makes the atoms move faster, but it also causes them to rotate more rapidly. For this reason, the specific heat of a diatomic gas is higher than the specific heat of a monatomic gas, as we see in Table 12.6. The table shows that water vapor, a triatomic gas, has an even higher specific heat, as we expect.

Heat Engines

Chapter 11 introduced the idea of a *heat engine,* a device that takes in heat energy from a hot reservoir, does useful work, and exhausts waste heat to a cold reservoir. Many real heat engines rely on gas processes, and we've now reached the point where we can examine heat engines more closely. A diagram of the (somewhat simplified) operation of a real heat engine, a two-stroke gasoline engine, is shown in Figure 12.28.

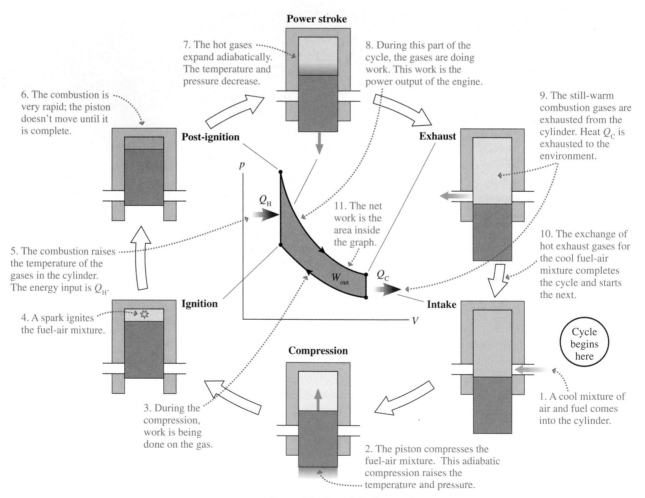

FIGURE 12.28 A simplified model of a gasoline engine.

NOTE ▶ It is common to use "gas" as shorthand for "gasoline." In this chapter, "gas" always refers to the state of matter. We will use "gasoline" or "fuel" to refer to the chemical energy source that you put in your tank. ◀

In this engine, the gases in the cylinder expand and contract in a repeating cycle as the piston moves back and forth. The cycle begins at the noted point with the introduction of gases (a mixture of air and fuel) into the cylinder. From this point, we can use our knowledge of gas processes to analyze the operation of the engine as it goes through its cycle. Figure 12.28 shows the connection between the operation of the engine and relevant points on the pV diagram of the gases in the engine's cylinder.

There are several important characteristics of this heat engine that apply generally to all heat engines:

- The sequence of processes forms a closed loop in the pV diagram. This must be true for the engine to go through its cycle over and over.
- Heat Q_H is taken in from a hot reservoir (in this case, from the burning fuel) at some point during the cycle. To bring the temperature back down at the end of the cycle, heat Q_C must be exhausted to a cold reservoir. This heat is carried out of the engine by the hot exhaust gases.
- The work done by the engine, W_{out}, is the area *inside* the closed loop on the pV diagram.
- Because the temperature returns to its starting point at the end of each cycle, there's no net change in thermal energy. Thus from energy conservation, $W_{out} = Q_H - Q_C$.

Work is the area inside the closed loop because of our earlier discovery that the work done in a single process is the area under the graph. The work is negative for the compression, because ΔV is negative, and positive for the expansion. Thus the net work is the area under the upper curve *minus* the area under the lower curve, which is just the area inside the closed loop. This is true for all heat engines, not just the example shown here.

Activ Physics ONLINE 8.12–8.14

EXAMPLE 12.14 Finding the pressure and temperature in a gasoline engine

The adiabatic compression stroke in an automobile engine raises the pressure to 17 atm and the temperature to 300°C. The ignition of the fuel-air mixture then provides 800 J of heat. For a typical engine, the cylinder contains 0.021 mol of gas. What are the pressure and temperature immediately after ignition?

PREPARE Ignition happens so fast there is no time for the piston to move, so the gas is heated in a constant-volume process. The gas is mostly air, so we'll use $C_V = 20.8$ J/mol · K as the molar specific heat at constant volume.

SOLVE The temperature increase during a constant-volume heating is

$$\Delta T = \frac{Q}{nC_V} = \frac{800 \text{ J}}{(0.021 \text{ mol})(20.8 \text{ J/mol} \cdot \text{K})} = 1800°C$$

Thus igniting the fuel raises the temperature from $T_i = 300°C$ at the end of the compression stroke to $T_f = 2100°C$. The volume doesn't change, so the new pressure is

$$p_f = p_i \frac{V_i}{V_f} \frac{T_f}{T_i} = (17 \text{ atm})(1)\frac{2100 + 273}{300 + 273} = 70 \text{ atm}$$

ASSESS The pressure and temperature are both very large, but that's to be expected following an explosion in a confined volume. When you drive your car, a series of such explosions in your engine's cylinders provide the necessary force to move your car down the road. In a typical car moving at highway speeds, there are over 100 explosions taking place under your hood each second.

12.8 Heat Transfer

You feel warmer when the sun is shining on you, colder when sitting on a cold concrete bench or when a stiff wind is blowing. This is due to the transfer of heat. Although we've talked about heat a lot in these last two chapters, we haven't said much about *how* heat is transferred from a hotter object to a colder object. There are four basic mechanisms, outlined in the table below, by which objects exchange heat with other objects or their surroundings. Evaporation was treated in an earlier section; in this section we will consider the other mechanisms.

Heat transfer mechanisms

When two objects are in direct physical contact, such as the soldering iron and the circuit board, heat is transferred by *conduction.* **Energy is transferred by direct contact.**

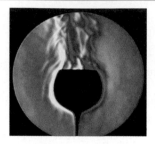

This special photograph shows air currents near a warm glass of water. Air near the glass is warmed and rises, taking thermal energy with it in a process known as *convection.* **Energy is transferred by the motion of molecules with high thermal energy.**

The lamp shines on the lambs huddled below, warming them. The energy is transferred by infrared *radiation,* a form of electromagnetic waves. **Energy is transferred by electromagnetic waves.**

As we saw in a previous section, the *evaporation* of liquid can carry away significant quantities of thermal energy. When you blow on a cup of cocoa, this increases the rate of evaporation, rapidly cooling it. **Energy is transferred by the removal of molecules with high thermal energy.**

Conduction

If you hold a metal spoon in a cup of hot coffee, the handle of the spoon soon gets warm. Thermal energy is transferred along the spoon from the coffee to your hand. The difference in temperature between the two ends drives this heat transfer by a process known as **conduction.** Conduction is the transfer of thermal energy directly through a physical material.

Figure 12.29 shows a copper rod placed between a hot reservoir (a fire) and a cold reservoir (a block of ice). We can use our atomic model to see how thermal energy is transferred along the rod by the interaction between atoms in the rod; fast-moving atoms at the hot end transfer energy to slower-moving atoms at the cold end.

Suppose we set up a series of experiments to measure the heat Q transferred through various rods. We would find the following trends in our data:

- Q increases if the temperature difference ΔT between the hot end and the cold end is increased.
- Q increases if the cross-section area A of the rod is increased.
- Q decreases if the length L of the rod is increased.
- Some materials (such as metals) have a high value of Q; a lot of energy is transferred. Other materials (such as wood) have a low value of Q; very little energy is transferred.

The final observation is one that is familiar to you; if you are stirring a pot of hot soup on the stove, you use a wood or plastic spoon rather than a metal one.

These experimental observations about heat conduction can be summarized in a single formula. If heat Q is transferred in a time interval Δt, the *rate* of heat

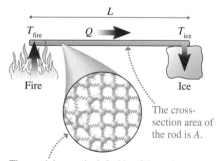

The particles on the left side of the rod are vibrating more vigorously than the particles on the right. The particles on the left transfer energy to the particles on the right via the bonds connecting them.

FIGURE 12.29 Conduction of heat in a solid rod.

transfer (joules per second) is $Q/\Delta t$. For a material of cross-section area A and length L, spanning a temperature difference ΔT, the rate of heat transfer is

$$\frac{Q}{\Delta t} = \left(\frac{kA}{L}\right)\Delta T \qquad (12.28)$$

Rate of conduction of heat across a temperature difference

The quantity k, which characterizes whether the material is a good conductor of heat or a poor conductor, is called the **thermal conductivity** of the material. Because the heat transfer rate J/s is a *power,* measured in watts, the units of k are W/m · K. Typical values of k for common materials are given in Table 12.7; a larger number for k means a material is a better conductor of heat.

Note that good conductors of electricity, such as silver and copper, are usually good heat conductors. One exception is diamond: Though diamond is a very poor electrical conductor, the strong bonds among atoms that make diamond such a hard material lead to a rapid transfer of thermal energy.

The weak bonds among molecules make most biological materials poor conductors of heat. Muscle is a better conductor than fat, so sea mammals have thick layers of fat for insulation. Land mammals insulate their bodies with fur, birds with feathers. Both fur and feathers trap a good deal of air, so their conductivity is similar to that of air. Air, like all gases, is a poor conductor of heat, as there are no bonds between adjacent atoms.

TABLE 12.7 Thermal conductivity values (measured at 20°C)

Material	k (W/m · K)
Diamond	1000
Silver	420
Copper	400
Iron	72
Stainless steel	14
Ice	1.7
Concrete	0.8
Plate glass	0.75
Skin	0.50
Muscle	0.46
Fat	0.21
Wood	0.2
Carpet	0.04
Fur, feathers	0.02–0.06
Air (27°C, 100 kPa)	0.026

EXAMPLE 12.15 Heat loss through a window

A 0.90 m by 1.2 m window (approximately 3 ft by 4 ft) in a house is made of a single pane of 3.0-mm-thick (approximately 1/8 in) glass. The temperature inside the house is 21°C; outside, it is a frosty −5°C. What power, in watts, must the furnace provide to compensate for heat loss through the window?

PREPARE The area through which heat is conducted is

$$A = (0.90 \text{ m})(1.2 \text{ m}) = 1.1 \text{ m}^2$$

The length of the conducting medium is the glass thickness, $L = 0.0030$ m. The temperature difference across the glass is

$$T = 21°C - (-5°C) = 26°C = 26 \text{ K}.$$

SOLVE Heat is transferred through the window to the outside. To keep the house at a constant temperature, the furnace must add heat energy at the same rate. The rate of conduction through the window is

$$\frac{Q}{\Delta t} = \left(\frac{(0.75 \text{ W/m} \cdot \text{K})(1.1 \text{ m}^2)}{0.0030 \text{ m}}\right)(26 \text{ K}) = 7100 \text{ W}$$

To compensate, the furnace must provide 7100 W of heating power.

ASSESS This is a great deal of power—three or four times the power of common electric space heaters. In modern construction, windows generally consist of two panes of glass with an air space in between. The low thermal conductivity of this air gap dramatically reduces the energy loss by conduction.

Carpet or tile? If you walk around your house barefoot, you will notice that tile floors feel much colder than carpeted floors. The tile floor isn't really colder—the temperatures of all the floors in your house are nearly the same. But tile has a much higher thermal conductivity than carpet, so more heat will flow from your feet into the tile than into carpet. This causes the tile to feel colder. Standing or lying on carpet, with its low thermal conductivity, is much more comfortable than on tile!

Warm water (colored) moves by convection.

A feather coat BIO Penguins are birds and thus have feathers, even though they do not fly. The feathers are more obvious on this penguin chick than on the adults. A penguin's short, dense feathers serve a different role than the flight feathers of other birds: They trap air to provide thermal insulation. The feathers are equipped with muscles at the base of the shafts to flatten them and eliminate these air pockets, otherwise the penguins would be too buoyant to swim underwater.

Convection

We noted that air is a poor conductor of heat. In fact, the data in Table 12.7 show that air has a thermal conductivity that is comparable to that of feathers. So why, on a cold day, will you be more comfortable if you are wearing a down jacket?

Thermal energy is easily transferred through air, water, and other fluids when the air and water can flow. A pan of water on the stove is heated at the bottom. This heated water expands and becomes less dense than the water above it, so it rises to the surface while cooler, denser water sinks to take its place. This transfer of thermal energy by the motion of a fluid is known as **convection.**

Convection is usually the main mechanism for heat transfer in fluid systems. On a small scale, convection mixes the pan of water that you heat on the stove; on a large scale, convection is responsible for making the wind blow and ocean currents circulate. Air is a very poor thermal conductor, but it is very effective at transferring energy by convection. To use air for thermal insulation, it is necessary to trap the air in small pockets to limit convection. And that's exactly what feathers, fur, double-paned windows, and fiberglass insulation do. Convection is much more rapid in water than in air, which is why people can die of hypothermia in 65°F water but can live quite happily in 65°F air.

Forced convection occurs when fluid is pushed by a pump or motor. In most buildings, forced hot-air heating transfers heat from the furnace to the rooms. Forced convection is also important for regulating body temperature. If your body is too hot, blood vessels near the cooler skin dilate to carry more blood. Blood flow transfers heat from the core of your body to the skin and then to your surroundings. Elephants use their ears for forced convection as well. The motion of air across the body from flapping ears is a very effective means of keeping cool.

Radiation

The sunlight that falls on the earth increases the thermal energy of the earth's surface each day. But the earth cools down at night; clearly, there must be some way for the earth to transfer energy back to space.

The earth and all other objects emit energy in the form of **radiation,** electromagnetic waves that are generated by oscillating electric charges in the atoms that form the object, an issue we'll explore further in later chapters on electricity and magnetism. These waves carry energy, transferring energy from the object that emits the radiation to the object that absorbs it. Electromagnetic waves carry energy from the sun; when this sunlight falls on your skin, the energy is absorbed, increasing your skin's thermal energy. Your skin also emits electromagnetic radiation, which reduces your skin's thermal energy. Radiation is a significant part of the energy balance that keeps your body at the proper temperature.

> **NOTE** ▶ The word "radiation" comes from "radiate," which means "to beam." You have likely used the word to refer to x-rays and other forms of *ionizing radiation.* This is not the sense we will use in this chapter. Here, we will use radiation to mean light and other forms of electromagnetic waves that can "beam" from an object. ◀

Visible light, the red-to-violet colors of the rainbow, is an electromagnetic wave. Electromagnetic waves of somewhat lower frequency are called *infrared radiation;* we can't see these waves, but we can sometimes sense them as "heat" as they are absorbed on our skin. All objects radiate; objects at room temperature and even somewhat higher temperatures emit virtually all of their radiation as infrared radiation. As the temperature increases, the charged particles oscillate more vigorously and an object becomes "red hot" and eventually "white hot"— some of the radiation is now in the form of visible light. The white light from an incandescent light bulb is radiation emitted by a thin wire filament that has been heated to a very high temperature by an electric current. The radiation that the earth emits, or that you emit, is no different, it is simply lower-frequency infrared radiation that you can't see.

Even though we can't see infrared radiation, some films and detectors are infrared sensitive and can record the infrared radiation from objects. The false-color thermal image of the elephant that opened this chapter shows the infrared emission as the elephant radiates energy into the cooler environment. Warmer patches, such as the trunk, emit more radiation and are easily distinguished from cooler patches that radiate less. Elephants do not sweat, so radiation from their trunk and ears is an important part of their temperature-regulation system.

If an object's temperature is increased, the increasingly vigorous oscillations radiate more energy. The energy radiated by any object thus shows a strong dependence on temperature. If heat energy Q is radiated in a time interval Δt by an object with surface area A and absolute temperature T, the *rate* of heat transfer $Q/\Delta t$ (joules per second) is found to be

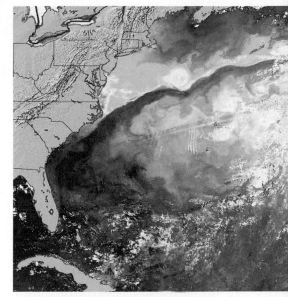

Global heat transfer This satellite image shows radiation emitted by the waters of the ocean off the east coast of the United States. You can clearly see the warm waters of the Gulf Stream, a large-scale convection that transfers heat to northern latitudes. The satellite "sees" the radiation from the earth, so this radiation must readily pass through the atmosphere into space. In fact, the *only* means by which the earth as a whole can reduce its thermal energy is by radiation, as there is no conduction or convection in the vacuum of space. This has consequences for the energy balance of the earth, as we consider in a section on the greenhouse effect at the end of Part III.

$$\frac{Q}{\Delta t} = e\sigma AT^4 \qquad (12.29)$$

Rate of heat transfer by radiation at temperature T (Stefan's Law)

Quantities in this equation are defined as follows:

- e is the **emissivity** of a surface, a measure of the effectiveness of radiation. The value of e ranges from 0 to 1. Human skin is a very effective radiator at body temperature, with $e = 0.97$, regardless of skin color.
- T is the absolute temperature in kelvin.
- A is the surface area in m^2.
- σ is a constant known as the Stefan-Boltzmann constant, with the value $\sigma = 5.67 \times 10^{-8}$ W/m$^2 \cdot$ K^4.

Notice the very strong fourth-power dependence on temperature. Doubling the absolute temperature of an object increases the radiant heat transfer by a factor of 16!

The amount of energy radiated by an animal can be surprisingly large. An adult human with bare skin in a room at a comfortable temperature has a skin temperature of approximately 33°C, or 306 K. A typical value for the skin's surface area is 1.8 m^2. With these values and the emissivity of skin noted above, we can calculate the rate of heat transfer via radiation from the skin:

$$\frac{Q}{\Delta t} = e\sigma AT^4 = (0.97)\left(5.67 \times 10^{-8} \frac{W}{m^2 \cdot K^4}\right)(1.8 \text{ m}^2)(306 \text{ K})^4 = 870 \text{ W}$$

As we learned in Chapter 11, the body at rest generates approximately 100 W of thermal energy. If the body radiated energy at 870 W, it would quickly cool. At this rate of emission, the body temperature would drop by 1°C every seven

The infrared radiation emitted by a dog is captured in this image. His cool nose and paws radiate much less energy than the rest of his body.

minutes! Clearly, there must be some mechanism to balance this emitted radiation: the radiation *absorbed* by the body.

When you sit in the sun, your skin warms due to the radiation you absorb. Even if you are not in the sun, you are absorbing the radiation emitted by the objects surrounding you. Suppose an object at temperature T is surrounded by an environment at temperature T_0. The *net* rate at which the object radiates heat energy—that is, radiation emitted minus radiation absorbed—is

$$\frac{Q_{\text{net}}}{\Delta t} = e\sigma A(T^4 - T_0^4) \tag{12.30}$$

This makes sense. An object should have no net energy transfer by radiation if it's in thermal equilibrium ($T = T_0$) with its surroundings. Note that the emissivity e appears for absorption as well; objects that are good emitters are good absorbers as well.

EXAMPLE 12.16 Determining energy loss by radiation for the body

A person with skin temperature of 33°C is in a room at 24°C. What is the net rate of heat transfer by radiation?

SOLVE Body temperature is $T = 33 + 273 = 306$ K; the temperature of the room is $T_0 = 24 + 273 = 297$ K. The net radiation rate, given by Equation 12.30, is

$$\frac{Q_{\text{net}}}{\Delta t} = e\sigma A(T^4 - T_0^4)$$

$$= (0.97)\left(5.67 \times 10^{-8}\,\frac{\text{W}}{\text{m}^2 \cdot \text{K}^4}\right)(1.8\,\text{m}^2)\left[(306\,\text{K})^4 - (297\,\text{K})^4\right] = 98\,\text{W}$$

ASSESS This is a reasonable value, roughly matching your resting metabolic power. When you are dressed (little convection) and sitting on wood or plastic (little conduction), radiation is your body's primary mechanism for dissipating the excess thermal energy of metabolism.

STOP TO THINK 12.6 Suppose you are an astronaut in space, hard at work in your sealed spacesuit. The only way that you can transfer heat to the environment is by

A. Conduction.   B. Convection.   C. Radiation.   D. Evaporation.

SUMMARY

The goal of Chapter 12 has been to use the atomic model of matter to explain and explore many macroscopic phenomena associated with heat, temperature, and the properties of matter.

GENERAL PRINCIPLES

Atomic Model

We model matter as being made of simple basic particles. The relationship of these particles to each other defines the phase.

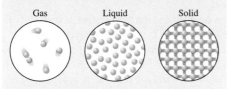

Gas   Liquid   Solid

The atomic model explains thermal expansion, specific heat, and heat transfer.

Kinetic Theory

Macroscopic properties of gases can be explained in terms of a microscopic view. The energy of motion of the particles of the gas is related to the temperature of the gas; the collisions of the particles with each other and the walls of the container are related to the pressure.

p, V, T

Ideal-Gas Law

The ideal-gas law can be derived from kinetic theory. It relates volume, pressure, and temperature in gases. The general law is

$$pV = nRT$$

For a gas process in a sealed container:

$$\frac{p_i V_i}{T_i} = \frac{p_f V_f}{T_f}$$

The gauge pressure of a gas is:

$$p_g = p - 1 \text{ atm}$$

IMPORTANT CONCEPTS

Effects of heat transfer

A system that is heated can either change temperature or change phase.

The **specific heat** c of a material is the heat required to raise 1 kg by 1 K.

$$Q = Mc\,\Delta T$$

The **heat of transformation** is the energy necessary to change the phase of 1 kg of a substance. Heat is added to change a solid to a liquid or a liquid to a gas; heat is removed to reverse these changes.

$$Q = \begin{cases} \pm ML_f \text{ (melt/freeze)} \\ \pm ML_v \text{ (boil/condense)} \end{cases}$$

The **molar specific heat** of a gas depends on the process.

For a constant-volume process: $\quad Q = nC_V\,\Delta T$

For a constant-pressure process: $\quad Q = nC_P\,\Delta T$

Mechanisms of heat transfer

An object can transfer heat to other objects or its environment:

Conduction is the transfer of heat by direct physical contact.

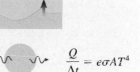

$$\frac{Q}{\Delta t} = \left(\frac{kA}{L}\right)\Delta T$$

Convection is the transfer of heat by the motion of a fluid.

Radiation is the transfer of heat by electromagnetic waves.

$$\frac{Q}{\Delta t} = e\sigma AT^4$$

A pV (pressure-volume) diagram is a useful means of looking at a process involving a gas.

A **constant-volume** process has no change in volume.	An **isobaric** process happens at a constant pressure.	An **isothermal** process happens at a constant temperature.	An **adiabatic** process involves no transfer of heat; the temperature changes.	The work done by a gas is the area under the graph.

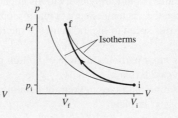

APPLICATIONS

Thermal expansion Objects experience an increase in volume and an increase in length when their temperature changes.

$$\Delta V = \beta V_i\,\Delta T \qquad \Delta L = \alpha L_i\,\Delta T$$

Calorimetry When two or more systems interact thermally, they come to a common final temperature determined by

$$Q_{net} = Q_1 + Q_2 + Q_3 + \cdots = 0$$

The number of **moles** is

$$n = \frac{M \text{ (in grams)}}{M_{mol}}$$

(MP) For instructor-assigned homework, go to
www.masteringphysics.com

Problem difficulty is labeled as I (straightforward) to IIII (challenging).

Problems labeled INT integrate significant material
from earlier chapters; BIO are of biological or medical
interest.

QUESTIONS

Conceptual Questions

1. You need to precisely measure the dimensions of a large wood panel for a construction project. Your metal tape measure was left outside for hours in the sun on a hot summer day, and now the tape is so hot it's painful to pick up. How will your measurements differ from those taken by your coworker, whose tape stayed in the shade? Explain.

2. If you buy a sealed bag of potato chips in Miami and drive with it to Denver, where the atmospheric pressure is lower, you will find that the bag gets very "puffy." Explain why.

3. If the pressure of a gas is really due to the *random* collisions of molecules with the walls of the container, why do pressure gauges—even very sensitive ones—give perfectly steady readings? Shouldn't the gauge be continually jiggling and fluctuating? Explain.

4. If you double the average speed of the molecules in a gas, by what factor does the pressure change? Give a simple explanation why the pressure changes by this factor.

5. Two gases have the same number of molecules per cubic meter (N/V) and the same rms speed. The molecules of gas 2 are more massive than the molecules of gas 1.
   a. Do the two gases have the same pressure? If not, which is larger?
   b. Do the two gases have the same temperature? If not, which is larger?

6. According to kinetic theory, the pressure of a gas depends on the number of molecules per cubic meter (N/V) and the rms speed of the gas molecules. Consider a sealed container of gas that is heated at constant volume.
   a. According to the ideal-gas law, does the pressure of the gas increase or stay the same? Explain.
   b. Does the number density of the gas increase or stay the same? Explain.
   c. What can you infer from these observations about a relationship between the gas temperature (a macroscopic parameter) and the rms speed of the molecules (a microscopic parameter)?

7. a. Which contains more particles, a mole of helium gas or a mole of oxygen gas? Explain.
   b. Which contains more particles, a gram of helium gas or a gram of oxygen gas? Explain.

8. You have 100 g of aluminum and 100 g of lead.
   a. Which contains a larger number of moles? Explain.
   b. Which contains more atoms? Explain.

9. Suppose you could suddenly increase the speed of every atom in a gas by a factor of 2.
   a. Does the thermal energy of the gas change? If so, by what factor? If not, why not?
   b. Does the molar specific heat change? If so, by what factor? If not, why not?

10. A gas is in a sealed container. By what factor does the gas pressure change if
    a. The volume is doubled and the temperature is tripled?
    b. The volume is halved and the temperature is tripled?

11. A gas is in a sealed container. By what factor does the gas temperature change if
    a. The volume is doubled and the pressure is tripled?
    b. The volume is halved and the pressure is tripled?

12. A gas is in a rigid container. The gas pressure is tripled and the temperature is doubled. What happens to the number of moles of gas in the container?

13. What is the maximum amount of work that a gas can do during a constant-volume process?

14. Materials A and B have equal densities, but A has a larger specific heat than B. You have 100 g cubes of each material. Cube A, initially at 0°C, is placed in good thermal contact with cube B, initially at 200°C. The cubes are inside a well-insulated container where they don't interact with their surroundings. Is their final temperature greater than, less than, or equal to 100°C? Explain.

15. Two containers hold equal masses of nitrogen gas at equal temperatures. You supply 10 J of heat to container A while not allowing its volume to change, and you supply 10 J of heat to container B while not allowing its pressure to change. Afterward, is temperature T_A greater than, less than, or equal to T_B? Explain.

16. You need to raise the temperature of a gas by 10°C. To use the smallest amount of heat energy, should you heat the gas at constant pressure or at constant volume? Explain.

17. The *rapid* compression of a gas by a fast-moving piston increases the gas temperature. For example, you likely have noticed that pumping up a bicycle tire causes the bottom of the pump to get warm. Consider a gas that is rapidly compressed by a piston.
    a. Does the thermal energy of the gas increase or stay the same? Explain.
    b. Is there a transfer of heat energy to the gas? Explain.
    c. Is work done on the gas? Explain.

18. A student is heating chocolate in a pan on the stove. He uses a cooking thermometer to measure the temperature of the chocolate and sees it varies as shown in Figure Q12.18. Describe what is happening to the chocolate in each of the three portions of the graph.

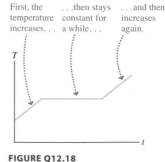

FIGURE Q12.18

19. In early spring, just as the ice on the surface of a lake finishes melting, a scientist measures the water temperature as a function of depth. Describe qualitatively how the temperature should change as the depth increases. Compare this result to the pattern that would be expected in the middle of summer.

20. If you bake a cake at high elevation, where atmospheric pressure is lower than at sea level, you will need to adjust the recipe. You will need to cook the cake for a longer time, and you will need to add less baking powder. (Baking powder is a leavening agent. As it heats, it releases gas bubbles that cause the cake to rise.) Explain why those adjustments are necessary.

21. The specific heat of aluminum is higher than that of iron. 1 kg blocks of iron and aluminum are heated to 100°C, and each is then dropped into its own 1 L beaker of 20°C water. Which beaker will end up with the warmer water? Explain.

22. A student is asked to sketch a pV diagram for a gas that goes through a cycle consisting of (a) an isobaric expansion, (b) a constant-volume reduction in temperature, and (c) an isothermal process that returns the gas to its initial state. The student draws the diagram shown in Figure Q12.22. What, if anything, is wrong with the student's diagram?

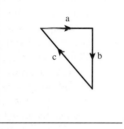

FIGURE Q12.22

23. If you have two spoons of the same size, one silver and one stainless steel, there is a quick test to tell which is which. Hold the end of a spoon in each hand, then lower them both into a cup of very hot water. One spoon will feel hot first. Is that the silver spoon or the stainless steel? Explain.

24. BIO People with body piercings are advised to remove their exposed body jewelry before engaging in cold-weather activities. The skin surrounding the metal in earrings, lip rings, nose rings, or eyebrow rings can get frostbite much more easily than other skin. Explain why.

25. If you live somewhere with cold, clear nights, you may have noticed some mornings when there was frost on open patches of ground but not under trees. This is because the ground under trees does not get as cold as open ground. Explain how tree cover keeps the ground under trees warmer.

Multiple-Choice Questions

26. | A tire is inflated to a gauge pressure of 35 psi. The absolute pressure in the tire is
    A. Less than 35 psi.
    B. Equal to 35 psi
    C. Greater than 35 psi.

27. | The number of atoms in a container is increased by a factor of 2 while the temperature is held constant. The pressure
    A. Decreases by a factor of 4.
    B. Decreases by a factor of 2.
    C. Stays the same.
    D. Increases by a factor of 2.
    E. Increases by a factor of 4.

28. | Suppose you do a calorimetry experiment to measure the specific heat of a penny. You take a number of pennies, measure their mass, heat them to a known temperature, and then drop them into a container of water at a known temperature. You then deduce the specific heat of a penny by measuring the temperature change of the water. Unfortunately, you didn't realize that you dropped one penny on the floor while transferring them to the water. This will
    A. Cause you to underestimate the specific heat.
    B. Cause you to overestimate the specific heat.
    C. Not affect your calculation of specific heat.

29. | A cup of water is heated with a heating coil that delivers 100 W of heat. In one minute, the temperature of the water rises by 20°C. What is the mass of the water?
    A. 72 g   B. 140 g
    C. 720 g   D. 1.4 kg

30. | Three identical beakers each hold 1000 g of water at 20°C. 100 g of liquid water at 0°C is added to the first beaker, 100 g of ice at 0°C is added to the second beaker, and the third beaker gets 100 g of aluminum at 0°C. The contents of which container end up at the lowest final temperature?
    A. The first beaker.
    B. The second beaker.
    C. The third beaker.
    D. All end up at the same temperature.

31. ‖ 100 g of ice at 0°C and 100 g of steam at 100°C interact thermally in a well-insulated container. The final state of the system is
    A. An ice-water mixture at 0°C.
    B. Water at a temperature between 0°C and 50°C.
    C. Water at 50°C.
    D. Water at a temperature between 50°C and 100°C.
    E. A water-steam mixture at 100°C.

32. | Suppose the 600 W of radiation emitted in a microwave oven is absorbed by a very lightweight cup holding 250 g of water. Approximately how long will it take to heat the water from 20°C to 80°C?
    A. 50 s   B. 100 s
    C. 150 s   D. 200 s

33. ‖ 40,000 J of heat is added to 1.0 kg of ice at −10°C. How much ice melts?
    A. 0.012 kg   B. 0.057 kg
    C. 0.12 kg   D. 1.0 kg

34. | Steam at 100°C causes worse burns than liquid water at
    BIO 100°C. This is because
    A. The steam is hotter than the water.
    B. Heat is transferred to the skin as steam condenses.
    C. Steam has a higher specific heat than water.
    D. Evaporation of liquid water on the skin causes cooling.

PROBLEMS

Section 12.1 The Atomic Model of Matter

1. | Which contains the most moles: 10 g of hydrogen gas, 100 g of carbon, or 50 g of lead?

2. |||| How many grams of water (H_2O) have the same number of oxygen atoms as 1.0 mol of oxygen gas?

3. || How many atoms of hydrogen are in 100 g of hydrogen peroxide (H_2O_2)?

4. | How many cubic millimeters (mm^3) are in 1 L?

5. | A box is 200 cm wide, 40 cm deep, and 3.0 cm high. What is its volume in m^3?

Section 12.2 Thermal Expansion

6. | A straight rod consists of a 1.2-cm-long piece of aluminum attached to a 2.0-cm-long piece of steel. By how much will the length of this rod change if its temperature is increased from 20°C to 40°C?

7. | The length of a steel beam increases by 0.73 mm when its temperature is raised from 22°C to 35°C. What is the length of the beam at 22°C?

8. | Older railroad tracks in the U.S. are made of 12-m-long pieces of steel. When the tracks are laid, gaps are left between the sections to prevent buckling when the steel thermally expands. If a track is laid at 16°C, how large should the gaps be if the track is not to buckle when the temperature is as high as 50°C?

9. | The temperature of an aluminum disk is increased by 120°C. By what percentage does its volume increase?

10. |||| If a container holds 0.865 L of ethyl alcohol at 12.0°C, what is the volume of the alcohol when its temperature is raised to 35.0°C?

Section 12.3 Pressure and the Kinetic Theory of an Ideal Gas

11. | What is the absolute pressure of the air in your car's tires, in psi, when your pressure gauge indicates they are inflated to 35.0 psi? Assume you are at sea level.

12. || Total lung capacity of a typical adult is approximately
BIO 5.0 L. Approximately 20% of the air is oxygen. At sea level and at an average body temperature of 37°C, how many moles of oxygen do the lungs contain at the end of an inhalation?

13. ||| Many cultures around the world still use a simple weapon
BIO called a blowgun, a tube with a dart that fits tightly inside. A sharp breath into the end of the tube launches the dart. When exhaling forcefully, a healthy person can supply air at a gauge pressure of 6.0 kPa. What force does this pressure exert on a dart in a 1.5-cm-diameter tube?

14. ||| When you stifle a sneeze, you can damage delicate tissues
BIO because the pressure of the air that is not allowed to escape may rise by up to 45 kPa. If this extra pressure acts on the inside of your 8.4-mm-diameter eardrum, what is the outward force?

15. ||| 7.5 mol of helium are in a 15 L cylinder. The pressure gauge on the cylinder reads 65 psi. What are (a) the temperature of the gas in °C and (b) the average kinetic energy of a helium atom?

16. | 3.0 mol of gas at a temperature of −120°C fills a 2.0 L container. What is the gas pressure?

17. || Determine the number of moles of oxygen gas in the following containers.
   a. Container A holds 0.30 m^3 at 25°C and an absolute pressure of 10 kPa.
   b. Container B holds 0.30 m^3 at 59°F and an absolute pressure of 12 kPa.
   c. Container C holds 0.10 m^3 at 25°C and a gauge pressure of 30 kPa.

18. || Determine the temperature (in °C) of the gases in the following containers.
   a. Container A, with volume 0.10 m^3, holds 2.0 mol of helium at an absolute pressure of 1.0 atm.
   b. Container B, with volume 0.20 m^3, holds 2.0 mol of argon at a gauge pressure of 0.50 atm.
   c. Container C, with volume 100 L holds 2.0 mol of neon at an absolute pressure of 3.0 atm.

19. |||| 10 g of liquid water is placed in a flexible bag, the air is excluded, and the bag is sealed. It is then placed in a microwave oven where the water is boiled to make steam at 100°C. What is the volume of the bag after all the water has boiled? Assume that the pressure inside the bag is equal to atmospheric pressure.

Section 12.4 Ideal-Gas Processes

20. || A cylinder contains 3.0 L of oxygen at 300 K and 2.4 atm. The gas is heated, causing a piston in the cylinder to move outward. The heating causes the temperature to rise to 600 K and the volume of the cylinder to increase to 9.0 L. What is the final gas pressure?

21. | A gas with initial conditions p_i, V_i, and T_i expands isothermally until $V_f = 2V_i$. What are (a) T_f and (b) p_f?

22. || 0.10 mol of argon gas is admitted to an evacuated 50 cm^3 container at 20°C. The gas then undergoes heating at constant volume to a temperature of 300°C.
   a. What is the final pressure of the gas?
   b. Show the process on a pV diagram. Include a proper scale on both axes.

23. | 0.10 mol of argon gas is admitted to an evacuated 50 cm^3 container at 20°C. The gas then undergoes an isobaric heating to a temperature of 300°C.
   a. What is the final volume of the gas?
   b. Show the process on a pV diagram. Include a proper scale on both axes.

24. || 0.10 mol of argon gas is admitted to an evacuated 50 cm^3 container at 20°C. The gas then undergoes an isothermal expansion to a volume of 200 cm^3.
   a. What is the final pressure of the gas?
   b. Show the process on a pV diagram. Include a proper scale on both axes.

25. | 0.0040 mol of gas undergoes the process shown in Figure P12.25.
   a. What type of process is this?
   b. What are the initial and final temperatures?

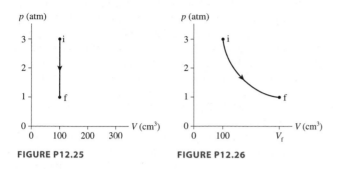

FIGURE P12.25   **FIGURE P12.26**

26. ||||| 0.0040 mol of gas follows the hyperbolic trajectory shown in Figure P12.26.
   a. What type of process is this?
   b. What are the initial and final temperatures?
   c. What is the final volume V_f?

27. || A gas with an initial temperature of 900°C undergoes the process shown in Figure P12.27.
   a. What type of process is this?
   b. What is the final temperature?
   c. How many moles of gas are there?

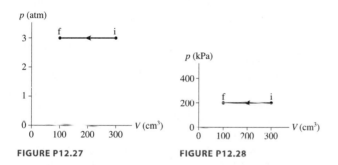

FIGURE P12.27   **FIGURE P12.28**

28. || How much work is done on the gas in the process shown in Figure P12.28?

29. ||||| It is possible to make a thermometer by sealing gas in a rigid container and measuring the pressure. Such a constant-volume gas thermometer is placed in an ice-water bath at 0.00°C. After reaching thermal equilibrium, the gas pressure is recorded as 55.9 kPa. The thermometer is then placed in contact with a sample of unknown temperature. After the thermometer reaches a new equilibrium, the gas pressure is 65.1 kPa. What is the temperature of this sample?

Section 12.5 Specific Heat and Heat of Transformation

30. | How much energy must be removed from a 200 g block of ice to cool it from 0°C to −30°C?

31. || How much heat is needed to change 20 g of mercury at 20°C into mercury vapor at the boiling point?

32. | a. 100 J of heat energy are transferred to 20 g of mercury. By how much does the temperature increase?
   b. How much heat is needed to raise the temperature of 20 g of water by the same amount?

33. ||| The maximum amount of water an adult in temperate climates can perspire in one hour is typically 1.8 L. However, after several weeks in a tropical climate the body can adapt, increasing the maximum perspiration rate to 3.5 L/hr. At what rate, in watts, is energy being removed when perspiring that rapidly? Assume all of the perspired water evaporates. At body temperature, the heat of vaporization of water is $L_v = 24 \times 10^5$ J/kg.

34. ||| When air is inhaled, it quickly becomes saturated with water vapor as it passes through the moist upper airways. When breathing dry air, about 25 mg of water are exhaled with each breath. At 12 breaths/min, what is the rate of energy loss due to evaporation? Express your answer in both watts and Calories per day. At body temperature, the heat of vaporization of water is $L_v = 24 \times 10^5$ J/kg.

35. || What minimum heat is needed to bring 100 g of water at 20°C to the boiling point and completely boil it away?

Section 12.6 Calorimetry

36. || 30 g of copper pellets are removed from a 300°C oven and immediately dropped into 100 mL of water at 20°C in an insulated cup. What will the new water temperature be?

37. | A copper block is removed from a 300°C oven and dropped into 1.00 kg of water at 20.0°C. The water quickly reaches 25.5°C and then remains at that temperature. What is the mass of the copper block?

38. ||||| A 750 g aluminum pan is removed from the stove and plunged into a sink filled with 10 kg of water at 20.0°C. The water temperature quickly rises to 24.0°C. What was the initial temperature of the pan?

39. ||| A 500 g metal sphere is heated to 300°C, then dropped into a beaker containing 4.08 kg of mercury at 20.0°C. A short time later the mercury temperature stabilizes at 99.0°C. Identify the metal.

40. ||||| Marianne really likes coffee, but on summer days she doesn't want to drink a hot beverage. If she is served 200 mL of coffee at 80°C in a well-insulated container, how much ice at 0°C should she add to obtain a final temperature of 30°C?

Section 12.7 Thermal Properties of Gases

41. | A container holds 1.0 g of argon at a pressure of 8.0 atm.
   a. How much heat is required to increase the temperature by 100°C at constant volume?
   b. How much will the temperature increase if this amount of heat energy is transferred to the gas at constant pressure?

42. | A container holds 1.0 g of oxygen at a pressure of 8.0 atm.
   a. How much heat is required to increase the temperature by 100°C at constant pressure?
   b. How much will the temperature increase if this amount of heat energy is transferred to the gas at constant volume?

43. | What is the temperature change of 1.0 mol of a monatomic gas if its thermal energy is increased by 1.0 J?

44. || The temperature of 2.0 g of helium is increased at constant volume by ΔT. What mass of oxygen can have its temperature increased by the same amount at constant volume using the same amount of heat?

45. ▌▌ How much work is done per cycle by a gas following the pV trajectory of Figure P12.45?

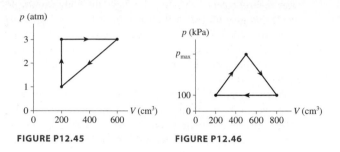

FIGURE P12.45   **FIGURE P12.46**

46. ▌▌ A gas following the pV trajectory of Figure P12.46 does 60 J of work per cycle. What is p_{max}?

Section 12.8 Heat Transfer

47. ▌ A 1.8-cm-thick wood floor covers a 4.0 m × 5.5 m room. The subfloor on which the flooring sits is at a temperature of 16.2°C, while the air in the room is at 19.6°C. What is the rate of heat conduction through the floor?

48. ▌▌ A copper-bottomed kettle, its bottom 24 cm in diameter and 3.0 mm thick, sits on a burner. The kettle holds boiling water, and energy flows into the water from the kettle bottom at 800 W. What is the temperature of the bottom surface of the kettle?

49. ▌▌ What is the rate of energy transfer by radiation from a metal cube 2.0 cm on a side that is at 700°C? Its emissivity is 0.20.

50. ▌ What is the greatest possible rate of energy transfer by radiation for a 5.0-cm-diameter sphere that is at 100°C?

51. ▌▌ Seals may cool themselves by
BIO using *thermal windows,* patches on their bodies with much higher than average surface temperature. Suppose a seal has a 0.030 m² thermal window at a temperature of 30°C. If the seal's surroundings are a frosty −10°C, what is the net rate of energy loss by radiation? Assume an emissivity equal to that of a human.

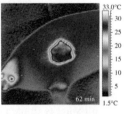

52. ▌ The glowing filament in a lamp is radiating energy at a rate of 60 W. At the filament's temperature of 1500°C, the emissivity is 0.23. What is the surface area of the filament?

General Problems

53. ▌ A rigid container holds 2.0 mol of gas at a pressure of 1.0 atm and a temperature of 30°C.
    a. What is the container's volume?
    b. What is the pressure if the temperature is raised to 130°C?

54. ▌ A 15-cm-diameter compressed-air tank is 50 cm tall. The pressure at 20°C is 150 atm.
    a. How many moles of air are in the tank?
    b. What volume would this air occupy at STP?

55. ▌▌▌▌ A 20-cm-diameter cylinder that is 40 cm long contains 50 g of oxygen gas at 20°C.
    a. How many moles of oxygen are in the cylinder?
    b. How many oxygen molecules are in the cylinder?
    c. What is the reading of a pressure gauge attached to the tank?

56. ▌▌▌▌ A 10-cm-diameter cylinder of helium gas is 30 cm long and at 20°C. The pressure gauge reads 120 psi.
    a. How many helium atoms are in the cylinder?
    b. What is the mass of the helium?

57. ▌▌▌▌ An empty room in a house near the Atlantic coast has a volume of 45 m³. What is the total kinetic energy of all of the gas molecules in the room on a 30°C summer day?

58. ▌▌▌ Suppose you take a deep breath on a chilly day, inhaling
BIO 4.0 L of air at −10°C. Assume that air pressure is 1.0 atm.
    a. How much heat must your body supply to warm the air to your internal body temperature of 37°C?
    b. How much does the volume of the air increase as it is warmed?

59. ▌ On average, each person in the industrialized world is responsible for the emission of 10,000 kg of carbon dioxide (CO_2) every year. This includes CO_2 that you generate directly, by burning fossil fuels to operate your car or your furnace, as well as CO_2 generated on your behalf by electric generating stations and manufacturing plants. CO_2 is a greenhouse gas that contributes to global warming. If you were to store your yearly CO_2 emissions in a cube at STP, how long would each edge of the cube be?

60. ▌▌▌▌ On a cool morning, when the temperature is 15°C, you measure the pressure in your car tires to be 30 psi. After driving 20 mi on the freeway, the temperature of your tires is 45°C. What pressure will your tire gauge now show?

61. ▌▌▌▌ Suppose you inflate your car tires to 35 psi on a 20°C day. Later, the temperature drops to 0°C. What is the pressure in your tires now?

62. ▌ The volume in a constant-*pressure* gas thermometer is directly proportional to the absolute temperature. A constant-pressure thermometer is calibrated by adjusting its volume to 1000 mL while it is in contact with a reference cell at 0.01°C. The volume increases to 1638 mL when the thermometer is placed in contact with a sample. What is the sample's temperature in °C?

63. ▌▌ A compressed-air cylinder is known to fail if the pressure exceeds 110 atm. A cylinder that was filled to 25 atm at 20°C is stored in a warehouse. Unfortunately, the warehouse catches fire and the temperature reaches 950°C. Does the cylinder explode?

64. ▌▌ A rigid sphere has a valve that can be opened or closed. The sphere with the valve open is placed in boiling water in a room where the air pressure is 1.0 atm. After a long period of time has elapsed, the valve is closed. What will be the pressure inside the sphere if it is then placed in (a) a mixture of ice and water and (b) an insulated box filled with dry ice, which is at −78.5°C?

65. ▌▌▌ 80 J of work are done on the gas in the process shown in Figure P12.65. What is V_f in cm³?

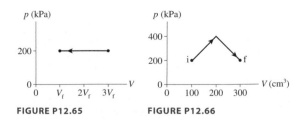

FIGURE P12.65   **FIGURE P12.66**

66. ‖ How much work is done by the gas in the process shown in Figure P12.66?

67. ‖ A container with 100 cm³ of gas is pressurized at constant volume until the gas pressure triples. It is then expanded at constant temperature until the pressure is one-half of the pressure the gas had before the experiment started.
    a. What is the final volume of the gas?
    b. Show this process on a pV diagram.

68. ‖ 0.10 mol of gas undergoes the process $1 \rightarrow 2$ shown in Figure P12.68.
    a. What are temperatures T_1 and T_2?
    b. What type of process is this?
    c. The gas undergoes constant-volume heating from point 2 until the pressure is restored to the value it had at point 1. What is the final temperature of the gas?

p (atm)

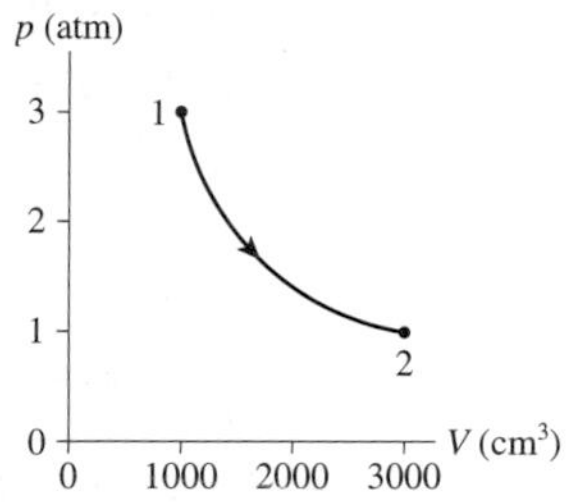

FIGURE P12.68

69. ‖ 10 g of dry ice (solid CO_2) is placed in a 10,000 cm³ container, then all the air is quickly pumped out and the container sealed. The container is warmed to 0°C, a temperature at which CO_2 is a gas.
    a. What is the gas pressure? Give your answer in atm.
       The gas then undergoes an isothermal compression until the pressure is 3.0 atm, immediately followed by an isobaric compression until the volume is 1000 cm³.
    b. What is the final temperature of the gas?
    c. Show the process on a pV diagram.

70. ‖ A 5.0-m-diameter garden pond holds 5.9×10^3 kg of water. Solar energy is incident on the pond at an average rate of 400 W/m². If the water absorbs all the solar energy and does not exchange energy with its surroundings, how many hours will it take to warm from 15°C to 25°C?

71. ‖ A typical nuclear reactor generates 1000 MW (1000 MJ/s) of electrical energy. In doing so, it produces 2000 MW of "waste heat" that must be removed from the reactor. Many reactors are sited next to large bodies of water so that they can use the water for cooling. Consider a reactor where the intake water is at 18°C. State regulations limit the temperature of the output water to 30°C so as not to harm aquatic organisms. How many kilograms of cooling water have to be pumped through the reactor each minute?

72. ‖ A 1200 kg car traveling at 60 mph quickly brakes to a halt. INT The kinetic energy of the car is converted to thermal energy of the disk brakes. The brake disks (one per wheel) are iron disks with a mass of 4.0 kg. Estimate the temperature rise in each disk as the car stops.

73. ‖ Suppose you drop a water balloon from a height of 10 m. If INT the balloon doesn't break on impact, its kinetic energy will be converted to thermal energy. Estimate the temperature rise of the water. Is this likely to be noticeable?

74. ‖ What is the maximum mass of lead you could melt with 1000 J of heat, starting from 20°C?

75. | An experiment measures the temperature of a 200 g substance while steadily supplying heat to it. Figure P12.75 shows the results of the experiment. What are (a) the specific heat of the liquid phase and (b) the heat of vaporization?

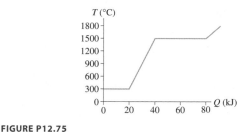

FIGURE P12.75

76. ‖‖ 10 g of aluminum at 200°C and 20 g of copper are dropped into 50 cm³ of ethyl alcohol at 15°C. The temperature quickly comes to 25°C. What was the initial temperature of the copper?

77. ‖ A 100 g ice cube at −10°C is placed in an aluminum cup whose initial temperature is 70°C. The system comes to an equilibrium temperature of 20°C. What is the mass of the cup?

78. ‖ A 50.0 g thermometer is used to measure the temperature of 200 g of water. The specific heat of the thermometer, which is mostly glass, is 750 J/kg · K, and it reads 20.0°C while lying on the table. After being completely immersed in the water, the thermometer's reading stabilizes at 71.2°C. What was the actual water temperature before it was measured?

79. ‖‖ Your 300 mL cup of coffee is too hot to drink when served at 90°C. What is the mass of an ice cube, taken from a −20°C freezer, that will cool your coffee to a pleasant 60°C?

80. ‖ A gas is compressed from 600 cm³ to 200 cm³ at a constant pressure of 400 kPa. At the same time, 100 J of heat energy is transferred out of the gas. What is the change in thermal energy of the gas during this process?

81. ‖ An expandable cube, initially 20 cm on each side, contains 3.0 g of helium at 20°C. 1000 J of heat energy are transferred to this gas. What are (a) the final pressure if the process is at constant volume and (b) the final volume if the process is at constant pressure?

82. ‖ 0.10 mol of a monatomic gas follows the process shown in Figure P12.82.
    a. How much heat energy is transferred to or from the gas during process $1 \rightarrow 2$?
    b. How much heat energy is transferred to or from the gas during process $2 \rightarrow 3$?
    c. What is the total change in thermal energy of the gas?

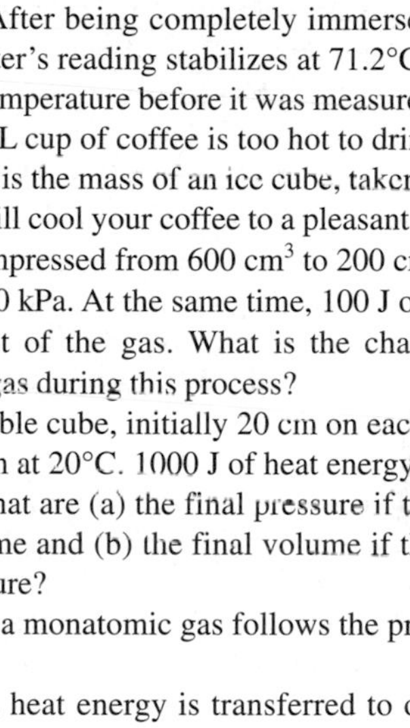

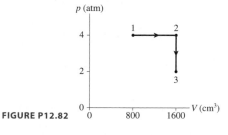

FIGURE P12.82

83. ‖ A monatomic gas follows the process $1 \rightarrow 2 \rightarrow 3$ shown in Figure P12.83. How much heat is needed for (a) process $1 \rightarrow 2$ and (b) process $2 \rightarrow 3$?

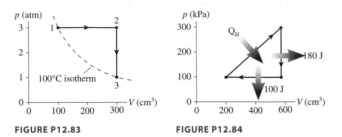

FIGURE P12.83   FIGURE P12.84

84. ‖‖ What are (a) the heat Q_H extracted from the hot reservoir INT and (b) the efficiency for a heat engine described by the pV diagram of Figure P12.84?

85. ‖‖ The top layer of your goose down sleeping bag has a thickness of 5.0 cm and a surface area of 1.0 m². When the outside temperature is −20°C, you lose 25 Cal/hr by heat conduction through the bag (which remains at a cozy 35°C inside). Assume that you're sleeping on an insulated pad that eliminates heat conduction to the ground beneath you. What is the thermal conductivity of the goose down?

86. ‖ Suppose you go outside in your fiber-filled jacket on a windless but very cold day. The thickness of the jacket is 2.5 cm, and it covers 1.1 m² of your body. The purpose of fiber- or down-filled jackets is to trap a layer of air, and it's really the air layer that provides the insulation. If your skin temperature is 34°C while the air temperature is −20°C, at what rate is heat being conducted through the jacket and away from your body?

87. ‖‖ Two thin, square copper plates are radiating energy at the same rate. The edge length of plate 2 is four times that of plate 1. What is the ratio of absolute temperatures T_1/T_2 of the plates?

88. ‖ The surface area of an adult human is about 1.8 m². Sup-
BIO pose a person with a skin temperature of 34°C is standing nude in a room where the air is 25°C but the walls are 17°C.
   a. There is a "dead-air" layer next to your skin that acts as insulation. If the dead-air layer is 5.0 mm thick, what is the person's rate of heat loss by conduction?
   b. What is the person's net radiation loss to the walls? The emissivity of skin is 0.97.
   c. Does conduction or radiation contribute more to the total rate of energy loss?
   d. If the person is metabolizing food at a rate of 155 W, does the person feel comfortable, chilly, or too warm?

In Problems 89 through 91 you are given the equation(s) used to solve a problem involving a container of gas. For each of these, you are to
a. Write a realistic problem for which this is the correct equation(s).
b. Draw a pV diagram.
c. Finish the solution of the problem.

89. ‖ $p_f = \dfrac{300 \text{ cm}^3}{100 \text{ cm}^3} \times 1 \times 2 \text{ atm}$

90. ‖ $(T_f + 273) \text{ K} = \dfrac{200 \text{ kPa}}{500 \text{ kPa}} \times 1 \times (400 + 273) \text{ K}$

91. ‖ $V_f = \dfrac{(400 + 273) \text{ K}}{(50 + 273) \text{ K}} \times 1 \times 200 \text{ cm}^3$

92. ‖ You are given an equation below used to solve a problem. You are to
   a. Write a realistic problem for which this is the correct equation.
   b. Finish the solution of the problem.

$$(2.72 \text{ kg})(140 \text{ J/kg} \cdot \text{K})(90°\text{C} - 15°\text{C})$$
$$+ (0.50 \text{ kg})(449 \text{ J/kg} \cdot \text{K})(90°\text{C} - T_i) = 0$$

Passage Problems

Thermal Properties of the Oceans

Seasonal temperature changes in the ocean only affect the top layer of water, to a depth of 500 m or so. This "mixed" layer is thermally isolated from the cold, deep water below. The average temperature of this top layer of the world's oceans, which has area 3.6×10^8 km², is approximately 17°C.

In addition to seasonal temperature changes, the oceans have experienced an overall warming trend over the last century that is expected to continue as the earth's climate changes. A warmer ocean means a larger volume of water; the oceans will

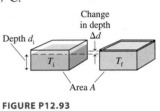

FIGURE P12.93

rise. Suppose the average temperature of the top layer of the world's oceans were to increase from a temperature T_i to a temperature T_f. The area of the oceans will not change, as this is fixed by the size of the ocean basin, so any thermal expansion of the water will cause the water level to rise, as shown in Figure P12.93. The original volume is the product of the original depth and the surface area, $V_i = A d_i$. The change in volume is given by $\Delta V = A \, \Delta d$.

93. ‖ If the top 500 m of ocean water increased in temperature from 17°C to 18°C, what would be the resulting rise in ocean height?
   A. 0.11 m   B. 0.22 m
   C. 0.44 m   D. 0.88 m

94. ‖ Approximately how much energy would be required to raise the temperature of the top layer of the oceans by 1°C? (1 m³ of water has a mass of 1000 kg.)
   A. 1×10^{24} J   B. 1×10^{21} J
   C. 1×10^{18} J   D. 1×10^{15} J

95. ‖ Water's coefficient of expansion varies with temperature. For water at 2°C, an increase in temperature of 1°C would cause the volume to
   A. Increase.   B. Stay the same.   C. Decrease.

96. ‖ The ocean is mostly heated from the top, by light from the sun. The warmer surface water doesn't mix much with the colder deep ocean water. This lack of mixing can be ascribed to a lack of
   A. Conduction.
   B. Convection.
   C. Radiation.
   D. Evaporation.

STOP TO THINK ANSWERS

Stop to Think 12.1: A. Both helium and neon are monatomic gases, where the basic particles are atoms. 5 mol of helium contain 5 times as many atoms as 1 mol of neon, though both samples have the same mass.

Stop to Think 12.2: A. The thermal expansion coefficients of aluminum are greater than those of iron. Heating the rod and the ring will expand the outer diameter of the rod and the inner diameter of the ring, but the ring's expansion will be greater.

Stop to Think 12.3: B. The product pV/T is constant. During the process, pV decreases by a factor of 2, so T must decrease by a factor of 2 as well.

Stop to Think 12.4: B. To solidify the lead, heat must be removed; this heat boils the water. The heat of vaporization of water is 10 times that of the heat of fusion of lead, so much less than 1 kg of water vaporizes as 1 kg of lead solidifies.

Stop to Think 12.5: A. The lead cools and the water warms as heat is transferred from the lead to the water. The specific heat of water is much larger than that of the lead, so the temperature change of the water is much less than that of the lead.

Stop to Think 12.6: C. With a sealed suit and no matter around you, there is no way to transfer heat to the environment except by radiation.

13

FLUIDS

This woman is floating in the Dead Sea, which has an exceptionally high density because of its salt content of about 35%, ten times that of the open sea. Why does she float so high in the water?

Looking Ahead ▶

The goal of Chapter 13 is to understand the static and dynamic properties of fluids. In this chapter you will learn to:

▶ Understand and use the concept of mass density.

▶ Understand pressure in liquids, and how pressure is measured.

▶ Use Archimedes' principle to understand buoyancy.

▶ Apply the equation of continuity to a moving fluid.

▶ Make qualitative predictions about pressures and forces in moving fluids.

▶ Apply Bernoulli's and Poiseuille's equations to problems of fluid flow.

Looking Back ◀

The material in this chapter depends on the concept of pressure in gases, on the conditions of equilibrium, and on how forces determine the acceleration of a particle. Please review:

◀ Sections 4.6 and 5.2 Newton's second law.

◀ Section 5.1 Static equilibrium.

◀ Sections 12.3–12.4 Pressure in gases.

414

This woman is floating on water, a fluid. The water itself is in motion. Surprisingly, we need no new laws of physics to understand how fluids flow or why some objects float while others sink. The physics of fluids, often called *fluid mechanics,* is an important application of Newton's laws and the law of conservation of energy—physics that you learned in Parts I and II.

We'll begin with *fluid statics,* situations in which the fluid remains at rest. We'll learn why submarines can be crushed by the ocean above, and why hot air balloons float. Then we'll turn to fluids in motion. We'll learn how pressure *differences* are responsible for accelerating fluids. These pressure differences can lead to large forces on objects in the fluids, and they explain how airplanes stay aloft and how blood pressure pushes blood through your circulatory system.

13.1 Fluids and Density

A **fluid** is simply a substance that flows. Because they flow, fluids take the shape of their container rather than retaining a shape of their own. You may think that gases and liquids are quite different, but both are fluids, and their similarities are often more important than their differences.

As you learned in Chapter 12, a gas, shown in Figure 13.1a, is a system in which each molecule moves freely through space until, on occasion, it collides with another molecule or with the wall of the container. The gas you are most familiar with is air, a mixture of mostly nitrogen and oxygen molecules. Gases are *compressible*. That is, the volume of a gas is easily increased or decreased, a consequence of the "empty space" between the molecules in a gas.

Liquids are more complicated than either gases or solids. Liquids, like solids, are nearly *incompressible*. This property tells us that the molecules in a liquid, as in a solid, are about as close together as they can get without coming into contact with each other. At the same time, a liquid flows and deforms to fit the shape of its container. The fluid nature of a liquid tells us that the molecules are free to move around.

Together, these observations suggest the model of a liquid shown in Figure 13.1b. Here you see a system in which the molecules are loosely held together by weak molecular bonds. The bonds are strong enough that the molecules never get far apart but not strong enough to prevent the molecules from sliding around each other.

Density

An important parameter that characterizes a macroscopic system is its *density*. Suppose you have several blocks of copper, each of different size. Each block has a different mass m and a different volume V. Nonetheless, all the blocks are copper, so there should be some quantity that has the *same* value for all the blocks, telling us, "This is copper, not some other material." The most important such parameter is the *ratio* of mass to volume, which we call the **mass density** ρ (lowercase Greek rho):

$$\rho = \frac{m}{V} \qquad (13.1)$$

Mass density of an object of mass m and volume V

Conversely, an object of density ρ and volume V has mass

$$m = \rho V \qquad (13.2)$$

The SI units of mass density are kg/m^3. Nonetheless, units of g/cm^3 are widely used. You need to convert these to SI units before doing most calculations. You must convert both the grams to kilograms and the cubic centimeters to cubic meters. The net result is the conversion factor

$$1 \text{ g/cm}^3 = 1000 \text{ kg/m}^3$$

The mass density is independent of the object's size. That is, mass and volume are parameters that characterize a *specific piece* of some substance—say copper—whereas the mass density characterizes the substance itself. All pieces of copper have the same mass density, which differs from the mass density of almost any other substance. Thus mass density allows us to talk about the properties of copper in general without having to refer to any specific piece of copper.

The mass density is usually called simply "the density" if there is no danger of confusion. However, we will meet other types of density as we go along, and sometimes it is important to be explicit about which density you are using. Table 13.1 provides a short list of mass densities of various fluids. Notice the enormous difference between the densities of gases and liquids. Gases have lower densities because the molecules in gases are farther apart than in liquids. Also, the density of a liquid varies only slightly with temperature, because its

(a) A gas

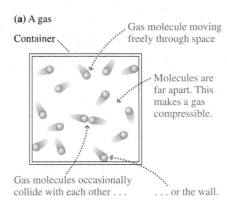

Container

Gas molecule moving freely through space

Molecules are far apart. This makes a gas compressible.

Gas molecules occasionally collide with each other . . .   . . . or the wall.

(b) A liquid

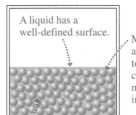

A liquid has a well-defined surface.

Molecules are about as close together as they can get. This makes a liquid incompressible.

Molecules make weak bonds with each other that keep them close together. But the molecules can slide around each other, allowing the liquid to flow and conform to the shape of its container.

FIGURE 13.1 Simple atomic-level models of gases and liquids.

TABLE 13.1 Densities of fluids at 1 atm pressure

Substance	ρ (kg/m^3)
Air (0°C)	1.28
Air (20°C)	1.20
Ethyl alcohol	790
Gasoline	680
Glycerin	1260
Helium gas (0°C)	0.179
Mercury	13,600
Oil (typical)	900
Water	1000
Seawater	1030
Dead Sea water	1240
Blood (whole)	1060

molecules are always nearly in contact. The density of a gas, such as air, has a stronger variation with temperature because it's easy to change the already large distance between the molecules.

What does it *mean* to say that the density of gasoline is 680 kg/m³? Recall in Chapter 1 we discussed the meaning of the word "per." We found that it meant "for each," so that 2 miles per hour means you travel 2 miles *for each* hour that passes. In the same way, saying that the density of gasoline is 680 kg per cubic meter means that there are 680 kg of gasoline *for each* one cubic meter of the liquid. If we have two m³ of gasoline, each will have a mass of 680 kg, so the total mass will be 2 × 680 kg = 1360 kg. The product ρV is the number of cubic meters times the mass of each cubic meter; that is, it is the total mass of the object.

EXAMPLE 13.1 Weighing the air in a living room

What is the mass of air in a living room having dimensions 4.0 m × 6.0 m × 2.5 m?

PREPARE Table 13.1 gives air density at a temperature of 20°C, which is about room temperature.

SOLVE The room's volume is

$$V = (4.0 \text{ m}) \times (6.0 \text{ m}) \times (2.5 \text{ m}) = 60 \text{ m}^3$$

The mass of the air is

$$m = \rho V = (1.20 \text{ kg/m}^3)(60 \text{ m}^3) = 72 \text{ kg}$$

ASSESS This is perhaps more mass—about that of an adult person—than you might have expected from a substance that hardly seems to be there. For comparison, a swimming pool this size would contain 60,000 kg of water.

STOP TO THINK 13.1 A piece of glass is broken into two pieces of different size. Rank in order, from largest to smallest, the mass densities of pieces 1, 2, and 3.

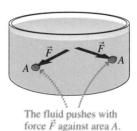

13.2 Pressure

In Chapter 12, you learned how a gas exerts a force on the walls of its container. Liquids also exert forces on the walls of their container, as shown in Figure 13.2, where a force $\vec{F}$ due to the liquid pushes against a small area A of the wall. Just as for a gas, we define the pressure at this point in the fluid to be the ratio of the force to the area on which the force is exerted:

$$p = \frac{F}{A} \tag{13.3}$$

This is the same as Equation 12.7 of Chapter 12. Recall also from Chapter 12 that the SI unit of pressure, the pascal, is defined as

$$1 \text{ pascal} = 1 \text{ Pa} = 1 \frac{\text{N}}{\text{m}^2}$$

The fluid pushes with force $\vec{F}$ against area A.

FIGURE 13.2 The fluid presses against area A with force $\vec{F}$.

It's important to realize that the force due to a fluid's pressure pushes not only on the walls of its container, but on *all* parts of the fluid itself. If you punch holes in a container of water, the water spurts out from the holes, as in Figure 13.3. It is the force due to the pressure of the water behind each hole that pushes the water forward though the holes.

To measure the pressure at any point within a fluid we can use the simple pressure-measuring device shown in Figure 13.4a. Because the spring constant k and the area A are known, we can determine the pressure by measuring the compression of the spring. Once we've built such a device, we can place it in various liquids and gases to learn about pressure. Figure 13.4b shows what we can learn from a series of simple experiments.

FIGURE 13.3 Water pushes the water *sideways,* out of the holes.

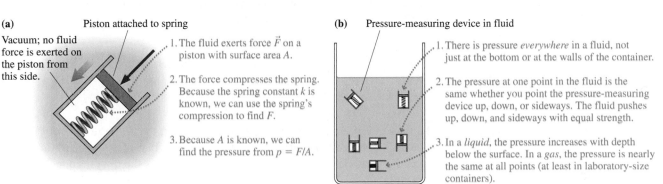

(a)   Piston attached to spring

Vacuum; no fluid force is exerted on the piston from this side.

1. The fluid exerts force $\vec{F}$ on a piston with surface area A.

2. The force compresses the spring. Because the spring constant k is known, we can use the spring's compression to find F.

3. Because A is known, we can find the pressure from $p = F/A$.

(b)   Pressure-measuring device in fluid

1. There is pressure *everywhere* in a fluid, not just at the bottom or at the walls of the container.

2. The pressure at one point in the fluid is the same whether you point the pressure-measuring device up, down, or sideways. The fluid pushes up, down, and sideways with equal strength.

3. In a *liquid*, the pressure increases with depth below the surface. In a *gas*, the pressure is nearly the same at all points (at least in laboratory-size containers).

FIGURE 13.4 Learning about pressure.

The first statement in Figure 13.4b emphasizes again that pressure exists at *all* points within a fluid, not just at the walls of the container. You may recall that tension exists at *all* points in a string, not only at its ends where it is tied to an object. We understood tension as the different parts of the string *pulling* against each other. Pressure is an analogous idea, except that the different parts of a fluid are *pushing* against each other.

Pressure in Liquids

Chapter 12 introduced our model of a gas. The molecules in a gas interact only when they collide with one another. Because of this, a gas almost instantaneously expands to fill the whole volume of a container. The pressure of the gas, which is the same at all points in the container, is then due to the collisions of the molecules with the container's walls.

In contrast to a gas, the molecules of a liquid interact by weak molecular bonds. These bonds tend to keep the molecules close together in a cohesive blob. If you introduce a liquid into a container, the force of gravity pulls the liquid down, causing it to fill the bottom of the container. It is this force of gravity—that is, the weight of the liquid—that is responsible for the pressure in a liquid. Pressure increases with depth in a liquid because the liquid below is being squeezed by all the liquid above, including any other liquid floating on the first liquid, and the weight of the air above the liquid.

We'd like to determine the pressure at a depth d below the surface of the liquid. We will assume that the liquid is at rest; flowing liquids will be considered later in this chapter. The shaded cylinder of liquid in Figure 13.5 extends from the surface to depth d. This cylinder, like the rest of the liquid, is in static equilibrium with $\vec{F}_{\text{net}} = \vec{0}$. Several forces act on this cylinder: its weight mg, a downward force $p_0 A$ due to the pressure p_0 at the surface of the liquid, an upward force pA due to the

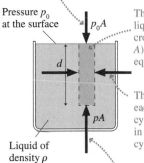

Whatever is above the liquid pushes down on the top of the cylinder.

Pressure p_0 at the surface

$p_0 A$

This cylinder of liquid (depth d, cross-section area A) is in static equilibrium.

d

The liquid on each side of the cylinder pushes in on the cylinder.

pA

Liquid of density ρ

The liquid beneath the cylinder pushes up on the cylinder. The pressure at depth d is p.

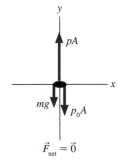

pA

mg $p_0 A$

$\vec{F}_{\text{net}} = \vec{0}$

Free-body diagram of the column of liquid. The horizontal forces cancel and are not shown.

FIGURE 13.5 Measuring the pressure at depth d in a liquid.

liquid beneath the cylinder pushing up on the bottom of the cylinder, and inward-directed forces due to the liquid pushing in on the sides of the cylinder. The forces due to the liquid pushing on the cylinder are a consequence of our earlier observation that different parts of a fluid push against each other. Pressure p, which is what we're trying to find, is the pressure at the bottom of the cylinder.

The horizontal forces cancel each other. The upward force balances the two downward forces, so

$$pA = p_0A + mg \qquad (13.4)$$

The liquid is a cylinder of cross-section area A and height d. Its volume is $V = Ad$ and its mass is $m = \rho V = \rho Ad$. Substituting this expression for the mass of the liquid into Equation 13.4, we find that the area A cancels from all terms. The pressure at depth d in a liquid is then

$$p = p_0 + \rho gd \qquad (13.5)$$

Pressure of a liquid with density ρ, at depth d

Because of our assumption that the fluid is at rest, the pressure given by Equation 13.5 is called the **hydrostatic pressure.** The fact that g appears in Equation 13.5 reminds us that the origin of this pressure is the gravitational force on the fluid.

As expected, $p = p_0$ at the surface, where $d = 0$. Pressure p_0 is often due to the air or other gas above the liquid. $p_0 = 1$ atm $= 101.3$ kPa for a liquid that is open to the air at sea level. In other situations, p_0 might be the pressure due to a piston or a closed surface pushing down on the top of the liquid.

NOTE ▶ Equation 13.5 assumes that the fluid is *incompressible;* that is, its density ρ doesn't increase with depth. This is an excellent assumption for liquids, but not a good one for a gas. Equation 13.5 should not be used for calculating the pressure of a gas. Gas pressure is found with the ideal-gas law. ◄

EXAMPLE 13.2 The pressure on a submarine

A submarine cruises at a depth of 300 m. What is the pressure at this depth? Give the answer both in pascals and atmospheres.

SOLVE The density of seawater, from Table 13.1, is $\rho = 1030$ kg/m³. At the surface, $p_0 = 1$ atm $= 101.3$ kPa. The pressure at depth $d = 300$ m is found from Equation 13.5 to be

$p = p_0 + \rho gd$

$= 1.013 \times 10^5$ Pa $+ (1030$ kg/m³$)(9.80$ m/s²$)(300$ m$)$

$= 3.13 \times 10^6$ Pa

Converting the answer to atmospheres gives

$p = 3.13 \times 10^6$ Pa $\times \dfrac{1 \text{ atm}}{1.013 \times 10^5 \text{ Pa}} = 30.9$ atm

ASSESS The pressure deep in the ocean is very great. The research submarine *Alvin,* shown in the left photo, can safely dive as deep as 4500 m, where the pressure is over 450 atm! Its viewports are over 3.5 inches thick to withstand this pressure. As shown in the right photo, each viewport is tapered, with its larger face toward the sea. The water pressure then pushes the viewports firmly into their conical seats, helping to seal them tightly.

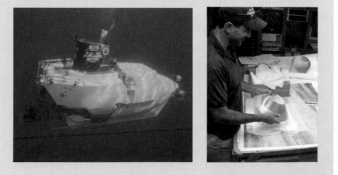

According to Equation 13.5, the hydrostatic pressure equation $p = p_0 + \rho g d$, the hydrostatic pressure in a liquid depends only on the depth and the pressure at the surface. This observation has some important implications. Figure 13.6a shows two connected tubes. It's certainly true that the larger volume of liquid in the wide tube weighs more than the liquid in the narrow tube. You might think that this extra weight would push the liquid in the narrow tube higher than in the wide tube. But it doesn't. If d_1 were larger than d_2, then, according to the hydrostatic pressure equation, the pressure at the bottom of the narrow tube would be higher than the pressure at the bottom of the wide tube. This *pressure difference* would cause the liquid to *flow* from right to left until the heights were equal.

Thus a first conclusion: **A connected liquid in hydrostatic equilibrium rises to the same height in all open regions of the container.** This is the familiar observation that "water seeks its own level."

Figure 13.6b shows two connected tubes of different shape. The conical tube holds more liquid above the dotted line, so you might think that $p_1 > p_2$. But it isn't. Both points are at the same depth, thus $p_1 = p_2$. You can arrive at the same conclusion by thinking about the pressure at the bottom of the tubes. If p_1 were larger than p_2, the pressure at the bottom of the left tube would be larger than the pressure at the bottom of the right tube. This would cause the liquid to flow until the pressures were equal. Thus a second conclusion: **The pressure is the same at all points on a horizontal line through a connected liquid in hydrostatic equilibrium.**

NOTE ▶ Both of these conclusions are restricted to liquids in hydrostatic equilibrium. The situation is entirely different for flowing fluids, as we'll see later in the chapter. ◀

(a)

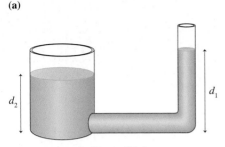

Is this possible?

(b)

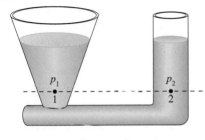

Is $p_1 > p_2$?

FIGURE 13.6 Some properties of a liquid in hydrostatic equilibrium are not what you might expect.

EXAMPLE 13.3 Pressure in a closed tube
Water fills the tube shown in Figure 13.7. What is the pressure at the top of the closed tube?

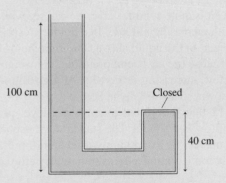

100 cm

Closed

40 cm

FIGURE 13.7 A bent tube closed at one end.

PREPARE This is a liquid in hydrostatic equilibrium. The closed tube is not an open region of the container, so the water cannot rise to an equal height. Nevertheless, the pressure is still the

same at all points on a horizontal line. In particular, the pressure at the top of the closed tube equals the pressure in the open tube at the height of the dotted line. Assume $p_0 = 1$ atm.

SOLVE A point 40 cm above the bottom of the open tube is at a depth of 60 cm. The pressure at this depth is

$$p = p_0 + \rho g d$$

$$= 1.013 \times 10^5 \text{ Pa} + (1000 \text{ kg/m}^3)(9.80 \text{ m/s}^2)(0.60 \text{ m})$$

$$= 1.072 \times 10^5 \text{ Pa}$$

$$= 1.06 \text{ atm}$$

The pressure at the top of the closed tube is thus 1.1 atm, to two significant figures.

ASSESS The water in the open tube *pushes* the water in the closed tube up against the top of the tube. Consequently, in accordance with Newton's third law, the top of the tube *presses down on the liquid* with a force of magnitude $F = pA$. This explains why the pressure at the top of the closed tube is greater than atmospheric pressure.

We can draw one more conclusion from the hydrostatic pressure equation $p = p_0 + \rho g d$. If we change the pressure p_0 at the surface to p_1, the pressure at depth d becomes $p' = p_1 + \rho g d$. The *change* in pressure $\Delta p = p_1 - p_0$ is the same at all points in the fluid, independent of the size or shape of the container.

This idea was first recognized by Blaise Pascal (the same Pascal for whom the pressure unit is named), and is called *Pascal's principle:*

> **Pascal's principle** If the pressure at one point in an incompressible fluid is changed, the pressure at every other point in the fluid changes by the same amount.

For example, if we compress the air above the open tube in Example 13.3 to a pressure of 1.5 atm, an increase of 0.5 atm, the pressure at the top of the closed tube will increase to 1.6 atm.

STOP TO THINK 13.2 Water is slowly poured into the container until the water level has risen into tubes 1, 2, and 3. The water doesn't overflow from any of the tubes. How do the water depths in the three columns compare to each other?

A. $d_1 > d_2 > d_3$.
B. $d_1 < d_2 < d_3$.
C. $d_1 = d_2 = d_3$.
D. $d_1 = d_2 > d_3$.
E. $d_1 = d_2 < d_3$.

Atmospheric Pressure

We live at the bottom of a "sea" of air that extends up many kilometers. As Figure 13.8 shows, there is no well-defined top to the atmosphere; it just gets less and less dense with increasing height until reaching zero in the vacuum of space. Nonetheless, 99% of the air in the atmosphere is below about 30 km.

If we recall that a gas like air is quite compressible, we can see why the atmosphere becomes less dense with increasing altitude. In a liquid, pressure increases with depth because of the weight of the liquid above. The same holds true for the air in the atmosphere, but because the air is compressible, the weight of the air above compresses the air below, increasing its density. At high altitudes there is very little air above to push down, so the density is less.

We learned in Chapter 12 that the global average sea-level pressure, the *standard atmosphere,* is 1 atm = 101,300 Pa. The standard atmosphere, usually referred to simply as "atmospheres," is a commonly used unit of pressure. But it is not an SI unit, so you must convert atmospheres to pascals before doing most calculations with pressure.

> **NOTE ▶** Unless you happen to live right at sea level, the atmospheric pressure around you is not exactly 1 atm. Pressure experiments use a barometer to determine the actual atmospheric pressure. For simplicity, this textbook will always assume that the pressure of the air is $p_{atmos} = 1$ atm unless stated otherwise. ◀

Atmospheric pressure varies not only with altitude, but also with changes in the weather. Large-scale regions of low-pressure air are created at the equator, where hot air rises and flows to the north and south temperate zones. There the air falls, creating high pressure zones. Local winds and weather are largely determined by presence and movement of air masses of differing pressure. You may have seen weather maps like the one shown in Figure 13.9 on the evening news. The letters H and L denote regions of high and low atmospheric pressure.

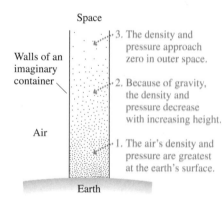

FIGURE 13.8 Atmospheric pressure and density.

Space

3. The density and pressure approach zero in outer space.

Walls of an imaginary container

2. Because of gravity, the density and pressure decrease with increasing height.

Air

1. The air's density and pressure are greatest at the earth's surface.

Earth

FIGURE 13.9 High and low pressure zones on a weather map.

Given that the pressure of the air at sea level is 101.3 kPa, you might wonder why the weight of the air doesn't crush your forearm when you rest it on a table. Your forearm has a surface area of ≈ 200 cm$^2 = 0.02$ m^2, so the force of the air pressing against it is ≈ 2000 N (≈ 450 pounds). How can you even lift your arm?

The reason, as Figure 13.10 shows, is that a fluid exerts pressure forces in *all* directions. There *is* a downward force of ≈ 2000 N on your forearm, but the air underneath your arm exerts an *upward* force of the same magnitude. The *net* force is just about zero.

The same arguments hold for deep-sea fish. Some fish have been observed at depths exceeding 8000 m. Here, the pressure is $\rho g h \approx 8 \times 10^7$ Pa ≈ 800 atm! But the pressure inside and outside the fish is the *same*, so there is no net force squeezing in on the fish.

Vacuum cleaners, suction cups, and other similar devices are powerful examples of how strong atmospheric pressure forces can be *if* the air is removed from one side of an object so as to produce an unbalanced force. The fact that we are *surrounded* by the fluid allows us to move around in the air or swim underwater, oblivious of these strong forces.

13.3 Measuring and Using Pressure

The pressure in a fluid is measured with a *pressure gauge,* which is often a device very similar to that in Figure 13.4. The fluid pushes against a spring, and the displacement of the spring is indicated on a scale. The familiar tire-pressure gauge shown in Figure 13.11 works in just this way. The pressure p_{tire} exerts a force $p_{tire}A$ on the front area A of the piston, while atmospheric pressure exerts a force $p_{atmos}A$ on the back of the piston. Thus the *net* pressure force on the piston is $(p_{tire} - p_{atmos})A$. This force compresses the spring until equilibrium is reached. Thus the movement of the scale depends not on the absolute pressure in the tire, but on the *difference* between the tire pressure and atmospheric pressure. This type of gauge measures the *gauge pressure* $p_g = p_{tire} - p_{atmos}$, an idea introduced in Chapter 12.

Solving Hydrostatic Problems

We now have enough information to formulate a set of rules for thinking about hydrostatic problems.

 TACTICS BOX 13.1 Hydrostatics   ✎ Exercise 5

❶ **Draw a picture.** Show open surfaces, pistons, boundaries, and other features that affect the pressure. Include height and area measurements and fluid densities. Identify the points at which you need to find the pressure.

❷ **Determine the pressure at surfaces.**
  ■ **Surface open to the air:** $p_0 = p_{atmos}$, usually 1 atm.
  ■ **Surface covered by a gas:** $p_0 = p_{gas}$.
  ■ **Closed surface:** $p = F/A$ where F is the force the surface, such as a piston, exerts on the fluid.

❸ **Use horizontal lines.** Pressure in a connected fluid is the same at any point along a horizontal line.

❹ **Allow for gauge pressure.** Pressure gauges read $p_g = p - 1$ atm.

❺ **Use the hydrostatic pressure equation.** $p = p_0 + \rho g d$.

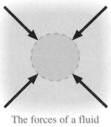

The forces of a fluid push in *all* directions.

FIGURE 13.10 Pressure forces in a fluid.

TRY IT YOURSELF

A crushing experience Heat a very small amount of water in a soda can until the water has boiled for a few seconds. Then, using tongs or gloves, quickly invert the can into a shallow pan of cold water. The can will dramatically implode! Boiling the water fills the can with water vapor at 1 atm, which rapidly condenses back into a liquid when the can is inverted into the cold water. Because liquid water occupies a much smaller volume than its vapor, this condensation creates a partial vacuum in the can. The greater pressure of the atmosphere outside then crushes it.

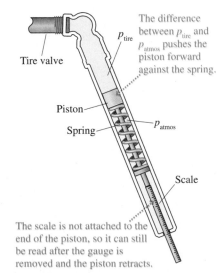

FIGURE 13.11 A tire gauge measures the difference between the tire's pressure and atmospheric pressure.

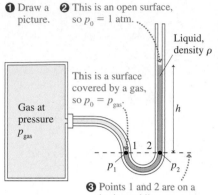

❶ Draw a picture. ❷ This is an open surface, so p_0 = 1 atm.

Liquid, density ρ

This is a surface covered by a gas, so $p_0 = p_{\text{gas}}$.

Gas at pressure p_{gas}

h

1   2

p_1 p_2

❸ Points 1 and 2 are on a horizontal line, so $p_1 = p_2$.

FIGURE 13.12 A manometer is used to measure gas pressure.

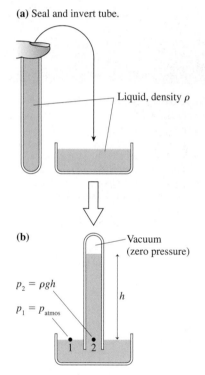

(a) Seal and invert tube.

Liquid, density ρ

(b)

Vacuum (zero pressure)

$p_2 = \rho g h$

$p_1 = p_{\text{atmos}}$

h

1   2

FIGURE 13.13 A barometer.

Manometers and Barometers

Gas pressure is sometimes measured with a device called a *manometer*. A manometer, shown in Figure 13.12, is a U-shaped tube connected to the gas at one end and open to the air at the other end. The tube is filled with a liquid—often mercury—of density ρ. The liquid is in static equilibrium. A scale allows the user to measure the height h of the right side above the left side.

Steps 1–3 from Tactics Box 13.1 lead to the conclusion that the pressures p_1 and p_2 must be equal. Pressure p_1, at the surface on the left, is simply the gas pressure: $p_1 = p_{\text{gas}}$. Pressure p_2 is the hydrostatic pressure at depth $d = h$ in the liquid on the right: $p_2 = 1$ atm $+ \rho g h$. Equating these two pressures gives

$$p_{\text{gas}} = 1 \text{ atm} + \rho g h \tag{13.6}$$

NOTE ▶ The height h has a *positive* value when the liquid is *higher* on the right than on the left ($p_{\text{gas}} > 1$ atm) and a *negative* value when the liquid is *lower* on the right than on the left ($p_{\text{gas}} < 1$ atm). ◀

Another important pressure-measuring instrument is the *barometer*, which is used to measure the atmospheric pressure p_{atmos}. Figure 13.13a shows a glass tube, sealed at the bottom, that has been completely filled with a liquid. If we temporarily seal the top end, we can invert the tube and place it in a beaker of the same liquid. When the temporary seal is removed, some, but not all, of the liquid runs out, leaving a liquid column in the tube that is a height h above the surface of the liquid in the beaker. This device, shown in Figure 13.13b, is a barometer. What does it measure? And why doesn't *all* the liquid in the tube run out?

We can analyze the barometer much as we did the manometer. Points 1 and 2 in Figure 13.13b are on a horizontal line drawn even with the surface of the liquid. The liquid is in hydrostatic equilibrium, so the pressure at these two points must be equal. Liquid runs out of the tube only until a balance is reached between the pressure at the base of the tube and the pressure of the air.

You can think of a barometer as rather like a seesaw. If the pressure of the atmosphere increases, it presses down on the liquid in the beaker. This forces liquid up the tube until the pressures at points 1 and 2 are equal. If the atmospheric pressure falls, liquid has to flow out of the tube to keep the pressures equal at these two points.

The pressure at point 2 is the pressure due to the weight of the liquid in the tube plus the pressure of the gas above the liquid. But in this case there is no gas above the liquid! Because the tube had been completely full of liquid when it was inverted, the space left behind when the liquid ran out is essentially a vacuum. Thus pressure p_2 is simply $p_2 = \rho g h$.

Equating p_1 and p_2, and noting that $p_1 = p_{\text{atmos}}$, gives

$$p_{\text{atmos}} = \rho g h \tag{13.7}$$

Thus we can measure the atmosphere's pressure by measuring the height of the liquid column in a barometer.

Equation 13.7 shows that the liquid height is $h = p_{\text{atmos}}/\rho g$. If a barometer were made using water, with $\rho = 1000 \text{ kg/m}^3$, the liquid column would be more than 10 m high, which is impractical. Instead, mercury, with its high density of 13,600 kg/m^3, is usually used. The average air pressure at sea level causes a column of mercury in a mercury barometer to stand 760 mm above the surface. We can then use Equation 13.7 to find that the average atmospheric pressure is

$$p_{\text{atmos}} = \rho_{\text{Hg}} g h = (13{,}600 \text{ kg/m}^3)(9.80 \text{ m/s}^2)(0.760 \text{ m})$$

$$= 1.013 \times 10^5 \text{ Pa} = 101.3 \text{ kPa}$$

This is the value given earlier as "one standard atmosphere." Now you can see that 1 atm = 101.3 kPa *because,* on average, a mercury barometer gives a reading of 760 mm.

Pressure Units

In practice, pressure is measured in a number of different units. This plethora of units and abbreviations has arisen historically as scientists and engineers working on different subjects (liquids, high-pressure gases, low-pressure gases, weather, etc.) developed what seemed to them the most convenient units. These units continue in use through tradition, so it is necessary to become familiar with converting back and forth between them. Table 13.2 gives the basic conversions.

TABLE 13.2 Pressure units

Unit	Abbreviation	Conversion to 1 atm	Uses
pascal	Pa	101.3 kPa	SI unit: $1 \text{ Pa} = 1 \text{ N/m}^2$ use in most calculations
atmosphere	atm	1 atm	general
millimeters of mercury	mm of Hg	760 mm of Hg	gases and barometric pressure
inches of mercury	in	29.92 in	barometric pressure in U.S. weather forecasting
pounds per square inch	psi	14.7 psi	U.S. engineering and industry

Blood Pressure

The last time you had a medical checkup, the doctor may have told you something like, "Your blood pressure is 120 over 80." What does that mean?

About every 0.8 s, assuming a pulse rate of 75 beats per minute, your heart "beats." The heart muscles contract and push blood out into your aorta. This contraction, like squeezing a balloon, raises the pressure in your heart. The pressure increase, in accordance with Pascal's principle, is transmitted through all your arteries.

Figure 13.14 is a pressure graph showing how blood pressure changes during one cycle of the heartbeat. The medical condition of *high blood pressure* usually means that your maximum (*systolic*) blood pressure is higher than necessary for blood circulation. The high pressure causes undue stress and strain on your entire circulatory system, often leading to serious medical problems. Low blood pressure can cause you to faint if you stand up quickly because the pressure isn't adequate to pump the blood up to your brain.

As shown in Figure 13.15, blood pressure is measured with a cuff that goes around your arm. The doctor or nurse pressurizes the cuff, places a stethoscope over the artery in your arm, then slowly releases the pressure while watching a pressure gauge. Initially, the cuff squeezes the artery shut and cuts off the blood flow. When the cuff pressure drops below the systolic pressure, the pressure pulse during each beat of your heart forces the artery open briefly and a squirt of blood goes through. You can feel this, and the doctor or nurse records the pressure when he or she hears the blood start to flow. This is your systolic pressure.

This pulsing of the blood through your artery lasts until the cuff pressure reaches the diastolic pressure. Then the artery remains open continuously and the blood flows smoothly. This transition is easily heard in the stethoscope, and the doctor or nurse records your base or *diastolic* pressure.

Blood pressure is measured in millimeters of mercury. And it is a gauge pressure, the pressure in excess of 1 atm. A fairly typical blood pressure of a healthy young adult is 120/80, meaning that the systolic pressure is $p_g = 120$ mm of Hg (absolute pressure $p = 880$ mm of Hg) and the diastolic pressure is 80 mm of Hg.

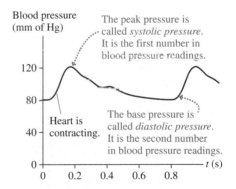

FIGURE 13.14 Blood pressure during one cycle of a heart beat.

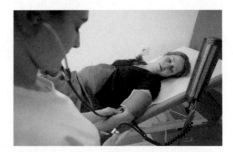

FIGURE 13.15 Measuring blood pressure with a manometer.

EXAMPLE 13.4 Measuring blood pressure

During a blood pressure measurement, a mercury-filled manometer is used to measure the pressure in the cuff. When the cuff pressure equals the systolic pressure, the height of the mercury in the open side of the manometer is 130 mm higher than on the side connected to the cuff. What is the patient's systolic pressure?

PREPARE We can use Equation 13.6 to find the pressure in the cuff, which is equal to the patient's blood pressure.

SOLVE Equation 13.6 gives the absolute pressure in the cuff as

$$p_{\text{cuff}} = 1 \text{ atm} + \rho g h$$

We want the blood pressure, which is the cuff pressure converted to gauge pressure. Thus

$$p_{\text{blood}} = p_{\text{cuff}} - 1 \text{ atm} = \rho g h$$

From Table 13.1, the density of mercury is $\rho = 13{,}600 \text{ kg/m}^3$. Using $h = 0.130 \text{ m}$ we have

$$p_{\text{blood}} = \rho g h = (13{,}600 \text{ kg/m}^3)(9.80 \text{ m/s}^2)(0.130 \text{ m})$$
$$= 17.3 \text{ kPa}$$

We can convert our result to millimeters of mercury, the proper unit for blood pressure, by using the conversion factor, $101.3 \text{ kPa} = 760 \text{ mm Hg}$. Thus

$$p_{\text{blood}} = 17.3 \text{ kPa}\left(\frac{760 \text{ mm Hg}}{101.3 \text{ kPa}}\right) = 130 \text{ mm of Hg}$$

ASSESS The blood pressure in mm of Hg is *equal* to the height difference of the mercury column in the manometer. This is what the units "mm of Hg" *mean:* a gauge pressure of 130 mm of Hg will push the open end of a mercury column 130 mm higher than the pressurized end.

CONCEPTUAL EXAMPLE 13.1 Blood pressure and the height of the arm

In Figure 13.15, the patient's arm is held at about the same height as her heart. Why is this?

REASON The hydrostatic pressure of a fluid varies with height. Although flowing blood is not in hydrostatic equilibrium, it is still true that blood pressure increases with the distance below the heart, and decreases above it. Because the upper arm when held beside the body is at the same height as the heart, the pressure here is the same as the pressure at the heart. If the patient held her arm straight up, the pressure cuff would be a distance $d \approx 25 \text{ cm}$ above her heart and the pressure would be *less* than the pressure at the heart by $\Delta p = \rho_{\text{blood}} g d \approx 20 \text{ mm of Hg}$.

ASSESS 20 mm of Hg is a substantial fraction of the average blood pressure. Measuring pressure above or below heart level could lead to a misdiagnosis of the patient's condition.

◀ **Pressure at the top** BIO A giraffe's head is some 2.5 m above its heart, compared to a distance of only about 30 cm for humans. To pump blood this extra height requires a blood pressure at the giraffe's heart that is some 170 mm of Hg higher than a human's, making its blood pressure more than twice as high as a human's.

13.4 Buoyancy

A rock, as you know, sinks like a rock. Wood floats on the surface of a lake. A penny with a mass of a few grams sinks, but a massive steel aircraft carrier floats. How can we understand these diverse phenomena?

An air mattress floats effortlessly on the surface of a swimming pool. But if you've ever tried to push an air mattress underwater, you know it is nearly impossible. As you push down, the water pushes up. This upward force of a fluid is called the **buoyant force.**

The basic reason for the buoyant force is easy to understand. Figure 13.16 shows a cylinder submerged in a liquid. The pressure in the liquid increases with depth, so the pressure at the bottom of the cylinder is larger than at the top. Both cylinder ends have equal area, so force $\vec{F}_{up}$ is larger than force $\vec{F}_{down}$. (Remember that pressure forces push in *all* directions.) Consequently, the pressure in the liquid exerts a *net upward force* on the cylinder of magnitude $F_{net} = F_{up} - F_{down}$. This is the buoyant force.

The submerged cylinder illustrates the idea in a simple way, but the result is not limited to cylinders or to liquids. Suppose we isolate a parcel of fluid of arbitrary shape and volume by drawing an imaginary boundary around it, as shown in Figure 13.17a. This parcel is in static equilibrium. Consequently, the parcel's weight force pulling it down must be balanced by an upward force. The upward force, which is exerted on this parcel of fluid by the surrounding fluid, is the buoyant force $\vec{F}_B$. The buoyant force matches the weight of the fluid: $F_B = w$.

Now imagine that we could somehow remove this parcel of fluid and instantaneously replace it with an object having exactly the same shape and size, as shown in Figure 13.17b. Because the buoyant force is exerted by the *surrounding* fluid, and the surrounding fluid hasn't changed, the buoyant force on this new object is *exactly the same* as the buoyant force on the parcel of fluid that we removed.

When an object (or a portion of an object) is immersed in a fluid, it *displaces* fluid that would otherwise fill that region of space. This fluid is called the **displaced fluid.** The displaced fluid's volume is exactly the volume of the portion of the object that is immersed in the fluid. Figure 13.17 leads us to conclude that the magnitude of the upward buoyant force matches the weight of this displaced fluid.

This idea was first recognized by the ancient Greek mathematician and scientist Archimedes, perhaps the greatest scientist of antiquity, and today we know it as *Archimedes' principle.*

> **Archimedes' principle** A fluid exerts an upward buoyant force $\vec{F}_B$ on an object immersed in or floating on the fluid. The magnitude of the buoyant force equals the weight of the fluid displaced by the object.

Suppose the fluid has density ρ_f and the object displaces volume V_f of fluid. The mass of the displaced fluid is $m_f = \rho_f V_f$ and so its weight is $w_f = \rho_f V_f g$. Thus Archimedes' principle in equation form is

$$F_B = \rho_f V_f g \qquad (13.8)$$

NOTE ▶ It is important to distinguish the density and volume of the displaced fluid from the density and volume of the object. To do so, we'll use subscript f for the fluid and o for the object. ◀

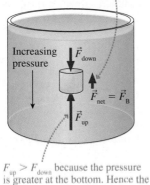

The net force of the fluid on the cylinder is the buoyant force $\vec{F}_B$.

$F_{up} > F_{down}$ because the pressure is greater at the bottom. Hence the fluid exerts a net upward force.

FIGURE 13.16 The buoyant force arises because the fluid pressure at the bottom of the cylinder is larger than at the top.

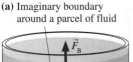

(a) Imaginary boundary around a parcel of fluid

These are equal because the parcel is in static equilibrium.

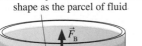

(b) Real object with same size and shape as the parcel of fluid

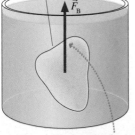

The buoyant force on the object is the same as on the parcel of fluid because the *surrounding* fluid has not changed.

FIGURE 13.17 The buoyant force on an object is the same as the buoyant force on the fluid it displaces.

EXAMPLE 13.5   Is the crown gold?

Legend has it that Archimedes was asked by King Hiero of Syracuse to determine whether a crown was of pure gold or if it had been adulterated with a lesser metal by an unscrupulous goldsmith. It was this problem that led him to the principle that bears his name. In a modern version of his method, a crown weighing 8.30 N in air is suspended underwater from a string. The tension in the string is measured to be 7.81 N. Is the crown pure gold?

PREPARE   To discover whether the crown is pure gold, we need to determine its density ρ_o. Because the crown is submerged, its volume V_o equals the volume V_f of displaced water, which we can find from Equation 13.8.

Figure 13.18 shows the forces acting on the crown. In addition to the familiar tension and weight forces, the water exerts an upward buoyant force on the crown. The size of the buoyant force is given by Archimedes' principle.

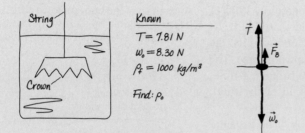

FIGURE 13.18   The forces acting on the submerged crown.

SOLVE   Because the crown's weight is $w_o = \rho_o V_o g$, its density is

$$\rho_o = \frac{w_o}{V_o g}$$

We know that $V_o = V_f$, and we can rearrange Equation 13.8 to give

$$V_o = V_f = \frac{F_B}{\rho_f g}$$

giving

$$\rho_o = \frac{w_o}{V_o g} = \frac{w_o}{\left(\dfrac{F_B}{\rho_f g}\right) g} = \frac{w_o \rho_f}{F_B}$$

We can find the magnitude F_B of the buoyant force from the fact that the crown is in static equilibrium, so the net force on it must be zero:

$$\sum F_y = F_B + T - w_o = 0$$

so that $F_B = w_o - T$. Putting this all together gives

$$\rho_o = \frac{w_o \rho_f}{F_B} = \frac{w_o \rho_f}{w_o - T}$$

$$= \frac{(8.30 \text{ N})(1000 \text{ kg/m}^3)}{(8.30 \text{ N}) - (7.81 \text{ N})} = 1.69 \times 10^4 \text{ kg/m}^3$$

The crown's density is considerably lower than that of pure gold, which is $1.93 \times 10^4 \text{ kg/m}^3$. The crown is not pure gold.

ASSESS   Our result shows that the closer the tension T (the *apparent weight*) is to w_o (the actual weight), the larger the density of the object. This makes sense, because the weight of a very dense object is much greater than the buoyant force on it, so that its true and apparent weights are nearly the same.

Float or Sink?

If you *hold* an object underwater and then release it, it either floats to the surface, sinks, or remains "hanging" in the water. How can we predict which it will do? Whether it heads for the surface or the bottom depends on whether the upward buoyant force F_B on the object is larger or smaller than the downward weight force w_o.

The magnitude of the buoyant force is $\rho_f V_f g$. The weight of a uniform object, such as a block of steel, is simply $\rho_o V_o g$. But a compound object, such as a scuba diver, may have pieces of varying density. If we define the **average density** to be $\rho_{avg} = m_o / V_o$, the weight of a compound object can be written $w_o = \rho_{avg} V_o g$.

Comparing $\rho_f V_f g$ to $\rho_{avg} V_o g$, and noting that $V_f = V_o$ for an object that is fully submerged, we see that an object floats or sinks depending on whether the

◄ **Submersible scales**   BIO   In Example 13.5, we saw how the density of an object could be determined by weighing it both underwater and in air. This idea is the basis of an accurate method for determining a subject's percentage of body fat. Fat has a lower density than lean muscle or bone, so a lower overall body density implies a greater proportion of body fat. To determine a subject's density, she is first weighed in air, then lowered completely into water and weighed again. Standard tables accurately relate body density to fat percentage.

fluid density ρ_f is larger or smaller than the object's average density ρ_{avg}. If the densities are equal, the object is in static equilibrium and hangs motionless. This is called **neutral buoyancy.** These conditions are summarized in Tactics Box 13.2.

(MP) TACTICS BOX 13.2 Finding whether an object floats or sinks   ✐ Exercise 8

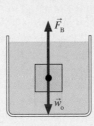

❶ Object sinks

❷ Object floats

❸ Object has neutral buoyancy

An object sinks if it weighs more than the fluid it displaces; that is, if its average density is greater than the density of the fluid:

$$\rho_{avg} > \rho_f$$

An object floats on the surface if it weighs less than the fluid it displaces; that is, if its average density is less than the density of the fluid:

$$\rho_{avg} < \rho_f$$

An object hangs motionless in the fluid if it weighs exactly the same as the fluid it displaces. It has neutral buoyancy if its average density equals the density of the fluid:

$$\rho_{avg} = \rho_f$$

As an example, steel is denser than water, so a chunk of steel sinks. Oil is less dense than water, so oil floats on water. Fish use *swim bladders* filled with air and scuba divers use weighted belts to adjust their average density to match the water. Both are examples of neutral buoyancy.

If you release a block of wood underwater, the net upward force causes the block to shoot to the surface. Then what? To understand floating, let's begin with a *uniform* object such as the block shown in Figure 13.19. This object contains nothing tricky, like indentations or voids. Because it's floating, it must be the case that $\rho_o < \rho_f$.

Now that the object is floating, it's in static equilibrium. The upward buoyant force, given by Archimedes' principle, exactly balances the downward weight of the object. That is

$$F_B = \rho_f V_f g = w_o = \rho_o V_o g \qquad (13.9)$$

In this case, the volume of the displaced fluid is *not* the same as the volume of the object. In fact, we can see from Equation 13.9 that the volume of fluid displaced by a floating object of uniform density is

$$V_f = \frac{\rho_o}{\rho_f} V_o \qquad (13.10)$$

which is *less* than V_o because $\rho_o < \rho_f$.

NOTE ▶ Equation 13.10 applies only to *uniform* objects. It does not apply to boats, hollow spheres, or other objects of nonuniform composition. ◀

An object of density ρ_o and volume V_o is floating on a fluid of density ρ_f.

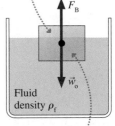

Fluid density ρ_f

The submerged volume of the object is equal to the volume V_f of displaced fluid.

FIGURE 13.19 A floating object is in static equilibrium.

◄**Hidden depths** You've probably heard it said that "90% of an iceberg is underwater." Equation 13.10 is the basis for that statement. Most icebergs break off glaciers and are fresh-water ice with a density of 917 kg/m³. The density of seawater is 1030 kg/m³. Thus

$$V_\text{f} = \frac{917 \text{ kg/m}^3}{1030 \text{ kg/m}^3} V_\text{o} = 0.89 V_\text{o}$$

V_f, the displaced water, is the volume of the iceberg that is underwater. You can see that, indeed, 89% of the volume of an iceberg is underwater.

CONCEPTUAL EXAMPLE 13.2 Which has the greater buoyant force?

A block of iron sinks to the bottom of a vessel of water while a block of wood of the *same size* floats. On which is the buoyant force greater?

REASON The buoyant force is equal to the volume of water displaced. The iron block is completely submerged, so it displaces a volume of water equal to its own volume. The wood block floats, so it displaces only the fraction of its volume that is under water, which is *less* than its own volume. The buoyant force on the iron block is therefore greater than on the wood one.

ASSESS This result may seem counterintuitive, but remember that the iron block sinks because of its high density, while the wood block floats because of its low density. A smaller buoyant force is sufficient to keep it floating.

EXAMPLE 13.6 Measuring the density of an unknown liquid

You need to determine the density of an unknown liquid. You notice that a block floats in this liquid with 4.6 cm of the side of the block submerged. When the block is placed in water, it also floats but with 5.8 cm submerged. What is the density of the unknown liquid?

PREPARE Assume that the block is an object of uniform composition. Figure 13.20 shows the block and defines the cross-section area A and submerged lengths h_u in the unknown liquid and h_w in water.

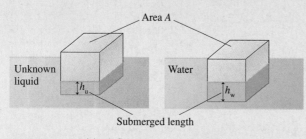

Area A

Unknown liquid Water

h_u h_w

Submerged length

FIGURE 13.20 A block floating in two liquids.

SOLVE The block is floating, so Equation 13.10 applies. The block displaces volume $V_\text{u} = Ah_\text{u}$ of the unknown liquid. Thus

$$V_\text{u} = Ah_\text{u} = \frac{\rho_\text{o}}{\rho_\text{u}} V_\text{o}$$

Similarly, the block displaces volume $V_\text{w} = Ah_\text{w}$ of the water, leading to

$$V_\text{w} = Ah_\text{w} = \frac{\rho_\text{o}}{\rho_\text{w}} V_\text{o}$$

Because there are two fluids, we've used subscripts w for water and u for the unknown in place of the fluid subscript f. The product $\rho_\text{o} V_\text{o}$ appears in both equations. In the first, $\rho_\text{o} V_\text{o} = \rho_\text{u} Ah_\text{u}$, and in the second $\rho_\text{o} V_\text{o} = \rho_\text{w} Ah_\text{w}$. Equating the right-hand sides gives

$$\rho_\text{u} Ah_\text{u} = \rho_\text{w} Ah_\text{w}$$

The unknown area A cancels, and the density of the unknown liquid is

$$\rho_\text{u} = \frac{h_\text{w}}{h_\text{u}} \rho_\text{w} = \frac{5.8 \text{ cm}}{4.6 \text{ cm}} 1000 \text{ kg/m}^3 = 1300 \text{ kg/m}^3$$

ASSESS Comparison with Table 13.1 shows that the unknown liquid is likely to be glycerin.

Boats and Balloons

A chunk of steel sinks, so how does a steel-hulled boat float? As we've seen, an object floats if the upward buoyant force—the weight of the displaced water—balances the weight of the object. A boat is really a large hollow shell whose weight is determined by the volume of steel in the hull. As Figure 13.21 shows, the volume of water displaced by a shell is *much* larger than the volume of the hull itself. As a boat settles into the water, it sinks until the weight of the displaced water exactly matches the boat's weight. It is then in static equilibrium, so it floats at that level.

The concept of buoyancy and flotation applies to all fluids, not just liquids. An object immersed in a gas such as air feels a buoyant force as well. Because the density of air is so small, this buoyant force is generally negligible. Nonetheless, even though the buoyant force due to air is small, an object will float in air if it weighs less than the air that it displaces. This is why a floating balloon cannot be filled with regular air. If it were, then the weight of the air inside it would equal the weight of the air it displaced, so that it would have no net upward force on it. Adding in the weight of the balloon itself would then lead to a net downward force. For a balloon to float, it must be filled with a gas that has a *lower* density than that of air. The following example illustrates how this works.

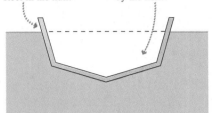

The weight of the boat is mostly the weight of the steel in the hull.

The buoyant force is equal to the weight of the water displaced by the hull.

Even though the density of water is much less than that of steel, the volume of water displaced by a hollow shell is very large. The boat will float if the weight of the displaced water matches the weight of the boat.

FIGURE 13.21 How a boat floats.

EXAMPLE 13.7 How big does a balloon need to be?
What diameter must a helium-filled balloon have to float with neutral buoyancy? The mass of the empty balloon is 2.0 g.

PREPARE We'll model the balloon as a sphere. The balloon will float when its weight—the weight of the empty balloon plus that of the helium—equals the weight of the air it displaces. The densities of air and helium are given in Table 13.1.

SOLVE The volume of the balloon is $V_{balloon}$. Its weight is

$$w_{balloon} = m_{balloon}g + \rho_{He}V_{balloon}g$$

where $m_{balloon}$ is the mass of the empty balloon. The weight of the displaced air is

$$w_{air} = \rho_{air}V_{air}g = \rho_{air}V_{ballon}g$$

where we noted that the volume of displaced air is the volume of the balloon. The balloon will just float when these two forces are equal, or when

$$\rho_{air}V_{balloon}g = m_{balloon}g + \rho_{He}V_{balloon}g$$

The g cancels, and we can solve for the volume of the balloon:

$$V_{balloon} = \frac{m_{balloon}}{\rho_{air} - \rho_{He}}$$

$$= \frac{2.0 \times 10^{-3} \text{ kg}}{1.3 \text{ kg/m}^3 - 0.18 \text{ kg/m}^3} = 1.8 \times 10^{-3} \text{ m}^3$$

A sphere has volume $V = (4\pi/3)r^3$. Thus the balloon's radius is

$$r = \left(\frac{3V_{balloon}}{4\pi}\right)^{\frac{1}{3}} = \left(\frac{3 \times (1.8 \times 10^{-3} \text{ m}^3)}{4\pi}\right)^{\frac{1}{3}} = 0.075 \text{ m}$$

The diameter of the balloon is twice this, or 15 cm.

ASSESS This example shows why a helium balloon will no longer float once its volume falls below a certain value.

Hot air rising A hot air balloon is filled with a low-density gas: Hot air! You learned in Chapter 12 that gases expand upon heating, thus lowering their density. The air at the top of a hot air balloon is surprisingly hot—about 100°C, the temperature of boiling water. Using the ideal-gas law, you should be able to show that the density of 100°C air is only 79% that of room-temperature air at 20°C. A burst of flame lowers the density of the air and thus the balloon's weight. The balloon rises when its weight becomes less than the weight of the cooler air it has displaced.

STOP TO THINK 13.3   An ice cube is floating in a glass of water that is filled entirely to the brim. When the ice cube melts, the water level will

A. Fall.   B. Stay the same.   C. Rise, causing the water to spill.

13.5 Fluids in Motion

The wind blowing through your hair, a white-water river, and oil gushing from an oil well are examples of fluids in motion. We've focused thus far on fluid statics, but it's time to turn our attention to fluid *dynamics*.

Fluid flow is a complex subject. Many aspects of fluid flow, especially turbulence and the formation of eddies, are still not well understood and are areas of current research. We will avoid these difficulties by using a simplified *model* of an *ideal* fluid. Our model can be expressed in three assumptions about the fluid:

1. The fluid is *incompressible*. This is a very good assumption for liquids, but it also holds reasonably well for a moving gas, such as air. For instance, even when a 100 mph wind slams into a wall, its density changes by only about one percent.
2. The flow is *steady*. That is, the fluid velocity at each point in the fluid is constant; it does not fluctuate or change with time. Flow under these conditions is called **laminar flow,** and it is distinguished from *turbulent flow*.
3. The fluid is *nonviscous*. Water flows much more easily than cold pancake syrup because the syrup is a very viscous liquid. Viscosity is resistance to flow, and assuming a fluid is nonviscous is analogous to assuming the motion of a particle is frictionless. Gases have very low viscosity, and even many liquids are well approximated as being nonviscous.

Later, in Section 13.7, we'll relax condition 3 and consider the effects of viscosity.

The rising smoke in the photograph of Figure 13.22 begins as laminar flow, recognizable by the smooth contours, but at some point undergoes a transition to turbulent flow. A laminar-to-turbulent transition is not uncommon in fluid flow. Our model of fluids can be applied to the laminar flow, but not to the turbulent flow.

FIGURE 13.22 Rising smoke changes from laminar flow to turbulent flow.

The Equation of Continuity

Consider a fluid flowing through a tube—oil through a pipe or blood through an artery. If the tube's diameter changes, as happens in Figure 13.23, what happens to the speed of the fluid?

When you squeeze a toothpaste tube, the volume of toothpaste that emerges matches the amount by which you reduce the volume of the tube. An *incompressible* fluid flowing through a rigid tube or pipe acts the same way. Fluid is neither created nor destroyed within the tube, and it cannot be stored. If volume V enters the tube during some interval of time Δt, then an equal volume of fluid must leave the tube.

To see the implications of this idea, suppose all molecules of the fluid in Figure 13.23 are moving forward with speed v_1 at a point where the cross-section area is A_1. Later, where the cross-section area is A_2, their speed is v_2. During an interval of time Δt, the first molecules move forward distance $\Delta x_1 = v_1 \Delta t$ and the second move $\Delta x_2 = v_2 \Delta t$. Because the fluid is incompressible, small volumes ΔV_1 and ΔV_2 must be equal. That is,

$$\Delta V_1 = A_1 \Delta x_1 = A_1 v_1 \Delta t = \Delta V_2 = A_2 \Delta x_2 = A_2 v_2 \Delta t \qquad (13.11)$$

Dividing both sides of the equation by Δt gives what is called the **equation of continuity:**

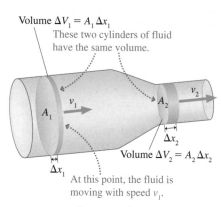

Volume $\Delta V_1 = A_1 \Delta x_1$
These two cylinders of fluid have the same volume.

Volume $\Delta V_2 = A_2 \Delta x_2$

At this point, the fluid is moving with speed v_1.

FIGURE 13.23 Flow speed changes through a tapered tube.

$$v_1A_1 = v_2A_2 \qquad (13.12)$$

The equation of continuity relating the speed v of an incompressible fluid to the cross-section area A of the tube in which it flows

(a) Garden hose

Garden hose with nozzle

Equations 13.11 and 13.12 say that **the volume of an incompressible fluid entering one part of a tube or pipe must be matched by an equal volume leaving downstream.**

An important consequence of the equation of continuity is that **flow is faster in narrower parts of a tube, slower in wider parts.** You're familiar with this conclusion from many everyday observations. The garden hose shown in Figure 13.24a squirts farther after you put a nozzle on it. This is because the narrower opening of the nozzle gives the water a higher exit speed. Water flowing from the faucet shown in Figure 13.24b picks up speed as it falls. As a result, the flow tube "necks down" to a smaller diameter.

The *rate* at which fluid flows through the tube—volume per second—is $\Delta V / \Delta t$. This is called the **volume flow rate** Q. We can see from Equation 13.11 that

$$Q = \frac{\Delta V}{\Delta t} = vA \qquad (13.13)$$

(b)

The SI units of Q are m^3/s, although in practice Q may be measured in cm^3/s, liters per minute, or, in the United States, gallons per minute and cubic feet per minute. Another way to express the meaning of the equation of continuity is to say that **the volume flow rate is constant at all points in a tube.**

NOTE ▶ An ideal fluid has no friction with the walls of its tube, so the speed is the same at all points on a line across the tube (i.e., at the wall and at the center). Real fluids have viscosity so that, as we'll see, the speed of the fluid is different at the tube's wall than at its center. But we can still use the equation of continuity for real fluids if we use the *average* speed of the fluid in Equations 13.12 and 13.13. ◀

FIGURE 13.24 The speed of the fluid changes as the flow diameter changes. This is a consequence of the equation of continuity.

EXAMPLE 13.8 Speed of water through a hose

A garden hose has an inside diameter of 16 mm. The hose can fill a 10 L bucket in 20 s.

a. What is the speed of the water out of the end of the hose?
b. What diameter nozzle would cause the water to exit with a speed 4 times greater than the speed inside the hose?

PREPARE Water is essentially incompressible, so the equation of continuity applies.

SOLVE

a. The volume flow rate is $Q = \Delta V/\Delta t = (10\text{ L})/(20\text{ s}) = 0.50\text{ L/s}$. To convert this to SI units, recall that $1\text{ L} =$

$1000\text{ mL} = 10^3\text{ cm}^3 = 10^{-3}\text{ m}^3$. Thus $Q = 5.0 \times 10^{-4}\text{ m}^3/\text{s}$. We can find the speed of the water from Equation 13.13:

$$v = \frac{Q}{A} = \frac{Q}{\pi r^2} = \frac{5.0 \times 10^{-4}\text{ m}^3/\text{s}}{\pi(0.0080\text{ m})^2} = 2.5\text{ m/s}$$

b. The quantity $Q = vA$ remains constant as the water flows through the hose and then the nozzle. To increase v by a factor of 4, A must be reduced by a factor of 4. The cross-section area depends on the square of the radius, so the area is reduced by a factor of 4 if the radius is reduced by a factor of 2. Thus the necessary nozzle diameter is 8 mm.

EXAMPLE 13.9 Blood flow in capillaries

The volume flow rate of blood leaving the heart to circulate throughout the body is about 5 L/min for a person at rest. All this blood eventually must pass through the smallest of blood vessels, the capillaries. Microscope measurements show that a

typical capillary is 6 μm in diameter, 1 mm long, and the blood flows through it at an average speed of 1 mm/s.

a. Estimate the total number of capillaries in the body.
b. Estimate the total surface area of all the capillaries.

Continued

The various lengths and areas are defined in Figure 13.25.

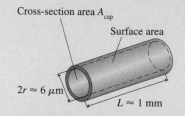

FIGURE 13.25 A capillary (not to scale).

PREPARE We can use the equation of continuity. The volume flow rate through all the capillaries must equal the volume flow rate leaving the heart.

SOLVE

a. We'll start by converting Q to SI units:

$$Q = 5\frac{L}{min} \times \frac{1\ m^3}{1000\ L} \times \frac{1\ min}{60\ s} = 8 \times 10^{-5}\ m^3/s$$

to one-significant-figure accuracy. Then the total area of all the capillaries is

$$A_{total} = \frac{Q}{v} = \frac{8 \times 10^{-5}\ m^3/s}{0.001\ m/s} = 0.08\ m^2$$

The cross-section area of each capillary is $A_{cap} = \pi r^2$, so the total number of capillaries is approximately

$$N = \frac{A_{total}}{A_{cap}} = \frac{0.08\ m^2}{\pi(3 \times 10^{-6}\ m)^2} = 3 \times 10^9$$

b. The surface area of one capillary is

$$A = \text{circumference of capillary} \times \text{length}$$

$$= 2\pi r L = 2\pi(3 \times 10^{-6}\ m)(0.001\ m) = 2 \times 10^{-8}\ m^2$$

so the total surface area of all the capillaries is about

$$A_{surface} = NA = (3 \times 10^9)(2 \times 10^{-8}\ m) = 60\ m^2$$

ASSESS The number of capillaries is huge, and the total surface area is about the floor area of a good-sized house. As we saw in Chapter 12, oxygen and nutrients move from the blood into cells by the slow process of diffusion. Only by having this large surface area available for diffusion can the required rate of gas and nutrient exchange be attained.

Representing Fluid Flow: Streamlines and Fluid Elements

To represent the flow of fluid is more complicated than representing the motion of a point particle because fluid flow is the collective motion of a vast number of particles. Figure 13.26 is an interesting photograph that gives us an idea of one possible fluid-flow representation. Here smoke is being used to help engineers visualize the air flow around a car in a wind tunnel. The smoothness of the flow tells us this is laminar flow. But notice also how the individual smoke trails retain their identity. They don't cross or get mixed together. Each smoke trail represents a *streamline* in the fluid.

Imagine that we could inject a tiny colored drop of water into a stream of water undergoing laminar flow. Because the flow is steady and the water is incompressible, this colored drop would maintain its identity as it flowed along. The path or trajectory followed by this "particle of fluid" is called a **streamline.** Smoke particles mixed with the air allow you to see the streamlines in the wind-tunnel photograph of Figure 13.26. Figure 13.27 illustrates three important properties of streamlines.

In studying the motion of a fluid, it is often also useful to consider a small *volume* of fluid, a volume containing many particles of fluid. Such a volume is called a **fluid element.** Figure 13.28 on the next page shows some of the important properties of a fluid element. Unlike a particle, a fluid element has an actual shape and volume. Although the shape of a fluid element may change as it moves, the equation of continuity requires that its volume remain constant. The progress of a fluid element as it moves along streamlines and changes shape is another very useful representation of fluid motion.

FIGURE 13.26 The laminar air flow around a car in a wind tunnel is made visible with smoke.

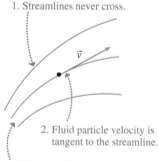

1. Streamlines never cross.

$\vec{v}$

2. Fluid particle velocity is tangent to the streamline.

3. The speed is higher where the streamlines are closer together.

◄ **FIGURE 13.27** Particles in a fluid move along streamlines.

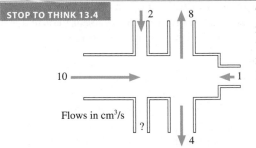

STOP TO THINK 13.4

Flows in cm³/s

The figure shows volume flow rates (in cm³/s) for all but one tube. What is the volume flow rate through the unmarked tube? Is the flow direction in or out?

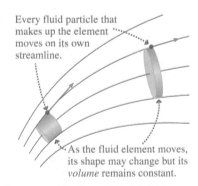

Every fluid particle that makes up the element moves on its own streamline.

As the fluid element moves, its shape may change but its *volume* remains constant.

FIGURE 13.28 Motion of a fluid element.

13.6 Fluid Dynamics

The equation of continuity describes a moving fluid but doesn't tell us anything about *why* the fluid is in motion. To understand the dynamics, consider the ideal fluid moving from left to right through the tube shown in Figure 13.29. The fluid moves at a steady speed v_1 in the wider part of the tube. In accordance with the equation of continuity, its speed is a higher, but steady, v_2 in the narrower part of the tube. If we follow a fluid element through the tube, we see that it undergoes an *acceleration* from v_1 to v_2 in the tapered section of the tube.

In the absence of friction, Newton's first law tells us that a particle will coast forever at a steady speed. Now in a fluid, viscosity is analogous to friction. In a viscous fluid, parts of the fluid moving at different speeds "rub" against each other, tending to slow the fluid down. But an ideal fluid is nonviscous and thus analogous to a particle moving without friction. Thus a fluid element moving through the constant-diameter sections of the tube in Figure 13.29 requires no force to "coast" at steady speed.

On the other hand, a fluid element moving through the neck of the tube accelerates, speeding up from v_1 to v_2. According to Newton's second law, for a particle to accelerate a net force must act on it. Similarly, for this fluid element to speed up there must be a net force acting on it.

What is the origin of this force? There are no external forces, and the horizontal motion rules out gravity. Instead, the fluid element is pushed from both ends by the *surrounding fluid*—that is, by *pressure forces*. The fluid element with cross-section area A in Figure 13.30 has a higher pressure on its left side than on its right. Thus the force $F_L = p_L A$ of the fluid pushing on its left side—a force pushing to the right—is greater than the force $F_R = p_R A$ from the fluid on its right. The net force, which points from the high pressure side of the element to the lower-pressure side, is

$$F_{net} = F_L - F_R = (p_L - p_R)A = A\,\Delta p$$

In other words, there's a net force on the fluid element, causing it to change speed, if and only if there's a pressure *difference* Δp between the two faces.

Thus, in order to accelerate the fluid elements in Figure 13.29 through the neck, the pressure p_1 in the wider section of the tube must be higher than the pressure p_2 in the narrower section. When the pressure is changing from one point in a fluid to another, we say that there is a **pressure gradient** in that region. We can also say that pressure forces are caused by pressure gradients, so **an ideal fluid accelerates wherever there is a pressure gradient.**

As a result, **the pressure is higher at a point along a streamline where the fluid is moving slower, lower where the fluid is moving faster.** This property of fluids was discovered in the 18th century by the Swiss scientist Daniel Bernoulli and is called the **Bernoulli effect.**

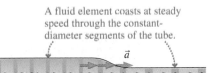

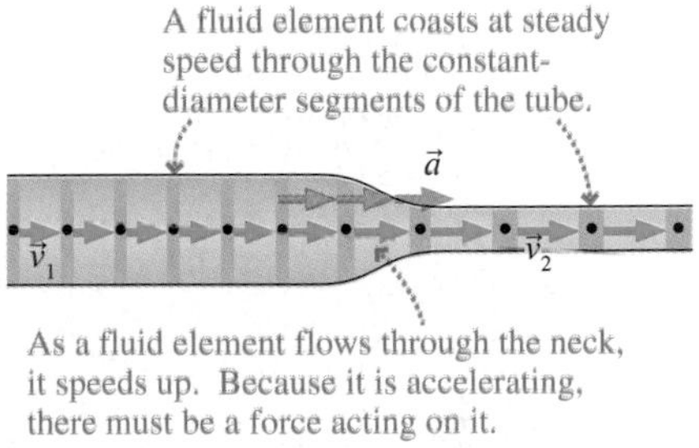

A fluid element coasts at steady speed through the constant-diameter segments of the tube.

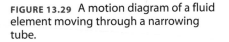

As a fluid element flows through the neck, it speeds up. Because it is accelerating, there must be a force acting on it.

FIGURE 13.29 A motion diagram of a fluid element moving through a narrowing tube.

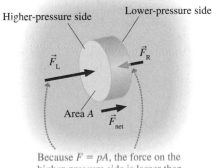

Higher-pressure side   Lower-pressure side

$\vec{F}_L$ $\vec{F}_R$

Area A $\vec{F}_{net}$

Because $F = pA$, the force on the higher-pressure side is larger than that on the lower-pressure side.

FIGURE 13.30 The net force on a fluid element due to pressure points from high to low pressure.

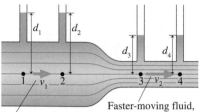

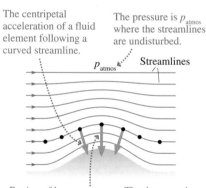

FIGURE 13.31 A Venturi tube measures flow speeds of a fluid.

The centripetal acceleration of a fluid element following a curved streamline.

The pressure is p_{atmos} where the streamlines are undisturbed.

p_{atmos} Streamlines

Region of lower pressure. The downward pressure gradient causes the centripetal force.

FIGURE 13.32 Centripetal acceleration of a fluid element is caused by a downward pressure gradient.

> **NOTE** ► It is important to realize that it is the change in pressure from high to low that *causes* the fluid to speed up. A high fluid speed doesn't *cause* a low pressure any more than a fast-moving particle causes the force that accelerated it. ◄

This relationship between pressure and fluid speed can be used to measure the speed of a fluid with a device called a *Venturi tube*. Figure 13.31 shows a simple Venturi tube suitable for a flowing liquid. The high pressure at point 1, where the fluid is moving slowly, causes the fluid to rise in the vertical pipe to a total height d_1. Because there's no vertical motion of the fluid, we can use the hydrostatic pressure equation to find that the pressure at point 1 is $p_1 = p_0 + \rho g d_1$. At point 3, on the same streamline, the fluid is moving faster and the pressure is lower, thus the fluid in the vertical pipe rises to a lower height d_3. The pressure *difference* is $\Delta p = \rho g (d_1 - d_3)$, so that the pressure difference across the neck of the pipe can be found by measuring the difference in heights of the fluid. We'll see later in this section how to relate this pressure difference to the increase of speed.

Note that the fluid height d_1 is the same as d_2, and d_3 is the same as d_4. For an ideal fluid, with no viscosity, no pressure difference is needed to keep the fluid moving at a constant speed, as it does in both the large- and small-diameter sections of the tube. In the next section we'll see how this result changes for a viscous fluid.

Lift

For a moving particle, a changing speed is just one way to accelerate. A particle also undergoes an acceleration if its *direction* of motion changes. Similarly, a moving fluid experiences an acceleration if it changes direction, requiring a net force to act on it. Figure 13.32 shows the wind blowing over a hill. Far from the hill, the airflow is undisturbed and the pressure is p_{atmos}. Near the ground, however, the streamlines have to *curve* around the hill, causing them to get bunched closer together.

You learned in Chapter 6 that curved motion requires a force pointing toward the center of curvature. As a fluid element follows the streamlines around the hill, the only force that can give it a centripetal acceleration is a pressure force due to the air above and below it. For the acceleration to point downward, the pressure near the ground, where the streamlines are closer together, must be *less* than the pressure p_{atmos} at slightly higher elevations.

Causing a fluid to follow a curved trajectory has some surprising but important consequences. To see this, try the simple experiment illustrated in Figure 13.33 before reading the next paragraph. Really, do try this!

TRY IT YOURSELF ►

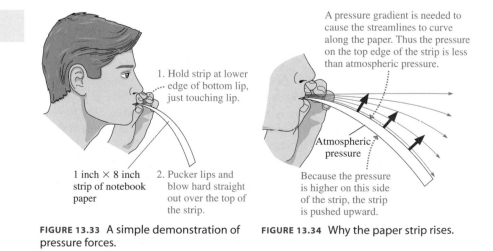

1. Hold strip at lower edge of bottom lip, just touching lip.

1 inch × 8 inch strip of notebook paper

2. Pucker lips and blow hard straight out over the top of the strip.

A pressure gradient is needed to cause the streamlines to curve along the paper. Thus the pressure on the top edge of the strip is less than atmospheric pressure.

Atmospheric pressure

Because the pressure is higher on this side of the strip, the strip is pushed upward.

FIGURE 13.33 A simple demonstration of pressure forces.

FIGURE 13.34 Why the paper strip rises.

What happened? You probably expected your breath to press the strip of paper down. Instead, the strip *rose*. In fact, the harder you blow, the more nearly the strip becomes parallel to the floor. As shown in Figure 13.34, the pressure near the strip must be less than atmospheric pressure to force the air to follow the curved path. This is just like the air flowing over the hill. Because there's a higher pressure beneath the strip, a net force pushes the strip upward. You can convince yourself that the curvature of the paper is important by placing the strip flat on a table. Blowing across it will not cause it to rise.

The upward motion of the paper strip is a simple example of the idea of *lift*. As you can imagine, an important application of this idea is the upward lift forces that act on airplane wings due to the air flowing past the wings. Figure 13.35 shows how this works. The top of a wing is curved, like the top of a gentle hill, while the bottom is flat. The air curves over the top of the wing, just like it does over the hill in Figure 13.32, which requires a lower pressure near the upper surface of the wing than in the surrounding air. Thus a *low pressure* zone develops just above the wing. The air moving *beneath* the wing is not strongly deflected, so the pressure there is close to p_{atmos}. The atmospheric pressure below pushes up; the low pressure above pushes down, but less strongly. The net result is an upward force—the lift.

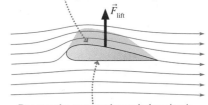

A pressure gradient is needed to accelerate the air around the curved upper surface of the wing. Thus the air just above the wing is a zone of low pressure.

$\vec{F}_{lift}$

Because the pressure beneath the wing is higher than the pressure above, there's a net upward force on the wing. This is lift.

FIGURE 13.35 Air flow over a wing generates lift by creating unequal pressures above and below.

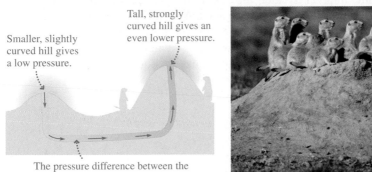

Smaller, slightly curved hill gives a low pressure.

Tall, strongly curved hill gives an even lower pressure.

The pressure difference between the two ends pushes air through the burrow.

Nature's air conditioning Prairie dogs ventilate their underground burrows with the same aerodynamic forces and pressures that give airplanes lift. The two entrances to their burrows are surrounded by mounds, one higher than the other. When the wind blows across these mounds the pressure is reduced at the top, just as for an airplane's wing. The taller mound, with its greater curvature, has the lower pressure of the two entrances. Air then is pushed through the burrow towards this lower-pressure side.

Bernoulli's Equation

We've seen that a pressure gradient causes a fluid to accelerate horizontally. Not surprisingly, gravity can cause a fluid to speed up or slow down if the fluid changes elevation. These are the key ideas of fluid dynamics. Now we would like to make these ideas quantitative by finding a numerical relationship between pressure, height, and the speed of a fluid. We can do so by applying the statement of conservation of mechanical energy you learned in Chapter 10,

$$\Delta K + \Delta U = W$$

where U is the gravitational potential energy and W is the work done by other forces, in this case pressure forces. Recall that we're still considering ideal fluids, with no friction or viscosity, so there's no dissipation of energy to thermal energy.

Figure 13.36 on the next page shows fluid flowing through a tube. The tube narrows from cross-section area A_1 to area A_2 as it bends uphill. Let's concentrate on the large segment of the fluid that is initially between the two planes, marked a and c, that are perpendicular to the flow. **This moving segment of fluid will be our *system* for the purpose of applying conservation of energy.**

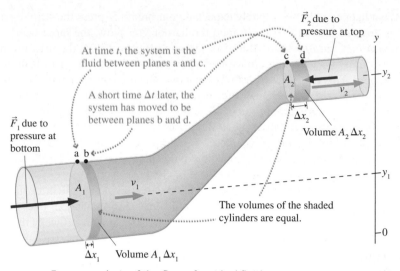

FIGURE 13.36 Energy analysis of the flow of an ideal fluid.

A short time interval Δt later, the system has moved a bit further along the tube, so that it is now between planes b and d. Because the tube is not of uniform diameter, the two ends of the fluid system do not move the same distance during Δt: The lower end moves a distance Δx_1 while the upper end moves Δx_2. These two distances are related by the continuity equation, Equation 13.12. We have

$$v_1 A_1 = \frac{\Delta x_1}{\Delta t} A_1 = v_2 A_2 = \frac{\Delta x_2}{\Delta t} A_2$$

or, canceling Δt,

$$A_1 \, \Delta x_1 = A_2 \, \Delta x_2 = V$$

Thus V is the volume of either of the small cylinders of fluid shown in Figure 13.36.

As the system moves, positive work is done on it by the force $\vec{F}_1$ due to the pressure p_1 of the fluid to the left of the system, while negative work is done on the system by the force $\vec{F}_2$ due to the pressure p_2 of the fluid to the right of the system. The positive work is

$$W_1 = F_1 \, \Delta x_1 = (p_1 A_1) \, \Delta x_1 = p_1 (A_1 \, \Delta x_1) = p_1 V$$

Similarly the negative work is

$$W_2 = -F_2 \, \Delta x_2 = -p_2 A_2 \, \Delta x_2 = -p_2 V$$

Thus the *net* work done on the system is

$$W = W_1 + W_2 = p_1 V - p_2 V = (p_1 - p_2) V$$

By conservation of energy, this work equals the change in the system's energy.

Initially the system had the kinetic and potential energy of all the fluid between points a and c; after time Δt has passed, it has the energy of all the fluid between b and d. The fluid between points b and c is common to both the initial and final systems, so there is *no* change in energy due to this part of the fluid. As the system moves, it *loses* the energy it had in cylinder 1 of length Δx_1, but *gains* energy in cylinder 2 of length Δx_2.

Let's find the kinetic energy contained in each of these cylinders. The mass of each of these cylinders is $m = \rho V$, where ρ is the density of the fluid. The kinetic energies of cylinders 1 and 2 are

$$K_1 = \frac{1}{2} \rho V v_1^2 \qquad \text{and} \qquad K_2 = \frac{1}{2} \rho V v_2^2$$

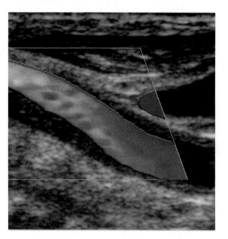

Living under pressure BIO When plaque builds up in major arteries, such as the carotid artery that supplies the head, dangerous drops in blood pressure can result. *Doppler ultrasound,* which you'll study in Chapter 15, uses sound waves to produce images of the interior of the body, and can detect the velocity of flowing blood. The image above shows the blood flow through a carotid artery with significant plaque buildup; yellow indicates a higher blood velocity than red. Once the velocities are known at two points along the flow, Bernoulli's equation can be used to deduce the corresponding pressure drop.

Thus the net *change* in kinetic energy is

$$\Delta K = K_2 - K_1 = \frac{1}{2}\rho V v_2^2 - \frac{1}{2}\rho V v_1^2$$

Similarly, the net change in the gravitational potential energy of our fluid system is

$$\Delta U = U_2 - U_1 = \rho V g y_2 - \rho V g y_1$$

We can use these expressions for ΔK, ΔU, and W to write the energy equation as

$$\underbrace{\frac{1}{2}\rho V v_2^2 - \frac{1}{2}\rho V v_1^2}_{\Delta K} + \underbrace{\rho V g y_2 - \rho V g y_1}_{\Delta U} = \underbrace{(p_1 - p_2)V}_{W}$$

The V cancels, and we can rearrange the remaining terms to get what is called **Bernoulli's equation:**

$$p_2 + \frac{1}{2}\rho v_2^2 + \rho g y_2 = p_1 + \frac{1}{2}\rho v_1^2 + \rho g y_1 \qquad (13.14)$$

Bernoulli's equation relating pressure p, speed v, and height y
at any two points in an ideal fluid

Equation 13.14, a quantitative statement of the ideas we developed earlier in this section, is really nothing more than a statement about work and energy. Using Bernoulli's equation is very much like using the law of conservation of energy. Rather than identifying a "before" and "after," you want to identify two points on a streamline. As the following example shows, Bernoulli's equation is often used in conjunction with the equation of continuity.

EXAMPLE 13.10 Pressure in an irrigation system

Water flows through the pipes shown in Figure 13.37. The water's speed through the lower pipe is 5.0 m/s and a pressure gauge reads 75 kPa. What is the reading of the pressure gauge on the upper pipe?

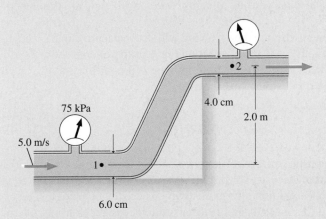

FIGURE 13.37 The water pipes of an irrigation system.

PREPARE Treat the water as an ideal fluid obeying Bernoulli's equation. Consider a streamline connecting point 1 in the lower pipe with point 2 in the upper pipe.

SOLVE Bernoulli's equation, Equation 13.14, relates the pressure, fluid speed, and heights at points 1 and 2. It is easily solved for the pressure p_2 at point 2:

$$p_2 = p_1 + \frac{1}{2}\rho v_1^2 - \frac{1}{2}\rho v_2^2 + \rho g y_1 - \rho g y_2$$

$$= p_1 + \frac{1}{2}\rho(v_1^2 - v_2^2) + \rho g(y_1 - y_2)$$

All quantities on the right are known except v_2, and that is where the equation of continuity will be useful. The cross-section areas and water speeds at points 1 and 2 are related by

$$v_1 A_1 = v_2 A_2$$

from which we find

$$v_2 = \frac{A_1}{A_2}v_1 = \frac{r_1^2}{r_2^2}v_1 = \frac{(0.030 \text{ m})^2}{(0.020 \text{ m})^2}(5.0 \text{ m/s}) = 11.25 \text{ m/s}$$

The pressure at point 1 is $p_1 = 75$ kPa $+ 1$ atm $= 176{,}300$ Pa. We can now use the above expression for p_2 to calculate $p_2 = 105{,}900$ Pa. This is the absolute pressure; the pressure gauge on the upper pipe will read

$$p_2 = 105{,}900 \text{ Pa} - 1 \text{ atm} = 4.6 \text{ kPa}$$

ASSESS Reducing the pipe size decreases the pressure because it makes $v_2 > v_1$. Gaining elevation also reduces the pressure.

Rank in order, from highest to lowest, the liquid heights h_1 to h_4 in tubes 1 to 4. The air flow is from left to right.

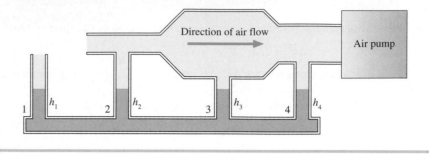

13.7 Viscosity and Poiseuille's Equation

The pressure drops between points 1 and 2.

In this region, the pressure gradient causes the fluid to speed up.

The pressure drops between points 3 and 4.

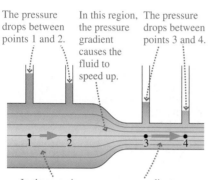

In these regions, a pressure gradient is needed simply to keep the fluid moving with constant speed.

FIGURE 13.38 A viscous fluid needs a pressure difference to keep it moving. Compare this to Figure 13.31, which showed an ideal fluid.

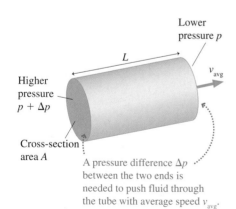

Lower pressure p

L

Higher pressure $p + \Delta p$

v_{avg}

Cross-section area A

A pressure difference Δp between the two ends is needed to push fluid through the tube with average speed v_{avg}.

FIGURE 13.39 The pressure difference needed to keep the fluid flowing is proportional to the fluid's viscosity.

Thus far, the only thing we've needed to know about a fluid is its density. It is the density that determines how the pressure in a static fluid increases with depth, and density appears in Bernoulli's equation for the motion of a fluid. But you know from everyday experience that another property of fluids is often crucial in determining how fluids flow. The densities of honey and water are not too different, but there's a huge difference in the way honey and water pour. The honey is much "thicker." This property of a fluid, which measures its resistance to flow, is called **viscosity.** A very viscous fluid flows slowly when poured and is difficult to force through a pipe or tube. The viscous nature of fluids is of key importance in understanding a wide range of real-world applications, from blood flow to the flight of birds to throwing a curveball.

An ideal fluid, with no viscosity, will "coast" at constant speed through a constant-diameter tube with no change in pressure. That's why the fluid heights are the same in the first and second pressure-measuring columns on the Venturi tube of Figure 13.31. But as Figure 13.38 shows, any real fluid requires a *pressure difference* between the ends of a tube to keep the fluid moving at constant speed. The size of the pressure difference depends on the viscosity of the fluid. Think about how much harder you have to suck on a straw while drinking a thick milkshake than you do to drink water or to pull air through the straw.

Figure 13.39 shows a viscous fluid flowing with a constant average speed v_{avg} through a tube of length L and cross-section area A. Experiments show that the pressure difference needed to keep the fluid moving is proportional to v_{avg} and to L and inversely proportional to A. We can write

$$\Delta p = 8\pi\eta\frac{Lv_{avg}}{A} \tag{13.15}$$

where $8\pi\eta$ is the constant of proportionality and η (lower-case Greek eta) is called the **coefficient of viscosity** (or just the *viscosity*). (The 8π enters the equation from the technical definition of viscosity, which need not concern us.)

Equation 13.15 makes sense. A more viscous fluid needs a larger pressure difference to push it through the tube; an ideal fluid, with $\eta = 0$, will keep flowing without any pressure difference. We can also see from Equation 13.15 that the units of viscosity are $N \cdot s/m^2$ or, equivalently, $Pa \cdot s$. In practice, a common unit of viscosity is the **poise** (pronounced "pwaz"), defined as 1 poise = $0.1 \text{ N} \cdot \text{s/m}^2$. However, the poise is not an SI unit and thus is not suitable for most calculations.

Table 13.3 gives values of η for some common fluids. Note that the viscosity of many liquids decreases *very* rapidly with temperature. Cold oil hardly flows at all, but hot oil pours almost like water.

Poiseuille's Equation

Viscosity has a profound effect on how a fluid moves through a tube. Figure 13.40a shows that in an ideal fluid, all fluid particles move with the same speed v, the speed that appears in the equation of continuity. For a viscous fluid, Figure 13.40b shows that the fluid moves fastest in the center of the tube. The speed decreases as you move away from the center of the tube until reaching zero on the walls of the tube. That is, the layer of fluid in contact with the tube doesn't move at all. Whether it be water through pipes or blood through arteries, the fact that the fluid at the outer edges "lingers" and barely moves allows deposits to build on the inside walls of a tube.

Although we can't characterize the flow of a viscous liquid by a single flow speed v, we can still define the *average* flow speed. Suppose a fluid with viscosity η flows through a circular pipe with radius R and cross-section area $A = \pi R^2$. From Equation 13.15, a pressure difference Δp between the ends of the pipe causes the fluid to flow with average speed

$$v_{avg} = \frac{R^2}{8\eta L} \Delta p \tag{13.16}$$

The average flow speed is directly proportional to the pressure difference; for the fluid to flow twice as fast, you would need to double the pressure difference between the ends of the pipe.

Equation 13.13 defined the volume flow rate $Q = \Delta V/\Delta t$ and found that $Q = vA$ for an ideal fluid. For viscous flow, where v isn't constant throughout the fluid, we simply need to replace v with the average speed v_{avg} found in Equation 13.16. Using $A = \pi R^2$ for a circular tube, we see that a pressure difference Δp causes a volume flow rate

$$Q = v_{avg}A = \frac{\pi R^4 \Delta p}{8\eta L} \tag{13.17}$$

Poiseuille's equation for viscous flow through a tube of radius R and length L

This result is called **Poiseuille's equation** after the French scientist Jean Poiseuille (1797–1869) who first performed this calculation.

One surprising result of Poiseuille's equation is the very strong dependence of the flow on the tube's radius; the volume flow rate is proportional to the *fourth* power of R. If you double the radius of a tube, the flow rate will increase by a factor of $2^4 = 16$. This strong radius dependence comes from two factors. First, the flow rate depends on the area of the tube, which is proportional to R^2; larger tubes carry more fluid. Second, the average speed is larger in a larger tube because the center of the tube, where the flow is fastest, is farther from the "drag" exerted by the wall. As we've seen, the average speed is also proportional to R^2. These two terms combine to give the R^4 dependence.

TABLE 13.3 Viscosities of fluids

Fluid	$\eta \ (\text{Pa} \cdot \text{s})$
Air (20°C)	1.8×10^{-5}
Water (20°C)	1.0×10^{-3}
Water (40°C)	0.7×10^{-3}
Water (60°C)	0.5×10^{-3}
Whole blood (37°C)	2.5×10^{-3}
Motor oil (−30°C)	3×10^5
Motor oil (40°C)	0.07
Motor oil (100°C)	0.01
Honey (15°C)	600
Honey (40°C)	20

(a) Ideal fluid

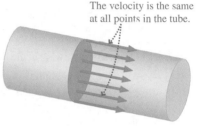

The velocity is the same at all points in the tube.

(b) Viscous fluid

The velocity is maximum at the center of the tube. It decreases away from center.

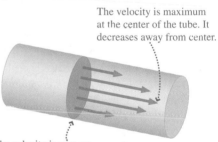

The velocity is zero on the walls of the tube.

FIGURE 13.40 Viscosity alters the velocities of the fluid particles.

CONCEPTUAL EXAMPLE 13.3 Blood pressure and cardiovascular disease

Cardiovascular disease is a narrowing of the arteries due to the buildup of plaque deposits on the interior walls. Magnetic resonance imaging, which you'll learn about in Chapter 24, can create exquisite three-dimensional images of the internal structure of the body. Shown are the carotid arteries that supply blood to the head, with a dangerous narrowing—a *stenosis*—indicated by the arrow.

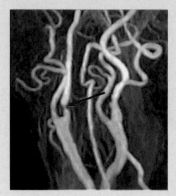

If a section of an artery has narrowed by 8%, not nearly as much as the stenosis shown, by what percent must the blood-pressure difference between the ends of the narrowed section increase to keep blood flowing at the same rate?

REASON According to Poiseuille's equation, the pressure difference Δp must increase to compensate for a decrease in the artery's radius R if the blood flow rate Q is to remain unchanged. If we write Poiseuille's equation as

$$R^4 \, \Delta p = 8\eta L Q/\pi$$

we see that the product $R^4 \, \Delta p$ must remain unchanged if the artery is to deliver the same flow rate. Let the initial artery radius and pressure difference be R_i and Δp_i. Disease decreases the radius by 8%, meaning that $R_f = 0.92 R_i$. The requirement

$$R_i^4 \, \Delta p_i = R_f^4 \, \Delta p_f$$

can be solved for the new pressure difference:

$$\Delta p_f = \frac{R_i^4}{R_f^4} \Delta p_i = \frac{R_i^4}{(0.92 R_i)^4} \Delta p_i = 1.4 \, \Delta p_i$$

That is, the pressure difference must increase by 40% to maintain the flow.

ASSESS Because the flow rate depends on R^4, even a small change in radius requires a large change in Δp to compensate. Either the person's blood pressure must increase, which is dangerous, or he or she will suffer a significant reduction in blood flow. For the stenosis shown in the image above the reduction in radius is much greater than 8%, and the pressure difference will be large and very dangerous.

EXAMPLE 13.11 Pressure drop along a capillary

In Example 13.9 we examined blood flow through a capillary. Using the numbers from that example, calculate the pressure "drop" from one end of a capillary to the other.

PREPARE Example 13.9 gives enough information to determine the flow rate through a capillary. We can then use Poiseuille's equation to calculate the pressure difference between the ends.

SOLVE The measured volume flow rate leaving the heart was given as about 5 L/min = 8×10^{-5} m³/s. This flow is divided among all the capillaries, which we found to number $N = 3 \times 10^9$. Thus the flow rate through each capillary is

$$Q_{cap} = \frac{Q_{heart}}{N} = \frac{8 \times 10^{-5} \text{ m}^3/\text{s}}{3 \times 10^9} = 2.7 \times 10^{-14} \text{ m}^3/\text{s}$$

Solving Poiseuille's equation for Δp, we get

$$\Delta p = \frac{8\eta L Q_{cap}}{\pi R^4}$$

$$= \frac{8(2.5 \times 10^{-3} \text{ Pa} \cdot \text{s})(0.001 \text{ m})(2.7 \times 10^{-14} \text{ m}^3/\text{s})}{\pi (3 \times 10^{-6} \text{ m})^4}$$

$$= 2100 \text{ Pa}$$

Converting to mm of mercury, the units of blood pressure, the pressure drop across the capillary is $\Delta p = 16$ mm of mercury.

ASSESS The average blood pressure provided by the heart (the average of systolic and diastolic pressure) is about 100 mm of mercury. A physiology textbook will tell you that the pressure has decreased to 35 mm by the time blood enters the capillaries, and it exits from capillaries into the veins at 17 mm. Thus the pressure drop across the capillaries is 18 mm of mercury. Our calculation, based on the laws of fluid flow and some simple estimates of capillary size, is in almost perfect agreement with measured values.

STOP TO THINK 13.6 A viscous fluid flows through the pipe shown. The three marked segments are of equal length. Across which segment is the pressure difference the greatest?

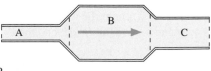

The goal of Chapter 13 has been to understand the static and dynamic properties of fluids.

GENERAL PRINCIPLES

Fluid Statics

Gases

- Freely moving particles
- Compressible
- Pressure mainly due to particle collisions with walls

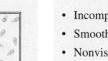

Liquids

- Loosely bound particles
- Incompressible
- Pressure due to the weight of the liquid
- Hydrostatic pressure at depth d is $p = p_0 + \rho g d$
- The pressure is the same at all points on a horizontal line through a liquid in hydrostatic equilibrium

Fluid Dynamics

Ideal-fluid model

- Incompressible
- Smooth, laminar flow
- Nonviscous

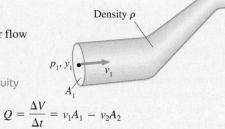

Density ρ

Equation of continuity

Volume flow rate $Q = \dfrac{\Delta V}{\Delta t} = v_1 A_1 = v_2 A_2$

Bernoulli's equation is a statement of energy conservation:

$$p_1 + \frac{1}{2}\rho v_1^2 + \rho g y_1 = p_2 + \frac{1}{2}\rho v_2^2 + \rho g y_2$$

Lift on a curved surface is generated when a fluid moves across the surface. It is due to the pressure gradients that cause the fluid's centripetal acceleration.

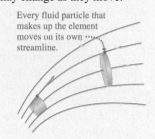

Poiseuille's equation governs viscous flow through a tube: $Q = v_{avg} A = \dfrac{\pi R^4\, \Delta p}{8 \eta L}$

IMPORTANT CONCEPTS

Density $\rho = m/V$, where m is mass and V is volume.

Pressure $p = F/A$, where F is force magnitude and A is the area on which the force acts.

- Exists at all points in a fluid
- Pushes equally in all directions
- Gauge pressure $p_g = p - 1$ atm

Viscosity η is the property of a fluid that makes it resist flowing.

Representing fluid flow

Streamlines are the paths of individual fluid particles.

The velocity of a fluid particle is tangent to its streamline.

$\vec{v}$

The speed is higher where the streamlines are closer together.

Fluid elements contain a fixed volume of fluid. Their shape may change as they move.

Every fluid particle that makes up the element moves on its own streamline.

APPLICATIONS

Buoyancy is the upward force of a fluid on an object immersed in the fluid.

Archimedes' principle The magnitude of the buoyant force equals the weight of the fluid displaced by the object.

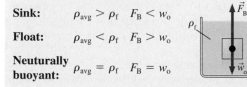

Sink:	$\rho_{avg} > \rho_f$ $F_B < w_o$
Float:	$\rho_{avg} < \rho_f$ $F_B > w_o$
Neutrally buoyant:	$\rho_{avg} = \rho_f$ $F_B = w_o$

Barometers measure atmospheric pressure. Atmospheric pressure is related to the height of the liquid column by $p_{atmos} = \rho g h$.

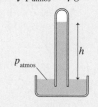

p_{atmos}

Manometers measure pressure. The pressure at the closed end of the tube is $p = 1$ atm $+ \rho g h$.

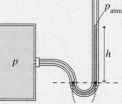

These two points are at the same pressure p.

For instructor-assigned homework, go to www.masteringphysics.com

Problem difficulty is labeled as | (straightforward) to ||||| (challenging).

Problems labeled INT integrate significant material from earlier chapters; BIO are of biological or medical interest.

QUESTIONS

Conceptual Questions

1. Which has the greater density, 1 g of mercury or 1000 g of water?
2. A 1×10^{-3} m^3 chunk of material has a mass of 3 kg.
   a. What is the material's density?
   b. Would a 2×10^{-3} m^3 chunk of the same material have the same mass? Explain.
   c. Would a 2×10^{-3} m^3 chunk of the same material have the same density? Explain.
3. You are given an irregularly shaped chunk of material and asked to find its density. List the *specific* steps that you would follow to do so.
4. Object 1 has an irregular shape. Its density is 4000 kg/m^3.
   a. Object 2 has the same shape and dimensions as object 1, but it is twice as massive. What is the density of object 2?
   b. Object 3 has the same mass and the same *shape* as object 1, but its size in all three dimensions is twice that of object 1. What is the density of object 3?
5. BIO When you get a blood transfusion the bag of blood is held above your body, but when you donate blood the collection bag is held below. Why is this?
6. BIO To explore the bottom of a 10-m-deep lake, your friend Tom proposes to get a long garden hose, put one end on land and the other in his mouth for breathing underwater, and descend into the depths. Susan, who overhears the conversation, reacts with horror and warns Tom that he will not be able to inhale when he is at the lake bottom. Why is Susan so worried?
7. Rank in order, from largest to smallest, the pressures at A, B, and C in Figure Q13.7. Explain.
8. Refer to Figure Q13.7. Rank in order, from largest to smallest, the pressures at D, E, and F. Explain.

FIGURE Q13.7

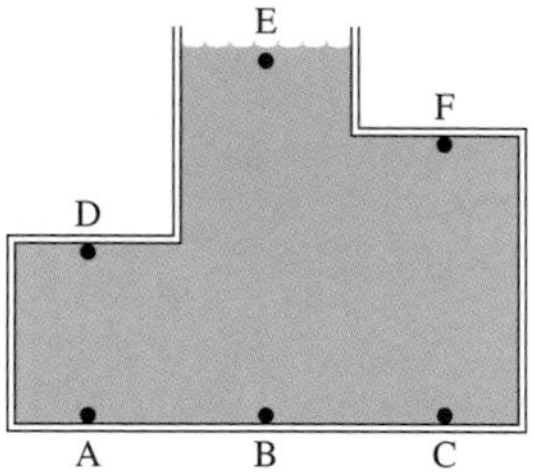

9. Cylinders A and B contain liquids. The pressure p_A at the bottom of A is higher than the pressure p_B at the bottom of B. Is the ratio p_A/p_B of the absolute pressures larger, smaller, or equal to the ratio of the gauge pressures? Explain.
10. In Figure Q13.10, A and B are rectangular tanks full of water. They have equal heights and equal side thicknesses (the dimension into the page), but different widths.

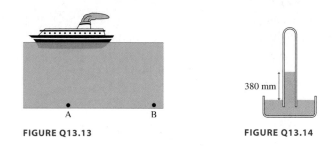

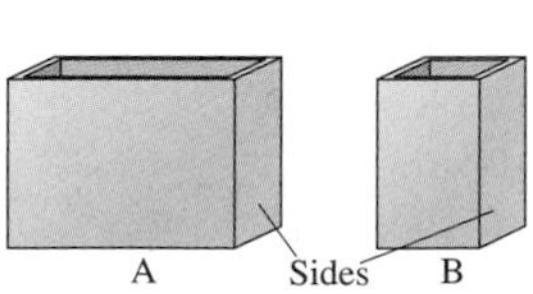

FIGURE Q13.10

   a. Compare the forces the water exerts on the bottoms of the tanks. Is F_A larger, smaller, or equal to F_B? Explain.
   b. Compare the forces the water exerts on the sides of the tanks. Is F_A larger, smaller, or equal to F_B? Explain.
11. Helium-filled weather balloons are spherical when they reach very high altitudes. However, they are only partially inflated with helium before they are released. Explain why this is done.

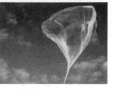

12. Water expands when heated. Suppose a beaker of water is heated from 10°C to 90°C. Does the pressure at the bottom of the beaker increase, decrease, or stay the same? Explain.
13. In Figure Q13.13, is p_A larger, smaller, or equal to p_B? Explain.

FIGURE Q13.13

FIGURE Q13.14

14. At sea level, the height of the mercury column in a sealed glass tube is 380 mm, as shown in Figure Q13.14. What can you say about the contents of the space above the mercury? Be as specific as you can.
15. Rank in order, from largest to smallest, the densities of objects A, B, and C in Figure Q13.15. Explain.

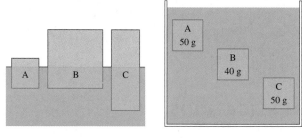

FIGURE Q13.15

FIGURE Q13.16

16. Objects A, B, and C in Figure Q13.16 have the same volume. Rank in order, from largest to smallest, the sizes of the buoyant forces F_A, F_B, and F_C on A, B, and C. Explain.
17. Refer to Figure Q13.16. Now A, B, and C have the same density, but still have the masses given in the figure. Rank in order, from largest to smallest, the sizes of the buoyant forces on A, B, and C. Explain.
18. When you stand on a bathroom scale, it reads 700 N. Suppose a giant vacuum cleaner sucks half the air out of the room, reducing the pressure to 0.5 atm. Would the scale reading increase, decrease, or stay the same? Explain.

19. Suppose you stand on a bathroom scale that is on the bottom of a swimming pool. The water comes up to your waist. Does the scale read more, less, or the same as your true weight? Explain.

20. When you place an egg in water, it sinks. If you add salt to the water, after some time the egg floats. Explain.

21. Submerged submarines contain tanks filled with water. To rise to the surface, compressed air is used to force the water out of the tanks. Explain why this works.

22. BIO Fish can adjust their buoyancy with an organ called the *swim bladder*. The swim bladder is a flexible gas-filled sac; the fish can increase or decrease the amount of gas in the swim bladder so that it stays neutrally buoyant—neither sinking nor floating. Suppose the fish is neutrally buoyant at some depth and then goes deeper. What needs to happen to the volume of air in the swim bladder? Will the fish need to add or remove gas from the swim bladder to maintain its neutral buoyancy?

23. Figure Q13.23 shows two identical beakers filled to the same height with water. Beaker B has a plastic sphere floating in it. Which beaker, with all its contents, weighs more? Or are they equal? Explain.

FIGURE Q13.23

24. A tub of water, filled to the brim, sits on a scale. Then a floating block of wood is placed in the tub, pushing some water over the rim. The water that overflows immediately runs off the scale. What happens to the reading of the scale?

25. Ships A and B have the same height and the same mass. Their cross-section profiles are shown in Figure Q13.25. Does one ship ride higher in the water (more height above the water line) than the other? If so, which one? Explain.

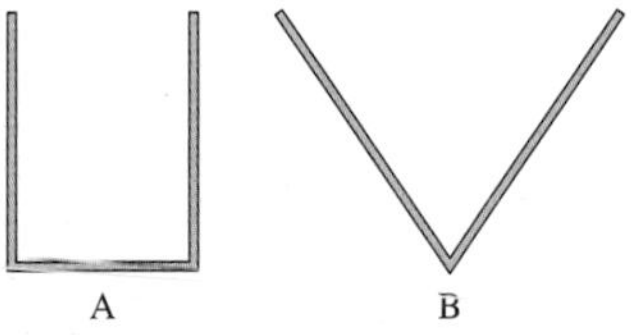

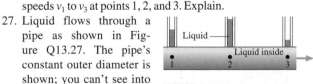

FIGURE Q13.25

26. Gas flows through a pipe, as shown in Figure Q13.26. The pipe's constant outer diameter is shown; you can't see into the pipe to know how the inner diameter changes. Rank in order, from largest to smallest, the gas speeds v_1 to v_3 at points 1, 2, and 3. Explain.

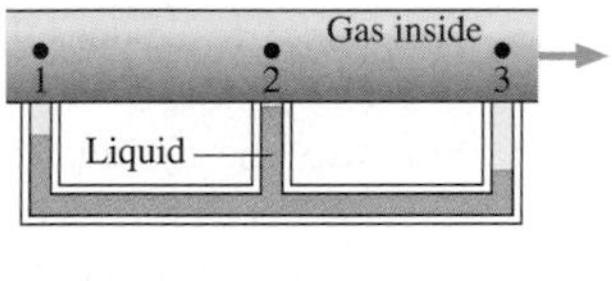

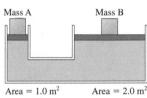

FIGURE Q13.26

27. Liquid flows through a pipe as shown in Figure Q13.27. The pipe's constant outer diameter is shown; you can't see into the pipe to know how the inner diameter changes. Rank in order, from largest to smallest, the flow speeds v_1 to v_3 at points 1, 2, and 3. Explain.

FIGURE Q13.27

28. A liquid with negligible viscosity flows through the pipe shown in Figure Q13.28. This is an overhead view.
    a. Rank in order, from largest to smallest, the flow speeds v_1 to v_4 at points 1 to 4. Explain.
    b. Rank in order, from largest to smallest, the pressures p_1 to p_4 at points 1 to 4. Explain.

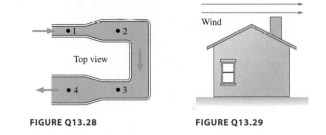

FIGURE Q13.28   **FIGURE Q13.29**

29. Wind blows over a house as shown in Figure Q13.29. A window on the ground floor is open. Is there an air flow through the house? If so, does the air flow in the window and out the chimney, or in the chimney and out the window? Explain.

30. Two pipes have the same inner cross-section area. One has a circular cross section and the other has a rectangular cross section with its height one-tenth its width. Through which pipe, if either, would it be easier to pump a viscous liquid? Explain.

Multiple-Choice Questions

31. | Figure Q13.31 shows a 100 g block of copper ($\rho = 8900$ kg/m^3) and a 100 g block of aluminum ($\rho = 2700$ kg/m^3) connected by a massless string that runs over two massless, frictionless pulleys. The two blocks exactly balance, since they have the same mass. Now suppose that the whole system is submerged in water. What will happen?

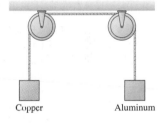

FIGURE Q13.31

   A. The copper block will fall, the aluminum block will rise.
   B. The aluminum block will fall, the copper block will rise.
   C. Nothing will change.
   D. Both blocks will rise.

32. | Masses A and B rest on very light pistons that enclose a fluid, as shown in Figure Q13.32. There is no friction between the pistons and the cylinders they fit inside. Which of the following is true?

FIGURE Q13.32

   A. Mass A is greater.   B. Mass B is greater.
   C. Mass A and mass B are the same.

33. ‖ If you dive underwater, you notice an uncomfortable pressure on your eardrums due to the increased pressure. The human eardrum has an area of about 70 mm^2 (7×10^{-5} m^2), and it can sustain a force of about 7 N without rupturing. If your body had no means of balancing the extra pressure (which, in reality, it does), what would be the maximum depth you could dive without rupturing your eardrum?
   A. 0.3 m   B. 1 m   C. 3 m   D. 10 m

34. | Suppose you take a basketball and submerge it in a swimming pool. To what minimum gauge pressure must you fill the ball for it to stay fully inflated at a depth of 2 m?
   A. 2 kPa   B. 4 kPa   C. 20 kPa   D. 40 kPa

35. ‖ A basketball has a mass of 0.50 kg and a volume of 8.0×10^{-3} m³. What is the magnitude of the net force on a basketball when it is fully submerged in water?
    A. 4.9 N   B. 74 N   C. 78 N   D. 83 N

36. | An object floats in water, with 75% of its volume submerged. What is its approximate density?
    A. 250 kg/m³   B. 750 kg/m³
    C. 1000 kg/m³   D. 1250 kg/m³

37. ‖ A syringe is being used to squirt water as shown in Figure Q13.37. The water is ejected from the nozzle at 10 m/s. At what speed is the plunger of the syringe being depressed?
    A. 0.01 m/s B. 0.1 m/s   C. 1 m/s   D. 10 m/s

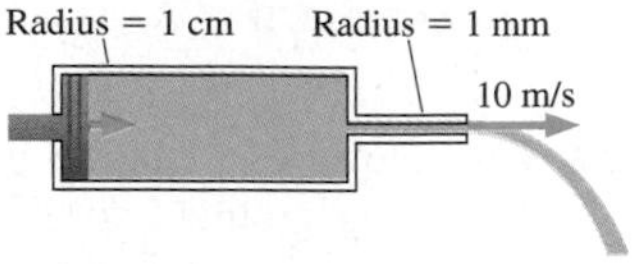

Radius = 1 cm   Radius = 1 mm

10 m/s

FIGURE Q13.37

38. ‖ Water flows through a 4.0-cm-diameter horizontal pipe at a speed of 1.3 m/s. The pipe then narrows down to a diameter of 2.0 cm. Ignoring viscosity, what is the pressure difference between the wide and narrow sections of the pipe?
    A. 850 Pa   B. 3400 Pa   C. 9300 Pa
    D. 12,700 Pa E. 13,500 Pa

39. ‖ A 15-m-long garden hose has an inner diameter of 2.5 cm. One end is connected to a spigot; 20°C water flows from the other end at a rate of 1.2 L/s. What is the gauge pressure at the spigot end of the hose?
    A. 1900 Pa   B. 2700 Pa   C. 4200 Pa
    D. 5800 Pa   E. 7300 Pa

PROBLEMS

Section 13.1 Fluids and Density

1. | A 100 mL beaker holds 120 g of liquid. What is the liquid's density in SI units?

2. | Containers A and B have equal volumes. Container A holds helium gas at 1.0 atm pressure and 0°C. Container B is completely filled with a liquid whose mass is 7000 times the mass of helium gas in container A. Identify the liquid in B.

3. | Air enclosed in a sphere has density $\rho = 1.4$ kg/m³. What will the density be if the radius of the sphere is halved, compressing the air within?

4. | Air enclosed in a cylinder has density $\rho = 1.4$ kg/m³.
    a. What will be the density of the air if the length of the cylinder is doubled while the radius is unchanged?
    b. What will be the density of the air if the radius of the cylinder is halved while the length is unchanged?

5. ‖ a. 50 g of gasoline are mixed with 50 g of water. What is the average density of the mixture?
    b. 50 cm³ of gasoline are mixed with 50 cm³ of water. What is the average density of the mixture?

6. ‖ Ethyl alcohol has been added to 200 mL of water in a container that has a mass of 150 g when empty. The resulting container and liquid mixture has a mass of 512 g. What volume of alcohol was added to the water?

7. | The average density of the body of a fish is 1080 kg/m³. To
BIO keep from sinking, the fish increases its volume by inflating an internal air bladder, known as a swim bladder, with air. By what percent must the fish increase its volume to be neutrally buoyant in fresh water? Use the Table 13.1 value for the density of air.

Section 13.2 Pressure

8. | The deepest point in the ocean is 11 km below sea level, deeper than Mt. Everest is tall. What is the pressure in atmospheres at this depth?

9. ‖ a. What volume of water has the same mass as 8.0 m³ of ethyl alcohol?
    b. If this volume of water completely fills a cubic tank, what is the pressure at the bottom?

10. ‖ A 1.0-m-diameter vat of liquid is 2.0 m deep. The pressure at the bottom of the vat is 1.3 atm. What is the mass of the liquid in the vat?

11. ‖ Find the force on the bottom of a 35-cm-tall, cylindrical beaker with a diameter of 5.0 cm. The beaker is filled to its brim with water.

12. ‖ The gauge pressure at the bottom of a cylinder of liquid is $p_g = 0.40$ atm. The liquid is poured into another cylinder with twice the radius of the first cylinder. What is the gauge pressure at the bottom of the second cylinder?

13. ‖‖ A research submarine has a 20-cm-diameter window 8.0 cm thick. The manufacturer says the window can withstand forces up to 1.0×10^6 N. What is the submarine's maximum safe depth? The pressure inside the submarine is maintained at 1.0 atm.

14. ‖ The highest that George can suck water up a very long straw is 2.0 m. (This is a typical value.) What is the lowest pressure that he can maintain in his mouth?

15. ‖‖ The two 60-cm-diameter cylinders in Figure P13.15, closed
BIO at one end, open at the other, are joined to form a single cylinder, then the air inside is removed.
    a. How much force does the atmosphere exert on the flat end of each cylinder?
    b. Suppose one cylinder is bolted to a sturdy ceiling. How many 100 kg football players would need to hang from the lower cylinder to pull the two cylinders apart?

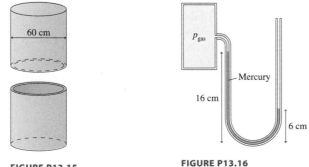

FIGURE P13.15

FIGURE P13.16

Section 13.3 Measuring and Using Pressure

16. | What is the gas pressure inside the box shown in Figure P13.16?

17. | The container shown in Figure P13.17 is filled with oil. It is open to the atmosphere on the left.
    a. What is the pressure at point A?
    b. What is the pressure difference between points A and B? Between points A and C?

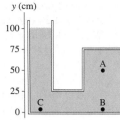

FIGURE P13.17

18. || Glycerin is poured into an open U-shaped tube until the height in both sides is 20 cm. Ethyl alcohol is then poured into one arm until the height of the alcohol column is 20 cm. The two liquids do not mix. What is the difference in height between the top surface of the glycerin and the top surface of the alcohol?

19. ||| A U-shaped tube, open to the air on both ends, contains mercury. Water is poured into the left arm until the water column is 10.0 cm deep. How far upward from its initial position does the mercury in the right arm rise?

20. | What is the height of a water barometer at atmospheric pressure?

21. | Postural hypotension is the occurrence of low (systolic)
BIO blood pressure when standing up too quickly from a reclined position, causing fainting or lightheadedness. For most people, a systolic pressure less than 90 mm of Hg is considered low. If the blood pressure in your brain is 120 mm when you are lying down, what would it be when you stand up? Assume that your brain is 40 cm from your heart and that $\rho = 1060$ kg/m^3 for your blood. Note: Normally, your blood vessels constrict and expand to keep your brain blood pressure stable when you change your posture.

Section 13.4 Buoyancy

22. || A 6.00-cm-diameter sphere with a mass of 89.3 g is neutrally buoyant in a liquid. Identify the liquid.

23. |||| A cargo barge is loaded in a saltwater harbor for a trip up a freshwater river. If the rectangular barge is 3.0 m by 20.0 m and sits 0.80 m deep in the harbor, how deep will it sit in the river?

24. | A 10 cm × 10 cm × 10 cm wood block with a density of 700 kg/m^3 floats in water.
    a. What is the distance from the top of the block to the water if the water is fresh?
    b. If it's seawater?

25. |||| What is the tension in the string in Figure P13.25?

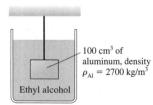

100 cm^3 of aluminum, density $\rho_{Al} = 2700$ kg/m^3

Ethyl alcohol

FIGURE P13.25

26. | A 10 cm × 10 cm × 10 cm block of steel ($\rho_{steel} = 7900$ kg/m^3) is suspended from a spring scale. The scale is in newtons.
    a. What is the scale reading if the block is in air?
    b. What is the scale reading after the block has been lowered into a beaker of oil and is completely submerged?

27. ||| Styrofoam has a density of 300 kg/m^3. What is the maximum mass that can hang without sinking from a 50-cm-diameter Styrofoam sphere in water? Assume the volume of the mass is negligible compared to that of the sphere.

28. || Calculate the buoyant force due to the surrounding air on a man weighing 800 N. Assume his average density is the same as that of water.

Section 13.5 Fluids in Motion

29. || River Pascal with a volume flow rate of 5.0×10^5 L/s joins with River Archimedes, which carries 10.0×10^5 L/s, to form the Bernoulli River. The Bernoulli River is 150 m wide and 10 m deep. What is the speed of the water in the Bernoulli River?

30. || Water flowing through a 2.0-cm-diameter pipe can fill a 300 L bathtub in 5.0 min. What is the speed of the water in the pipe?

31. | A pump is used to empty a 6000 L wading pool. The water exits the 2.5-cm-diameter hose at a speed of 2.1 m/s. How long will it take to empty the pool?

32. || A 1.0-cm-diameter pipe widens to 2.0 cm, then narrows to 0.50 cm. Liquid flows through the first segment at a speed of 4.0 m/s.
    a. What are the speeds in the second and third segments?
    b. What is the volume flow rate through the pipe?

Section 13.6 Fluid Dynamics

33. | What does the top pressure gauge in Figure P13.33 read?

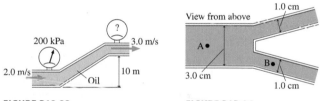

FIGURE P13.33   **FIGURE P13.34**

34. | The 3.0-cm-diameter water line in Figure P13.34 splits into two 1.0-cm-diameter pipes. All pipes are circular and at the same elevation. At point A, the water speed is 2.0 m/s and the gauge pressure is 50 kPa. What is the gauge pressure at point B?

35. |||| A rectangular trough, 2.0 m long, 0.60 m wide, and 0.45 m deep, is completely full of water. One end of the trough has a small drain plug right at the bottom edge. When you pull the plug, at what speed does water emerge from the hole?

Section 13.7 Viscosity and Poiseuille's Equation

36. | What pressure difference is required between the ends of a 2.0-m-long, 1.0-mm-diameter horizontal tube for 40°C water to flow through it at an average speed of 4.0 m/s?

37. ||| Water flows at 0.25 L/s through a 10-m-long garden hose 2.5 cm in diameter that is lying flat on the ground. The temperature of the water is 20°C. What is the gauge pressure of the water where it enters the hose?

38. || Figure P13.38 shows a water-filled syringe with a 4.0-cm-long needle. What is the gauge pressure of the water at the point P, where the needle meets the wider chamber of the syringe?

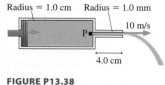

Radius = 1.0 cm   Radius = 1.0 mm
10 m/s
P
4.0 cm

FIGURE P13.38

General Problems

39. | The density of gold is 19,300 kg/m³. 197 g of gold is shaped into a cube. What is the length of each edge?

40. ||| The density of copper is 8920 kg/m³. How many moles are
INT in a 2.0 cm × 2.0 cm × 2.0 cm cube of copper?

41. || The density of aluminum is 2700 kg/m³. How many atoms
INT are in a 2.0 cm × 2.0 cm × 2.0 cm cube of aluminum?

42. || A 50-cm-thick layer of oil floats on a 120-cm-thick layer of water. What is the pressure at the bottom of the water layer?

43. || An oil layer floats on 85 cm of water in a tank. The pressure at the bottom of the tank is 112.0 kPa. How thick is the oil?

44. | The little Dutch boy saved Holland by sticking his finger in the leaking dike. If the water level was 2.5 m above his finger, *estimate* the force of the water on his finger.

45. |||| a. In Figure P13.45, how much force does the fluid exert on the end of the cylinder at A?
    b. How much force does the fluid exert on the end of the cylinder at B?

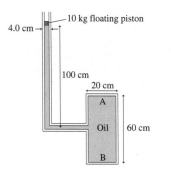

FIGURE P13.45

46. ||| A friend asks you how much pressure is in your car tires. You know that the tire manufacturer recommends 30 psi, but it's been a while since you've checked. You can't find a tire gauge in the car, but you do find the owner's manual and a ruler. Fortunately, you've just finished taking physics, so you tell your friend, "I don't know, but I can figure it out." From the owner's manual you find that the car's mass is 1500 kg. It seems reasonable to assume that each tire supports one-fourth of the weight. With the ruler you find that the tires are 15 cm wide and the flattened segment of the tire in contact with the road is 13 cm long. What answer will you give your friend?

47. ||| A diver 50 m deep in 10°C fresh water exhales a 1.0-cm-
INT diameter bubble. What is the bubble's diameter just as it reaches the surface of the lake, where the water temperature is 20°C?
    Hint: Assume that the air bubble is always in thermal equilibrium with the surrounding water.

48. || A 6.0-cm-tall cylinder floats in water with its axis perpendicular to the surface. The length of the cylinder above water is 2.0 cm. What is the cylinder's mass density?

49. | A sphere completely submerged in water is tethered to the bottom with a string. The tension in the string is one-third the weight of the sphere. What is the density of the sphere?

50. | You need to determine the density of a ceramic statue. If you suspend it from a spring scale, the scale reads 28.4 N. If you then lower the statue into a tub of water so that it is completely submerged, the scale reads 17.0 N. What is the density?

51. | A 5.0 kg rock whose density is 4800 kg/m³ is suspended by a string such that half of the rock's volume is under water. What is the tension in the string?

52. |||| The bottom of a steel "boat" is a 5.0 m × 10 m × 2.0 cm piece of steel ($\rho_{\text{steel}} = 7900$ kg/m³). The sides are made of 0.50-cm-thick steel. What minimum height must the sides have for this boat to float in perfectly calm water?

53. ||| A 2.0 mL syringe has an inner diameter of 6.0 mm, a needle
BIO inner diameter of 0.25 mm, and a plunger pad diameter (where you place your finger) of 1.2 cm. A nurse uses the syringe to inject medicine into a patient whose blood pressure is 140/100. Assume the liquid is an ideal fluid.
    a. What is the minimum force the nurse needs to apply to the syringe?
    b. The nurse empties the syringe in 2.0 s. What is the flow speed of the medicine through the needle?

54. || A child's water pistol shoots water through a 1.0-mm-
INT diameter hole. If the pistol is fired horizontally 70 cm above the ground, a squirt hits the ground 1.2 m away. What is the volume flow rate during the squirt? Ignore air resistance.

55. | A long horizontal tube has a square cross section with sides of width L. A fluid moves through the tube with speed v_1. The tube then changes to a circular cross section with diameter L. What is the fluid's speed in the circular part of the tube?

56. ||| A hurricane wind blows across a 6.00 m × 15.0 m flat roof at a speed of 130 km/hr.
    a. Is the air pressure above the roof higher or lower than the pressure inside the house? Explain.
    b. What is the pressure difference?
    c. How much force is exerted on the roof? If the roof cannot withstand this much force, will it "blow in" or "blow out"?

57. || Water flows from the pipe shown in Figure P13.57 with a speed of 4.0 m/s.
    a. What is the water pressure as it exits into the air?
    b. What is the height h of the standing column of water?

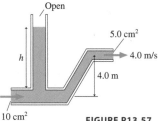

FIGURE P13.57

58. || Air flows through the tube shown in Figure P13.58. Assume that air is an ideal fluid.
    a. What are the air speeds v_1 and v_2 at points 1 and 2?
    b. What is the volume flow rate?

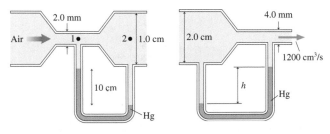

FIGURE P13.58   **FIGURE P13.59**

59. ||| Air flows through the tube shown in Figure P13.59 at a rate of 1200 cm³/s. Assume that air is an ideal fluid. What is the height h of mercury in the right side of the U-tube?

60. ‖ Water flows at 5.0 L/s through a horizontal pipe that narrows smoothly from 10 cm diameter to 5.0 cm diameter. A pressure gauge in the narrow section reads 50 kPa. What is the reading of a pressure gauge in the wide section?

61. ‖‖‖ The mercury manometer
INT shown in Figure P13.61 is attached to a gas cell. The mercury height h is 120 mm when the cell is placed in an ice-water mixture. The mercury height drops to 30 mm when the device is carried into an industrial freezer. What is the freezer temperature?
Hint: The right tube of the manometer is much narrower than the left tube. What reasonable assumption can you make about the gas volume?

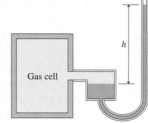

FIGURE P13.61

62. ‖ Figure P13.62 shows a section of a long tube that narrows near its open end to a diameter of 1.0 mm. Water at 20°C flows out of the open end at 0.020 L/s. What is the gauge pressure at point P, where the diameter is 4.0 mm?

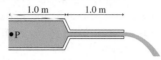

FIGURE P13.62

63. ‖ Smoking tobacco is bad for your circulatory health. In an
BIO attempt to maintain the blood's capacity to deliver oxygen, the body increases its red blood cell production, and this increases the viscosity of the blood. In addition, nicotine from tobacco causes arteries to constrict.

For a nonsmoker, with blood viscosity of 2.5×10^{-3} Pa · s, normal blood flow requires a pressure difference of 8.0 mm of Hg between the two ends of an artery. If this person were to smoke regularly, his blood viscosity would increase to 2.7×10^{-3} Pa · s, and the arterial diameter would constrict to 90% of its normal value. What pressure difference would be needed to maintain the same blood flow?

64. ‖ A stiff, 10-cm-long tube with an inner diameter of 3.0 mm is attached to a small hole in the side of a tall beaker. The tube sticks out horizontally. The beaker is filled with 20°C water to a level 45 cm above the hole, and it is continually topped off to maintain that level. What is the volume flow rate through the tube?

Passage Problems

Blood Pressure and Blood Flow BIO

The blood pressure at your heart is approximately 100 mm of Hg. As blood is pumped from the left ventricle of your heart, it flows through the aorta, a single large blood vessel with a diameter of about 2.5 cm. The speed of blood flow in the aorta is about 60 cm/s. Any change in pressure as blood flows in the aorta is due to the change in height: the vessel is large enough that viscous drag is not a major factor. As the blood moves through the circulatory system, it flows into successively smaller and smaller blood vessels until it reaches the capillaries. Blood flows in the capillaries at the much lower speed of approximately 0.7 mm/s. The diameter of capillaries and other small blood vessels is so small that viscous drag is a major factor.

65. ‖ There is a limit to how long your neck can be. If your neck were too long, no blood would reach your brain! What is the maximum height a person's brain could be above his heart, given the noted pressure and assuming that there are no valves or supplementary pumping mechanisms in the neck? The density of blood is 1060 kg/m³.
A. 0.97 m   B. 1.3 m   C. 9.7 m   D. 13 m

66. ‖ Because the flow speed in your capillaries is much less than in the aorta, the total cross section area of the capillaries considered together must be much larger than that of the aorta. Given the flow speeds noted, the total area of the capillaries considered together is equivalent to the cross-section area of a single vessel of approximately what diameter?
A. 25 cm   B. 50 cm   C. 75 cm   D. 100 cm

67. ‖ Suppose that in response to some stimulus a small blood vessel narrows to 90% its original diameter. If there is no change in the pressure across the vessel, what is the ratio of the new volume flow rate to the original flow rate?
A. 0.66   B. 0.73   C. 0.81   D. 0.90

68. ‖ Sustained exercise can increase the blood flow rate of the heart by a factor of five with only a modest increase in blood pressure. This is a large change in flow. Although several factors come into play, which of the following physiological changes would most plausibly account for such a large increase in flow with a small change in pressure?
A. A decrease in the viscosity of the blood.
B. Dilation of the smaller blood vessels to larger diameters.
C. Dilation of the aorta to larger diameter.
D. An increase in the oxygen carried by the blood.

STOP TO THINK ANSWERS

Stop to Think 13.1: $\rho_1 = \rho_2 = \rho_3$. Density depends only on what the object is made of, not how big the pieces are.

Stop to Think 13.2: C. These are all open tubes, so the liquid rises to the same height in all three despite their different shapes.

Stop to Think 13.3: B. The weight of the displaced water equals the weight of the ice cube. When the ice cube melts and turns into water, that amount of water will exactly fill the volume that the ice cube is now displacing.

Stop to Think 13.4: 1 cm³/s out. The fluid is incompressible, so the sum of what flows in must match the sum of what flows out.

13 cm³/s is known to be flowing in while 12 cm³/s flows out. An additional 1 cm³/s must flow out to achieve balance.

Stop to Think 13.5: $h_2 > h_4 > h_3 > h_1$. The liquid level is higher where the pressure is lower. The pressure is lower where the flow speed is higher. The flow speed is highest in the narrowest tube, zero in the open air.

Stop to Think 13.6: A. All three segments have the same volume flow rate Q. According to Poiseuille's equation, the segment with the smallest radius R has the greatest pressure difference Δp.

Properties of Matter

The goal of Part III has been to understand macroscopic systems and their properties. We've introduced no new fundamental laws in these chapters. Instead, we've broadened and extended the scope of Newton's law and the law of conservation of energy. In many ways, Part III has focused on *applications* of the general principles introduced in Parts I and II.

That matter exists in three phases—solids, liquids, gases—is perhaps the most basic fact about macroscopic systems. We looked at an atomic-level picture of the three phases of matter, and we also found a connection between *heat* and *phase changes*. Both liquids and gases are *fluids,* and we spent quite a bit of time investigating the properties of fluids, ranging from pressure and the ideal-gas law to buoyancy, fluid flow, and Bernoulli's equation.

Along the way, we introduced *mechanical properties* of matter, such as density and viscosity, and *thermal prop-*

erties, such as thermal-expansion coefficients and specific heats. You should now be able to use these properties to figure out what happens when you heat a system, whether an object will float, and how fast a fluid flows. These are all very practical, useful things to know, with many applications to the world around us.

Last, but not least, we discovered that many *macroscopic* properties are determined by the *microscopic* motions of atoms and molecules. For example, we found that the ideal-gas law is based on the idea that gas pressure is due to the collisions of atoms with the walls of the container. Similarly, regularities in the specific heats of gases were explained on the basis of how atoms and molecules share the thermal energy of a system. Atoms are important, even if we can't see them or follow their individual motions.

KNOWLEDGE STRUCTURE III **Properties of Matter**

BASIC GOAL How can we describe the macroscopic flows of energy and matter in heat transfer and fluid flow?

GENERAL PRINCIPLES **Phases of matter**

Gas: Freely moving particles. Thermal energy is kinetic energy of motion of particles. Pressure due to collisions of particles with walls of container.

Liquid: Particles are loosely bound, but flow is possible. Heat is transferred by convection and conduction. Pressure is due to gravity.

Solid: Particles are joined by spring-like bonds. Increasing temperature causes expansion. Heat is transferred by conduction only.

Heat must be added to change a solid to a liquid and a liquid to a gas. Heat must be removed to reverse these changes.

Ideal-gas processes

We can graph processes on a pV diagram. Work by the gas is the area under the graph.

For a gas process in a sealed container, the ideal gas law relates values of pressure, volume, and temperature:

$$\frac{p_i V_i}{T_i} = \frac{p_f V_f}{T_f}$$

- Constant-volume process $V = $ constant
- Isothermal process $T = $ constant
- Isobaric process $p = $ constant
- Adiabatic process $Q = 0$

Heat and heat transfer

The **specific heat** c of a material is the heat required to raise the temperature of 1 kg by 1 K.

$$Q = Mc\,\Delta T$$

When two ore more systems interact thermally, they come to a common final temperature determined by

$$Q_{net} = Q_1 + Q_2 + Q_3 + \cdots = 0$$

Conduction transfers heat by direct physical contact.
Convection transfers of heat by the motion of a fluid.
Radiation transfers heat by electromagnetic waves.

Fluid statics

Archimedes' principle: The magnitude of the buoyant force equals the weight of the fluid displaced by the object.

An object sinks if it is more dense than the fluid in which it is submerged; if it is less dense, it floats.

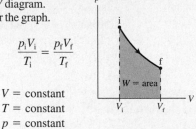

Pressure increases with depth in a fluid. The pressure at depth d is

$$p = p_0 + \rho g d$$

Fluid flow
Ideal fluid flow is laminar, incompressible, and nonviscous. The **equation of continuity** is

$$v_1 A_1 = v_2 A_2$$

Bernoulli's equation is

$$p_1 + \tfrac{1}{2}\rho v_1^2 + \rho g y_1 =$$
$$p_2 + \tfrac{1}{2}\rho v_2^2 + \rho g y_2$$

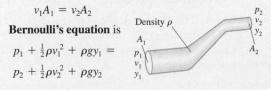

Size and Life

Physicists look for simple models and general principles that underlie and explain diverse physical phenomena. In the first 13 chapters of this book, you've seen that just a handful of general principles and laws can be used to solve a wide range of problems. Can this approach have any relevance to a subject like biology? It may seem surprising, but there *are* general "laws of biology" that apply, with quantitative accuracy, to organisms as diverse as elephants and mice.

Let's look at an example. An elephant uses more metabolic power than a mouse. This is not surprising, as an elephant is quite a bit bigger. But recasting the data shows an interesting trend. When we looked at the energy required to raise the temperature of different substances, we considered *specific heat*. The "specific" meant that we considered the heat required for one kilogram. For animals, rather than metabolic rate, we can look at the *specific metabolic rate,* the metabolic power used *per kilogram* of tissue. If we factor out the mass difference between a mouse and an elephant, are their specific metabolic powers the same?

In fact, the specific metabolic rate varies quite a bit among mammals, as the graph of specific metabolic rate versus mass shows. But there is an interesting trend: all of the data points do lie on a single smooth curve. In other words, there really is a biological *law* we can use to predict a mammal's metabolic rate knowing only its mass M. In particular, the specific metabolic rate is proportional to $M^{-0.25}$. Because a 4000 kg elephant is 160,000 times more massive than a 25 g mouse, the mouse's specific metabolic power is $(160,000)^{0.25} = 20$ times that of the elephant. A law that shows how a property scales with the size of a system is called a *scaling law*.

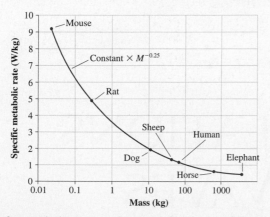

Specific metabolic rate as a function of body mass follows a simple scaling law.

A similar scaling law holds for birds, reptiles—and even bacteria. Why should a single, simple relationship hold true for organisms that range in size from a single cell to a 100 ton blue whale? Interestingly, no one knows for sure. It is a matter of current research to find out just what this and other scaling laws tell us about the nature of life.

Perhaps the metabolic-power scaling law is a result of heat transfer. In Chapter 12, we noted that all metabolic energy used by an animal ends up as heat, which must be transferred to the environment. A 4000 kg elephant has 160,000 times the mass of a 25 g mouse, but it only has about 3,000 times the surface area. The heat transferred to the environment depends on the surface area; the more surface area, the greater the rate of heat transfer. An elephant with a mouse-sized metabolism simply wouldn't be able to dissipate heat fast enough—it would quickly overheat and die.

If heat dissipation were the only factor limiting metabolism, we can show that the specific metabolic rate should scale as $M^{-0.33}$, quite different from the $M^{-0.25}$ scaling observed. Clearly, another factor is at work. Exactly what underlies the $M^{-0.25}$ scaling is still a matter of debate, but some recent analysis suggests the scaling is due to limitations not of heat transfer but of fluid flow. Cells in mice, elephants, and all mammals receive nutrients and oxygen for metabolism from the blood stream. Because the minimum size of a capillary is about the same for all mammals, the structure of the circulatory system must vary from animal to animal. The human aorta has a diameter of about 1 inch; in a mouse, the diameter is approximately $1/20^{th}$ of this. Thus a mouse has fewer levels of branching to smaller and smaller blood vessels as we move from the aorta to the capillaries. The smaller blood vessels in mice mean that viscosity is more of a factor throughout the circulatory system. The circulatory system of a mouse is quite different from that of an elephant.

A model of specific metabolic rate based on blood-flow limitations predicts a $M^{-0.25}$ law, exactly as observed. The model also makes other testable predictions. For example, the model predicts that the smallest possible mammal should have a body mass of about 1 gram—exactly the size of the smallest shrew. Even smaller animals have different types of circulatory systems; in the smallest animals, nutrient transport is by diffusion alone. But the model can be extended to predict that the specific metabolic rate for these animals will follow a scaling law similar to that for mammals, exactly as observed. It is too soon to know if this model will ultimately prove to be correct, but it's indisputable that there are large-scale regularities in biology that follow mathematical relationships based on the laws of physics.

OSCILLATIONS AND WAVES

Motion That Repeats Again and Again

Up to this point in the book, we have generally considered processes that have a clear starting and ending point, such as a car accelerating from rest to a final speed, or a solid being heated from an initial to a final temperature. In Part IV, we begin to consider processes that are *periodic*—they repeat. A child on a swing, a boat bobbing on the water, and even the repetitive bass beat of a rock song are *oscillatory motions* that happen over and over without a starting or ending point. The *period*, the time for one cycle of the motion, will be a key parameter for us to consider as we look at oscillatory motion.

Our first goal will be to develop the language and tools needed to describe oscillations, ranging from the swinging of the bob of a pendulum clock to the bouncing of a car on its springs. Once we understand oscillations, we will extend our analysis to consider oscillations that travel—*waves*.

The Wave Model

We've had great success modeling the motion of complex objects as the motion of one or more particles. We were even able to explain the macroscopic properties of matter, such as pressure and temperature, in terms of the motion of the atomic particles that comprise all matter.

Now it's time to explore another way of looking at nature, the *wave model*. Familiar examples of waves include

- Ripples on a pond.
- The sound of thunder.
- The swaying ground of an earthquake.
- A vibrating guitar string.
- The colors of a rainbow.

Despite the great diversity of types and sources of waves, there is a single, elegant physical theory that is capable of describing them all. Our exploration of wave phenomena will call upon water waves, sound waves, and light waves for examples, but our goal will be to emphasize the unity and coherence of the ideas that are common to *all* types of waves. As was the case with the particle model, we will use the wave model to explain a wide range of phenomena.

When Waves Collide

The collision of two particles is a dramatic event. Energy and momentum are transferred as the two particles head off in different directions. Something much gentler happens when two waves come together—the two waves pass through each other unchanged. Where they overlap, we get a *superposition* of the two waves. We will finish our discussion of waves by analyzing the standing waves that result from the superposition of two waves traveling in opposite directions. The physics of standing waves will allow us to understand how your vocal tract can produce such a wide range of sounds, and how your ears are able to analyze them.

Wolves are social animals, and they howl to communicate over distances of several miles with other members of their pack. How are such sounds made? How do they travel through the air? And how are other wolves able to hear these sounds from such a great distance?

14

OSCILLATIONS

A gibbon moves in a series of circular arcs as it swings through the forest canopy, pivoting about one handhold and then the next. How does an understanding of oscillatory motion allow us to analyze the swinging motion of a gibbon—and the walking motion of other animals?

Looking Ahead ▶▶

The goal of Chapter 14 is to understand systems that oscillate with simple harmonic motion. In this chapter you will learn to:

▶ Describe oscillatory motion and the type of force that leads to it.

▶ Use graphical and mathematical representations of oscillatory motion.

▶ Understand how conservation of energy applies in simple harmonic motion.

▶ Apply concepts of damping and resonance in oscillating systems.

Looking Back ◀◀

This chapter brings together concepts from many previous chapters. Oscillatory motion is produced by a linear restoring force like that of a spring. We will use the concepts of circular motion and conservation of energy to describe and analyze oscillatory motion. Please review:

◀ Section 3.8 Period and frequency.

◀ Section 6.2 Velocity and acceleration in circular motion.

◀ Section 8.3 Springs and restoring forces.

◀ Sections 10.2 and 10.3 Energy transformations and the conservation of energy.

A marble rolling back and forth in the bottom of a bowl, a car bouncing up and down on its springs, and a pendulum bob keeping time in a clock are all examples of **oscillatory motion,** a repetitive motion back and forth about an equilibrium position. Swinging motions and vibrations of all kinds are oscillatory motions. All oscillatory motion is **periodic:** the motion repeats, over and over, with a fixed *period,* the time for one cycle of the motion.

Examples of oscillatory motion abound, including pieces of vibrating machinery, ringing bells, and the current in electric circuits used to drive antennas. A vibrating loudspeaker cone pushes air molecules back and forth to send out a sound wave, showing that oscillations are closely related to waves.

Our goal in this chapter is to study the physics of oscillations. Much of our analysis will be focused on the most basic form of oscillatory motion, *simple harmonic motion.* After we explore simple harmonic motion, we will use it to help us understand more complex motions.

14.1 Equilibrium and Oscillation

Consider a marble that is free to roll inside a spherical bowl, as shown in Figure 14.1. The marble has an **equilibrium position** at the bottom of the bowl where it will rest with no net force on it. If you push the marble away from equilibrium, the marble's weight leads to a net force directed back toward the equilibrium position. We call this a **restoring force** because it acts to restore equilibrium. The magnitude of this restoring force increases if the marble is moved farther away from the equilibrium position.

If you pull the marble to the side and release it, it doesn't just roll back to the bottom of the bowl and stay put. It keeps on moving, rolling up and down each side of the bowl, repeatedly moving through its equilibrium position, as we see in Figure 14.2. We call such repetitive motion an **oscillation.** This oscillation is a result of an interplay between the restoring force and the marble's inertia, something we will see in all of the oscillations we consider.

The key characteristic of an oscillation is the *period,* the time for the motion to repeat. We have seen the concepts of frequency and period when we studied circular motion. We will directly connect the ideas of simple harmonic motion and circular motion later in the chapter. For now, we'll review the concepts of frequency and period and see how they apply to oscillatory motion.

Frequency and Period

An electrocardiogram (ECG), such as the one shown in Figure 14.3, is a record of the electrical signals of the heart as it beats. We will explore the ECG in some detail in Chapter 21. Although the shape of a typical ECG is rather complex, notice that it has a *repeating pattern.* For this or any oscillation, the time to complete one full cycle is called the **period** of the oscillation. Period is given the symbol T.

An equivalent piece of information is the number of cycles, or oscillations, completed per second. If the period is $\frac{1}{10}$ s, then the oscillator can complete 10 cycles in one second. Conversely, an oscillation period of 10 s allows only $\frac{1}{10}$ of a cycle to be completed per second. In general, T seconds per cycle implies that $1/T$ cycles will be completed each second. The number of cycles per second is called the **frequency** f of the oscillation. The relationship between frequency and period is therefore

$$f = \frac{1}{T} \quad \text{or} \quad T = \frac{1}{f} \tag{14.1}$$

The units of frequency are **hertz,** abbreviated Hz. By definition,

$$1 \text{ Hz} = 1 \text{ cycle per second} = 1 \text{ s}^{-1}$$

We will frequently deal with very rapid oscillations and make use of the units shown in Table 14.1.

NOTE ▶ Uppercase and lowercase letters *are* important. 1 MHz is 1 megahertz = 10^6 Hz, but 1 mHz would be 1 millihertz = 10^{-3} Hz! ◀

EXAMPLE 14.1 Frequency and period of a radio station

An FM radio station broadcasts at a frequency of 100 MHz. What is the period?

SOLVE The frequency f of oscillations in the radio transmitter is 100 MHz = 1.0×10^8 Hz. The period is the inverse of the frequency, hence

$$T = \frac{1}{f} = \frac{1}{1.0 \times 10^8 \text{ Hz}} = 1.0 \times 10^{-8} \text{ s} = 10 \text{ ns}$$

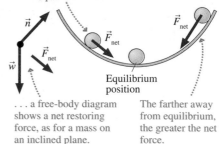

When the ball is displaced from equilibrium . . .

. . . a free-body diagram shows a net restoring force, as for a mass on an inclined plane.

The farther away from equilibrium, the greater the net force.

FIGURE 14.1 Equilibrium and restoring forces for a ball in a bowl.

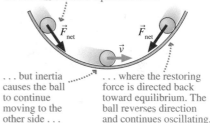

When moved from equilibrium and released, a restoring force pulls the ball back toward equilibrium . . .

. . . but inertia causes the ball to continue moving to the other side . . .

. . . where the restoring force is directed back toward equilibrium. The ball reverses direction and continues oscillating.

FIGURE 14.2 The motion of a ball rolling in a bowl.

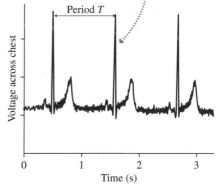

Successive beats of the heart produce approximately the same signal.

Period T

FIGURE 14.3 An electrocardiogram.

TABLE 14.1 Common units of frequency

Frequency	Period
10^3 Hz = 1 kilohertz = 1 kHz	1 ms
10^6 Hz = 1 megahertz = 1 MHz	1 μs
10^9 Hz = 1 gigahertz = 1 GHz	1 ns

Oscillatory Motion

Let's return to the marble in the bowl and describe its motion in more detail. We start with a position-versus-time graph for the marble, with positions to the right of equilibrium positive and those to the left negative. Figure 14.4 shows a graph of the motion, with important points noted. This graph has the form of a *cosine function*. A graph or a function that has the form of a sine or cosine function is called **sinusoidal.** A sinusoidal oscillation is called **simple harmonic motion,** often abbreviated SHM.

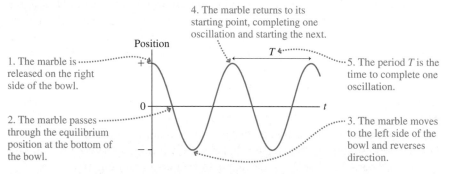

4. The marble returns to its starting point, completing one oscillation and starting the next.

1. The marble is released on the right side of the bowl.

2. The marble passes through the equilibrium position at the bottom of the bowl.

5. The period *T* is the time to complete one oscillation.

3. The marble moves to the left side of the bowl and reverses direction.

FIGURE 14.4 A position-versus-time graph for a marble rolling in a bowl.

A marble rolling in the bottom of a bowl undergoes simple harmonic motion, as does a car bouncing up and down on its springs. SHM is very common, but we'll find that most cases of SHM can be modeled as one of two simple systems: a mass oscillating on a spring or a pendulum swinging back and forth. The following table shows two examples.

Examples of simple harmonic motion

Oscillating system		Related real-world example	
Mass on a spring 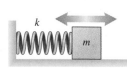	The mass oscillates back and forth due to the restoring force of the spring. The period depends on the mass and the stiffness of the spring.	Vibrations in the ear 	Sound waves entering the ear cause the oscillation of a membrane in the cochlea. The vibration can be modeled as a mass on a spring. The period of oscillation of a segment of the membrane depends on mass (the thickness of the membrane) and stiffness (the rigidity of the membrane.)
Pendulum 	The mass oscillates back and forth due to the restoring gravitational force. The period depends on the length of the pendulum and the acceleration of gravity.	Motion of legs while walking	The motion of a walking animal's legs can be modeled as pendulum motion. The rate at which the legs swing depends on the length of the legs and the acceleration of gravity.

STOP TO THINK 14.1   Two oscillating systems have periods T_1 and T_2, with $T_1 < T_2$. How are the frequencies of the two systems related?

A. $f_1 < f_2$ B. $f_1 = f_2$ C. $f_1 > f_2$

14.2 Linear Restoring Forces and Simple Harmonic Motion

Simple harmonic motion can occur when a system has a restoring force that pushes it back toward equilibrium. We will begin our study with the very simple system of a mass and a spring, analyzing it in detail before moving on to other oscillatory systems.

Figure 14.5 shows a glider that rides with very little friction on an air track. There is a spring connecting the glider to the end of the track. When the spring is neither stretched nor compressed, the net force on the glider is zero. The glider just sits there—this is the equilibrium position.

If the glider is now displaced from this equilibrium position by Δx, the spring exerts a force back toward equilibrium—a restoring force. In Section 8.3, we found that the spring force is given by Hooke's law: $(F_{sp})_x = -k\,\Delta x$. If we set the origin of our coordinate system at the equilibrium position, the displacement from equilibrium Δx is equal to x, thus the spring force can be written as $(F_{sp})_x = -kx$.

The net force on the glider is simply the spring force, so we can write

$$(F_{net})_x = -kx \qquad (14.2)$$

LINEAR
p.38

This is a **linear restoring force.** That is, **the net force is toward the equilibrium position and is linearly proportional to the distance from equilibrium.**

Motion of a Mass on a Spring

If we pull the air track glider of Figure 14.5 a short distance to the right and release it, it will oscillate back and forth. Figure 14.6 shows actual data from an experiment in which a computer was used to measure the position of a glider 20 times every second. This is a position-versus-time graph that has been rotated 90° from its usual orientation in order for the x-axis to match the motion of the glider.

The object's maximum displacement from equilibrium is called the **amplitude** A of the motion. The object's position oscillates between $x = -A$ and $x = +A$.

> **NOTE ▶** When interpreting a graph, notice that the amplitude is the distance from equilibrium to the maximum, *not* the distance from the minimum to the maximum. ◀

The graph of the position is sinusoidal, so this is simple harmonic motion. The motion is "simple" because the restoring force has a simple, linear form. **Oscillation about an equilibrium position with a linear restoring force is always simple harmonic motion.** There are many other circumstances in which a linear restoring force exists, and we will see simple harmonic motion in these cases as well.

Vertical Mass on a Spring

Figure 14.7 on the following page shows a block of mass m hanging from a spring with spring constant k. Will this mass-spring system have the same simple harmonic motion as the horizontal system we just saw, or will gravity add an additional complication? An important fact to notice is that the equilibrium position of

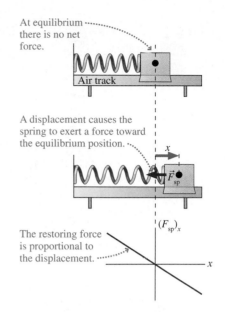

At equilibrium there is no net force.

Air track

A displacement causes the spring to exert a force toward the equilibrium position.

$\vec{F}_{sp}$

The restoring force is proportional to the displacement.

$(F_{sp})_x$

x

FIGURE 14.5 The restoring force on an air track glider attached to a spring.

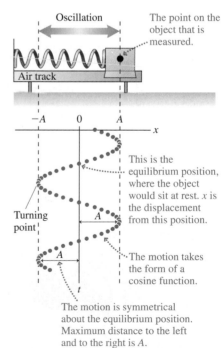

Oscillation

The point on the object that is measured.

Air track

$-A$ 0 A x

This is the equilibrium position, where the object would sit at rest. x is the displacement from this position.

Turning point

A

The motion takes the form of a cosine function.

t

The motion is symmetrical about the equilibrium position. Maximum distance to the left and to the right is A.

FIGURE 14.6 An experiment showing the oscillation of an air-track glider.

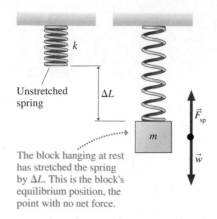

Unstretched spring

ΔL

$\vec{F}_{sp}$

m

$\vec{w}$

The block hanging at rest has stretched the spring by ΔL. This is the block's equilibrium position, the point with no net force.

FIGURE 14.7 The equilibrium position of a mass on a vertical spring.

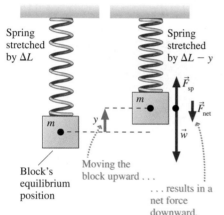

Spring stretched by ΔL

Spring stretched by $\Delta L - y$

m

m

y

$\vec{F}_{sp}$

$\vec{F}_{net}$

$\vec{w}$

Block's equilibrium position

Moving the block upward . . .

. . . results in a net force downward.

FIGURE 14.8 Displacing the block from its equilibrium position produces a restoring force.

Pendulum motion at the playground.

the block is *not* where the spring is at its unstretched length. At the equilibrium position of the block, where it hangs motionless, the spring has stretched by ΔL.

Finding ΔL is a static-equilibrium problem in which the upward spring force balances the downward weight force of the block. The y-component of the spring force is given by Hooke's law,

$$(F_{sp})_y = -k\,\Delta y = +k\,\Delta L \tag{14.3}$$

Equation 14.3 makes a distinction between ΔL, which is simply a *distance* and is a positive number, and the displacement Δy. The block is displaced downward, so $\Delta y = -\Delta L$. Newton's first law for the block in equilibrium is

$$(F_{net})_y = (F_{sp})_y + w_y = k\,\Delta L - mg = 0 \tag{14.4}$$

from which we can find

$$\Delta L = \frac{mg}{k} \tag{14.5}$$

This is the distance the spring stretches when the block is attached to it.

Suppose we now displace the block from this equilibrium position, as shown in Figure 14.8. We've placed the origin of the y-axis at the block's equilibrium position in order to be consistent with our analyses of oscillations throughout this chapter. If the block moves upward, as in the figure, the spring gets shorter compared to its equilibrium length, but the spring is still *stretched* compared to its unstretched length in Figure 14.7. When the block is at position y, the spring is stretched by an amount $\Delta L - y$ and hence exerts an *upward* spring force $F_{sp} = k(\Delta L - y)$. The net force on the block at this point is

$$(F_{net})_y = (F_{sp})_y + w_y = k(\Delta L - y) - mg = (k\,\Delta L - mg) - ky \tag{14.6}$$

But $k\,\Delta L - mg$ is zero, from Equation 14.4, so the net force on the block is simply

$$(F_{net})_y = -ky \tag{14.7}$$

LINEAR
p. 38

Equation 14.7 for a mass hung from a spring has the same form as Equation 14.2 for the horizontal spring, where we found $(F_{net})_x = -kx$. That is, the restoring force for vertical oscillations is identical to the restoring force for horizontal oscillations. **The role of gravity is to determine where the equilibrium position is, but it doesn't affect the restoring force for displacement from the equilibrium position.** Because it has a linear restoring force, **a mass on a vertical spring oscillates with simple harmonic motion.** The motion has the same form as that of the air-track glider.

The Pendulum

The chapter opened with a picture of a gibbon, whose body swings back and forth below a tree branch. The motion of the gibbon is essentially that of a **pendulum,** a mass hung from a pivot point by a light string or rod. A pendulum oscillates about its equilibrium position, but is this simple harmonic motion? To answer this question, we need to examine the restoring force on the pendulum. If the restoring force is linear, the motion will be simple harmonic.

Figure 14.9a shows a mass m attached to a string of length L and free to swing back and forth. The pendulum's position can be described either by the arc of length s or the angle θ, both of which are zero when the pendulum hangs straight down. Because angles are measured counterclockwise, s and θ are positive when the pendulum is to the right of center, negative when it is to the left.

Two forces are acting on the mass: the string tension $\vec{T}$ and the weight $\vec{w}$. The motion is along a circular arc. We choose a coordinate system on the mass with one axis along the radius of the circle and the other tangent to the circle. We divide the forces into tangential components, parallel to the motion (denoted with a subscript "t"), and components directed toward the center of the circle. These are shown on the free-body diagram of Figure 14.9b.

The mass must move along a circular arc, as noted. As Figure 14.9b shows, the net force in this direction is the tangential component of the weight:

$$(F_{\text{net}})_t = \sum F_t = w_t = -mg\sin\theta \tag{14.8}$$

This is the restoring force pulling the mass back toward the equilibrium position.

Suppose we restrict the pendulum's oscillations to *small angles* of less than about 10°, or 0.17 rad. For such small angles, if the angle θ is in radians, it turns out that

$$\sin\theta \approx \theta$$

This result is called the **small-angle approximation.**

> **NOTE** ▶ You should check this approximation on your calculator. Put your calculator in radian mode, enter some values that are less than 0.17 rad, take the sine, and see what you get. ◀

Recall from Chapter 6 that the angle is related to the arc length by $\theta = s/L$. Using this and the small-angle approximation, we can write the restoring force as

$$(F_{\text{net}})_t = -mg\sin\theta \approx -mg\theta = -mg\frac{s}{L} = -\left(\frac{mg}{L}\right)s \tag{14.9}$$

The net force is directed toward the equilibrium position, and it is linearly proportional to the displacement s from equilibrium. In other words, **the force on a pendulum is a linear restoring force for small angles, so the pendulum will undergo simple harmonic motion.**

Whenever there is a linear restoring force, there can be simple harmonic motion. There are many examples beyond the few we have considered here, such as sloshing water in a cup or the vibration of a bridge or a building swaying in the wind. We will see these and other examples in the coming sections.

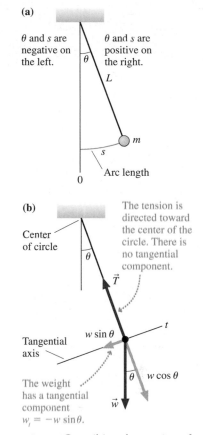

(a)
θ and s are negative on the left.
θ and s are positive on the right.
θ
L
m
s
Arc length
0

(b)
Center of circle
The tension is directed toward the center of the circle. There is no tangential component.
θ
$\vec{T}$
t
Tangential axis
$w\sin\theta$
θ $w\cos\theta$
The weight has a tangential component $w_t = -w\sin\theta$.
$\vec{w}$

FIGURE 14.9 Describing the motion of and force on a pendulum.

STOP TO THINK 14.2 A ball is hung from a rope, making a pendulum. When it is pulled 5° to the side, the restoring force is 1.0 N. What will be the magnitude of the restoring force if the ball is pulled 10° to the side?

A. 0.5 N
B. 1.0 N
C. 1.5 N
D. 2.0 N

14.3 Describing Simple Harmonic Motion

Now that we know what *causes* simple harmonic motion, we can begin to develop a mathematical description. Once we do this for one system, the mass on a spring, we can adapt the description to *any* system. The details will vary, but the basic form of the motion will be exactly the same.

Let's look again at the oscillating air-track glider of Figure 14.5. If we plot data for the position versus time, we get the graph of Figure 14.10a. You can see that the amplitude in this case is $A = 0.17$ m, or 17 cm. You can also measure the period to be $T = 1.60$ s. Thus the oscillation frequency is $f = 1/T = 0.625$ Hz. You can get a sense of the speed of the glider by noting the spacing of the points—where they are close together, the speed is low; where they are far apart, the speed is high. The speed is instantaneously zero at the peaks and valleys of the position-versus-time graph.

Figure 14.10b is a velocity-versus-time graph for the oscillating mass. The velocity graph also has a sinusoidal shape, oscillating between $-v_{max}$ (maximum speed to the left) and v_{max} (maximum speed to the right). As the figure shows,

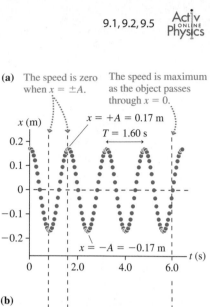

(a) The speed is zero when $x = \pm A$.   The speed is maximum as the object passes through $x = 0$.

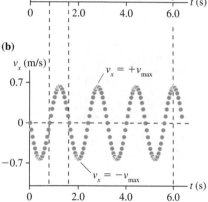

(b)

FIGURE 14.10 Position and velocity graphs of experimental data for an air-track cart on a spring.

- The instantaneous velocity is zero at the points where $x = \pm A$. These are the *turning points* of the motion.
- The maximum speed v_{max} is reached as the object passes through the equilibrium position at $x = 0$ m. The *velocity* is positive as the object moves to the right but *negative* as it moves to the left.

If we did a series of experiments on this system, we would find that **the frequency does not depend on the amplitude of the motion;** small oscillations and large oscillations have the same period. This is a key feature of simple harmonic motion that we will explore in more detail later in the chapter. Keep in mind that all these features of the motion of a mass on a spring also apply to other examples of simple harmonic motion.

Kinematics of Simple Harmonic Motion

Figure 14.11 redraws the position-versus-time and velocity-versus-time graphs of Figure 14.10 as smooth curves. An acceleration-versus-time graph is introduced as well. The motion divides naturally into phases lasting one-quarter of a cycle:

- At $t = 0$, the mass starts out with $x = A$ and $v_x = 0$. The acceleration a_x is negative. As soon as the mass is released, v_x becomes negative and the mass begins moving back toward equilibrium.
- At $t = \frac{1}{4}T$, the mass passes through its equilibrium position, $x = 0$. The speed is a maximum, with $v_x = -v_{max}$.
- At $t = \frac{1}{2}T$, the mass is at the other end of its motion, with $x = -A$ and $v_x = 0$. The acceleration a_x is positive, so the mass now begins moving in the positive x-direction.
- At $t = \frac{3}{4}T$, the mass again passes through its equilibrium position, $x = 0$, in the opposite direction. The speed is again a maximum, with $v_x = v_{max}$.
- At $t = T$, the mass is back to its starting position, and the cycle begins again.

NOTE ▶ The graphs of Figure 14.11 are for an oscillation in which the object just happened to be at $x = A$ at $t = 0$. You can certainly imagine a different set of *initial conditions,* with the object at $x = -A$, or even somewhere in the middle of an oscillation. But our choice of $t = 0$ is arbitrary. Even if an oscillation begins at a different position, it will certainly pass through the position $x = A$, and we can simply choose to set $t = 0$ at this instant. ◀

The position-versus-time graph in Figure 14.11 is a cosine curve. We can write the object's position as

$$x(t) = A\cos\left(\frac{2\pi t}{T}\right) \tag{14.10}$$

where the notation $x(t)$ indicates that the position x is a *function* of time t. Because $\cos(2\pi \text{ rad}) = \cos(0 \text{ rad})$, it's easy to see that the position at time $t = T$ is the same as the position at $t = 0$. In other words, this is a cosine function with period T. We can write Equation 14.10 in an alternative form. Because the oscillation frequency is $f = 1/T$, we can write

$$x(t) = A\cos(2\pi ft) \tag{14.11}$$

NOTE ▶ The argument $2\pi ft$ of the cosine function is in *radians.* That will be true throughout this chapter. Don't forget to set your calculator to radian mode before working oscillation problems. ◀

Just as the position graph was a cosine function, the velocity graph shown in Figure 14.11 is an "upside-down" sine function with the same period T. The velocity v_x, which is a function of time, can be written

$$v_x(t) = -v_{max}\sin\left(\frac{2\pi t}{T}\right) = -v_{max}\sin(2\pi ft) \tag{14.12}$$

NOTE ▶ v_{max} is the maximum *speed* and thus is inherently a *positive* number. The minus sign in Equation 14.12 is needed to turn the sine function upside down. ◀

We've seen the position and velocity for a mass on a spring; how about the acceleration? As we saw earlier, the restoring force that causes the mass to oscillate with simple harmonic motion, Equation 14.2 is $(F_{net})_x = -kx$. Using Newton's second law, we see that this force causes an acceleration

$$a_x = \frac{(F_{net})_x}{m} = -\frac{k}{m}x \tag{14.13}$$

Acceleration is proportional to the position x, but with a minus sign. Consequently we expect the acceleration-versus-time graph to be an inverted form of the position-versus-time graph. This is, indeed, what we find; the acceleration-versus-time graph in Figure 14.11 is clearly an "upside-down" cosine function with the same period T. We can write the acceleration as

$$a_x(t) = -a_{max}\cos\left(\frac{2\pi t}{T}\right) = -a_{max}\cos(2\pi ft) \tag{14.14}$$

In this and coming chapters, we will use sine and cosine functions extensively, so we will summarize some of the key aspects of mathematical relationships using these functions.

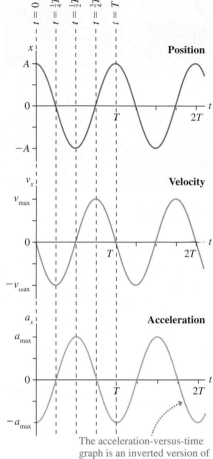

The acceleration-versus-time graph is an inverted version of the position-versus-time graph.

FIGURE 14.11 The position, velocity, and acceleration graphs for simple harmonic motion.

Small bird, fast wings BIO As this rufous hummingbird flaps its wings, the motion of the wing tips is approximately simple harmonic. The frequency of the motion is about 45 Hz, so the period of the wingbeat is only a few hundredths of a second. As we'll see in Equation 14.15, this short period means that the tips of the hummingbird's tiny wings are moving at a maximum speed of over 15 m/s—nearly 35 mph!

Sinusoidal relationships   ✎ Exercise 6   (MP)

A quantity that oscillates in time and can be written

$$x = A\sin\left(\frac{2\pi t}{T}\right)$$

or

$$x = A\cos\left(\frac{2\pi t}{T}\right)$$

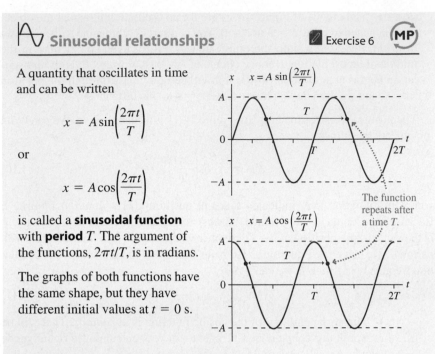

is called a **sinusoidal function** with **period** T. The argument of the functions, $2\pi t/T$, is in radians.

The graphs of both functions have the same shape, but they have different initial values at $t = 0$ s.

LIMITS If x is a sinusoidal function, then x is

- *bounded*—it can only take values between A and $-A$.
- *periodic*—it repeats the same sequence of values over and over again. Whatever value x has at time t, it has the same value at $t + T$.

SPECIAL VALUES The function x has special values at certain times:

	$t = 0$	$t = \frac{1}{4}T$	$t = \frac{1}{2}T$	$t = \frac{3}{4}T$	$t = T$
$x = A\sin(2\pi t/T)$	0	A	0	$-A$	0
$x = A\cos(2\pi t/T)$	A	0	$-A$	0	A

EXAMPLE 14.2 Motion of a glider on a spring

An air-track glider oscillates on a spring at a frequency of 0.50 Hz. Suppose the glider is pulled to the right of its equilibrium position by 12 cm and then released. Where will the glider be 1.0 s after its release? What is its velocity at this point?

PREPARE The glider undergoes simple harmonic motion with amplitude 12 cm. The frequency is 0.50 Hz, so the period is $T = 1/f = 2.0$ s. The glider is released at maximum extension from the equilibrium position, meaning that we can take this point to be $t = 0$.

SOLVE 1.0 s is exactly half the period. As the graph of the motion in Figure 14.12 shows, half a cycle brings the glider to its left turning point, 12 cm to the left of the equilibrium position. The velocity at this point is zero.

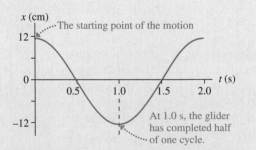

FIGURE 14.12 Position-versus-time graph for the glider.

We were able to determine the velocity in Example 14.2 because $t = 0$ was a turning point where the instantaneous velocity was zero. To determine the velocity at other points, we need to know v_{max} and use Equation 14.12. How does v_{max} depend on other variables of the motion? Basic reasoning about the motion tells us a few things. A large amplitude implies a high speed because the glider moves a large distance. And a small period implies a high speed as well because the glider must complete its motion in a short time. Later in the chapter we will show that both of these assumptions are correct, and that the maximum speed is

$$v_{max} = 2\pi f A = \frac{2\pi A}{T} \tag{14.15}$$

EXAMPLE 14.3 Analyzing the motion of a spring toy

A classic wooden toy hangs from a spring that lets it bob up and down. If the toy is raised up 20 cm and released, it makes 15 complete oscillations in 10 seconds.

a. What is the period of oscillation?
b. What is the toy's maximum speed?
c. What are the position and velocity at $t = 0.80$ s?

PREPARE The toy is a mass on a vertical spring, and so will undergo simple harmonic motion. The toy begins its motion at its maximum positive displacement.

SOLVE a. The oscillation frequency is

$$f = \frac{15 \text{ oscillations}}{10 \text{ s}} = 1.5 \text{ oscillations/s} = 1.5 \text{ Hz}$$

Thus the period is $T = 1/f = 0.667$ s.

b. The oscillation amplitude is $A = 0.20$ m. The maximum speed, given by Equation 14.15, is

$$v_{max} = \frac{2\pi A}{T} = \frac{2\pi(0.20 \text{ m})}{0.667 \text{ s}} = 1.9 \text{ m/s}$$

c. The object starts at $y = +A$ at $t = 0$. This is exactly the oscillation described by Equations 14.10 and 14.12. The position at $t = 0.80$ s is

$$y = A\cos\left(\frac{2\pi t}{T}\right) = (0.20 \text{ m})\cos\left(\frac{2\pi(0.80 \text{ s})}{0.667 \text{ s}}\right)$$

$$= (0.20 \text{ m})\cos(7.54 \text{ rad}) = 0.062 \text{ m} = 6.2 \text{ cm}$$

The velocity at this instant of time is

$$v_y = -v_{max}\sin\left(\frac{2\pi t}{T}\right) = -(1.9 \text{ m/s})\sin\left(\frac{2\pi(0.80 \text{ s})}{0.667 \text{ s}}\right)$$

$$= -(1.9 \text{ m/s})\sin(7.54 \text{ rad}) = -1.8 \text{ m/s}$$

ASSESS At $t = 0.80$ s, which is slightly more than one period, the object is 6.2 cm above the equilibrium position and moving down at 1.8 m/s. Notice the use of radians in the calculations.

Connection to Uniform Circular Motion

Both circular motion and simple harmonic motion are motions that repeat. Many of the concepts we have used for describing simple harmonic motion were introduced in our study of circular motion. We will use the close connection between the two to extend our knowledge of the kinematics of circular motion to that of simple harmonic motion.

We can demonstrate the relationship between circular and simple harmonic motion with a simple experiment. Suppose a turntable has a small ball glued to the edge. As Figure 14.13a shows, we can make a "shadow movie" of the ball by projecting a light past the ball and onto a screen. The ball's shadow oscillates back and forth as the turntable rotates.

If you place a real object on a real spring directly below the shadow, as shown in Figure 14.13b, and if you adjust the turntable to have the same period as the spring, you will find that the shadow's motion exactly matches the simple harmonic motion of the object on the spring. **Uniform circular motion projected onto one dimension is simple harmonic motion.**

We can show why the projection of circular motion is simple harmonic motion by considering the particle in Figure 14.14 on the next page. As in Chapter 6, we can locate the particle by the angle ϕ measured counterclockwise from the x-axis. Projecting the ball's shadow onto a screen in Figure 14.13 is

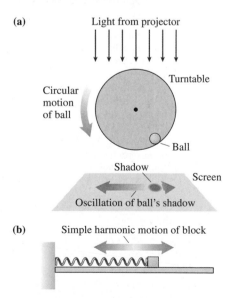

FIGURE 14.13 A projection of the circular motion of a rotating ball matches the SHM of an object on a spring.

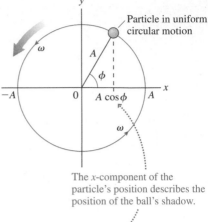

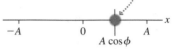

The x-component of the particle's position describes the position of the ball's shadow.

FIGURE 14.14 A particle in uniform circular motion with radius A and angular velocity ω.

TRY IT YOURSELF

SHM in your microwave It's possible to do something like the turntable demonstration right at home. Place a tall (microwave safe) glass or cup on the outside edge of the turntable in your microwave oven. Start the oven. The turntable will rotate, moving the cup in a circle. Stand in front of the oven with your eyes level with the cup. Watch the cup, paying attention to the side-to-side motion. This motion is the horizontal component of the circular motion, and so it is simple harmonic motion.

equivalent to observing just the x-component of the particle's motion. Figure 14.14 shows that the x-component, when the particle is at angle ϕ, is

$$x = A\cos\phi$$

If the particle starts from $\phi_0 = 0$ at $t = 0$, its angle at a later time t is simply

$$\phi = \omega t$$

where ω is the particle's *angular velocity,* as defined in Chapter 6. Recall that the angular velocity is related to the frequency f by

$$\omega = 2\pi f$$

The particle's x-component can therefore be expressed as

$$x(t) = A\cos(2\pi ft) \qquad (14.16)$$

This is identical to Equation 14.11 for the position of a mass on a spring! **The x-component of a particle in uniform circular motion is simple harmonic motion.**

We can use this correspondence to deduce more details. Figure 14.15 is the same motion we looked at in Figure 14.14, but here we've shown the velocity vector (tangent to the circle) and the acceleration vector (a centripetal acceleration pointing to the center of the circle). The magnitude of the velocity vector is the particle's speed. Recall from Chapter 6 that the speed of a particle in circular motion with radius A and frequency f is $v = 2\pi fA$. Thus the x-component of the velocity vector, which is pointing in the negative x-direction, is

$$v_x = -v\sin\phi = -(2\pi f)A\sin(2\pi ft)$$

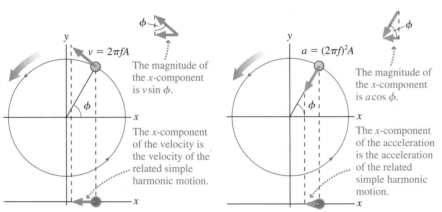

FIGURE 14.15 Projection of the velocity and acceleration vectors.

According to the correspondence between circular motion and simple harmonic motion, this is the velocity of an object in simple harmonic motion. This is, indeed, exactly Equation 14.12, which we deduced from the graph of the motion, if we define the maximum speed, as we did in Equation 14.15, to be

$$v_{max} = 2\pi fA = \frac{2\pi A}{T}$$

Similarly, the magnitude of the acceleration vector is the centripetal acceleration $a = v^2/A = (2\pi f)^2 A$. The x-component of the acceleration vector, which is the acceleration for simple harmonic motion, is

$$a_x = -a\cos\phi = -(2\pi f)^2 A\cos(2\pi ft)$$

The maximum acceleration is

$$a_{max} = (2\pi f)^2 A \qquad (14.17)$$

We can now summarize our findings for the position, velocity, and acceleration of an object in simple harmonic motion:

$$x(t) = A\cos(2\pi ft)$$
$$v_x(t) = -(2\pi f)A\sin(2\pi ft) \qquad (14.18)$$
$$a_x(t) = -(2\pi f)^2 A\cos(2\pi ft)$$

SINUSOIDAL
p. 460

Position, velocity, and acceleration for an object in simple harmonic motion with frequency f and amplitude A

Any simple harmonic motion will follow these equations. If you know the amplitude and the frequency, the motion is completely specified.

EXAMPLE 14.4 Measuring the sway of a tall building

The John Hancock Center in Chicago is 110 stories high. Strong winds can cause the building to sway, as is the case with all tall buildings. On particularly windy days, the top of the building is known to oscillate with an amplitude of 40 cm ($\approx$16 in) and a period of 7.7 s. What are the maximum speed and acceleration of the top of the building?

PREPARE We will assume that the oscillation of the building is simple harmonic motion with amplitude $A = 0.40$ m. The frequency can be computed from the period:

$$f = \frac{1}{T} = \frac{1}{7.7\ s} = 0.13\ Hz$$

SOLVE We can use Equations 14.15 and 14.17 for the maximum velocity and acceleration to compute:

$$v_{max} = 2\pi fA = 2\pi(0.13\ Hz)(0.40\ m) = 0.33\ m/s$$
$$a_{max} = (2\pi f)^2 A = [2\pi(0.13\ Hz)]^2(0.40\ m) = 0.27\ m/s^2$$

In terms of the acceleration of gravity, the maximum acceleration is

$$a_{max} = 0.027g$$

ASSESS The acceleration is quite small, as you would expect; if it were large, building occupants would certainly complain! Even if they don't notice the motion directly, office workers on high floors of high buildings may experience a bit of nausea when the oscillations are large because the acceleration affects the equilibrium organ in the inner ear.

STOP TO THINK 14.3 The figure shows four identical oscillators at different points in their motion. Which is moving fastest at the time shown?

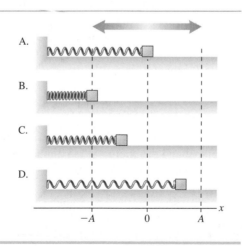

Elastic cords lead to the up-and-down motion of a bungee jump.

14.4 Energy in Simple Harmonic Motion

A bungee jumper falls, increasing in speed, until the elastic cords attached to his ankles start to stretch. The kinetic energy of his motion is transformed into the elastic potential energy of the cords. Once the cords reach their maximum stretch, his velocity reverses. He rises as the elastic potential energy of the cords is transformed back into kinetic energy. If he keeps bouncing (simple harmonic motion), this transformation happens again and again.

This interplay between kinetic and potential energy is very important to understanding simple harmonic motion. The five diagrams in Figure 14.16a show the position and velocity of a mass on a spring at successive points in time. Let's look at the potential and kinetic energy of the object as it moves through these points.

The object begins at rest, with the spring at a maximum extension; the kinetic energy is zero, and the potential energy is a maximum. As the spring contracts, the object speeds up until it reaches the center point of its oscillation, the equilibrium point. At this point, the potential energy is zero and the kinetic energy is a maximum. As the object continues to move, it slows down as it compresses the spring. Eventually, it reaches the turning point, where its instantaneous velocity is zero. At this point, the kinetic energy is zero and the potential energy is again a maximum.

Now we'll specify that the object has mass m, the spring has spring constant k, and the motion takes place on a frictionless surface. You learned in Chapter 10 that the elastic potential energy of a spring stretched by a distance x from its equilibrium position is

$$U = \frac{1}{2}kx^2 \qquad (14.19)$$

The potential energy is zero at the equilibrium position and is a maximum when the spring is at its maximum extension or compression. The mechanical energy of the object is the sum of its potential energy and its kinetic energy. Because there is no friction and there are no other forces to do work on the system, the mechanical energy is conserved. We can write

$$E = K + U = \frac{1}{2}mv^2 + \frac{1}{2}kx^2 = \text{constant} \qquad (14.20)$$

Figure 14.16b shows a graph of the potential energy, kinetic energy, and total energy for the object as it moves. You can see that, as the object goes through its motion, energy is transformed from potential to kinetic and then back to potential. In particular, you can see that **the object has purely potential energy at** $x = \pm A$ **and purely kinetic energy as it passes through the equilibrium point at** $x = 0$. At maximum displacement, with $x = \pm A$ and $v_x = 0$ the energy is

$$E(\text{at } x = \pm A) = U_{\text{max}} = \frac{1}{2}kA^2 \qquad (14.21)$$

At $x = 0$, where $v_x = \pm v_{\text{max}}$, the energy is

$$E(\text{at } x = 0) = K_{\text{max}} = \frac{1}{2}mv_{\text{max}}^2 \qquad (14.22)$$

The system's mechanical energy is conserved, so the energy at maximum displacement and the energy at maximum speed, Equations 14.22 and 14.23, must be equal. That is

$$\frac{1}{2}m(v_{\text{max}})^2 = \frac{1}{2}kA^2 \qquad (14.23)$$

By solving Equation 14.23 for the maximum speed we can see that it is related to the amplitude by

$$v_{\text{max}} = \sqrt{\frac{k}{m}}A \qquad (14.24)$$

Earlier we found that

$$v_{\text{max}} = 2\pi f A \qquad (14.25)$$

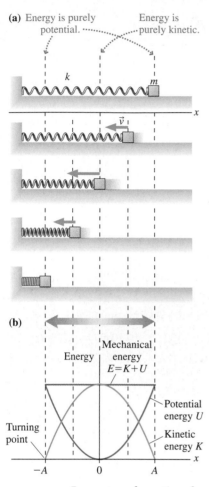

(a) Energy is purely potential. Energy is purely kinetic.

x

$\vec{v}$

(b)

Energy

Mechanical energy $E = K + U$

Turning point

Potential energy U

Kinetic energy K

$-A \qquad 0 \qquad A \qquad x$

FIGURE 14.16 Energy transformations for a mass on a spring.

Comparing Equations 14.24 and 14.25, we see that the frequency, and thus the period, of an oscillating mass on a spring is determined by the spring constant k and the object's mass m:

$$f = \frac{1}{2\pi}\sqrt{\frac{k}{m}} \quad \text{and} \quad T = 2\pi\sqrt{\frac{m}{k}} \qquad (14.26)$$

SQUARE-ROOT
p. 346

Frequency and period of SHM for
mass m on a spring with spring constant k

We make two observations about this equation:

- **The frequency of simple harmonic motion is determined by the physical properties of the oscillator.** The frequency of a mass on a spring is determined by (1) the mass and (2) the stiffness of the spring, as outlined in Figure 14.17. This dependence of frequency on a force term and an inertia term will also apply to other oscillators.
- **The frequency of simple harmonic motion does not depend on the amplitude A.** A small oscillation and a large oscillation have the same frequency.

CONCEPTUAL EXAMPLE 14.1   The period of a bobbing bottle

A cylindrical bottle is floating in still water. A push on the top of the bottle causes the bottle to bob up and down. How does the period change as the amplitude of the bobbing slowly decreases?

REASON The motion of the bottle is simple harmonic. The period does not depend on the amplitude, so it will stay constant as the motion dies down.

ASSESS The fact that the period doesn't depend on the amplitude may seem counterintuitive, but it turns out to be a general property of every type of simple harmonic motion. As time goes on, the speed of the bottle decreases, but so does the amplitude—the distance it travels. The bottle is moving more slowly, but it is not going as far. The time for one cycle, the period, stays the same.

Now that we have a complete description of simple harmonic motion in one system, we will summarize the details in a Tactics Box, showing how to use this information to solve oscillation problems.

 TACTICS BOX 14.1   Identifying and analyzing
simple harmonic motion   Exercise 11

❶ If the net force acting on a particle is a linear restoring force, the motion is simple harmonic motion around the equilibrium position.

❷ The position, velocity, and acceleration as a function of time are given in Equation 14.18. The equations are given here in terms of x, but they can be written in terms of y, θ, or some other variable if the situation calls for it.

❸ The amplitude A is the maximum value of the displacement from equilibrium. The maximum speed and the maximum magnitude of the acceleration are $v_{max} = (2\pi f)A$ and $a_{max} = (2\pi f)^2 A$.

❹ The frequency f (and hence the period $T = 1/f$) depends on the physical properties of the particular oscillator, but f does *not* depend on A.

For a mass on a spring, the frequency is given by $f = \frac{1}{2\pi}\sqrt{\frac{k}{m}}$.

❺ Mechanical energy is conserved. As the oscillation proceeds, energy is transformed from kinetic to potential energy and then back again.

High frequency
Stiff spring,
low mass

Increasing the stiffness of the spring will increase the restoring force. This *increases* the frequency.

$$f = \frac{1}{2\pi}\sqrt{\frac{k}{m}}$$

Low frequency
Soft spring,
high mass

Increasing the mass will increase the inertia of the system. This *decreases* the frequency.

FIGURE 14.17 Frequency dependence on mass and spring stiffness.

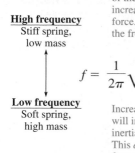

Measuring mass in space Astronauts on Skylab, a US space station in the 1970's, monitored their weight to determine the effects of prolonged space flight on the body. But the astronauts' weightlessness in space meant they couldn't just hop on a scale! Instead, they became the moving mass in a mass-spring system, holding on tightly to a platform that oscillated back and forth on a spring. By measuring the period of this motion, they could deduce their mass.

Activ
ONLINE
Physics   9.3, 9.4, 9.6–9.9

EXAMPLE 14.5 Finding the frequency of an oscillator

A spring has an unstretched length of 10.0 cm. A 25 g mass is hung from the spring, stretching it to a length of 15.0 cm. If the mass is pulled down and released so that it oscillates, what will be the frequency of the oscillation?

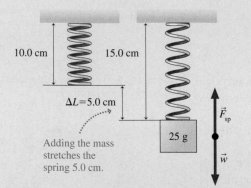

10.0 cm   15.0 cm

$\Delta L = 5.0$ cm

Adding the mass stretches the spring 5.0 cm.

25 g

$\vec{F}_{sp}$

$\vec{w}$

FIGURE 14.18 Visual overview of a mass suspended from a spring.

PREPARE The spring provides a linear restoring force, so the motion will be simple harmonic, as noted in Tactics Box 14.1. The oscillation frequency depends on the spring constant, which we can determine from the stretch of the spring. Figure 14.18 gives a visual overview of the situation.

SOLVE When the mass hangs at rest, after stretching the spring to 15 cm, the net force on it must be zero. Thus the magnitude of the upward spring force equals the downward weight, giving $k\Delta L = mg$. The spring constant is thus

$$k = \frac{mg}{\Delta L} = \frac{(0.025 \text{ kg})(9.8 \text{ m/s}^2)}{0.050 \text{ m}} = 4.9 \text{ N/m}$$

Now that we know the spring constant, we can compute the oscillation frequency:

$$f = \frac{1}{2\pi}\sqrt{\frac{k}{m}} = \frac{1}{2\pi}\sqrt{\frac{4.9 \text{ N/m}}{0.025 \text{ kg}}} = 2.2 \text{ Hz}$$

ASSESS 2.2 Hz is 2.2 oscillations per second. This seems like a reasonable frequency for a mass on a spring. A frequency in the kHz range (thousands of oscillations per second) would have been suspect!

EXAMPLE 14.6 Weighing DNA molecules

It has recently become possible to "weigh" individual DNA molecules by measuring the influence of their mass on a nanoscale oscillator. Figure 14.19 shows a thin rectangular cantilever etched out of silicon (density 2300 kg/m³). If pulled down and released, the end of the cantilever

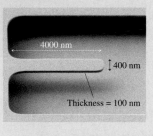

4000 nm

400 nm

Thickness = 100 nm

FIGURE 14.19 A nano-scale cantilever.

vibrates with simple harmonic motion, moving up and down like a diving board after a jump. When the end of the cantilever is bathed with DNA molecules whose ends have been modified to bind to a surface, one or more molecules may attach to the end of the cantilever. The addition of their mass causes a very slight—but measurable—decrease in the oscillation frequency.

A vibrating cantilever of mass M can be modeled as a simple block of mass $\frac{1}{3}M$ attached to a spring. (The factor of $\frac{1}{3}$ arises from the moment of inertia of a bar pivoted at one end: $I = \frac{1}{3}ML^2$.) Neither the mass nor the spring constant can be determined very accurately—perhaps only to two significant figures—but the oscillation frequency can be measured with very high precision simply by counting the oscillations. In one experiment, the cantilever was initially vibrating at exactly 12 MHz. Attachment of a DNA molecule caused the frequency to decrease by 50 Hz. What was the mass of the DNA molecule?

PREPARE We determine the mass of the cantilever as the product of its volume and its density ($M = \rho V$), giving:

$$M = (2300 \text{ kg/m}^3)(4000 \times 10^{-9} \text{ m} \times 400 \times 10^{-9} \text{ m} \times 100 \times 10^{-9} \text{ m})$$

$$= 3.7 \times 10^{-16} \text{ kg}$$

We will model the cantilever as a block of mass $m = \frac{1}{3}M = 1.2 \times 10^{-16}$ kg oscillating on a spring with spring constant k. When the mass increases to $m + m_{DNA}$, the oscillation frequency decreases from $f_0 = 12,000,000$ Hz to $f_1 = 11,999,950$ Hz.

SOLVE The oscillation frequency of a mass on a spring is given by Equation 14.26. Addition of mass doesn't change the spring constant, so solving this equation for k allows us to write

$$k = m(2\pi f_0)^2 = (m + m_{DNA})(2\pi f_1)^2$$

The 2π terms cancel, and we can rearrange this equation to give

$$\frac{m + m_{DNA}}{m} = 1 + \frac{m_{DNA}}{m} = \left(\frac{f_0}{f_1}\right)^2 = \left(\frac{12,000,000 \text{ Hz}}{11,999,950 \text{ Hz}}\right)^2$$

$$= 1.0000083$$

Subtracting 1 from both sides gives

$$\frac{m_{DNA}}{m} = 0.0000083$$

and thus

$$m_{DNA} = 0.0000083m = (0.0000083)(1.2 \times 10^{-16} \text{ kg})$$

$$= 1.0 \times 10^{-21} \text{ kg} = 1.0 \times 10^{-18} \text{ g}$$

ASSESS This is a reasonable mass for a DNA molecule. It's a remarkable technical achievement to be able to measure a mass this small. With a slight further improvement in sensitivity, scientists will be able to determine the number of base pairs in a strand of DNA simply by weighing it!

EXAMPLE 14.7 Slowing a mass with a spring collision

A 1.5 kg mass slides across a horizontal, frictionless surface at a speed of 2.0 m/s until it collides with and sticks to the free end of a spring with spring constant 50 N/m. The spring's other end is anchored to a wall. How far has the spring compressed when the mass is, at least for an instant, at rest? How much time does it take for the spring to compress to this point?

PREPARE This is a collision problem, but we can solve it using the tools of simple harmonic motion. Figure 14.20 gives a visual overview of the problem. The motion is along the x-axis. We have set $x = 0$ at the uncompressed end of the spring, which is the point of collision. Once the mass hits and sticks, it is going to start oscillating with simple harmonic motion. The position-versus-time graph shows that the motion of the mass until it stops is $\frac{1}{4}$ of a cycle of simple harmonic motion—though the starting point is different from our usual choice. During this quarter cycle of motion, the kinetic energy of the mass is transformed to potential energy of the compressed spring.

SOLVE Conservation of energy tells us that potential energy of the spring when fully compressed is equal to the initial kinetic energy of the mass, so we write

$$\frac{1}{2}m(v_x)_i^2 = \frac{1}{2}kx_f^2$$

The final position of the mass is

$$x_f = (v_x)_i\sqrt{\frac{m}{k}} = (2.0 \text{ m/s})\sqrt{\frac{1.5 \text{ kg}}{50 \text{ N/m}}} = 0.35 \text{ m}$$

This is the compression of the spring.

Because the compression is $\frac{1}{4}$ of a cycle of simple harmonic motion, the time needed to stop the mass is $t_f = \frac{1}{4}T$. The period is computed using Equation 14.26, so we write:

$$t_f = \frac{1}{4}T = \frac{1}{4}2\pi\sqrt{\frac{m}{k}} = \frac{\pi}{2}\sqrt{\frac{1.5 \text{ kg}}{50 \text{ N/m}}} = 0.27 \text{ s}$$

This is the time required for the spring to compress.

ASSESS Interestingly, the time needed to stop the mass does not depend on its initial velocity or on the distance the spring compresses. The motion is simple harmonic, thus the period depends only on the stiffness of the spring and the mass, not on the details of the motion.

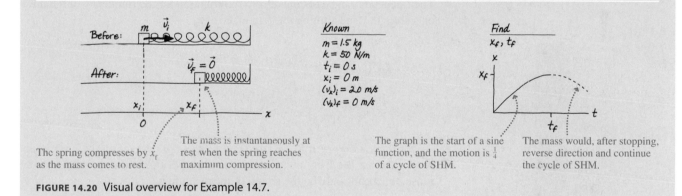

Before: m $\vec{v_i}$ k

After: $\vec{v_f} = \vec{0}$

x_i x_f

0

Known
$m = 1.5$ kg
$k = 50$ N/m
$t_i = 0$ s
$x_i = 0$ m
$(v_x)_i = 2.0$ m/s
$(v_x)_f = 0$ m/s

Find
x_f, t_f

x

x_f

t_f

The spring compresses by x_f as the mass comes to rest.

The mass is instantaneously at rest when the spring reaches maximum compression.

The graph is the start of a sine function, and the motion is $\frac{1}{4}$ of a cycle of SHM.

The mass would, after stopping, reverse direction and continue the cycle of SHM.

FIGURE 14.20 Visual overview for Example 14.7.

Many real-world collisions (such as a pole vaulter landing on a foam pad) involve a moving object that is stopped by an elastic object. In other cases (such as a bouncing rubber ball) the object itself is elastic. The time of the collision is reasonably constant, independent of the speed of the collision, for the reasons noted in the above example.

▶ **Automobile collision times** When a car hits a stationary barrier, it takes approximately 0.1 s to come to rest, regardless of the initial speed, something we can understand with a simple model of the collision. The crumpling of the front of a car during a collision is quite complex, but for many cars the force is approximately proportional to the displacement during the compression of the front of the car. As long as we only consider this initial compression, we can model the body of the car as a mass and the front of the car as a spring. With this model, the collision is similar to that of Example 14.7, and, as in the example, the time for the car to come to rest does not depend on the initial speed.

STOP TO THINK 14.4 Four oscillators have masses and spring constants as shown in the figure. Rank in order, from highest to lowest, the frequencies of the oscillators.

A. k — $4m$

B. $\frac{1}{2}k$ — m

C. k — $2m$

D. $2k$ — m

14.5 Pendulum Motion

9.10–9.12

As we've already seen, a simple pendulum—a mass at the end of a string or a rod that is free to pivot—is another system that exhibits simple harmonic motion. Everything we have learned about the mass on a spring can be applied to the pendulum as well.

In the first part of the chapter, we looked at the restoring force in the pendulum. For a pendulum of length L displaced by an arc length s, as in Figure 14.21, the tangential restoring force is

$$(F_{net})_t = -\frac{mg}{L}s \qquad (14.27)$$

NOTE ▶ Recall that this equation holds only for small angles. ◀

This linear restoring force has exactly the same form as the net force in a mass-spring system, but with the constants mg/L in place of the constant k. Given this, we can quickly deduce the essential features of pendulum motion by replacing k, wherever it occurs in the oscillating spring equations, with mg/L.

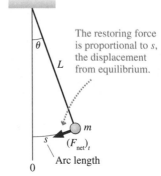

The restoring force is proportional to s, the displacement from equilibrium.

FIGURE 14.21 A simple pendulum.

- The oscillation of a pendulum is simple harmonic motion; the equations of motion can be written for the arc length or the angle:

$$s(t) = A\cos(2\pi f t) \qquad \text{or} \qquad \theta(t) = \theta_{max}\cos(2\pi f t)$$

- The frequency can be obtained from the equation for the frequency of the mass on a spring by substituting mg/L in place of k:

$$f = \frac{1}{2\pi}\sqrt{\frac{g}{L}} \qquad \text{and} \qquad T = 2\pi\sqrt{\frac{L}{g}} \qquad (14.28)$$

Frequency of a pendulum of length L, with acceleration due to gravity g

- As for a mass on a spring, the frequency does not depend on the amplitude. Note also that **the frequency, and hence the period, is independent of the mass.** It depends only on the length of the pendulum.

Galileo was the first person to study the pendulum in detail. He realized that the pendulum's fixed frequency would serve as the basis of an accurate clock. Pendulum clocks were the most accurate timepieces available until well into the 20th century.

Pendulum prospecting The period of a pendulum clock does not depend on the amplitude, but it does depend on the strength of gravity. Soon after Christian Huygens built an accurate pendulum clock in 1656 (the photo shows a replica), Jean Richer discovered that the clock ran more slowly near the equator. Richer correctly surmised that this was due to the weaker gravity near the equator because of the greater distance from the center of the earth. In later years, more accurate pendulums were built that could detect much smaller variations in gravity—small enough that they could sense the presence of dense mineral deposits or low-density strata containing petroleum.

EXAMPLE 14.8 Designing a pendulum for a clock
A grandfather clock is designed so that one swing of the pendulum in either direction takes 1.00 s. What is the length of the pendulum?

PREPARE One period of the pendulum is two swings, so the period is $T = 2.00$ s.

SOLVE The period is independent of the mass and depends only on the length. From Equation 14.28,

$$T = \frac{1}{f} = 2\pi\sqrt{\frac{L}{g}}$$

Solving for L, we find

$$L = g\left(\frac{T}{2\pi}\right)^2 = 0.993 \text{ m}$$

ASSESS A pendulum clock with a "tick" or "tock" each second requires a long pendulum of about 1 m—which is why these clocks were originally known as "tall case clocks."

Physical Pendulums and Locomotion

In Chapter 6, we computed maximum walking speed using the ideas of circular motion. We can also model the motion of your legs during walking as pendulum motion. When you walk, you push off with your rear leg and then let it swing forward for the next stride. At normal, comfortable walking speeds, you use very little force to bring your leg forward. Your leg swings forward under the influence of gravity—like a pendulum.

Try this: Stand on one leg, and gently swing your free leg back and forth. There is a certain frequency at which it will naturally swing. This is your leg's pendulum frequency. Now, try swinging your leg at twice this frequency. You can do it, but it is very difficult. The muscles that move your leg back and forth aren't very strong because under normal circumstances they don't need to apply much force.

A pendulum, like your leg, whose mass is distributed along its length is known as a **physical pendulum.** The motion of a physical pendulum is similar to that of a simple pendulum, but its frequency depends on the distribution of mass.

Figure 14.22 shows a simple pendulum and a physical pendulum of the same length. The position of the center of gravity of the physical pendulum is at a distance d from the pivot.

What will be the frequency of a physical pendulum? This is really a rotational motion problem similar to those we considered in Chapter 7. We would expect the frequency to depend on the moment of inertia and the distance to the center of gravity as follows:

- The moment of inertia I is a measure of an object's resistance to rotation. Increasing the moment of inertia while keeping other variables equal should cause the frequency to decrease. In an expression for the frequency of the physical pendulum, we would expect I to appear in the denominator.
- When the pendulum is pushed to the side, a gravitational torque pulls it back. The greater the distance d of the center of gravity from the pivot point, the greater the torque. Increasing this distance while keeping the other variables constant should cause the frequency to increase. In an expression for the frequency of the physical pendulum, we would expect d to appear in the numerator.

A careful analysis of the motion of the physical pendulum produces a result for the frequency that matches these expectations:

$$f = \frac{1}{2\pi}\sqrt{\frac{mgd}{I}} \tag{14.29}$$

Frequency of a physical pendulum of mass m,
moment of inertia I, with center of gravity distance d from the pivot

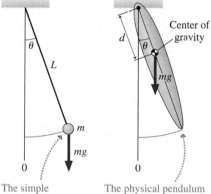

The simple pendulum is a small mass m at the end of a light rod of length L.

The physical pendulum is an extended object with mass m, length L, and moment of inertia I.

FIGURE 14.22 A simple pendulum and a physical pendulum of equal length.

EXAMPLE 14.9 Finding the frequency of a swinging leg
A student in a biomechanics lab measures the length of his leg, from hip to heel, to be 0.90 m. What is the frequency of the pendulum motion of the student's leg? What is the period?

PREPARE We can model a human leg reasonably well as a rod of uniform cross section, pivoted at one end (the hip). Recall from Chapter 7 that the moment of inertia of a rod pivoted about its end is $\frac{1}{3}mL^2$. The center of gravity of a uniform leg is at the midpoint, so $d = L/2$.

Continued

How do you hold your arms? You maintain your balance when walking or running by moving your arms back and forth opposite the motion of your legs. You hold your arms so that the natural period of their pendulum motion matches that of your legs. At a normal walking pace, your arms are extended and naturally swing at the same period as your legs. When you run, your gait is more rapid. To decrease the period of the pendulum motion of your arms to match, you bend them at the elbows, shortening their effective length and increasing the natural frequency of oscillation. To test this for yourself, try running fast with your arms fully extended. It's quite awkward!

SOLVE The frequency of a physical pendulum is given by Equation 14.29. Before we put in numbers, we will use symbolic relationships and simplify:

$$f = \frac{1}{2\pi} \sqrt{\frac{mgd}{I}} = \frac{1}{2\pi} \sqrt{\frac{mg(L/2)}{\frac{1}{3}mL^2}} = \frac{1}{2\pi} \sqrt{\frac{3}{2}\frac{g}{L}}$$

The expression for the frequency is similar to that for the simple pendulum, but with an additional numerical factor of 3/2 inside the square root. The numerical value of the frequency is

$$f = \frac{1}{2\pi} \sqrt{\left(\frac{3}{2}\right)\left(\frac{9.8 \text{ m/s}^2}{0.90 \text{ m}}\right)} = 0.64 \text{ Hz}$$

The period is

$$T = \frac{1}{f} = 1.6 \text{ s}$$

ASSESS Notice that we didn't need to know the mass of the leg to find the period. The period of a physical pendulum does not depend on the mass, just as it doesn't for the simple pendulum. The period depends only on the *distribution* of mass. When you walk, swinging your free leg forward to take another stride corresponds to half a period of this pendulum motion. For a period of 1.6 s, this is 0.80 s. For a normal walking pace, one stride in just under one second sounds about right.

As you walk, your legs do swing as physical pendulums as you bring them forward. The frequency is fixed by the length of your legs and their distribution of mass; it doesn't depend on amplitude. Consequently, you don't increase your walking speed by taking more rapid steps—changing the frequency is quite difficult. You simply take longer strides, changing the amplitude, but not the frequency.

Gibbons and other apes move through the forest canopy by a hand-over-hand swinging motion called *brachiation*. In this motion, the body swings under a pivot point where a hand grips a tree branch, a clear example of pendulum motion. A brachiating ape will increase its speed by taking bigger swings; it does this by "pumping" the swinging motion, much as you do when increasing your amplitude on a playground swing. But because the period of the pendulum motion is fixed, a brachiating gibbon can only go so fast. At some point a maximum speed is reached, and gibbons and other apes break into a different gait, launching themselves from branch to branch through the air.

STOP TO THINK 14.5   A pendulum clock is made with a metal rod. It keeps perfect time at a temperature of 20°C. At a higher temperature, the metal rod expands. How will this change the clock's timekeeping?

A. The clock will run fast.
B. The clock will keep perfect time.
C. The clock will run slow.

14.6 Damped Oscillations

A real pendulum clock must have some energy input, otherwise the oscillation of the pendulum would slowly decrease in amplitude due to air resistance. If you strike a bell, the oscillation will soon die away as energy is lost to sound waves in the air and dissipative forces within the metal of the bell.

All real oscillators do run down—some very slowly but others quite quickly—as their mechanical energy is transformed into the thermal energy of the oscillator and its environment. An oscillation that runs down and stops is called a **damped oscillation**.

For a pendulum, the main energy loss is due to air resistance, which we called the *drag force* in Chapter 4. When we learned about the drag force, we noted that it depends on velocity: the faster the motion, the bigger the drag force. For this reason, the decrease in amplitude of an oscillating pendulum will be fastest at the start of the motion. For a pendulum or other oscillator with modest damping, we end up with a graph of motion like that in Figure 14.23a. The maximum displacement, x_{max}, decreases with time. As the oscillation decays, the *rate* of the decay decreases; the difference between successive peaks is less.

If we plot a smooth curve that connects the peaks of successive oscillations (we call such a curve an *envelope*) we get the dotted line shown in Figure 14.23b. It's possible, using calculus, to show that x_{max} decreases with time as

$$x_{max}(t) = Ae^{-t/\tau} \tag{14.30}$$

where $e \approx 2.718$ is the base of the natural logarithm and A is the *initial* amplitude. This steady decrease of x_{max} with time is called an **exponential decay.**

The constant τ (lowercase Greek tau) in Equation 14.30 is called the **time constant.** After one time constant has elapsed—that is, at $t = \tau$—the maximum displacement x_{max} has decreased to

$$x_{max}(\text{at } t = \tau) = Ae^{-1} = A/e \approx 0.37A$$

In other words, the oscillation has decreased after one time constant to about 37% of its initial value. The time constant τ measures the "characteristic time" during which damping causes the amplitude of the oscillation to decay away. An oscillation that decays quickly has a small time constant whereas a "lightly damped" oscillator, which decays very slowly, has a large time constant.

Because exponential decay is something we will see again, we will look at it in more detail.

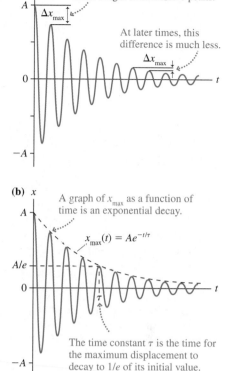

FIGURE 14.23 The motion of a damped oscillator.

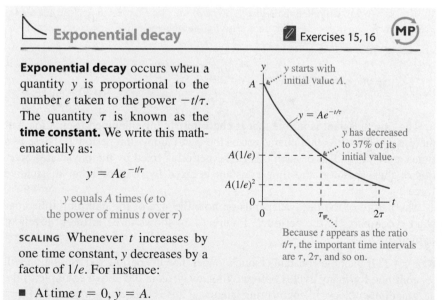

Exponential decay   Exercises 15, 16   (MP)

Exponential decay occurs when a quantity y is proportional to the number e taken to the power $-t/\tau$. The quantity τ is known as the **time constant.** We write this mathematically as:

$$y = Ae^{-t/\tau}$$

y equals A times (e to the power of minus t over τ)

SCALING Whenever t increases by one time constant, y decreases by a factor of $1/e$. For instance:

- At time $t = 0$, $y = A$.
- Increasing time to $t = \tau$ reduces y to A/e.
- A further increase to $t = 2\tau$ reduces y by another factor of $1/e$ to A/e^2.

Generally, we can say that

At $t = n\tau$, y has the value A/e^n.

LIMITS As t becomes large, y becomes very small and approaches zero.

Damping smoothes the ride A car's wheels are attached to the car's body with springs so that the wheels can move up and down as the car moves over an uneven road. Because the car-spring system is a simple harmonic oscillator, a shock absorber is added to damp unwanted oscillations. The shock absorber is designed to give *critical damping* (see next page) so that, after a bump, the wheel quickly and smoothly returns to its equilibrium position.

Different Amounts of Damping

The damped oscillation shown in Figure 14.23 continues for a long time. The amplitude isn't zero after one time constant, or two, or three. . . . Mathematically, the oscillation never ceases, though the amplitude will eventually be so small as to be undetectable. For practical purposes, we can speak of the time constant τ as the *lifetime* of an oscillation—a measure of about how long it takes to decay. An oscillation with a large time constant will persist for a long time; one with a small time constant will decay quickly.

EXAMPLE 14.10 Finding a clock's decay time

The pendulum in a grandfather clock has a period of 1.00 s. If the clock's driving spring is allowed to run down, damping due to friction will cause the pendulum to slow to a stop. If the time constant for this decay is 300 s, how long will it take for the pendulum's swing to be reduced to half its initial amplitude?

SOLVE Equation 14.30 gives an expression for the decay of the amplitude of a damped harmonic oscillator:

$$x_{max}(t) = Ae^{-t/\tau}$$

As noted in Tactics Box 14.1, we can equally well write this equation in terms of the pendulum's angle in a straightforward manner:

$$\theta_{max}(t) = \theta_i e^{-t/\tau}$$

where θ_i is the initial angle of swing. At some time t, the time we wish to find, the amplitude has decayed to half its initial value. At this time,

$$\theta_{max}(t) = \theta_i e^{-t/\tau} = \frac{1}{2}\theta_i$$

The θ_i cancels, giving

$$e^{-t/\tau} = \frac{1}{2}$$

To solve this for t, first take the natural logarithm of both sides and use the logarithm property $\ln(e^a) = a$. Doing so gives

$$\ln(e^{-t/\tau}) = -\frac{t}{\tau} = \ln\left(\frac{1}{2}\right) = -\ln 2$$

In the last step we used the property $\ln(1/b) = -\ln b$. Now we can solve for t, finding

$$t = \tau \ln 2$$

The time constant was specified as $\tau = 300$ s, so

$$t = (300 \text{ s})(0.693) = 208 \text{ s}$$

It will take 208 s, or about 3.5 min, for the oscillations to decay by half after the spring has run down.

ASSESS The time is less than the time constant, which makes sense. The time constant is the time for the amplitude to decay to 37% of its initial value; we are looking for the time to decay to 50% of its initial value, which should be a shorter time. The time to decay to $\frac{1}{2}$ of the initial amplitude, $t = \tau \ln 2$, could be called the *half-life*. We will see this expression again when we work with radioactivity, another example of an exponential decay.

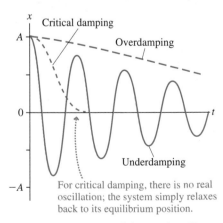

FIGURE 14.24 Motion of damped oscillator systems.

The time constant is a time that is characteristic of a damped oscillation, but there is another time that characterizes the oscillation—the period T. These two times are independent of each other. The period is fixed by the physical properties of the oscillator; the time constant is fixed by drag or other dissipative forces.

The nature of the decay depends on how the size of the damping time constant τ compares to the period T. Figure 14.24 shows three different levels of damping:

- $\tau \gg T$: If the time constant is much longer than the period, the oscillation will continue for many cycles before dying away, as in the above example of the grandfather clock. In engineering language, this is *underdamping*.
- $\tau \sim T$: If the time constant is just about the same size as the period, the system is *critically damped*. The system doesn't really oscillate. It simply moves smoothly to its equilibrium position in the shortest possible time. A shock absorber in a car is designed to have critical damping.
- $\tau \ll T$: If the time constant is much shorter than the period, the system is so "sticky" that the return to equilibrium is very slow. This is *overdamping*.

STOP TO THINK 14.6   Rank in order, from largest to smallest, the time constants τ_A to τ_D of the decays shown in the figure. The scales on all of the graphs are the same.

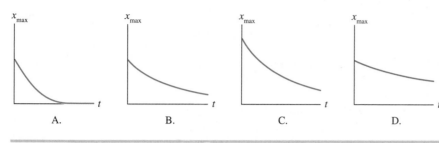

A.   B.   C.   D.

14.7 Driven Oscillations and Resonance

If you jiggle a cup of water, the water sloshes back and forth. This is an example of an oscillator (the water in the cup) subjected to a periodic external force (from your hand). This motion is called a **driven oscillation.**

We can give many examples of driven oscillations. The electromagnetic coil on the back of a loudspeaker cone provides a periodic magnetic force to drive the cone back and forth, causing it to send out sound waves. Earthquakes cause the surface of the earth to move back and forth; this motion causes buildings to oscillate, possibly producing damage or collapse.

Consider an oscillating system that, when left to itself, oscillates at a frequency f_0. We will call this the **natural frequency** of the oscillator. f_0 is simply the frequency of the system if it is displaced from equilibrium and released.

Suppose that this system is now subjected to a *periodic* external force of frequency f_{ext}. This frequency, which is called the **driving frequency,** is completely independent of the oscillator's natural frequency f_0. Somebody or something in the environment selects the frequency f_{ext} of the external force, causing the force to push on the system f_{ext} times every second. The external force causes the oscillation of the system, so it will oscillate at f_{ext}, the driving frequency, not at its natural frequency f_0.

Let's return to the example of the cup of water. If you nudge the cup, you will notice that the water sloshes back and forth at a particular frequency; this is the natural frequency f_0. Now shake the cup at some frequency; this is the driving frequency, f_{ext}. As you shake the cup, the oscillation amplitude of the water depends very sensitively on the frequency f_{ext} of your hand. If the driving frequency is near the natural frequency of the system, the oscillation amplitude may become so large that water splashes out of the cup.

Any driven oscillator will show a similar dependence of amplitude on the driving frequency. Suppose a mass on a spring has a natural frequency $f_0 = 2$ Hz. We can use an external force to push and pull on the mass at frequency f_{ext}, measure the amplitude of the resulting oscillation, and then repeat this over and over for many different driving frequencies. A graph of amplitude versus driving frequency, such as the one in Figure 14.25, is called the oscillator's **response curve.**

At the right and left edges of Figure 14.25, the driving frequency is substantially different from the oscillator's natural frequency. The system oscillates, but its amplitude is very small. The system simply does not respond well to a driving frequency that differs much from f_0. As the driving frequency gets closer and closer to the natural frequency, the amplitude of the oscillation rises dramatically. After all, f_0 is the frequency at which the system "wants" to oscillate, so it is quite happy to respond to a driving frequency near f_0. Hence the amplitude reaches a maximum when the driving frequency matches the system's natural frequency:

Serious sloshing Water in a cup has a natural frequency at which it will slosh back and forth. The same is true of larger bodies of water. Water in Canada's Bay of Fundy would naturally move into or out of the bay with a period of 12 hours. This is nearly equal to the period of the tidal force of the moon; the two daily high tides are 12.5 hours apart. This *resonance,* a close match between the bay's natural frequency and the moon's driving frequency, produces a huge tidal amplitude. Low tide can be as much as 16 m below high tide, leaving boats high and dry.

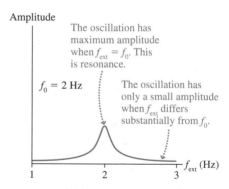

FIGURE 14.25 The response curve shows the amplitude of a driven oscillator at frequencies near its natural frequency $f_0 = 2$ Hz.

Amplitude

$f_0 = 2$ Hz

A lightly damped system has a very tall and very narrow response curve.

$\tau = 50T$

A system with significant damping has a much smaller response.

$\tau = 20T$

$\tau = 5T$

f_{ext} (Hz)

1   2   3

FIGURE 14.26 The response curve becomes higher and narrower as the damping is reduced.

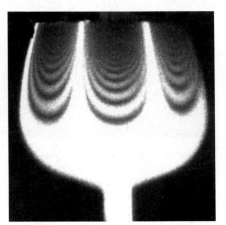

An optical technique called interferometry reveals the oscillations of a wine glass.

$f_{ext} = f_0$. This large-amplitude response to a driving force whose frequency matches the natural frequency of the system is a phenomenon called **resonance.** Within the context of driven oscillations, the natural frequency f_0 is often called the **resonance frequency.**

The amplitude can become exceedingly large when the frequencies match, especially if there is very little damping. Figure 14.26 shows the response curve of the oscillator of Figure 14.25 with different amounts of damping. Three different graphs are plotted, each with a different time constant for damping. The three graphs have damping that ranges from $\tau = 50T$ (light damping) to $\tau = 5T$ (heavy damping).

A classic example of resonance is the use of sound to break a glass. If you have ever tapped on the edge of a wine glass, you know that it has a certain natural frequency at which it will vibrate. If a sound wave of just the right frequency vibrates the glass, it is possible to increase the amplitude of the oscillation to such a degree that the glass will shatter.

Breaking a glass this way isn't easy. First, you need a glass that has very little damping. As you can see from Figure 14.25, the amplitude of a driven oscillation at resonance is higher if the damping is lower. Second, you need very good pitch control; the sound must be at precisely the right frequency, so that it is an excellent match to the natural frequency of the glass. Third, the sound has to be very, very loud, so that the amplitude of the driving force is quite large. With a very nice crystal wine glass, a signal generator set at just the right frequency, and a very powerful amplifier, it is possible to break a glass via resonance with sound.

CONCEPTUAL EXAMPLE 14.2 **Fixing an unwanted resonance**

Railroad cars have a natural frequency at which they rock side to side. This can lead to problems on certain stretches of track that have bumps where the rails join. If the joints alternate sides, with a bump on the left rail and then on the right, a train car moving down the track is bumped one way and then the other. For a train moving at just the right speed, the bumps can cause rocking with amplitude large enough to derail the train. One solution is to replace the track with longer rails with greater distance between bumps. How does this solve the problem?

REASON The large amplitude of oscillation is produced by a resonance between the frequency at which the train car rocks and the frequency at which the car hits the bumps. Lengthening the distance between bumps will decrease the driving frequency of the bumps, eliminating the resonance.

ASSESS Unintended resonances can cause serious difficulties. If the resonance frequency of a building matches the frequency of an earthquake, serious structural damage or failure may occur.

Resonance and Hearing

Resonance in a system means that certain frequencies produce a large response and others do not. This can be very useful. For instance, resonance is responsible for the frequency discrimination of the ear.

As we will see in the next chapter, sound is a vibration in air. Like any vibration, it has a certain frequency. Your ear provides a very sensitive measurement of this frequency—and a resonance in tissues in your ear is responsible. Figure 14.27 on the next page provides an overview of the structures by which sound waves that enter the ear produce vibrations in the cochlea, the coiled, fluid-filled, sound-sensing organ of the inner ear.

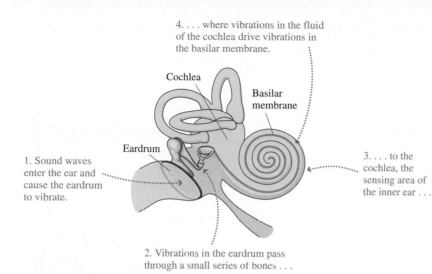

4. . . . where vibrations in the fluid of the cochlea drive vibrations in the basilar membrane.

Cochlea

Basilar membrane

Eardrum

1. Sound waves enter the ear and cause the eardrum to vibrate.

3. . . . to the cochlea, the sensing area of the inner ear . . .

2. Vibrations in the eardrum pass through a small series of bones . . .

FIGURE 14.27 The structures of the ear.

Figure 14.28 shows a very simplified model of the cochlea. As a sound wave travels down the cochlea, it causes a large-amplitude vibration of the basilar membrane at the point where the membrane's natural oscillation frequency matches the sound frequency—a resonance. Lower-frequency sound causes a response farther from the stapes. Sensitive hair cells on the membrane sense the vibration and send nerve signals to your brain. The point on the membrane generating the largest signal tells your brain the frequency of the sound. In this way, your ear and brain can discriminate between two sounds that are very close together in frequency.

We now know a bit about how your ear responds to the vibration of a sound wave—but how does this vibration get from a source to your ear? This is a topic we will consider in the next chapter, when we look at *waves,* oscillations that travel.

▶ **FIGURE 14.28** Resonance plays a role in determining the frequencies of sounds we hear.

To analyze the cochlea, we imagine the spiral structure unrolled, with the basilar membrane separating two fluid-filled chambers.

The stapes, the last of the small bones, transfers vibrations into fluid in the cochlea.

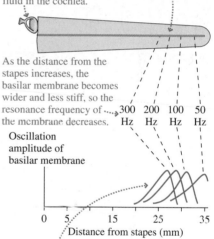

As the distance from the stapes increases, the basilar membrane becomes wider and less stiff, so the resonance frequency of the membrane decreases.

300   200   100   50
Hz   Hz   Hz   Hz

Oscillation amplitude of basilar membrane

0   5   15   25   35
Distance from stapes (mm)

Sounds of different frequencies cause different responses in the basilar membrane.

The goal of Chapter 14 has been to understand systems that oscillate with simple harmonic motion.

GENERAL PRINCIPLES

Restoring Forces

SHM occurs when a **linear restoring force** acts to return a system to an equilibrium position.

Mass on spring

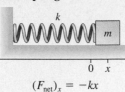

$$(F_{net})_x = -kx$$

The frequency of a mass on a spring depends on the mass and the spring constant.

$$f = \frac{1}{2\pi}\sqrt{\frac{k}{m}}$$

Pendulum

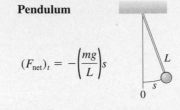

$$(F_{net})_t = -\left(\frac{mg}{L}\right)s$$

The frequency of a pendulum depends on the length and the acceleration due to gravity.

$$f = \frac{1}{2\pi}\sqrt{\frac{g}{L}}$$

Energy

If there is no friction or dissipation, kinetic and potential energy are alternately transformed into each other in SHM, but the total mechanical energy $E = K + U$ is conserved.

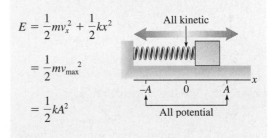

$$E = \frac{1}{2}mv_x^2 + \frac{1}{2}kx^2$$

$$= \frac{1}{2}mv_{max}^2$$

$$= \frac{1}{2}kA^2$$

IMPORTANT CONCEPTS

Oscillation

An **oscillation** is a repetitive motion about an equilibrium position. The **amplitude** A is the maximum displacement from equilibrium. The period T is the time for one cycle. We may also characterize an oscillation by its frequency f.

Simple Harmonic Motion (SHM)

SHM is an oscillation that is described by a sinusoidal function. All systems that undergo SHM can be described by the same functional forms.

Position-versus-time is a cosine function.

Velocity-versus-time is an inverted sine function.

Acceleration-versus-time is an inverted cosine function.

$$x(t) = A\cos(2\pi ft)$$

$$v_x(t) = -v_{max}\sin(2\pi ft)$$
$$v_{max} = 2\pi fA$$

$$a_x(t) = -a_{max}\cos(2\pi ft)$$
$$a_{max} = (2\pi f)^2 A$$

APPLICATIONS

Damping

Simple harmonic motion with damping (due to drag) decreases in amplitude over time. The **time constant** τ determines how quickly the amplitude decays.

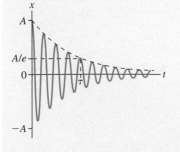

Resonance

A system that oscillates has a **natural frequency** of oscillation f_0. **Resonance** occurs if the system is driven with a frequency f_{ext} that matches this natural frequency. This may produce a large amplitude of oscillation.

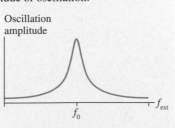

Physical pendulum

A **physical pendulum** is a pendulum with mass distributed along its length. The frequency depends on the position of the center of gravity and the moment of inertia.

The motion of legs during walking can be described using a physical pendulum model.

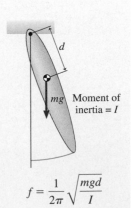

$$f = \frac{1}{2\pi}\sqrt{\frac{mgd}{I}}$$

For instructor-assigned homework, go to
www.masteringphysics.com

Problem difficulty is labeled as I (straightforward) to II (challenging).

Problems labeled INT integrate significant material from earlier chapters; BIO are of biological or medical interest.

QUESTIONS

Conceptual Questions

1. Give three examples of *oscillatory* motion. (Note that circular motion is similar to, but not the same as oscillatory motion.)
2. A person's heart rate is given in beats per minute. Is this a period or a frequency?
3. Figure Q14.3 shows the position-versus-time graph of a particle in SHM.
   a. At what time or times is the particle moving to the right at maximum speed?
   b. At what time or times is the particle moving to the left at maximum speed?
   c. At what time or times is the particle instantaneously at rest?

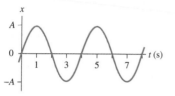

FIGURE Q14.3

4. A block oscillating on a spring has an amplitude of 20 cm. What will be the amplitude if its total mechanical energy is doubled? Explain.
5. A block oscillating on a spring has a maximum speed of 20 cm/s. What will be the block's maximum speed if its total mechanical energy is doubled? Explain.
6. A block oscillating on a spring has 2.0 J of mechanical energy. What will be the energy if its amplitude is doubled? Explain.
7. A block oscillating on a spring has a maximum speed of 30 cm/s. What will be the block's maximum speed if the initial elongation of the spring is doubled?
8. For the graph in Figure Q14.8, determine the frequency f and the oscillation amplitude A.

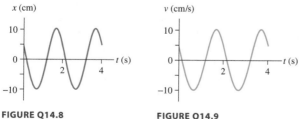

FIGURE Q14.8   **FIGURE Q14.9**

9. For the graph in Figure Q14.9, determine the frequency f and the oscillation amplitude A.

10. A block oscillating on a spring has period $T = 2.0$ s.
    a. What is the period if the block's mass is doubled? Explain.
    b. What is the period if the value of the spring constant is quadrupled?
    c. What is the period if the oscillation amplitude is doubled while m and k are unchanged?
    Note: You do not know values for either m or k. Do *not* assume any particular values for them. The required analysis involves thinking about ratios.
11. A pendulum on Planet X, where the value of g is unknown, oscillates with a period of 2.0 s. What is the period of this pendulum if:
    a. Its mass is doubled?
    b. Its length is doubled?
    c. Its oscillation amplitude is doubled?
    Note: You do not know the values of m, L, or g, so do not assume any specific values.
12. As we saw in Chapter 6, the acceleration due to gravity is slightly less in Denver than in Miami. If a pendulum clock keeps perfect time in Miami, will it run fast or slow in Denver? Explain.
13. Sprinters push off from the ball of
    BIO their foot, then bend their knee to bring their foot up close to the body as they swing their leg forward for the next stride. Why is this an effective strategy for running fast?

14. Gibbons move through the trees by
    BIO swinging from successive hand-holds, as we have seen. To increase their speed, gibbons may bring their legs close to their bodies. How does this help them move more quickly?

15. Describe the difference between τ and T. Don't just *name* them; say what is different about the physical concepts that they represent.
16. What is the difference between the driving frequency and the natural frequency of an oscillator?
17. A person driving a truck on a "washboard" road, one with regularly spaced bumps, notices an interesting effect: When the truck travels at low speed, the amplitude of the vertical motion of the car is small. If the truck's speed is increased, the amplitude of the vertical motion also increases, until it becomes quite unpleasant. But if the speed is increased yet further, the amplitude decreases, and at high speeds the amplitude of the vertical motion is small again. Explain what is happening.

Multiple-Choice Questions

18. | A spring has an unstretched length of 20 cm. A 100 g mass hanging from the spring stretches it to an equilibrium length of 30 cm.
    a. Suppose the mass is pulled down to where the spring's length is 40 cm. When it is released, it begins to oscillate. What is the amplitude of the oscillation?
       A. 5.0 cm   B. 10 cm   C. 20 cm   D. 40 cm
    b. For the data given above, what is the frequency of the oscillation?
       A. 0.10 Hz   B. 0.62 Hz   C. 1.6 Hz   D. 10 Hz
    c. Suppose this experiment were done on the moon, where the acceleration of gravity is approximately 1/6 of that on the earth. How would this change the frequency of the oscillation?
       A. The frequency would decrease.
       B. The frequency would increase.
       C. The frequency would stay the same.

19. ‖ Figure Q14.19 represents the motion of a mass on a spring.

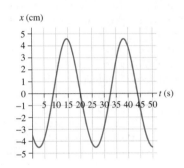

FIGURE Q14.19

    a. What is the period of this oscillation?
       A. 12 s   B. 24 s   C. 36 s
       D. 48 s   E. 50 s
    b. What is the amplitude of the oscillation?
       A. 1.0 cm   B. 2.5 cm   C. 4.5 cm
       D. 5.0 cm   E. 9.0 cm
    c. What is the position of the mass at time $t = 30$ s?
       A. −4.5 cm   B. −2.0 cm   C. 0.0 cm
       D. 4.5 cm   E. 30 cm
    d. When is the first time the velocity of the mass is zero?
       A. 0 s   B. 2 s   C. 8 s
       D. 10 s   E. 13 s
    e. At which of these times does the kinetic energy have its maximum value?
       A. 0 s   B. 8 s   C. 13 s
       D. 26 s   E. 30 s

20. | A ball of mass m oscillates on a spring with spring constant $k = 200$ N/m. The ball's position is $x = (0.350 \text{ m}) \cos 15.0t$, with t measured in seconds.
    a. What is the amplitude of the ball's motion?
       A. 0.175 m   B. 0.350 m   C. 0.700 m
       D. 7.50 m   E. 15.0 m
    b. What is the frequency of the ball's motion?
       A. 0.35 Hz   B. 2.39 Hz   C. 5.44 Hz
       D. 6.28 Hz   E. 15.0 Hz

    c. What is the value of the mass m?
       A. 0.45 kg   B. 0.89 kg   C. 1.54 kg
       D. 3.76 kg   E. 6.33 kg
    d. What is the total mechanical energy of the oscillator?
       A. 1.65 J   B. 3.28 J   C. 6.73 J
       D. 10.1 J   E. 12.2 J
    e. What is the ball's maximum speed?
       A. 0.35 m/s   B. 1.76 m/s   C. 2.60 m/s
       D. 3.88 m/s   E. 5.25 m/s

21. ‖ Figure Q14.21 shows a block on a frictionless surface attached to a spring. The block is pulled out to position $x_i = 20$ cm, then given a "kick" so that it moves to the right with speed v_i. The block then oscillates with an amplitude of 50 cm.

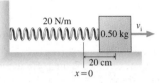

FIGURE Q14.21

    a. What is the oscillation frequency?
       A. 0.025 Hz   B. 0.32Hz   C. 0.81 Hz
       D. 1.0 Hz   E. 6.4 Hz
    b. What is the total mechanical energy of the oscillator?
       A. 1.7 J   B. 2.5 J   C. 4.6 J
       D. 6.0 J   E. 12 J
    c. What was the initial speed v_i?
       A. 2.0 m/s   B. 2.9 m/s   C. 3.3 m/s
       D. 5.1 m/s   E. 12 m/s

22. | A heavy brass ball is used to make a pendulum with a period of 5.5 s. How long is the cable that connects the pendulum ball to the ceiling?
    A. 4.7 m   B. 6.2 m   C. 7.5 m   D. 8.7 m

23. | Suppose you travel to the moon, and you take with you two timepieces: a pendulum clock and a wristwatch that runs with a wheel and a mainspring. (The wheel and spring works, essentially, like a mass on a spring, but the wheel rotates back and forth rather than moving up and down.) Which will keep good time on the moon?
    A. Only the pendulum clock.
    B. Only the wristwatch.
    C. Both timepieces.
    D. Neither timepiece.

24. ‖| Very loud sounds can damage hearing by injuring the
BIO vibration-sensing hair cells on the basilar membrane. Suppose a person has injured hair cells on a segment of the basilar membrane close to the stapes. What type of sound is most likely to have produced this particular pattern of damage?
    A. Loud music with a mix of different frequencies
    B. A very loud, high-frequency sound
    C. A very loud, low-frequency sound

PROBLEMS

Section 14.1 Equilibrium and Oscillation

Section 14.2 Linear Restoring Forces and Simple Harmonic Motion

1. | When a guitar string plays the note "A," the string vibrates at 440 Hz. What is the period of the vibration?

2. || In the aftermath of an intense earthquake, the earth as a whole "rings" with a period of 54 minutes. What is the frequency (in Hz) of this oscillation?

3. | You hear a faucet dripping steadily in your bathroom sink and decide to measure the frequency of the drips. You start a stopwatch on hearing a drip. Calling that the first drip, you count succeeding drips and stop the watch on the fifteenth drip. The watch reads 50 s. What is the frequency?

4. | In taking your pulse, you count 75 heartbeats in 1 min.
BIO What are the period (in s) and frequency (in Hz) of your heart's oscillations?

5. | Make a table with 3 columns and 8 rows. In row 1, label the columns θ (°), θ (rad), and $\sin\theta$. In the left column, starting in row 2, write 0, 2, 4, 6, 8, 10, and 12.
   a. Convert each of these angles, in degrees, to radians. Put the results in column 2. Show four decimal places.
   b. Calculate the sines. Put the results, showing four decimal places, in column 3.
   c. What is the first angle for which θ and $\sin\theta$ differ by more than 0.0010?
   d. Over what range of angles does the small-angle approximation appear to be valid?

Section 14.3 Describing Simple Harmonic Motion

6. || An air-track glider attached to a spring oscillates between the 10 cm mark and the 60 cm mark on the track. The glider completes 10 oscillations in 33 s. What are the (a) period, (b) frequency, (c) amplitude, and (d) maximum speed of the glider?

7. ||| An air-track glider is attached to a spring. The glider is pulled to the right and released from rest at $t = 0$ s. It then oscillates with a period of 2.0 s and a maximum speed of 40 cm/s.
   a. What is the amplitude of the oscillation?
   b. What is the glider's position at $t = 0.25$ s?

8. | What are the (a) amplitude and (b) frequency of the oscillation shown in Figure P14.8?

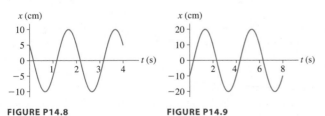

FIGURE P14.8   **FIGURE P14.9**

9. | What are the (a) amplitude and (b) frequency of the oscillation shown in Figure P14.9?

10. | An object in simple harmonic motion has an amplitude of 4.0 cm and a frequency of 2.0 Hz. Draw a position graph showing two cycles of the motion.

11. | An object in simple harmonic motion has an amplitude of 6.0 cm and a frequency of 0.50 Hz. Draw a position graph showing two cycles of the motion.

12. || During an earthquake, the top of a building oscillates with an amplitude of 30 cm at 1.2 Hz. What are the magnitudes of (a) the maximum displacement, (b) the maximum velocity, and (c) the maximum acceleration of the top of the building?

13. | Some passengers on an ocean cruise may suffer from motion sickness as the ship rocks back and forth on the waves. At one position on the ship, passengers experience a vertical motion of amplitude 1 m with a period of 15 s.
   a. To one significant figure, what is the maximum acceleration of the passengers during this motion?
   b. What fraction is this of g?

14. || A child standing at the edge of a 6.0-m-diameter merry-go-
INT round travels counterclockwise with a period of 10 s. Draw graphs showing just the x-component of (a) the child's position, (b) the child's velocity, and (c) the acceleration of the child, as functions of time for two times around the ride. Be sure to supply appropriate scales on the vertical axes.

Section 14.4 Energy in Simple Harmonic Motion

15. ||| a. When the displacement of a mass on a spring is $\frac{1}{2}A$, what fraction of the mechanical energy is kinetic energy and what fraction is potential energy?
    b. At what displacement, as a fraction of A, is the energy half kinetic and half potential?

16. ||| A 1.0 kg block is attached to a spring with spring constant 16 N/m. While the block is sitting at rest, a student hits it with a hammer and almost instantaneously gives it a speed of 40 cm/s. What are
    a. The amplitude of the subsequent oscillations?
    b. The block's speed at the point where $x = \frac{1}{2}A$?

17. || A block attached to a spring with unknown spring constant oscillates with a period of 2.00 s. What is the period if
    a. The mass is doubled?
    b. The mass is halved?
    c. The amplitude is doubled?
    d. The spring constant is doubled?
    Parts a to d are independent questions, each referring to the initial situation.

18. || A 200 g air-track glider is attached to a spring. The glider is pushed in 10.0 cm and released. A student with a stopwatch finds that 10 oscillations take 12.0 s. What is the spring constant?

19. | The position of a 50 g oscillating mass is given by $x(t) = (2.0 \text{ cm}) \cos 10t$, where t is in seconds. Determine:
    a. The amplitude.   b. The period.
    c. The spring constant.   d. The maximum speed.
    e. The total energy.   f. The velocity at $t = 0.40$ s.

20. || A 200 g mass attached to a horizontal spring oscillates at a frequency of 2.0 Hz. At one instant, the mass is at $x = 5.0$ cm and has $v_x = -30$ cm/s. Determine:
    a. The period.   b. The amplitude.
    c. The maximum speed.   d. The total energy.

21. ‖‖ A 507 g mass oscillates with an amplitude of 10.0 cm on a spring whose spring constant is 20.0 N/m. Determine:
    a. The period.
    b. The maximum speed.
    c. The total energy.
22. ‖‖ A 300 g oscillator has a speed of 95.4 cm/s when its displacement is 3.00 cm and 71.4 cm/s when its displacement is 6.00 cm. What is the oscillator's maximum speed?

Section 14.5 Pendulum Motion

23. | A mass on a string of unknown length oscillates as a pendulum with a period of 4.00 s. What is the period if
    a. The mass is doubled?
    b. The string length is doubled?
    c. The string length is halved?
    d. The amplitude is doubled?
    Parts a to d are independent questions, each referring to the initial situation.
24. ‖ A 200 g ball is tied to a string. It is pulled to an angle of 8.00° and released to swing as a pendulum. A student with a stopwatch finds that 10 oscillations take 12.0 s. How long is the string?
25. ‖‖ The angle of a pendulum is given by $\theta(t) = (0.10 \text{ rad})\cos 5t$, where t is in seconds. Determine:
    a. The amplitude.   b. The frequency.
    c. The length of the string.   d. The angle at $t = 2.0$ s.
26. | It is said that Galileo discovered a basic principle of the pendulum—that the period is independent of the amplitude— by using his pulse to time the period of swinging lamps in the cathedral as they swayed in the breeze. Suppose that one oscillation of a swinging lamp takes 5.5 s. How long is the lamp chain?
27. ‖ The acceleration due to gravity on the moon is 1.62 m/s². What is the length of a pendulum whose period on the moon matches the period of a 2.00-m-long pendulum on the earth?
28. | Astronauts on the first trip to Mars take along a pendulum that has a period on earth of 1.50 s. The period on Mars turns out to be 2.45 s. What is the Martian acceleration due to gravity?
29. ‖ A building is being knocked down with a wrecking ball, which is a big metal sphere that swings on a 10-m-long cable. You are (unwisely!) standing directly beneath the point from which the wrecking ball is hung when you notice that the ball has just been released and is swinging directly toward you. How much time do you have to move out of the way?
30. | Interestingly, there have been several studies using cadav-
BIO ers to determine the moment of inertia of human body parts by letting them swing as a pendulum about a joint. In one study, the center of gravity of a 5.0 kg lower leg was found to be 18 cm from the knee. When pivoted at the knee and allowed to swing, the oscillation frequency was 1.6 Hz. What was the moment of inertia of the lower leg?
31. ‖ A pendulum clock keeps time by the swinging of a uniform solid rod pivoted at one end. The angular position of the rod is given by $\theta(t) = (0.175 \text{ rad})\sin(\pi t)$, where t is in seconds.
    a. What is the angular position of the rod at $t = 0.250$ s?
    b. What is the period of oscillation?
    c. How long is the rod?
32. | A thin, circular hoop with a radius of 0.22 m is hanging from its rim on a nail. When pulled to the side and released,

the hoop swings back and forth as a physical pendulum. The moment of inertia of a hoop for a rotational axis passing through its edge is $I = 2MR^2$. What is the period of oscillation of the hoop?

Section 14.6 Damped Oscillations

33. ‖ The amplitude of an oscillator decreases to 36.8% of its initial value in 10.0 s. What is the value of the time constant?
34. ‖ A gust of wind hits a flagpole and starts it oscillating. After 10 s, the amplitude of its motion has decreased to 25% of its initial value. What is the value of the time constant?
35. ‖ Calculate and draw an accurate displacement graph from $t = 0$ s to $t = 10$ s of a damped oscillator having a frequency of 1.0 Hz and a time constant of 4.0 s.
36. ‖ A small earthquake starts a lamppost vibrating back and forth. The amplitude of the vibration of the top of the lamppost is 6.5 cm at the moment the quake stops, and 8.0 s later it is 1.8 cm.
    a. What is the time constant for the damping of the oscillation?
    b. What was the amplitude of the oscillation 4.0 s after the quake stopped?

Section 14.7 Driven Oscillations and Resonance

37. | A 25 kg child sits on a 2.0-m-long rope swing. You are going to give the child a small, brief push at regular intervals. If you want to increase the amplitude of her motion as quickly as possible, how much time should you wait between pushes?
38. ‖ Vision is blurred if the head is vibrated at 29 Hz because the
BIO vibrations are resonant with the natural frequency of the eyeball held by the musculature in its socket. If the mass of the eyeball is 7.5 g, a typical value, what is the effective spring constant of the musculature attached to the eyeball?

General Problems

39. | A spring has an unstretched length of 12 cm. When an 80 g ball is hung from it, the length increases by 4.0 cm. Then the ball is pulled down another 4.0 cm and released.
    a. What is the spring constant of the spring?
    b. What is the period of the oscillation?
    c. Draw a position-versus-time graph showing the motion of the ball for three cycles of the oscillation. Let the equilibrium position of the ball be $y = 0$. Be sure to include appropriate units on the axes so that the period and the amplitude of the motion can be determined from your graph.
40. | A 0.40 kg ball is attached to a spring with spring constant 12 N/m. If the ball is pulled down 0.20 m from the equilibrium position and released, what is its maximum speed while it oscillates?
41. | A spring is hanging from the ceiling. Attaching a 500 g physics book to the spring causes it to stretch 20.0 cm in order to come to equilibrium.
    a. What is the spring constant?
    b. From equilibrium, the book is pulled down 10.0 cm and released. What is the period of oscillation?
    c. What is the book's maximum speed? At what position or positions does it have this speed?

42. ⫴ A spring with spring constant 15.0 N/m hangs from the ceiling. A ball is attached to the spring and allowed to come to rest. It is then pulled down 6.00 cm and released. If the ball makes 30 oscillations in 20.0 s, what are its (a) mass and (b) maximum speed?

43. ⫴ A spring is hung from the ceiling. When a coffee mug is attached to its end, the spring stretches 2.0 cm before reaching its new equilibrium length. The mug is then pulled down slightly and released. What is the frequency of oscillation?

44. ⫴ On your first trip to Planet X you happen to take along a 200 g mass, a 40.0-cm-long spring, a meter stick, and a stopwatch. You're curious about the acceleration due to gravity on Planet X, where ordinary tasks seem easier than on earth, but you can't find this information in your Visitor's Guide. One night you suspend the spring from the ceiling in your room and hang the mass from it. You find that the mass stretches the spring by 31.2 cm. You then pull the mass down 10.0 cm and release it. With the stopwatch you find that 10 oscillations take 14.5 s. Can you now satisfy your curiosity?

45. | An object oscillating on a spring has the position graph shown in Figure P14.45. Draw a position graph if the following changes are made.
    a. The amplitude is halved and the frequency is halved.
    b. The mass is quadrupled.
    Parts a and b are independent questions, each starting from the graph shown.

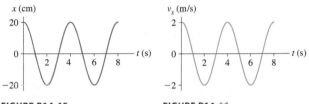

FIGURE P14.45   **FIGURE P14.46**

46. | An object oscillating on a spring has the velocity graph shown in Figure P14.46. Draw a velocity graph if the following changes are made.
    a. The amplitude is doubled and the frequency is halved.
    b. The mass is quadrupled.
    Parts a and b are independent questions, each starting from the graph shown.

47. ‖ The two graphs in Figure P14.47 are for two different vertical mass/spring systems.
    a. What is the frequency of system A? What is the first time at which the mass has maximum speed while traveling in the upward direction?
    b. What is the period of system B? What is the first time at which the mechanical energy is all potential?
    c. If both systems have the same mass, what is the ratio k_A/k_B of their spring constants?

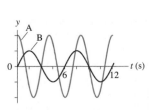

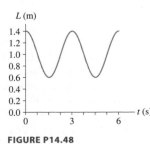

FIGURE P14.47   **FIGURE P14.48**

48. | Astronauts in space cannot weigh themselves by standing on a bathroom scale. Instead, they determine their mass by oscillating on a large spring, as we saw on page 465. Suppose an astronaut attaches one end of a large spring, with spring constant 240 N/m, to her belt and the other end to a hook on the wall of the space capsule. A fellow astronaut then pulls her away from the wall and releases her. The spring's length as a function of time is shown in Figure P14.48. What is her mass?

49. ⫴ A 100 g ball attached to a spring with spring constant 2.50 N/m oscillates horizontally on a frictionless table. Its velocity is 20.0 cm/s when $x = -5.00$ cm.
    a. What is the amplitude of oscillation?
    b. What is the speed of the ball when $x = 3.00$ cm?

50. ‖ The ultrasonic transducer used in a medical ultrasound imaging device is a very thin disk ($m = 0.10$ g) driven back and forth in SHM at 1.0 MHz by an electromagnetic coil.
    a. The maximum restoring force that can be applied to the disk without breaking it is 40,000 N. What is the maximum oscillation amplitude that won't rupture the disk?
    b. What is the disk's maximum speed at this amplitude?

51. | A 2 g spider is dangling at the end of a silk thread. You can make the spider bounce up and down on the thread by tapping lightly on his feet with a pencil. You soon discover that you can give the spider the largest amplitude on his little bungee cord if you tap exactly once every 3 seconds. What is the approximate spring constant of the spider's silk thread?

52. ‖ A compact car has a mass of 1200 kg. Assume that the car has one spring on each wheel, that the springs are identical, and that the mass is equally distributed over the four springs.
    a. What is the spring constant of each spring if the empty car bounces up and down 2.0 times each second?
    b. What will be the car's oscillation frequency while carrying four 70 kg passengers?

53. | A 500 g air-track glider attached to a spring with spring constant 10 N/m is sitting at rest on a frictionless air track. A 250 g glider is pushed toward it from the far end of the track at a speed of 120 cm/s. It collides with and sticks to the 500 g glider. What are the amplitude and period of the subsequent oscillations?

54. ⫼ A 1.00 kg block is attached to a horizontal spring with spring constant 2500 N/m. The block is at rest on a frictionless surface. A 10.0 g bullet is fired into the block, in the face opposite the spring, and sticks.
    a. What was the bullet's speed if the subsequent oscillations have an amplitude of 10.0 cm?
    b. Could you determine the bullet's speed by measuring the oscillation frequency? If so, how? If not, why not?

55. ⫼ Figure P14.55 shows two springs, each with spring constant 20 N/m, connecting a 2.5 kg block to two walls. The block slides on a frictionless surface. If the block is displaced from equilibrium, it will undergo simple harmonic motion. What is the frequency of that motion?

FIGURE P14.55

56. ‖ Bungee Man is a superhero who does super deeds with the help of Super Bungee cords. The Super Bungee cords act like ideal springs no matter how much they are stretched. One day, Bungee Man stopped a school bus that had lost its brakes by hooking one end of a Super Bungee to the rear of the bus as it passed him, planting his feet, and holding on to the other end of the Bungee until the bus came to a halt. (Of course, he then had to quickly release the Bungee before the bus came flying back at him.) The mass of the bus, including passengers, was 12,000 kg, and its speed was 21.2 m/s. The bus came to a stop in 50.0 m.
   a. What was the spring constant of the Super Bungee?
   b. How much time after the Super Bungee was attached did it take the bus to stop?

57. ‖‖‖ Two 50 g blocks are held 30 cm above a table. As shown in Figure P14.57, one of them is just touching a 30-cm-long spring. The blocks are released at the same time. The block on the left hits the table at exactly the same instant as the block on the right first comes to an instantaneous rest. What is the spring constant?

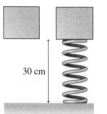

30 cm

FIGURE P14.57

58. ‖ The earth's acceleration due to gravity varies from 9.78 m/s² at the equator to 9.83 m/s² at the poles. A pendulum whose length is precisely 1.000 m can be used to measure g. Such a device is called a *gravimeter*.
   a. How long do 100 oscillations take at the equator?
   b. How long do 100 oscillations take at the north pole?
   c. Suppose you take your gravimeter to the top of a high mountain peak near the equator. There you find that 100 oscillations take 201 s. What is g on the mountain top?

59. ‖‖‖ A pendulum clock has a heavy bob supported on a very thin steel rod that is 1.00000 m long at 20°C.
   a. To 6 significant figures, what is the clock's period? Assume that g is 9.80 m/s² exactly.
   b. To 6 significant figures, what is the period if the temperature increases by 10°C?
   c. The clock keeps perfect time at 20°C. At 30°C, after how many hours will the clock be off by 1.0 s?

60. ‖‖‖ A pendulum consists of a massless, rigid rod with a mass at one end. The other end is pivoted on a frictionless pivot so that it can turn through a complete circle. The pendulum is inverted, so the mass is directly above the pivot point, then released. The speed of the mass as it passes through the lowest point is 5.0 m/s. If the pendulum later undergoes small-amplitude oscillations at the bottom of the arc, what will the frequency be?

61. ‖ Two side-by-side pendulum clocks have heavy bobs at the ends of rigid, very lightweight arms. One pendulum has a 38.8-cm-long rod, the other a 24.8-cm-long rod. Each clock makes one tick for each complete swing of its pendulum.
   a. Determine the frequencies and periods of the two clocks.
   b. Because the two pendulums have different frequencies, their ticks are usually "out of step." However, you notice that they do get back into step (tick at the same instant) at regular intervals. How much time elapses between one such event and the next?
   c. The getting-into-step phenomenon is, itself, periodic. What is the frequency of this phenomenon? Can you see a relationship between its frequency and the frequencies of the two clocks?

62. ‖‖‖ The 15 g head of a bobble-head doll oscillates in SHM at a frequency of 4.0 Hz.
   a. What is the spring constant of the spring on which the head is mounted?
   b. Suppose the head is pushed 2.0 cm against the spring, then released. What is the head's maximum speed as it oscillates?
   c. The amplitude of the head's oscillations decreases to 0.50 cm in 4.0 s. What is the head's time constant?

63. ‖ An oscillator with a mass of 500 g and a period of 0.50 s has an amplitude that decreases by 2.0% during each complete oscillation. If the initial amplitude is 10 cm, what will be the amplitude after 25 oscillations?

64. ‖ An infant's toy has a 120 g wooden animal hanging from a spring. If pulled down gently, the animal oscillates up and down with a period of 0.50 s. His older sister pulls the spring a bit more than intended. She pulls the animal 30 cm below its equilibrium position, then lets go. The animal flies upward and detaches from the spring right at the animal's equilibrium position. If the animal does not hit anything on the way up, how far above its equilibrium position will it go?

65. ‖‖‖‖ A 200 g oscillator in a vacuum chamber has a frequency of 2.0 Hz. When air is admitted, the oscillation decreases to 60% of its initial amplitude in 50 s. How many oscillations will have been completed when the amplitude is 30% of its initial value?

66. ‖ While seated on a tall bench, extend your lower leg a small amount and then let it swing freely about your knee joint, with no muscular engagement. It will oscillate as a damped pendulum. Figure P14.66 is a graph of the lower leg angle versus time in such an experiment.
   a. Is this an overdamped, underdamped, or critically damped oscillation?
   b. Estimate the time constant for this oscillation.

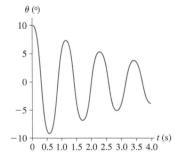

FIGURE P14.66

67. ‖ A 2.0 kg block oscillates up and down on a spring with spring constant 220 N/m. Its initial amplitude is 15 cm. If the time constant for damping of the oscillation is 3.0 s, how much mechanical energy has been dissipated from the block-spring system after 6.0 s?

Passage Problems

Web Spiders and Oscillations BIO

All spiders have special organs that make them exquisitely sensitive to vibrations. Web spiders detect vibrations of their web to determine what has landed in their web, and where.

In fact, spiders carefully adjust the tension of strands to "tune" their web. Suppose an insect lands and is trapped in a web. The silk of the web serves as the spring in a spring-mass system while the body of the insect is the mass. The frequency of oscillation depends on the restoring force of the web and the mass of the insect. Spiders respond more quickly to larger—and therefore more valuable—prey, which they can distinguish by the web's oscillation frequency.

Suppose a 12 mg fly lands in the center of a horizontal spider's web, causing the web to sag by 3.0 mm.

68. | Assuming that the web acts like a spring, what is the spring constant of the web?
   A. 0.039 N/m   B. 0.39 N/m   C. 3.9 N/m   D. 39 N/m

69. | Modeling the motion of the fly on the web as a mass on a spring, at what frequency will the web vibrate when the fly hits it?
   A. 0.91 Hz   B. 2.9 Hz   C. 9.1 Hz   D. 29 Hz

70. | If the web were vertical rather than horizontal, how would he frequency of oscillation be affected?
   A. The frequency would be higher.
   B. The frequency would be lower.
   C. The frequency would be the same.

71. | Spiders are more sensitive to oscillations at higher frequencies. For example, a low-frequency oscillation at 1 Hz can be detected for amplitudes down to 0.1 mm, but a high-frequency oscillation at 1 kHz can be detected for amplitudes as small as 0.1 μm. For these low- and high-frequency oscillations, we can say that
   A. The maximum acceleration of the low-frequency oscillation is greater.
   B. The maximum acceleration of the high-frequency oscillation is greater.
   C. The maximum accelerations of the two oscillations are approximately equal.

STOP TO THINK ANSWERS

Stop to Think 14.1: C. The frequency is inversely proportional to the period, so a shorter period implies a larger frequency.

Stop to Think 14.2: D. The restoring force is proportional to the displacement. If the displacement from equilibrium is doubled, the force is doubled as well.

Stop to Think 14.3: A. The maximum speed occurs when the mass passes through its equilibrium position.

Stop to Think 14.4: $f_D > f_C = f_B > f_A$. The frequency is determined by the ratio of k to m.

Stop to Think 14.5: C. The increase in length will cause the frequency to decrease and thus the period will increase. The time between ticks will increase, so the clock will run slow.

Stop to Think 14.6: $\tau_D > \tau_B = \tau_C > \tau_A$. The time constant is the time to decay to 37% of the initial height. The time constant is independent of the initial height.

15

TRAVELING WAVES AND SOUND

This bat's ears are much more prominent than its eyes. It would appear that hearing is a much more important sense than sight for bats. How does a bat use sound waves to locate prey?

Looking Ahead ▶▶

The goal of Chapter 15 is to learn the basic properties of traveling waves. In this chapter you will learn to:

- ▶ Understand how a wave travels through a medium.
- ▶ Recognize the properties of sinusoidal waves.
- ▶ Discern the important characteristics of sound and light waves.
- ▶ Apply energy and power concepts to waves.
- ▶ Understand the decibel scale for loudness of sound.
- ▶ Use the Doppler effect to find the speed of wave sources and observers.

Looking Back ◀◀

The material in this chapter is closely related to the concept of simple harmonic motion. Please review:

- ◀ Section 14.2 The properties and mathematical description of simple harmonic motion.

The world is full of waves, although you may not always be aware of them. The "waviness" of a water wave is readily apparent, from the ripples on a pond to ocean waves large enough to surf. It's less apparent that sound and light are also waves because their wave properties are discovered only by careful observations and experiments.

Waves are closely related to oscillations, which we learned about in the previous chapter. If you hold your hand on your throat while you speak, you can feel the vibrations of your vocal cords. This vibration produces a *traveling wave*, a sound wave that moves through the air to the ears of those around you.

In our treatment of waves we will pay special attention to sound waves, but we will consider other traveling waves as well. Light waves travel from the sun to the earth. A sudden fracture in the earth's crust sends out a shock wave that is felt far away as an earthquake. To understand phenomena such as these we need both new models and new mathematics. Our overarching goal in the next two chapters is to understand the properties and characteristics that are common to waves of all types. In other words, we want to find the "essence of waviness" that all waves possess, to have a single model that explains all of the different types of waves we will encounter.

15.1 The Wave Model

The *particle model* that we have been using since Chapter 1 allowed us to simplify the treatment of motion of complex objects by considering them to be particles. Balls, cars, and rockets obviously differ from one another, but the general features of their motions are well described by treating them as particles. As we saw in Chapter 3, a ball or a rock or a car flying through the air will undergo the same motion. The particle model helps us understand this underlying simplicity.

In this chapter we will introduce the basic properties of waves with a **wave model** that emphasizes those aspects of wave behavior common to all waves. Although sound waves, water waves, and radio waves are clearly different, the wave model will allow us to understand many of the important features they share.

The wave model is built around the idea of a **traveling wave,** which is an organized disturbance that travels with a well-defined wave speed. This definition seems straightforward, but several new terms must be understood to gain a complete understanding of the concept of a traveling wave.

Mechanical Waves

Mechanical waves are waves that involve the motion of a substance through which they move, the **medium.** For example, the medium of a water wave is the water, the medium of a sound wave is the air, and the medium of a wave on a stretched string is the string.

As a wave passes through a medium, the atoms that make up the medium are displaced from equilibrium, much like the initial perturbation that begins an oscillation. This is a **disturbance** of the medium. The water ripples of Figure 15.1 are a disturbance of the water's surface.

A wave disturbance is created by a *source*. The source of a wave might be a rock thrown into water, your hand plucking a stretched string, or an oscillating loudspeaker cone pushing on the air. Once created, the disturbance travels outward through the medium at the **wave speed** v. This is the speed with which a ripple moves across the water or a pulse travels down a string.

The disturbance propagates through the medium, and a wave does transfer *energy,* but **the medium as a whole does not travel!** The ripples on the pond (the disturbance) move outward from the splash of the rock, but there is no outward flow of water. Likewise, the particles of a string oscillate up and down but do not move in the direction of a pulse traveling along the string. **A wave transfers energy, but it does not transfer any material or substance outward from the source.**

Electromagnetic and Matter Waves

Mechanical waves require a medium, but there are waves that do not. Two important types of such waves are electromagnetic waves and matter waves.

Electromagnetic waves are waves of an *electromagnetic field*. Electromagnetic waves are very diverse, including visible light, radio waves, microwaves, and x rays. Electromagnetic waves require no material medium and can travel through a vacuum; light can travel through space, though sound cannot. At this point, we have not defined what an "electromagnetic field" is, so we won't worry about the precise nature of what is "waving" in electromagnetic waves. The wave model can describe many of the important aspects of these waves without a detailed description of their exact nature. We'll look more closely at electromagnetic waves in Chapter 25, once we have a full understanding of electric and magnetic fields.

One of the most significant discoveries of the 20th century was that material particles, such as electrons and atoms, have wave-like characteristics. We will learn how a full description of matter at an atomic scale requires an understanding of such **matter waves** in Chapter 28.

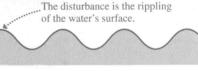
The disturbance is the rippling of the water's surface.

The water is the medium.

FIGURE 15.1 Ripples on a pond are a traveling wave.

You may have been at a sporting event in which spectators do "The Wave." The wave moves around the stadium, but the spectators (the medium, in this case) stay right where they are! This is a clear example of the principle that a wave does not transfer any material.

Transverse and Longitudinal Waves

Most waves fall into two general classes, *transverse* and *longitudinal*. For mechanical waves, these terms describe the relationship between the motion of the particles that carry the wave and the motion of the wave itself.

Two types of wave motion

A transverse wave

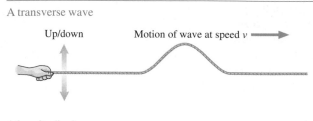

For mechanical waves, a **transverse wave** is a wave in which the particles in the medium move *perpendicular* to the direction in which the wave travels. Shaking the end of a stretched string up and down creates a wave that travels along the string in a horizontal direction while the particles that make up the string oscillate vertically.

A longitudinal wave

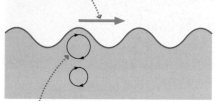

In a **longitudinal wave,** the particles in the medium move *parallel* to the direction in which the wave travels. Here we see a chain of masses connected by springs. If you give the first mass in the chain a sharp push, a disturbance travels down the chain by compressing and expanding the springs.

The motion of the water wave.

The motion of the molecules of water carrying the wave.

FIGURE 15.2 A water wave exhibits both transverse and longitudinal characteristics.

Water waves have characteristics of both transverse and longitudinal waves. When a water wave passes, individual water molecules describe a circular path as in Figure 15.2, exhibiting both transverse and longitudinal motion.

The rapid motion of the earth's crust during an earthquake can produce a disturbance that travels through the entire earth. The detection and analysis of these waves is the most important tool we have for analyzing the nature of the earth's interior. The two most important types of earthquake waves are S waves (which are transverse) and P waves (which are longitudinal). The longitudinal P waves are faster, but the transverse S waves are more destructive. Residents of a city a few hundred kilometers from an earthquake will feel the resulting P waves as much as a minute before the S waves, giving them a crucial early warning.

STOP TO THINK 15.1 Spectators at a sporting event do "The Wave," as shown in the photo on the previous page. Is this a transverse or longitudinal wave?

15.2 Traveling Waves

When you drop a pebble in a pond, waves travel outward. But how does this happen? How does a mechanical wave travel through a medium? In answering this question, we must be careful to distinguish the motion of the wave from the motion of the particles that make up the medium. The wave itself is not a particle, so we cannot apply Newton's laws to the wave. However, we can use Newton's laws to examine how the medium responds to a disturbance.

Waves on a String

Figure 15.3 on the next page shows a transverse *wave pulse* traveling to the right along a stretched string. Imagine watching a little dot on the string as a wave pulse passes by. As the pulse approaches from the left, the string near the dot begins to curve. This is the **leading edge** of the pulse. Once the string curves, the tension forces pulling on a small segment of string no longer cancel each other. Instead, as you can see in the first part of the figure, the tension in the string exerts an upward net force on this segment of the string, causing it to accelerate upward.

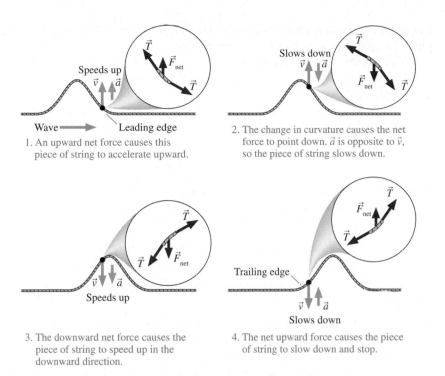

1. An upward net force causes this piece of string to accelerate upward.

2. The change in curvature causes the net force to point down. $\vec{a}$ is opposite to $\vec{v}$, so the piece of string slows down.

3. The downward net force causes the piece of string to speed up in the downward direction.

4. The net upward force causes the piece of string to slow down and stop.

FIGURE 15.3 The motion of a small segment of string as a wave pulse travels to the right.

The string's curvature reverses near the top of the pulse, causing the net force and the acceleration to point downward. Thus this piece of string slows until reaching a turning point at the highest point of the pulse, then it speeds up in the downward direction. This is much like a mass on a spring as the mass approaches the highest point and then reverses direction. Finally, the force and acceleration turn upward on the **trailing edge** of the pulse. The piece of string decelerates and comes to rest when the pulse has completely passed by.

No new physical principles are required to understand how a wave moves. The motion of a pulse along a string is a direct consequence of the tension acting on the segments of the string. An external force may have been required to create the pulse, but **once started the pulse continues to move because of the internal dynamics of the medium.**

Sound Waves

Let's see how a sound wave in air is created using a loudspeaker. When the loudspeaker cone in Figure 15.4a moves forward, it compresses the air in front of it, as shown in Figure 15.4b. The *compression* is the disturbance that travels forward through the air. This is much like the sharp push on the end of the chain of springs on the previous page; thus **a sound wave is a longitudinal wave.** We usually think of sound waves as traveling in air, but sound can travel through any gas, through liquids, and even through solids. A wave similar to that in Figure 15.4b is produced if you hit the end of a metal rod with a hammer.

The motion of a wave on a string is determined by the internal dynamics of the string. Similarly, the motion of a sound wave in air is determined by the physics of gases that we explored in Chapter 12. Once created, the wave in Figure 15.4b will propagate forward; its motion is entirely determined by the properties of the air.

Wave Speed Is a Property of the Medium

The above discussions of waves on a string and sound waves in air show that the motion of these waves is due to the properties of the medium. A careful analysis of these waves would lead to the important and somewhat surprising conclusion

(a) The loudspeaker cone moves in and out in response to electrical signals.

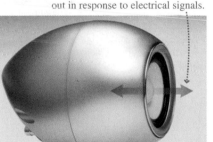

(b) The loudspeaker receives an electrical pulse, and pushes out sharply. This creates a compression of the air.

Compression

v_{sound}

Speaker

Molecules

This compression is the disturbance. It travels at the wave speed v_{sound}.

FIGURE 15.4 A sound wave produced by a loudspeaker.

that **the wave speed does not depend on the shape or size of the pulse, how the pulse was generated, or how far it has traveled.**

What properties of a string determine the speed of waves traveling along the string? The only likely candidates are the string's mass, length, and tension. Because a pulse doesn't travel faster on a longer and thus more massive string, neither the total mass m nor the total length L is important. Instead, the speed depends on the mass-to-length *ratio,* which is called the **linear density** μ of the string:

$$\mu = \frac{m}{L} \tag{15.1}$$

Linear density characterizes the *type* of string we are using. A fat string has a larger value of μ than a skinny string made of the same material. Similarly, a steel wire has a larger value of μ than a plastic string of the same diameter.

NOTE ▶ We'll assume that strings are *uniform,* meaning that the linear density is the same everywhere along the length of the string. ◀

How does the speed of a wave on a string vary with the tension and the linear density?

Sensing water waves BIO The African clawed frog has a hunting strategy similar to that of many spiders: It sits and waits for prey to come to it. Like a spider, the frog detects prey animals by the vibrations they cause—not in a web, but in the water. The frog has an array of sensors called the lateral line organ on each side of its body. This organ, highlighted in the above photo, detects oscillations of the water due to passing waves. The frog can determine where the waves come from and what type of animal made them, and thus whether a strike is called for.

10.2 Activ Physics

- A string with a greater tension will exert larger forces on the segments of the string shown in Figure 15.3. The string will respond more rapidly, so the wave will move at a higher speed. **Wave speed increases with increasing tension.**
- A string with a greater linear density has more inertia. It will respond less rapidly, so the wave will move at a lower speed. **Wave speed decreases with increasing linear density.**

A full analysis of the motion of the string would lead to an expression for the speed of a wave that shows both of these trends:

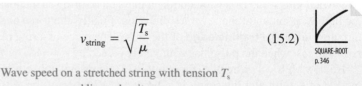

$$v_{\text{string}} = \sqrt{\frac{T_s}{\mu}} \tag{15.2}$$

SQUARE-ROOT
p.346

Wave speed on a stretched string with tension T_s and linear density μ

The subscript s on the symbol T_s for the string's tension will distinguish it from the symbol T for the *period* of oscillation.

Every point on a wave pulse travels with the speed given by Equation 15.2. You can increase the wave speed either by *increasing* the string's tension (make it tighter) or by *decreasing* the string's linear density (make it skinnier). We'll examine the implications for stringed musical instruments in Chapter 16.

EXAMPLE 15.1 When does the spider sense its lunch?

All spiders are very sensitive to vibrations. An orb spider will sit at the center of its large, circular web and monitor radial threads for vibrations created when an insect lands. Assume that these threads are made of silk, with a linear density of 1.0×10^{-5} kg/m, and are under a tension of 0.40 N. If an insect lands in the web 30 cm from the spider, how long will it take for the spider to find out?

PREPARE When the insect hits the web, a wave pulse will be transmitted along the silk fibers. The speed of the wave depends on the properties of the silk.

SOLVE First, determine the speed of the wave:

$$v = \sqrt{\frac{T_s}{\mu}} = \sqrt{\frac{0.40 \text{ N}}{1.0 \times 10^{-5} \text{ kg/m}}} = 200 \text{ m/s}$$

The time for the wave to travel a distance $d = 30$ cm to reach the spider is

$$\Delta t = \frac{d}{v} = \frac{0.30 \text{ m}}{200 \text{ m/s}} = 1.5 \text{ ms}$$

ASSESS Spider webs are made of very light strings under significant tension, so the wave speed is quite high. The time for signals to travel across the web is very short.

What properties of a gas determine the speed of a sound wave traveling through the gas? It seems plausible that the speed of a sound pulse is related to the speed with which the molecules of the gas move—faster molecules should mean a faster sound wave. In Chapter 11, we found that the typical speed of an atom of mass m, the root-mean-square speed, is

$$v_{\text{rms}} = \sqrt{\frac{3k_{\text{B}}T}{m}}$$

where k_{B} is Boltzmann's constant and T the absolute temperature in kelvin. A thorough analysis finds that the sound speed is slightly less than this rms speed, but has the same dependence on the temperature and the molecular mass:

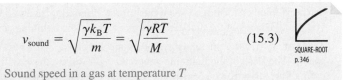

$$v_{\text{sound}} = \sqrt{\frac{\gamma k_{\text{B}}T}{m}} = \sqrt{\frac{\gamma RT}{M}} \qquad (15.3)$$

Sound speed in a gas at temperature T

SQUARE-ROOT
p. 346

In Equation 15.3 M is the molar mass (kg per mol). γ is a constant that depends on the gas. $\gamma = 1.67$ for monatomic gases such as helium, and $\gamma = 1.40$ for diatomic gases such as nitrogen and oxygen.

Certain trends in Equation 15.3 are worth mentioning:

Actıv
ONLINE
Physıcs   10.3

TABLE 15.1 The speed of sound

Medium	Speed (m/s)
Air (0°C)	331
Air (20°C)	343
Helium (0°C)	970
Ethyl alcohol	1170
Water	1480
Human tissue (ultrasound)	1540
Lead	1200
Aluminum	5100
Granite	6000
Diamond	12,000

- At a given temperature, the speed of sound in a gas varies significantly with the molecular mass. At room temperature, the speed of sound in helium is much greater than that in air.
- The speed of sound in air (and other gases) varies with temperature. **For calculations in this chapter, use the speed of sound in air at 20°C, 343 m/s, unless otherwise specified.**

Table 15.1 gives the speed of sound in various materials. The greater forces between molecules in liquids and solids result in higher sound speeds than in gases. Generally, sound waves travel faster in liquids than in gases, and faster in solids than in liquids. The speed of sound in a solid depends on its density and stiffness. Light, stiff solids (such as diamond) transmit sound at very high speeds. The sound speed is much less in dense solids such as lead.

EXAMPLE 15.2 Finding the speed of sound in a cold gas
What is the speed of sound through the gas boiling off a flask of liquid nitrogen? The boiling point of nitrogen is $-196°C$.

PREPARE The speed of sound in a gas depends on the temperature and molar mass of the gas. Because there's no temperature change in a phase transition, the gas boiling off the liquid is at the boiling point temperature, $T = -196°C = 77$ K. The molar mass of molecular nitrogen (N_2) is $M = 0.028$ kg/mol, and $\gamma = 1.40$ for a diatomic gas.

SOLVE The speed of sound in a gas, given by Equation 15.3, is

$$v_{\text{sound}} = \sqrt{\frac{\gamma RT}{M}} = \sqrt{\frac{(1.40)(8.31 \text{ J/mol} \cdot \text{K})(77 \text{ K})}{0.028 \text{ kg/mol}}}$$

$$= 180 \text{ m/s}$$

ASSESS This is much less than the speed of sound in air (which is mostly nitrogen) at 0°C. But that's not unexpected at such a low temperature.

Electromagnetic waves, such as light, travel at a much greater speed than do mechanical waves. As we'll discuss in Chapter 25, all electromagnetic waves travel at the same speed in a vacuum. We call this speed the **speed of light,** which we represent with the symbol c. The value of the speed of light in a vacuum is

$$v_{\text{light}} = c = 3.00 \times 10^8 \text{ m/s} \qquad (15.4)$$

This is almost one million times the speed of sound in air! At this speed, light could circle the earth 7.5 times in one second.

NOTE ▶ The speed of electromagnetic waves is lower when they travel through a material, but this value for the speed of light in vacuum is also good for electromagnetic waves traveling through air. ◀

TRY IT YOURSELF

Distance to a lightning strike Sound travels approximately 1 km in 3 s, or 1 mi in 5 s. When you see a lightning flash, start counting seconds. When you hear the thunder, stop counting. Divide the result by 5, and you will have the approximate distance to the lightning strike in miles.

EXAMPLE 15.3 How far away was the lightning?

During a thunderstorm, you see a flash from a lightning strike. 8.0 seconds later, you hear the crack of the thunder. How far away did the lightning strike?

PREPARE Two different kinds of waves are involved, with very different wave speeds. The flash of the lightning generates light waves; these will travel from the point of the strike to your position essentially instantaneously. (It takes light about 5 μs to travel 1 mile—not something you will notice.) The strike also generates sound waves that you hear as thunder; these travel much more slowly. The delay between the flash and the thunder is due to the time it takes for the sound wave to travel.

SOLVE We will assume that the speed of sound has its room temperature (20°C) value of 343 m/s. During the time between seeing the flash and hearing the thunder, the sound travels a distance

$$d = v \, \Delta t = (343 \text{ m/s})(8.0 \text{ s}) = 2.7 \times 10^3 \text{ m} = 2.7 \text{ km}$$

ASSESS You can use the time between the flash and the thunder to estimate the distance to a lightning strike, as in the "Try It Yourself" activity.

STOP TO THINK 15.2   Suppose you shake the end of a stretched string to produce a wave. Which of the following actions would increase the speed of the wave down the string? There may be more than one correct answer; if so, give all that are correct.

A. Move your hand up and down more quickly as you generate the wave.
B. Move your hand up and down a greater distance as you generate the wave.
C. Use a heavier string of the same length, under the same tension.
D. Use a lighter string of the same length, under the same tension.
E. Stretch the string tighter to increase the tension.
F. Loosen the string to decrease the tension.

15.3 Graphical and Mathematical Descriptions of Waves

Now that we know a bit about waves and how they travel, it's time to develop our understanding further by describing waves with graphs and equations.

Describing waves and their motion takes a bit more thought than describing particles and their motion. Until now, we have been concerned with quantities, such as position and velocity, that depend only on time. We can write these, as we did in Chapter 14, as $x(t)$ or $v(t)$, indicating that x and v are *functions* of the time variable t. Functions of the single variable t are all right for a particle, because a particle is in only one place at a time, but a wave is not localized. It is spread out through space at each instant of time. To describe a wave mathematically requires a function that specifies what the medium is doing not only at an instant of time (when) but also at a point in space (where); that is, a function of *two* variables.

NOTE ▶ The analysis that follows is for a wave on a string, which is easy to visualize, but the results apply to any traveling wave. ◀

Snapshot and History Graphs

It is helpful to think about waves graphically before we discuss the mathematical details. Consider the wave pulse shown moving along a stretched string in Figure 15.5. (We will consider somewhat artificial triangular and square-shaped wave pulses in this section to clearly show the edges of the pulse.) The graph shows the string's displacement y at a particular instant of time t_1 as a function of position x along the string. This is a "snapshot" of the wave, much like what you might make with a camera whose shutter is opened briefly at t_1. A graph that shows the wave's displacement as a function of position at a single instant of time is called a **snapshot graph.** For a wave on a string, a snapshot graph is literally a picture of the wave at this instant.

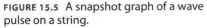

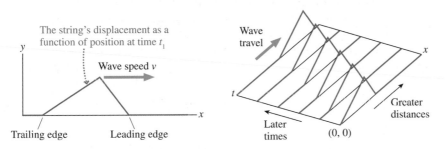

FIGURE 15.5 A snapshot graph of a wave pulse on a string.

FIGURE 15.6 The traveling wave of Figure 15.5.

As the wave moves, we can take further snapshots. If we place these next to each other, we get a clear picture of the wave's motion, as in Figure 15.6. Note that the wave's shape stays the same as it moves.

We will see this type of picture again, but we will more often simply draw several snapshot graphs one below the other, as in Figure 15.7. This figure shows a sequence of snapshot graphs as the wave of Figure 15.5 continues to move. These are like successive frames from a movie, and reminiscent of the sequences of pictures we saw in Chapter 1. The wave pulse moves forward distance $\Delta x = v\,\Delta t$ during the time interval Δt. That is, the wave moves with constant speed. Notice also the motion of the point on the string; as we noted, the motion of the medium is distinct from that of the wave.

A snapshot graph tells only half the story. It tells us *where* the wave is and how it varies with position, but only at one instant of time. It gives us no information about how the wave *changes* with time. As a different way of portraying the wave, suppose we follow the dot marked on the string in Figure 15.7 and produce a graph showing how the displacement of this dot changes with time. The result, shown in Figure 15.8, is a displacement-versus-time graph at a single position in space. This is called a **history graph,** as it tells the history of that particular point in the medium.

You might think we have made a mistake; the graph of Figure 15.8 is reversed compared to Figure 15.7. It is not a mistake, but it requires careful thought to see why. As the wave moves toward the dot, the steep leading edge causes the dot to rise quickly. On the displacement-versus-time history graph, *earlier* times

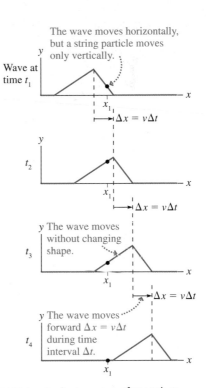

FIGURE 15.7 A sequence of snapshot graphs shows the wave in motion.

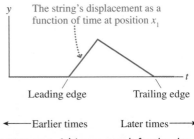

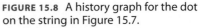

FIGURE 15.8 A history graph for the dot on the string in Figure 15.7.

(smaller values of *t*) are to the *left* and later times (larger *t*) to the right. Thus the leading edge of the wave is on the *left* side of the Figure 15.8 history graph. As you move to the right on Figure 15.8 you see the slowly falling trailing edge of the wave as it moves past the dot at later times.

Sinusoidal Waves

Waves can come in many different shapes, but for the mathematical description of wave motion we will focus on a particular shape, the **sinusoidal wave.** This is the type of wave produced by a source that oscillates with simple harmonic motion. For example, a loudspeaker cone that oscillates in SHM radiates a sinusoidal sound wave. The sinusoidal electromagnetic waves broadcast by television and FM radio stations are generated by electrons oscillating back and forth in the antenna wire with SHM.

Figure 15.9 shows a sinusoidal wave moving through a medium. The source of the wave, which is undergoing vertical SHM, is located at $x = 0$. Notice how the wave crests move with steady speed toward larger values of x at later times t.

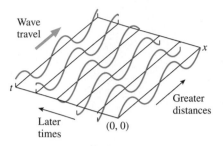

FIGURE 15.9 A sinusoidal wave moving along the *x*-axis.

Figure 15.10 shows a history graph for a point in a string as a sinusoidal wave passes by. This graph is identical to the graphs you worked with in Chapter 14, meaning each point in the medium oscillates with simple harmonic motion. The *period* of the wave, shown on the graph, is the time interval for one cycle of the motion. The period is related to the wave *frequency f* by $T = 1/f$, exactly as in SHM.

Because each point on the string oscillates up and down in SHM with period T, we can describe the motion of a point on the string with the familiar expression

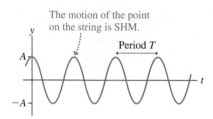

FIGURE 15.10 History graph for one point on a string carrying a sinusoidal wave.

$$y(t) = A\cos\left(2\pi\frac{t}{T}\right) \tag{15.5}$$

NOTE ▶ As it was in Chapter 14, the argument of the cosine function is in radians. Make sure that your calculator is in radian mode before starting a calculation with the trigonometric equations in this chapter. ◀

Figure 15.11 shows a snapshot graph for the sinusoidal wave of Figure 15.10, a record of the displacement of the medium at one point in time. Here we see the wave stretched out in space, moving to the right with speed *v*. The **amplitude** *A* of the wave is the maximum value of the displacement. The crests of the wave—the high points—have displacement $y_{\text{crest}} = A$ and the troughs—the low points—have displacement $y_{\text{trough}} = -A$.

An important characteristic of a sinusoidal wave is that it is periodic *in space* as well as in time. As you move from left to right along the "frozen" wave in the snapshot graph of Figure 15.11, the disturbance repeats itself over and over. The distance spanned by one cycle of the motion is called the **wavelength** of the wave. Wavelength is symbolized by λ (lowercase Greek lambda) and, because it is a length, it is measured in units of meters. The wavelength is shown in Figure 15.11 as the distance between two crests, but it could equally well be the distance between two troughs.

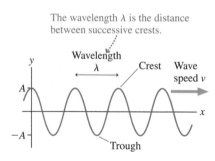

FIGURE 15.11 Snapshot graph of a string carrying a sinusoidal wave.

NOTE ▶ Wavelength is the spatial analog of period. The period *T* is the *time* in which the disturbance at a single point in space repeats itself. The wavelength λ is the *distance* in which the disturbance at one instant of time repeats itself. ◀

The snapshot graph shows that the wave, at a frozen instant in time, is also a sinusoidal function of the distance x along the wave, with wavelength λ. At this particular time, the displacement is given by:

$$y(x) = A\cos\left(2\pi\frac{x}{\lambda}\right) \tag{15.6}$$

Equation 15.5 gives the displacement as a function of time at one point in space, and Equation 15.6 gives the displacement as a function of position at one instant of time. How can we combine these two to form a single expression that is a complete description of the wave? As we'll verify, a wave traveling to the right is described by the equation

$$y(x, t) = A\cos\left(2\pi\left(\frac{x}{\lambda} - \frac{t}{T}\right)\right) \tag{15.7}$$

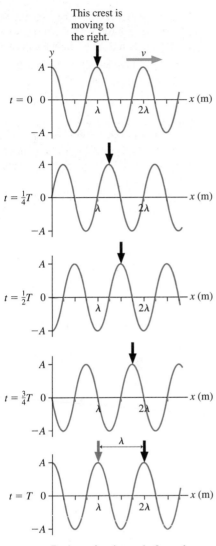

SINUSOIDAL
p.460

Displacement of a traveling wave moving to the right with amplitude A, wavelength λ, and period T

The notation $y(x, t)$ indicates that the displacement y is a function of the *two* variables x and t. We must specify both where (x) and when (t) before we can calculate the displacement of the wave.

We can understand why this expression describes a wave traveling to the right by looking at Figure 15.12. In the figure, we have graphed Equation 15.7 at five instants of time, each separated by one-quarter of the period T, to make five snapshot graphs. One full period has elapsed between the first graph and the last. By following the crest marked with the arrow, you can see that this is a sinusoidal wave moving to the right. As the wave has passed, each point in the medium has undergone exactly one complete oscillation.

For a wave traveling to the left, we have a slightly different form:

$$y(x, t) = A\cos\left(2\pi\left(\frac{x}{\lambda} + \frac{t}{T}\right)\right) \tag{15.8}$$

Displacement of a traveling wave moving to the left

NOTE ▶ A wave moving to the right (the $+x$-direction) has a $-$ in the expression while a wave moving to the left (the $-x$-direction) has a $+$. Remember that the sign is *opposite* the direction of travel. ◀

This crest is moving to the right.

During a time interval of exactly one period, the crest has moved forward exactly one wavelength.

FIGURE 15.12 A series of snapshot graphs at time increments of one-quarter of the period T. This is a traveling wave moving to the right.

EXAMPLE 15.4 Determining the rise and fall of a boat

A boat is moving to the right at 5.0 m/s against a sinusoidal ocean wave that is traveling to the left. The waves have 2.0 m between the top of the crests and the bottom of the troughs. The period of the waves is 8.3 s, and their wavelength is 110 m. At one instant, the boat sits on a crest of the wave. 20 s later, what is the vertical displacement of the boat?

PREPARE We begin with a visual overview. Let $t = 0$ be the instant the boat is on the crest, and draw a snapshot graph of the traveling wave at that time. Because the wave is traveling to the left, we will use Equation 15.8 to represent the wave. The boat begins at a crest of the wave, so we see that the boat can start the problem at $x = 0$.

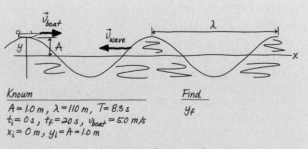

FIGURE 15.13 Visual overview for Example 15.4.

Continued

The distance between the high and low points of the wave is 2.0 m; the amplitude is half this, so $A = 1.0$ m. The wavelength and period are given in the problem.

SOLVE The boat is moving to the right at 5.0 m/s. At $t_f = 20$ s the boat is at position

$$x_f = (5.0 \text{ m/s})(20 \text{ s}) = 100 \text{ m}$$

We need to find the wave's displacement at this position and time. Substituting known values for amplitude, wavelength, and period into Equation 15.8, for a wave traveling to the left, we obtain the following equation for the wave:

$$y(x, t) = (1.0 \text{ m}) \cos\left(2\pi\left(\frac{x}{110 \text{ m}} + \frac{t}{8.3 \text{ s}}\right)\right)$$

At $t_f = 20$ s and $x_f = 100$ m, the boat's displacement on the wave is

$$y_f = y(\text{at } 100 \text{ m}, 20 \text{ s}) = (1.0 \text{ m}) \cos\left(2\pi\left(\frac{100 \text{ m}}{110 \text{ m}} + \frac{20 \text{ s}}{8.3 \text{ s}}\right)\right)$$

$$= -0.42 \text{ m}$$

ASSESS The final displacement is negative—meaning the boat is in a trough of a wave, not underwater. Don't forget that your calculator must be in radian mode when you make your final computation!

The Fundamental Relationship for Sinusoidal Waves

10.1   Actįv
ONLINE
Physįcs

In Figure 15.12, one critical observation is that the wave crest marked by the arrow has moved one full wavelength between the first graph and the last. That is, **during a time interval of exactly one period T, each crest of a sinusoidal wave travels forward a distance of exactly one wavelength λ.** Because speed is distance divided by time, the wave speed must be

$$v = \frac{\text{distance}}{\text{time}} = \frac{\lambda}{T} \qquad (15.9)$$

Using $f = 1/T$, it is customary to write Equation 15.9 in the form

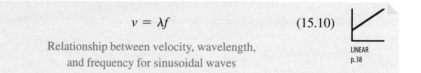

$$v = \lambda f \qquad (15.10)$$

Relationship between velocity, wavelength, and frequency for sinusoidal waves

LINEAR
p. 38

Although Equation 15.10 has no special name, it is *the* fundamental relationship for sinusoidal waves. When using it, keep in mind the *physical* meaning that a wave moves forward a distance of one wavelength during a time interval of one period.

We will frequently see problems in which we are given two of the variables in Equation 15.10 but we need to find the third, as in the following example.

EXAMPLE 15.5 Writing the equation for a wave

A sinusoidal wave with an amplitude of 1.5 cm and a frequency of 100 Hz travels at 200 m/s in the positive x-direction. Write the equation for the wave's displacement as it travels.

PREPARE The problem statement gives the following characteristics of the wave:

$$A = 1.5 \text{ cm} = 0.015 \text{ m} \qquad v = 200 \text{ m/s} \qquad f = 100 \text{ Hz}$$

SOLVE To write the equation for the wave, we need the amplitude, the wavelength, and the period. The amplitude was given in the problem statement. We can use the fundamental relationship for sinusoidal waves to find the wavelength:

$$\lambda = \frac{v}{f} = \frac{200 \text{ m/s}}{100 \text{ Hz}} = 2.0 \text{ m}$$

The period can be calculated from the frequency:

$$T = \frac{1}{f} = \frac{1}{100 \text{ Hz}} = 0.010 \text{ s}$$

With these values in hand, we can write the equation for the wave's displacement using Equation 15.7:

$$y(x, t) = (0.015 \text{ m}) \cos\left(2\pi\left(\frac{x}{2.0 \text{ m}} - \frac{t}{0.010 \text{ s}}\right)\right)$$

Three waves travel to the right with the same speed. Which wave has the highest frequency?

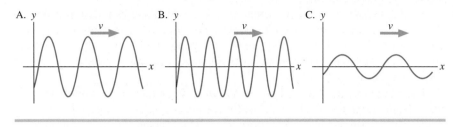

15.4 Sound and Light Waves

Although there are many kinds of waves in nature, two are especially significant for us as humans. These are sound waves and light waves, the basis of hearing and seeing.

Sound Waves

We saw in Figure 15.4 how a loudspeaker creates a sound wave. If the loudspeaker cone moves with simple harmonic motion, it will create a sinusoidal sound wave, as illustrated in Figure 15.14. Each time the cone moves forward, it collides with air molecules and pushes them closer together, creating a region of higher pressure. A half cycle later, as the cone moves backward, the air has room to expand and the pressure decreases. These regions of higher and lower pressure are called **compressions** and **rarefactions.**

As Figure 15.14 suggests, it is often most convenient and informative to think of a sound wave as a pressure wave. As the graph of the pressure shows, the pressure oscillates sinusoidally around the atmospheric pressure p_{atm}. This is a snapshot graph of the wave at one instant of time, and the distance between two adjacent crests (two points of maximum compression) is the wavelength λ.

When the wave reaches your ear, the oscillating pressure causes your eardrums to vibrate. This vibration is transferred through your inner ear to the cochlea, where it is sensed, as we learned in Chapter 14. Humans with normal hearing are able to detect sinusoidal sound waves with frequencies between about 20 Hz and about 20,000 Hz, or 20 kHz. Low frequencies are perceived as a "low pitch" bass note while high frequencies are heard as a "high pitch" treble note.

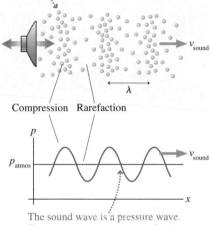

The loudspeaker cone moves back and forth, creating regions of higher and lower pressure—compressions and rarefactions.

The sound wave is a pressure wave. Compressions are crests; rarefactions are troughs.

FIGURE 15.14 A sound wave is a pressure wave.

EXAMPLE 15.6 Range of wavelengths of sound
What are the wavelengths of sound waves at the limits of human hearing and at the midrange frequency of 500 Hz? Notes sung by human voices are near 500 Hz, as are notes played by striking keys near the center of a piano keyboard.

PREPARE We will do our calculation at room temperature, 20°C, so we will use $v = 343$ m/s for the speed of sound.

SOLVE We can solve for the wavelengths given the fundamental relationship $v = f\lambda$:

$f = 20$ Hz: $\lambda = \dfrac{v}{f} = \dfrac{343 \text{ m/s}}{20 \text{ Hz}} = 17 \text{ m}$

$f = 500$ Hz: $\lambda = \dfrac{v}{f} = \dfrac{343 \text{ m/s}}{500 \text{ Hz}} = 0.69 \text{ m}$

$f = 20$ kHz: $\lambda = \dfrac{v}{f} = \dfrac{343 \text{ m/s}}{20 \times 10^3 \text{ Hz}} = 0.017 \text{ m} = 1.7 \text{ cm}$

ASSESS The wavelength of a 20 kHz note is a small 1.7 cm. At the other extreme, a 20 Hz note has a huge wavelength of 17 m! A wave moves forward one wavelength during a time interval of one period, and a wave traveling at 343 m/s can move 17 m during the $\frac{1}{20}$ s period of a 20 Hz note.

TABLE 15.2 Range of hearing for animals

Animal	Range of hearing (Hz)
Elephant	<5–12,000
Owl	200–12,000
Human	20–20,000
Dog	30–45,000
Mouse	1,000–90,000
Bat	2,000–100,000
Porpoise	75–150,000

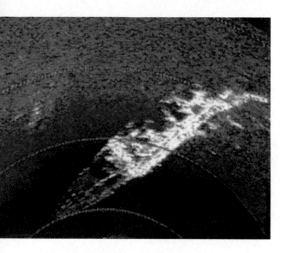

Short wavelengths mean sharper images
In the dark depths of the ocean, research submarines use sonar (which stands for "**so**und **n**avigation **a**nd **r**anging") to navigate through and investigate their surroundings. At short distances, high-frequency sound waves can create images of remarkable detail. This sonar image of the Titanic was made from 60 m away with 675 kHz ultrasound; notice the sharp outline, and the small structures visible on the deck.

It is well known that dogs are sensitive to high frequencies that we cannot hear. Other animals also have quite different ranges of hearing than humans; some examples are shown in Table 15.2.

Elephants communicate over great distances with vocalizations at frequencies far too low for us to hear. These low-frequency sounds are transmitted with less energy loss than high-frequency sounds, allowing elephants to hear each other from over 6 km away—important for these social animals that may forage over very large areas.

There are animals that use frequencies well above the range of our hearing as well. High-frequency sounds are useful for *echolocation*—emitting a pulse of sound and listening for its reflection. Bats, which generally feed at night, rely much more on their hearing than their sight. They find and catch insects by echolocation, emitting loud chirps whose reflections are detected by their large, sensitive ears. The frequencies that they use are well above the range of our hearing; we call such sound **ultrasound.** Why do bats and other animals use high frequencies for this purpose?

In Chapter 19, we will look at the *resolution* of optical instruments. The finest detail that your eye—or any optical instrument—can detect is limited by the wavelength of light. Smaller wavelengths allow for the imaging of smaller details.

The same limitations apply to the acoustic image of the world made by bats. In order to sense fine details of their surroundings, bats must use sound of very short wavelength (and thus high frequency). A 50 kHz chirp from a little brown bat has a wavelength of just over half a centimeter, allowing the bat to precisely locate an insect that reflects it. Other animals that use echolocation, such as porpoises, also produce and sense high-frequency sounds.

Sound travels very well though tissues in the body and so can be a useful tool for making images of the body's interior. Again, the fine details necessary for a clinical diagnosis require the short wavelengths of ultrasound. X rays create very good images of bones; ultrasound is used for creating images of soft tissues. It is also useful in cases where the radiation exposure of x rays should be avoided. You have certainly seen ultrasound images taken during pregnancy, where the use of x rays is clearly undesirable.

EXAMPLE 15.7 Ultrasonic frequencies in medicine

To make a sufficiently detailed ultrasound image of a fetus in its mother's uterus, a physician has decided that a wavelength of 0.50 mm is needed. What frequency is required?

PREPARE The speed of ultrasound in the body is given in Table 15.1 as 1540 m/s.

SOLVE We can use the familiar relationship among speed, wavelength, and frequency to calculate:

$$f = \frac{v}{\lambda} = \frac{1540 \text{ m/s}}{0.50 \times 10^{-3} \text{ m}} = 3.1 \times 10^6 \text{ Hz} = 3.1 \text{ MHz}$$

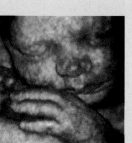

Computer processing of an ultrasound image shows fine detail.

ASSESS This is a reasonable result. Clinical ultrasound uses frequencies in the range of 1–20 MHz. Lower frequencies have greater penetration; higher frequencies (and thus shorter wavelengths) show finer detail.

Light and Other Electromagnetic Waves

A light wave is an electromagnetic wave, an oscillation of the electromagnetic field. Other electromagnetic waves, such as radio waves, microwaves, and ultraviolet light, have the same physical characteristics as light waves even though we cannot sense them with our eyes. As we saw earlier, all electromagnetic waves, regardless of wavelength or frequency, travel through vacuum (or air) with the same speed: $c = 3.00 \times 10^8$ m/s.

The wavelengths of light are extremely small. Visible light is an electromagnetic wave with a wavelength (in air) of roughly 400 nm ($=400 \times 10^{-9}$ m) to 700 nm ($=700 \times 10^{-9}$ m). Each wavelength is perceived as a different color. Longer wavelengths, in the 600–700 nm range, are seen as orange or red light; shorter wavelengths, in the 400–500 nm range, are seen as blue or violet light.

Figure 15.15 shows that the visible spectrum is a small slice out of the much broader **electromagnetic spectrum.** We will have much more to say about light and the rest of the electromagnetic spectrum in future chapters.

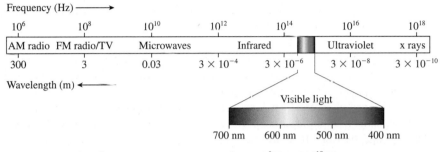

Frequency (Hz) ⟶

| 10^6 | 10^8 | 10^{10} | 10^{12} | 10^{14} | | 10^{16} | 10^{18} |

| AM radio | FM radio/TV | Microwaves | | Infrared | | Ultraviolet | x rays |

| 300 | 3 | 0.03 | 3×10^{-4} | 3×10^{-6} | | 3×10^{-8} | 3×10^{-10} |

Wavelength (m) ⟵

Visible light

700 nm   600 nm   500 nm   400 nm

FIGURE 15.15 The electromagnetic spectrum from 10^6 Hz to 10^{18} Hz.

EXAMPLE 15.8 Finding the frequency of microwaves

The wavelength of microwaves in a microwave oven is 12 cm. What is the frequency of the waves?

PREPARE Microwaves are electromagnetic waves, so their speed is the speed of light, $c = 3.00 \times 10^8$ m/s.

SOLVE Using the fundamental relationship for sinusoidal waves, we find

$$f = \frac{3.00 \times 10^8 \text{ m/s}}{0.12 \text{ m}} = 2.5 \times 10^9 \text{ Hz} = 2.5 \text{ GHz}$$

ASSESS Many modern cordless phones work at frequencies of 2.4 GHz, and cell phones at just a bit under 2.0 GHz. These frequencies are not very different from those of the waves that heat your food in a microwave oven, though the power is much less.

NOTE ▶ The wave model we've been developing applies equally well to string, sound, water, and electromagnetic waves. Features such as wavelength, frequency, and speed are characteristics of waves in general. So while we don't yet know much about the details of electromagnetic fields, we can use the wave model to say some significant things about the properties of electromagnetic waves. ◀

15.5 Energy and Intensity

A traveling wave transfers energy from one point to another. The sound wave from a loudspeaker sets your eardrum into motion. Light waves from the sun warm the earth and, if focused with a lens, can start a fire. The *power* of a wave is the rate, in joules per second, at which the wave transfers energy. As you learned

▶ **The better to hear you with** BIO The great grey owl has its ears on the front of its face, hidden behind its facial feathers. Its round face works like a radar dish, collecting the energy of sound waves and "funneling" it into the ears. This focusing of the sound allows these owls to sense very quiet sounds. Having ears on the front of the face allows them to precisely determine the source of sounds as well, an asset for a bird of prey. These owls can hear—and locate—mice moving underneath a thick blanket of snow.

(a)

Wave fronts are the crests of the wave. They are spaced one wavelength apart.

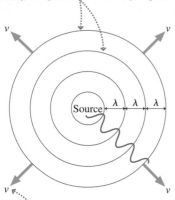

The circular wave fronts move outward from the source at speed v.

(b)

Very far away from the source, small sections of the wave fronts appear to be straight lines.

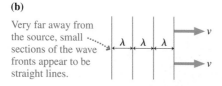

FIGURE 15.16 The wave fronts of a circular or spherical wave.

Very far from the source, small segments of spherical wave fronts appear to be planes. The wave is cresting at every point in these planes.

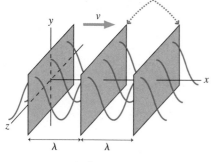

FIGURE 15.17 A plane wave.

The wave intensity at this surface is $I = P/a$.

Plane waves of power P

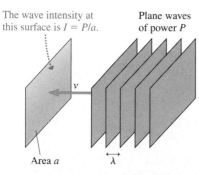

Area a

FIGURE 15.18 Plane waves of power P impinge on area a.

in Chapter 10, power is measured in watts. A person singing or shouting as loud as possible is emitting energy in the form of sound waves at a rate of about 1 W, or 1 J/s. A 25 W light bulb emits about 1 W of visible light, with the other 24 W of power being emitted as heat, or infrared radiation, rather than as visible light. In this section, we will learn how to characterize the power of waves. A first step in doing so is to understand how waves change as they spread out.

Circular, Spherical, and Plane Waves

Suppose you were to take a photograph of ripples spreading on a pond. If you mark the location of the *crests* on the photo, your picture would look like Figure 15.16a. The lines that locate the crests are called **wave fronts,** and they are spaced precisely one wavelength apart. A wave like this is called a **circular wave.** It is a two-dimensional wave that spreads across a surface.

Although the wave fronts are circles, you would hardly notice the curvature if you observed a small section of the wave front very far away from the source. The wave fronts would appear to be parallel lines, still spaced one wavelength apart and traveling at speed v as in Figure 15.16b.

Many waves of interest, such as sound waves or light waves, move in three dimensions. For example, loudspeakers and light bulbs emit **spherical waves.** The crests of the wave form a series of concentric spherical shells separated by the wavelength λ. In essence, the waves are three-dimensional ripples. It will still be useful to draw wave-front diagrams such as Figure 15.16a, but now the circles are slices through the spherical shells locating the wave crests.

If you observe a spherical wave very far from its source, the small piece of the wave front that you can see is a little patch on the surface of a very large sphere. If the radius of the sphere is sufficiently large, you will not notice the curvature and this little patch of the wave front appears to be a plane. Figure 15.17 illustrates the idea of a **plane wave.**

Power, Energy, and Intensity

Imagine doing two experiments with a light bulb that emits 2 W of visible light. In the first, you hang the bulb in the center of a room and allow the light to illuminate the walls. In the second experiment, you use mirrors and lenses to "capture" the bulb's light and focus it onto a small spot on one wall. (This is what a slide projector does.) The energy emitted by the bulb is the same in both cases, but, as you know, the light is much brighter when focused onto a small area. We would say that the focused light is more *intense* than the diffuse light that goes in all directions. Similarly, a loudspeaker that beams its sound forward into a small area produces a louder sound in that area than a speaker of equal power that radiates the sound in all directions. Quantities such as brightness and loudness depend not only on the rate of energy transfer, or power, but also on the *area* that receives that power.

Figure 15.18 shows a wave impinging on a surface of area a. The surface is perpendicular to the direction in which the wave is traveling. This might be a real, physical surface, such as your eardrum or a solar cell, but it could equally well be a mathematical surface in space that the wave passes right through. If the wave has power P, we define the **intensity** I of the wave to be

$$I = \frac{P}{a} \tag{15.11}$$

The SI units of intensity are W/m². Because intensity is a power-to-area ratio, a wave focused into a small area has a larger intensity than a wave of equal power that is spread out over a large area.

EXAMPLE 15.9 The intensity of a laser beam
A bright, tightly focused laser pointer emits 1.0 mW of light power into a beam that is 1.0 mm in diameter. What is the intensity of the laser beam?

SOLVE The light waves of the laser beam pass through a circle of diameter 1.0 mm. The intensity of the laser beam is

$$I = \frac{P}{a} = \frac{P}{\pi r^2} = \frac{0.0010 \text{ W}}{\pi (0.00050 \text{ m})^2} = 1300 \text{ W/m}^2$$

ASSESS This intensity is roughly equal to the intensity of sunlight at noon on a summer day. The difference between the sun and a small laser is not their intensities, which are about the same, but their powers. The laser has a small power of 1 mW. It produces a very intense wave only because the area through which the wave passes is very small. In contrast, the sun radiates a total power of about 4×10^{26} W, but this immense power is spread throughout a very large area. Because the beam of the laser is about as intense as the direct, unfiltered light of the sun, you do not want the beam to shine into your eye.

Sound from a loudspeaker and light from a light bulb become less intense as you get farther from the source. This is because spherical waves spread out to fill larger and larger volumes of space. To conserve energy, the wave's amplitude has to decrease with increasing distance r.

If a source of spherical waves radiates uniformly in all directions, then, as Figure 15.19 shows, the power at distance r is spread uniformly over the surface of a sphere of radius r. The surface area of a sphere is $a = 4\pi r^2$, so the intensity of a uniform spherical wave is

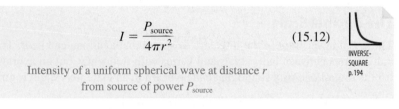

$$I = \frac{P_{\text{source}}}{4\pi r^2} \qquad (15.12)$$

INVERSE-
SQUARE
p. 194

Intensity of a uniform spherical wave at distance r
from source of power P_{source}

The inverse-square dependence of r is really just a statement of energy conservation. The source emits energy at the rate P joules per second. The energy is spread over a larger and larger area as the wave moves outward. Consequently, the *energy per area* must decrease in proportion to the surface area of a sphere.

If the intensity at distance r_1 is $I_1 = P_{\text{source}}/4\pi r_1^2$ and the intensity at r_2 is $I_2 = P_{\text{source}}/4\pi r_2^2$, then you can see that the intensity *ratio* is

$$\frac{I_1}{I_2} = \frac{r_2^2}{r_1^2} \qquad (15.13)$$

You can use Equation 15.13 to compare the intensities at two distances from a source without needing to know the power of the source.

NOTE ▶ Wave intensities are strongly affected by reflections and absorption. Equations 15.12 and 15.13 apply to situations such as the light from a star or the sound from a firework exploding high in the air. Indoor sound does *not* obey a simple inverse-square law because of the many reflecting surfaces. ◀

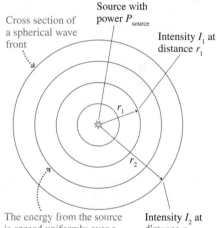

The energy from the source is spread uniformly over a spherical surface of area $4\pi r^2$.

FIGURE 15.19 A source emitting uniform spherical waves.

STOP TO THINK 15.4   A plane wave, a circular wave, and a spherical wave each have the same intensity. Each of the waves travels the same distance. Afterward, which wave has the largest intensity?

A. The plane wave.   B. The circular wave.   C. The spherical wave.

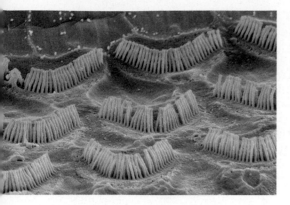

Hearing hairs BIO This electron microscope picture shows the hair cells in the cochlea of the ear that are responsible for sensing sound. Each cell has a curved row of tiny hairs, which give your ears their remarkable sensitivity. Even very tiny vibrations transmitted into the fluid of the cochlea deflect the hairs, triggering a response in the cells. Motion of the basilar membrane by as little as 0.5 nm, about 5 atomic diameters, can produce an electrical response in the hair cells. Large vibrations can damage these sensitive structures, which is why exposure to loud sounds diminishes hearing.

The loudest animal in the world BIO The blue whale is the largest animal in the world, up to 30 m (about 100 ft) long, weighing 150,000 kg or more. It is also the loudest. At close range in the water, the 10–to–30 second calls of the blue whale would be intense enough to damage tissues in your body. Blue whales produce sound with a larynx, much as humans do, but at much lower frequencies of 10–40 Hz that can travel great distances in water. Their loud, low-frequency calls can be heard by other whales hundreds of miles away.

15.6 Loudness of Sound

Suppose you are listening to one person singing. If nine more people join in, so a total of 10 people are singing in unison, you will certainly perceive the sound of the group as being louder, but it won't be 10 times as loud as the solo singer, even though the intensity is 10 times as great. In fact, you will perceive the sound of 10 people singing together to be approximately twice as loud as the sound of one person.

Now, let's add another 90 people to make a choir of 100 people. This will increase the intensity by another factor of 10, but it will seem to you only twice as loud as the 10-person ensemble, or four times as loud as the original soloist.

Generally, **increasing the sound intensity by a factor of 10 results in an increase in perceived loudness by approximately a factor of 2.** This allows your ears to be sensitive over a very wide range of intensities. The difference in intensity between the quietest sound you can detect and the loudest you can safely hear is a factor of 1,000,000,000,000! A normal conversation has 10,000 times the sound intensity of a whisper, but it sounds only about 16 times as loud.

The loudness of sound is measured by a quantity called the **sound intensity level.** Because of the wide range of intensities we can hear, and the fact that the difference in perceived loudness is much less than the actual difference in intensity, the sound intensity level is measured on a *logarithmic scale*. The next section will explain what this means. The units of sound intensity level (i.e., of loudness) are decibels, a word you have likely heard.

The Decibel Scale

There is a lower limit to the intensity of sound that a human can hear. The exact value varies between individuals and varies with frequency, but an average value for the lowest intensity sound that can be heard in an extremely quiet room is

$$I_0 = 1.0 \times 10^{-12} \text{ W/m}^2$$

This intensity is called the *threshold of hearing*. A sound wave can have less intensity than this, but you won't be able to perceive it.

It's convenient and logical to place the zero of our loudness scale at the threshold of hearing. All other sounds can then be referenced to this intensity. To create a loudness scale, we define the *sound intensity level,* expressed in **decibels** (dB) as

$$\beta = (10 \text{ dB}) \log_{10}\left(\frac{I}{I_0}\right) \tag{15.14}$$

Sound intensity level in decibels for a sound of intensity I

The decibel is named after Alexander Graham Bell, inventor of the telephone. Sound intensity level is actually dimensionless, since it's formed from the ratio of two intensities, so decibels are actually just a *name* to remind us that we're dealing with an intensity *level* rather than a true intensity.

Equation 15.14 takes the base-10 logarithm of the intensity ratio I/I_0. As a reminder, logarithms work like this:

If you express a number as a power of 10 . . . . . . the logarithm is the exponent.

$$\log_{10}(1000) = \log_{10}(10^3) = 3$$

Right at the threshold of hearing, where $I = I_0$, the sound intensity level is

$$\beta = (10 \text{ dB}) \log_{10}\left(\frac{I_0}{I_0}\right) = (10 \text{ dB}) \log_{10}(1) = (10 \text{ dB}) \log_{10}(10^0) = 0 \text{ dB}$$

The threshold of hearing corresponds to 0 dB, as we wanted. 0 dB doesn't mean no sound; it means that, for most people, no sound is heard. Dogs have more sensitive hearing than humans, and most dogs can easily perceive a 0 dB sound.

A comparison between sound intensity in W/m² and the associated sound intensity level in dB is shown in Figure 15.20. The major point to notice is that the sound intensity level increases by 10 dB each time the actual intensity increases by a *factor* of 10. For example, the sound intensity level increases from 70 dB to 80 dB when the sound intensity increases from 10^{-5} W/m² to 10^{-4} W/m². We found earlier that sound is perceived as "twice as loud" when the intensity increases by a factor of 10. In terms of decibels, we can say that the loudness of a sound doubles with each increase in the sound intensity level by 10 dB.

Table 15.3 lists the intensities and sound intensity levels for a number of sounds. Although 130 dB is the threshold of pain, quieter sounds can damage your hearing. A fairly short exposure to 120 dB can cause damage to the hair cells in the ear, but lengthy exposure to sound intensity levels of over 85 dB can produce damage as well.

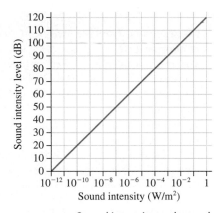

FIGURE 15.20 Sound intensity and sound intensity level scales compared.

TABLE 15.3 Intensity and sound intensity levels of common environmental sounds

Sound	β (dB)	I (W/m²)
Threshold of hearing	0	1.0×10^{-12}
Person breathing, at 3 m	10	1.0×10^{-11}
A whisper, at 1 m	20	1.0×10^{-10}
Classroom during test, no talking	30	1.0×10^{-9}
Residential street, no traffic	40	1.0×10^{-8}
Quiet restaurant	50	1.0×10^{-7}
Normal conversation, at 1 m	60	1.0×10^{-6}
Busy traffic	70	1.0×10^{-5}
Vacuum cleaner, for user	80	1.0×10^{-4}
Niagara Falls, at viewpoint	90	1.0×10^{-3}
Pneumatic hammer, at 2 m	100	0.010
Home stereo at max volume	110	0.10
Rock concert	120	1.0
Threshold of pain	130	10

EXAMPLE 15.10 Finding the loudness of a shout

A person shouting at the top of his lungs emits about 1.0 W of energy as sound waves. What is the sound intensity level 1.0 m from such a person?

PREPARE We will assume that the shouting person emits a spherical sound wave, in which case the intensity decreases according to Equation 15.12.

SOLVE At a distance of 1.0 m, the sound intensity is

$$I = \frac{P}{4\pi r^2} = \frac{1.0 \text{ W}}{4\pi (1.0 \text{ m})^2} = 0.080 \text{ W/m}^2$$

Thus the sound intensity level is

$$\beta = (10 \text{ dB}) \log_{10}\left(\frac{0.080 \text{ W/m}^2}{1.0 \times 10^{-12} \text{ W/m}^2}\right) = 110 \text{ dB}$$

ASSESS This is quite loud (compare with values in Table 15.3), as you might expect. In calculating sound intensity level, be sure to use the $\log_{10}$ button on your calculator, not the natural logarithm button. On many calculators, $\log_{10}$ is labeled LOG and the natural logarithm is labeled LN.

Equation 15.14 allows us to compute the sound intensity level β from the intensity I. We can do the reverse, finding an intensity from the sound intensity level, by taking the inverse of the $\log_{10}$ function. Recall, from the definition of the base-10 logarithm, that $10^{\log(x)} = x$. Applying this to Equation 15.14, we find

$$I = (1.0 \times 10^{-12} \text{ W/m}^2) 10^{(\beta/10 \text{ dB})} \tag{15.15}$$

EXAMPLE 15.11 **How far away can you hear a conversation?**

The sound intensity level 1.0 m from a person talking in a normal conversational voice is 60 dB. Suppose you are outside, 100 m from the person speaking. If it is a very quiet day with very minimal background noise, will you be able to hear him or her?

PREPARE We know how sound intensity changes with distance, but not sound intensity level, so we need to break this problem into several steps. First, we will convert the sound intensity level at 1.0 m into intensity, using Equation 15.15. Second, we will compute the intensity at a distance of 100 m, using Equation 15.13. Finally, we will convert this result back to a sound intensity level, using Equation 15.14, so that we can judge the loudness.

SOLVE The sound intensity level of a normal conversation at 1.0 m is 60 dB. The intensity is

$$I(1\text{ m}) = (1.0 \times 10^{-12}\text{ W/m}^2)10^{(60\text{ dB}/10\text{ dB})}$$

$$= 1.0 \times 10^{-6}\text{ W/m}^2$$

The intensity at 100 m can be found using Equation 15.13:

$$\frac{I(100\text{ m})}{I(1\text{ m})} = \frac{(1\text{ m})^2}{(100\text{ m})^2}$$

$$I(100\text{ m}) = I(1\text{ m})(1.0 \times 10^{-4}) = 1.0 \times 10^{-10}\text{ W/m}^2$$

This intensity corresponds to a sound intensity level

$$\beta = (10\text{ dB})\log_{10}\left(\frac{1.0 \times 10^{-10}\text{ W/m}^2}{1.0 \times 10^{-12}\text{ W/m}^2}\right) = 20\text{ dB}$$

This is above the threshold for hearing—about the level of a whisper—so it could be heard on a very quiet day.

ASSESS The sound is well within what the ear can detect. However, normal background noise is rarely less than 40 dB, which would make the conversation difficult to decipher.

15.7 The Doppler Effect and Shock Waves

In this final section, we will look at sounds from moving objects and for moving observers. You've likely noticed that the pitch of an ambulance's siren drops as it goes past you. A higher pitch suddenly becomes a lower pitch. This change in frequency, which is due to the motion of the ambulance, is called the *Doppler effect*. A more dramatic effect happens when an object moves faster than the speed of sound. The crack of a whip is a *shock wave* produced when the tip moves at such a *supersonic* speed. These examples—and much of this section—concern sound waves, but the phenomena we will explore apply generally to all waves.

Sound Waves from a Moving Source

Figure 15.21a on the next page shows a source of sound waves moving away from Pablo and toward Nancy at a steady speed v_s. The subscript s indicates that this is the speed of the source, not the speed of the waves. The source is emitting sound waves of frequency f_0 as it travels. Part a of the figure is a motion diagram showing the position of the source at times $t = 0$, T, $2T$, and $3T$, where $T = 1/f_0$ is the period of the waves.

Nancy measures the frequency of the wave emitted by the *approaching source* to be f_+. At the same time, Pablo measures the frequency of the wave emitted by the *receding source* to be f_-. Our task is to relate these measured frequencies to the frequency f_0 and speed v_s of the source.

After a wave crest leaves the source, its motion is governed by the properties of the medium. The motion of the source cannot affect a wave that has already been emitted. Thus each circular wave front in Figure 15.21b is centered on the point from which it was emitted. You can see that the wave crests are bunched up in the direction in which the source is moving and are stretched out behind it. The distance between one crest and the next is one wavelength, so the wavelength λ_+ that Nancy measures is *less* than the wavelength $\lambda_0 = v/f_0$ that would be emitted if the source were at rest. Similarly, λ_- behind the source is larger than λ_0.

These crests move through the medium at the wave speed v. Consequently, the frequency $f_+ = v/\lambda_+$ detected by the observer whom the source is approaching is

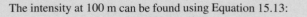

Catching a wave on the sun SOHO (the **So**lar and **H**eliospheric **O**bservatory) is a satellite studying the sun. The instrument on the satellite responsible for images like this measures the Doppler effect of light emitted from the surface of the sun. If the surface is rising toward the satellite, the light is shifted to higher frequencies; these positions are shown darker on the image. If the surface is falling, the light is shifted lower and is shown lighter. This series of images of the sun's surface shows a wave produced by the disruption of a solar flare. The waves reached speeds of 400,000 kilometers per hour and carried an energy 40,000 times that of a large earthquake.

(a) Motion of the source

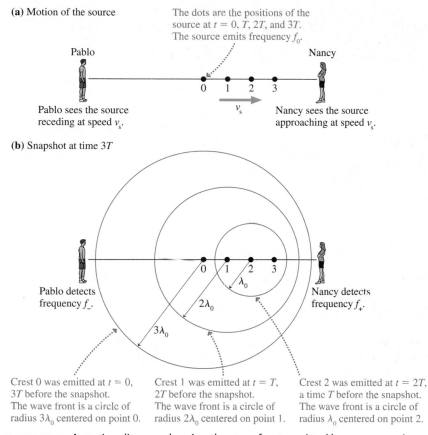

The dots are the positions of the source at $t = 0$, T, $2T$, and $3T$. The source emits frequency f_0.

Pablo

Nancy

Pablo sees the source receding at speed v_s.

Nancy sees the source approaching at speed v_s.

(b) Snapshot at time $3T$

Pablo detects frequency f_-.

Nancy detects frequency f_+.

Crest 0 was emitted at $t = 0$, $3T$ before the snapshot. The wave front is a circle of radius $3\lambda_0$ centered on point 0.

Crest 1 was emitted at $t = T$, $2T$ before the snapshot. The wave front is a circle of radius $2\lambda_0$ centered on point 1.

Crest 2 was emitted at $t = 2T$, a time T before the snapshot. The wave front is a circle of radius λ_0 centered on point 2.

FIGURE 15.21 A motion diagram showing the wave fronts emitted by a source as it moves to the right at speed v_s.

higher than the frequency f_0 emitted by the source. Similarly, $f_- = v/\lambda_-$ detected behind the source is *lower* than frequency f_0. This change of frequency when a source moves relative to an observer is called the **Doppler effect.** A quantitative analysis based on this information would show that the frequency heard by a stationary observer depends on whether the observer sees the source approaching or receding:

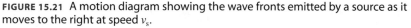

Actlv Physics
ONLINE   10.8, 10.9

$$f_+ = \frac{f_0}{1 - v_s/v}$$

Observed frequency of a wave of speed v emitted from a source approaching at speed v_s

(15.16)

$$f_- = \frac{f_0}{1 + v_s/v}$$

Observed frequency of a wave of speed v emitted from a source receding at speed v_s

As expected, $f_+ > f_0$ (the frequency is higher) for an approaching source, because the denominator is less than 1, and $f_- < f_0$ (the frequency is lower) for a receding source.

EXAMPLE 15.12 How fast are the police driving?

A police siren has a frequency of 550 Hz as the police car approaches you, 450 Hz after it has passed you and is moving away. How fast are the police traveling? Assume that the temperature is 20°C.

PREPARE The siren's frequency is altered by the Doppler effect. The frequency is f_+ as the car approaches and f_- as it moves away. We can write down two equations for these frequencies and solve for the speed of the police car, v_s.

SOLVE Because our goal is to find v_s, we rewrite Equations 15.16 as

$$f_0 = (1 + v_s/v)f_- \quad \text{and} \quad f_0 = (1 - v_s/v)f_+$$

Subtracting the second equation from the first, we get

$$0 = f_- - f_+ + \frac{v_s}{v}(f_- + f_+)$$

Now we can solve for the speed v_s:

$$v_s = \frac{f_+ - f_-}{f_+ + f_-}v = \frac{100 \text{ Hz}}{1000 \text{ Hz}} \, 343 \text{ m/s} = 34 \text{ m/s}$$

ASSESS This is pretty fast (about 75 mph), but is reasonable for a police car speeding with the siren on.

A Stationary Source and a Moving Observer

Suppose the police car in Example 15.12 is at rest while you drive toward it at 34 m/s. You might think that this is equivalent to having the police car move toward you at 34 m/s, but there is an important difference. Mechanical waves move through a medium, and the Doppler effect depends not just on how the source and the observer move with respect to each other but also how they move with respect to the medium. The frequency heard by an observer moving at speed v_0 relative to a stationary source emitting frequency f_0 is

$$f_+ = (1 + v_0/v)f_0$$

Doppler effect for an observer approaching a source

(15.17)

$$f_- = (1 - v_0/v)f_0$$

Doppler effect for an observer receding from a source

A quick calculation shows that the frequency of the police siren as you approach it at 34 m/s is 545 Hz, not the 550 Hz you heard as it approached you at 34 m/s.

The Doppler Effect for Light Waves

If a source of light waves is receding from you, the wavelength λ_- that you detect is longer than the wavelength λ_0 emitted by the source. Because the wavelength is shifted toward the red end of the visible spectrum, the longer wavelengths of light, this effect is called the **red shift.** Similarly, the light you detect from a source moving toward you is **blue shifted** to shorter wavelengths.

In the 1920s, an analysis of the spectra of many distant galaxies showed that they *all* had a distinct red shift—all distant galaxies are moving away from us. How can we make sense of this observation? The astronomer Edwin Hubble concluded that the galaxies of the universe are *all* moving apart from each other. Extrapolating backward in time brings us to a point when all the matter of the universe—and even space itself, according to the theory of relativity—began rushing out of a primordial fireball. Many observations and measurements since have given support to the idea that the universe began in a *Big Bang* about 14 billion years ago.

Frequency Shift on Reflection from a Moving Object

A wave striking a barrier or obstacle can *reflect* and travel back toward the source of the wave, a process we will examine more closely in the next chapter. For sound, the reflected wave is called an *echo*. A bat is able to determine the distance to a flying insect by measuring the time between the emission of an ultrasonic

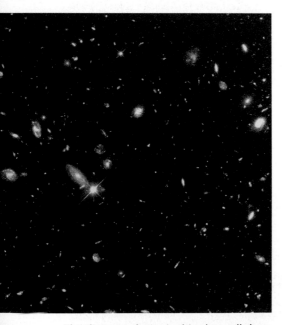

The distant galaxies in this photo all show significant red shifts. The farther away the galaxy, the greater the shift.

chirp and the detection of the echo from the insect. But if the bat is to catch the insect, it's just as important to know where the insect is going—its velocity. The bat can figure this out by noting the *frequency shift* in the reflected wave, another application of the Doppler effect.

Suppose a sound wave of frequency f_0 travels toward a moving object. The object would see the wave's frequency Doppler shifted to a higher frequency f_+, as given by Equation 15.17. The wave reflected back toward the sound source by the moving object is also Doppler shifted to a higher frequency because, to the source, the reflected wave is coming from a moving object. Thus the echo from a moving object is "double Doppler shifted." If the object's speed v_o is much less than the wave speed v ($v_0 \ll v$), the frequency *shift* for waves reflected from a moving object is

$$\Delta f = \pm 2f_0 \frac{v_o}{v} \tag{15.18}$$

Frequency shift of waves reflected from an object moving at speed v_o

The $+$ case is for objects moving toward the source of sound, the $-$ for objects moving away. Notice that there is no shift (i.e., the reflected wave has the same frequency as the emitted wave) if the reflecting object is at rest.

The *Doppler blood flow meter* is an important application of the frequency shift of waves reflected from moving objects. If an ultrasound emitter is pressed against the skin, the sound waves reflect off tissues in the body. In most cases, the frequency of the reflected wave is the same as that of the emitted wave. However, some of the sound waves reflect from blood cells moving through arteries toward or away from the emitter. The moving blood cells produce a frequency shift Δf in the reflected wave. By measuring Δf, doctors can determine blood flow speeds in an entirely noninvasive manner.

If we add a frequency-shift measurement to an ultrasound imaging unit, like the one in Example 15.7, we have a device—called *Doppler ultrasound*—that can show not only structure but also motion. *Doppler ultrasound* is a very valuable tool in cardiology because an image can reveal the motion of the heart muscle and the blood in the chambers of the heart in addition to the structure of the heart itself.

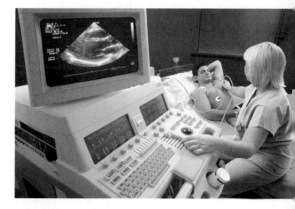

A Doppler ultrasound image can show the motion of the tissues of the heart and of blood within the heart.

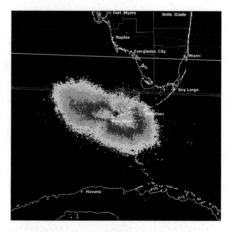

Keeping track of wildlife with radar
Doppler radar is tuned to measure only those reflected radio waves that have a frequency shift, eliminating reflections from stationary objects and showing only objects in motion. Televised Doppler radar images of storms are made using radio waves reflected from moving water droplets. But this Doppler radar image of an area off the tip of Florida was made on a clear night with no rain. The blue and green patch isn't a moving storm, it's a moving flock of birds. Migratory birds move at high altitudes at night, so Doppler radar makes an excellent tool for analyzing their movements.

EXAMPLE 15.13 Ultrasound frequency to measure blood flow

A biomedical engineer is designing a Doppler blood flow meter to measure blood flow in an artery where a typical flow speed is known to be 0.60 m/s. What ultrasound frequency should she use to produce a frequency shift of 1500 Hz when this flow is detected?

SOLVE We can rewrite Equation 15.18 to calculate the frequency of the emitter:

$$f_0 = \left(\frac{\Delta f}{2}\right)\left(\frac{v}{v_{cell}}\right)$$

The values on the right side are all known. Thus the required ultrasound frequency is

$$f_0 = \left(\frac{1500 \text{ Hz}}{2}\right)\left(\frac{1540 \text{ m/s}}{0.60 \text{ m/s}}\right) = 1.9 \text{ MHz}$$

ASSESS Doppler units to measure blood flow in deep tissues actually work at about 2.0 MHz, so our answer is reasonable.

Frequency shift on reflection is observed for all types of waves, not just sound waves. Radar (which stands for "**ra**dio **d**etection **a**nd **r**anging") units work by emitting radio waves and observing the waves that are reflected from objects. The

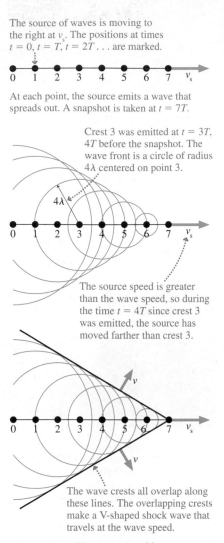

The source of waves is moving to the right at v_s. The positions at times $t = 0$, $t = T$, $t = 2T$ . . . are marked.

At each point, the source emits a wave that spreads out. A snapshot is taken at $t = 7T$.

Crest 3 was emitted at $t = 3T$, $4T$ before the snapshot. The wave front is a circle of radius 4λ centered on point 3.

4λ

The source speed is greater than the wave speed, so during the time $t = 4T$ since crest 3 was emitted, the source has moved farther than crest 3.

v

v

The wave crests all overlap along these lines. The overlapping crests make a V-shaped shock wave that travels at the wave speed.

FIGURE 15.22 Waves emitted by a source traveling faster than the speed of the waves in a medium.

time between the emission of a pulse and its return gives an object's position. The change in frequency of the returned pulse gives the object's speed. This is the principle behind the radar guns used by traffic police, as well as the Doppler radar images you have seen in weather reports.

Shock Waves

When sound waves are emitted by a moving source, the frequency is shifted, as we have seen. Now, let's look at what happens when the source speed v_s exceeds the wave speed v and the source "outruns" the waves it produces.

Earlier in this section, in Figure 15.21, we looked at a motion diagram of waves emitted by a moving source. Figure 15.22 is the same diagram, but in this case the source is moving faster than the waves. This motion causes the waves to overlap. (Compare Figure 15.22 to Figure 15.21.) The amplitudes of the overlapping waves add up to produce a very large amplitude wave—a **shock wave.**

Anything that moves faster than the speed of sound in air will create such a shock wave. The speed of sound was broken on land in 1997 by a British team driving the Thrust SCC in Nevada's Black Rock Desert. Figure 15.23a shows the shock waves being produced during a **supersonic** (faster than the speed of sound) run. You can see distortion of the landscape behind the car where the waves add to produce regions of high pressure. This shock wave travels along with the car. If the car went by you, the passing of the shock wave would produce a **sonic boom,** the distinctive loud sound you may have heard when a supersonic aircraft passes. The crack of a whip is a sonic boom as well, though on a much smaller scale.

FIGURE 15.23 Extreme and everyday examples of shock waves.

Shock waves are not necessarily an extreme phenomenon. A boat (or even a duck!) can easily travel faster than the speed of the water waves it creates; the resulting wake, with its characteristic "V" shape as shown in Figure 15.23b, is really a shock wave.

STOP TO THINK 15.5   Amy and Zack are both listening to the source of sound waves that is moving to the right. Compare the frequencies each hears.

A. $f_{Amy} > f_{Zack}$
B. $f_{Amy} = f_{Zack}$
C. $f_{Amy} < f_{Zack}$

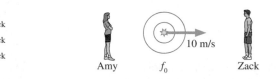

The goal of Chapter 15 has been to learn the basic properties of traveling waves.

GENERAL PRINCIPLES

The Wave Model

This model is based on the idea of a traveling wave, which is an organized disturbance traveling at a well-defined **wave speed** v.

- In transverse waves the particles of the medium move *perpendicular* to the direction in which the wave travels.

- In longitudinal waves the particles of the medium move *parallel* to the direction in which the wave travels.

A wave transfers energy, but there is no material or substance transferred.

Mechanical waves require a material **medium.** The speed of the wave is a property of the medium, not the wave. The speed does not depend on the size or shape of the wave.

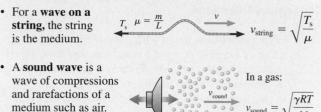

- For a **wave on a string,** the string is the medium.

$$T_s \quad \mu = \frac{m}{L} \qquad v_{string} = \sqrt{\frac{T_s}{\mu}}$$

- A **sound wave** is a wave of compressions and rarefactions of a medium such as air.

In a gas:

$$v_{sound} = \sqrt{\frac{\gamma RT}{M}}$$

Electromagnetic waves are waves of the electromagnetic field. They do not require a medium. All electromagnetic waves travel at the same speed in a vacuum, $c = 3.00 \times 10^8$ m/s.

IMPORTANT CONCEPTS

Graphical representation of waves

A snapshot graph is a picture of a wave at one instant in time. For a periodic wave, the **wavelength** λ is the distance between crests.

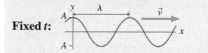

Fixed t:

A history graph is a graph of the displacement of one point in a medium versus time. For a periodic wave, the **period** T is the time between crests.

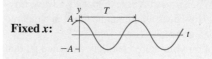

Fixed x:

Mathematical representation of waves

Sinusoidal waves are produced by a source moving with simple harmonic motion. The equation for a sinusoidal wave is a function of position and time.

$$y(x, t) = A \cos\left(2\pi\left(\frac{x}{\lambda} \pm \frac{t}{T}\right)\right)$$

+: wave travels to left
−: wave travels to right

For sinusoidal and other periodic waves:

$$T = \frac{1}{f} \qquad v = f\lambda$$

The intensity of a wave is the ration of the power to the area.

$$I = P/A$$

For a **spherical wave** the power decreases with the surface area of the spherical **wave fronts:**

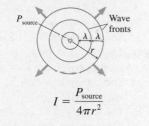

$$I = \frac{P_{source}}{4\pi r^2}$$

APPLICATIONS

The loudness of a sound is given by the sound intensity level. This is a logarithmic function of intensity and is in units of **decibels.**

- The usual **reference level** is the quietest sound that can be heard:

$$I_0 = 1.0 \times 10^{-12} \text{ W/m}^2$$

- The sound intensity level in dB is computed relative to this value:

$$\beta = (10 \text{ dB}) \log_{10}\left(\frac{I}{I_0}\right)$$

- A sound at the reference level corresponds to 0 dB.

The Doppler effect is a shift in frequency when there is relative motion of a wave source (frequency f_0, wave speed v) and an observer.

Moving source, stationary observer:

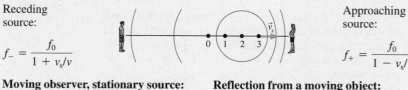

Receding source:

$$f_- = \frac{f_0}{1 + v_s/v}$$

Approaching source:

$$f_+ = \frac{f_0}{1 - v_s/v}$$

Moving observer, stationary source:

Approaching the source:

$$f_+ = (1 + v_o/v)f_0$$

Moving away from the source:

$$f_- = (1 - v_o/v)f_0$$

Reflection from a moving object:

For $v_o \ll v$, $\Delta f = \pm 2f_0 \dfrac{v_o}{v}$

When an object moves faster than the wave speed in a medium, a shock wave is formed.

QUESTIONS

Conceptual Questions

1. a. In your own words, define what a *transverse wave* is.
   b. Give an example of a wave that, from your own experience, you know is a transverse wave. What observations or evidence tells you this is a transverse wave?
2. a. In your own words, define what a *longitudinal wave* is.
   b. Give an example of a wave that, from your own experience, you know is a longitudinal wave. What observations or evidence tells you this is a longitudinal wave?
3. The wave pulses shown in Figure Q15.3 travel along the same string. Rank in order, from largest to smallest, their wave speeds v_1, v_2, and v_3. Explain.

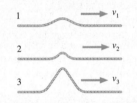

FIGURE Q15.3

4. Is it ever possible for one sound wave in air to overtake and pass another? Explain.
5. A wave pulse travels along a string at a speed of 200 cm/s. What will be the speed if:
   a. The string's tension is doubled?
   b. The string's mass is quadrupled (but its length is unchanged)?
   c. The string's length is quadrupled (but its mass is unchanged)?
   d. The string's mass and length are both quadrupled?
   Note that parts a–d are independent and refer to changes made to the original string.
6. An ultrasonic range finder sends out a pulse of ultrasound and measures the time between the emission of the pulse and the return of an echo from an object. This time is used to determine the distance to the object. To get good accuracy from the device, a user must enter the air temperature in the room. Why is this?
7. When water freezes, the density decreases and the bonds between molecules become stronger. Do you expect the speed of sound to be greater in liquid water or in water ice?
8. Figure Q15.8 is a history graph showing the displacement as a function of time at one point on a string. Did the displacement at this point reach its maximum of 2 mm *before* or *after* the interval of time when the displacement was a constant 1 mm? Explain how you interpreted the graph to answer this question.

   y (mm)

   2

   1

   0.00   0.04   0.08   t (s)

   FIGURE Q15.8

9. Figure Q15.9 shows a history graph *and* a snapshot graph for a wave pulse on a string. They describe the same wave from two perspectives.
   a. In which direction is the wave traveling? Explain.
   b. What is the speed of this wave?

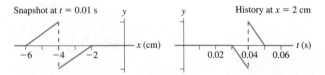

Snapshot at $t = 0.01$ s   History at $x = 2$ cm

FIGURE Q15.9

10. Rank in order, from largest to smallest, the wavelengths λ_1 to λ_3 for sound waves having frequencies $f_1 = 100$ Hz, $f_2 = 1000$ Hz, and $f_3 = 10,000$ Hz. Explain.
11. Explain why there is a factor of 2π in Equation 15.5.
12. A laser beam has intensity I_0.
    a. What is the intensity, in terms of I_0, if a lens focuses the laser beam to 1/10 its initial diameter?
    b. What is the intensity, in terms of I_0, if a lens defocuses the laser beam to 10 times its initial diameter?
13. Sound wave A delivers 2 J of energy in 2 s. Sound wave B delivers 10 J of energy in 5 s. Sound wave C delivers 2 mJ of energy in 1 ms. Rank in order, from largest to smallest, the sound powers P_A, P_B, and P_C of these three waves. Explain.
14. When you want to "snap" a towel, the best way to wrap the towel is so that the end that you hold and shake is thick, and the far end is thin. When you shake the thick end, a wave travels down the towel. How does wrapping the towel in a tapered shape help make for a good snap?
15. The volume control on your stereo is likely designed so that each time you turn it by one click, the loudness increases by a certain number of dB. Does each click increase the output power by a fixed amount as well?
16. You are standing at $x = 0$ m, listening to seven identical sound sources described by Figure Q15.16. At $t = 0$ s, all seven are at $x = 343$ m and moving as shown below. The sound from all seven will reach your ear at $t = 1$ s. Rank in order, from highest to lowest, the seven frequencies f_1 to f_7 that you hear at $t = 1$ s. Explain.

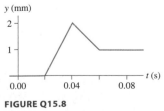

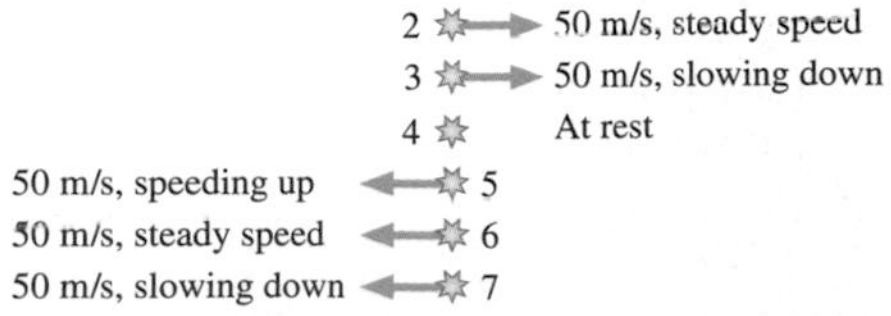

FIGURE Q15.16

17. A bullet can travel at speeds of over 1000 m/s. When a bullet is fired from a rifle, the actual firing makes a distinctive sound, but people at a distance may hear a second, different sound that is even louder. Explain the source of this sound.

Multiple-Choice Questions

18. | Denver, Colorado has an oldies station that calls itself "KOOL 105." This means that they broadcast radio waves at a frequency of 105 MHz. Suppose that they decide to describe their station by its wavelength (in meters), instead of by its frequency. What name would they now use?
    A. KOOL 0.35   B. KOOL 2.85
    C. KOOL 3.5   D. KOOL 285

19. | What is the frequency of blue light with a wavelength of 400 nm?
    A. 1.33×10^3 Hz   B. 7.50×10^{12} Hz
    C. 1.33×10^{14} Hz   D. 7.50×10^{14} Hz

20. | Ultrasound can be used to deliver energy to tissues for ther-
    BIO apy. It can penetrate tissue to a depth approximately 200 times its wavelength. What is the approximate depth of penetration of ultrasound at a frequency of 5.0 MHz?
    A. 0.29 mm   B. 1.4 cm
    C. 6.2 cm   D. 17 cm

21. ‖ A sinusoidal wave traveling on a string has a period of 0.20 s, a wavelength of 32 cm, and an amplitude of 3 cm. The speed of this wave is
    A. 0.60 cm/s   B. 6.4 cm/s
    C. 15 cm/s   D. 160 cm/s

22. ‖ Two strings of different linear density are joined together and pulled taut. A sinusoidal wave on these strings is traveling to the right, as shown in Figure Q15.22. When the wave goes across the boundary from string 1 to string 2:

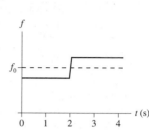
FIGURE Q15.22

String 1   String 2

    A. The velocity increases.
    B. The velocity stays the same.
    C. The velocity decreases.

23. ‖ You stand at $x = 0$ m, listening to a sound that is emitted at frequency f_0. Figure Q15.23 shows the frequency you hear during a four-second interval. Which of the following describes the motion of the sound source?

FIGURE Q15.23

    A. It moves from left to right and passes you at $t = 2$ s.
    B. It moves from right to left and passes you at $t = 2$ s.
    C. It moves toward you but doesn't reach you. It then reverses direction at $t = 2$ s.
    D. It moves away from you until $t = 2$ s. It then reverses direction and moves toward you but doesn't reach you.

PROBLEMS

Section 15.1 The Wave Model

Section 15.2 Traveling Waves

1. | The wave speed on a string under tension is 200 m/s. What is the speed if the tension is doubled?

2. | The wave speed on a string is 150 m/s when the tension is 75.0 N. What tension will give a speed of 180 m/s?

3. | A wave travels along a string at a speed of 280 m/s. What will be the speed if the string is replaced by one made of the same material and under the same tension but having twice the radius?

4. | The back wall of an auditorium is 26.0 m from the stage. If you are seated in the middle row, how much time elapses between a sound from the stage reaching your ear directly and the same sound reaching your ear after reflecting from the back wall?

5. ‖‖ A hammer taps on the end of a 4.00-m-long metal bar at room temperature. A microphone at the other end of the bar picks up two pulses of sound, one that travels through the metal and one that travels through the air. The pulses are separated in time by 11.0 ms. What is the speed of sound in this metal?

6. | In an early test of sound propagation through the ocean, an underwater explosion of 1 pound of dynamite in the Bahamas was detected 3200 km away on the coast of Africa. How much time elapsed between the explosion and the detection?

Section 15.3 Graphical and Mathematical Descriptions of Waves

7. | Figure P15.7 is a snapshot graph of a wave at $t = 0$ s. Draw the history graph for this wave at $x = 6$ m, for $t = 0$ s to 6 s.

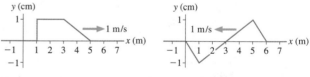

FIGURE P15.7   **FIGURE P15.8**

8. | Figure P15.8 is a snapshot graph of a wave at $t = 2$ s. Draw the history graph for this wave at $x = 0$ m, for $t = 1$ s to 8 s.

9. | Figure P15.9 is a history graph at $x = 0$ m of a wave moving to the right at 1 m/s. Draw a snapshot graph of this wave at $t = 1$ s.

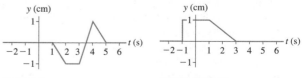

FIGURE P15.9   **FIGURE P15.10**

10. | Figure P15.10 is a history graph at $x = 2$ m of a wave moving to the left at 1 m/s. Draw the snapshot graph of this wave at $t = 0$ s.

11. | A sinusoidal wave has period 0.20 s and wavelength 2.0 m. What is the wave speed?

12. | A sinusoidal wave travels with speed 200 m/s. Its wavelength is 4.0 m. What is its frequency?

13. ‖ The motion detector used in a physics lab sends and receives 40 kHz ultrasonic pulses. A pulse goes out, reflects off the object being measured, and returns to the detector. The lab temperature is 20°C.
    a. What is the wavelength of the waves emitted by the motion detector?
    b. How long does it take for a pulse that reflects off an object 2.5 m away to make a round trip?

14. | The displacement of a wave traveling in the positive x-direction is $y(x, t) = (3.5 \text{ cm}) \cos(2.7x - 124t)$, where x is in m and t is in s. What are the (a) frequency, (b) wavelength, and (c) speed of this wave?

15. ‖ The displacement of a wave traveling in the negative x-direction is $y(x, t) = (5.2 \text{ cm}) \cos(5.5x + 72t)$, where x is in m and t is in s. What are the (a) frequency, (b) wavelength, and (c) speed of this wave?

16. | Figure P15.16 is a snapshot graph at $t = \frac{1}{4}$ s of a traveling wave with displacement $y(x,t) = (2.0 \text{ cm}) \times \cos(2\pi x - 4\pi t)$, where x is in m and t is in s.

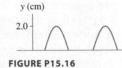
FIGURE P15.16

    a. Reproduce this graph on your page, then use a *dotted* line to show the snapshot graph at $t = 0$ s and a *dashed* line to show the snapshot graph at $t = \frac{1}{8}$ s.
    b. What is the speed of this wave?

17. | Figure P15.17 is a snapshot graph of a wave at $t = 0$ s. What are the amplitude, wavelength, and frequency of this wave?

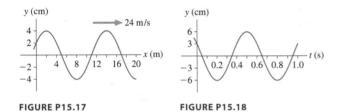

FIGURE P15.17   **FIGURE P15.18**

18. | Figure P15.18 is a history graph at $x = 0$ m of a wave moving to the right at 2 m/s. What are the amplitude, frequency, and wavelength of this wave?

Section 15.4 Sound and Light Waves

19. | A dolphin emits ultrasound at 100 kHz and uses the timing
BIO of reflections to determine the position of objects in the water. What is the wavelength of this ultrasound?

20. ‖ a. What is the wavelength of a 2.0 MHz ultrasound wave traveling through aluminum?
    b. What frequency of electromagnetic wave would have the same wavelength as the ultrasound wave of part a?

21. | a. At 20°C, what is the frequency of a sound wave in air with a wavelength of 20 cm?
    b. What is the frequency of an electromagnetic wave with a wavelength of 20 cm?
    c. What would be the wavelength of a sound wave in water that has the same frequency as the electromagnetic wave of part b?

22. | a. What is the frequency of blue light that has a wavelength of 450 nm?
    b. What is the frequency of red light that has a wavelength of 650 nm?

23. ‖ a. Telephone signals are often transmitted over long distances by microwaves. What is the frequency of microwave radiation with a wavelength of 3.0 cm?
    b. Microwave signals are beamed between two mountaintops 50 km apart. How long does it take a signal to travel from one mountaintop to the other?

24. ‖ a. An FM radio station broadcasts at a frequency of 101.3 MHz. What is the wavelength?
    b. What is the frequency of a sound source that produces the same wavelength in 20°C air?

Section 15.5 Energy and Intensity

25. ‖ Sound is detected when a sound wave causes the eardrum
BIO to vibrate (see Figure 14.27). Typically, the diameter of the
INT eardrum is about 8.4 mm in humans. How much energy is delivered to your eardrum each second when someone whispers (20 dB) a secret in your ear?

26. ‖ A fan at a rock concert is 30 m from the stage, where the
BIO sound intensity level is 110 dB. Her eardrums have a diameter
INT of 8.4 mm. How much energy is transferred to each eardrum in one second?

27. ‖ The intensity of electromagnetic waves from the sun is 1.4 kW/m² just above the earth's atmosphere. Eighty percent of this reaches the surface at noon on a clear summer day. Suppose you model your back as a 30 cm × 50 cm rectangle. How many joules of solar energy fall on your back as you work on your tan for 1.0 hr?

28. | The sun emits electromagnetic waves with a power of 4.0×10^{26} W. Determine the intensity of electromagnetic waves from the sun just outside the atmospheres of (a) Venus, (b) Mars, and (c) Saturn. Refer to the table of astronomical data on the inside back cover.

29. | On the other side of the Milky Way, Planet M orbits a star at a distance of 1.31×10^{11} m. Planet N orbits at a distance of 1.49×10^{12} m. Light from the star has an intensity of 1840 W/m² at Planet M. What is the intensity of the light at Planet N?

Section 15.6 Loudness of Sound

30. | What is the sound intensity level of a sound with an intensity of 3.0×10^{-6} W/m²?

31. ‖ What is the sound intensity of a whisper at a distance of 2.0 m, in W/m²? What is the corresponding sound intensity level in dB?

32. | The sound intensity from a jack hammer breaking concrete is 2.0 W/m² at a distance of 2.0 m from the point of impact. This is sufficiently loud to cause permanent hearing damage if the operator doesn't wear ear protection. What are (a) the sound intensity and (b) the sound intensity level for a person watching from 50 m away?

33. ‖ A concert loudspeaker suspended high off the ground emits 35 W of sound power. A small microphone with a 1.0 cm² area is 50 m from the speaker. What are (a) the sound intensity and (b) the sound intensity level at the position of the microphone?

34. | A rock band playing an outdoor concert produces sound at 120 dB 5.0 m away from their single working loudspeaker. What is the sound intensity level 35 m from the speaker?

Section 15.7 The Doppler Effect and Shock Waves

35. | An opera singer in a convertible sings a note at 600 Hz while cruising down the highway at 90 km/hr. What is the frequency heard by
    a. A person standing beside the road in front of the car?
    b. A person standing beside the road behind the car?

36. || An osprey's call is a distinct whistle at 2200 Hz. An osprey
    BIO calls while diving at you, to drive you away from her nest. You hear the call at 2300 Hz. How fast is the osprey approaching?

37. | A whistle you use to call your hunting dog has a frequency of 21 kHz, but your dog is ignoring it. You suspect the whistle may not be working, but you can't hear sounds above 20 kHz. To test it, you ask a friend to blow the whistle, then you hop on your bicycle. In which direction should you ride (toward or away from your friend) and at what minimum speed to know if the whistle is working?

38. | A friend of yours is loudly singing a single note at 400 Hz while driving toward you at 25.0 m/s on a day when the speed of sound is 340 m/s.
    a. What frequency do you hear?
    b. What frequency does your friend hear if you suddenly start singing at 400 Hz?

39. || While anchored in the middle of a lake, you count exactly three waves hitting your boat every 10 s. You raise anchor and start motoring slowly in the same direction the waves are going. When traveling at 1.5 m/s, you notice that exactly two waves are hitting the boat from behind every 10 s. What is the speed of the waves on the lake?

40. || A Doppler blood flow unit emits ultrasound at 5.0 MHz.
    BIO What is the frequency shift of the ultrasound reflected from blood moving in an artery at a speed of 0.20 m/s?

General Problems

41. | You're watching a carpenter pound a nail. He hits the nail twice a second, but you hear the sound of the strike when his hammer is fully raised. What is the minimum distance from you to the carpenter? Assume the air temperature is 20°C.

42. |||| A 2.50 kHz sound wave is transmitted through an aluminum rod.
    a. What is its wavelength in the aluminum?
    b. What is the sound wave's frequency when it passes into the air?
    c. What is its wavelength in air?

43. ||| Oil explorers set off explosives to make loud sounds, then listen for the echoes from underground oil deposits. Geologists suspect that there is oil under 500-m-deep Lake Physics. It's known that Lake Physics is carved out of a granite basin. Explorers detect a weak echo 0.94 s after exploding dynamite at the lake surface. If it's really oil, how deep will they have to drill into the granite to reach it?

44. || A 2.0-m-long string is under 20 N of tension. A pulse travels the length of the string in 50 ms. What is the mass of the string?

45. || Andy (mass 80 kg) uses a 3.0-m-long rope to pull Bob
    INT (mass 60 kg) across the floor ($\mu_k = 0.20$) at a constant speed of 1.0 m/s. Bob signals to Andy to stop by "plucking" the rope, sending a wave pulse forward along the rope. The pulse reaches Andy 150 ms later. What is the mass of the rope?

46. | If a bungee cord is stretched horizontally to a length of 2.5 m, the tension in the cord is 2.1 N. A transverse pulse on the cord makes a round trip in 1.6 s. A pulse makes the round trip in 1.6 s if the cord is stretched to a length of 3.5 m. What is the tension in the 3.5-m-long cord?

47. || String 1 in Figure P15.47 has linear density 2.0 g/m and string 2 has linear density 4.0 g/m. A student sends pulses in both directions by quickly pulling up on the knot, then releasing it. What should the string lengths L_1 and L_2 be if the pulses are to reach the ends of the strings simultaneously?

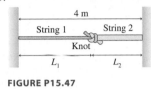

FIGURE P15.47

48. | Ships measure the distance to the ocean bottom with *sonar*. A pulse of sound waves is aimed at the ocean bottom, then sensitive microphones listen for the echo. The graph in Figure P15.48 shows the delay time as a function of the ship's position as it crosses 60 km of ocean. Draw a graph of the ocean bottom. Let the ocean surface define $y = 0$ and ocean bottom have negative values of y. This way your graph will be a picture of the ocean bottom. The speed of sound in ocean water varies slightly with temperature, but you can use 1500 m/s as an average value.

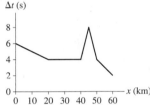

FIGURE P15.48

49. ||| A long string is shaken at its left end at 5.0 Hz, sending a sinusoidal wave down the string at 2.0 m/s. After the wave has traveled 1.0 m, the linear density of the string suddenly decreases by a factor of four. Draw a snapshot graph of the first 3.0 m of the string at an instant when the left end is at its equilibrium position.

50. || A coyote can locate a sound source with good accuracy by
    BIO comparing the arrival times of a sound wave at its two ears. Suppose a coyote is listening to a bird whistling at 1000 Hz. The bird is 3.0 m away, directly in front of the coyote's right ear. The coyote's ears are 15 cm apart.
    a. What is the distance between the bird and the coyote's left ear?
    b. What is the difference in the arrival time of the sound at the left ear and the right ear?
    c. What is the ratio of this time difference to the period of the sound wave?
    Hint: You are looking for the difference between two numbers that are nearly the same. What does this near equality imply about the necessary precision during intermediate stages of the calculation?

51. | A 256 Hz sound wave in 20°C air propagates into the water of a swimming pool. What are the water-to-air ratios of the wave's frequency, wave speed, and wavelength?

52. ‖ Earthquakes produce seismic waves that travel through the earth. The speed of longitudinal seismic waves, called P waves, is 8000 m/s. Transverse seismic waves, called S waves, travel at a slower 4500 m/s. A seismograph records the two waves from a distant earthquake. If the S wave arrives 2.0 min after the P wave, how far away was the earthquake? You can assume that the waves travel in straight lines, although actual seismic waves follow more complex routes.

53. ‖‖ One way to monitor global warming is to measure the average temperature of the ocean. Researchers are doing this by measuring the time it takes sound pulses to travel underwater over large distances. At a depth of 1000 m, where ocean temperatures hold steady near 4°C, the average sound speed is 1480 m/s. It's known from laboratory measurements that the sound speed increases 4.0 m/s for every 1.0°C increase in temperature. In one experiment, where sounds generated near California are detected in the South Pacific, the sound waves travel 8000 km. If the smallest time change that can be reliably detected is 1.0 s, what is the smallest change in average temperature that can be measured?

54. ‖ Figure P15.54 shows two snapshot graphs taken 10 ms apart, with the blue curve being the first snapshot. What are the (a) wavelength, (b) speed, (c) frequency, and (d) amplitude of this wave?

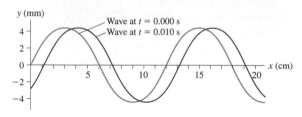

FIGURE P15.54

55. | Low-frequency vertical oscillations are one possible cause BIO of motion sickness, with 0.3 Hz having the strongest effect. Your boat is bobbing in place at just the right frequency to cause you the maximum discomfort.
    a. How much time elapses between two waves hitting the ship?
    b. If the wave crests appear to be about 30 m apart, what would you estimate to be the speed of the waves?

56. | A wave on a string is described by $y(x, t) = (3.0 \text{ cm}) \times \cos[2\pi(x/(2.4 \text{ m}) + t/(0.20 \text{ s}))]$, where x is in m and t in s.
    a. In what direction is this wave traveling?
    b. What are the wave speed, frequency, and wavelength?
    c. At $t = 0.50 \text{ s}$, what is the displacement of the string at $x = 0.20 \text{ m}$?

57. | Write the y-equation for a wave traveling in the negative x-direction with wavelength 50 cm, speed 4.0 m/s, and amplitude 5.0 cm.

58. | Write the y-equation for a wave traveling in the positive x-direction with frequency 200 Hz, speed 400 m/s, and amplitude 0.010 mm.

59. | A wave is described by the expression $y(x, t) = (3.0 \text{ cm}) \times \cos(1.5x - 50t)$, where x is in m and t is in s.
    a. Make an accurate snapshot graph of this wave.
    b. What is the speed of the wave and in what direction is it traveling?

60. ‖‖ A point on a string undergoes simple harmonic motion as a INT sinusoidal wave passes. When a sinusoidal wave with speed 24 m/s, wavelength 30 cm, and amplitude of 1.0 cm passes, what is the maximum speed of a point on the string?

61. | A simple pendulum is INT made by attaching a small cup of sand with a hole in the bottom to a 1.2 m long string. The pendulum is mounted on the back of a small motorized car. As the car drives forward, the pendulum swings from side to side and leaves a trail of sand as shown in Figure P15.61. How fast was the car moving?

FIGURE P15.61

62. ‖ a. A typical 100 W lightbulb produces 4.0 W of visible light. (The other 96 W are dissipated as heat and infrared radiation.) What is the light intensity on a wall 2.0 m away from the lightbulb?
    b. A krypton laser produces a cylindrical red laser beam 2.0 mm in diameter with 2.0 W of power. What is the light intensity on a wall 2.0 m away from the laser?

63. ‖ An AM radio station broadcasts with a power of 25 kW at a frequency of 920 kHz. Estimate the intensity of the radio wave at a point 10 km from the broadcast antenna.

64. ‖‖ The earth's average distance from the sun is 1.50×10^{11} m. At this distance, the intensity of radiation from the sun is 1.38 kW/m². The earth's radius is 6.37×10^6 m. What is the total solar power received by the earth? (For comparison, total human power consumption is roughly 10^{13} W.)

65. ‖‖ Lasers can be used to drill or cut material. One such laser generates a series of high-intensity pulses rather than a continuous beam of light. Each pulse contains 500 mJ of energy and lasts 10 ns. The laser fires 10 such pulses per second.
    a. What is the *peak power* of the laser light? The peak power is the power output during one of the 10 ns pulses.
    b. What is the average power output of the laser? The average power is the total energy delivered per second.
    c. A lens focuses the laser beam to a 10-μm-diameter circle on the target. During a pulse, what is the light intensity on the target?
    d. The intensity of sunlight at the surface of the earth at midday is about 1100 W/m². What is the ratio of the laser intensity on the target to the intensity of the midday sun?

66. ‖‖ The sound intensity 50 m from a wailing tornado siren is 0.10 W/m².
    a. What is the sound intensity level?
    b. In a noisy neighborhood, the weakest sound likely to be heard over background noise is 60 dB. Estimate the maximum distance at which the siren can be heard.

67. ‖ A stereo is claimed to put out 150 W per channel, for a total of 300 W. This is electrical power to the speakers; the actual sound output is much less. Suppose a stereo loudspeaker actually did put out 300 W of sound power, radiated equally in all directions. What would be the sound intensity level 5.0 m from the speaker?

68. ‖ Five trumpet players have a sound intensity level of 75 dB. What would be the sound intensity level of 20 such trumpet players?

69. ‖ A speaker at an open-air concert emits 600 W of sound power, radiated equally in all directions.
    a. What is the intensity of the sound 5.00 m from the speaker?
    b. What sound intensity level would you experience there if you did not have any protection for your ears?
    c. Earplugs you can buy in the drugstore have a noise reduction rating of 23 decibels. If you are wearing those earplugs but your friend Phil is not, how far from the speaker should Phil stand to experience the same loudness as you?

70. | A bat locates insects by emitting ultrasonic "chirps" and
    BIO then listening for echoes. The lowest-frequency chirp of a big brown bat is 26 kHz. How fast would the bat have to fly, and in what direction, for you to just barely be able to hear the chirp at 20 kHz?

71. | A physics professor demonstrates the Doppler effect by
    INT tying a 600 Hz sound generator to a 1.0-m-long rope and whirling it around her head in a horizontal circle at 100 rpm. What are the highest and lowest frequencies heard by a student in the classroom? Assume the room temperature is 20°C.

72. | Ocean waves with wavelength 1.2 m and period 1.5 s are moving past a pier. A boy runs along the pier, in the direction opposite to the motion of the wave, at 3.5 m/s. How many wave crests pass the boy each second?

73. ‖ A source of sound moves toward you at speed v_s and away from Jane, who is standing on the other side of it. You hear the sound at twice the frequency as Jane. What is the speed of the source? Assume that the speed of sound is 340 m/s.

Passage Problems

Echolocation BIO

As discussed in the chapter, many species of bats find flying insects by emitting pulses of ultrasound and listening for the reflections. This technique is called **echolocation**. Bats possess several adaptations that allow them to echolocate very effectively.

74. | Although we can't hear them, the ultrasonic pulses are very loud. In order not be deafened by the sound they emit, bats can temporarily turn off their hearing. Muscles in the ear cause the bones in their middle ear to separate slightly, so that they don't transmit vibrations to the inner ear. After an ultrasound pulse ends, a bat can hear an echo from an object 1 m away. Approximately how much time after a pulse is emitted is the bat ready to hear its echo?
    A. 0.5 ms   B. 1 ms   C. 3 ms   D. 6 ms

75. | Bats are sensitive to very small changes in frequency of the reflected waves. What information does this allow them to determine about their prey?
    A. Size   B. Speed   C. Distance   D. Species

76. | Some bats have specially shaped noses that they use to focus the ultrasound pulses in the forward direction. Why is this useful?
    A. They are not distracted by echoes from several directions.
    B. The energy of the pulse is concentrated in a smaller area, so the intensity is larger.
    C. The pulse goes forward only, so it doesn't affect the bat's hearing.

77. | Some bats utilize a sound pulse with a rapidly decreasing frequency. A decreasing-frequency pulse has
    A. Decreasing wavelength.
    B. Decreasing speed.
    C. Increasing wavelength.
    D. Increasing speed.

STOP TO THINK ANSWERS

Stop to Think 15.1: Transverse. The wave moves horizontally through the crowd, but individual spectators move up and down, transverse to the motion of the wave.

Stop to Think 15.2: D, E. Shaking your hand faster or farther will change the shape of the wave, but this will not change the wave speed; the speed is a property of the medium. Changing the linear density of the string or its tension will change the wave speed. To increase the speed, you must decrease the linear density or increase the tension.

Stop to Think 15.3: B. All three waves have the same speed, so the frequency is highest for the wave with the shortest wave-

length. (Imagine the three waves moving to the right; the one with the crests closest together has the crests passing by most rapidly.)

Stop to Think 15.4: A. The plane wave does not spread out, so its intensity will be constant. The other two waves spread out, so their intensity will decrease.

Stop to Think 15.5: C. The source is moving toward Zack, so he observes a higher frequency. The source is moving away from Amy, so she observes a lower frequency.

16

SUPERPOSITION AND STANDING WAVES

The didgeridoo is a simple musical instrument played by aboriginal tribes in Australia. It consists of a hollow tube that is open at both ends. How can the player get a wide range of notes out of such a simple device?

A person playing a musical instrument such as a trumpet, flute, or didgeridoo makes sound waves. Listeners hear the traveling sound waves that the instrument emits, but the more important sound waves are the ones that stay inside the instrument's tube. These *standing waves* determine the sound that the instrument produces. Standing waves also exist in laser cavities and microwave ovens, and they are responsible for determining the sound of your voice.

Before we can talk about standing waves, we have to understand how two or more traveling waves interact with each other. What happens when two waves overlap? This combination is called a *superposition*. As with other wave concepts, the principle of superposition will apply to all waves that we study, from water waves to sound waves to light waves.

16.1 The Principle of Superposition

Figure 16.1a shows two baseball players, Alan and Bill, at batting practice. Unfortunately, someone has turned the pitching machines so that pitching machine A throws baseballs toward Bill while machine B throws toward Alan. If two baseballs are launched at the same time, and with the same speed, they collide at the crossing point and bounce away. Two baseballs cannot occupy the same point of space at the same time.

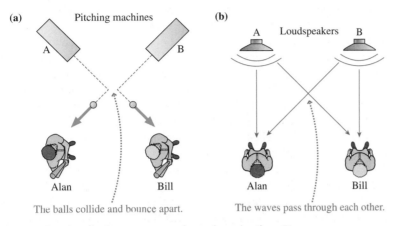

(a) Pitching machines   **(b)** Loudspeakers

Alan   Bill   Alan   Bill

The balls collide and bounce apart.   The waves pass through each other.

FIGURE 16.1 Two baseballs cannot pass through each other. Two waves can.

But unlike baseballs, sound waves *can* pass directly through each other. In Figure 16.1b, Alan and Bill are listening to the stereo system in the locker room after practice. Both hear the music quite well, without distortion or missing sound, so the sound wave that travels from speaker A toward Bill must pass through the wave traveling from speaker B toward Alan, with no effect on either. This is a basic property of waves.

What happens to the medium at a point where two waves are present simultaneously? What is the displacement of the medium at this point? Figure 16.2 shows a sequence of photos of two wave pulses traveling along a stretched string. In the first photo, the waves are approaching each other. In the second, the waves overlap, and the displacement of the string is larger than it was for either of the individual waves. A careful measurement would reveal that the displacement is the sum of the displacements of the two individual waves. In the third frame, the waves have passed through each other and continue on as if nothing had happened.

This result is not limited to stretched strings; the outcome is the same whenever two waves of any type pass through each other. This is known as the *principle of superposition.*

> **Principle of superposition** When two or more waves are *simultaneously* present at a single point in space, the displacement of the medium at that point is the sum of the displacements due to each individual wave.

To use the principle of superposition you must know the displacement that each wave would cause if it traveled through the medium alone. Then you go through the medium *point by point* and add the displacements due to each wave *at that point* to find the net displacement at that point. The outcome will be different at each and every point in the medium because the displacements are different at each point.

Let's illustrate this principle with an idealized example. Figure 16.3 shows five snapshot graphs taken 1 s apart of two waves traveling at the same speed (1 m/s) in opposite directions along a string. The principle of superposition comes into play wherever the waves overlap. The displacement of each wave is shown as a dotted line. The solid line is the sum *at each point* of the two displacements at that point. This is the displacement that you would actually observe as the two waves pass through each other.

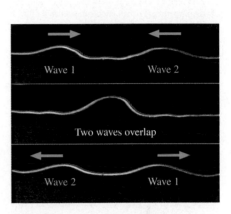

FIGURE 16.2 Two wave pulses on a stretched string pass through each other.

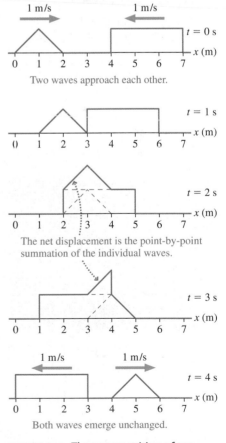

FIGURE 16.3 The superposition of two waves on a string as they pass through each other.

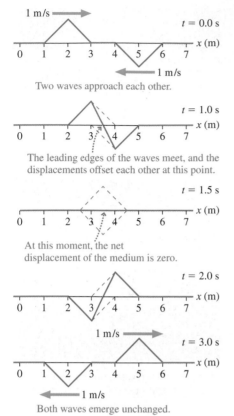

1 m/s

$t = 0.0$ s

1 m/s

Two waves approach each other.

$t = 1.0$ s

The leading edges of the waves meet, and the displacements offset each other at this point.

$t = 1.5$ s

At this moment, the net displacement of the medium is zero.

$t = 2.0$ s

$t = 3.0$ s

1 m/s

Both waves emerge unchanged.

FIGURE 16.4 Two waves with opposite displacements produce destructive interference.

Constructive and Destructive Interference

The superposition of two waves is often called **interference.** The displacements of the waves in Figure 16.3 are both positive, so the total displacement of the medium where they overlap is larger than it would be due to either of the waves separately. We call this **constructive interference.**

Figure 16.4 shows another series of snapshot graphs of two counterpropagating waves, but this time one has a negative displacement. The principle of superposition still applies, but now the displacements are opposite each other. The displacement of the medium where the waves overlap is *less* than it would be due to either of the waves separately. We call this **destructive interference.**

In the series of graphs in Figure 16.4 the displacement of the medium at $x = 3.5$ m is always zero. The positive displacement of the wave traveling to the left and the negative displacement of the wave traveling to the right always exactly cancel at this spot. The complete cancellation at one point of two waves traveling in opposite directions is something we will see again. When the displacements of the waves cancel, where does the energy of the wave go? We know that the waves continue on unchanged after their interaction, so no energy is dissipated. Consider the graph at $t = 1.5$ s. There is no net displacement at any point of the medium at this instant, but the *string is moving rapidly.* The energy of the waves hasn't vanished—it is in the form of the kinetic energy of the medium.

STOP TO THINK 16.1 Two pulses on a string approach each other at speeds of 1 m/s. What is the shape of the string at $t = 6$ s?

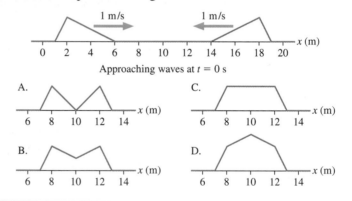

Approaching waves at $t = 0$ s

A.   B.   C.   D.

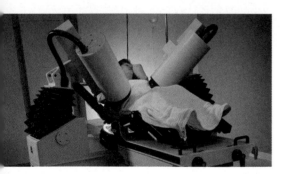

◀ **Breaking stones with sound** BIO As we saw in Chapter 15, waves carry energy. The energy of high-intensity ultrasonic waves can be used to break up kidney stones so that they can be cleared from the body, a technique known as *shock wave lithotripsy.* This machine uses two generators, each of which produces ultrasonic waves. The two waves, which enter the body at different points, are directed so that they overlap and produce constructive interference at the position of a stone. This allows the individual waves to have lower intensity, minimizing tissue damage as they pass through the body, while still providing a region of high intensity right where it is needed.

16.2 Standing Waves

When you pluck a guitar string or a rubber band stretched between your fingers, you create waves. But how is this possible? There isn't really anywhere for the waves to go, because the string or the rubber band is held between two fixed ends. Figure 16.5 on the next page shows a photograph of waves on a stretched elastic cord. This is a wave, though it may not look like one, because it doesn't "travel" either right or left. Waves that are "trapped" between two boundaries, like those in the photo or on a guitar string, are what we call *standing waves.* **Individual points on the string oscillate up and down, but the wave itself does not travel.** It is called a **standing wave** because the crests and troughs "stand in place" as it oscillates. As we'll see, a standing wave isn't a totally new kind of wave; it is simply the superposition of two traveling waves moving in opposite directions.

Superposition Creates a Standing Wave

Suppose we have a string on which two sinusoidal waves of equal wavelength and amplitude travel in opposite directions, as in Figure 16.6a. When the waves meet, the displacement of the string will be a superposition of these two waves. Figure 16.6b shows nine snapshot graphs, at intervals of $\frac{1}{8}T$, of the two waves as they move through each other. The red and green dots identify particular crests on each of the waves to help you see that the red wave is traveling to the right and the green wave to the left. At *each point,* the net displacement of the medium is found by adding the red displacement and the green displacement. The resulting blue wave is the superposition of the two traveling waves.

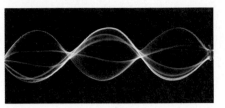

FIGURE 16.5 The motion of a standing wave on a string.

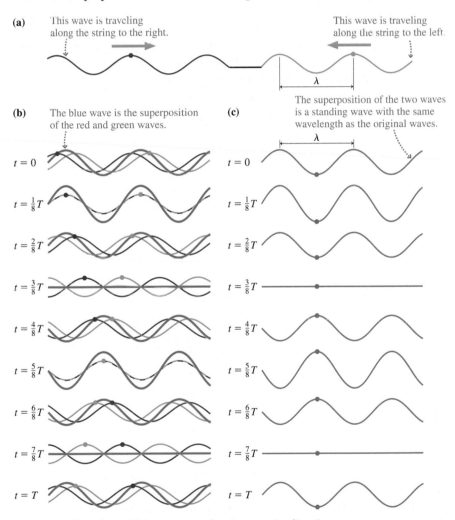

(a) This wave is traveling along the string to the right.

This wave is traveling along the string to the left.

λ

(b) The blue wave is the superposition of the red and green waves.

(c) The superposition of the two waves is a standing wave with the same wavelength as the original waves.

λ

$t = 0$

$t = \frac{1}{8}T$

$t = \frac{2}{8}T$

$t = \frac{3}{8}T$

$t = \frac{4}{8}T$

$t = \frac{5}{8}T$

$t = \frac{6}{8}T$

$t = \frac{7}{8}T$

$t = T$

FIGURE 16.6 Two sinusoidal waves traveling in opposite directions.

Figure 16.6b is rather complicated, so Figure 16.6c shows just the blue superposition of the two waves. This is the wave that you would actually observe in the medium. Interestingly, the blue dot shows that the wave in Figure 16.6c is moving neither right nor left. The superposition of the two counterpropagating traveling waves is a standing wave. Notice that the wavelength of the standing wave, the distance between two crests or two troughs, is the same as the wavelengths of the two traveling waves that combine to produce it.

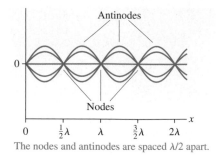

The nodes and antinodes are spaced λ/2 apart.

FIGURE 16.7 Standing waves are often represented as they would be seen in a strobe photograph, which makes the nodes and antinodes clearly visible.

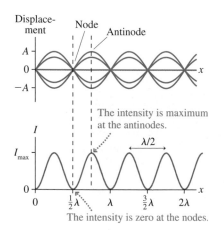

FIGURE 16.8 Intensity of a standing wave.

Nodes and Antinodes

In Figure 16.7 we have superimposed the nine snapshot graphs of Figure 16.6c into a single graphical representation of this standing wave. Compare this to Figure 16.5 and you can see why we called the vibrating string a standing wave. A striking feature of a standing-wave pattern is the existence of points that *never move!* These points, which are spaced λ/2 apart, are called **nodes.** Halfway between the nodes are points where the particles in the medium oscillate with maximum displacement. These points of maximum amplitude are called **antinodes,** and you can see that they are also spaced λ/2 apart. This means that **the wavelength of a standing wave is *twice* the distance between successive nodes or successive antinodes.**

It seems surprising and counterintuitive that some particles in the medium have no motion at all. This happens for the same reason we saw in Figure 16.4: The two waves exactly offset each other at that point. Look carefully at the two traveling waves in Figure 16.6b. You will see that the nodes occur at points where at *every instant* of time the displacements of the two traveling waves have equal magnitudes but *opposite signs.* Thus the superposition of the displacements at these points is always zero—they are points of destructive interference. The antinodes have large displacements. They correspond to points where the two displacements have equal magnitudes and the *same sign* at all times. Constructive interference at these points gives a displacement twice that of each individual wave.

The intensity of a wave is largest at points where it oscillates with maximum amplitude. You can see in Figure 16.8 that the points of maximum intensity along the standing wave occur at the antinodes; the intensity is zero at the nodes. If this is a standing sound wave, the loudness of the sound varies from zero (no sound) at the nodes to a maximum at the antinodes and then back to zero. The key idea is that **the intensity is maximum at points of constructive interference and zero at points of destructive interference.**

EXAMPLE 16.1 Setting up a standing wave
Two children hold an elastic cord at each end. Each child shakes the end of the cord 2.0 times per second, sending waves at 3.0 m/s toward the middle of the cord, where the two waves combine to create a standing wave. What is the distance between adjacent nodes?

SOLVE The distance between adjacent nodes is λ/2. The wavelength, frequency, and speed are related as $v = f\lambda$, as we saw in Chapter 15, so the wavelength is

$$\lambda = \frac{v}{f} = \frac{3.0 \text{ m/s}}{2.0 \text{ Hz}} = 1.5 \text{ m}$$

The distance between adjacent nodes is λ/2 and thus is 0.75 m.

10.4–10.6   Act|v ONLINE Physics

16.3 Standing Waves on a String

The oscillation of a guitar string is a standing wave. A standing wave is naturally produced on a string when both ends are fixed (i.e., tied down), as in the case of a guitar string or the string in the photo of Figure 16.5. We also know that a standing wave is produced when there are two counterpropagating traveling waves. But you don't shake both ends of a guitar string to produce the standing wave! How do we actually get two traveling waves on a string with both ends fixed? Before we can answer this question, we need a brief explanation of what happens when a traveling wave encounters a boundary or a discontinuity.

Reflections

We know that light reflects from mirrors; it can also reflect from the surface of a pond or from a pane of glass. As we saw in Chapter 15, sound waves reflect as well; that's how an echo is produced. To understand reflections, we'll look at waves on a string, but the results will be completely general and can be applied to other waves as well.

Suppose we have a string that is attached to a wall or other fixed support, as in Figure 16.9. The wall is what we will call a *boundary*—it's the end of the medium. When the pulse reaches this boundary, it reflects, moving away from the wall. *All* the wave's energy is reflected, hence **the amplitude of a wave reflected from a boundary is unchanged.**

Figure 16.9 shows that the amplitude doesn't change when the pulse reflects, but the pulse is inverted. We can understand why by looking at the forces at the boundary. When the pulse reaches the wall, the piece of string right at the wall feels an upward force from the approaching pulse. But this piece of the string is fixed in place, so there must be an opposite force from the wall on this piece of string. This downward force leads to an inverted reflected pulse.

Waves also reflect from what we will call a *discontinuity,* a point where there is a change in the properties of the medium. Figure 16.10a shows a discontinuity between a string with a larger linear density and one with a smaller linear density. The tension is the same in both strings, so the wave speed is slower on the left, faster on the right. Whenever a wave encounters a discontinuity, some of the wave's energy is *transmitted* forward and some is reflected. Because energy must be conserved, both the transmitted and the reflected pulse have a smaller amplitude than the initial pulse.

In Figure 16.10b, an incident wave encounters a discontinuity at which the wave speed decreases. Once again, some of the wave's energy is transmitted and some is reflected.

In Figure 16.10a, the reflected pulse is right-side up. The string on the right is quite light and provides little resistance, so the junction moves up and down as the pulse passes. This motion of the string is just like the original "snap" of the string that started the pulse, so the reflected pulse has the same orientation as the original pulse. In Figure 16.10b, the string on the right is more massive, so it looks more like the fixed boundary in Figure 16.9 and the reflected pulse is again inverted.

Creating a Standing Wave

Now that we understand reflections, let's create a standing wave. Figure 16.11 shows a string of length L that is tied at $x = 0$ and $x = L$. This string has *two* boundaries where reflections can occur. If you wiggle the string in the middle, sinusoidal waves travel outward in both directions and soon reach the boundaries, where they reflect. The reflections at the ends of the string cause two waves of *equal amplitude and wavelength* to travel in opposite directions along the string. As we've just seen, these are the conditions that cause a standing wave!

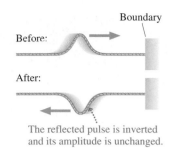

The reflected pulse is inverted and its amplitude is unchanged.

FIGURE 16.9 A wave reflects when it encounters a boundary.

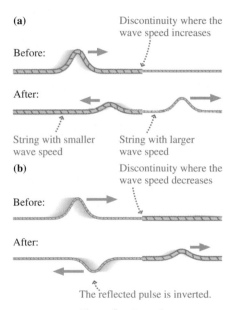

The reflected pulse is inverted.

FIGURE 16.10 The reflection of a wave at a discontinuity.

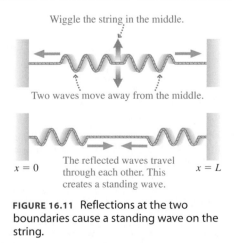

FIGURE 16.11 Reflections at the two boundaries cause a standing wave on the string.

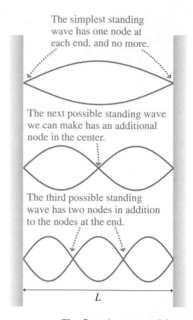

The simplest standing wave has one node at each end, and no more.

The next possible standing wave we can make has an additional node in the center.

The third possible standing wave has two nodes in addition to the nodes at the end.

L

FIGURE 16.12 The first three possible standing waves on a string of length L.

What kind of standing waves might develop on the string? There are two conditions that must be met:

- Because the string is tied down at the ends, the displacements at $x = 0$ and $x = L$ must be zero at all times. Stated another way, we require nodes at both ends of the string.
- We know that standing waves have a spacing of $\lambda/2$ between nodes. This means that the nodes must be equally spaced.

Figure 16.12 sketches the first few possible waves that meet these conditions. These are called the standing wave **modes** of the string. To help quantify the possible waves, we can assign a **mode number** m to each. The first wave in Figure 16.12, with a node at each end, has a mode number $m = 1$. The next wave is $m = 2$, and so on.

NOTE ▶ Figure 16.12 shows only the first three modes, for $m = 1$, $m = 2$, and $m = 3$. But there are many more modes, for all possible values of m. ◀

The distance between adjacent nodes is $\lambda/2$, so the different modes have different wavelengths. For the first mode in Figure 16.12, the distance between nodes is the length of the string, so we can write

$$\lambda_1 = 2L$$

The subscript identifies the mode number; in this case $m = 1$. For $m = 2$, the distance between nodes is $L/2$; this means that $\lambda_2 = L$. Generally, for any mode m the wavelength is given by the following equation:

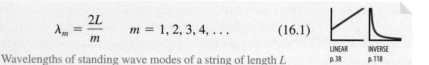

$$\lambda_m = \frac{2L}{m} \qquad m = 1, 2, 3, 4, \ldots \qquad (16.1)$$

Wavelengths of standing wave modes of a string of length L

LINEAR
p. 38

INVERSE
p. 118

These are the only possible wavelengths for standing waves on the string. **A standing wave can exist on the string *only* if its wavelength is one of the values given by Equation 16.1.**

NOTE ▶ Other wavelengths, which would be perfectly acceptable wavelengths for a traveling wave, cannot exist as a *standing* wave of length L because they do not meet the constraint of having a node at each end of the string. ◀

If standing waves are possible only for certain wavelengths, then only specific oscillation frequencies are allowed. Because $\lambda f = v$ for a sinusoidal wave, the oscillation frequency corresponding to wavelength λ_m is

$$f_m = \frac{v}{\lambda_m} = \frac{v}{2L/m} = m\left(\frac{v}{2L}\right) \qquad m = 1, 2, 3, 4, \ldots \qquad (16.2)$$

Frequencies of standing wave modes of a string of length L

TRY IT YOURSELF

◀**Through the glass darkly** A piece of window glass is a discontinuity to a light wave, so it both transmits and reflects light. To verify this, look at the windows in a brightly lit room at night. The small percentage of the interior light that reflects from windows is more intense than the light coming in from outside, so reflection dominates and the windows show a mirror-like reflection of the room. Now turn out the lights. With no more reflected interior light you will be able to see the transmitted light from outside.

Figure 16.13 shows the first three modes with their wavelengths and frequencies. You can see that **the mode number *m* is equal to the number of antinodes of the standing wave.** You can therefore tell a string's mode of oscillation by counting the number of antinodes (*not* the number of nodes).

In Chapter 14, we looked at the concept of *resonance*. A mass on a spring has a certain frequency at which it "wants" to oscillate. If the system is driven at its resonance frequency, it will develop a large amplitude of oscillation. A stretched string will support standing waves, meaning it has a series of frequencies at which it "wants" to oscillate: the frequencies of the different standing wave modes. We can call these **resonant modes,** or more simply, just **resonances.** A small oscillation of a stretched string at a frequency near one of its resonant modes will cause it to develop a standing wave with a large amplitude. In Figure 16.14, a small amplitude oscillation at the *m* = 4 mode frequency of a stretched string produces a large amplitude standing wave.

> **NOTE** ▶ When we draw standing wave modes, as in Figure 16.13, we usually only show the *envelope* of the wave, the greatest extent of the motion of the string. The string's motion is actually continuous and goes through all intermediate positions as well, as we see from the time-exposure photograph of a standing wave in Figure 16.14. ◀

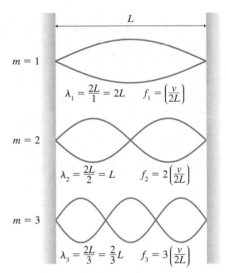

FIGURE 16.13 Possible standing waves of a string fixed at both ends.

The Fundamental and the Higher Harmonics

The sequence of possible frequencies for a standing wave on a string has an interesting pattern that is worth exploring. The first mode has frequency:

$$f_1 = \frac{v}{2L} \tag{16.3}$$

We call this the **fundamental frequency** of the string. All of the other modes have frequencies that are multiples of this fundamental frequency. We can rewrite Equation 16.2 in terms of the fundamental frequency as

$$f_m = m f_1 \qquad m = 1, 2, 3, 4, \ldots \tag{16.4}$$

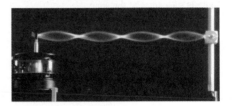

FIGURE 16.14 A resonant mode of a stretched string.

The allowed standing-wave frequencies are all integer multiples of the fundamental frequency. This sequence of possible frequencies is called a set of **harmonics.** The fundamental frequency f_1 is also known as the *first harmonic,* the *m* = 2 wave at frequency f_2 is called the *second harmonic,* the *m* = 3 wave is called the *third harmonic,* and so on. The frequencies above the fundamental frequency, the harmonics with *m* = 2, 3, 4, . . . , are referred to as the **higher harmonics.**

EXAMPLE 16.2 Identifying harmonics on a string

A 2.50-m-long string vibrates as a 100 Hz standing wave with nodes 1.00 m and 1.50 m from one end of the string and at no points in between these two. Which harmonic is this, and what is the string's fundamental frequency?

PREPARE We begin with the visual overview in Figure 16.15, in which we sketch the appearance of this particular standing wave. We set up an *x*-axis with one end of the string at *x* = 0 m and the

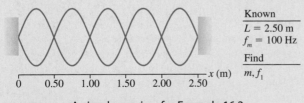

Known
$L = 2.50$ m
$f_m = 100$ Hz

Find
m, f_1

FIGURE 16.15 A visual overview for Example 16.2.

other end at *x* = 2.50 m. These must be nodes. We also know there are nodes at 1.00 m and 1.50 m, but there are no nodes in between these. This means that the spacing between nodes is 0.50 m. Figure 16.15 is a sketch of a standing wave with a node every 0.50 m.

SOLVE We can simply count the number of antinodes of the standing wave to deduce the mode number; this is mode *m* = 5. The frequencies of the harmonics are given by $f_m = m f_1$, hence the fundamental frequency is

$$f_1 = \frac{f_5}{5} = \frac{100 \text{ Hz}}{5} = 20 \text{ Hz}$$

ASSESS Let's look at this another way to check the value for the mode number. The distance between nodes is 0.50 m. In a 2.5 m length of string, we can fit five such segments, giving us five antinodes, so this is *m* = 5, as we found.

Standing waves on a bridge This photograph shows the Tacoma Narrows suspension bridge on the day in 1940 when it experienced a catastrophic oscillation that led to its collapse. Aerodynamic forces caused the amplitude of a particular resonant mode of the bridge to increase dramatically until the bridge failed. In this photo, the red line shows the original line of the deck of the bridge. You can clearly see the large amplitude of the oscillation and the node at the center of the span.

Stringed Musical Instruments

Think about stringed musical instruments, such as the guitar, the piano, and the violin. These instruments all have strings that are fixed at both ends and tightened to create tension. A disturbance is generated on the string by plucking, striking, or bowing. Regardless of how it is generated, the disturbance creates standing waves on the string. Understanding the sound of a stringed musical instrument means understanding standing waves.

In Chapter 15, we saw that the speed of a wave on a stretched string depended on T_s, the tension in the string and μ, the linear density, as $v = \sqrt{T_s/\mu}$. Combining this with Equation 16.3, we find that the fundamental frequency of a stretched string is

$$f_1 = \frac{v}{2L} = \frac{1}{2L}\sqrt{\frac{T_s}{\mu}} \tag{16.5}$$

When you pluck or bow a string, you initially excite a wide range of frequencies. However, resonance sees to it that the only frequencies to persist are those of the possible standing waves. The musical note you hear when the string is sounded is the fundamental frequency f_1. The higher harmonics determine the *tone quality,* something we will explore later in the chapter.

> **NOTE ▶** In Equation 16.5, v is the wave speed *on the string,* not the speed of sound in air. ◀

For instruments like the guitar or the violin, the strings are all the same length and under approximately the same tension. Were that not the case, the neck of the instrument would tend to twist toward the side of higher tension. The strings have different frequencies because they differ in linear density. The lower-pitched strings are "fat" while the higher-pitched strings are "skinny." This difference changes the frequency by changing the wave speed. Small adjustments are then made in the tension to bring each string to the exact desired frequency.

CONCEPTUAL EXAMPLE 16.1 Tuning and playing a guitar

A guitar has strings of a fixed length. Plucking a string makes a particular musical note. A player can make other notes by pressing the string against frets, metal bars on the neck of the guitar, as shown in Figure 16.16. The fret becomes the new end of the string, making the effective length shorter.

FIGURE 16.16 Guitar frets.

a. A guitar player plucks a string to play a note. He then presses down on a fret to make the string shorter. Does the new note have a higher or lower frequency?

b. The frequency of one string is too low. (Musically, we say the note is "flat.") How must the tension be adjusted to bring the string to the right frequency?

REASON

a. The fundamental frequency, the note we hear, is $f_1 = (1/2L)\sqrt{T_s/\mu}$. Because f_1 is inversely proportional to L, decreasing the string length increases the frequency.

b. Because f is proportional to the square root of T_s, the player must increase the tension to increase the fundamental frequency.

ASSESS If you watch someone play a guitar, you can see that he or she plays higher notes by moving the fingers to shorten the strings.

EXAMPLE 16.3 Setting the tension in a guitar string

The fifth string on a guitar plays the musical note A, at a frequency of 110 Hz. On a typical guitar, this string is stretched between two fixed points 0.640 m apart, and this length of string has a mass of 2.86 g. What is the tension in the string?

PREPARE Strings sound at their fundamental frequency, so 110 Hz is f_1.

SOLVE The linear density of the string is

$$\mu = (2.86 \times 10^{-3}\,\text{kg})/(0.640\,\text{m}) = 4.47 \times 10^{-3}\,\text{kg/m}$$

We can rearrange Equation 16.5 for the fundamental frequency to solve for the tension in terms of the other variables:

$$T_s = (2Lf_1)^2\mu = (2(0.640\,\text{m})(110\,\text{Hz}))^2(4.47 \times 10^{-3}\,\text{kg/m})$$
$$= 88.6\,\text{N}$$

ASSESS A guitar has six strings. If each has approximately the same tension, the total force on the neck of the guitar is six times this tension, a bit more than 500 N. This is quite a bit of force.

Standing Electromagnetic Waves

The standing wave descriptions we've found for a vibrating string are actually valid for any transverse wave, including an electromagnetic wave. For example, standing light waves can be established between two parallel mirrors that reflect the light back and forth. The mirrors are boundaries, analogous to the boundaries at the ends of a string. In fact, this is exactly how a laser works. The two facing mirrors in Figure 16.17 form what is called a *laser cavity.*

Because the mirrors act exactly like the points to which a string is tied, the light wave must have a node at the surface of each mirror. (To allow some of the light to escape the laser cavity and form the *laser beam,* one of the mirrors lets some of the light through. This doesn't affect the node.)

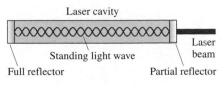
Laser cavity

Standing light wave

Laser beam

Full reflector Partial reflector

FIGURE 16.17 A laser contains a standing light wave between two parallel mirrors.

▶ **Microwave modes** A microwave oven uses a type of electromagnetic wave—microwaves, with a wavelength of about 12 cm—to heat food. The inside walls of a microwave oven are reflective to microwaves, so we have the correct conditions to set up a standing wave. This isn't a good thing! A standing wave has high intensity at the antinodes and low intensity at the nodes, so your oven has hot spots and cold spots, as we see in this thermal image showing the interior of an oven with a thin layer of water that has been "cooked" for a short time. A turntable in a microwave oven keeps the food moving so that no part of your dinner remains at a node or an antinode.

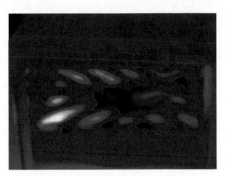

EXAMPLE 16.4 Finding the mode number for a laser
A helium-neon laser (such as the laser in supermarket barcode scanners) emits light of wavelength $\lambda = 633$ nm. A typical cavity for such a laser is 15.0 cm long. What is the mode number of the standing wave in this cavity?

PREPARE Because a light wave is a transverse wave, Equation 16.1 for λ_m applies to a laser as well as a vibrating string.

SOLVE The standing light wave in a laser cavity has a mode number m that is approximately

$$m = \frac{2L}{\lambda} = \frac{2 \times 0.150 \text{ m}}{633 \times 10^{29} \text{ m}} = 474{,}000$$

ASSESS The standing light wave inside a laser cavity may have half a million anti-nodes! This is a consequence of the very small wavelength of light.

STOP TO THINK 16.2 A standing wave on a string is shown. Which of the modes shown below (on the same string) has twice the frequency of the original wave?

Original standing wave

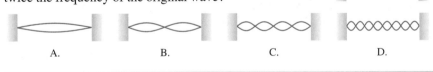

A. B. C. D.

16.4 Standing Sound Waves

Wind instruments like flutes, trumpets, and didgeridoos work very differently from stringed instruments. The player blows air into one end of a tube, producing standing waves of sound that make the notes we hear. In this section, we will look at the properties of such standing sound waves.

Recall that a sound wave is a longitudinal pressure wave. The air molecules oscillate back and forth parallel to the direction in which the wave is traveling, creating *compressions* (regions of higher pressure) and *rarefactions* (regions of lower pressure). Consider a sound wave confined to a long, narrow column of air, such as the air in a tube. A wave traveling down the tube eventually reaches the end, where it encounters the atmospheric pressure of the surrounding environment. This is a discontinuity, much like the small rope meeting the big rope in Figure 16.10b. Part

Fiery interference In this apparatus, a speaker at one end of the metal tube emits a sinusoidal wave. The wave reflects from the other end, which is closed, to make a counter-propagating wave and set up a standing sound wave in the tube. The tube is filled with propane gas that exits through small holes on top. The burning propane allows us to easily discern the nodes and the antinodes of the standing sound wave. An exciting demonstration—but one you shouldn't try yourself!

of the wave's energy is transmitted out into the environment, allowing you to hear the sound, and part is reflected back into the tube. Reflections at both ends of the tube create waves traveling both directions inside the tube, and their superposition, like that of the reflecting waves on a string, is a standing wave.

We start by looking at a sound wave in a tube open at both ends. What kind of standing waves can exist in such a tube? Because the ends of the tube are open to the atmosphere, the pressure at the ends is "fixed" at atmospheric pressure and cannot vary. This is analogous to a stretched string that is fixed at the end. As a result, **the open end of a column of air must be a node of the pressure wave.**

Figure 16.18 shows a column of air open at both ends. We call this an *open-open tube*. Whereas the antinodes of a standing wave on a string are points where the string oscillates with maximum displacement, the antinodes of a standing sound wave are where the pressure has the largest variation, creating, alternately, the maximum compression and the maximum rarefaction. In Figure 16.18, the air molecules squeeze together on the left side of the tube. Then, in Figure 16.18b, half a cycle later, the air molecules squeeze together on the right side. The varying density creates a variation in pressure across the tube, as the graphs show. Figure 16.18c combines the information of Figures 16.18a and 16.18b into a graph of the pressure of the standing sound wave in the tube.

As the standing wave oscillates, the air molecules "slosh" back and forth along the tube with the wave frequency, alternately pushing together (maximum pressure at the antinode) and pulling apart (minimum pressure at the antinode). This makes sense, because sound is a longitudinal wave in which the air molecules oscillate parallel to the tube.

NOTE ▶ The variation in pressure from atmospheric pressure in a real standing sound wave is much smaller than Figure 16.18 implies. When we display graphs of the pressure in sound waves, we won't generally graph the pressure p. Instead, we will graph Δp, the variation from atmospheric pressure. ◀

Many musical instruments, such as a flute, can be modeled as open-open tubes. The flutist blows across one end to create a standing wave inside the tube, and a note of this frequency is emitted from both ends of the flute. The possible standing waves in tubes, like standing waves on strings, are resonances of the system. A gentle puff of air across the mouthpiece of a flute can cause large standing waves at these resonant frequencies.

Other instruments work differently from the flute. A trumpet or a clarinet is a column of air open at the bell end but *closed* by the player's lips at the mouthpiece. To be complete in our treatment of sound waves in tubes, we need to consider tubes that are closed at one or both ends. At a closed end, the air molecules can alternately rush toward the wall, creating a compression, and then rush away from the wall, leaving a rarefaction. Thus **a closed end of an air column is an antinode of pressure.**

Figure 16.19 shows graphs of the first three standing-wave modes of a tube open at both ends (an *open-open tube*), a tube closed at both ends (a *closed-closed tube*), and a tube open at one end but closed at the other (an *open-closed tube*), all with the same length L. These are graphs of the pressure wave, with a node at open ends and an antinode at closed ends. The standing wave in the closed-closed tube looks like the wave in the open-open tube except that the positions of the nodes and antinodes are interchanged. In both cases there are m half-wavelength segments between the ends, thus the wavelengths and frequencies of an open-open tube and a closed-closed tube are the same as those of a string tied at both ends:

(a) At one instant

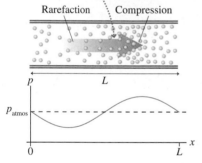

(b) Half a cycle later

The shift between compression and rarefaction means a motion of molecules along the tube.

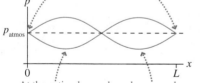

(c) At the ends of the tube, the pressure is equal to atmospheric pressure. These are nodes.

At the antinodes, each cycle sees a change from compression to rarefaction and back to compression.

FIGURE 16.18 The $m = 2$ standing sound wave inside an open-open column of air.

$$\begin{cases} \lambda_m = \dfrac{2L}{m} \\[2ex] f_m = m\left(\dfrac{v}{2L}\right) = mf_1 \end{cases} \qquad m = 1, 2, 3, 4, \ldots \qquad (16.6)$$

Wavelengths and frequencies of standing sound wave modes
in an open-open or closed-closed tube

(a) Open-open **(b)** Closed-closed **(c)** Open-closed

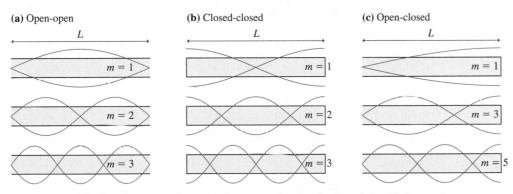

FIGURE 16.19 The first three standing sound wave modes in columns of air with different ends. These graphs show the pressure variation in the tube.

The open-closed tube is different, as we can see from Figure 16.19. The $m = 1$ mode has a node at one and an antinode at the other, and so has only one-quarter of a wavelength in a tube of length L. The $m = 1$ wavelength is $\lambda_1 = 4L$, twice the $m = 1$ wavelength of an open-open or a closed-closed tube. Consequently, **the fundamental frequency of an open-closed tube is half that of an open-open or a closed-closed tube of the same length.**

The wavelength of the next mode of the open-closed tube is $4L/3$. Because this is $1/3$ of λ_1, we assign $m = 3$ to this mode. The wavelength of the subsequent mode is $4L/5$, so this is $m = 5$. In other words, an open-closed tube allows only odd-numbered modes. Consequently, the possible wavelengths and frequencies are

$$\begin{cases} \lambda_m = \dfrac{4L}{m} \\[2ex] f_m = m\dfrac{v}{4L} = mf_1 \end{cases} \qquad m = 1, 3, 5, 7, \ldots \qquad (16.7)$$

Wavelengths and frequencies of standing sound wave modes
in an open-closed tube

NOTE ▶ Because sound is a pressure wave, the graphs of Figure 16.19 are *not* "pictures" of the wave as they are for a string wave. The graphs show the pressure variation versus position x. The tube itself is shown merely to indicate the location of the open and closed ends, but the diameter of the tube is *not* related to the amplitude of the wave. ◀

We are now in a position to suggest the following problem-solving strategy, not just for sound waves, but any standing wave.

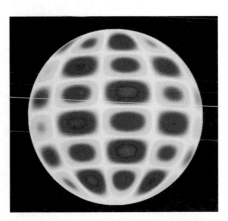

Standing waves in the sun The sun is a three-dimensional ball of hot gas. Violent convection in the gas as heat emerges from the sun's core can set up standing waves that can be seen in the motion of the sun's surface. This computer model shows one standing-wave mode. The red areas are moving up at this instant while the blue areas are moving down; they'll reverse during the next half cycle. These are the antinodes. The yellow areas in between, which are not moving, are the nodes. This is a more complex standing wave than a one-dimensional gas in a tube, but the basic ideas are the same.

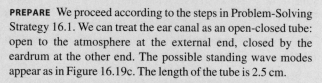

PROBLEM-SOLVING
STRATEGY 16.1 Standing waves

PREPARE

- For sound waves, determine what sort of pipe or tube you have: open-open, closed-closed, or open-closed.
- For string or light waves, the ends will be fixed points.
- Determine known values: length of the tube or string, frequency, wavelength, positions of nodes or antinodes.
- It may be useful to sketch a visual overview, including a picture of the relevant mode.

SOLVE For a string, the allowed frequencies and wavelengths are given by Equations 16.1 and 16.2. For sound waves in an open-open or closed-closed tube, the allowed frequencies and wavelengths are given by Equation 16.6; for an open-closed tube, by Equation 16.7.

ASSESS Does your final answer seem reasonable? Is there another way to check on your results? For example, the number of antinodes gives the mode. Does this agree?

EXAMPLE 16.5 Resonances of the ear canal

The eardrum, which transmits vibrations to the sensory organs of your ear, lies at the end of the ear canal. As Figure 16.20 shows, the ear canal in adults is about 2.5 cm in length. What frequency standing waves can occur within the ear canal that are within the range of human hearing? The speed of sound in the warm air of the ear canal is 350 m/s.

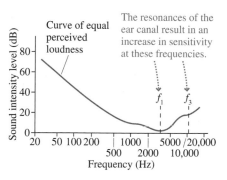

FIGURE 16.20 The anatomy of the ear.

PREPARE We proceed according to the steps in Problem-Solving Strategy 16.1. We can treat the ear canal as an open-closed tube: open to the atmosphere at the external end, closed by the eardrum at the other end. The possible standing wave modes appear as in Figure 16.19c. The length of the tube is 2.5 cm.

SOLVE Equation 16.7 gives the allowed frequencies in an open-closed tube. We are looking for the frequencies in the range 20 Hz–20,000 Hz. The fundamental frequency is

$$f_1 = \frac{v}{4L} = \frac{350 \text{ m/s}}{4(0.025 \text{ m})} = 3500 \text{ Hz}$$

The higher harmonics are odd multiples of this frequency:

$$f_3 = 3(3500 \text{ Hz}) = 10,500 \text{ Hz}$$
$$f_5 = 5(3500 \text{ Hz}) = 17,500 \text{ Hz}$$

These three modes lie within the range of human hearing; higher modes are greater than 20,000 Hz.

ASSESS These standing wave modes are resonances of the ear canal. There will be an increased amplitude of oscillation within the ear canal in response to sounds at these frequencies.

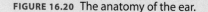

FIGURE 16.21 A curve of equal perceived loudness.

How important are the resonances of the ear canal calculated in Example 16.5? In Chapter 15, you learned about the decibel scale for measuring loudness of sound. In fact, your ears have varying sensitivity to sounds of different frequencies. Figure 16.21 shows a curve of *equal perceived loudness,* the sound intensity level (in dB) required to give the *impression* of equal loudness for sinusoidal waves of the noted frequency. Lower values mean that your ear is more sensitive at that frequency. In general, the curve decreases to about 1000 Hz, the frequency at which your hearing is most acute, then slowly rises at higher frequencies. However, this general trend is punctuated by two dips in the curve, showing two frequencies at which a quieter sound produces the same perceived loudness. As you can see, these two dips correspond to the resonances f_1 and f_3 of the ear canal. The resonances are frequencies at which the air in the ear canal "wants" to oscillate. Incoming sounds at these frequencies produce a larger oscillation, resulting in an increased sensitivity to these frequencies.

Wind Instruments

With a wind instrument, blowing into the mouthpiece creates a standing sound wave inside a tube of air. The player changes the notes by using her fingers to cover holes or open valves, changing the effective length of the tube. The first open hole becomes a node because the tube is open to the atmosphere at that point. The fact that the holes are on the side, rather than literally at the end, makes very little difference. The length of the tube determines the standing wave resonances, and thus the musical note that the instrument produces.

Many wind instruments have a "buzzer" at one end of the tube, such as a vibrating reed on a saxophone or clarinet, or the musician's vibrating lips on a trumpet or trombone. Buzzers like these generate a continuous range of frequencies rather than single notes, which is why they sound like a "squawk" if you play on just the mouthpiece without the rest of the instrument. When connected to the body of the instrument, most of those frequencies cause little response. But the frequencies from the buzzer that match the resonant frequencies of the instrument cause the buildup of large amplitudes at these frequencies—standing-wave resonances. The combination of these frequencies makes the musical note that we hear.

Truly classical music The oldest known musical instruments are bone flutes from burials in central China. The flutes in the photo are up to 9000 years old, and are made from naturally hollow bones from crowned cranes. The positions of the holes determine the frequencies that the flutes can produce. Soon after the first flutes were created, the design was standardized so that different flutes would play the same notes—including the notes in the modern Chinese musical scale.

CONCEPTUAL EXAMPLE 16.2 Comparing the flute and the clarinet

A flute and a clarinet have about the same length, but the lowest note that can be played on the clarinet is much lower than the lowest note that can be played on the flute. Why is this?

REASON A flute is an open-open tube; the frequency of the fundamental mode is $f_1 = v/2L$. A clarinet is open at one end, but the player's lips and the reed close it at the other end. The clarinet is thus an open-closed tube with a fundamental frequency $f_1 = v/4L$. This is about half the fundamental frequency of the flute, so the lowest note on the clarinet has a much lower pitch. In musical terms, it's about an octave lower than the flute.

ASSESS The possible standing wave modes depend on whether the end of an air column is open or closed. This changes the notes that can be played, as we have seen, and it also affects the way the instrument sounds. A flute and a clarinet sound very different to your ear; we will have more to say about this difference later in the chapter.

EXAMPLE 16.6 The importance of warming up

Players of wind instruments know that it is important to warm up their instruments in order for them to play in tune—that is, at the desired frequency. A "cold" flute plays the note A at 440 Hz when the air temperature is 20°C.

a. How long is the tube? At 20°C, the speed of sound in air is 343 m/s.
b. As the player blows air through the flute, the air inside the instrument warms up. Once the air temperature inside the flute has risen to 32°C, increasing the speed of sound to 350 m/s, what is the frequency?
c. At the higher temperature, how much must the length of the tube be changed to bring the frequency back to 440 Hz?

SOLVE A flute is an open-open tube with fundamental frequency $f_1 = v/2L$.

a. At 20°C, the length corresponding to 440 Hz is

$$L = \frac{v}{2f} = \frac{343 \text{ m/s}}{2(440 \text{ Hz})} = 0.390 \text{ m}$$

b. As the speed of sound increases, the frequency changes to

$$f_1(\text{at } 32°C) = \frac{350 \text{ m/s}}{2(0.390 \text{ m})} = 449 \text{ Hz}$$

c. To bring the flute back into tune, the length must be increased to give a frequency of 440 Hz with a speed of 350 m/s. The new length is

$$L = \frac{v}{2f_1} = \frac{350 \text{ m/s}}{2(440 \text{ Hz})} = 0.398 \text{ m}$$

Thus the flute must be increased in length by 8 mm.

ASSESS The dependence of the frequency on temperature means that wind instruments must have an adjustable joint to change the tube length so that players can keep them in tune as the temperature changes.

STOP TO THINK 16.3   A tube that is open at both ends supports a standing wave with harmonics at 300 Hz and 400 Hz, with no harmonics between. What is the fundamental frequency of this tube?

A. 50 Hz   B. 100 Hz   C. 150 Hz   D. 200 Hz   E. 300 Hz

16.5 Speech and Hearing

When you hear a particular note played on a guitar, it sounds very different from the same note played on a trumpet. And you have perhaps been to a lecture in which the speaker talked at essentially the same pitch the entire time—but you could still understand what was being said. Clearly, there is more to your brain's perception of sound than pitch alone. How do you tell the difference between a guitar and a trumpet? How do you distinguish between an "oo" vowel sound and an "ee" vowel sound at the same pitch?

The Frequency Spectrum

To this point, we have pictured sound waves as sinusoidal waves, with a well-defined frequency. In fact, most of the sounds that you hear are not pure sinusoidal waves. Most sounds are a mix, or superposition, of different frequencies. For example, we have seen how certain standing-wave modes are possible on a stretched string. When you pluck a string on a guitar, you generally don't excite just one standing-wave mode—you simultaneously excite many different modes.

If you play the note "middle C" on a guitar, the fundamental frequency is 262 Hz. There will be a standing wave at this frequency, but there will also be standing waves at the frequencies 524 Hz, 786 Hz, 1048 Hz, . . . all the higher harmonics predicted by Equation 16.4.

Figure 16.22a is a bar chart showing all the frequencies present in the sound of the vibrating guitar string. The height of each bar shows the relative intensity of that harmonic. The fundamental frequency has the highest intensity, but many other harmonics have significant intensities as well. A bar chart showing the relative intensities of the different frequencies is called the **frequency spectrum** of the sound.

When your brain interprets the mix of frequencies from the guitar in Figure 16.22a, it identifies the fundamental frequency as the *pitch*. 262 Hz corresponds to a middle C, so you will identify the pitch as a middle C, even though the sound consists of many different frequencies. Your brain uses the higher harmonics to determine the **tone quality,** which is also called the *timbre*. The tone quality—and therefore the higher harmonics—is what makes a middle C played on a guitar sound quite different from a middle C played on a trumpet. The frequency spectrum of a trumpet would show a very different pattern of the relative intensities of the higher harmonics, and this different mix of higher harmonics gives the trumpet a different sound.

The actual sound wave produced by a guitar playing middle C is shown in Figure 16.22b. The sound wave is periodic, with a period of 3.81 ms that corresponds to the 262 Hz fundamental frequency. But the wave doesn't have a simple sinusoidal shape; it is more complex. **The higher harmonics don't change the period of the sound wave, they only change its shape.** The sound wave of a trumpet playing middle C would also have a 3.81 ms period, but its shape would be entirely different.

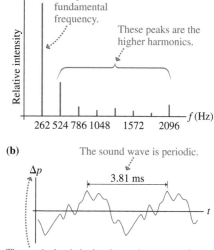

(a)

This peak is the fundamental frequency.

These peaks are the higher harmonics.

Relative intensity

262 524 786 1048   1572   2096 f (Hz)

(b)   The sound wave is periodic.

Δp

3.81 ms

t

The vertical axis is the change in pressure from atmospheric pressure due to the sound wave.

FIGURE 16.22 The frequency spectrum and a graph of the sound wave of a guitar playing a note with fundamental frequency 262 Hz.

CONCEPTUAL EXAMPLE 16.3 Playing the didgeridoo

The didgeridoo, a musical instrument developed by aboriginal Australians, is deceptively simple. It consists of a tube (a eucalyptus stem or branch hollowed out by termites) of $1\frac{1}{2}$ m or more in length. The player presses his lips against the end and blows air through his lips as with a trumpet. He may also make sounds with his vocal cords. Skilled players can make a wide variety of sounds. How is this possible with such a simple instrument?

REASON Because the lips seal one end, a didgeridoo has the resonances of an open-closed tube, given by Equation 16.7. The instrument can therefore produce many different frequencies. Changing the vibration of the lips and the sounds from the vocal cords can change the mix of standing-wave modes that are produced, leading to very different sounds.

One instrument, many sounds A synthesizer can be adjusted to sound like a flute, a clarinet, a trumpet, a piano—or any other musical instrument. The keys on a synthesizer determine what fundamental frequency to produce. The other controls adjust the mix of higher harmonics to match the frequency spectrum of various musical instruments, effectively mimicking their sounds.

Vowels and Formants

Try this: Keep your voice at the same pitch, and say the "ee" sound, as in "beet," then the "oo" sound as in "boot". Pay attention to how you reshape your mouth as you move back and forth between the two sounds. The two vowel sounds are at the same pitch, and thus have the same period, but they sound quite different. The difference in sound arises from the difference in the higher harmonics, just as for musical instruments. As you speak, you adjust the properties of your vocal tract to produce different mixes of harmonics that make the "ee," "oo," "ah," and other vowel sounds.

Speech begins with the vibration of your vocal cords, stretched bands of tissue in your throat. The vibration is similar to that of a wave on a stretched string. In ordinary speech, the average fundamental frequency for adult males and females is about 150 Hz and 250 Hz, respectively, but you can change the vibration frequency by changing the tension of your vocal cords. That's how you make your voice higher or lower as you sing. If you have a cold, your vocal cords can become inflamed and swollen, which increases their linear density. This makes the pitch of your voice lower.

As is the case with a guitar string, your vocal cords produce a mix of different frequencies as they vibrate—the fundamental frequency and a rich mixture of higher harmonics. If you were to put a microphone in your throat and measure the sound waves right at your vocal cords, you would measure a frequency spectrum similar to that in Figure 16.23a.

There is more to the story, though. Before reaching the opening of your mouth, sound from your vocal cords must pass through your vocal tract—a series of hollow cavities including your throat, your mouth, and your nose. The vocal tract acts much like a series of tubes, and, as in any tube, certain frequencies will set up standing-wave resonances. The rather broad standing-wave resonances of the vocal tract are called **formants**. Harmonics of your vocal cords at or near the formant frequencies will be amplified; harmonics that are far from a formant will be suppressed. Figure 16.23b shows the formants of an adult male making an "ee" vowel sound. The filtering of the vocal cord harmonics by the formants is clear.

You can change the shape and frequencies of the formants, and thus the sound you make, by changing the shape and length of your vocal tract. You do this by changing your mouth opening and the shape and position of your tongue. The first two formants for an "ee" sound are at roughly 270 Hz and 2300 Hz, but for an "oo" sound they become 300 Hz and 870 Hz. The much lower second formant of the "oo" emphasizes midrange frequencies, making a "calming" sound, while the more strident sound of "ee" comes from enhancing the higher frequencies.

(a) Frequencies from the vocal cords

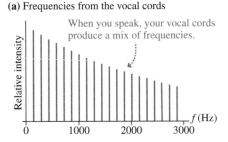

When you speak, your vocal cords produce a mix of frequencies.

(b) Actual spoken frequencies (Vowel sound "ee")

When you form your vocal tract to make a certain vowel sound it increases the amplitudes of certain frequencies and suppresses others.

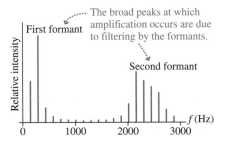

The broad peaks at which amplification occurs are due to filtering by the formants.

First formant

Second formant

FIGURE 16.23 The frequency spectrum from the vocal cords, and after passing through the vocal tract.

◄ **Saying "ah"** BIO Why, during a throat exam, does a doctor ask you to say "ah"? This particular vowel sound is formed by opening the mouth and the back of the throat wide—giving a clear view of the tissues of the throat.

CONCEPTUAL EXAMPLE 16.4 **High-frequency hearing loss**

As you age, your hearing sensitivity will decrease. This decrease is not uniform; for most people, the loss of sensitivity is greater for higher frequencies. The loss of sensitivity at high frequencies may make it difficult to understand what others are saying. Why is this?

REASON It is the high-frequency components of speech that allow us to distinguish different vowel sounds. A decrease in sensitivity to these higher frequencies makes it more difficult to make such distinctions.

ASSESS For high-frequency hearing loss, simple amplification of sound will not help; this will amplify all frequencies equally. A good hearing aid should only amplify the portions of the audio spectrum for which sensitivity is reduced.

STOP TO THINK 16.4   These sound waves represent notes played on different musical instruments. Which has the highest pitch?

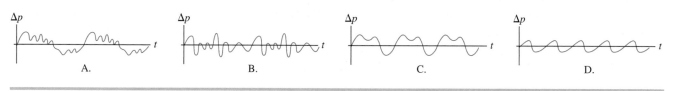

16.6 The Interference of Waves from Two Sources

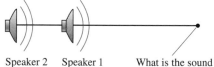

(a) Two sound waves overlapping along a line

Speaker 2   Speaker 1   What is the sound at this point?

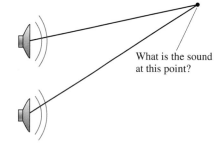

(b) Two overlapping spherical sound waves

What is the sound at this point?

FIGURE 16.24 Interference of waves from two sources.

Perhaps you have seen headphones that offer "active noise reduction." When you turn on the headphones, they produce sound that somehow *cancels* noise from the external environment. How does adding sound to a system make it quieter? This is one of the questions we will answer in this section, in which we look at the interference of waves from two sources.

We began the chapter by noting that waves, unlike particles, can pass through each other. Where they do, the principle of superposition tells us that the displacement of the medium is the sum of the displacements due to each wave acting alone. Consider the two loudspeakers in Figure 16.24, both emitting sound waves with the same frequency. In Figure 16.24a, sound from loudspeaker 2 passes loudspeaker 1, then two overlapped sound waves travel to the right along the *x*-axis. What sound is heard at the point indicated with the dot? And what about at the dot in Figure 16.24b, where the speakers are side by side? These are the two cases we will consider in this section. Although we'll use sound waves for our discussion, the results are completely general and apply to all waves. In Chapter 17, we will use these ideas to study the interference of light waves.

Interference Along a Line

Figure 16.25 on the next page shows two loudspeakers sending traveling waves along a line. The graphs are slightly displaced from each other so that you can see what each wave is doing, but the *physical situation* is one in which the waves are traveling *on top of* each other. We assume that the two speakers emit sound waves of identical frequency f, wavelength λ, and amplitude A.

At every point along the line, the net sound pressure wave will be the sum of the pressures from the individual waves. That's the principle of superposition. Because the two speakers are separated by one wavelength, the two waves are aligned crest-to-crest and trough-to-trough. Waves aligned this way are said to be **in phase;** waves that are in phase march along "in step" with each other. If we add the two waves point by point, we see that their superposition is a traveling wave with wavelength λ and twice the amplitude of the individual waves. This is constructive interference.

> **NOTE** ▶ Textbook pictures can be misleading because they're frozen in time. The net sound wave in Figure 16.25 is a *traveling* wave, moving to the right with the speed of sound v. It differs from the two individual waves only by having twice the amplitude. This is not a standing wave with nodes and anti-nodes that remain in one place. ◀

If d_1 and d_2 are the distances from loudspeakers 1 and 2 to a point at which we want to know the sound wave, their difference $\Delta d = d_2 - d_1$ is called the **path-length difference.** It is the *extra* distance traveled by wave 2 on the way to the point where the two waves are combined. In Figure 16.25, we see that constructive interference results from a path-length difference $\Delta d = \lambda$. But increasing Δd by an additional λ would produce exactly the same result, so we will also have constructive interference for $\Delta d = 2\lambda$, $\Delta d = 3\lambda$, and so on. In other words, **two waves will be in phase and will produce constructive interference any time their path-length difference is a whole number of wavelengths.**

In Figure 16.26, the two speakers are separated by half a wavelength. In this case, the crests of one wave align with the troughs of the other and the waves march along "out of step" with each other. We say that the two waves are **out of phase.** When two waves are out of phase, they are equal and opposite at every point. Consequently, the sum of the two waves is zero *at every point*. The superposition of two out-of-phase waves to produce a wave with zero amplitude is destructive interference.

The destructive interference of Figure 16.26 results from a path-length difference $\Delta d = \frac{1}{2}\lambda$. Again, increasing Δd by an additional λ would produce a picture that looks exactly the same, so we will also have destructive interference for $\Delta d = 1\frac{1}{2}\lambda$, $\Delta d = 2\frac{1}{2}\lambda$, and so on. That is, **two waves will be out of phase and will produce destructive interference any time their path-length difference is a whole number of wavelengths plus half a wavelength.**

Summing up, for two identical sources of waves, constructive interference occurs when the path-length difference is

$$\Delta d = m\lambda \qquad m = 0, 1, 2, 3, \ldots \qquad (16.8)$$

and destructive interference occurs when the path-length difference is

$$\Delta d = \left(m + \frac{1}{2}\right)\lambda \qquad m = 0, 1, 2, 3, \ldots \qquad (16.9)$$

> **NOTE** ▶ The path-length difference needed for constructive or destructive interference depends on the wavelength, and hence the frequency. If one particular frequency interferes destructively, another may not. ◀

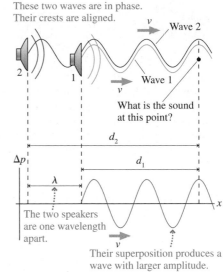

These two waves are in phase. Their crests are aligned.

The two speakers are one wavelength apart.

Their superposition produces a wave with larger amplitude.

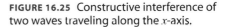

FIGURE 16.25 Constructive interference of two waves traveling along the x-axis.

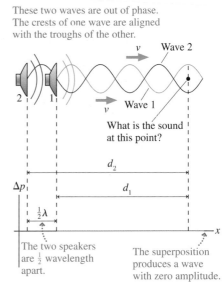

These two waves are out of phase. The crests of one wave are aligned with the troughs of the other.

The two speakers are $\frac{1}{2}$ wavelength apart.

The superposition produces a wave with zero amplitude.

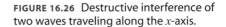

FIGURE 16.26 Destructive interference of two waves traveling along the x-axis.

The path-length difference is not necessarily the distance between the speakers, as we see in the next example. It is simply the difference in the distances traveled by the two waves.

EXAMPLE 16.7 **Is the sound loud or quiet? Part I**

Two loudspeakers 42.0 m apart and facing each other emit identical 115 Hz sinusoidal sound waves. Susan is walking along a line between the speakers. As she walks, she finds herself moving through loud and quiet spots. If Susan stands 19.5 m from one speaker, is she standing at a quiet spot or a loud spot? Assume that the speed of sound is 345 m/s.

PREPARE As Susan walks along the line between the speakers, she moves between points of constructive interference (loud spots) and destructive interference (quiet spots). Is her current position one of constructive or destructive interference? This will depend on the path-length difference. We start with a visual overview of the situation in Figure 16.27.

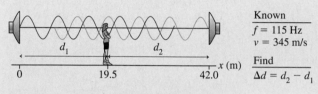

Known
$f = 115$ Hz
$v = 345$ m/s

Find
$\Delta d = d_2 - d_1$

FIGURE 16.27 Visual overview for Example 16.7

SOLVE At Susan's position, the distances the two waves travel to reach her are

$$d_1 = 19.5 \text{ m} \qquad d_2 = 42.0 \text{ m} - 19.5 \text{ m} = 22.5 \text{ m}$$

At the point where the two waves reach Susan and interfere, their path-length difference is

$$\Delta d = d_2 - d_1 = 3.0 \text{ m}$$

To know if we have constructive or destructive interference, we need to compare this with the wavelength:

$$\lambda = \frac{v}{f} = \frac{345 \text{ m/s}}{115 \text{ Hz}} = 3.0 \text{ m}$$

Because the path-length difference is exactly one wavelength, Susan is standing at a point of constructive interference. That is, she is standing at a loud spot.

ASSESS As you'll recall from earlier in the chapter, two counterpropagating waves create a standing wave. In this case, the conditions for constructive and destructive interference are the same as the conditions for antinodes and nodes, respectively, that we found earlier. Susan is at an antinode of the standing wave. The ideas of interference give us a different perspective on standing waves.

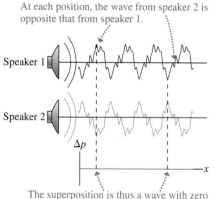

At each position, the wave from speaker 2 is opposite that from speaker 1.

The superposition is thus a wave with zero amplitude at all points.

FIGURE 16.28 Destructive interference of two waves.

The analysis above assumed that the two loudspeakers were emitting identical waves. Another interesting and important case of interference, illustrated in Figure 16.28, occurs when one loudspeaker emits a sound wave that is *the exact inverse* of the wave from the other speaker. If the speakers are side by side, so that $\Delta d = 0$, the superposition of these two waves will result in destructive interference; they will completely cancel. This destructive interference does not require the waves to have any particular frequency or any particular shape. As long as the waves are exact inverses of each other, there will be destructive interference.

Headphones with *active noise reduction* use this technique. A microphone on the outside of the headphones measures ambient sound. A circuit inside the headphones produces an inverted version of the microphone signal and sends it to the headphone speakers. The real ambient sound and the inverted version of the ambient sound coming from the speakers arrive at the ears together and interfere destructively, thus significantly reducing the sound intensity. In this case, *adding* sound results in a *lower* overall intensity inside the headphones!

Interference of Spherical Waves

Interference along a line illustrates the idea of interference, but it's not very realistic. In practice, sound waves from a loudspeaker or light waves from a light bulb spread out as spherical waves. Figure 16.29 on the next page shows a wave-front diagram for a spherical wave. Recall that the wave fronts represent the *crests* of the wave and are spaced by the wavelength λ. Halfway between two wave fronts is a trough of the wave. What happens when two spherical waves overlap? For example, imagine two loudspeakers emitting identical waves radiating sound in all directions. Figure 16.30 shows the wave fronts of the two waves.

This is a static picture, of course, so you have to imagine the wave fronts spreading out as new circular rings are born at the speakers. The waves overlap as they travel, and, as was the case in one dimension, this causes interference.

Consider a particular point like that marked by the red dot in Figure 16.30. The two waves each have a crest at this point, so there is constructive interference here. But at other points, such as that marked by the black dot, a crest overlaps a trough, so this is a point of destructive interference.

Notice—simply by counting the wave fronts—that the red dot is three wavelengths from speaker 2 ($r_2 = 3\lambda$) but only two wavelengths from speaker 1 ($r_1 = 2\lambda$). The path-length difference of the two waves arriving at the red dot is $\Delta r = r_2 - r_1 = \lambda$. That is, the wave from speaker 2 has to travel one full wavelength more than the wave from speaker 1, so the waves are in phase (crest aligned with crest) and interfere constructively. You should convince yourself that Δr is a *whole number of wavelengths* at every point where two wave fronts intersect.

Similarly, the path length difference at the black dot, where the interference is destructive, is $\Delta r = \frac{1}{2}\lambda$. As with interference along a line, destructive interference results when the path-length difference is a whole number of wavelengths plus half a wavelength.

Thus the general rule for determining whether there is constructive or destructive interference at any point is the same for spherical waves as for waves traveling along a line. For identical sources, constructive interference occurs when the path-length difference is

$$\Delta r = m\lambda \qquad m = 0, 1, 2, 3, \ldots \qquad (16.10)$$

Destructive interference occurs when the path-length difference is

$$\Delta r = \left(m + \frac{1}{2}\right)\lambda \qquad m = 0, 1, 2, 3, \ldots \qquad (16.11)$$

The conditions for constructive and destructive interference are the same for spherical waves as for waves along a line. And the treatment we have seen for sound waves can be applied to any wave, as we have noted. For any two wave sources, the following Tactics Box sums up how to determine if the interference at a point is constructive or destructive.

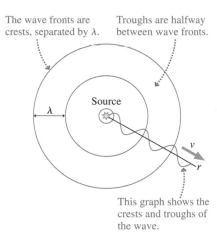

The wave fronts are crests, separated by λ.   Troughs are halfway between wave fronts.

This graph shows the crests and troughs of the wave.

FIGURE 16.29 A spherical wave.

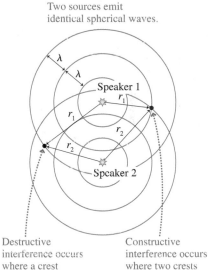

Two sources emit identical spherical waves.

Destructive interference occurs where a crest overlaps a trough.   Constructive interference occurs where two crests overlap.

FIGURE 16.30 The overlapping wave patterns of two sources.

TACTICS BOX 16.1 **Identifying constructive and**   ✎ Exercise 9
destructive interference

❶ Identify the path length from each source to the point of interest. Compute the path-length difference $\Delta r = |r_2 - r_1|$.
❷ Find the wavelength, if it is not specified.
❸ If the path-length difference is a whole number of wavelengths ($\lambda, 2\lambda, 3\lambda \ldots$), crests are aligned with crests and there is constructive interference.
❹ If the path-length difference is a whole number of wavelengths plus a half wavelength $\left(1\frac{1}{2}\lambda, 2\frac{1}{2}\lambda, 3\frac{1}{2}\lambda \ldots\right)$, crests are aligned with troughs and there is destructive interference.

NOTE ▶ Keep in mind that interference is determined by Δr, the path-length *difference*, not by r_1 or r_2. ◀

EXAMPLE 16.8 Is the sound loud or quiet? Part II

Two speakers are 3.0 m apart, and play identical tones of frequency 170 Hz. Sam stands directly in front of one speaker at a distance of 4.0 m. Is this a loud spot or a quiet spot? Assume that the speed of sound in air is 340 m/s.

PREPARE Figure 16.31 shows a visual overview of the situation, showing the positions of and path lengths from each speaker.

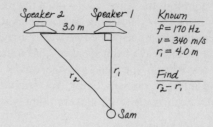

FIGURE 16.31 Visual overview for Example 16.8.

SOLVE Following the steps in Tactics Box 16.1, we first compute the path-length difference. r_1, r_2, and the distance between the speakers form a right triangle, so we can use the Pythagorean theorem to find

$$r_2 = \sqrt{(4.0 \text{ m})^2 + (3.0 \text{ m})^2} = 5.0 \text{ m}$$

Thus the path-length difference is

$$\Delta r = r_2 - r_1 = 1.0 \text{ m}.$$

Next, we compute the wavelength:

$$\lambda = \frac{v}{f} = \frac{340 \text{ m/s}}{170 \text{ Hz}} = 2.0 \text{ m}$$

The path-length difference is $\frac{1}{2}\lambda$, so this is a point of destructive interference. Sam is at a quiet spot.

So far, we have only looked at interference at particular points. What can we say about the overall pattern of points at which we have constructive or destructive interference? For instance, the red dot in Figure 16.30 is only one point where $\Delta r = \lambda$; you should be able to locate several more. Taken together, all the points with $\Delta r = \lambda$ form a curved line along which constructive interference is occurring. Another curved line of constructive interference connects all the points at which $\Delta r = 2\lambda$, another connects the $\Delta r = 3\lambda$ points, and so on. These lines, shown as red lines in Figure 16.32 on the next page, are called **antinodal lines.** They are analogous to the antinodes of a standing wave, hence the name. An antinode is a *point* of constructive interference; for spherical waves, oscillation at maximum amplitude occurs along a continuous *line*. To understand this idea better, imagine the static picture of Figure 16.32 evolving with time. As the wave fronts expand, the *intersection point* of two rings moves outward along one of the red lines.

Similarly, we can connect together points where Δr is a multiple of λ plus $\frac{1}{2}\lambda$. The black dot in Figure 16.30 is just one of many points with $\Delta r = \frac{1}{2}\lambda$. Together, these points of destructive interference form a **nodal line** along which the displacement is always zero. The nodal lines in Figure 16.32 are shown in black.

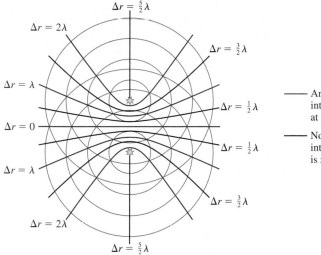

FIGURE 16.32 The points of constructive and destructive interference fall along antinodal and nodal lines.

You are regularly exposed to sound from two separated sources: stereo speakers. When you walk across a room in which a stereo is playing, why don't you hear a pattern of loud and soft sounds as you cross antinodal and nodal lines? First, we don't listen to single frequencies. Music is a complex sound wave, with many frequencies, but only one frequency at a time satisfies the condition for constructive or destructive interference. Most of the sound frequencies are not affected. Second, reflections of sound waves from walls and furniture make the situation much more complex than the idealized two-source picture in Figure 16.32. Sound wave interference can be heard, but it takes careful selection of a pure tone and a room with no hard, reflecting surfaces. But interference that's rather tricky to demonstrate with sound waves is easy to produce with light waves. We'll return to this example in the next chapter.

The two water waves overlap, leading to patterns of constructive and destructive interference.

STOP TO THINK 16.5 These speakers emit identical sound waves with a wavelength of 1.0 m. At the point indicated, is the interference constructive, destructive, or something in between?

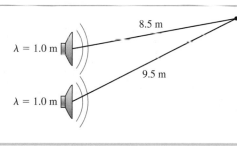

16.7 Beats

Thus far we have looked at the superposition of waves from sources having the same wavelength and frequency. We can also use the principle of superposition to investigate a phenomenon that is easily demonstrated with two sources of slightly *different* frequency and wavelength.

Suppose two sinusoidal waves are traveling toward your ear, as shown in Figure 16.33 on the next page. The two waves have the same amplitude but slightly different frequencies: The red wave has a slightly higher frequency (and thus a slightly shorter wavelength) than the green wave. This slight difference

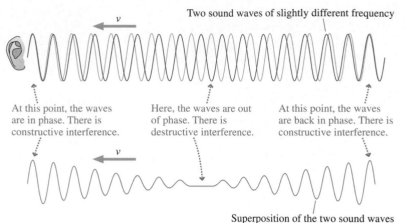

At this point, the waves are in phase. There is constructive interference.

Here, the waves are out of phase. There is destructive interference.

At this point, the waves are back in phase. There is constructive interference.

Superposition of the two sound waves

FIGURE 16.33 The superposition of two sound waves with slightly different frequencies.

causes the waves to combine in a manner that alternates between constructive and destructive interference. Their superposition, drawn in blue below the two waves, is a wave whose amplitude shows a periodic variation. As the waves reach your ear, you will hear a single tone whose intensity is *modulated*. That is, the sound goes up and down in volume, loud, soft, loud, soft, . . . , making a distinctive sound pattern called **beats.**

Suppose the two waves have frequencies f_1 and f_2 that differ only slightly, so that $f_1 \approx f_2$. A complete mathematical analysis would show that the air oscillates against your eardrum at frequency

$$f_{osc} = \frac{1}{2}(f_1 + f_2)$$

This is the *average* of f_1 and f_2, and it differs little from either since the two frequencies are nearly equal. Further, the intensity of the sound is modulated at a frequency called the *beat frequency:*

$$f_{beat} = |f_1 - f_2| \tag{16.12}$$

The beat frequency is simply the *difference* between the two individual frequencies.

Figure 16.34 is a history graph of the wave at the position of your ear. You can see both the sound wave oscillation at frequency f_{osc} and the much slower intensity oscillation at frequency f_{beat}. Frequency f_{osc} determines the pitch you hear while f_{beat} determines the frequency of the loud-soft-loud modulations of the sound intensity.

Musicians can use beats to tune their instruments. If one flute is properly tuned at 440 Hz but the other plays 438 Hz, the flutists will hear two loud-soft-loud beats per second. The second flutist is "flat" and needs to shorten her flute very slightly to bring the frequency up to 440 Hz.

Many measurement devices use beats to determine an unknown frequency by comparing it to a known frequency. For example, Chapter 15 described a Doppler blood-flow meter that used the Doppler shift of ultrasound reflected from moving blood to determine its speed. The meter determines this very small frequency shift by combining the emitted wave and the reflected wave and measuring the resulting beat frequency. The beat frequency is equal to the shift in frequency on reflection, exactly what is needed to determine the blood speed. Another example of using beats to make a measurement is given in the following example.

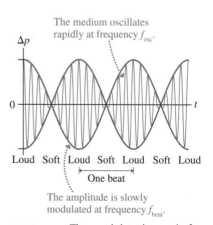

The medium oscillates rapidly at frequency f_{osc}.

The amplitude is slowly modulated at frequency f_{beat}.

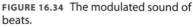

FIGURE 16.34 The modulated sound of beats.

EXAMPLE 16.9 **Detecting bats using beats**

The little brown bat is a common bat species in North America. It emits echolocation pulses at a frequency of 40 kHz, well above the range of human hearing. To
allow observers to "hear" these bats, the bat detector shown in Figure 16.35 combines the bat's sound wave at frequency f_1 with a wave of frequency f_2 from a tunable oscillator. The resulting beat frequency is isolated with a filter, then amplified
and sent to a loudspeaker. To what frequency should the tunable oscillator be set to
produce an audible beat frequency of 3 kHz?

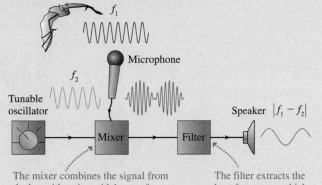

The mixer combines the signal from
the bat with a sinusoidal wave from an
oscillator. The result is a modulated wave.

The filter extracts the
beat frequency, which
is sent to the speaker.

FIGURE 16.35 The operation of a bat detector.

SOLVE The beat frequency is $f_{\text{beat}} = |f_1 - f_2|$, so the oscillator frequency and the
bat frequency need to *differ* by 3 kHz. An oscillator frequency of either 37 kHz or
43 kHz will work nicely.

STOP TO THINK 16.6 You hear three beats per second when two sound tones are
generated. The frequency of one tone is known to be 610 Hz. The frequency of
the other is

A. 604 Hz.
B. 607 Hz.
C. 613 Hz.
D. 616 Hz.
E. Either A or D.
F. Either B or C.

The goal of Chapter 16 has been to use the idea of superposition to understand the phenomena of interference and standing waves.

GENERAL PRINCIPLES

Principle of Superposition

The displacement of a medium when more than one wave is present is the sum of the displacements due to each individual wave.

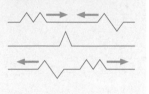

Interference

In general, the superposition of two or more waves into a single wave is called interference.

Constructive interference occurs when crests are aligned with crests and troughs with troughs. We say the waves are in phase. It occurs when the path-length difference $\Delta d = d_2 - d_1$ is a whole number of wavelengths.

Destructive interference occurs when crests are aligned with troughs. We say the waves are out of phase. It occurs when the path-length difference $\Delta d = d_2 - d_1$ is a whole number of wavelengths plus half a wavelength.

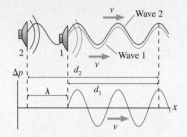

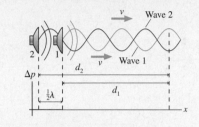

IMPORTANT CONCEPTS

Standing Waves

Two identical traveling waves moving in opposite directions create a standing wave.

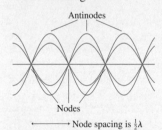

The boundary conditions determine which standing wave frequencies and wavelengths are allowed. The allowed standing waves are **modes** of the system.

A standing wave on a string has a node at each end. Possible modes:

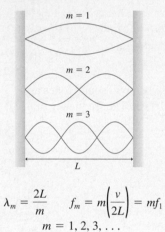

$$\lambda_m = \frac{2L}{m} \qquad f_m = m\left(\frac{v}{2L}\right) = mf_1$$
$$m = 1, 2, 3, \ldots$$

A standing sound wave in a tube can have different boundary conditions: open-open, closed-closed or open-closed.

Open-open
$$f_m = m\left(\frac{v}{2L}\right)$$
$$m = 1, 2, 3, \ldots$$

Closed-closed
$$f_m = m\left(\frac{v}{2L}\right)$$
$$m = 1, 2, 3, \ldots$$

Open-closed
$$f_m = m\left(\frac{v}{4L}\right)$$
$$m = 1, 3, 5, \ldots$$

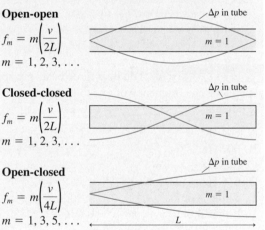

APPLICATIONS

Beats (loud-soft-loud-soft modulations of intensity) are produced when two waves of slightly different frequency are superimposed.

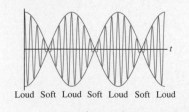

Loud Soft Loud Soft Loud Soft Loud

$$f_{\text{beat}} = |f_1 - f_2|$$

Standing waves are multiples of a **fundamental frequency,** the frequency of the lowest mode. The higher modes are the higher **harmonies.**

For sound, the fundamental frequency determines the perceived **pitch;** the higher harmonics determine the **tone quality.**

Our vocal cords create a range of harmonics. The mix of higher harmonics is changed by our vocal tract to create different vowel sounds.

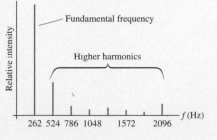

QUESTIONS

Conceptual Questions

1. Light can pass easily through water and through air, but light will reflect from the surface of a lake. What does this tell you about the speed of light in air and in water?

2. Ocean waves are partially reflected from the entrance to a harbor, where the depth of the water is suddenly less. What does this tell you about the speed of waves in water of different depths?

3. A string has an abrupt change in linear density at its midpoint so that the speed of a pulse on the left side is 2/3 of that on the right side.
   a. On which side is the linear density greater? Explain.
   b. From which side would you start a pulse so that its reflection from the midpoint would not be inverted? Explain.

4. A guitarist finds that the frequency of one of her strings is too low by 1.4%. Should she increase or decrease the tension of the string? Explain.

5. Figure Q16.5 shows a standing wave on a string that is oscillating at frequency f_0. How many antinodes will there be if the frequency is doubled to $2f_0$? Explain.

FIGURE Q16.5

6. Figure Q16.6 shows a standing sound wave in a tube of air that is open at both ends.
   a. Which mode (value of m) standing wave is this?
   b. Is the air vibrating horizontally or vertically?

FIGURE Q16.6

7. If you take snapshots of a standing wave on a string at different times, you will find certain times at which the wave is totally flat. What has happened to the energy of the wave at those times?

8. A typical flute is about 66 cm long. A piccolo is a very similar instrument, though it is smaller, with a length of about 32 cm. How does the pitch of a piccolo compare to that of a flute?

9. Some pipes on a pipe organ are open at both ends, others are closed at one end. For pipes that play low-frequency notes, there is an advantage to using pipes that are closed at one end. What is the advantage?

10. An organ pipe is tuned to exactly 384 Hz when the room temperature is 20°C. Later, the room temperature increases to 25°C. What happens to the frequency of the pipe's note? Does it increase, decrease, or stay the same? Explain.

11. BIO A friend's voice sounds different over the telephone than it does in person. This is because telephones do not transmit frequencies over about 3000 Hz. 3000 Hz is well above the normal frequency of speech, so why does eliminating these high frequencies change the sound of a person's voice?

12. Suppose you were to play a trumpet after breathing helium, in which the speed of sound is much greater than in air. Would the pitch of the instrument be higher or lower than normal, or would it be unaffected by being played with helium inside the tube rather than air?

13. If you pour liquid in a tall, narrow glass, you may hear sound with a steadily rising pitch. What is the source of the sound, and why does the pitch rise as the glass fills?

14. BIO When you speak after breathing helium, in which the speed of sound is much greater than in air, your voice sounds quite different. The frequencies emitted by your vocal cords do not change since they are determined by the mass and tension of your vocal cords. So what *does* change when your vocal tract is filled with helium rather than air?

15. BIO Sopranos can sing notes at very high frequencies—over 1000 Hz. When they sing such high notes, it can be difficult to understand the words they are singing. Use the concepts of harmonics and formants to explain this.

16. BIO If a cold gives you a stuffed-up nose, it changes the way your voice sounds, even if your vocal cords are not affected. Explain why this is so.

17. BIO Figure Q16.17 shows wave fronts of a circular wave. Are the displacements at the following pairs of positions *in phase* or *out of phase*? Explain.
    a. A and B   b. C and D   c. E and F

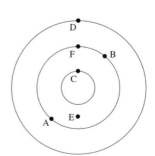

FIGURE Q16.17

Multiple-Choice Questions

Questions 18 through 20 refer to the snapshot graph Figure Q16.18.

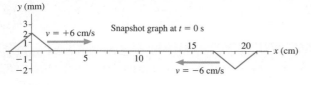

FIGURE Q16.18

18. | At $t = 1$ s, what is the displacement y of the string at $x = 7$ cm?
   A. −1.0 mm   B. 0 mm   C. 0.5 mm
   D. 1.0 mm   E. 2.0 mm

19. | At $x = 3$ cm, what is the earliest time that y will equal 2 mm?
   A. 0.5 s   B. 0.7 s   C. 1.0 s
   D. 1.5 s   E. 2.5 s

20. | The pulses overlap at $t = 1.5$ s. At $x = 10$ cm, what is the value of y at this time?
   A. −2.0 mm   B. −1.0 mm   C. −0.5 mm
   D. 0 mm   E. 1.0 mm

21. || Two sinusoidal waves with the same amplitude A and frequency f travel in opposite directions along a long string. You stand at one point and watch the string. The maximum displacement of the string at that point is
   A. A B. $2A$ C. 0
   D. There is not enough information to decide.

22. | A student in her physics lab measures the standing wave modes of a tube. The lowest frequency that makes a resonance is 20 Hz. As the frequency is increased, the next resonance is at 60 Hz. What will be the next resonance after this?
   A. 80 Hz   B. 100 Hz   C. 120 Hz   D. 180 Hz

23. || Two guitar strings made of the same type of wire have the same length. String 1 has a higher pitch than string 2. Which of the following is true?
   A. The wave speed of string 1 is greater than that of string 2.
   B. The tension in string 2 is greater than that in string 1.
   C. The wavelength of the lowest standing wave mode on string 2 is longer than that on string 1.
   D. The wavelength of the lowest standing wave mode on string 1 is longer than that on string 2.

24. | The frequency of the lowest standing wave mode on a 1.0-m-long string is 20 Hz. What is the wave speed on the string?
   A. 10 m/s   B. 20 m/s   C. 30 m/s   D. 40 m/s

25. | Suppose you pluck a string on a guitar and it produces the note A at a frequency of 440 Hz. Now you press your finger down on the string against one of the frets, making this point the new end of the string. The newly shortened string has 4/5 the length of the full string. When you pluck the string, its frequency will be:
   A. 350 Hz   B. 440 Hz   C. 490 Hz   D. 550 Hz

PROBLEMS

Section 16.1 The Principle of Superposition

1. | Figure P16.1 is a snapshot graph at $t = 0$ s of two waves on a taut string approaching each other at 1 m/s. Draw six snapshot graphs, stacked vertically, showing the string at 1 s intervals from $t = 1$ s to $t = 6$ s.

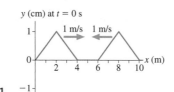

FIGURE P16.1

2. | Figure P16.2 is a snapshot graph at $t = 0$ s of two waves approaching each other at 1 m/s. Draw six snapshot graphs, stacked vertically, showing the string at 1 s intervals from $t = 1$ s to $t = 6$ s.

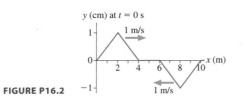

FIGURE P16.2

3. || Figure P16.3 is a snapshot graph at $t = 0$ s of two waves approaching each other at 1 m/s. Draw four snapshot graphs, stacked vertically, showing the string at $t = 2, 4, 6$, and 8 s.

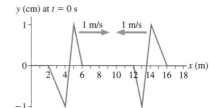

FIGURE P16.3

4. || Figure P16.4 is a snapshot graph at $t = 0$ s of two waves approaching each other at 1 m/s. Draw four snapshot graphs, stacked vertically, showing the string at $t = 2, 4, 6$, and 8 s.

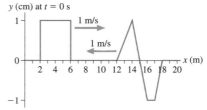

FIGURE P16.4

5. ‖ Figure P16.5a is a snapshot graph at $t = 0$ s of two waves on a string approaching each other at 1 m/s. At what time was the snapshot graph in Figure P16.5b taken?

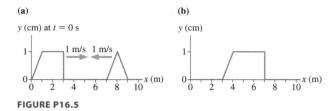

FIGURE P16.5

Section 16.2 Standing Waves

6. ‖ Figure P16.6 is a snapshot graph at $t = 0$ s of two waves moving to the right at 1 m/s. The string is fixed at $x = 8$ m. Draw four snapshot graphs, stacked vertically, showing the string at $t = 2, 4, 6$, and 8 s.

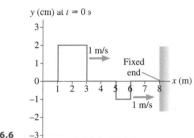

FIGURE P16.6

7. ‖ Figure P16.7 is a snapshot graph at $t = 0$ s of a pulse on a string moving to the right at 1 m/s. The string is fixed at $x = 5$ m. Draw a history graph spanning the time interval $t = 0$ s to $t = 10$ s for the location $x = 3$ m on the string.

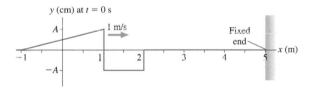

FIGURE P16.7

8. | At $t = 0$ s, a small "upward" (positive y) pulse centered at $x = 6.0$ m is moving to the right on a string with fixed ends at $x = 0.0$ m and $x = 10.0$ m. The wave speed on the string is 4.0 m/s. At what time will the string next have the same appearance that it did at $t = 0$ s?

Section 16.3 Standing Waves on a String

9. ‖ A 2.0-m-long string is fixed at both ends and tightened until the wave speed is 40 m/s. What is the frequency of the standing wave shown in Figure P16.9?

FIGURE P16.9

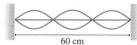

60 cm

FIGURE P16.10

10. | Figure P16.10 (below left) shows a standing wave oscillating at 100 Hz on a string. What is the wave speed?

11. | A bass guitar string is 89 cm long with a fundamental frequency of 30 Hz. What is the wave speed on this string?

12. ‖ The fundamental frequency of a guitar string is 384 Hz. What is the fundamental frequency if the tension in the string is reduced by half?

13. | a. What are the three longest wavelengths for standing waves on a 240-cm-long string that is fixed at both ends?
    b. If the frequency of the second-longest wavelength is 50.0 Hz, what is the frequency of the third-longest wavelength?

14. ‖‖ Standing waves on a 1.0-m-long string that is fixed at both ends are seen at successive frequencies of 24 Hz and 36 Hz.
    a. What are the fundamental frequency and the wave speed?
    b. Draw the standing-wave pattern when the string oscillates at 36 Hz.

15. | A 121-cm-long, 4.00 g string oscillates in its $m = 3$ mode with a frequency of 180 Hz and a maximum amplitude of 5.00 mm. What are (a) the wavelength and (b) the tension in the string?

16. ‖ A guitar string with a linear density of 2.0 g/m is stretched between supports that are 60 cm apart. The string is observed to form a standing wave with three antinodes when driven at a frequency of 420 Hz. What are (a) the frequency of the fifth harmonic of this string and (b) the tension in the string?

17. ‖‖ A violin string has a standard length of 32.8 cm. It sounds the musical note A (440 Hz) when played without fingering. How far from the end of the string should you place your finger to play the note C (523 Hz)?

18. ‖ The lowest note on a grand piano has a frequency of 27.5 Hz. The entire string is 2.00 m long and has a mass of 400 g. The vibrating section of the string is 1.90 m long. What tension is needed to tune this string properly?

19. | A carbon dioxide laser emits infrared. A CO_2 laser with a cavity length of 53.00 cm oscillates in the $m = 100,000$ mode. What are the wavelength and frequency of the laser beam?

Section 16.4 Standing Sound Waves

20. | The lowest frequency in the audible range is 20 Hz. (a) What is the length of (a) the shortest open-open tube and (b) the shortest open-closed tube needed to produce this frequency?

21. | The contrabassoon is the wind instrument capable of sounding the lowest pitch in an orchestra. It is folded over several times to fit its impressive 18 ft length into a reasonable size instrument.
    a. If we model the instrument as an open-closed tube, what is its fundamental frequency? The sound speed inside is 350 m/s because the air is warmed by the player's breath.
    b. The actual fundamental frequency of the contrabassoon is 27.5 Hz, which should be different from your answer in part a. This means the model of the instrument as an open-closed tube is a bit too simple. But if you insist on using that model, what is the "effective length" of the instrument?

22. | Figure P16.22 shows a standing sound wave in an 80-cm-long tube. The tube is filled with an unknown gas. What is the speed of sound in this gas?

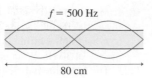

$f = 500$ Hz

80 cm

FIGURE P16.22

23. | What are the three longest wavelengths for standing sound waves in a 121-cm-long tube that is (a) open at both ends and (b) open at one end, closed at the other?

24. | The lowest pedal note on a large pipe organ has a fundamental frequency of 16 Hz. This extreme bass note is more felt as a rumble than heard with the ears. What is the length of the open-closed pipe that makes that note? Assume a speed of sound of 340 m/s.

25. ‖ The fundamental frequency of an open-open tube is 1500 Hz when the tube is filled with 0°C helium. What is its frequency when filled with 0°C air?

26. ‖‖ A drainage pipe running under a freeway is 30.0 m long. Both ends of the pipe are open, and wind blowing across one end causes the air inside to vibrate.
    a. If the speed of sound on a particular day is 340 m/s, what will be the fundamental frequency of air vibration in this pipe?
    b. What is the frequency of the lowest harmonic that would be audible to the human ear?
    c. What will happen to the frequency in the later afternoon as the air begins to cool?

27. ‖‖ You've decided to install a pipe organ in your living room. Unfortunately, your ceiling is only 3.2 m high. To reach the lowest notes possible, you've decided to use pipes closed at the floor end and open at the top, where they'll be sounded. You need 50 cm clearance at the top end to install the compressed-air box and mount the pipes in the box. If you keep your living room at 20°C, what is the frequency of the lowest note that you will be able to obtain from your organ?

28. ‖‖ A child has an ear canal that is 1.3 cm long. At what sound
BIO frequencies in the audible range will the child have increased hearing sensitivity?

29. | Although the vocal tract is quite complicated, we can make
BIO a simple model of it as an open-closed tube extending from the opening of the mouth to the diaphragm, the large muscle separating the abdomen and the chest cavity. What is the length of this tube if its fundamental frequency equals a typical speech formant frequency of 200 Hz? Assume a sound speed of 350 m/s. Does this result for the tube length seem reasonable, based on observations on your own body?

Section 16.6 The Interference of Waves from Two Sources

30. ‖‖‖ Two loudspeakers in a 20°C room emit 686 Hz sound waves along the x-axis. What is the smallest distance between the speakers for which the interference of the sound waves is destructive?

31. ‖ Two loudspeakers emit sound waves along the x-axis. The sound has maximum intensity when the speakers are 20 cm apart. The sound intensity decreases as the distance between the speakers is increased, reaching zero at a separation of 30 cm.
    a. What is the wavelength of the sound?
    b. If the distance between the speakers continues to increase, at what separation will the sound intensity again be a maximum?

32. ‖ Two identical loudspeakers separated by distance d emit 170 Hz sound waves along the x-axis. As you walk along the axis, away from the speakers, you don't hear anything even though both speakers are on. What are three possible values for d? Assume a sound speed of 340 m/s.

33. ‖ Figure P16.33 shows the circular wave fronts emitted by two sources. Make a table with rows labeled P, Q, and R and columns labeled r_1, r_2, Δr, and C/D. Fill in the table for points P, Q, and R, giving the distances as multiples of λ and indicating, with a C or a D, whether the interference at that point is constructive or destructive.

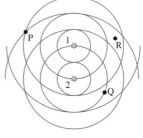

FIGURE P16.33

34. ‖‖‖ Two identical loudspeakers 2.0 m apart are emitting 1800 Hz sound waves into a room where the speed of sound is 340 m/s. Is the point 4.0 m directly in front of one of the speakers, perpendicular to the plane of the speakers, a point of maximum constructive interference, perfect destructive interference, or something in between?

Section 16.7 Beats

35. | Two strings are adjusted to vibrate at exactly 200 Hz. Then the tension in one string is increased slightly. Afterward, three beats per second are heard when the strings vibrate at the same time. What is the new frequency of the string that was tightened?

36. | A flute player hears four beats per second when she compares her note to a 523 Hz tuning fork (the note C). She can match the frequency of the tuning fork by pulling out the "tuning joint" to lengthen her flute slightly. What was her initial frequency?

General Problems

37. | The fundamental frequency of a standing wave on a 1.0-m-long string is 440 Hz. What would be the wave speed of a pulse moving along this string?

38. | A guitarist plucks the D-string of his instrument, causing it to vibrate at 147 Hz. If he now presses his finger down on the string, creating a new effective end for the string at the string's exact midpoint, what will be the frequency of the string?

39. ‖ An 80-cm-long steel string with a linear density of 1.0 g/m
INT is under 200 N tension. It is plucked and vibrates at its fundamental frequency. What is the wavelength of the sound wave that reaches your ear in a 20°C room?

40. | A string, stretched between two fixed posts, forms standing wave resonances at 325 Hz and 390 Hz. What is the greatest possible value of its fundamental frequency?

41. ‖‖ A violinist places her finger so that the vibrating section of
INT a 1.0 g/m string has a length of 30 cm, then she draws her bow across it. A listener nearby in a 20°C room hears a note with a wavelength of 40 cm. What is the tension in the string?

42. ‖ A particularly beautiful note reaching your ear from a rare
INT Stradivarius violin has a wavelength of 39.1 cm. The room is slightly warm, so the speed of sound is 344 m/s. If the string's linear density is 0.60 g/m and the tension is 150 N, how long is the vibrating section of the violin string?

43. ‖ A heavy piece of hanging sculpture is suspended by a 90-
INT cm-long, 5.0 g steel wire. When the wind blows hard, the
wire hums at its fundamental frequency of 80 Hz. What is
the mass of the sculpture?

44. │ An experiment finds that standing waves on a 0.80-m-long
string, fixed at both ends, occur at 24 Hz and 32 Hz, but at no
frequencies in between.
a. What is the fundamental frequency?
b. What is the wave speed on the string?
c. Draw the standing wave pattern for the string at 32 Hz.

45. ‖‖‖ Astronauts visiting Planet X have a 2.5-m-long string
INT whose mass is 5.0 g. They tie the string to a support, stretch it
horizontally over a pulley 2.0 m away, and hang a 1.0 kg mass
on the free end. Then the astronauts begin to excite standing
waves on the string. Their data show that standing waves exist
at frequencies of 64 Hz and 80 Hz, but at no frequencies in
between. What is the value of g, the acceleration due to gravity, on Planet X?

46. ‖‖ A 75 g bungee cord has an equilibrium length of 1.2 m. The
INT cord is stretched to a length of 1.8 m, then vibrated at 20 Hz.
This produces a standing wave with two antinodes. What is
the spring constant of the bungee cord?

47. ‖‖‖ A 2.5-cm-diameter steel cable
(with density 7900 kg/m³) that
is part of the suspension system
for a footbridge stretches 14 m
between the tower and the
ground). After walking over
the bridge, a hiker finds that the
cable is vibrating in its fundamental mode with a period of
0.40 s. What is the tension in the cable?

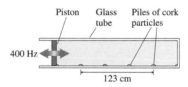

48. ‖ A steel wire is used to
INT stretch a spring, as shown in
Figure P16.48. An oscillat-
ing magnetic field drives the

Steel wire   Pull
Spring

FIGURE P16.48

steel wire back and forth. A standing wave with three antinodes is created when the spring is stretched 8.0 cm. What
stretch of the spring produces a standing wave with two
antinodes?

49. ‖ Just as you are about to step into a nice hot bath, a small
earthquake rattles your bathroom. Immediately afterward, you
notice that the water in the tub is oscillating. The water in the
center seems to be motionless while the water at the two ends
alternately rises and falls, like a seesaw. You happen to know
that your bathtub is 1.4 m long, and you count 10 complete
oscillations of the water in 20 s.
a. What is the wavelength of this standing wave?
b. What is the speed of the waves that are reflecting back and
forth inside the tub to create the standing wave?

50. ‖ A microwave generator
can produce microwaves
at any frequency between
10 GHz and 20 GHz. As
Figure P16.50 shows, the

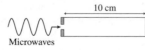

10 cm

Microwaves

FIGURE P16.50

microwaves are aimed, through a small hole, into a "microwave cavity" that consists of a 10-cm-long cylinder with
reflective ends.
a. Which frequencies will create standing waves in the
microwave cavity?
b. For which of these frequencies is the cavity midpoint an
antinode?

51. ‖ An open-open organ pipe is 78.0 cm long. An open-closed
pipe has a fundamental frequency equal to the third harmonic of the open-open pipe. How long is the open-closed
pipe?

52. │ A narrow column of air is found to have standing waves at
frequencies of 390 Hz, 520 Hz, and 650 Hz and at no frequencies in between these. The behavior of the tube at frequencies
less than 390 Hz or greater than 650 Hz is not known.
a. Is this an open-open tube or an open-closed tube? Explain.
b. How long is the tube?

53. ‖ An open-open aluminum pipe has an $m = 1$ standing wave
at 262 Hz. It is also possible to set up a longitudinal standing
sound wave in the aluminum itself by striking it on the end.
What will be the frequency of the $m = 1$ mode in the
aluminum?

54. ‖ In 1866, the German scientist Adolph Kundt developed a
technique for accurately measuring the speed of sound in various gases. A long glass tube, known today as a Kundt's tube,
has a vibrating piston at one end and is closed at the other.
Very finely ground particles of cork are sprinkled in the bottom of the tube before the piston is inserted. As the vibrating
piston is slowly moved forward, there are a few positions that
cause the cork particles to collect in small, regularly spaced
piles along the bottom. Figure P16.54 shows an experiment in
which the tube is filled with pure oxygen and the piston is driven at 400 Hz.

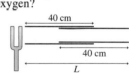

Piston   Glass   Piles of cork
          tube   particles

400 Hz

123 cm

FIGURE P16.54

a. Do the cork particles collect at standing wave nodes or
antinodes? (Hint: consider the appearance of the ends of
the tube.)
b. What is the speed of sound in oxygen?

55. ‖ A 40-cm-long tube has a
40-cm-long insert that can be
pulled in and out, as shown in
Figure P16.55. A vibrating tuning fork is held next to the tube.
As the insert is slowly pulled out,

40 cm

40 cm

L

FIGURE P16.55

the sound from the tuning fork creates standing waves in the
tube when the total length L is 42.5 cm, 56.7 cm, and 70.9 cm.
What is the frequency of the tuning fork? The air temperature
is 20°C.

56. ‖‖ A 1.0-m-tall vertical tube is filled with 20°C water. A tuning fork vibrating at 580 Hz is held just over the top of the
tube as the water is slowly drained from the bottom. At what
water heights, measured from the bottom of the tube, will
there be a standing wave in the tube?

57. ‖ A 50-cm-long wire with a mass of 1.0 g and a tension of
440 N passes across the open end of an open-closed tube of
air. The wire, which is fixed at both ends, is bowed at the center so as to vibrate at its fundamental frequency and generate a
sound wave. Then the tube length is adjusted until the fundamental frequency of the tube is heard. What is the length of
the tube? Assume the speed of sound is 340 m/s.

58. ▐▐▐ A 25-cm-long wire with a linear density of 20 g/m passes across the open end of an 85-cm-long open-closed tube of air. If the wire, which is fixed at both ends, vibrates at its fundamental frequency, the sound wave it generates excites the second vibrational mode of the tube of air. What is the tension in the wire? Assume the speed of sound is 340 m/s.

59. ▐▐▐▐ Two loudspeakers located along the x-axis as shown in Figure P16.59 produce sounds of equal frequency. Speaker 1 is at the origin, while the location of speaker 2 can be varied by a remote control wielded by the listener. He notices maxima in the sound intensity when speaker 2 is located at $x = 0.75$ m and 1.00 m, but at no points in between. What is the frequency of the sound? Assume the speed of sound is 340 m/s.

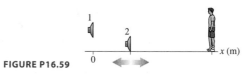

FIGURE P16.59

60. ▐▐▐▐ You are standing 2.50 m directly in front of one of the two loudspeakers shown in Figure P16.60. They are 3.00 m apart and both are playing a 686 Hz tone in phase. As you begin to walk directly away from the speaker, at what distances from the speaker do you hear a *minimum* sound intensity? The room temperature is 20°C.

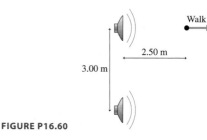

FIGURE P16.60

61. ▐▐▐ FM station KCOM ("All commercials, all the time") transmits simultaneously, at a frequency of 99.9 MHz, from two broadcast towers placed precisely 31.5 m apart along a north-south line.
    a. What is the wavelength of KCOM's transmissions?
    b. Suppose you stand 90.0 m due east of the point halfway between the two towers with your portable FM radio. Will you receive a strong or weak signal at this position? Why?
    c. You then stand 90.0 m due north of the northern tower with your radio. Will you receive a strong or weak signal at this position? Why?

62. ▐▐▐ Two loudspeakers, 4.0 m apart and facing each other, play identical sounds of the same frequency. You stand halfway between them, where there is a maximum of sound intensity. Moving from this point toward one of the speakers, you encounter a minimum of sound intensity when you have moved 0.25 m.
    a. What is the frequency of the sound?
    b. If the frequency is then increased while you remain 0.25 m from the center, what is the first frequency for which that location will be a maximum of sound intensity?

63. ▐▐▐▐ Two radio antennas are separated by 2.0 m. Both broadcast identical 750 MHz waves. If you walk around the antennas in a circle of radius 10 m, how many maxima will you detect?

64. ▐▐▐ Two stones, one large and one small, hit water simultaneously and generate ripples. The large stone creates ripples twice as high as the small stone. At the midpoint between the splashes, what is the ratio of the maximum possible ripple amplitude to the minimum possible ripple amplitude?

65. ▐ Certain birds produce vocalizations consisting of two distinct frequencies that are not harmonically related—that is, the two frequencies are not harmonics of a common fundamental frequency. These two frequencies must be produced by two different vibrating structures in the bird's vocal tract.
    a. Wood ducks have been observed to make a call with approximately equal intensities at 850 Hz and 1200 Hz. The membranes that produce the vocalizations do not seem to vibrate at frequencies less than 500 Hz. Given this limitation, could the two frequencies noted be higher harmonics of a lower-frequency fundamental?
    b. If we model the duck's vocal tract as an open-closed tube, what length corresponds to a fundamental frequency equal to the lower of the two frequencies in a?

66. ▐ Piano tuners tune pianos by listening to the beats between the harmonics of two different strings. When properly tuned, the note A should have the frequency 440 Hz and the note E should be at 659 Hz. The tuner can determine this by listening to the beats between the third harmonic of the A and the second harmonic of the E.
    a. A tuner first tunes the A string very precisely by matching it to a 440 Hz tuning fork. She then strikes the A and E strings simultaneously and listens for beats between the harmonics. What beat frequency indicates that the E string is properly tuned?
    b. The tuner starts with the tension in the E string a little low, then tightens it. What is the frequency of the E string when she hears four beats per second?

67. ▐▐▐ A flutist assembles her flute in a room where the speed of sound is 342 m/s. When she plays the note A, it is in perfect tune with a 440 Hz tuning fork. After a few minutes, the air inside her flute has warmed to where the speed of sound is 346 m/s.
    a. How many beats per second will she hear if she now plays the note A as the tuning fork is sounded?
    b. How far does she need to extend the "tuning joint" of her flute to be in tune with the tuning fork?

68. ▐▐ A student waiting at a stoplight notices that her turn signal, which has a period of 0.85 s, makes one blink exactly in sync with the turn signal of the car in front of her. The blinker of the car ahead then starts to get ahead, but 17 s later the two are exactly in sync again. What is the period of the blinker of the other car?

69. ▐▐ Two loudspeakers emit 400 Hz notes. One speaker sits on the ground. The other speaker is in the back of a pickup truck. You hear eight beats per second as the truck drives away from you. What is the truck's speed?

70. ▐ A Doppler blood flow meter emits ultrasound at a frequency of 5.0 MHz. What is the beat frequency between the emitted waves and the waves reflected from blood cells moving away from the emitter at 0.15 m/s?

71. ▐ An ultrasound unit is being used to measure a patient's heartbeat by combining the emitted 2.0 MHz signal with the sound waves reflected from the moving tissue of one point on the heart. The beat frequency between the two signals has a maximum value of 520 Hz. What is the maximum speed of the heart tissue?

Passage Problems

Harmonics and Harmony

You know that certain musical notes sound good together—harmonious—whereas others do not. This harmony is related to the various harmonics of the notes.

The musical notes C (262 Hz) and G (392 Hz) make a pleasant sound when played together; we call this consonance. As Figure P16.72 shows, the harmonics of the two notes are either far from each other or very close to each other (within a few Hz). This is the key to consonance: harmonics that are spaced either far apart or very close. The close harmonics have a beat frequency of a few Hz that is perceived as pleasant. If the harmonics of two notes are close but not too close, the rather high beat frequency between the two is quite unpleasant. This is what we hear as dissonance. Exactly how much a difference is maximally dissonant is a matter of opinion, but harmonic separations of 30 or 40 Hz seem to be quite unpleasant for most people.

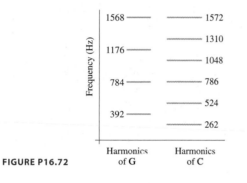

FIGURE P16.72

72. | What is the beat frequency between the second harmonic of G and the third harmonic of C?
    A. 1 Hz   B. 2 Hz   C. 4 Hz   D. 6 Hz
73. | Would a G-flat (frequency 370 Hz) and a C played together be consonant or dissonant?
    A. Consonant
    B. Dissonant
74. | An organ pipe open at both ends is tuned so that its fundamental frequency is a G. How long is the pipe?
    A. 43 cm   B. 87 cm   C. 130 cm   D. 173 cm
75. | If the C were played on an organ pipe that was open at one end and closed at the other, which of the harmonic frequencies in Figure 16.72 would be present?
    A. All of the harmonics in the figure would be present.
    B. 262, 786, and 1310 Hz
    C. 524, 1048, and 1572 Hz
    D. 262, 524, and 1048 Hz

STOP TO THINK ANSWERS

Stop to Think 16.1: C. The figure shows the two waves at $t = 6$ s and their superposition. The superposition is the *point-by-point* addition of the displacements of the two individual waves.

$$0 \quad 2 \quad 4 \quad 6 \quad 8 \quad 10 \quad 12 \quad 14 \quad 16 \quad 18 \quad 20 \quad x\,(\text{m})$$

Stop to Think 16.2: C. Standing-wave frequencies are $f_m = mf_1$. The original wave has frequency $f_2 = 2f_1$ because it has two antinodes. The wave with frequency $2f_2 = 4f_1$ is the $m = 4$ mode with four antinodes.

Stop to Think 16.3: B. 300 Hz and 400 Hz are not f_1 and f_2 because 400 Hz $\neq$ 2 × 300 Hz. Instead, both are multiples of the fundamental frequency. Because the difference between them is 100 Hz, we see that $f_3 = 3 \times 100$ Hz and $f_4 = 4 \times 100$ Hz. Thus $f_1 = 100$ Hz.

Stop to Think 16.4: D. Highest pitch, or highest frequency, corresponds to the shortest period. For a complex wave, period is the time required for the entire wave pattern to repeat.

Stop to Think 16.5: Constructive interference. The path-length difference is $\Delta r = 1.0$ m $= \lambda$. Interference is constructive when the path-length difference is a whole number of wavelengths.

Stop to Think 16.6: F. The beat frequency is the difference between the two frequencies.

Oscillations and Waves

As we have studied oscillations and waves, one point we have emphasized is the *unity* of the basic physics. The mathematics of oscillation describes the motion of a mass on a spring or the motion of a gibbon swinging from a branch. The same theory of waves works for string waves and sound waves and light waves. A few basic ideas enable us to understand a wide range of physical phenomena.

The physics of oscillations and waves is not quite as easily summarized as the physics of particles. Newton's laws and the conservation laws are two very general sets of principles about particles, principles that allowed us to develop the powerful problem-solving strategies of Parts I and II. The knowledge structure of oscillations and waves, shown below, rests more heavily on *phenomena* than on general principles. This knowledge structure doesn't contain a problem-solving strategy or a wide range of general principles, but is instead a logical grouping of the major topics you studied. This is a different way of structuring knowledge, but it still provides you with a mental framework for analyzing and thinking about wave problems.

The physics of oscillations and waves will be with us for the rest of the book. Part V is an exploration of optics, beginning with a detailed study of light as a wave. In Part VI, we will study the nature of electromagnetic waves in more detail. Finally, in Part VII, we will see that matter has a wave nature. There, the wave ideas we have developed in Part IV will lead us into the exciting world of quantum physics, finishing with some quite remarkable insights about the nature of light and matter.

KNOWLEDGE STRUCTURE IV Oscillations and Waves

BASIC GOALS
How can we describe oscillatory motion?
How does a wave travel through a medium?
What are the distinguishing features of waves?
What happens when two waves meet?

GENERAL PRINCIPLES
Simple harmonic motion occurs when a linear restoring force acts to return a system to equilibrium.
The period of simple harmonic motion depends on physical properties of the system but not the amplitude.
All sinusoidal waves, whether water waves, sound waves, or light waves, have the same functional form.
When two waves meet, they pass through each other, combining where they overlap by superposition.

Simple harmonic motion

$$x(t) = A\cos(2\pi ft)$$

$$v_x(t) = -v_{max}\sin(2\pi ft)$$

$$v_{max} = 2\pi fA$$

$$a_x(t) = -a_{max}\cos(2\pi ft)$$

$$a_{max} = (2\pi f)^2 A$$

The frequency of a mass on a spring depends on the spring constant and the mass:

$$f = \frac{1}{2\pi}\sqrt{\frac{k}{m}}$$

The frequency of a pendulum depends on the length and the acceleration of gravity:

$$f = \frac{1}{2\pi}\sqrt{\frac{g}{L}}$$

Traveling waves

All traveling waves are described by the same equation:

$$y = A\cos\left(2\pi\left(\frac{x}{\lambda} \pm \frac{t}{T}\right)\right)$$

+: wave travels to left
−: wave travels to right

The wave speed depends on the properties of the medium.

$$v_{string} = \sqrt{\frac{T_s}{\mu}}$$

The speed, wavelength, period and frequency of a wave are related.

$$T = \frac{1}{f} \qquad v = f\lambda$$

Superposition and interference

The superposition of two waves is the sum of the displacements of the two individual waves.

Constructive interference occurs when crests are aligned with crests and troughs with troughs. Constructive interference occurs when the path-length difference is a whole number of wavelengths.

Destructive interference occurs when crests are aligned with troughs. Destructive interference occurs when the path-length difference is a whole number of wavelengths plus half a wavelength.

Standing waves

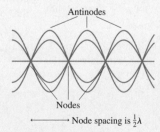
Antinodes
Nodes
Node spacing is $\frac{1}{2}\lambda$

Standing waves are due to the superposition of two traveling waves moving in opposite directions.

The boundary conditions determine which standing wave frequencies and wavelengths are allowed. The allowed standing waves are modes of the system.

For a string of length L, the modes have wavelength and frequency:

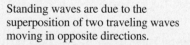

$$\lambda_m = \frac{2L}{m} \qquad f_m = m\left(\frac{v_{string}}{2L}\right) = mf_1 \qquad m = 1, 2, 3, 4, \ldots$$

Waves in the Earth and the Ocean

In December 2004, a large earthquake off the coast of Indonesia produced a devastating water wave, called a *tsunami*, that caused tremendous destruction thousands of miles away from the earthquake's epicenter. The tsunami was a dramatic illustration of the energy carried by waves.

It was also a call to action. Many of the communities hardest hit by the tsunami were struck hours after the waves were generated, long after seismic waves from the earthquake that passed through the earth had been detected at distant recording stations, long after the possibility of a tsunami was first discussed. With better detection and more accurate models of how a tsunami is formed and how a tsunami propagates, the affected communities could have received advance warning. The study of physics may seem an abstract undertaking with few practical applications, but on this day a better scientific understanding of these waves could have averted tragedy.

Let's use our knowledge of waves to explore the properties of a tsunami. In Chapter 15, we saw that a vigorous shake of one end of a rope causes a pulse to travel along it, carrying energy as it goes. The earthquake that produced the Indian Ocean tsunami of 2004 caused a sudden upward displacement of the sea floor that produced a corresponding rise in the surface of the ocean. This was

the *disturbance* that produced the tsunami, very much like a quick shake on the end of a rope. The resulting wave propagated through the ocean, as we see in the figure.

This simulation of the tsunami looks much like the ripples that spread when you drop a pebble into a pond. But there is a big difference—the scale. The fact that you can see the individual waves on this diagram that spans 5000 km is quite revealing. To show up so clearly, the individual wave pulses must be very wide—up to hundreds of kilometers from front to back.

A tsunami is actually a "shallow water wave," even in the deep ocean, because the depth of the ocean is much less than the width of the wave. Consequently, a tsunami travels differently than normal ocean waves. In Chapter 15 we learned that wave speeds are fixed by the properties of the medium. That is true for normal ocean waves, but the width of the wave causes a tsunami to "feel the bottom." Its wave speed is determined by the depth of the ocean: the greater the depth, the greater the speed. In the deep ocean, a tsunami travels at hundreds of kilometers per hour, much faster than a typical ocean wave. Near shore, as the ocean depth decreases, so does the speed of the wave.

The height of the tsunami in the open ocean was about half a meter. Why should such a small wave—one that ships didn't even notice as it passed—be so fearsome? Again, it's the *width* of the wave that matters. Because a tsunami is the wave motion of a considerable mass of water, great energy is involved. As the front of a tsunami wave nears shore, its speed decreases, and the back of the wave moves faster than the front. Consequently, the width decreases. The water begins to pile up, and the wave dramatically increases in height.

The Indian Ocean tsunami had a height of up to 15 m when it reached shore, with a width of up to several kilometers. This tremendous mass of water was still moving at high speed, giving it a great deal of energy. A tsunami reaching the shore isn't like a typical wave that breaks and crashes. It is a kilometers-wide wall of water that moves onto the shore and just keeps on coming. In many places, the water reached 2 km inland.

The impact of the Indian Ocean tsunami was devastating, but it was the first tsunami for which scientists were able to use satellites and ocean sensors to make planet-wide measurements. An analysis of the data has helped us better understand the physics of these ocean waves. We won't be able to stop future tsunamis, but with a better knowledge of how they are formed and how they travel, we will be better able to warn people to get out of their way.

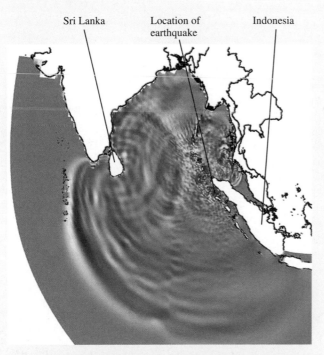

Sri Lanka   Location of earthquake   Indonesia

One frame from a computer simulation of the Indian Ocean tsunami three hours after the earthquake that produced it. The disturbance propagating outward from the earthquake is clearly seen, as are wave reflections from the island of Sri Lanka.

Mathematics Review

Algebra

Using exponents:
$$a^{-x} = \frac{1}{a^x} \qquad a^x a^y = a^{(x+y)} \qquad \frac{a^x}{a^y} = a^{(x-y)} \qquad (a^x)^y = a^{xy}$$

$$a^0 = 1 \qquad a^1 = a \qquad a^{1/n} = \sqrt[n]{a}$$

Fractions:
$$\left(\frac{a}{b}\right)\left(\frac{c}{d}\right) = \frac{ac}{bd} \qquad \frac{a/b}{c/d} = \frac{ad}{bc} \qquad \frac{1}{1/a} = a$$

Logarithms: Natural (base e) logarithms: If $a = e^x$, then $\ln(a) = x$ $\qquad \ln(e^x) = x \qquad e^{\ln(x)} = x$

Base 10 logarithms: If $a = 10^x$, then $\log_{10}(a) = x \qquad \log_{10}(10^x) = x \qquad 10^{\log_{10}(x)} = x$

The following rules hold for both natural and base 10 algorithms:

$$\ln(ab) = \ln(a) + \ln(b) \qquad\qquad \ln\left(\frac{a}{b}\right) = \ln(a) - \ln(b) \qquad\qquad \ln(a^n) = n\ln(a)$$

The expression $\ln(a + b)$ cannot be simplified.

Linear equations: The graph of the equation $y = ax + b$ is a straight line. a is the slope of the graph. b is the y-intercept.

Proportionality: To say that y is proportional to x, written $y \propto x$, means that $y = ax$, where a is a constant. Proportionality is a special case of linearity. A graph of a proportional relationship is a straight line that passes through the origin. If $y \propto x$, then

$$\frac{y_1}{y_2} = \frac{x_1}{x_2}$$

Slope $a = \dfrac{\text{rise}}{\text{run}} = \dfrac{\Delta y}{\Delta x}$; y-intercept $= b$

Quadratic equation: The quadratic equation $ax^2 + bx + c = 0$ has the two solutions $x = \dfrac{-b \pm \sqrt{b^2 - 4ac}}{2a}$.

Geometry and Trigonometry

Area and volume:

Rectangle
$$A = ab$$

Triangle
$$A = \tfrac{1}{2}ab$$

Circle
$$C = 2\pi r$$
$$A = \pi r^2$$

Rectangular box
$$V = abc$$

Right circular cylinder
$$V = \pi r^2 l$$

Sphere
$$A = 4\pi r^2$$
$$V = \tfrac{4}{3}\pi r^3$$

Arc length and angle:　The angle θ in radians is defined as $\theta = s/r$.

The arc length that spans angle θ is $s = r\theta$.

2π rad $= 360°$

Right triangle:　Pythagorean theorem　$c = \sqrt{a^2 + b^2}$ or $a^2 + b^2 = c^2$

$$\sin \theta = \frac{b}{c} = \frac{\text{far side}}{\text{hypotenuse}} \qquad \theta = \sin^{-1}\left(\frac{b}{c}\right)$$

$$\cos \theta = \frac{a}{c} = \frac{\text{adjacent side}}{\text{hypotenuse}} \qquad \theta = \cos^{-1}\left(\frac{a}{c}\right)$$

$$\tan \theta = \frac{b}{a} = \frac{\text{far side}}{\text{adjacent side}} \qquad \theta = \tan^{-1}\left(\frac{b}{a}\right)$$

In general, if it is known that sine of an angle θ is x, so $x = \sin \theta$, then we can find θ by taking the *inverse sine* of x, denoted $\sin^{-1} x$. Thus $\theta = \sin^{-1} x$. Similar relations apply for cosines and tangents.

General triangle:　$\alpha + \beta + \gamma = 180° = \pi$ rad

Identities:

$$\tan \alpha = \frac{\sin \alpha}{\cos \alpha} \qquad\qquad \sin^2 \alpha + \cos^2 \alpha = 1$$

$$\sin(-\alpha) = -\sin \alpha \qquad\qquad \cos(-\alpha) = \cos \alpha$$

$$\sin(2\alpha) = 2\sin \alpha \cos \alpha \qquad\qquad \cos(2\alpha) = \cos^2 \alpha - \sin^2 \alpha$$

Expansions and Approximations

Binomial approximation:　$(1 + x)^n \approx 1 + nx$　if　$x \ll 1$

Small-angle approximation:　If $\alpha \ll 1$ rad, then $\sin \alpha \approx \tan \alpha \approx \alpha$ and $\cos \alpha \approx 1$.

The small-angle approximation is excellent for $\alpha < 5°$ (≈ 0.1 rad) and generally acceptable up to $\alpha \approx 10°$.

Periodic Table of Elements

Symbol key:
27
Co
58.9
(Atomic number / Symbol / Atomic mass)

Period	1	2		Transition elements										13	14	15	16	17	18
1	1 H 1.0																		2 He 4.0
2	3 Li 6.9	4 Be 9.0												5 B 10.8	6 C 12.0	7 N 14.0	8 O 16.0	9 F 19.0	10 Ne 20.2
3	11 Na 23.0	12 Mg 24.3												13 Al 27.0	14 Si 28.1	15 P 31.0	16 S 32.1	17 Cl 35.5	18 Ar 39.9
4	19 K 39.1	20 Ca 40.1	21 Sc 45.0	22 Ti 47.9	23 V 50.9	24 Cr 52.0	25 Mn 54.9	26 Fe 55.8	27 Co 58.9	28 Ni 58.7	29 Cu 63.5	30 Zn 65.4		31 Ga 69.7	32 Ge 72.6	33 As 74.9	34 Se 79.0	35 Br 79.9	36 Kr 83.8
5	37 Rb 85.5	38 Sr 87.6	39 Y 88.9	40 Zr 91.2	41 Nb 92.9	42 Mo 95.9	43 Tc 96.9	44 Ru 101.1	45 Rh 102.9	46 Pd 106.4	47 Ag 107.9	48 Cd 112.4		49 In 114.8	50 Sn 118.7	51 Sb 121.8	52 Te 127.6	53 I 126.9	54 Xe 131.3
6	55 Cs 132.9	56 Ba 137.3	57 La 138.9	72 Hf 178.5	73 Ta 180.9	74 W 183.9	75 Re 186.2	76 Os 190.2	77 Ir 192.2	78 Pt 195.1	79 Au 197.0	80 Hg 200.6		81 Tl 204.4	82 Pb 207.2	83 Bi 209.0	84 Po 209.0	85 At 210.0	86 Rn 222.0
7	87 Fr 223.0	88 Ra 226.0	89 Ac 227.0	104 Rf 261	105 Db 262	106 Sg 263	107 Bh 264	108 Hs 269	109 Mt 268	110 Ds 271	111 Rg 272	112 285							

Inner transition elements

Lanthanides 6

58 Ce 140.1	59 Pr 140.9	60 Nd 144.2	61 Pm 144.9	62 Sm 150.4	63 Eu 152.0	64 Gd 157.3	65 Tb 158.9	66 Dy 152.5	67 Ho 164.9	68 Er 167.3	69 Tm 168.9	70 Yb 173.0	71 Lu 175.0

Actinides 7

90 Th 232.0	91 Pa 231.0	92 U 238.0	93 Np 237.0	94 Pu 239.1	95 Am 241.1	96 Cm 244.1	97 Bk 249.1	98 Cf 252.1	99 Es 257.1	100 Fm 257.1	101 Md 258.1	102 No 259.1	103 Lr 262.1

Answers

Chapter 1

Answers to odd-numbered multiple-choice questions
21. C
23. A
25. B
27. D

Answers to odd-numbered problems

1.
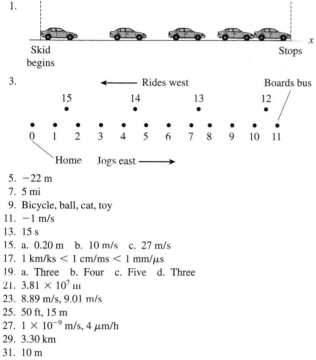

Skid begins ... Stops

3. Rides west — Boards bus

Home   Jogs east →

5. -22 m
7. 5 mi
9. Bicycle, ball, cat, toy
11. -1 m/s
13. 15 s
15. a. 0.20 m   b. 10 m/s   c. 27 m/s
17. 1 km/ks $<$ 1 cm/ms $<$ 1 mm/μs
19. a. Three   b. Four   c. Five   d. Three
21. 3.81×10^7 m
23. 8.89 m/s, 9.01 m/s
25. 50 ft, 15 m
27. 1×10^{-9} m/s, 4 μm/h
29. 3.30 km
31. 10 m
33. (100 m, 61° north of east)

37.
Pictorial representation   **Motion diagram**
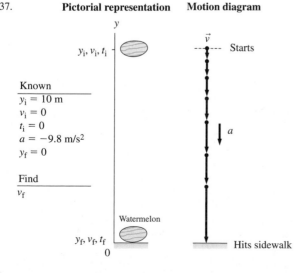

y

y_i, v_i, t_i

Known
$y_i = 10$ m
$v_i = 0$
$t_i = 0$
$a = -9.8$ m/s^2
$y_f = 0$

Find
v_f

Watermelon

y_f, v_f, t_f
0

$\vec{v}$ --- Starts

a

Hits sidewalk

41.
Known
$\theta = 20°$
$x_i = 0$ $v_0 = 10$ m/s
$t_i = 0$ $a < 0$
$v_f = 0$

Find
$h = x_f \sin \theta$

Pictorial representation
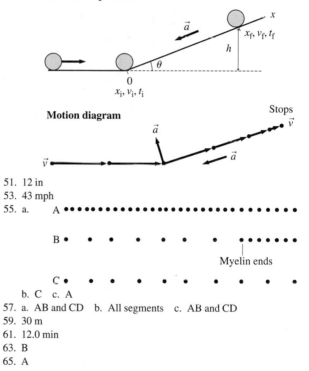

$\vec{a}$ x
x_f, v_f, t_f
h
θ
0
x_i, v_i, t_i

Motion diagram

Stops
$\vec{a}$
$\vec{v}$
$\vec{a}$
$\vec{v}$

51. 12 in
53. 43 mph
55. a.
A •••••••••••••••••••••••••••

B • • • • • •   • ••••••••

Myelin ends

C • • • • • •   • • • • •

b. C   c. A
57. a. AB and CD   b. All segments   c. AB and CD
59. 30 m
61. 12.0 min
63. B
65. A

Chapter 2

Answers to odd-numbered multiple-choice questions
17. C
19. D
21. B
23. A
25. D

Answers to odd-numbered problems

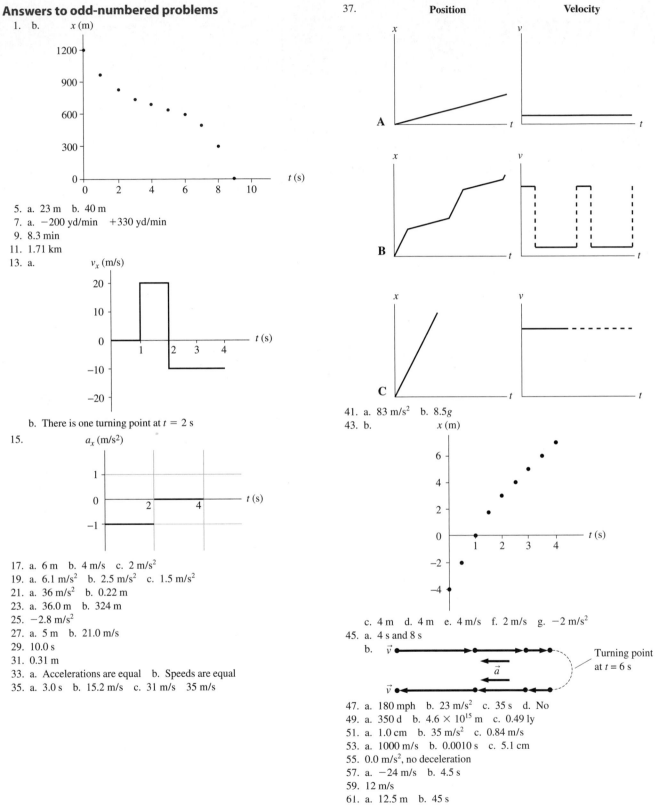

1. b. x (m)

5. a. 23 m   b. 40 m
7. a. −200 yd/min   +330 yd/min
9. 8.3 min
11. 1.71 km
13. a. v_x (m/s)

   b.   There is one turning point at $t = 2$ s
15. a_x (m/s²)

17. a. 6 m   b. 4 m/s   c. 2 m/s²
19. a. 6.1 m/s²   b. 2.5 m/s²   c. 1.5 m/s²
21. a. 36 m/s²   b. 0.22 m
23. a. 36.0 m   b. 324 m
25. −2.8 m/s²
27. a. 5 m   b. 21.0 m/s
29. 10.0 s
31. 0.31 m
33. a. Accelerations are equal   b. Speeds are equal
35. a. 3.0 s   b. 15.2 m/s   c. 31 m/s   35 m/s

37.   Position   Velocity

41. a. 83 m/s²   b. 8.5g
43. b. x (m)

   c. 4 m   d. 4 m   e. 4 m/s   f. 2 m/s   g. −2 m/s²
45. a. 4 s and 8 s
   b. $\vec{v}$ Turning point
        $\vec{a}$ at $t = 6$ s
        $\vec{v}$
47. a. 180 mph   b. 23 m/s²   c. 35 s   d. No
49. a. 350 d   b. 4.6 × 10¹⁵ m   c. 0.49 ly
51. a. 1.0 cm   b. 35 m/s²   c. 0.84 m/s
53. a. 1000 m/s   b. 0.0010 s   c. 5.1 cm
55. 0.0 m/s², no deceleration
57. a. −24 m/s   b. 4.5 s
59. 12 m/s
61. a. 12.5 m   b. 45 s
63. −1.0 m/s²
65. a. 4.1 s   b. Both same
67. a. 4.0 s   b. 16 m

69. a. 100 m   b.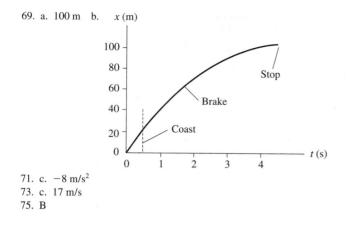

71. c. -8 m/s^2
73. c. 17 m/s
75. B

Chapter 3

Answers to odd-numbered multiple-choice questions

17. D
19. a. D   b. A   c. C   d. A
21. a. C   b. D
23. B
25. D

Answers to odd-numbered problems

3. b. Greater
5. The acceleration vector is toward the center in each case.

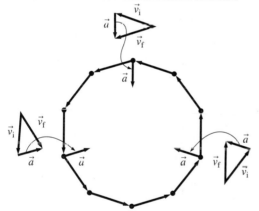

7. 12 m/s
9. 87 m/s
11. a. $d_x = 70.7$ m, $d_y = -70.7$ m   b. $v_x = 282$ m/s, $v_y = 103$ m/s
    c. $a_x = 0.0$ m/s^2, $a_y = -5.0$ m/s^2
13. a. $v = 45$ mph, $\theta = 63°$ b. $a = 6.3$ m/s^2, $\theta = -72°$
15. 5.1 km
17. 530 m
19. 1.3 s
21. Ball 1: 5 m/s Ball 2: 15 m/s
23. 2.0 km/h
25. 0.4 m

27. a. 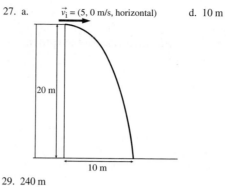   d. 10 m

29. 240 m
31. a. 0.064 s   b. 780 m/s
33. a. 15 s   b. 330 m   c. 280 m
35. 4.7 cm
37. a. 4.0 m/s^2 b. 32 m/s^2
39. Magnitude $= 0.004$ m/s^2, direction toward center of circle
41. a. $D_x = 8$ $D_y = 7$
    b.

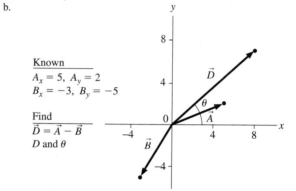

    c. $D = 11$ $\theta = 41°$ north of east
43. a. $B_x = -2$ $B_y = 2$ b. $B = 2.8$ $\theta = 45°$ north of west
45. a. 100 m, lower
    b. (500 m, east) $+$ (5000 m, north) $-$ (100 m, vertical)
47. 7.5 m
49. 31° south of east for 20 min
51. a. 63.5 m   b. 7.05 s
53. No
55. a. 39 mi   b. 20 mph
57. 49° west of south
59. a. 19 mph, 27° south of west   b. 8.5 mph, 3.3° east of north
61. a. $(v_x)_0 = 2.0$ m/s $(v_y)_0 = 4.0$ m/s
    $(v_x)_2 = 2.0$ m/s $(v_y)_2 = 0.0$ m/s
    $(v_x)_3 = 2.0$ m/s $(v_y)_3 = -2.0$ m/s
    b. 2.0 m/s   c. 63.4° north of east
63. a. $v = L\sqrt{g/2h}$ b. 6.0 m/s,   c. Yes, it is reasonable
65. 6.0 m
67. 15° and 75°
69. No
71. a. 6.1 m/s   b. 1.8 m
73. b. 235 m
75. C
77. B
79. B

Chapter 4

Answers to odd-numbered multiple-choice questions
21. C
23. D
25. A
27. B
29. C

Answers to odd-numbered problems
1. First is rear-end; second is head-on
3. People in back can fly forward and injure people in front
5.

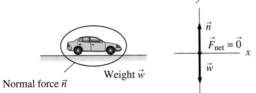

7. Weight, tension force by rope
9. Weight, normal force by ground, kinetic friction force by ground
11. Weight, normal force by ground, kinetic friction force by ground
13. $m_1 = 0.08$ kg, $m_3 = 0.50$ kg
15. a. 2.4 m/s^2 b. 0.6 m/s^2
17. a. 16 m/s^2 b. 4 m/s^2 c. 8 m/s^2 d. 32 m/s^2
19. 2.5 m/s^2
21. 0.025 kg
23. a. 25 N b. 100 N c. 70 N
25. 0.02 m/s^2
29.

Force identification   **Free-body diagram**

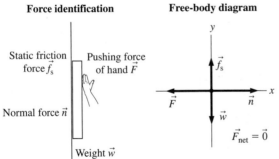

35.

Force identification   **Free-body diagram**

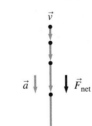

37. The pairs are identified on the diagram.

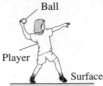

(i) Earth

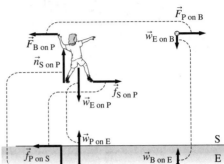

(ii)

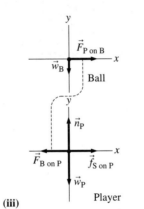

(iii)

39. **Motion diagram**   41. **Motion diagram**

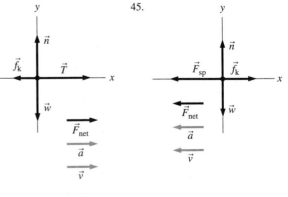

43.   45.

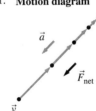

47.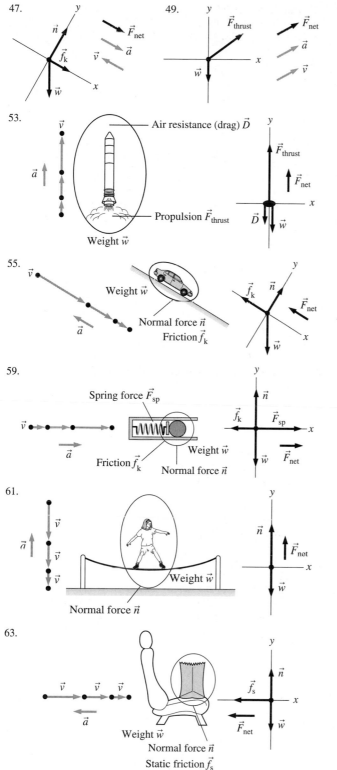

49.

53.

Air resistance (drag) $\vec{D}$

Propulsion $\vec{F}_{thrust}$

Weight $\vec{w}$

55.

Weight $\vec{w}$

Normal force $\vec{n}$

Friction $\vec{f}_k$

59.

Spring force $\vec{F}_{sp}$

Friction $\vec{f}_k$

Weight $\vec{w}$

Normal force $\vec{n}$

61.

Weight $\vec{w}$

Normal force $\vec{n}$

63.

Weight $\vec{w}$

Normal force $\vec{n}$

Static friction $\vec{f}_s$

67. a.

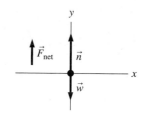

b. Greater

69. C

71. C

Chapter 5

Answers to odd-numbered multiple-choice questions

25. a. C   b. D
27. a. C   b. B
29. C
31. C
33. D

Answers to odd-numbered problems

1. $T_1 = 87$ N $T_2 = 50$ N
3. 110 N each
5. 510 N each
7. F_x (N)

9. a_x (m/s²)

11. $a_x = 1.0$ m/s² $a_y = 0.0$ m/s²
13. a. 0.0 N   b. 0.0 N   c. 250 N
15. 9800 N toward the rear
17. a. 540 N   b. 89 N
19. a. 780 N   b. 1600 N
21. a. 780 N   b. 1100 N
23. a. 5.9 N   b. 5.1 N
25. 0.25
27. 136 m
29. a. 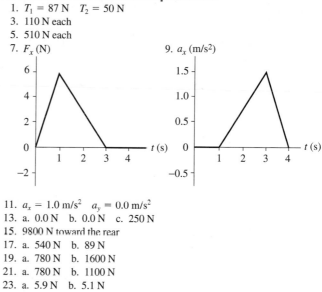   b.

Direction of motion

$\vec{F}_{net} = \vec{0}$

$\vec{F}_{net}$

c. 4.9 m/s²   d. 2.9 m/s²

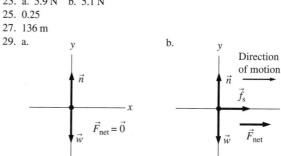

31. 170 m/s
33. a. 5.2×10^{-8} m   b. Much less
35. a. 6 N   b. 10 N
37. a. 20 N   b. 21 N
39. a. 530 N   b. 5300 N
41. At $t = 1$ s, $F_{net} = 8$ N; at $t = 4$ s, $F_{net} = 0$ N; at $t = 7$ s, $F_{net} = -12$ N
43. a. 490 N   b. 740 N
45. a. 20000 m/s²   b. 2800 N   c. 2800 N   d. Forehead no, cheek yes
47. a. 230 N   b. 0.20 m/s
49. a. 3.96 N   b. 2.32 N
51. a. 1.3 m/s²   b. 2.0 m/s²
53. a. 9.8 m/s²   b. -9.8 m/s²   c. 20 m/s²   d. 2.5 m/s   e. 0.31 m
55. a. 17 m/s   b. 150 m
57. Stay at rest
59. 1.9×10^5 N
61. a. $-5g$ b. $3g$
63. 9800 N
65. b. 0.36 s
67. c. 100 m
69. c. 2.8 m/s²
71. A
73. D

47. 22 m/s
49. a. 1.7 m/s²
    b.   c. 0.012 N, up   d. 0.0098 N

$\vec{F}$

$\vec{w}$

51. 54 N
53. a. 9.0 N
55. 2400 m
57. a. 3.0×10^{24} kg   b. 0.90 m/s²
59. a. 3.8 m/s²   b. 1.6 m/s
61. (12 cm, 0 cm)
63. 6.5×10^{23} kg
65. 0.48 m/s
67. b. 19.8 m/s
69. b. 1.00×10^8 m
71. B
73. A

Chapter 6

Answers to odd-numbered multiple-choice questions
23. C
25. B
27. A
29. D
31. C

Answers to odd-numbered problems
1. 1.7×10^{-3} rad/s
3. a. 1.3 rad   72°   b. 3.9×10^{-5} rad/s
5. 3.0 rad
7. 3.9 m/s
9. a. 20 s   b. 2.5 m/s
11. $v = 5.7$ m/s $a = 110$ m/s²
13. $T_3 > T_1 = T_4 > T_2$
15. 9400 N, toward center, static friction
17. a. 1.7×10^3 m/s²   b. 240 N
19. 12 m/s
21. 20 m/s
23. 1.58 m/s²
25. 1/2
27. 6.0×10^{-4}
29. a. 3.5×10^{22} N   b. 2.0×10^{20} N   c. 0.56%
31. 1600 earth days
33. 92 min   7700 m/s
35. 4.2 h
39. North pole, by 2.5 N
41. 5.5 m/s
43. a. Two; upward normal force and inward normal force   b. 0.24 N
45. a. 5.0 N   b. 30 rpm

Chapter 7

Answers to odd-numbered multiple-choice questions
17. C
19. A
21. A
23. C

Answers to odd-numbered problems
1. 11 m/s
3. a. 157 rad/s²   b. 50 rev
5. $\tau_1 < \tau_2 = \tau_3 < \tau_4$
7. -0.20 N · m
9. 5.7 N
11. a. 0.036 N · m   b. 100 N · m
13. 5.5 N · m, counterclockwise
15. a. 1.1 m   b. 1.1 m
17. 12 N · m
19. -98 N · m
21. a. 34 N · m   b. 24 N · m
23. a. 1.7 m   b. 833 N · m
25. 7.2×10^{-7} kg · m²
27. 120 kg · m²
29. 32
31. a. 6.9 kg · m²   b. 0.75 rad/s²
33. 1.6 kg · m²
35. 0.047 N · m
37. 17 rad/s²
39. 0.50 s
41. 2.2 m/s²
43. 16.8 m/s

45.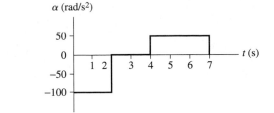

47. $-0.94 \, \text{N} \cdot \text{m}$
49. 7.5 cm
51. a. (8.7 cm, 8.2 cm)   c. 43°
53. $0.018 \, \text{kg} \cdot \text{m}^2$
55. $1.1 \, \text{kg} \cdot \text{m}^2$
57. $6.0 \times 10^{-4} \, \text{kg} \cdot \text{m}^2$
59. a. 0.039 N   b. 38 rpm
61. 3.4 m/s
63. $-0.28 \, \text{N} \cdot \text{m}$
65. 1.1 s
67. B
69. B

Chapter 8

Answers to odd-numbered multiple-choice questions
15. B
17. B
19. C

Answers to odd-numbered problems
1. Right 470 N; left 160 N
3. 140 N
5. No, there is a net torque
7. 590 N
9. 1
11. 8.9°
13. 15 cm
15. 830 N/m
17. a. 13 cm   b. 8 cm
19. 18 cm
21. $1.5 \, \text{m/s}^2$
23. a. 2.0 mm   b. 0.25 mm
25. 7900 N
27. a. 4000 N   b. 4.0 mm
29. 15 cm
31. 0.0078%
33. 1.4 mm
35. a. $F_1 = 750 \, \text{N}$ b. 1000 N
37. 94 N
39. 350 N
41. $81 \, \text{N} \cdot \text{m}$
43. 4.0 m
45. a. 49 N   b. 1500 N/m   c. 3.4 cm
47. 17.4 N
49. a. 1.0 cm   b. 4.5 N, to the left
51. 0.31 m
53. a. 8.2 m/s   b. 30 N
55. 2.7 cm
57. 25%
59. 30%
61. a. $4.8 \times 10^{-2} \, \text{m}^2$ b. 70000 kg   c. 14%
63. D
65. A

Chapter 9

Answers to odd-numbered multiple-choice questions
19. D
21. D
23. B
25. B

Answers to odd-numbered problems
1. 75 m/s
3. 6.0 N
5. a. $1.5 \times 10^4 \, \text{kg} \cdot \text{m/s}$ b. $8.0 \, \text{kg} \cdot \text{m/s}$
7. a. 1.5 m/s, to the right   b. 0.5 m/s, to the right
9. 110 N, directed opposite the original motion
11. a. 19 000 N, directed opposite the original motion,
    b. 280 000 N, directed opposite the original motion
13. 0.205 m/s
15. 0.081 m/s, no
17. 0.143 m/s
19. 0.200 m/s
21. 4.8 m/s
23. 2.3 m/s
25. 300 m/s
27. $(p_{1x})_f = -2 \, \text{kg} \cdot \text{m/s}$, $(p_{1y})_f = 4 \, \text{kg} \cdot \text{m/s}$
29. (14 m/s, 45°)
31. $510 \, \text{kg} \cdot \text{m}^2/\text{s}$
33. $(0.025 \, \text{kg} \cdot \text{m}^2/\text{s}$, into page)
35. 1.3 rev/s
37. 0.20 s
39. 940 N
41. 3.8 m/s
43. a.

	p_x	p_y
After launch	$1.08 \, \text{kg} \cdot \text{m/s}$	$0.625 \, \text{kg} \cdot \text{m/s}$
At top	$1.08 \, \text{kg} \cdot \text{m/s}$	$0.00 \, \text{kg} \cdot \text{m/s}$
Before landing	$1.08 \, \text{kg} \cdot \text{m/s}$	$-0.625 \, \text{kg} \cdot \text{m/s}$

   b. p_x is constant because no forces act on the ball in the x-direction
45. $7.5 \times 10^{-10} \, \text{kg} \cdot \text{m/s}$
47. a. 6.4 m/s   b. 360 N
49. 3.6 m/s
51. 30 m/s
53. 2.0 m/s
55. a. $6.7 \times 10^{-8} \, \text{m/s}$ b. $2 \times 10^{-10}\%$
57. 13 s
59. 440 m/s
61. 20.0 m/s, downward
63. 200 m/s
65. $1.5 \times 10^7 \, \text{m/s}$, to the right
67. 0.81 m/s, 73° below $+x$-axis
69. 4.5 rpm
71. a. 2.8 m/s   b. 2.2 m/s, 25 cm
73. 50 rpm
75. 18 rev/s, clockwise
77. B
79. C

Chapter 10

Answers to odd-numbered multiple-choice questions
29. C
31. C
33. C

Answers to odd-numbered problems

1. 0 J
3. 12,500 J by the weight, -7920 J by $\vec{T}_1$, -4580 J by $\vec{T}_2$
5. a. 0 J   b. 2600 J   c. -2600 J
7. The bullet
9. 2
11. 0 J
13. 1.8 kg $\cdot$ m^2
15. a. 6.8×10^5 J   b. 46 m   c. No
17. a. 13 m/s   b. 14 m/s
19. 31 m/s
21. a. Yes   b. 14 m/s
23. 0.63 m
25. 3.0 m/s
27. 9.7 J
29. 17 m/s
31. a. Potential energy is transformed to kinetic and thermal energy.
    b. 550 J
33. 0.86 m/s and 2.9 m/s
35. Proton: 1.7×10^7 m/s to left; Carbon: 3.1×10^6 m/s to right
39. a. 180 J   b. 59 W
41. 45 kW
43. a. 30 N   b. 45 W
45. 2.0×10^4 W
47. a. 7.7 m/s   b. 6.6 m/s
49. 2.3 m/s
51. a. 0.20 kJ   b. 98 N   c. 2.0 m   d. 0.20 kJ
53. 15 m/s
55. 51 cm
57. a. 15 m/s   b. They will go hungry.
59. 3.8 m, not dependent on mass
61. a. $\sqrt{\dfrac{(m + M)kd^2}{m^2}}$ b. 200 m/s
    c. 0.9975 kinetic energy transformed to thermal energy
63. 7.9 m/s
65. a. $0.048v_i$ b. 95%
67. a. 100 N   b. 0.20 kW   c. 1.2 kW
69. 5.5×10^4 L
71. B
73. C
75. A
77. B
79. C

Chapter 11

Answers to odd-numbered multiple-choice questions

21. A
23. B
25. C

Answers to odd-numbered problems

1. 8.4×10^5 J
3. 3.3%
5. a. 4.2×10^6 J   b. 43 km
7. 230,000 J = 55,000 cal = 55 Cal
9. 1.4 km
11. 710 m
13. a. 200 J   b. 16,000 J/day   c. 0.0095 donuts/day
15. 1.0 K
17. 30 K
19. -700 J

21. -150 J
23. a. 15 kJ   b. 27%
25. 25%
27. a. 7°C   b. -4°C
29. 1.5
31. a. 200 J   b. 250 J
33. a. (b) only   b. (a) only
35. b. 0.125
37. 0.011 kg
39. a. 1.4×10^6 J   b. 1.4×10^6 J
41. a. 190 kJ   b. 760 kJ   c. 180 Cal   d. 140 Cal   e. 1200 W
43. 180 m/s, 1×10^{-22} J
45. 1.2×10^4 m
47. a. 3400 J   b. 3400 J
49. 320 K
51. 2/3
53. 30°C
55. 1/2
57. a. 110 m   b. 32%
59. 8.0%
61. C
63. A
65. C

Chapter 12

Answers to odd-numbered multiple-choice questions

27. D
29. A
31. E
33. D.

Answers to odd-numbered problems

1. Carbon
3. 3.54×10^{24}
5. 0.024 m^3
7. 4.7 m
9. 0.83%
11. 49.7 psi
13. 1.1 N
15. a. -141°C   b. 2.7×10^{-21} J
17. a. 1.2 mol   b. 1.5 mol   c. 5.3 mol
19. 17 L
21. a. $T_2 = T_1$ b. $p_2 = p_1/2$
23. a. 98 cm^3
    b.

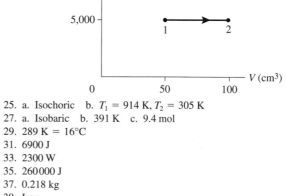

25. a. Isochoric   b. $T_1 = 914$ K, $T_2 = 305$ K
27. a. Isobaric   b. 391 K   c. 9.4 mol
29. 289 K = 16°C
31. 6900 J
33. 2300 W
35. 260 000 J
37. 0.218 kg
39. Iron
41. a. 31 J   b. 60°C
43. 0.080°C or K

45. 41 J
47. 830 W
49. 24 W
51. 6.0 W
53. a. 0.050 m³   b. 1.3 atm
55. a. 1.6 mol   b. 9.4 × 10²³   c. 200 kPa
57. 6.8 × 10⁶ J
59. 17 m
61. 32 psi
63. No
65. 200 cm³
67. a. 600 cm³
    b.

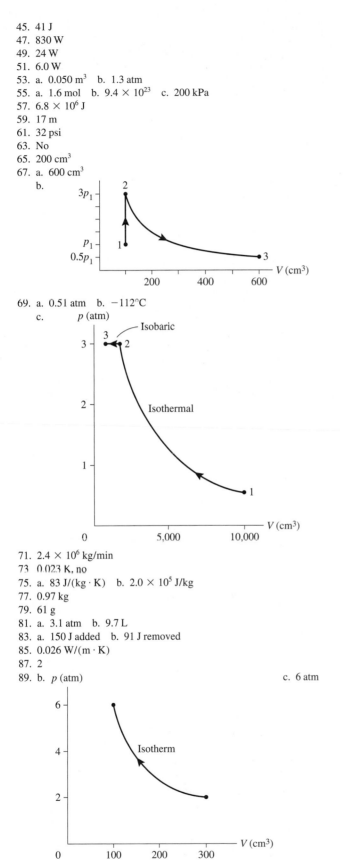

69. a. 0.51 atm   b. −112°C
    c. p (atm)

71. 2.4 × 10⁶ kg/min
73. 0.023 K, no
75. a. 83 J/(kg · K)   b. 2.0 × 10⁵ J/kg
77. 0.97 kg
79. 61 g
81. a. 3.1 atm   b. 9.7 L
83. a. 150 J added   b. 91 J removed
85. 0.026 W/(m · K)
87. 2
89. b. p (atm)   c. 6 atm

91. b. p c. 420 cm³

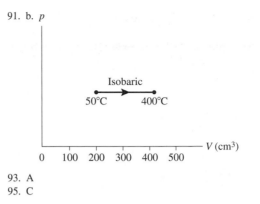

93. A
95. C

Chapter 13

Answers to odd-numbered multiple-choice questions

31. A
33. D
35. B
37. B
39. A

Answers to odd-numbered problems

1. 1200 kg/m³
3. 11 kg/m³
5. a. 810 kg/m³   b. 840 kg/m³
7. 8.0%
9. a. 6.3 m³   b. 1.2 × 10⁵ Pa
11. 210 N
13. 3.2 km
15. a. 2.9 × 10⁴ N   b. 30
17. a. 110 kPa   b. 4400 Pa for both A–B and A–C
19. 3.7 mm
21. 89 mm Hg
23. 2 cm
25. 1.9 N
27. 46 kg
29. 1.0 m/s
31. 97 min
33. 110 kPa
35. 3.0 m/s
37. 260 Pa
39. 2.17 cm
41. 4.8 × 10²³
43. 27 cm
45. a. 5800 N   b. 6000 N
47. 1.8 cm
49. 750 kg/m³
51. 44 N
53. a. 0.38 N   b. 20 m/s
55. $\dfrac{4}{\pi}v_1 \approx 1.3v_1$
57. a. p_{atmos} b. 4.6 m
59. 4.4 cm
61. −28°C
63. 13 mm Hg
65. B
67. A

Chapter 14

Answers to odd-numbered multiple-choice questions
19. a. B   b. C   c. B   d. B   e. B
21. a. D   b. B   c. B
23. B

Answers to odd-numbered problems
1. 2.3 ms
3. 0.30 Hz
5. c. 12°   d. 0° to 10°
7. a. 0.13 m   b. 9.0 cm
9. a. 20 cm   b. 0.25 Hz
11.

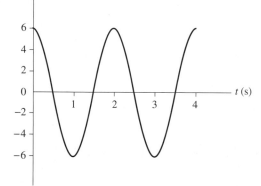

13. a. 0.2 m/s^2 b. $\frac{1}{50}g$
15. a. $\frac{1}{4}$ is potential; $\frac{3}{4}$ is kinetic   b. $A/\sqrt{2}$
17. a. 2.83 s   b. 1.41 s   c. 2.00 s   d. 1.41 s
19. a. 2.0 cm   b. 0.63 s   c. 5.0 N/m   d. 20 cm/s   e. 1.0×10^{-3} J
    f. 15 cm/s
21. a. 1.00 s   b. 0.628 m/s   c. 0.100 J
23. a. 4.00 s   b. 5.66 s   c. 2.83 s   d. 4.00 s
25. a. 0.20 rad   b. 0.80 Hz   c. 0.39 m   d. −0.084 rad
27. 33.1 cm
29. 1.6 s
31. a. 0.124 rad   b. 2.00 s   c. 1.49 m
33. 10.0 s
37. 2.8 s
39. a. 20 N/m   b. 0.40 s
41. a. 24.5 N/m   b. 0.898 s   c. 0.700 m/s through equilibrium position
43. 3.5 Hz

45.
x (cm)

(a)

x (cm)

(b)

47. a. 0.25 Hz; 3.0 s   b. 6.0 s; 1.5 s   c. 9/4
49. a. 6.40 cm   28.3 cm/s
51. 0.08 N/m
53. 0.11 m; 1.7 s
55. 0.64 Hz
57. 8.1 N/m
59. a. 2.00709 s   b. 2.00721 s   c. 4.6 h
61. a.

	f (Hz)	*T* (s)
24.8 cm rod	1.00	1.00
38.8 cm rod	0.800	1.25

   b. 5.00 s   c. 0.200 Hz
63. 6.0 cm
65. 250 oscillations
67. 0.0453 J
69. C
71. B

Chapter 15

Answers to odd-numbered multiple-choice questions
19. D
21. D
23. D

Answers to odd-numbered problems
1. 280 m/s
3. 140 m/s
5. 6060 m/s

7.

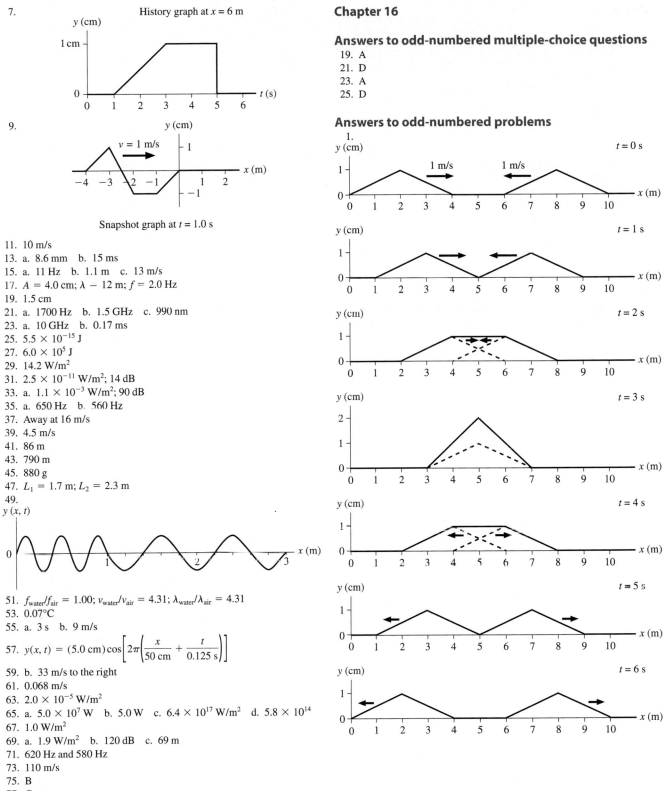

History graph at $x = 6$ m

9.

Snapshot graph at $t = 1.0$ s

11. 10 m/s
13. a. 8.6 mm   b. 15 ms
15. a. 11 Hz   b. 1.1 m   c. 13 m/s
17. $A = 4.0$ cm; $\lambda = 12$ m; $f = 2.0$ Hz
19. 1.5 cm
21. a. 1700 Hz   b. 1.5 GHz   c. 990 nm
23. a. 10 GHz   b. 0.17 ms
25. 5.5×10^{-15} J
27. 6.0×10^5 J
29. 14.2 W/m^2
31. 2.5×10^{-11} W/m^2; 14 dB
33. a. 1.1×10^{-3} W/m^2; 90 dB
35. a. 650 Hz   b. 560 Hz
37. Away at 16 m/s
39. 4.5 m/s
41. 86 m
43. 790 m
45. 880 g
47. $L_1 = 1.7$ m; $L_2 = 2.3$ m
49.

51. $f_{water}/f_{air} = 1.00$; $v_{water}/v_{air} = 4.31$; $\lambda_{water}/\lambda_{air} = 4.31$
53. 0.07°C
55. a. 3 s   b. 9 m/s
57. $y(x, t) = (5.0\ \text{cm})\cos\left[2\pi\left(\dfrac{x}{50\ \text{cm}} + \dfrac{t}{0.125\ \text{s}}\right)\right]$
59. b. 33 m/s to the right
61. 0.068 m/s
63. 2.0×10^{-5} W/m^2
65. a. 5.0×10^7 W   b. 5.0 W   c. 6.4×10^{17} W/m^2 d. 5.8×10^{14}
67. 1.0 W/m^2
69. a. 1.9 W/m^2 b. 120 dB   c. 69 m
71. 620 Hz and 580 Hz
73. 110 m/s
75. B
77. C

Chapter 16

Answers to odd-numbered multiple-choice questions
19. A
21. D
23. A
25. D

Answers to odd-numbered problems
1.

5. 4 s

7.

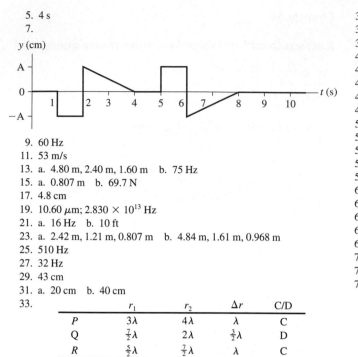

9. 60 Hz

11. 53 m/s

13. a. 4.80 m, 2.40 m, 1.60 m   b. 75 Hz

15. a. 0.807 m   b. 69.7 N

17. 4.8 cm

19. 10.60 μm; 2.830 × 10^{13} Hz

21. a. 16 Hz   b. 10 ft

23. a. 2.42 m, 1.21 m, 0.807 m   b. 4.84 m, 1.61 m, 0.968 m

25. 510 Hz

27. 32 Hz

29. 43 cm

31. a. 20 cm   b. 40 cm

33.

	r_1	r_2	Δr	C/D
P	3λ	4λ	λ	C
Q	$\frac{7}{2}\lambda$	2λ	$\frac{3}{2}\lambda$	D
R	$\frac{5}{2}\lambda$	$\frac{7}{2}\lambda$	λ	C

35. 203 Hz

37. 880 m/s

39. 1.2 m

41. 260 N

43. 12 kg

45. 8.2 m/s^2

47. 19 kN

49. a. 2.8 m   b. 1.4 m/s

51. 13.0 cm

53. 3800 Hz

55. 1210 Hz

57. 18 cm

59. 1400 Hz

61. a. 3.00 m   b. Strong   c. Weak

63. 20

65. a. No   b. 10 cm

67. a. 5 beats/s   b. 4.6 mm

69. 7.0 m/s

71. 0.40 m/s

73. B

75. B

Credits

Index